Inner Space/Outer Space

D0813842

Theoretical Astrophysics

David N. Schramm, series editor

Inner Space/Outer Space

The Interface between Cosmology and Particle Physics

Edited by
Edward W. Kolb, Michael S. Turner, David Lindley, Keith Olive, and David Seckel

The University of Chicago Press

Chicago and London

Edward W. Kolb, Michael S. Turner, David Lindley, Keith Olive, and David Seckel are members of the Theoretical Astrophysics Group at the Fermi National Accelerator Laboratory. Kolb is also professor of astronomy and astrophysics at the University of Chicago. Turner is professor of physics and of astronomy and astrophysics at the University of Chicago.

For the sake of a more affordable book, this volume has been printed from camera-ready copy prepared under the supervision of the volume's editors.

The University of Chicago Press, Chicago 60637
The University of Chicago Press, Ltd., London

© 1986 by The University of Chicago
All rights reserved. Published 1986
Printed in the United States of America

95 94 93 92 91 90 89 88 87 86 5 4 3 2 1

Library of Congress Cataloging-in-Publication Data
Main entry under title:

Inner space/outer space.

(Theoretical astrophysics)
Proceedings of a conference hosted at the Fermi National Accelerator Laboratory in May 1984.
Bibliography: p.
Includes index.
1. Cosmology—Congresses. 2. Astrophysics—Congresses. 3. Particles (Nuclear physics)—Congresses. I. Kolb, Edward W. II. Fermi National Accelerator Laboratory. III. Series.
QB980.I56 1985 523.01 85-24510
ISBN 0-226-45032-5
ISBN 0-226-45033-3 (pbk.)

Contents

PREFACE

This book is the proceedings of Inner Space/Outer Space, a conference held at Fermi National Accelerator Laboratory during the first week of May 1984. The conference was attended by over 200 scientists from throughout the world. The attendees represented a very diverse group of scientists: astronomers, theoretical astrophysicists, cosmologists, high-energy experimentalists, high-energy theorists, low-temperature physicists, relativists, and cosmic-ray physicists. The unifying theme that brought this diverse group of people together is the connection between physics of the microworld (Inner Space) and physics of the macroworld (Outer Space). The connection between particle physics and cosmology links together the smallest scales probed by man, the realm of elementary particle physics, and physics of the largest scale imaginable (the entire Universe), the realm of cosmology.

The interdisciplinary field of particle physics and cosmology has now reached a level of maturity where it is possible to present a book that summarizes the present status of the field. In this book we have attempted to present a comprehensive review of particle physics and cosmology by collecting into a single volume the contributions of leading experts in the field. The papers are grouped under nine subject headings: Standard models of particle physics and cosmology; Microwave background radiation; Origin and evolution of large scale structure; Inflationary Universe; Massive magnetic monopoles; Supersymmetry, supergravity and quantum gravity; Cosmological constraints on particle physics; Cosmology in extra dimensions; and Directions and connections in particle physics and cosmology. In each chapter there are one or two long review articles summarizing the status of each subject, and several shorter contributions that report on recent, more specialized developments.

The conference was organized by the editors of this book, who at the time comprised the Fermilab Astrophysics Group. Chris Quigg and David Schramm also served on the organizing committee. The Group was started in 1983, jointly funded by the Innovative Research Program of the NASA Office of Space Science and Applications and the Department of Energy through Fermilab. In planning the conference, the organizers had the help and cooperation of many people. In particular, the pre-conference planning and organization of Marilyn Paul, Ann Burwell, and Olga Zdanovics insured that the meeting ran smoothly. The conference photographs were taken by Rick Fenner. Finally, conference registration was handled by Sue Winchester, Teri Martin, Michelle Gleason, Sam Yarborough, Raeburn Wheeler and Pat Oleck. All papers were re-typed at Fermilab under the supervision of Olga Zdanovics, with assistance from Pat Oleck, Sue Grommis, and Sam Yarborough. Their careful attention to detail is reflected throughout the book. The meeting and book would not have been possible without the financial support of Fermilab, NASA, the Department of Energy, and the National Science Foundation (under agreement No. PHY-8411472: any opinion, findings and conclusions or recommendations expressed in this publication are those of the authors and do not necessarily reflect the views of the National Science Foundation). Finally we would like to thank everyone who participated in the conference or contributed to this volume.

INTRODUCTION

Important progress in astrophysics has often been spurred by the infusion of new ideas in physics. At least three infusions of new physics have taken place already this century -- the introduction of atomic physics, nuclear physics, and general relativity to astrophysics, and all three have resulted in revolutions in astrophysics. A fourth such revolution seems to be taking place at present, in the application of modern particle theory to astrophysics. This infusion of new physical principles has already led to a renaissance in cosmology and holds forth the possibility of sorting out the history of the Universe back to times as early as 10^{-43} seconds after the bang or even earlier!

Before the current revolution began some five or so years ago, our understanding of the Universe was embodied in the so-called standard hot big bang cosmology. This very successful model provides us with a reliable description of the history of the Universe from about 10^{-2} sec after the bang until today, some 15 billion years later. According to the standard cosmology, the time and temperature of the Universe are related by

$$t(\mathrm{sec}) \simeq T(\mathrm{MeV})^{-2}$$

valid for $t \lesssim 10^{10}$ sec.

At about 10^{-2} sec after the bang, the temperature was about 10 MeV and conditions were becoming suitable for the synthesis of light nuclei from the free neutrons and protons present. The standard cosmology, supplemented by now well-understood nuclear physics, makes definite predictions about the elements synthesized: ^{4}He - about 25% by mass; D and ^{3}He - each about a few parts in 10^5; and ^{7}Li - about a part in 10^9. At present these 'predictions' (more precisely, 'postdictions') agree quite well with the primordial abundances of these elements inferred by present observation, giving us confidence that the standard cosmology is reliable at least back to 10^{-2} sec after the bang.

At about 1000 yrs after the bang, matter rather than radiation began to dominate the energy density of the Universe, at which time structure formation in the Universe began. Small primordial density perturbations began to grow via the Jeans (or gravitational) instability. Shortly thereafter, about 300,000 yrs after the bang when the temperature was about 3000K, matter and radiation decoupled, the free nuclei and electrons combined to form neutral atoms and the radiation scattered for the last time off the free electrons. By the time the Universe was a few billion years old, the primordial perturbations have grown into galaxies. Current observations probe the Universe at three very different epochs and give evidence to the credibility of the standard cosmology -- (i) the observations of galaxies and QSO's with redshifts approaching 4; (ii) measurements of the 2.7 K microwave background radiation, both the spectrum and anisotropy; (iii) the primordial abundances of the light elements.

As successful as the standard cosmology is, it leaves a number of very fundamental facts unexplained (although it can accommodate all of them!). These puzzling facts include: the large-scale isotropy and homogeneity; the flatness of the observed Universe; the precise nature and origin of the primordial density perturbations; the net baryon number of the Universe and matter-to-radiation ratio, quantified together as the baryon number-to-photon number ratio, which is about $(3-7) \times 10^{-10}$; the nature of the dark matter which pervades the cosmos and dominates the present mass density; how, in detail, the structure which is so conspicuous today evolved; and finally, the smallness of the cosmological constant, which on dimensional grounds one would expect to be at least 120 orders-of-magnitude larger than we know it can be.

The application of modern particle theory to cosmology allows us to begin to bring the earliest history of the Universe into focus and begin to understand some of these puzzling facts. To understand earlier and earlier times we must understand physics at higher and higher energies. Our knowledge of Quantum Chromo Dynamics, the SU(3) part of the standard SU(3)×SU(2)×U(1) low energy theory, implies that at about 10^{-5} sec the Universe underwent a transition from an unconfined quark phase to the present phase where quarks are confined to be in color singlet hadrons. Our understanding of the electroweak interaction, the SU(2)×U(1) part of the theory, leads us to believe that the Universe should have undergone a phase transition at about 10^{-11} sec, in which the full symmetry of the electroweak interaction present at high temperature is broken.

Speculations about even earlier events are necessarily based upon physical principles at energies which have not yet been probed by experiment. Nonetheless, these speculations have proven to be even more interesting and promising. Interactions predicted by Grand Unified Theories which violate baryon-number conservation as well as C and CP, lead to the Universe developing a net baryon number when the Universe was of order 10^{-34} sec old. Once The Grand Unified Theory is known, the net baryon number of the Universe can be calculated in much the same way that the primordial abundances of the light elements is calculated. Phase transitions in the very early Universe, possibly associated with spontantous symmetry breaking, can lead to the Universe expanding exponentially for a period of time; this exponential growth or inflation may be able to account for the large-scale isotropy and homogeneity and flatness of the Universe, and the origin of the density inhomogeneities necessary for structure formation. Some theories of the very early Universe predict that various relics, including particles -- e.g., massive neutrinos, photinos, and axions; topological structures -- e.g., monopoles, strings, and domain walls, should be left over in a noticeable abundance. It is attractive to speculate that the dark matter may be comprised of relic particles from the early Universe, or that cosmic strings may play an important role in galaxy formation, or even that enough relic monopoles remain to be detected. Kaluza-Klein and Superstring theories predict that there are more than the three space dimensions that we observe today and that while the additional dimensions today are virtually inaccessible because of their small spatial extent, early on, times less than 10^{-43} sec, all of the spatial dimensions were on equal footing, implying that to understand the evolution of the early Universe we will need to study cosmological models with 'extra dimensions' and understand why only three spatial dimensions are accessible to us today. If one can indeed describe the Universe, space-time manifold and all by a wave function, then even the initial state of the Universe may be subject to calculation.

In the "History of the Universe" (see Figure) we present our best guess of the early history of the Universe. Although the earliest history of the Universe is only now starting to come into focus, the potential revolutionary implications are very apparent. We may very

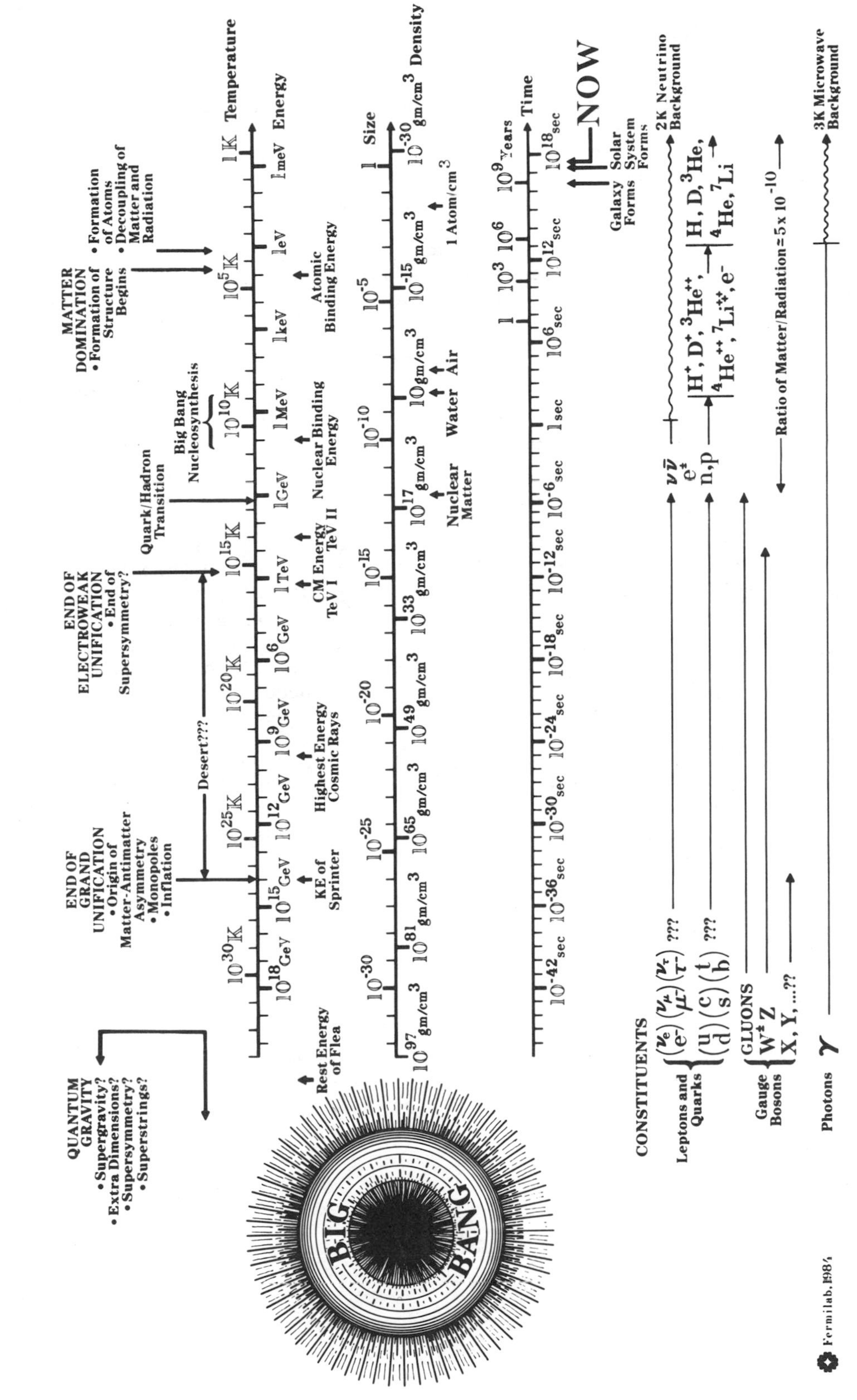

QUANTUM GRAVITY
• Supergravity?
• Extra Dimensions?
• Supersymmetry?
• Superstrings?
END OF GRAND UNIFICATION
• Origin of Matter-Antimatter Asymmetry
• Monopoles
• Inflation
Desert???
END OF ELECTROWEAK UNIFICATION
• End of Supersymmetry?
Quark/Hadron Transition
Big Bang Nucleosynthesis
MATTER DOMINATION
• Formation of Structure Begins
• Formation of Atoms
• Decoupling of Matter and Radiation
10^{30}K
10^{25}K
10^{20}K
10^{15}K
10^{10}K
10^{5}K
1 K
Temperature
10^{18}GeV
10^{15}GeV
10^{12}GeV
10^{9}GeV
10^{6}GeV
1 TeV
1 GeV
1 MeV
1 keV
1 eV
1 meV
Energy
Rest Energy of Flea
KE of Sprinter
Highest Energy Cosmic Rays
CM Energy TeV I TeV II
Nuclear Binding Energy
Atomic Binding Energy
10^{-30}
10^{-25}
10^{-20}
10^{-15}
10^{-10}
10^{-5}
1
Size
10^{97} gm/cm^3
10^{81} gm/cm^3
10^{65} gm/cm^3
10^{49} gm/cm^3
10^{33} gm/cm^3
10^{17} gm/cm^3
10 gm/cm^3
10^{-15} gm/cm^3
10^{-30} gm/cm^3
Density
Nuclear Matter
Water
Air
1 Atom/cm^3
10^{-42}sec
10^{-36}sec
10^{-30}sec
10^{-24}sec
10^{-18}sec
10^{-12}sec
10^{-6}sec
1 sec
10^{6}sec
10^{12}sec
10^{18}sec
1
10^{3}
10^{6}
10^{9} Years
Time
Galaxy Forms
Solar System Forms
NOW
BIG BANG
CONSTITUENTS
Leptons and Quarks
$\binom{\nu_e}{e^-}\binom{\nu_\mu}{\mu^-}\binom{\nu_\tau}{\tau^-}$???
$\binom{u}{d}\binom{c}{s}\binom{t}{b}$???
$\nu\bar{\nu}$
$e^\pm$
n, p
$H^+, D^+, {}^3He^{++}$,
${}^4He^{++}, {}^7Li^{+++}, e^-$
H, D, ^{3}He,
^{4}He, ^{7}Li
2K Neutrino Background
Gauge Bosons
GLUONS
$W^\pm$ Z
X, Y, ...??
Ratio of Matter/Radiation ≈ 5 x 10^{-10}
Photons
γ
3K Microwave Background
Fermilab, 1984

well be close to understanding many, if not all of the cosmological facts left unexplained by the standard cosmology. At the very least it is by now clear the answers to some of our most pressing questions lie in the earliest moments of the Universe.

One of the lessons of modern particle physics seems to be that Nature reveals more of her simplicity and symmetry at higher and higher energies. The unification of the weak and electromagnetic interactions at an energy of a few hundred GeV has wet our appetite for more -- the unification of the strong interactions with the weak and electromagnetic interactions, and a quantum mechanical description of gravity and unification with the other interactions. Grand Unified Theories do the former, while Supergravity or Superstring theories seem to offer the best hope of doing the latter. Unfortunately, the energy scales involved exceed our present technical capabilities by many orders of magnitude. The early Universe it appears had no such limitations, and so it has become a 'laboratory' for studying physics at very high energies, and the cosmological, and sometimes astrophysical, implications of such theories are of the greatest significance, because at present they are the testing grounds.

The Inner Space/Outer Space connection is still in its scientific infancy, but already it has begun to revolutionize our thinking about the origin and evolution of the cosmos and the nature of the fundamental laws. This volume serves to summarize the present status of the interface between particle physics and cosmology, beginning with a review of the well-established -- the standard models of cosmology and of particle physics, then continuing on to the speculations -- structure formation in a Universe dominated by exotic debris, inflation, the status for experimental verification of physics beyond the standard $SU(3)\times SU(2)\times U(1)$ model, theoretical possibilities for the unification of gravity with the other forces, cosmological constraints on particle physics, cosmology with extra dimensions, and finally, a chapter which deals with future directions in both particle physics and cosmology.

PART I

STANDARD MODELS OF PARTICLE PHYSICS AND COSMOLOGY

In the first chapter the standard models of particle physics and cosmology are presented. These standard models represent our current understanding of particle physics, cosmology, and the connection between them. Paul Langacker gives an overview of particle physics starting with the observed "low-energy" world of SU(3)×SU(2)×U(1), which is a good description of physics below energies of 100 GeV (i.e., sizes greater than 10^{-16}cm.) and concluding with a review of the status of grand unification, which is an attempt to model physics at energies of 10^{15} GeV (10^{-28} cm.). The standard particle physics model described by Langacker provides the input for modelling the Universe at very early times.

In "The big bang, the Universe, and everything", Gary Steigman gives the big picture in a summary of our present understanding of cosmology by reviewing the successes and shortcomings of the big-bang model of the Universe. He pays particular attention to primordial nucleosynthesis, because it provides the best probe of the conditions in the Universe as early as a fraction of a second after the bang. Steigman gives a theorist's view of the basic cosmological parameters, such as the Hubble constant, the mass density of the Universe, and the abundance of the light elements. Sandage and Tammann, Huchra, and Pagel describe the observational details in the determination of these crucial parameters.

The possibility of an early generation of stars, Population III stars, is reviewed in a cosmological context by B. J. Carr. The existence of such a population would have a multitude of cosmological and astrophysical implications. The remainder of the chapter is devoted to reports of recent work with important implications in our picture of what we consider to be the "Standard Models".

INNER SPACE/OUTER SPACE
FERMILAB

The Present Status of Grand Unification and Proton Decay

Paul Langacker

INTRODUCTION

In the last fifteen years or so there has been a tremendous advance in our understanding of the elementary particles and their interactions. We now have a mathematically consistent field theory--the standard model--which successfully describes the elementary particles at all energy scales that have been probed at existing accelerators or under known astrophysical or terrestrial conditions. Nevertheless, few physicists consider the standard model to be a serious candidate for the ultimate theory of nature. It is not that there is anything inconsistent or incorrect about the standard model, it is just that it is too complicated and arbitrary (the simplest version has 19 free parameters!).

One attempt to constrain the arbitrary features of the standard model are grand unified theories (GUTs), in which the strong, weak, and electromagnetic interactions are all parts of a more fundamental underlying theory. In addition to their simplicity and elegance, grand unified theories have a number of practical implications. These include: (a) an explanation of charge quantization (the equality of the magnitudes of the electron and proton electric charges), which is an input rather than a prediction in the standard model. (b) The weak angle $\sin^2\theta_W$ and the b quark mass are successfully predicted by many GUTs. (c) Essentially all but the simplest GUT predict nonzero neutrino masses. Massive neutrinos could dominate the energy density of the universe and would be a candidate for the dark matter in galaxies and clusters. (d) All GUTs predict the existence of superheavy (e.g., 10^{16} GeV) magnetic monopoles, which may have been produced prolifically during phase transitions in the very early universe. (e) Grand unified theories predict the existence of extremely weak new interactions which typically violate both baryon and lepton number. These interactions could have generated the observed cosmological excess of baryons (protons and neutrons) over anti-baryons dynamically during the first 10^{-35} sec. after the Big Bang. (f) Perhaps the most dramatic prediction of most GUTs is proton (and bound neutron) decay. Despite the extremely long lifetime prediction ($\gtrsim 10^{30}$yr), several experiments in various parts of the world are searching for proton decay. The present lower limit on the lifetime for decays into the $e^+\pi^0$ mode is 2×10^{32}yr, which probably rules out the simplest GUT. However, many variations and modifications of the simplest model allow longer lifetimes and/or different decay modes (e.g., supersymmetric models often favor $\bar{\nu}K$ decays). In fact, there are several candidate events for decays into more complicated decay modes, but alternate interpretations (e.g., the interaction of cosmic ray produced neutrinos in the detector) cannot be excluded at present.

THE STANDARD MODEL

The standard model is a gauge field theory, which means that the equations of motion are invariant under a group of (gauge) transformations which can be performed independently at different space-time points. Gauge invariance implies the existence of spin-1 vector (gauge) bosons, one for each generator of the gauge group, which can mediate interactions between the particles in the theory. Furthermore, the amplitude for emission or absorption of a gauge boson or for the self-interaction between gauge bosons is prescribed by the gauge invariance up to a coupling constant that must be determined experimentally.

The standard model is based on the complicated direct product group $G_S = SU_3 \times SU_2 \times U_1$, where the three factors have coupling constants g_S, g, and g', respectively. To see how G_S acts, consider the 15 left-handed (i.e., negative helicity) fermions of the first family (which accounts for ordinary terrestrial and astrophysical matter):

$$
\begin{array}{ccccc}
 & \multicolumn{3}{c}{G(SU_3)} & \\
 & \longleftarrow\!\!\longrightarrow & \longleftarrow\!\!\longrightarrow & & \\
W^{\pm}(SU_2)\updownarrow & \begin{pmatrix} u_R \\ d_R \end{pmatrix}_L & \begin{pmatrix} u_G \\ d_G \end{pmatrix}_L & \begin{pmatrix} u_B \\ d_B \end{pmatrix}_L & \begin{pmatrix} \nu_e \\ e^- \end{pmatrix}_L \\
 & u^c_{RL} & u^c_{GL} & u^c_{BL} & e^+_L \\
 & d^c_{RL} & d^c_{GL} & d^c_{BL} & -
\end{array}
\qquad (1)
$$

u(d) represent up (down) quarks of electric charge 2e/3 (-e/3), where e is the proton charge; R, G, B refer to the three values (red, green, blue) of the color quantum number, the superscript c refers to antiparticle, and L refers to left-handed. The right-handed particles are related by CPT ($\psi_R \underset{CPT}{\longleftrightarrow} \psi^c_L$) and are therefore not independent.

The strong interactions are believed to be due to the SU_3 color gauge group, in which different colors of quarks (or antiquarks) are rotated into each other. Associated with each of the eight SU_3 generators is a massless gluon (G), which can be emitted or absorbed by quarks, as in Figure 1a. This theory (quantum chromo-dynamics[1] or QCD) has been extremely successful in describing the qualitative features of the strong interactions. However, it is very difficult to test the model quantitatively because of the large value of the gauge coupling constant $g_S = O(1)$.

The electromagnetic and weak (electroweak) interactions are associated with the $SU_2 \times U_1$ group.[2] The left-handed quarks and leptons transform as doublets under SU_2. Two of the SU_2 generators are associated with massive electrically charged bosons $W^{\pm}$, which mediate the ordinary charged current weak interactions (Figure 1b). The gauge coupling constant g and the W mass are related to the phenomenological Fermi constant by $G_F/\sqrt{2} = g^2/8M_W^2$. From G_F and the value of g determined from the electromagnetic and neutral current data one can predict[3] $M_W = 83.0^{+2.9}_{-2.7}$ GeV. The W was discovered at CERN last year[4] by the UA1 and UA2 groups with masses (80.9 ± 1.5 ± 2.4) GeV and (81.0 ± 2.5 ± 1.3) GeV, respectively, in excellent agreement with the $SU_2 \times U_1$ prediction.

A linear combination of the third SU_2 generator and the U_1 generator is associated with a neutral Z boson, which mediates "neutral current" processes such as shown in Figure 1c. The neutral current interaction depends on g and the parameter $\sin^2\theta_W = g'^2/(g^2 + g'^2)$. The

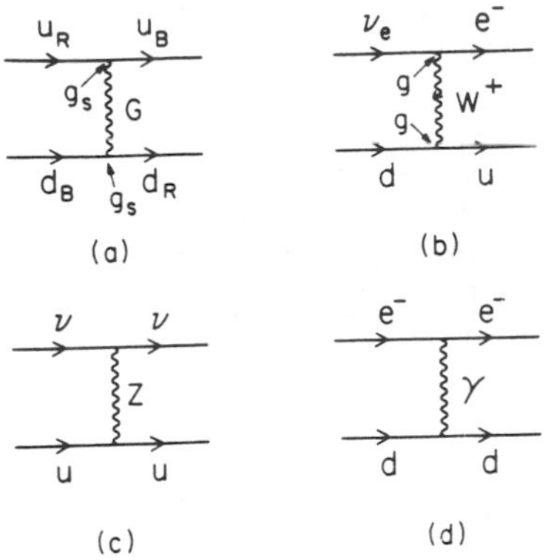

Figure 1: Standard model processes; (a) gluon exchange between quarks, (b) charged current weak process, (c) neutral current process, (d) photon exchange.

neutral current interaction has by now been observed and studied for a variety of processes (e.g., neutrino scattering, atomic parity violation, polarized eD asymmetries, and e^+e^- annihilation asymmetries), and the $SU_2 \times U_1$ predictions for the form of the interaction have been verified in detail.[5] In particular, the Z was also discovered at CERN in 1983, with the masses $(95.6 \pm 1.4 \pm 2.9)$ GeV and $(91.9 \pm 1.3 \pm 1.4)$ GeV determined by the UA1 and UA2 groups[5] in excellent agreement with the predicted value[3] $93.8^{+2.4}_{-2.2}$ GeV.

The orthogonal combination of SU_2 and U_1 generators generates the ordinary electromagnetic interaction, mediated by the massless photon (Figure 1d). The electric charge is $e = g \sin\theta_W$.

The standard $SU_3 \times SU_2 \times U_1$ model has been spectacularly successful. It is a mathematically consistent renormalizable field theory that is compatible with every known fact in particle physics. Moreover, it successfully predicted the existence and detailed form of the neutral current interaction, the W and Z boson masses, and the charmed quark (in order to avoid strangeness changing neutral currents). Nevertheless, it cannot be considered the final story--it is simply too complicated and arbitrary and leaves too many fundamental questions unanswered.

These difficulties can be summarized under four headings.

a) The gauge problem: the standard model gauge group is a complicated direct product of three groups with three distinct coupling constants. Furthermore, the U_1 factor renders electric charge assignments arbitrary except for constraints from the cancellation of anomalies. Hence, there is no explanation of the quantization of fermion and boson charges in multiples of e/3.

b) The fermion problem: the fermions are assigned to a complicated reducible representation of the $SU_3 \times SU_2 \times U_1$ group. No fundamental explanation is given for the existence of the heavy fermion families (c, s, ν_μ, μ) and (t, b, ν_τ, τ^-) (the t and ν_τ have not yet been directly observed), which are essentially identical to the ordinary fermion family (u, d, ν_e, e^-) except for mass. Although well studied in the laboratory and in cosmic ray interactions, these particles appear to play little role in ordinary matter. Furthermore, the fermion masses and mixing angles are not predicted by the model.

c) The Higgs problem: $SU_2 \times U_1$ symmetry is broken by the introduction of fundamental Higgs fields ϕ, which should have mass^2 values μ_ϕ^2 that are not too much larger in magnitude than M_W^2. However, μ_ϕ^2 receives quadratically divergent corrections

$$\delta\mu_\phi^2 = O(g^2, \lambda, h^2)\, \Lambda^2 \tag{2}$$

from the one loop diagram of Figure 2, where Λ is the mass scale at which the integral is cut off. Λ is typically the Planck mass $m_P = G_N^{-1/2} \approx 10^{19}$ GeV. Hence, an incredibly accurate cancellation

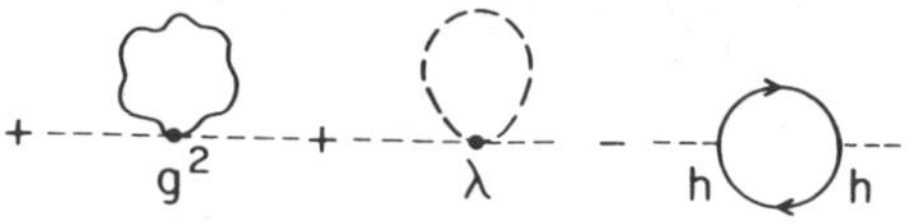

Figure 2: Corrections to μ_ϕ^2 from gauge, Higgs and fermion loops.

(fine-tuning) between the bare value of μ_ϕ^2 and the correction is needed.

d) The graviton problem: the standard model does not incorporate quantum gravity.

Another characterization of many of these problems is that the standard model with massless neutrinos involves nineteen free parameters (three gauge couplings, two θ parameters, nine fermion masses, three KM mixing angles, one CP phase, and the W and Higgs masses, minus one overall mass scale). If one includes classical gravity, one must add m_P and the (observationally tiny) cosmological constant to the list.

GRAND UNIFIED THEORIES[6-8]

Grand unified theories constrain some of the arbitrary features of the standard model by embedding G_S in a larger simple group G with a simple coupling constant g_G. Hence, the strong, weak, and electromagnetic interactions are unified: they are simply different parts of a larger underlying gauge interaction. Similarly, quarks (q), antiquarks ($\bar{q}$), leptons (ℓ), and antileptons ($\bar{\ell}$) are fundamentally similar in most GUTs: the extra symmetries in G typically transform q, $\bar{q}$, ℓ, and $\bar{\ell}$ into each other. This will lead to charge quantization. Furthermore, the gauge bosons associated with the new symmetry generators can usually mediate baryon and lepton number violating processes such as proton (or bound neutron) decay.

Typical diagrams shown in Figure 3 imply such decays as $p \to e^+\pi^0$ (or $e^+\rho^0$, $e^+\pi^+\pi^-$, etc.), $p \to \bar{\nu}\pi^+$ (or $\bar{\nu}\pi^+\pi^0$, etc.), $n \to e^+\pi^-$, $n \to \bar{\nu}\pi^0$, etc. The lifetime is expected to be of order

$$\tau_p \approx \frac{1}{\alpha^2}\frac{M_X^4}{m_p^5}, \tag{3}$$

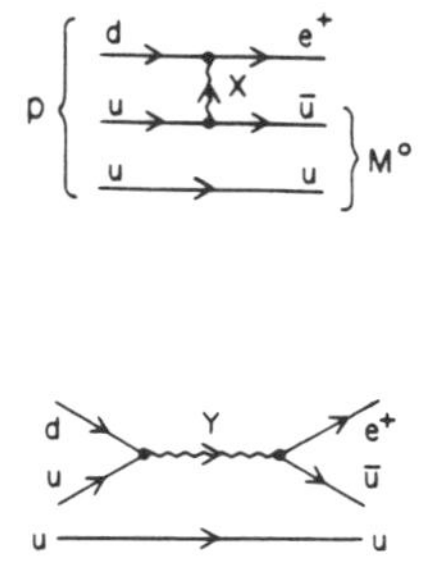

Figure 3: Typical proton decay diagrams. Similar diagrams imply bound neutron decay or decays into $\bar{\nu}$ + mesons.

where $\alpha = e^2/4\pi$ is the fine structure constant. It is known experimentally that if the proton does decay, the lifetime must be $\gtrsim 10^{31}$yr, implying $M_X \gtrsim 5 \times 10^{14}$GeV, more than twelve orders of magnitude larger than the W and Z masses! Fortunately, however, M_X can be independently predicted from the observed ratios of strong and electroweak coupling constants: the prediction is $M_X \sim 2 \times 10^{14}$GeV ($\tau_p \sim 10^{29\pm2}$yr), in the general range but slightly lower than the experimental constraint.

The simplest and most popular GUT is the Georgi-Glashow SU_5 model.[6] The fermion representations are still rather complicated in SU_5: each family is assigned to a reducible 5* + 10 dimensional representation:

$$\underset{X,Y}{\overset{5^*}{W^\pm \updownarrow \begin{pmatrix} \nu_e & \\ & d^c \\ e^- & \end{pmatrix}_L}} \qquad \underset{X,Y}{\overset{10}{\begin{pmatrix} & u & \\ e^+ & & u^c \\ & d & \\ & \longleftrightarrow \quad \longleftrightarrow & \end{pmatrix}_L}} \qquad (4)$$

where the color indices have been suppressed. In addition to the twelve generators and corresponding gauge bosons of the standard model, there are twelve new generators associated with transformations between adjacent columns in Eq. (4). These are X^i, i = R, G, B, which carries color and has electric charge 4/3 e; Y^i, i = R, G, B, with electric charge 1/3 e, and their antiparticles $\bar{X}_i$, $\bar{Y}_i$. The X and Y (which are nearly degenerate) can mediate proton and bound neutron decay as in Figure 3.

IMPLICATIONS OF SU_5 AND SIMILAR MODELS

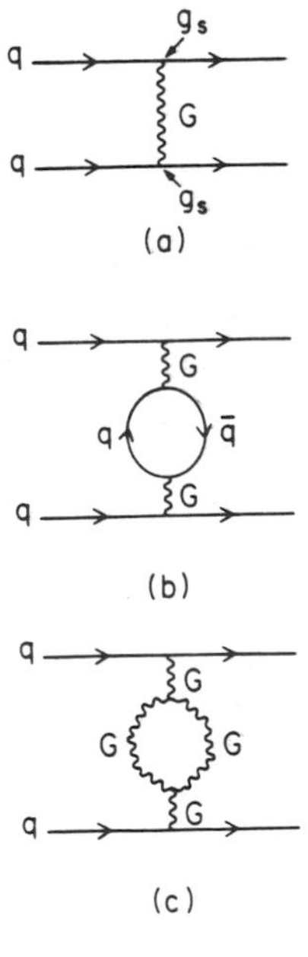

Figure 4: (a) Single gluon exchange, (b) vacuum polarization from a virtual $\bar{q}q$ pair (screening), (c) vacuum polarization from virtual gluons (antiscreening).

Charge quantization follows from the fact that quarks and leptons are combined in multiplets and because the electric charge operator Q is a generator of SU_5. Hence, the sum of charges of all particles in a multiplet must vanish. For the 5* in Eq. (4) this implies $Q_{\nu_e} + Q_{e^-} + 3\, Q_{d^c} = 0$, where the 3 is due to the three colors of d^c. For $Q_{\nu_e} = 0$ this implies $Q_d = + 1/3\, Q_{e^-} = - 1/3e$, from which one easily obtains $Q_p = e = -Q_{e^-}$.

Another implication concerns the coupling constants g_s, g, and g'. These are not actually constants, but functions of the typical momentum--squared Q^2 of the interacting particles. This is due to higher order processes such as the vacuum polarization diagrams of Figure 4 which imply momentum dependent screening or antiscreening of the gauge interaction. The properly normalized couplings $g_3(Q^2) \equiv g_s(Q^2)$, $g_2(Q^2) \equiv g(Q^2)$, and $g_1(Q^2) \equiv \sqrt{5/3}\, g'(Q^2)$ of the SU_3, SU_2, and U_1 subgroups are predicted to approach the common value $g_5(Q^2)$ for momenta $Q^2 \gg M_X^2$ (for which symmetry breaking effects can be ignored) as shown in Figure 5.

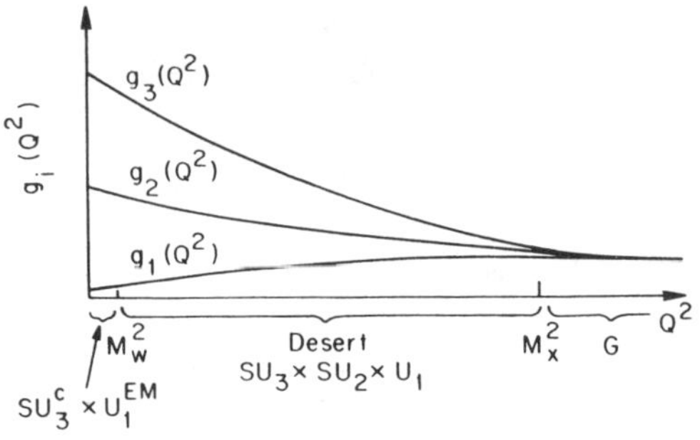

Figure 5: The Q^2 evolution of the coupling constants. The threshold-free region between M_W and M_X is referred to as the desert. M_X is smaller than the Planck mass in sensible GUTs.

From the observed ratios of coupling constants at low (accelerator) energies and the theoretical Q^2 dependence, it is possible to predict M_X. In principle, M_X is determined independently from $\alpha/\alpha_s = e^2/g_s^2$ and from $\sin^2\theta_W = g'^2/(g^2 + g'^2)$. In practice, the uncertainties in $\sin^2\theta_W$ do not allow an accurate determination, so one inverts the logic to use α/α_s to predict both M_X and $\sin^2\theta_W$ (the latter prediction is then a consistency condition that the three couplings all meet at the same point). The major uncertainty in the prediction is the value of the QCD parameter $\Lambda_{\overline{MS}}$, related to α_s by

$$\alpha_s(Q^2) = \frac{12\pi}{25\ \ln(Q^2/\Lambda^2_{\overline{MS}})} + \text{H.O.T.} \tag{5}$$

A plausible range,

$$\Lambda_{\overline{MS}} = 160^{+100}_{-80}\ \text{MeV}, \tag{6}$$

was given by Buras[9] in 1981 and is still valid. However, each determination of $\Lambda_{\overline{MS}}$ involves serious uncertainties.[10] The determinations from deep inelastic scattering suffer from serious higher twist ambiguities,[11] and one phenomenological analysis[12] obtains considerably larger (typically O(300 MeV)) values. The values derived from $e^+e^- \rightarrow$ jets depend very sensitively on the Monte Carlo program used to describe the hadronization.[13] Finally, the ratio of upsilon widths $B(\mu^+\mu^-)/B(3g)$ is experimentally precise. However, the $O(\alpha_s)$ correction (relative to the leading α_s^3 term) to the theoretical formula is large,[10] casting doubt on the validity of the expansion. The leading correction to the $B(\Upsilon gg)/B(3g)$ ratio is small because of cancellations[10] (leading to the value $\Lambda_{\overline{MS}} = 100 \pm 50$ MeV) but that unfortunately does not guarantee that higher order terms are unimportant.

I will utilize the plausible range (Eq. (6)) for $\Lambda_{\overline{MS}}$, but because of the serious uncertainties involved, I will also consider the more conservative limit

$$\Lambda_{\overline{MS}} < 400\ \text{MeV} \tag{7}$$

The results are[14]

$$M_X(\text{GeV}) = 2.1 \times 10^{14} \times (1.5)^{\pm 1} \times \left[\frac{\Lambda_{\overline{MS}}}{0.16\ \text{GeV}}\right] \times \left[\frac{1}{1.5}\right]^{n_H - 1} \times (1.2)^{F-3} \tag{8}$$

$$\sin^2\hat{\theta}_W(M_W) = 0.214 \pm 0.003 + 0.006\ \ell n \left[\frac{0.16\ \text{GeV}}{\Lambda_{\overline{MS}}}\right] + 0.004\ (n_H - 1) - 0.0004\ (F-3) \tag{9}$$

where n_H is the number of light Higgs doublets, F is the number of fermion families, and the quoted errors are from higher order terms, thresholds, and m_t (<50 GeV). The result for $\sin^2\hat{\theta}_W(M_W)$ is in spectacular agreement with the world average[5] 0.215 ± 0.014. (Some recent determinations give slightly larger central values of $\sin^2\hat{\theta}_W(M_W)$ but agree with the stated errors.)

Some models, including SU_5 with only 5 and 24 dimensional Higgs representations, make predictions[15] for the b mass (since the b and τ are in the same multiplet). The prediction $m_b \approx (4.7 - 5.5)$ GeV for F = 3, m_t = 20 GeV, and $\Lambda_{\overline{MS}}$ given by Eq. (6) is in excellent agreement with the experimental value (4.5 - 5.5) GeV (there is some ambiguity[8] as to the relation between the experimental and theoretical values for m_b). The agreement would be spoiled for too large a top mass. One has $m_t < 180$, 145, 20 GeV for $\Lambda_{\overline{MS}}$ = 80, 160, 260 MeV, respectively. A fourth fermion family is essentially ruled out for this class of model unless $\Lambda_{\overline{MS}}$ is small ($\lesssim$ 80 MeV).

Unfortunately these models also predict $m_s \approx 500$ MeV, which is larger than most phenomenological determinations (150 - 300 MeV). Even worse, one has

$$\frac{m_s}{m_d} = \frac{m_\mu}{m_e} \sim 200 , \tag{10}$$

while current algebra determinations typically yield $m_s/m_d \approx 20$. These problems could be fixed by adding a Higgs 45, but then the prediction for m_b would also be lost. It has been suggested[16] that small effective interactions associated with the Planck scale could shift the quark masses by a few MeV, which would suffice for m_s/m_d.

Most grand unified theories other than the minimal SU_5 model predict nonzero neutrino masses. They are typically Majorana (lepton number violating) in the range $(10^{-9} - 10^{+2})$ eV. The direct laboratory limits on neutrino masses are not very stringent:[17]

$$m_{\nu_\tau} < 250\ \text{MeV} \qquad (\tau\ \text{decay})$$

$$m_{\nu_\mu} < 0.49\ \text{MeV} \qquad (\pi\ \text{decay})$$

$$m_{\nu_e} < 46\ \text{eV} \qquad (H^3 \to He^3 e^- \bar{\nu}_e) \tag{11}$$

Lubimov et al.[18] have reported a nonzero ν_e mass in the range (20-40) eV. This has not yet been confirmed, but a number of sensitive experiments are under way.[17] There is no positive evidence for neutrinoless double beta decay ($\beta\beta_{0\nu}$) or for neutrino oscillations, but these depend not only on the neutrino masses but also on mixing angles and, for $\beta\beta_{0\nu}$, on the Majorana or Dirac nature of the neutrinos and on their CP parities.

Stable neutrinos with masses in the 10 eV range could dominate the energy density of the universe. One has $\sum_i m_{\nu_i} < 100$ eV from limits on the large scale deceleration. Massive neutrinos are also candidates for the dark matter in galaxies and clusters, although that view runs into difficulties in detail.[19] Even tiny neutrino masses (e.g., $\gtrsim 10^{-5}$ eV) could, for large enough mixing angles, lead to oscillations which would reduce the predicted flux of solar neutrinos.

All grand unified theories predict the existence of superheavy magnetic monopoles,[20] typically with masses

$$M_m \sim \frac{M_X}{\alpha} \sim 10^{16} \text{ GeV}. \tag{12}$$

These may have been prolifically produced during GUT phase transitions in the very early universe. For many models the predicted mass density of relic monopoles is too large by many orders of magnitude.[20] This monopole problem was one of the primary original motivations for the inflationary universe models.[21]

The baryon number violating interactions in GUTs allow the observed cosmological baryon asymmetry

$$\frac{n_B}{n_\gamma} \approx 10^{-10}, \quad n_{\bar{B}} \approx 0 , \tag{13}$$

where n_B, $n_{\bar{B}}$, n_γ are the present average number densities of baryons, anti-baryons, and microwave photons, respectively, to be generated dynamically in the first 10^{-35}s after the Big Bang.[22,8] The simplest version of SU_5 yields n_B/n_γ about ten orders of magnitude too small. However, this is easily remedied in non-minimal models with a second Higgs 5 (which can be superheavy) or with a fourth fermion family.

PROTON DECAY EXPERIMENTS[23]

One ton of matter contains $\approx 6 \times 10^{29}$ nucleons. If the nucleon lifetime were 10^{32} years, for example, one would need 1700 tons of matter to observe 10 decays per year (assuming 100% detection efficiency).

Most large proton decay experiments have used either water or iron as their primary material. The water detectors are surrounded or interspersed with photomultiplier tubes which can detect the Cerenkov light that would be emitted by charged decay products. For example, for

$$p \to e^+\pi^0$$
$$\quad\quad \hookrightarrow 2\gamma$$

each of the photons would convert to an electromagnetic shower, implying three Cerenkov cones. The largest detector (of the Irvine-Michigan-Brookhaven (IMB) collaboration) utilizes 7000 tons (3300 tons fiducial) of water in an underground tank of dimension 23m × 18m × 17m. It is surrounded by 2048 5" phototubes. Several other experiments use iron as their source of nucleons, with the iron sheets separated by various types of charged particle counters.

Natural radioactivity is not a serious background for proton decay experiments, because the typical energies are much less than the proton mass. Much more serious are muons produced by cosmic-ray interactions in the atmosphere. All existing experiments have been performed deep underground in mines or tunnels in order to reduce this muon flux. Probably the most serious background is due to neutrinos produced in cosmic-ray interactions. Reactions such as $\bar{\nu}_e p \to e^+\pi^0 n$ with the neutron unobserved can simulate proton decay. After applying kinematic cuts and taking Fermi motion, etc., into account, the neutrino background rate would equal the proton decay signal for a true lifetime of 10^{33} yr. It would therefore be very difficult (though not impossible) for terrestrial experiments to detect proton decay for a lifetime much longer than 10^{33} yr.

Recent, current, and planned proton decay experiments are listed in Table I, along with a partial list of results.

The IMB group,[28] with 1844 ton-years of running, has set a lower limit of 10^{31} yr for the partial lifetime τ_i ($\equiv \tau/B_i$, where B_i is the branching ratio for mode i) for 32 decay modes. In particular, $\tau_{e^+\pi^0} > 2 \times 10^{32}$ yr (for 2260 ton-years) and $\tau_{\bar{\nu}K^+} > 1 \times 10^{31}$ yr (these modes are expected to be important in SU_5-like models and in supersymmetric GUTs, respectively). The IMB group does have several events that could be considered as candidates for such modes as μ^+K^0, but the number is consistent with the expectations from neutrino background.

Table I. Status of proton decay experiments. All lifetimes are in years.

Group [location]	Depth (meters water equivalent)	mass (tons) [Fiducial]	Type	Status/Results
Pennsylvania(24) Homestake mine, [N. Dakota]	4200	300 [150]	Cereknov	$\tau_{\mu+X} > 3 \times 10^{31}$ (completed)
Minnesota-ANL(25) [Soudan mine, Minnesota]	1800	30	Proportional tubes	$\tau(SU_5) > 0.8 \times 10^{30}$ (completed)
Tata, Osaka, Tokyo(26) Kolar Gold Field, [S. India]	7600	140	Proportional tubes	3 contained candidates ($e^+\pi^0$, $\bar{\nu}K^+$, $e^+\pi^-$ or μ^+K^0) $\tau_i \sim 1.1 \times 10^{31}$(?)
Frascati, Milano, Turin(27) [M. Blanc Tunnel]	5000	150	Limited streamer	225 ton-years $\tau_{e^+\pi^0} > 1.5 \times 10^{31}$ $\tau_{\mu^+K^0} > 0.6 \times 10^{31}$ [1 candidate; 0.2 background]
Irvine, Michigan, Brookhaven(28) [Morton Mine, Ohio]	1570	7000 [3300]	Cerenkov	1844 ton-years $\tau_i > 10^{31}$, 32 modes $\tau_{e^+\pi^0} > 2 \times 10^{32}$ (2260 t-y) $\tau_{\mu^+K^0} > 4.4 \times 10^{31}$ $\tau_{\mu^+\eta} > 9 \times 10^{31}$ $\tau_{\bar{\nu}K^+} > 1 \times 10^{31}$

Table I. Status of proton decay experiments. All lifetimes are in years. (Concl.)

Group location	Depth (meters water equivalent)	mass (tons) Fiducial	Type	Status/Results
KEK, Tokyo, Tsukuba[29] [Kamioka, Japan]	2700	3000 [1000]	Cerenkov	485 ton-years, $\tau_i > 10^{31}$, 32 modes $\tau_{e^+\pi^o} > 3.2 \times 10^{31}$ $\tau_{\bar{\nu}K^+} > 1.1 \times 10^{31}$ $\tau_{\mu^+\eta} > 1.2 \times 10^{31}$ [1 candidate; .01 bkg] $\tau_{e^+\omega^o} > 1.1 \times 10^{31}$ [1 candidate; 0.6 bkg]
Harvard, Purdue, Wisconsin[30] [Silver King Mine, Utah]	1500	800 [500]	Cerenkov	480 ton-years $N \to \mu\mu X$ $\tau_i > 1.3 \times 10^{31}$ [2 candidates]
Orsay, Ecole Pol., Saclay, Wuppertal [Frejus Tunnel]	4500	1500	Flash chamber	800 tons by end 1984
Argonne, Minn., Oxford, Rutherford, Tufts [Soudan II]	2000	1000-5000	Drift chamber	300 tons (summer 1986) 1000 tons (summer 1987)
Kolar II	-	1000	-	400 tons by end 1984

The Kamioka experiment[29] utilizes specially designed 20" diameter phototubes which yield excellent energy resolution. After 485 ton-years of running, they have observed two serious proton decay candidate events (as well as a lower limit of 10^{31} years for the partial lifetimes into 32 decay modes). The favored interpretations of the two events are into $\mu^+\eta$ and $e^+\omega^o$, respectively, for which the expected backgrounds are 0.01 event and 0.6 events, respectively. It should perhaps be emphasized that no theory has predicted either of these modes as dominant.

The Mont Blanc experiment[27] has one candidate for a μ^+K^o decay (expected background 0.2 events), but it should be recalled that the IMB group, with a factor of 8 more exposure, has no signal above background for this mode.

The Harvard-Purdue-Wisconsin experiment[30] has concentrated on dimuon signals (i.e., $N \to \mu\mu X$, where X represents unobserved particles), where one of the muons could be from the decay of a pion or kaon. They have two- candidate events which are compatible with decays into μ^+K^o, $\mu^+\rho^o$, $\mu^{\pm}K^+$ etc., one of the two in a kinematic region where little background is expected.

The Kolar Gold Field experiment[26] has a number of nucleon decay candidates, three of which are completely contained in the detector. The latter are candidates for $p \to e^+\pi^o$, $p \to \bar{\nu}K^+$, and $n \to e^+\pi^-$ or $p \to \mu^+K^o$, respectively. If true nucleon decays, the partial lifetimes for these modes would be $\approx 1.1 \times 10^{31}$ yr. However, it is difficult to accept these events given the much more stringent lower limits from the larger Cerenkov detectors.

To summarize, proton decay has not been established at present. The partial lifetime $\tau_{e^+\pi^o}$ is greater than 2×10^{32} yr and $\tau_i > 10^{31}$ yr for $\approx$ 32 decay modes. However, a lifetime of order (few) $\times 10^{31}$ yr for many modes more complicated than $e^+\pi^o$ is still possible, and in fact there are several candidate events for decays into such modes as μ^+K^0, $\mu^+\eta$, and $e^+\omega^0$.

PROTON DECAY IN SU_5 AND SIMILAR MODELS

The uncertainties in relating the quark lines in Figure 3 to physical nucleon and meson states make a precise theoretical estimate of the proton lifetime very difficult.

There are three basic classes of diagrams, as shown in Figure 6. The quark decay diagram (Figure 6a) is unimportant due to phase space suppression. Most of the early estimates[8] of the proton decay rate considered only two quark fusion, on the grounds that the three quark fusion diagram requires the overlap of three quarks at one point. However, a number of recent estimates[31-44] suggest that the second and third diagrams may be of the same sign and comparable in magnitude (typically, the ratio of the two amplitudes is found to be 1/3 and 3 for the $e^+\pi^0$ mode).

The results of these recent estimates for the $e^+\pi^0$ and two body lifetimes are listed in Table II.

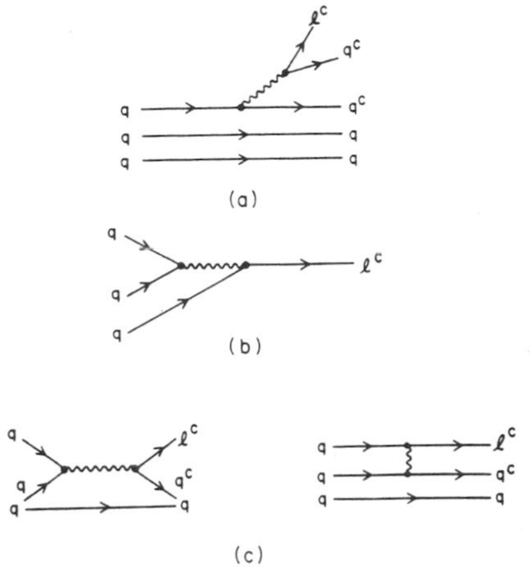

Figure 6: (a) Quark decay diagram, (b) three quark fusion diagram (a meson or gluon must be radiated from one or more of the initial quark lines), (c) two quark fusion diagrams.

Table II. Recent estimates of the proton lifetime in years for SU_5 with $F = 3$, $n_H = 1$, and $M_X = 2.1 \times 10^{14}$ GeV.

Author	Method	$\tau_p \to$ 2 body	$\tau_p \to e^+\pi^0$
Berezinsky et al.(31)	PCAC, QCD sum rules	3.2×10^{28}	6.4×10^{28}
Tomozawa(32	PCAC, VMD	$(0.9\text{-}6)\times 10^{29}$	$(1.5\text{-}10.4)\times10^{29}$
Isgur, Wise(33)	quark	-	4.6×10^{29}
Lucha, Stretmnitzer(34)	Bethe-Salpeter	2.8×10^{29}	6.0×10^{29}
Donoghue, Golowich(35)	bag	$(2.6) \times 10^{29}$	6.0×10^{29}
Thomas, McKellar(36)	cloudy bag	2.5×10^{28}	4.2×10^{28}
Meljanac et al.(37)	bag	$(4.5\text{-}64) \times 10^{28}$	$(7\text{-}180) \times 10^{28}$
Dupont et al.(38)	PCAC, VMD	7.8×10^{27}	2×10^{28}
Brodsky et al.(39)	Light cone	-	8×10^{27}
Mitra, Ramanathan(40)	Bethe-Salpeter	-	6.9×10^{31}
Hamber, Parisi(41)	Lattice	1.5×10^{28}	3.0×10^{28}
Chang, Wu(42)	Bethe-Salpeter	1.1×10^{30}	8.1×10^{30}
Kaymakcalan et al.(43)	chiral Lag. bag, quark	$(1\text{-}16) \times 10^{29}$	$(1.7\text{-}28) \times 10^{29}$
Goldhaber et al.(44)	semi-classical	-	possible supression by 10-100

From these results I will take

$$\tau_{p\to e^+\pi^0}(yr) = 4.5 \times 10^{29 \pm 0.7} \left(\frac{M_X}{2.1\times10^{14}\text{GeV}}\right)^4 \qquad (14)$$

The theoretical uncertainty $10^{\pm 0.7} \sim (5)^{\pm 1}$ is a guess based loosely on the spread of values in Table II. Unfortunately, the hadronic uncertainties in the calculations are large and difficult to estimate. The true uncertainty could therefore be larger.

Combining this with Eq. (8), we find

$$\tau_{p\to e^+\pi^0}(yr) = 4.5 \times 10^{29 \pm 0.9} \left(\frac{\Lambda_{\overline{MS}}}{160 \text{ MeV}}\right)^4$$

$$\longrightarrow \begin{cases} 4.5 \times 10^{29 \pm 1.7} < 2.3 \times 10^{31} & \text{for } \Lambda_{\overline{MS}} = 160^{+100}_{-80} \\ < 1.4 \times 10^{32} & \text{for } \Lambda_{\overline{MS}} < 400 \text{ MeV} \end{cases} \qquad (15)$$

The present limit, $\tau_{p\to e^+\pi^0} > 2.0 \times 10^{32}$ yr from the IMB experiment[28] therefore essentially rules out minimal SU_5. Unfortunately, it is difficult to make an absolute statement to this effect because of the possibility that the uncertainties in the matrix elements and in $\Lambda_{\overline{MS}}$ may have been underestimated.

For completeness, the recent calculations[31-44] also yield

$$\tau_{p\to 2\text{ body}} \sim 2 \times 10^{29\pm1.7} \text{ yr}$$

$$\tau_{n\to e^+\pi^-} = \frac{1}{2}\,\tau_{p\to e^+\pi^0}$$

$$\tau_{n\to 2\text{ body}} \approx 1.5 \times 10^{29\pm1.7} \text{ yr} \qquad (16)$$

Typical branching ratios are given in Table III.

Table **III.** Typical proton decay branching ratios in SU_5.

Mode	B(%)	Mode	B(%)
$e^+\pi^0$	40-60	$\bar{\nu}_e\ \pi^+$	$\frac{2}{5}B\ (e^+\pi^0)$
$e^+\rho^0$	1-10	$\bar{\nu}_e\ \rho^+$	$\frac{2}{5}B\ (e^+\rho^0)$
$e^+\eta$	<2	$\mu^+\ K^0$	1-20
$e^+\omega$	5-20	$\bar{\nu}_\mu\ K^+$	<2

MODIFICATIONS AND UNCERTAINTIES

The results in Eqs. (14-16) are for SU_5 with $n_H = 1$, $F = 3$, and $m_t \lesssim 50$ GeV. It is important to consider the sensitivity of the τ_p and $\sin^2\theta_W$ predictions to small and not so small modifications of the model. It will be seen that many GUTs are still compatible with observations (although they typically have less predictive power than minimal SU_5). The basic ideas of grand unification are sufficiently attractive that searches for proton decay and determinations of $\sin^2\theta_W$ should be pushed as far as possible.

Changes to the lifetime prediction relative to the minimal SU_5 prediction τ_p^{min} of Eqs. (15,16) and to $\sin^2\theta_W$ (relative to Eq. (9)) are shown in Table IV. It is seen that additional light Higgs multiplets reduce the lifetime by a factor of 5 while extra fermion generations have little effect.

Table IV. Changes in the minimal SU_5 predictions (9), (15)-(16) due to various modifications.

Modification	τ_p/τ_p^{min}	$\Delta \sin^2\theta_W$
Light Particles[45]		
$n_H > 1$	$(1/5)^{n_H-1}$	$+0.004\ (n_H-1)$
m_t	10%	-
$F = 4$	1.3	-0.0004
$F = 5$	2.6	-0.0008
Multiplet Splitting[46]	$\lesssim 10^3$	$\lesssim 0.01$
Higgs Color Triplet[47]		
$(m_H > 10^{10})$	$\lesssim 1.8$	-0.0013-0
New Heavy Higgs[47]	$\lesssim 150$	~0.001
Mixing Effects[48,49]	$\lesssim \frac{1}{\sin^2\theta_c} \sim 20$	-
Other Groups[8]	$\lesssim 1$	-
3 Scale Models[51]	$1-10^4$	0=0.01 (input)
Effective nonrenormalizable interactions[52,53]	small effects	[For $M \sim m_P$[52]]
	10^2	-.01 [For $M \sim 10^{17}$ GeV[53]]

Multiplet splitting[46] refers to the addition of fermion or Higgs multiplets which split into light ($O(M_W)$) and heavy ($O(M_X)$) sectors. These can affect τ_p and $\sin^2\theta_W$ in either direction. For example, a desired pattern can be achieved by the addition of a $10 + \overline{10}$ of fermions to the theory, which contains the new fermions (U,D), $\overline{C}$, and E^+, where U has the same charge and color as the u quark, etc. The results in Table IV are for a superheavy E^+ and light U, D, $\overline{C}$. It should be emphasized that such a model is extremely ad hoc. In particular, it is hard to understand how the mass splitting comes about.

The Higgs 5 contains a color triplet H^{α} in addition to the ordinary $SU_{2} \times U_{1}$ doublet. It is seen that if $m_{H} > 10^{10}$ GeV, the effects on $\sin^2\theta_W$ and the ordinary (gauge boson) contributions to τ_p are small. Other representations of new heavy Higgs bosons can have far more significant effects. The results listed in Table IV assume that the Higgs masses are within an order of magnitude of M_X.

Models with relatively simple Higgs structure (e.g., SU_5 with only Higgs 5's and 24) automatically imply that the lightest q, $\bar{q}$, ℓ and $\bar{\ell}$ are all associated together in multiplets. In models with more compicated Higgs structure[48] it is conceivable that, for example, heavy leptons and light quarks could be associated. In this case, the dominant amplitude would be into energetically forbidden channels such as $p \to \tau^+\pi^0$. In most cases, the lifetime is expected to increase by no more than $\sin^{-2}\theta_c \sim 20$, however. One exception to this limitation is models in which the proton is made absolutely stable by imposing a new quantum number of the theory.[49] Such models generally involve a doubling of fermions and can lead to the dramatic signal of baryon number violation at accelerators.

Other gauge groups G, such as SO_{10}, SO_{22}, E_6, etc., give identical predictions as SU_5 for $\sin^2\theta_W$ provided they break directly to $SU_3 \times SU_2 \times U_1$ at M_X. The proton lifetime is generally reduced slightly because of the additional gauge bosons. For example, in SO_{10} with $M_{X'} = M_X$, one has $\tau_{p\to e^+X} = (5/8)\,\tau_{p\to e^+X}(SU_5)$. The extra currents in still larger groups generally do not induce transitions between light fermions and are therefore not relevant to proton decay.[50] The relative branching ratios for $p \to e^+M_i^0$, where $M_i^0 = \pi^0, \rho^0, \eta$, or ω, or for $\bar{\nu}\pi^+/\bar{\nu}\rho^+$ are the same for all gauge boson mediated models in this class (but are very sensitive to the poorly known hadronic matrix elements), while the ratios $e^+/\bar{\nu}$, e^+/μ^+, and $(\Delta S = 1)/(\Delta S = 0)$ are very model dependent. The latter would therefore be useful for distinguishing between models if proton decay is observed.

Theories with 3 or more symmetry breaking scales can lead to longer lifetimes. For example,[51]

$$SO_{10} \xrightarrow[M_X \gg 10^{14}\ \text{GeV}]{} SU_3 \times SU_{2L} \times SU_{2R} \times U_1' \xrightarrow[M_R \gtrsim 10^{9}\ \text{GeV}]{} SU_3 \times SU_2 \times U_1 \xrightarrow[M_W]{} SU_3 \times U_1^{EM} \qquad (17)$$

allows[51] $M_X \sim M_5\ (M_5/M_R)^{1/2} \gg M_5$, where M_5 is the SU_5 scale $\sim 2 \times 10^{14}$ GeV. $\sin^2\theta_W$ depends on mass ratios in such schemes, so one has no real prediction. If one requires $\Delta \sin^2\theta_W < 0.01$ for the pattern in Eq. (17) then $M_R > 10^{12}$ GeV and $M_X \lesssim 2 \times 10^{15}$ GeV.

Effective non-renormalizable operators such as $Tr(F_{\mu\nu}F^{\mu\nu}\phi^{adj})/M$, where ϕ^{adj} is the adjoint Higgs representation, may be generated if a GUT is embedded in a larger theory (e.g., super-gravity or Kaluza Klein) with mass scale M.

Proton decay can also be mediated by the exchange of Higgs bosons, such as the color triplet H^{α} in the 5 of SU_5. Because of the weaker Higgs couplings, such mechanisms are only experimentally relevant if the

Higgs mass happens to be in the approximate range 10^{10} - 10^{11} GeV. (There is no particular reason for m_H to actually be in this range.) Also, the ordinary gauge mediated decays would still have to be suppressed by some other mechanism. Since Higgs Yukawa couplings tend to be proportional to the fermion mass, decays involving muons and kaons should dominate. In SU_5 with a single 5, the branching ratio for μ^+K^0 is in the range[54-55] 35-80%.

Most of the protons studied in existing experiments are found in nuclei. In addition to Fermi motion, there is the additional problem that emitted mesons may be absorbed or scattered in the nucleus[8] (this effect has been included in the IMB analysis). Also, the decay rate may be enhanced[8] by ≈ 2 due to the exchange of a virtual π^0 (which would lead to a continuous e^+ spectrum).

SUPERSYMMETRY

Supersymmetries[56] (SUSY) are generated by fermionic charges Q which transform bosons (B) into fermions (F)

$$Q|F\rangle = |B\rangle, \quad Q|B\rangle = |F\rangle . \tag{18}$$

If unbroken, supersymmetry implies that bosons and fermions are combined in degenerate supermultiplets with relations between their couplings.

There is currently not a shred of experimental evidence for supersymmetry. However, there are several strong theoretical motivations. The first is purely esthetic: SUSY theories are very elegant. Secondly, local supersymmetry (supergravity) is probably the most promising approach to quantum gravity. Finally, the strongest motivation is the Higgs problem: in an exactly supersymmetric theory the boson and fermion loops in Figure 2 cancel, preventing any renormalization of the Higgs mass. The recent interest in supersymmetric grand unified theories (SUSY-GUTs) was largely due to the hope of solving the hierarchy problem (the incredibly tiny ratio of μ_ϕ^2 to M_X^2). SUSY has unfortunately not succeeded in explaining the origin of this tiny ratio, but at least it can prevent the ratio from being renormalized from its tree level value.

Since we do not observe degenerate fermions and bosons, SUSY--if present at all--must be broken so that $m_B \neq m_F$. The quadratic divergences in Figure 2 still cancel for soft or spontaneous SUSY breaking, so one is left with $\delta\mu_\phi^2 = O(g^2,\lambda)\,|m_F^2 - m_B^2|$. Hence, if SUSY is to solve the Higgs problem ($\delta\mu_\phi^2 \lesssim O(M_W^2)$) we must have $|m_F - m_B| \lesssim O(1\ \text{TeV})$ for the scale of observed SUSY breaking.

For the simplest (N = 1) supersymmetry, to be considered here, the supermultiplets consist of boson-fermion pairs. For a variety of reasons,[56] no known particle pair can be interpreted as SUSY partners. Therefore, one must double the number of fields in the theory. Also the supersymmetry requires the existence of two light Higgs doublets ϕ_1 and ϕ_2 to give mass to both the u and d quarks. The minimal field content is shown in Table V.

The phenomenological implications of these new particles include direct production at accelerators, rare transitions and decays (e.g., $\mu \to e\gamma$ and the K_L-K_S mass difference), CP violation (ε, ε' and the neutron electric dipole moment), strong parity violation, and cosmology. These have been discussed in several excellent reviews,[56] so I will only consider the implications for $\sin^2\phi_W$ and proton decay.

Table V. The minimal field content for low energy SUSY. The physical (mass eigenstate) fields may be mixtures of fields with the same spin, charge, and color.

$J = 1$	$J = \frac{1}{2}$	$J = 0$
$W^{\pm}$	$\tilde{w}^{\pm}$ (wino)	
Z	$\tilde{z}$ (zino)	
γ	$\tilde{\gamma}$ (photino)	
g	$\tilde{g}$ (gluino)	
	q_L, q_R	$\tilde{q}_L$, $\tilde{q}_R$ (scalar quarks)
	ℓ_L, ℓ_R	$\tilde{\ell}_L$, $\tilde{\ell}_R$ (scalar leptons)
	$\tilde{\phi}_1$, $\tilde{\phi}_2$ (Higgsinos)	ϕ_1, ϕ_2

PROTON DECAY

The combination of N = 1 supersymmetry with a grand unified group G requires the existence of fermionic partners for the superheavy gauge and Higgs field (e.g., $\tilde{x}$, $\tilde{y}$, $\tilde{H}^{\alpha}$, and adjoint fermions in SUSY-SU_5). Assuming that

$$G \underset{M_X^{SUSY}}{\rightarrow} SU_3 \times SU_2 \times U_1 \underset{M_W}{\rightarrow} SU_3 \times U_1^{EM} , \qquad (19)$$

with SUSY spontaneously broken at a scale F such that $|m_F - m_B| \lesssim 0$ (1 TeV), one has to consider at least four mass scales (m_P, M_X^{SUSY}, F, M_W). One can consider many special cases, such as $F \approx M_X^{SUSY}$ (presumably with SUSY and G broken in the same sector) or $M_X^{SUSY} \approx m_P$. I will simply assume a hidden sector supergravity scheme with $F \approx \sqrt{M_W m_P} \sim 10^{10}$-$10^{11}$ GeV and with M_X^{SUSY} determined by the renormalization group equations (RGE) for the coupling constants.[57,58] Similar results occur in SUSY-GUTs with explicit (soft) SUSY breaking.[57]

The supersymmetry modifies the usual GUT predictions for proton decay in two ways. The new SUSY partners of the ordinary light particles modify the RGE, implying larger values for $\sin^2\phi_W$ and the unification mass, as shown in Table VI. For two light Higgs doublets (the minimum number) M_X^{SUSY} is considerably larger than the non-SUSY value ($\approx 2.1 \times 10^{14}$ GeV), and the proton lifetime due to X, Y exchange becomes much longer (and probably unobservable). Also the prediction for $\sin^2\phi_W$ is slightly larger than the experimental value 0.215 ± 0.015. $n_H = 4$ implies a shorter lifetime but an unacceptable value for $\sin^2\phi_W$. The predictions[59] for m_b are slightly larger than in ordinary SU_5 and strongly suggest $F < 4$. A number of authors[60] have suggested that predictions similar to ordinary GUTs may be obtained by adding extra multiplets to the theory. For example, a split 10 + 10* chiral multiplet can be used to obtain lower M_X^{SUSY} and $\sin^2\phi_W$ and an observable

Table VI. Predictions of Marciano and Senjanovic[58] for M_X^{SUSY}, $\tau_p^{X,Y}$ (the proton lifetime due to ordinary X, Y exchange) and $\sin^2\hat{\theta}_W(M_W)$. n_H is the number of light Higgs doublets.

n_H	$\Lambda_{\overline{MS}}$(MeV)	M_X^{SUSY} (GeV)	$\tau_p^{X,Y}$ (yr)	$\sin^2\hat{\theta}_W(M_W)$
2	100	4.8×10^{15}	2×10^{34}	0.239
2	200	1.1×10^{16}	4×10^{35}	0.235
4	100	2.6×10^{14}	1×10^{29}	0.260
4	200	5.5×10^{14}	3×10^{30}	0.258

lifetime for $\tau_{p\to e^+\pi^0}$. Another possibility[61] is for the color triplet Higgs to have a low mass $m_H \sim 10^{10-11}$ GeV $<< M_X^{SUSY}$ (which was motivated by cosmological arguments). In this case, one has a lower $\sin^2\phi_W$ and an observable $p \to \mu^+K^0$ rate may be mediated by H exchange. (It is necessary in this case to forbid or highly suppress the dimension 5 operators discussed below.)

The other effect on proton decay is the existence of new diagrams[62] involving the SUSY partners (Figure 7).

The dimension 4 diagram of Figure 7a involves the exchange of a light scalar quark. It is not suppressed by any power of M_X^{SUSY} and, if present, would lead to a disastrously short proton lifetime. Fortunately, it can be forbidden[62] by imposing discrete symmetries on the Lagrangian.

The colored Higgsino exchange in the left half of the box diagram in Figure 7b leads to the dimension 5 operator

$$L \sim \frac{h^2}{M_{\tilde{H}}} qq\tilde{q}\tilde{\ell} \qquad (20)$$

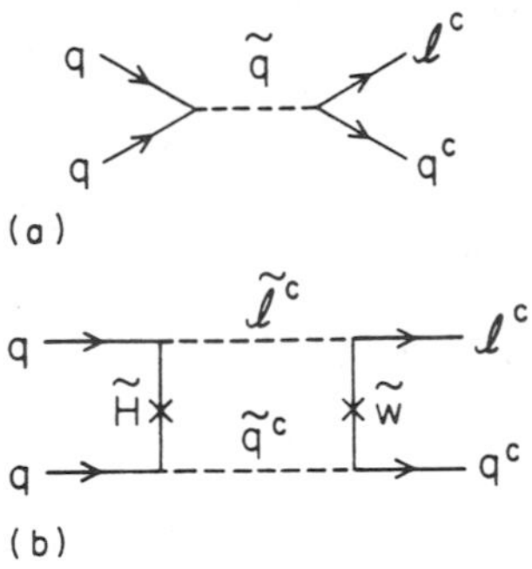

Figure 7: New proton decay diagrams that can occur in SUSY-GUTs; (a) a dimension 4 diagram, which must be forbidden (e.g., by extra discrete symmetries) to avoid a disastrously short proton lifetime, (b) a dimension 5 diagram.

where h is a Yukawa coupling. The dimension 5 operators can be forbidden[62] by extra symmetries, but if they are allowed, they lead--after exchanging a (light) wino[63]--to proton decay with a lifetime $\tau_p \sim (M_{\tilde{H}})^2$ (rather than the usual M_X^4). This appears disastrous (since one expects $M_{\tilde{H}} \sim O(M_X^{SUSY})$), but fortunately there are a number of suppressions compared to the ordinary SU_5 case.[64] These include (a) $M_X^{SUSY} > M_X^{SU_5}$; (b) a $1/16\pi^2$ factor from the loop; (c) two (small) Yukawa couplings; (d) smaller higher order enhancement factors; and (e) possible mass suppressions.

The operator is antisymmetric in flavor indices, so one expects $\Delta S = 1$ to dominate. Also, the Yukawa vertices favor μ^+, ν_μ^c and ν_τ^c decays. Assuming $\tilde{q}$ mixing angles similar to the KM angles, $p\to\nu_\tau^c K^+$, $n\to\nu_\tau^c K^0$ are expected to dominate, with branching ratios[64,65] of 90%, 10%, 10^{-3}, 10^{-4}, 10^{-5}, and 10^{-6} for $N\to\nu_\tau^c K$, $\nu^c\pi$, μ^+K, $\mu^+\pi$, e^+K and $e^+\pi$, respectively.

The actual lifetime calculation suffers from the same kinds of hadronic uncertainties as for ordinary SU_5. Detailed calculations[64,65] yield

$$\tau_{N\to\nu K}(yr) \approx 10^{26\pm2}\left(\frac{M_{\tilde{H}}}{10^{16}\ \mathrm{GeV}}\right)^2\left(\frac{K(m_{\tilde{q}},m_{\tilde{\chi}},m_{\tilde{w}})}{1\ \mathrm{GeV}}\right)^2 \tag{21}$$

where K is a complicated function which is of order $m_{\tilde{q}}^2/m_{\tilde{w}}$ for $m_{\tilde{q}} \approx m_{\tilde{\chi}} > m_{\tilde{w}}$. For $\tau_N > 10^{30}$ yr one requires

$$\left(\frac{M_{\tilde{H}}}{10^{16}}\right)\frac{m_{\tilde{q}}^2}{m_{\tilde{w}}} > 10^{2\pm1}\ \mathrm{GeV}^2 , \tag{22}$$

which is just satisfied for $M_{\tilde{H}} \sim 10^{16}$ GeV, $m_{\tilde{w}} \approx M_W$, and $m_{\tilde{q}} \approx 100$ GeV. Although there is some freedom to vary $M_{\tilde{H}}$ and other masses, proton decay should be observable if theories of this type are correct.

An alternative set of models, in which the supersymmetry and grand unification symmetry are broken in the same sector, has a unification mass close to the Planck scale because of new light supermultiplets.[66] Typically, $\tau_p \sim 10^{34}$yr from dimension 5 operators in such models.[66]

CONCLUSION

The standard $SU_3 \times SU_2 \times U_1$ gauge model is a remarkably successful theory of the ordinary strong, weak, and electromagnetic interactions up to an energy of ≈ 100 GeV. However, the standard model is too complicated and arbitrary to be taken seriously as the ultimate theory of nature.

One possible extension is grand unification, which is motivated mainly by the gauge problem (GUTs have shed little light on the fermion and Higgs problems.) Minimal SU_5 is probably ruled out by the non-observation of proton decay, although it is hard to be absolutely sure because of the large (and difficult to quantify) uncertainties in the hadronic matrix elements and in the determination of $\Lambda_{\overline{MS}}$. However, the basic motivations for GUTs are still strong and attractive, and it is likely that at least some aspects of grand unification are correct. In partiuclar, there are many modifications of and alternatives to minimal SU_5 which have longer proton lifetimes. Such models generally have less predictive power, but $\tau_p \sim 10^{34}$yr seems like a reasonable (and an ambitious) goal for future proton decay experiments.

Another approach is supersymmetry. The ideas of SUSY are very elegant, but unfortunately the viable models are extremely complicated. The Higgs problem suggests that the new particles predicted by SUSY should be lighter than ≈ 1 TeV. In addition to their direct production, they may have observable consequences for rare transitions and decays, CP violation, electroweak parameters, parity violation, and proton decay.

It seems very likely that some form of grand unification exists, probably associated with a mass scale of $\approx 10^{14}$ GeV. Also, the Higgs problem strongly indicates the existence of new physics at the TeV scale. (Supersymmetry, technicolor, or something entirely different.) Intermediate scales, which could lead to phenomena such as neutron-antineutron oscillations,[67] are also possible, but the motivations are less compelling. We desperately need experimental clues as to what lies beyond the standard model. Evidence for TeV and 10^{14} GeV scales is very hard to come by, so every type of experiment (new particle searches at existing and future accelerators, baryon and lepton number violation, (e.g., neutrinoless double β decay), neutrino mass, flavor changing and other rare decays and transitions, CP violation, precision electroweak tests, monopole searches, cosmological and astrophysical implications) should be pushed as far as possible.

REFERENCES

[1]For a review, see W. Marciano and H. Pagels, Phys. Rep. **36**, 137 (1978).
[2]S. Weinberg, Phys. Rev. Lett. **19**, 1264 (1967); A. Salam in Elementary Particle Theory, ed. N. Svartholm (Almquist and Wiksells, Stockholm, 1969) 367; S. L. Glashow, Nucl. Phys. **22**, 579 (1961); S. L. Glashow, J. Iliopoulos, and L. Maiani, Phys. Rev. **D2**, 1285 (1970).
[3]W. J. Marciano and A. Sirlin, Phys. Rev. **D29**, 945 (1984); M. Consoli, S. Lo Presti, L. Maiani, Nucl. Phys. **B223**, 474 (1983).
[4]UA1: G. Arnison et al., Phys. Lett. **129B**, 273 (1983);
UA2: P. Bagnaia et al., Phys. Lett. **129B**, 130 (1983).
[5]J. E. Kim et al., Rev. Mod. Phys. **53**, 211 (1981); W.J. Marciano and A. Sirlin, Phys. Rev. Lett. **46**, 163 (1981) and Nucl. Phys. **B189**, 442 (1981); C. H. Llewellyn Smith and J. R. Wheater, Phys. Rev. Lett. **B105**, 486 (1981) and Nucl. Phys. **B208**, 27 (1982); M. Davier, Journ de Physique, Suppl. **43**, C3-471 (1982); P. Q. Hung and J. J. Sakurai, Ann. Rev. Nucl. Part. Sci. **31**, 375 (1981); L. M. Sehgal, Aachen preprint 84/03.
[6]H. Georgi and S. L. Glashow, Phys. Rev. Lett. **32**, 438 (1974).
[7]J. C. Pati and A. Salam, Phys. Rev. Lett. **31**, 661 (1973), Phys. Rev. **D8**, 1240 (1973).
[8]For reviews, see P. Langacker, Phys. Rep. **72**, 185 (1981); 1981 Int. Symp. on Lepton and Photon Interactions, ed. W. Pfeil (Bonn, 1981) p. 823; Particles and Fields - 1983, ed. A. Abashian (AIP, N.Y., 1984) p. 251; J. Ellis, 21st Scottish Univ. Summer School, ed. K. C. Bowler and D. G. Sutherland (Edinburgh, 1981) p. 201; 1983 Int. Symp. on Lepton and Photon Interactions (Cornell, 1983); W. J. Marciano, Fourth Workshop on Grand Unification, ed. H. A. Weldon, et al., (Birkhäuser, Boston, 1983) p. 13, J. C. Pati, Maryland preprint 83-188, to be published in Proceedings of ICOMAN, Frascati, 1983; M. Gell-Mann, P. Ramond, and R. Slansky, Rev. Mod. Phys. **50**, 721 (1978).
[9]A. Buras, 1981 Int. Symp. on Lepton and Photon Interactions, p. 636.
[10]G. P. Lepage, 1983 Int. Symp. on Lepton and Photon Interactions, ed. D. G. Cassel and D. L. Kreinick (Cornell, Ithaca, N.Y. 1983), p. 565, and references therein. P. B. MacKenzie and G. P. Lepage, Phys. Rev. Lett. **47**, 1244 (1981).
[11]See, for example, R. M. Barnett, Phys. Rev. **D27**, 98 (1983).
[12]A. Devoto et al., Phys. Rev. **D27**, 508 (1983).
[13]R. Field, 1983 Int. Symp. on Lepton and Photon Interactions, p. 593.
[14]W. J. Marciano, ref. 8, and references therein.
[15]D. V. Nanopoulos and D. A. Ross, Phys Lett. **108B**, 351 (1982); J. Oliensis and M. Fischler, Phys. Rev. **D28**, 194 (1983).
[16]J. Ellis and M. K. Gaillard, Phys. Lett. **88B**, 315 (1979).

[17]For reviews, see F. Sciulli, these proceedings; F. Boehm, Particles and Fields - 1983, p. 1 and Fourth Workshop on Grand Unification, p. 163; M. H. Shaevitz, 1983 Int. Symp. on Lepton and Photon Interactions, p. 132.
[18]V. A. Lubimov et al., Phys. Lett. **94B**, 266 (1980) and Proc. of the Int. Europhysics Conf., Brighton, eds. J. Guy and C. Costain (Rutherford Lab, 1983) p. 386.
[19]See, for example, G. Steigman, these proceedings.
[20]See the talks by J. Preskill and J. Stone, these proceedings.
[21]A. Guth, these proceedings.
[22]For a recent review, see E. W. Kolb and M. S. Turner, Ann. Rev. Nucl. Part. Sci. **33**, 645 (1983).
[23]For reviews, see D. H. Perkins, CERN/EP 84-7; L. R. Sulak, Particles and Fields - 1983, p. 21; E. Fiorini, 1983 Int. Symp. of Lepton and Photon Interactions, p. 405; M. Goldhaber, P. Langacker and R. Slansky, Science **210**, 851 (1980).
[24]M. L. Cherry et al., Phys. Rev. Lett **47**, 1507 (1981).
[25]J. Bartelt et al., Phys. Rev. Lett. **50**, 651 (1983); E. Peterson, Fourth Workshop on Grand Unification, p. 35.
[26]M. R. Krishnaswamy et al., Phys. Lett. **106B**, 339 (1981), **115B**, 349 (1982); B. V. Sreekantan, Fourth Workshop on Grand Unification, p. 25.
[27]G. Battistoni et al., Phys. Lett. **133B**, 454 (1983).
[28]S. Errede, Washington APS meeting, April 1984; B. G. Cortez, et al., Phys. Rev. Lett. **52**, 1092 (1984); R. M. Bionta et al., Phys. Rev. Lett. **51**, 27 (1983).
[29]M. Koshiba, Washington APS meeting, April 1984.
[30]D. Cline, Washington APS Meeting, April 1984.
[31]V. S. Berezinsky, B. L. Ioffe, and Ya. I. Kogan, Phys. Lett. **105B**, 33 (1981).
[32]Y. Tomozawa, Phys. Rev. Lett. **46**, 463 (1981) and Michigan Preprint UMHE 82-45.
[33]N. Isgur and M. B. Wise, Phys. Lett. **117B**, 179 (1982).
[34]W. Lucha, Phys. Lett. **122B**, 381 (1983); W. Lucha and H. Stremnitzer, Z. Phys. **C17**, 229 (1983).
[35]J. F. Donoghue and E. Golowich, Phys. Rev. **D26**, 3092 (1982). The quoted result is from three quark fusion only.
[36]A. W. Thomas and B.H.J. McKellar, Nucl. Phys. **B227**, 206 (1983).
[37]S. Meljanac et al., Nucl. Phys. **B206**, 298 (1982); **B228**, 56 (1983).
[38]Y. Dupont, T. N. Pham and T. N. Truong, Nucl. Phys. **B228**, 77 (1983).
[39]S. J. Brodsky et al., SLAC preprint 3141.
[40]A. N. Mitra and R. Ramanathan, Phys. Lett. **128B**, 381 (1983) and ICTP preprint IC/83/109.
[41]H. Hamber and G. Parisi, Phys. Rev. **D27**, 208 (1983).
[42]C. H. Chang and Y. S. Wu, Phys. Lett. **132B**, 363 (1983).
[43]O. Kaymakcalan, C. H. Lo, and K. C. Wali, Phys. Rev. **D29**, 1962 (1984).
[44]A. S. Goldhaber, T. Goldman, and S. Nussinov, LA-UR-83-3008.
[45]W. J. Marciano, ref. 8, and references therein.
[46]Fermion splitting has been considered by A. Yu. Smirnov, JETP Lett. **31**, 737 (1980) and by P. Frampton and S. L. Glashow, Phys. Lett. **131B**, 340 (1983) and E (to appear); higher order effects and m_b are discussed by S. Nandi, Austin preprint CPT DOE-ER-03992-545. Higgs splitting is considered in K. Hagiwara et al., Wisconsin preprint MAD/PH/142 and ref. 8.
[47]W. J. Marciano, ref. 8. See also G. Lazarides and Q. Shafi, Phys. Rev. **D26**, 1188 (1982); T. Hubsch, S. Pallua, Phys. Lett. **138B**, 279 (1984) and ref. 8.

[48] Recent examples include V. Berezinsky and A. Smirnov, CERN preprint TH-3786; A. Nelson, Phys. Lett. **134B**, 422 (1984); D. Altschüler et al., Phys. Lett. **119B**, 351 (1982); S. Nandi et al., Phys. Lett. **113B**, 165 (1982), and Y. Achiman and J. Keymer, Wuppertal preprint WU B83-16. See also V. Berezinsky and A. Smirnov, Phys. Lett. **97B**, 371 (1980 and ref. 8).

[49] P. Langacker and D. Sahdev, Phys. Rev. **D28**, 2248 (1983); V. A. Kuzmin and M. E. Shaposhnikov, Phys. Lett. **125B**, 449 (1983), and references therein.

[50] T. Schücker, Phys. Lett. **B101**, 321 (1981); S. Nandi, A. Stern, and E.C.G. Sudarshan, Phys. Rev. **D26**, 1653 (1982).

[51] See reference 8 and R. Robinett and J. L. Rosner, Phys. Rev. **D26**, 2388 (1982); Y. Tosa, G. C. Branco, and R. E. Marshak, Phys. Rev. **D28**, 1731 (1983); E. Ma and K. Whisnant, Phys. Lett. **131B**, 343 (1983); D. P. Derendinger et al., CERN preprint TH3770; T. W. Kephart and N. Nakagawa, Purdue preprint TH 84-5.

[52] C. T. Hill, Phys. Lett. **135B**, 47 (1984). Additional solutions and supersymmetric models are also discussed.

[53] Q. Shafi and C. Wetterich, Phys. Rev. Lett. **52**, 875 (1984).

[54] E. Golowich, Phys. Rev. **D24**, 2899 (1981); W. B. Rolnick, Phys. Rev. **D24**, 1434 (1981).

[55] B. A. Campbell, J. Ellis, D. V. Nanopoulos, TH-3787-CERN preprint.

[56] For recent reviews, see J. Polchinski, this proceedings and Fourth Workshop on Grand Unification, p. 325; H. E. Haber and G. L. Kane, Michigan preprint UM-HE-TH 83-17; H. P. Nilles, Geneva preprint UGVA-DPT 1983/12-412; S. Dawson, E. Eichten, and C. Quigg, Fermilab-Pub-83/82-THY.

[57] S. Dimopoulos and H. Georgi, Nucl. Phys. **B193**, 150 (1981); N. Sakai, Zeit. Phys. **C11**, 153 (1981).

[58] L. E. Ibanez and G. G. Ross, Phys. Lett. **105B**, 439 (1981); M. B. Einhorn and D.R.T. Jones, Nucl. Phys. **B196**, 475 (1982); W. Marciano and G. Senjanovic, Phys. Rev. **D25**, 3092 (1982).

[59] D. V. Nanopoulos and D. Ross, Phys. Lett. **118B**, 99 (1982).

[60] V. Berezinsky and A. Smirnov, INR preprint; A. Masiero et al., Phys. Lett. **115B**, 298 (1982); Y. Igarishi et al., Phys. Lett. **116B**, 349 (1982).

[61] C. Kounas, A. B. Lahanas, D. V. Nanopoulos and K. Tamvakis, Phys. Lett. **118B**, 91 (1982) and references therein.

[62] S. Weinberg, Phys. Rev. **D26**, 287 (1982); N. Sakai and T. Yanagida, Nucl. Phys. **B197**, 533 (1982). Four boson operators have been considered by J. P. Derendinger and C. A. Savoy, Phys. Lett. **118B**, 347 (1982) and N. Sakai, Phys. Lett. **121B**, 130 (1983).

[63] There may be comparable contributions from gluino exchange if $m_d \neq m_u$. See J. Milutinovic, P. Pal and G. Senjanovic, Santa Barbara Inst. for Th. Physics preprint, and R. Arnowitt, Fifth Workshop on Grand Unification, Brown Univ., April 1984.

[64] J. Ellis, D. V. Nanopoulos and S. Rudaz, Nucl. Phys. **B202**, 43 (1982); S. Dimopoulos, S. Raby and F. Wilczek, Phys. Lett. **112B**, 133 (1983); W. Lang, Nucl. Phys. **B203**, 277 (1982).

[65] V. M. Belyaev and M. I. Vysotski, Phys. Lett. **127B**, 215 (1983); P. Salati and C. Wallet, Nucl. Phys. **B209**, 389 (1982); W. Lucha, Nucl. Phys. **B221**, 300 (1983); S. Chadha and M. Daniel, Nucl. Phys. **B229**, 105 (1983), Phys. Lett. **137B**, 374 (1984) and Rutherford preprint RL-83-099 (1983); J. Ellis et al., Phys. Lett. **121B**, 123 (1983), and **124B**, 484 (1983); S. J. Brodsky et al., ref. 39; B. A. Campbell et al., ref. 55.

[66] B. Ovrut and S. Raby, Phys. Lett. **138B**, 72 (1984).
[67] For a review, see G. Fidecaro, _Fourth Workshop on Grand Unification_, p. 196.

The Big Bang, the Universe and Everything

Gary Steigman

INTRODUCTION

The size, scope and enthusiastic high quality of this workshop is a testimony to the importance and timeliness of the interdisciplinary research at the frontier of particle physics and cosmology. The melding together of the microscopic with the macroscopic has led to a synergistic assault on fundamental problems yielding new insights and significant progress. As the first speaker at this workshop, it is my intent to survey the cosmic connections between the physics of the very, very small and the physics of the very, very large.

In the first part of my talk I will examine those "Cosmological Puzzles" which have been a focus for our endeavor. I will also explore the "solutions" proposed for many of these puzzles. It is a tribute to our boldness (or, perhaps, a measure of our naivete!) that we've had the courage to address such fundamental issues as, "Why is the hot big bang hot?", "Why is there a Universal baryon asymmetry?", "How did the Universe get to be so smooth, to live so long and to have so much entropy?", "Where are the relic monopoles?", "Why is the cosmological constant so small?"

For the second part of my talk I will concentrate on Primordial Nucleosynthesis. The remarkable success of the predictions of Big Bang Nucleosynthesis provides valuable support (perhaps, the only support!) for our ambitious enterprise. From a comparison between the predicted abundances of the light elements and current observational data we justify our extrapolation to early epochs in the evolution of the Universe and set constraints on quantities of importance in particle physics and cosmology.

COSMOLOGICAL PUZZLES

The following is a selection of those problems on the frontier between particle physics and cosmology whose solution may provide the "Rosetta Stone" of our cooperative quest. To set the stage, I will briefly define the various puzzles. Then, to provide a basis for a clearer appreciation of their significance, I will make a short digression into some fundamental--and very simple--cosmography and cosmology. Having set forth the puzzles, in the following section I will outline some of the proposed solutions.

(i) Baryon Asymmetry: The physics of the low energy world we inhabit and that which we explore at our highest energy accelerators is remarkably symmetric between particles and antiparticles, yet all our astrophysical observations clearly reveal an overwhelmingly asymmetric Universe.[1] On the cosmological scale, no significant amounts of antimatter exist. Equally important, nucleons are rare; averaged over the Universe there is only one nucleon for every one to ten billion microwave (blackbody) photons. How did the Universe get to be asymmetric and why are there so many relic photons for each nucleon?

(ii) Cosmological Constant: In the Einstein equations describing the dynamical evolution of the Universe there is a term involving the cosmological constant Λ. Λ behaves as if it were the energy density of the vacuum. This term is an embarrassment. In many field theories the energy density of the vacuum is infinite and, is ignored. In general relativity, however, all energy gravitates; the vacuum energy density cannot so easily be swept under the rug. Even if the vacuum energy density is formally infinite, the difference in energy density between the various vacua in many spontaneously broken gauge theories is finite--but very large. It may not be unreasonable to identify Λ with such energy density differences. Observations, however, lead to the conclusion that, if not identically zero, Λ is some 50 to 120 orders of magnitude smaller than its "natural" value. Why is Λ so very, very small? Is Λ identically zero?

(iii) Horizon-Homogeneity Puzzle: On the largest scales the Universe is very smooth as is revealed by the distribution of galaxies and radio sources and by the isotropy of the radio, X-ray and microwave background radiations. Yet, in the conventional models for the dynamical evolution of the Universe, the radiation observed today was emitted from regions which had never been in contact with each other. The finite speed of light leads to causal horizons which guarantee that if the large-scale Universe had been inhomogeneous early on, no physical processes could have produced the observed homogeneity. How, then, did the Universe arrange itself to be so very smooth on large scales?

(iv) Age-Flatness-Entropy Puzzle: The Universe is very old ($t_0 \gtrsim 10 \times 10^9$ yr). To put this in some perspective it is useful to recall that Newton's gravitational constant may be combined with Planck's constant and the speed of light to obtain a "natural" timescale for cosmology: $t_P = (Gh/c^5)^{1/2} \approx 10^{-43}$s. In "natural" units then, $t_0 \gtrsim 10^{60} t_P$. This old age for the Universe would, perhaps, not be so remarkable were it not for another aspect of the equation describing the dynamical evolution of the Universe. There is a term in that equation which depends on the 3-space curvature. For times later than t_{CD} or temperatures smaller than T_{CD}, the Universe becomes "Curvature-Dominated". It might be expected that $t_{CD} \approx t_P$ ($T_{CD} \approx T_P$). However, even the present Universe may not yet have become Curvature-Dominated: $t_{CD} \gtrsim 10^{60} t_P$, $T_{CD} \lesssim 10^{-30} T_P$. How, then, is it that our Universe has lived so long and remained so "flat"?

Intimately related to the Age-Flatness Puzzle is the question of the enormous entropy of the Universe. We have already noted that the entropy per baryon (which is, up to numerical factors of order unity, the photon-to-nucleon ratio) is large ($\sim 10^9$-10^{10}). A dimensionless measure of the universal entropy which--in the absence of entropy generating processes--is epoch independent is the number of relic photons per unit co-moving volume $\mathcal{N}$. A lower bound to $\mathcal{N}$ comes from counting up all the microwave photons in the presently observable Universe ($\ell_0 \approx ct_0 \approx 10^{28}$cm): $\mathcal{N} \gtrsim 10^{90}$. A "natural" guess at the expected value of this dimensionless, scale independent constant might have been $\mathcal{N} \approx 0(1)$. Why is $\mathcal{N}$ so very, very large?

(v) Monopole Problem: In spontaneously broken gauge theories whenever a U(1) factor appears, superheavy magnetic monopoles exist.[2,3] In conventional Grand Unified Theories (GUTs), a U(1) factor appears at the GUT scale ($T_{GUT} \approx 10^{-4} T_P$) leading to monopoles whose mass is $M_M \approx \alpha_{GUT}^{-1} T_{GUT} \approx 10^{-3} T_P \approx 10^{16} M_N$ (where M_N is the nucleon mass). Estimates of the relic abundance of such superheavy monopoles[4,5,6] are large, roughly one monopole for every thousand or so nucleons. The energy density in such superheavy monopoles is so large that very early on the Universe becomes monopole dominated. Such a Universe would have evolved very rapidly and would be only ~ 30,000 years old when the temperature of the relic photons fell to the presently observed value of ~3K. Clearly, the relic monopoles have gone away or, perhaps, they never came. Where are the relic monopoles?

To better appreciate the significance of some of the problems outlined above, it will be worthwhile to review some very simple results concerning the geometry and dynamics of the Universe. The observed isotropy and homogeneity of the Universe leads to a unique choice of metric, the Robertson-Walker metric.

$$ds^2 = (cdt)^2 - a^2(t)\,[\frac{dr^2}{1-kr^2} + r^2\,(d\theta^2 + \sin^2\theta d\phi^2)]. \qquad (1)$$

In Eq. (1), r, θ, ϕ are co-moving co-ordinates; they expand along with the general expansion of the Universe which is described by the evolution of the scale factor, a(t); k is the dimensionless, 3-space curvature. It is convenient to introduce a new co-moving radial co-ordinate $\Theta = \Theta(r)$, where $d\Theta^2 = |k|dr^2\,(1-kr^2)^{-1}$. Along with Θ, we may define a renormalized scale factor $R = R(t) = a(t)|k|^{-1/2}$ so that the metric may be written equivalently as,

$$ds^2 = (cdt)^2 - R^2(t)[d\Theta^2+\ldots]. \qquad (1')$$

From the form of the metric it follows that a photon emitted at t = 0 will, by time t, have travelled a co-moving radial distance,

$$\Theta_H(t) = \int_0^t \frac{cdt}{R(t)} \qquad (2)$$

Θ_H is the co-moving radial co-ordinate of the "causal" horizon; physical events cannot, by time t, have had their effects propagate over larger distances.

It also follows from the metric that the "physical" volume at time t out (from the origin at Θ = 0) to co-moving coordinate Θ is of the form,

$$V(\Theta,t) = R^3(t)\;\Phi(\Theta). \qquad (3)$$

Φ(Θ) depends on the sign of the curvature k; for $\Theta \ll 1$, $\Phi \to 4\pi\Theta^3/3$ independent of k. The volume within the causal horizon is: $V_H(t) = V(\Theta_H(t),\, t)$.

If the Robertson-Walker metric is used in the Einstein equations, those equations governing the dynamics of the evolution of the Universe (i.e., of the scale factor, R(t)) emerge.

$$H^2 = [\frac{1}{R}(\frac{dR}{dt})]^2 = \frac{8\pi G\rho}{3} \pm (\frac{c}{R})^2 + \frac{\Lambda}{3}. \qquad (4)$$

In Eq. (4), H = H(t) is the Hubble parameter, $\rho = \rho(t)$ is the mass density and Λ is the cosmological constant. The second term on the right hand side of Eq. (4) is the effect of the 3-space curvature. Since there are two unknown functions of time in Eq. (4), R(t) and ρ(t), another equation is required. That equation, which follows from the conservation of the stress-energy tensor, may be written in a familiar form.

$$S = \left(\frac{P + \varepsilon}{T}\right)V = \left(\frac{P + \varepsilon}{T}\right) R^3(t)\ \Phi\ (\Theta). \tag{5}$$

$S = S(\Theta,t)$ is the entropy in the volume V; P is the pressure and $\varepsilon = \rho c^2$ is the energy density; T is the temperature. S is clearly extensive: $S \propto V$. Now, for a Universe dominated by extremely relativistic (ER) particles--a "Radiation-Dominated" Universe: $P + \varepsilon = 4\varepsilon/3 \approx 4Tn$, where n is the number density of the ER particles (e.g., photons). It follows from Eq. (5) that

$$S \approx 4\ N \equiv 4\mathcal{N}\,\Phi(\Theta)\ , \tag{6a}$$

and

$$\mathcal{N} \equiv [R(t)T(t)]^3 \tag{6b}$$

N is the number of ER particles in the physical volume V; $\mathcal{N}$ is the number of ER particles per unit co-moving volume. From conservation of the stress-energy tensor, S = constant. Therefore, $\mathcal{N}$ is independent of Θ and t; it is a dimensionless number which defines the cosmological model. For example, in a closed (k>0) model, $\Theta_{MAX} = \pi$ and $\Phi_{TOT} = 2\pi^2$ so that, up to factors of order unity (or, ten), $\mathcal{N}$ is the total number of ER particles in the Universe.

It is useful to rewrite Eq. (4) using $\mathcal{N}$ and "Planck" units t_P and T_P (where $kT_P = M_Pc^2 = (\hbar c^5/G)^{1/2} \approx 1.2 \times 10^{19}$GeV). Ignoring some numerical factors of order unity,

$$(Ht_P)^2 \approx \left(\frac{T}{T_P}\right)^4 \pm \mathcal{N}^{-2/3} \left(\frac{T}{T_P}\right)^2 + \left(\frac{T_V}{T_P}\right)^4 . \tag{7}$$

In Eq. (7) we have used $\rho_{ER} \propto T^4$ and have written $\Lambda \equiv t_\Lambda^{-2} = t_P^{-2}\ (T_V/T_P)^4$.

It is now easy to see why it was claimed earlier that Λ is "unnaturally" small. Current data suggests that $\Lambda \lesssim H_0^2$ where H_0 is the present value of the Hubble parameter ($H_0^{-1} \approx$ 10-20 billion years) so that $t_\Lambda \gtrsim 10^{60}\ t_P$ or, $\Lambda t_P^2 \lesssim 10^{-120}$. The "temperature" of the vacuum $T_V \lesssim 10^{-30}T_P$. In the subsequent discussion the last term on the right hand side of Eq. (7), the Λ-term, will be ignored.

Comparing the first two terms on the right hand side of Eq. (7), it is clear that as the Universe expands (R increases) and cools (T decreases), the Universe becomes "Curvature-Dominated" (CD) for $T \lesssim T_{CD} = \mathcal{N}^{-1/3}\ T_P$. Current data are inadequate to determine whether or not by the present epoch ($T_0 \approx 3K$) the Universe has become CD so that, $T_{CD} \lesssim 3K$. It therefore follows that $\mathcal{N} \gtrsim 10^{95}$; $\mathcal{N}$ or, equivalently, the entropy is enormous.

The effect of this enormous entropy is to guarantee that our Universe is "old" and "flat". To obtain a dimensionless measure of the curvature, compare the "Hubble distance" c/H with the scale factor R,

$$\left(\frac{c/H}{R}\right)^2 = |\Omega - 1| = [\mathcal{N}^{2/3} \left(\frac{T}{T_P}\right)^2 \pm 1]^{-1} , \tag{8a}$$

$$\Omega = \rho/\rho_c \; ; \; \rho_c = \frac{3H^2}{8\pi G} . \tag{8b}$$

ρ_c is the "critical" density and Ω is the dimensionless density, the ratio of the actual density to its critical value (which depends, through H, on how fast the Universe is expanding). If we evaluate this dimensionless measure of the curvature at the Planck epoch,

$$\left(\frac{c/H}{R}\right)^2_P = |\Omega_P - 1| = (\mathcal{N}^{2/3} \pm 1)^{-1} \approx (t_{CD}/t_P \pm 1)^{-1} . \tag{9}$$

These simple algebraic manipulations permit us to see more clearly the intimate connection among the age-flatness-entropy puzzles; they are several aspects of one puzzle. Since the entropy is so very large ($\mathcal{N} \gtrsim 10^{95}$), the Universe is very old ($t_0 \gtrsim t_{CD} \gtrsim 10^{60}\, t_P$). The early Universe was incredibly flat $(c/HR)_P \lesssim 10^{-30}$; equivalently, the density at that epoch differed from the critical value by, at most, an incredibly tiny amount: $\Omega_P - 1 \lesssim \pm 10^{-60}$.

Finally let us return briefly to the horizon-homogeneity puzzle. Consider a source which emits radiation at t_e ($< t_0$) which is received today; the co-moving distance to the source is

$$\Theta = \int_{t_e}^{t_0} \frac{c\,dt}{R(t)} . \tag{10}$$

Two such sources in opposite directions on the sky are separated (in <u>co-moving</u> coordinates) by $\Delta\Theta \approx 2\Theta$. We are interested in knowing if information could have been exchanged by such sources <u>prior</u> to the time the radiation we've received was emitted. The co-moving distance over which information could have been transmitted is the horizon size at t_e.

$$\Theta_H\,(t_e) = \int_0^{t_e} \frac{c\,dt}{R(t)} . \tag{11}$$

If $\Delta\Theta > \Theta_H\,(t_e)$ then no information could have been exchanged prior to emission. A simple example, adequate for illustrative purposes, will suffice. Consider a "Matter-Dominated" ($\rho \propto R^{-3}$), Einstein-deSitter ($\rho = \rho_c$) Universe.

$$\frac{R}{R_0} = \left(\frac{t}{t_0}\right)^{2/3} \; ; \; \frac{R}{R_0} = \frac{1}{1+z} . \tag{12}$$

z is the redshift of the emitting source. It is then easy to see that

$$\Delta\Theta/\Theta_H = 2[(1+z)^{1/2} - 1]. \tag{13}$$

For $z \gtrsim 5/4$, $\Delta\Theta > \Theta_H$ so that information could not have been exchanged between large redshift QSOs or the sites of the last scattering of the relic photons ($z \approx 10^3$). Indeed, roughly 3×10^4 horizon volumes from the epoch of recombination could fit within the present horizon. With $\Theta_H(t_e) << \Delta\Theta_H$, how did the Universe get to be so smooth?

COSMOLOGICAL SOLUTIONS

If the current concerted assault on problems of particle physics and cosmology is to be taken seriously, the puzzles outlined above must be tackled, if not solved. To date, the reviews are mixed concerning the success of our interdisciplinary enterprise. First, the good news.

(i) Baryon Asymmetry: During the early evolution of the Universe nucleons and antinucleons (or, their constituent quarks and antiquarks) were abundant: $n_N \approx n_{\bar{N}} \approx n_\gamma$; in every cubic centimeter were roughly equal numbers of nucleons, antinucleons and blackbody photons. Today, however, $n_N << n_{\bar{N}} \approx n_\gamma$. How did the baryon asymmetry develop and, why are there so few nucleons (equivalently, why is the big bang hot)? The solution was proposed in a prescient article by Sakharov.[7] Sakharov's recipe requires three ingredients: Baryon Number conservation must be violated (clearly!); C and CP conservation must be violated (even if baryon number is not conserved, reaction rates or decay rates of particles and antiparticles may exactly compensate); departure from equilibrium is required for baryon number violating interactions (lest, by detailed balance, "effective" baryon conservation be restored). The union of particle physics and cosmology provides the necessary ingredients. In most GUTs, baryon number is not conserved (in unifying the weak, electromagnetic and strong interactions, transformations of baryons into leptons and, vice-versa, become possible). GUTs also provide for C and CP nonconservation; here it should be noted that in many GUTs, the <u>amount</u> of C and CP violation is insufficient to account for the observed baryon asymmetry. Non-equilibrium is guaranteed by the evolution of the Universe. For example, in the "out-of-equilibrium decay scenario",[8] the baryon asymmetry is generated by the baryon nonconserving decays of a superheavy gauge (or Higgs) boson, X. For $T >> M_X$, the Xs are in equilibrium ($n_X \approx n_\gamma$) while for $T << M_X$, the reaction (and/or decay) rates are too slow compared to the universal expansion rate for equilibrium to be maintained ($\exp(-M_X/kT) << n_X/n_\gamma << 1$).

Through the marriage of microscopic particle physics with macroscopic cosmology there emerges a solution to the baryon asymmetry puzzle. The Universe is not matter-antimatter symmetric because the basic particle physics is not. The evolution of the Universe ensures that this fact will be exploited when the appropriate reactions fall out of equilibrium. Qualitatively, this solution is a spectacular success. Quantitatively, the situation is less clear. The baryon-to-photon ratio, $\eta = n_N/n_\gamma$ is determined locally, by microscopic physics (i.e., η depends on the amount of C and CP violation, the mass and lifetime of the X, etc.). The observed value of $\eta \approx 10^{-10} - 10^{-9}$ can be "postdicted" in several GUTs but, there are at present, too many adjustable parameters to claim predictive power for η. For an excellent review of these and related issues, see Kolb and Turner[9] and references therein.

(iii) - (v) Horizon/Homogeneity - Age/Flatness/Entropy - Monopole Puzzles: In one fell swoop, Alan Guth[10] proposed the solution to all these puzzles: Inflation. The basic idea is that very early on, when the GUT symmetry is broken, the Universe temporarily remains trapped in the "symmetric" phase and its energy density is dominated by the energy density of the "false" vacuum. Since ρ_V = constant, the Universe expands exponentially (inflates!); see Eq. (4). Eventually (after many e-foldings) the Universe makes the transition to the "true" vacuum and the vacuum energy ("latent heat") is released, reheating the Universe to $T \approx O(T_{GUT})$.

Inflation solves the horizon/homogeneity puzzle since the exponential growth of the scale factor permits a tiny, causally connected ($\Theta << \Theta_H$) region to grow sufficiently to encompass the entire, presently observable Universe. Furthermore after reheating, when $T \approx O(T_{GUT})$, since the scale factor is exponentially larger than it was when the Universe first cooled below T_{GUT}, the entropy, $\mathcal{N} = (RT)^3$, is enormous; see Eq. (6b). As discussed earlier, the age/flatness/entropy puzzles are different aspects of one puzzle. By providing a mechanism for the generation of huge amounts of entropy, Guth's inflationary scenario automatically solves the age/flatness puzzles as well.

At this point, a cautionary note is called for. As attractive as the inflationary scenario is, it is not the only possible solution to the above puzzles. Any mechanism which, sufficiently early on, releases enormous amounts of entropy will--de facto--"solve" the age/flatness/entropy puzzles. For example, it would not be surprising to learn (if a satisfactory theory of quantum gravity is ever developed) that the Universe emerges from the Planck epoch ($t \lesssim t_P$; $T \gtrsim T_P$) with $\mathcal{N} \gtrsim 10^{90}$. Furthermore, if the mechanism for generating the large entropy is a local one, depending only on the microphysics, then $\mathcal{N}$ will be the same everywhere in the Universe. With the entropy ($\mathcal{N}$) and the entropy per baryon ($\propto \eta^{-1}$) both determined by the microphysics, the horizon/homogeneity puzzle is solved. Inflation is one--but, perhaps not the only--solution to these puzzles.

Finally, to the monopole problem. If inflation occurs after the monopoles exist, the enormous production of entropy dilutes the relic abundance of monopoles. Indeed, if the presently observable Universe inflated from one, causally connected region, then there will be O(1) monopoles in the entire Universe!

It is, by now, well known that the inflationary model originally proposed by Guth[10] fails.[11,12] In its place, we have "New Inflation".[13,14] Even this scenario has its problems but, since a significant part of this workshop is devoted to "Inflation", it would be inappropriate for me to linger here any longer. Suffice it to say that the successes of the inflationary scenario are so striking that it is generally accepted that some variation on the same theme will underlie the "true theory of the world".

(ii) Cosmological Constant: And now for the bad news. There is, at present, no satisfactory explanation of the extreme smallness of the cosmological constant. The inflationary models exploit the energy difference between various vacua but do not determine the energy density of the $T \approx 0$ vacuum. Why it is that ρ_V (at $T \approx 0$) is so many orders of magnitude smaller than any of the $\Delta\rho_V$, remains a puzzle. To be sure, there have been explanations which appeal to quantum gravity or to supersymmetry/supergravity. At present, these seem equivalent to appeals to the "Tooth Fairy" or to "Asparagus".

PRIMORDIAL NUCLEOSYNTHESIS

The first part of this talk has been devoted to a general overview of several of the issues of fundamental importance to particle physics and cosmology. For the remainder of the talk I intend to concentrate, in some detail, on one specific issue: Big Bang Nucleosynthesis. The synthesis of the light elements during the early evolution of the Universe provides a (the only!) probe of the early Universe. Through a step-by-step comparison of the predictions of big bang nucleosynthesis with the abundances of the light elements inferred from observational data, several goals may be accomplished. Most fundamental is the goal of establishing the validity of the standard model. The observed abundances (relative to hydrogen) of deuterium, helium-3, helium-4 and lithium-7 range over nearly ten orders of magnitude. That the standard model predicts abundances in agreement with those observed is a remarkable achievement and lends credence to our extrapolations from present conditions (age ~ 10 billion years, temperature ~ 3K, density ~ 10^{-30}gcm^{-3}) to those at the time of element synthesis (age ~ minutes, temperature ~ 10^{10}K, density ~ 10^{-2}gcm^{-3}) and beyond to the conditions of the very early Universe of most interest to particle physics. Having established, qualitatively, the success of the standard model, we may proceed to a detailed quantitative comparison from which we may derive limits to the nucleon abundance ($\eta \equiv N/\gamma$), the number of light, weakly interacting particles (N_ν) and, to other quantities of interest to particle physics and cosmology (e.g., anisotropy, inhomogeneity, neutrino degeneracy, etc.).

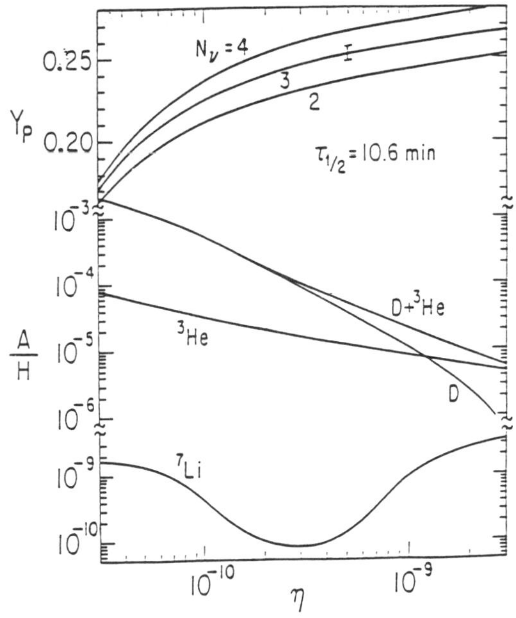

Figure 1: The predicted abundances[20] of D, ^{3}He and ^{7}Li (by number relative to H) and ^{4}He (by mass Y_p) as a function of nucleon-to-photon ratio η for $\tau_{1/2} = 10.6$ min.; for ^{4}He the predictions for $N_\nu = 2, 3, 4$ are shown and the size of the "error" bar shows the range in Y_p which corresponds to $10.4 < \tau_{1/2} < 10.8$ min.

Since all of the material germane to this discussion has appeared elsewhere in print, I will provide a somewhat detailed overview of the results and conclusions; I urge those interested to find the details and further references in the relevant articles.[15-21]

The predictions of primordial nucleosynthesis in the standard model are summarized in Figure 1 and Table I taken from Yang et al.[20] It is seen that, for the range of nucleon abundance of interest (η ~ few × 10^{-11} to a few × 10^{-9}), the predicted abundance of the light elements cover a range of roughly nine to ten orders of magnitude (as do the observed abundances!). The predicted abundance of ^{7}Li (^{7}Li/H ~ 10^{-10} - 10^{-9}) has the characteristic "valley" shape because, as η increases at low η ($\lesssim 2$-3×10^{-10}), ^{7}Li is

TABLE I: Primordial Abundances*of D, ^{3}He, ^{4}He and ^{7}Li

$10^{10}\eta$	10^5(D/H)	10^5(D+^{3}He)/H	10^{10}(^{7}Li/H)	Y_p
1	49	53	4.4	0.225
1.5	25	28	1.8	0.233
2	16	18	1.1	0.237
3	8.1	9.7	0.76	0.243
4	5.1	6.5	1.0	0.246
5	3.6	4.8	1.7	0.248
6	2.7	3.8	2.7	0.250
7	2.1	3.1	3.9	0.252
8	1.7	2.6	5.3	0.253
9	1.4	2.3	6.9	0.254
10	1.1	2.0	8.6	0.255
15	0.48	1.2	17	0.258
20	0.23	0.87	25	0.261

* For N_ν=3 and $\tau_{1/2}$ = 10.6 min.

being burned away whereas at higher η, ^{7}Be is being formed and, before being destroyed, e$^-$ captures to ^{7}Li. The abundances of D and ^{3}He ($\sim 10^{-5}$ - 10^{-3}) decrease with increasing nucleon abundance because D and ^{3}He are being burned to ^{4}He. Note that the abundances of D and ^{3}He differ from that of ^{7}Li by some 5($\pm$1) orders of magnitude over the interesting range of η (as do the observed abundances!).

In contrast to the abundances of the other light elements, that of ^{4}He varies much more slowly with changing nucleon abundance. Over some two orders of magnitude in η, the helium-4 mass fraction (Y_P) ranges from ~0.2 to ~0.3. The predicted abundance of ^{4}He depends on the neutron-to-proton ratio at the time nucleosynthesis begins in earnest. Y_P, then, is a probe of a competition between the early expansion rate (which depends on the total mass-energy density) and the weak interaction rate (which depends on the neutron half-life, $\tau_{1/2}$). The early energy density is dominated by relativistic particles (e.g., light (<<MeV) or massless neutrinos) so that the expansion rate at the time of nucleosynthesis will increase with increasing numbers of relativistic particles (N_ν). The faster the Universe expands, the more neutrons survive to be incorporated in ^{4}He. The effect of 2-, 3- and 4-neutrino "flavors" is illustrated in Figure 1; the results in Table I are for the "standard" model (N_ν = 3). Since it is the competition between the expansion rate and the weak interaction rate which is of importance, uncertainties in $\tau_{1/2}$ are reflected in uncertainties in the predicted helium-4 mass fraction. The "error" bar in Figure 1 shows the effect of

an uncertainty in $\tau_{1/2}$ of ±0.2 min.; the entries in Table I are for $\tau_{1/2}$ = 10.6 min. Note that (as observed!) there are some four orders of magnitude between the ^4He and D abundances.

After presenting a brief summary of the best estimates--at present--of the abundances of D, ^3He and ^7Li, I will turn to a comparison of the inferred primordial abundances with the predictions of the standard model. From the overall, qualitative success of this comparison I will proceed to derive quantitative constraints on the nucleon abundance (η) which will lead to constraints on the predicted helium-4 mass fraction (Y_P). Comparing Y_P with the primordial abundance inferred from recent data will provide a check on the concordance of the standard model and constraints on deviations from the standard model (e.g., a limit to N_ν).

Abundances of D, ^3He, ^7Li: From studies of solar wind, meteoritic and lunar material, it is possible to infer the presolar abundances of deuterium, helium-3 and lithium-7. Of greatest cosmological value are the derived abundances of D and ^3He. The gas rich meteorites and the solar wind provide a sample of material in which the presolar deuterium has been burned to helium-3. In such material, the measured abundance of ^3He reflects the presolar abundance of D plus ^3He. From the best data,[22-24]

$$\left(\frac{D + {}^3He}{{}^4He}\right)_\odot = (4.03 \pm 0.19) \times 10^{-4}. \tag{14}$$

In contrast, the carbonaceous chondrites are such primitive material that, it is generally believed, the lowest observed $^3He/^4He$ ratio represents the presolar abundance of ^3He, uncontaminated by deuterium burned to helium-3. The most accurate data[25,26] suggests,

$$({}^3He/{}^4He)_\odot = (1.54 \pm 0.05) \times 10^{-4}. \tag{15}$$

If Eq. (15) is subtracted from Eq. (14), an indirect estimate of the presolar abundance of deuterium is obtained.

$$(D/{}^4He)_\odot = (2.5 \pm 0.2) \times 10^{-4}. \tag{16}$$

To convert the above ratios relative to ^4He to abundances relative to H requires a knowledge of the solar abundance of ^4He. Unfortunately, at present there are large uncertainties in $y_\odot \equiv (^4He/H)_\odot$. Spectroscopically, Heasley and Milkey[27] find $y_\odot = 0.10 \pm 0.025$ ($Y_\odot = 0.28 \pm 0.05$). Indirect estimates follow from comparisons of solar models with solar observations; Bahcall et al.[28] derive $Y_\odot = 0.25 \pm 0.01$ ($0.081 \leq y_\odot \leq 0.090$) and Gough[29] finds $Y_\odot = 0.25 \pm 0.02$ ($0.076 \leq y_\odot \leq 0.095$). If we adopt $0.08 \lesssim y_\odot \lesssim 0.10$, then we derive the following lower bound to the deuterium abundance and upper bound to the $D+^3He$ abundance.

$$(D/H)_\odot \gtrsim (2.0 \pm 0.2) \times 10^{-5} \tag{17a}$$

$$[(D + {}^3He)/H]_\odot \lesssim (4.0 \pm 0.2) \times 10^{-5}. \tag{17b}$$

These abundances are consistent with those derived from observations of the giant planets[30,31] and spectroscopic data on ^3He in the Sun.[32] At the "3σ" level it seems likely that $(D/H)_{\odot} \gtrsim 1 \times 10^{-5}$ and $[(D + {}^3He)/H]_{\odot} \lesssim 5 \times 10^{-5}$.

In passing, I note that the abundance of lithium derived from meteoritic data[33] is comparable to that found in Pop I stars[34,35] and in the interstellar gas.[36] To within a factor of 2 or so, $({}^7Li/H)_{\odot} \approx ({}^7Li/H)_{PopI} \approx ({}^7Li/H)_{ISM} \approx 1 \times 10^{-9}$. The lithium abundance in the Sun is much below that in the meteorites since, in the last 4.5 billion years, much of the presolar lithium has been burned away. Since some stars destroy lithium and others produce it, the "Pop I" abundance may provide more information about galactic chemical evolution and less about the primordial abundance of lithium; more about this issue shortly.

The status of the interstellar abundances of D and ^3He are, at present, confused. Current data on deuterium in the relatively local ISM spans a range too large to be accounted for by the claimed statistical accuracy of the data: $(D/H)_{ISM} \approx (1/4 - 4) \times 10^{-5}$. The "best" data suggests $\langle D/H\rangle_{ISM} = 2 \times 10^{-5}$ (±25%).[37,38] Having noted time variations in some of the data, Vidal-Madjar et al.[39] suggest that much of the previously "observed" deuterium has, in fact, been hydrogen at the "wrong" velocity. Vidal-Madjar et al.[39] suggest, therefore, that $(D/H)_{ISM} \lesssim 1/2 \times 10^{-5}$. As for ^3He, it is expected[40] that $({}^3He/H)_{ISM} > ({}^3He/H)_{\odot}$ due to ^3He production in stars of ~ 1-2 $M_{\odot}$. Rood et al.[41] have detected ^3He in three HII regions and (W3, W43, W51) and derive abundances which cover a surprisingly large range: $4 \lesssim 10^5 \, ({}^3He/H) \lesssim 20$. For these regions the helium-3 abundance exceeds that in the presolar nebula by factors ranging from ~2 to an order of magnitude (see Eq. (17)). However, in two HII regions (W49, M17), Rood et al.[41] can only set upper limits to the ^3He abundance ($^3He/H \lesssim 2 \times 10^{-5}$) which are comparable to the presolar abundance. Further observations are required to help clarify this issue (has the abundance of ^3He increased or decreased since the formation of the solar system ...the galaxy?).

Since the primordial abundances of D, ^3He and ^7Li have been affected by subsequent processing of material during the course of galactic evolution, it is our task to use the current data to infer the primordial abundances. It is virtually certain that deuterium has decreased during the history of the galaxy, so that $(D/H)_P \gtrsim (D/H)_{\odot} \gtrsim (D/H)_{ISM}$. From the solar system and interstellar data discussed earlier, it is likely that

$$(D/H)_P \gtrsim (1\text{-}2) \times 10^{-5}. \tag{18}$$

The situations regarding ^3He and ^7Li are less clear since they are burned away in some stars and produced in others. Yang et al.[20] have noted that since deuterium is burned to helium-3 and, not all ^3He is destroyed by stars, the presolar abundances of D and ^3He provide an upper bound to the sum of the primordial abundances of D and ^3He.

$$\left(\frac{D + {}^3He}{H}\right)_P \leq \left(\frac{D + {}^3He}{H}\right)_{\odot} + \left(\frac{1}{g} - 1\right)\left(\frac{{}^3He}{H}\right)_{\odot} . \tag{19}$$

In Eq. (19), g is the fraction of ^3He which survives stellar processing. Yang et al.[20] estimate that $g \gtrsim 1/4 - 1/2$ so that the likely upper bounds to the sum of the abundances are

$$\left(\frac{D + {}^3He}{H}\right)_P \lesssim (6\text{-}10) \times 10^{-5}. \tag{20}$$

Concerning the primordial abundance of lithium-7, very interesting --and important--developments have occurred recently. Spite and Spite[42,43] have observed lithium in the atmospheres of halo and old disk stars - Pop II stars. From their data they derive

$$({}^7Li/H)_P \approx ({}^7Li/H)_{PopII} \approx (1.1 \pm 0.4) \times 10^{10}. \tag{21}$$

Their observations suggest a primordial abundance which is an order of magnitude below that derived for the presolar nebula and/or the present interstellar gas; the bulk of the "Pop I" abundance of lithium must have been synthesized during the course of galactic evolution.

Comparison with the Standard Model: Our best estimates for the primordial abundances (or, bounds to them) for D, ^3He and ^7Li are contained in Eqs. (18), (20) and (21). Let us now compare these estimates with the abundances predicted by the standard model ($N_\nu = 3$, $\tau_{1/2} = 10.6$ min) as summarized in Figure 1 and Table I. From the lower bound to the deuterium abundance, we derive an upper bound to η, the nucleon-to-photon ratio: $\eta_{10} \lesssim 7\text{-}10$. Notice, however, that for $\eta_{10} \gtrsim 5$, too much lithium-7 is produced during the big bang. Now, it may be the case that $({}^7Li/H)_P > ({}^7Li/H)_{PopII}$ if there was some destruction of lithium prior to the formation of the stars observed by the Spites.[42,43] However, since deuterium is burned away at temperatures lower than those which destroy lithium, it is clear that $({}^7Li/D)_P \leq ({}^7Li/D)_{PopII}$. Of course, the Pop II abundance of deuterium is unknown, but $(D/H)_{PopII} \geq (D/H)_\odot$ so that $({}^7Li/D)_P \lesssim 2 \times 10^{-5}$. From Table I this suggests that $\eta_{10} \lesssim 7$. The Pop II abundance of lithium leads to estimates of the primordial abundance which are consistent with the standard model for $1.5 \lesssim \eta_{10} \lesssim 5\text{-}7$. Now, however, there is a problem at the lower end of the range for η. For low η, the primordial abundances of D and ^3He are very large. Primordial deuterium would, today, manifest itself as deuterium and/or helium-3. To avoid overproducing D plus ^3He requires: $\eta_{10} \gtrsim 3\text{-}4$.

To summarize, the standard model can account for the primordial abundances of D, ^3He and ^7Li as inferred from current observational data. There is a relatively narrow range of nucleon abundance which is consistent with the derived abundances

$$\text{D:} \quad \eta_{10} \lesssim 7\text{-}10, \tag{22a}$$

$$\text{D} + {}^3\text{He:} \quad \eta_{10} \gtrsim 3\text{-}4, \tag{22b}$$

$${}^7\text{Li:} \quad \eta_{10} \lesssim 5\text{-}7. \tag{22c}$$

If the lithium data is ignored, $3\text{-}4 \lesssim \eta_{10} \lesssim 7\text{-}10$. Including lithium narrows the range, $3\text{-}4 \lesssim \eta_{10} \lesssim 5\text{-}7$. Note that for η in these ranges, the standard model produces helium-4 in an amount whose mass fraction is $Y_P \approx 0.24\text{-}0.25$.

Consistency with ^4He?: The current status of the primordial abundance of ^4He as derived from recent observational data has been reviewed by Pagel,[15-17] by Steigman,[18,19] and by Yang et al.[20] and a comprehensive discussion of the crucial issues may be found in the Proceedings of the ESO Workshop (1983) and in Kinman and Davidson.[44] A primordial mass fraction $Y_P = 0.24 \pm 0.02$ is consistent with all current high quality data. For the highest quality data, the uncertainty in Y_P may be reduced by, perhaps, a factor of two: $0.22\text{-}0.23 \lesssim Y_P \lesssim 0.25\text{-}0.26$.

In comparing with Table I, it is clear that $Y_P \lesssim 0.25$ is consistent with the standard model for $N_\nu = 3$ and $\tau_{1/2} \gtrsim 10.4$ min. for $\eta_{10} \lesssim 7$ (for $\tau_{1/2} \gtrsim 10.4$ min., $\lesssim 0.003$ should be subtracted from the entries in Table I). Indeed, given the uncertainty in the third decimal place, even $\eta_{10} \lesssim 10$ may be consistent with the inferred primordial abundance. To the accuracy of current data, there is complete concordance between the predictions of the standard model ($N_\nu = 3$, $\tau_{1/2} = 10.6 \pm 0.2$ min.) and the observed abundances of D, ^{3}He, ^{7}Li and ^{4}He provided that $3\text{-}4 \lesssim \eta_{10} \lesssim 7\text{-}10$. The Spites'[42,43] observations of ^{7}Li as well as the best data on ^{4}He suggest that the upper limit on the nucleon abundance should be reduced somewhat: $\eta_{10} \lesssim 5\text{-}7$.

The sensitivity of the predicted abundance of ^{4}He to the number of neutrino flavors (see Figure 1) offers a challenge to the standard model; the model is potentially falsifiable. Although the τ-neutrino could, in principle, be heavy (>> 1 MeV), it is already known that the e- and μ-neutrinos are light (<< 1 Mev) so that $N_\nu \geqq 2$. For $N_\nu \geqq 2$, $\tau_{1/2} \gtrsim 10.4$ min. and $\eta_{10} \gtrsim 3$, the predicted primordial mass fraction of ^{4}He exceeds 0.23. If future observations should ever establish that the primordial abundance were less than 0.23, the standard model would be in trouble. At present, no such contradiction between theory and observation exists. However, such a conflict would materialize if there are additional flavors of neutrinos (or, other light "inos"). For $N_\nu = 4$ (and $\tau_{1/2} \gtrsim 10.4$ min., $\eta_{10} \gtrsim 3$), the predicted abundance of ^{4}He increases to $Y_P > 0.252$. Given the uncertainty (observationally) in the third decimal place, this is not inconsistent with current data. Thus, $N_\nu \lesssim 4$ is (barely) permitted.

Cosmological Constraints: We have seen that the predictions of primordial nucleosynthesis in the standard model ($N_\nu = 3 \pm 1$, $\tau_{1/2} = 10.6 \pm 0.2$ min.) are completely consistent with the primordial abundances of the light elements as inferred from current observational data provided that the nucleon abundance lies in the range $3\text{-}4 \lesssim \eta_{10} \lesssim 7\text{-}10$. This narrow range for η firmly establishes the result that the universal mass density in nucleons is small.

$$\Omega_N = \rho_N/\rho_c = 0.00353 h_0^{-2} \theta^3 \eta_{10}, \tag{23}$$

where h_0 is the present value of the Hubble parameter in units of 100 kms^{-1}Mpc^{-1} ($1/2 \lesssim h_0 \lesssim 1$) and θ is the present temperature of the microwave radiation in units of 2.7K ($1 \lesssim \theta \lesssim 1.1$). For the allowed ranges of values for h_0, θ and η_{10}, Ω_N is bounded by: $0.01 \lesssim \Omega_N \lesssim 0.14\text{-}0.19$. The lower end of this range is comparable to the mass density inferred from the dynamics of the luminous parts of galaxies ($\Omega_N \gtrsim 0.01 \approx \Omega_{gal}$). It is of more than trivial significance that the nucleon density as inferred from the study of primordial nucleosynthesis is consistent with that obtained from an application of Newtonian dynamics to the visible matter in the Universe. It is, of course, well known that the universal density of luminous matter is greatly exceeded by the mass density of a dark component.[45] The clustered component of the total mass density is currently estimated to be $\Omega_{cl} \approx 0.2 \pm 0.1$.[46-48] Nucleons are capable of accounting for some--perhaps all--of this dark matter ($\Omega_N \lesssim 0.1$-0.2, $\Omega_{cl} \approx 0.1$-0.3).

It is, however, very clear that nucleons alone cannot "close" the Universe ($\Omega_N \neq 1$). Indeed, from Eq. (23) with $h_0 \gtrsim 1/2$ and $\theta \lesssim 1.1$, $\Omega_N \lesssim 0.019\eta_{10}$ so that $\Omega_N = 1$ would require $\eta_{10} \gtrsim 52$. For such a large nucleon abundance the predicted primordial abundance of deuterium would fall short of that observed by some two orders of magnitude while the pregalactic abundance of lithium would exceed that observed by one to two orders of magnitude; a ^{4}He mass fraction $Y_P \gtrsim 0.27$ would also be predicted. A nucleon dominated Universe must be a low density Universe: $\Omega_N \lesssim 0.1$-0.2; an Einstein-deSitter Universe ($\Omega_0 = 1$) must be dominated by exotic matter.

SUMMARY

Several of the most profound and fundamental problems in cosmology have, in recent years, been subject to a concerted assault by particle physicists and astrophysicists. The blending of GUTs and the early evolution of the Universe appears to provide the recipe for producing the universal baryon asymmetry. Inflation, apparently, is the panacea for the ills of the horizon/homogeneity, age/flatness/entropy and monopole problems. The solution to the puzzle of the magnitude of the vacuum energy density--the cosmological constant problem--remains elusive.

In contrast to the ambitious endeavors outlined in the first part of my talk, big bang nucleosynthesis rests on a much more substantial foundation. As a window on the early Universe which can be confronted with observational data, primordial nucleosynthesis occupies a unique position. A comparison of theory and observation establishes the consistency of the standard model, limits the number of light neutrinos ($N_\nu \lesssim 4$) and bounds the universal density of nucleons (3-$4 \lesssim \eta_{10} \lesssim 7$-$10$, $0.01 \lesssim \Omega_N \lesssim 0.14$-$0.19$). This success and these modest achievements should give courage--as if any were needed--to the more tenuous exploits we will hear described in the course of this workshop.

ACKNOWLEDGEMENTS

The nucleosynthesis results described here have been obtained in the course of a very fruitful series of collaborations with K. A. Olive, D. N. Schramm, M. S. Turner and J. Yang. I am grateful to J. Felten for calling attention to an error in a preliminary version of this manuscript. This research is supported at Bartol by DOE Grant DE-AC02-78ER05007.

REFERENCES

[1] G. Steigman, Ann. Rev. Astron. Astrophys. **14**, 339 (1976).
[2] G. 't Hooft, Nucl. Phys. **B79**, 276 (1974).
[3] A. M. Polyakov, JETP Lett. **20**, 194 (1974).
[4] T. Kibble, J. Phys. **A9**, 1387 (1976).
[5] Ya. B. Zeldovich and M. Yu. Khlopov, Phys. Lett. **79B**, 239 (1978).
[6] J. Preskill, Phys. Rev. Lett. **43**, 1365 (1979).
[7] A. Sakharov, JETP Lett. **5**, 24 (1967).

[8]S. Weinberg, Phys. Rev. Lett. **42**, 850 (1979).
[9]E. W. Kolb and M. S. Turner, Ann. Rev. Nucl. Part. Sci. **33**, 645 (1983).
[10]A. Guth, Phys. Rev. **D23**, 347 (1981).
[11]S. W. Hawking, I. Moss and J. Stewart, Phys. Rev. **D26**, 1681 (1982).
[12]A. Guth and E. Weinberg, Nucl. Phys. **B212**, 321 (1983).
[13]A. D. Linde, Phys. Lett. **108B**, 389 (1982).
[14]A. Albrecht and P. J. Steinhardt, Phys. Rev. Lett. **48**, 1220 (1982).
[15]B.E.J. Pagel, Phil. Trans. Roy. Soc. Lond. **A307**, 19 (1982).
[16]B.E.J. Pagel, Proc. of the ESO Workshop on Primordial Helium (Kunth, Shaver and Kjar, eds.), 1983.
[17]B.E.J. Pagel, Proceedings this Workshop, (1984).
[18]G. Steigman, Proc. of the ESO Workshop on Primordial Helium (Kunth, Shaver and Kjar, eds.), p. 13 (1983).
[19]G. Steigman, Proc. of the Yerkes Workshop on Challenges and Developments in Nucleosynthesis (W. D. Arnett and J. W. Truran, eds.), In Press (1984).
[20]J. Yang, M. S. Turner, G. Steigman, D. N. Schramm and K. A. Olive, Ap. J. **281**, 493 (1984).
[21]G. Steigman and A. Boesgaard, Ann. Rev. Astron. Astrophys. **23**, (In Press, 1985).
[22]P. M. Jeffrey and E. Anders, Geochim. Cosmochim. Acta **34**, 1175 (1970).
[23]D. C. Black, Geochim. Cosmochim. Acta **36**, 347 (1972).
[24]J. Geiss and H. Reeves, Astron. Astrophys. **18**, 126 (1972).
[25]U. Frick and R. K. Moniot, Proc. 8th Lunar Sci. Conf., 229 (1977).
[26]P. Eberhardt, Proc. 9th Lunar Sci. Conf., 1027 (1978).
[27]J. N. Heasley and R. W. Milkey, Ap. J. **221**, 677 (1978).
[28]J. N. Bahcall, W. F. Huebner, S. H. Lubow, P. D. Parker and R. K. Ulrich, Rev. Mod. Phys. **54**, 767 (1982).
[29]D. O. Gough, Proc. of the ESO Workshop on Prim. Helium (Shaver, Kunth and Kjar, eds.), 117 (1983).
[30]T. Encrenaz and M. Combes, Icarus **52**, 54 (1982).
[31]V. Kunde, R. Hanel, W. Maguire, D. Gautier, J. P. Baluteau, A. Marten, A. Chedin, N. Husson and N. Scott, Ap. J. **263**, 443 (1982).
[32]D.N.B. Hall, Ap. J. **197**, 509 (1975).
[33]B. Mason, Cosmochemistry, Part I. Meteorites (1979).
[34]R. A. Zappala, Ap. J. **172**, 57 (1972).
[35]D. K. Duncan, Ap. J. **248**, 651 (1981).
[36]P. A. Vanden Bout, R. L. Snell, S. S. Vogt and R. G. Tull, Ap. J. **221**, 598 (1978).
[37]P. Bruston, J. Audouze, A. Vidal-Madjar and C. Laurent, Ap. J. **243**, 161 (1981).
[38]D. G. York, Private Communication (1983).
[39]A. Vidal-Madjar, C. Laurent, C. Gry, P. Bruston, R. Ferlet and D. G. York, Astron. Astrophys. **120**, 58 (1983).
[40]R. T. Rood, G. Steigman and B. M. Tinsley, Ap. J. (Lett.) **207**, L57 (1976).
[41]R. T. Rood, T. M. Bania and T. L. Wilson, Ap. J. **280**, 629 (1984).
[42]F. Spite and M. Spite, Astron. Astrophys. **115**, 357 (1982a).
[43]F. Spite and M. Spite, Nature **297**, 483 (1982b).
[44]T. D. Kinman and K. Davidson, Ap. J., In Press (1984).
[45]S. M. Faber and J. S. Gallagher, Ann. Rev. Astron. Astrophys. **17**, 135 (1979).
[46]W. Press and M. Davis, Ap. J. **259**, 449 (1982).

[47]M. Aaronson, J. Huchra, J. Mould, P. L. Schechter and R. B. Tully, Ap. J. **258**, 64 (1982).
[48]M. Davis and P.J.E. Peebles, Ann. Rev. Astron. Astrophys. **21**, 109 (1983).

The Dynamical Parameters of the Expanding Universe as They Constrain Homogeneous World Models With and Without Λ

Allan Sandage and G.A. Tammann

ABSTRACT

The only three known reliable methods to obtain the Hubble constant by direct observation are reviewed, with the result that $H_0 = 50 \pm 7$ km s^{-1} Mpc^{-1}, or $H_0^{-1} = (19.5\pm3) \times 10^9$ years. The three known methods of merit to measure the density parameter directly (counting the luminosity density, measuring the local perturbation of the velocity field due to the Virgo cluster over-density, and using the results of big bang nucleosynthesis and the observed abundance of D, He^3, He^4, and Li^7) are consistent with $\Omega_0 = 0.1$ within a factor of 2. The age of the Universe, determined by adding the galaxy formation time to the age of our galaxy, is taken as 18.3×10^9 years.

The standard Friedmann models with $\Lambda = 0$ are specified completely by the values of H_0, T_0, and Ω_0. The values $H_0 = 50$, $T_0 = 18$ by, $\Omega_0 \simeq 0.1$ are consistent with a standard $\Lambda = 0$ model with negative spatial curvature (k<0) that will expand forever. A model with $H_0 = 100$ km s^{-1} M_{pc}^{-1} is not possible with $\Lambda = 0$.

Conditions for an exactly flat (k≡0) model are explored, consistent with the flatness condition of the inflationary Universe of Grand Unification. The condition for this is $\Lambda c^2 = 3H_0^2\ (1-\Omega_0)$. Within the observational constraints that $H_0 = 50 \pm$ say 15 and $T_0 = 18 \pm 5$, such flat models exist if $\Omega_0 \simeq 0.1$ and $\Lambda c^2 \sim 3H_0^2$.

I. INTRODUCTION

On scales larger than ~300 Mpc (i.e. over regions separated by expansion velocities of at least 15,000 km s^{-1}), the Universe is observed to be homogeneous;[1,2] the mean density averaged over volumes of this size is much the same from cell to cell no matter which cell is considered.

As the scale is decreased, structure begins to appear,[3-5] and at ~100 Mpc (v~5000 km s^{-1}) superclusters appear whose form is either pancakes or simple strings of galaxies connecting the cores of dense clusters.[6-9]

On scales of ~20 Mpc the principal structures are the dense envelopes surrounding the cluster cores such as that of our own Virgo-complex supercluster.[10] Between the scale at ~100 Mpc and this scale at ~20 Mpc, voids appear, first discovered convincingly by Gregory and Thompson.[11]

On still smaller scales of ~2 Mpc, the cores of the clusters are the principal structures, appearing at the intersections of the supercluster pancakes.[7] These are to be identified with the clusters cataloged by Abell[12] and by Zwicky et. al.[13] Smaller groups of galaxies exist with higher frequency, also at this scale. The Local Group is the best known example. The greatest density fluctuations on the smallest scale are the galaxies themselves, with typical scale-lengths of ~0.1 Mpc for the largest, ranging down to at least one hundredth of that.

One of the principal problems of cosmology is to understand the formation of such a complex hierarchy, embedded in the uniform large-scale Universe required by the very high order isotropy ($\Delta T/T \leq 10^{-4}$ over angles ≥ 1 arc minute) of the 3°K radiation.[14]

Or perhaps it is more important to look at the point the other way around -- to understand the large scale homogeneity and isotropy in the presence of the bewildering hierarchy of density contrasts.

In the first approach, concentrating on the structure problem, we accept the homogeneity as given (in much the same way as Le Comte accepted the Universe), and attempt to calculate the structures from the observed dynamical parameters that define the homogeneous Friedmann-Lemaître-Robertson-Walker models. There are three: (1) the expansion rate now -- the Hubble constant H_0, (2) the ratio of the actual mean density ρ_0 to ρ_{crit} (defined as $3H_0^2/8\pi G$ required for closure, but only if $\Lambda=0$); the ratio ρ_0/ρ_{crit} is named Ω_0, and (3) the cosmological constant Λ. Armed with the values of these parameters together with an assumed spectrum of the density fluctuations $\delta\rho/\rho$, we attempt to calculate the growth of $\delta\rho/\rho$ with time such that the observed hierarchical structure of the moment is reproduced (see Silk 1984 for a review).[8]

The second way of thinking is more complex; it attempts to answer three questions raised by the observations themselves. (1) The Universe appears to be very nearly flat in its geometry i.e. $k/R^2 \simeq 0$. Why? This requires the special condition that the kinetic energy-density of the expansion now be very close (to within a factor of ~10) to the self gravitational energy-density. The observations show this to be true, i.e. $\Omega_0 = 0(1)$. Why this should be so nearly so at the present epoch is a mystery called the flatness problem. (2) Why is the Universe at large so homogeneous over the enormous scale of 10^{24} km, which is many orders of magnitude larger than the size of regions that were causally connected at the time of last photon scattering (the horizon problem)? (3) What is the cause of the initial $\delta\rho/\rho$ fluctuations that grow into structure, given the observed homogeneity and isotropy (i.e. the paradox of Lifshitz 1946)?[15]

These are the principal subjects of much of this conference. It is our duty here only to approach the problem by the first way, reviewing the observations that provide the present estimates of the Hubble constant, the value of Ω_0 via $\rho_0 \equiv (3H_0^2/8\pi G)\Omega_0$, and the need (or not) for $\Lambda \neq 0$ for a model on the largest scale.

Part of the data that leads to our preferred parameter-set for the Universe at large was reviewed in the ESO/CERN conference on the same topic (Sandage and Tammann 1984a,[16] hereafter ST). New developments since then on the local streaming motions within the Virgo supercluster, and relative to the microwave background (MWB), and on the recalibration of various distance indicators that came with the completion of the Virgo cluster survey,[17] make the observational situation somewhat more secure. We do not discuss here in much detail those points made in the ESO/CERN report, but do discuss with a somewhat different emphasis the value of H_0, the infall velocity to Virgo leading to a small value of Ω_0, and the consequences of the $k/R^2 \simeq 0$ flatness condition required by inflation.

The following sections treat the Hubble constant (Section II), ways to obtain Ω_0 and the consequences of the data concerning conditions on the type of dark matter permitted (Section III), the age of the Universe as a crucial constraint on the models (Section IV), the standard model with $\Lambda = 0$ (Section V), limits on the models with $\Lambda \neq 0$ (Section VI), and the conditions for $k/R^2 = 0$ (Section VII).

II. THE HUBBLE CONSTANT

A. Form of the Hubble Law

The most secure fact of observational cosmology is that the velocity-distance relation is linear, i.e. $n = 1$ in $v = Hr^n$. The result is based on observations of first ranked cluster galaxies (i.e. objects from a highly restricted region of the luminosity function -- the standard candles) with redshift that range from $z \equiv \Delta\lambda/\lambda_0 = 0.003$ to ~ 1.

The work was carried out at Palomar from 1950 through ~1975[18-24] and continued with different methods elsewhere.[25,26] Two tests that give $n = 1 \pm 0.02$ are the decline in the apparent luminosity of standard candles as z^2, and the decline of their angular diameter as z; both tests with very high accuracy (op. cit.). To test for the linearity of the Hubble law, the data are restricted to $z \lesssim 0.5$ where the effects of look-back time [i.e. (1) the changing of the World picture into the World map due to the light travel-time, and (2) the effects of evolution] are small.

That the observed redshift-distance relation is linear is taken to be highly significant. It is the only type of velocity field that (a) looks identical in all directions to all observers no matter what their position, and (b) permits a singularity (all points together in a finite past) when all points in the manifold are at every other point. These are the necessary (but by no means sufficient) conditions for the "big bang" to have occurred, assumed throughout this conference.

Although these facts of observation and the linear property do not prove that the redshift is due to expansion, no viable alternative to the many manifestations of the redshift has successfully survived ~50 years of scrutiny since Hubble's announcement in 1929. Tests of the expansion hypothesis have been attempted via the surface brightness =f(redshift) relation. True expansion requires $SB \sim (1+z)^{-4}$ for standard surfaces, whereas most non-expanding models require $SB \sim (1+z)^{-1}$. The observational test is difficult for technical reasons, and is not yet satisfactory.

Nevertheless, in view of the otherwise extraordinary near-coincidence of the three time scales given by H_0^{-1}, the age of the Galaxy, and the age of the elements, the true expansion hypothesis, bizzare as it is, is apparently the only "likely" possibility.

B. Distances to Galaxies

The problem of determining the Hubble constant breaks into two quite separate pieces. (1) Precise distances to nearby galaxies must be found using all parts of the arsenal of fundamental distance indicators that one considers reliable. As these distance indicators are relatively faint (none brighter than $M_B \simeq -6$) we have not been able to push beyond distance moduli of $(m-M_B) \cong 29.0$ (i.e. $D \simeq 6$ Mpc) using them, although, to be sure, secondary indicators exist which hold promise to $D \simeq 20$ Mpc even with ground-based telescopes. (2) It is from these primary calibrating galaxies with (precise) local distances that secondary indicators must be calibrated to carry the distance scale to $D > 20$ Mpc into the general Hubble expansion field. If D can be found to ≳50 Mpc, the Hubble constant can be found directly. If D can only be obtained to distances of the Virgo Cluster at ~22 Mpc, then knowledge of the local velocity perturbation due to the Virgo over-density must also be had so as to tie the local value of v/D to the global value of H_0. Much recent attention has been given to this problem (see Ref. 27 for a review with recent references).

This is not the place to enter into a tedious discussion of the technical problems of absolute magnitude calibration of primary indicators, and especially into the effects of severe observational bias on the final result according to the way objects for study are chosen (i.e. if picked from a magnitude-limited or a distance-limited sample). Our purpose is merely to list what now seems to us to be the three most promising ways to carry the local distance scale (defined by Local Group, M81, NGC2403, NGC300, and M101 Groups) into the D > 20 Mpc regime where the expansion dominates over the peculiar motions. We do not dwell upon the local scale using Cepheids to D ≲ 6 Mpc, as there is general agreement among most workers on these distances (to within the uncertainty of the zero-point of the Cepheid P-L relation itself). A review of this problem has been given elsewhere.[27] We emphasize that it is not within this local sphere of 6 Mpc where we find the 1.5 mag difference that is required between the two scales that give H_0 = 50 or 100 km s^{-1} Mpc^{-1} for the global Hubble constant, hence most of the factor of 4 difference in luminosity of the candles must be sought after the leap has been made to the more remote realm, and it is this problem that we concentrate on.

C. Three Routes to Remote Distances

In this section we describe the methods of merit in the inverse relation to their power. For those readers who do not find their favorite method here, we refer to the more detailed discussion mentioned earlier[27] for explanations.

1. The total luminosity of ScI type galaxies was hoped to be a superior indicator of distance after van den Bergh's[28] introduction and first calibration of his luminosity classes, where only a small dispersion in absolute magnitude was suggested, of order $\sigma(M) \simeq 0.4$ mag. We used the precepts of the method to obtain $H_0 = 50 \pm 5$ km s^{-1} Mpc^{-1}, calibrated via M101,[29] but later we claimed to have destroyed our own result[30] by showing that the van den Bergh luminosity classes overlap greatly in luminosity. We now believe we were too hasty in this rejection, based on new data in the Virgo Cluster. We argue here that a probable lower limit to H_0 can be inferred from the data.

The van den Bergh ScI galaxies in the Virgo Cluster have a near Gaussian luminosity function of dispersion $\sigma = 0.36$ mag. However, there are only 5 cases, but these five show a clean difference from all other Virgo Spirals (see the photographic Atlas in Sandage, Binggeli, and Tammann).[31] The evidence is shown in Fig. 1 which is the distribution of apparent magnitudes for current star-producing galaxies in the Virgo Cluster core. Among the 6 examples of ScI shown, the faintest is an ScI-II. Its elimination gives a mean apparent magnitude for the remaining 5 of $\langle B_T \rangle = 10.45$, again with $\sigma(M) = 0.36$ mag.

Note that the luminosity function for all types and classes is approximately Gaussian -- at the very least going to zero at both the bright and faint ends. The ScI group of 5 has the smallest dispersion.

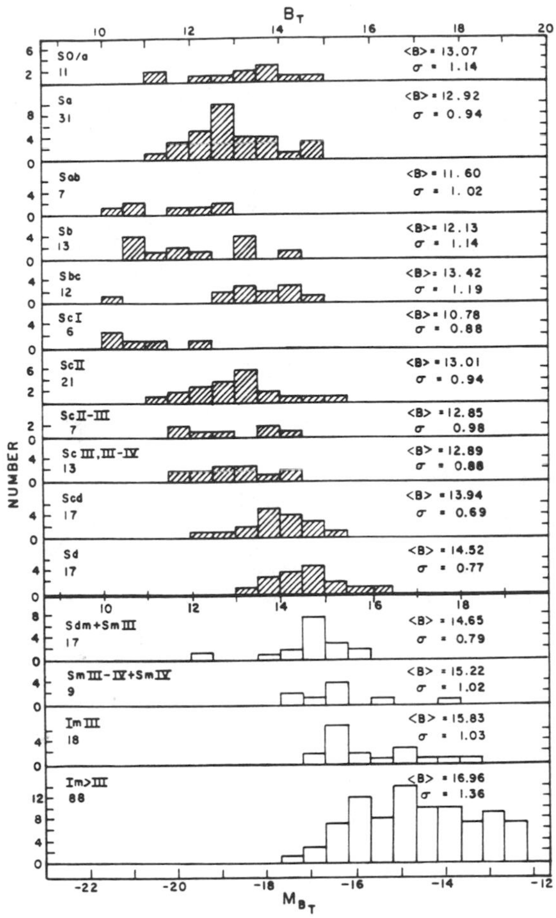

Fig. 1. Distribution of apparent magnitudes of current star-producing galaxies in the Virgo Cluster 6° core, binned into Hubble morphological types and van den Bergh luminosity classes.

We found a similar result earlier using field ScI galaxies,[29] the brighter ones taken from a magnitude limited sample of $B_T = 13$, the fainter from a sample with $B_T = 15.7$. The Hubble diagram for this group is shown in Fig. 2. The theoretical linear velocity -- distance line of slope 5 has been put through the points, fitting the data well. Some evidence of selection (Malmquist) bias was present in the original data[29] making our determination of H_0 from them too high, all other things being equal.

Figure 1 could be used to determine H_0 via the distance to Virgo if $\langle M_{B_T} \rangle$ were known for a distance-limited sample of ScI galaxies, and if the streaming motions connected with Virgo were also known. Figure 2 could be used to determine H_0 directly, independent of such motions, but again only if $\langle M_{B_T} \rangle$ were known.

The only ScI galaxy whose (precise) distance is known is M101 at $m - M = 29.2$ (D=6.9 Mpc) from Cepheids.[32,33] Using this modulus and its observed apparent magnitude gives $M^{0,i}_{B_T} = -21.31$ for M101 itself. If this is the required mean value of $\langle M_{B_T} \rangle$ for ScI's in a distance limited sample, Fig. 2, using $B - m_{pg} = 0.1$ mag., requires $H_0 = 51$ km s^{-1} Mpc^{-1}.

The second method uses Fig. 1. With $\langle B_T \rangle = 10.45$ and $M_{B_T} = -21.31$, the distance to Virgo is $m - M = 31.76$ or D = 22.5 Mpc. Correction for the streaming motions (our infall to Virgo, and Virgo's infall to the MWB: Tammann and Sandage[27]) requires $H_0 = 48 \pm 7$ via this alternate route.

The principle uncertainty with this method is equating M_B for M101 alone to $\langle M_B \rangle_{ScI}$. We are swimming within $\sigma(M_B)$ for ScI. But we note that, because M101 is the nearest ScI, the probability is that $M_B = -21.31$ is fainter than $\langle M_{B_T} \rangle$ (the Malmquist bias), hence $H_0 = 50$

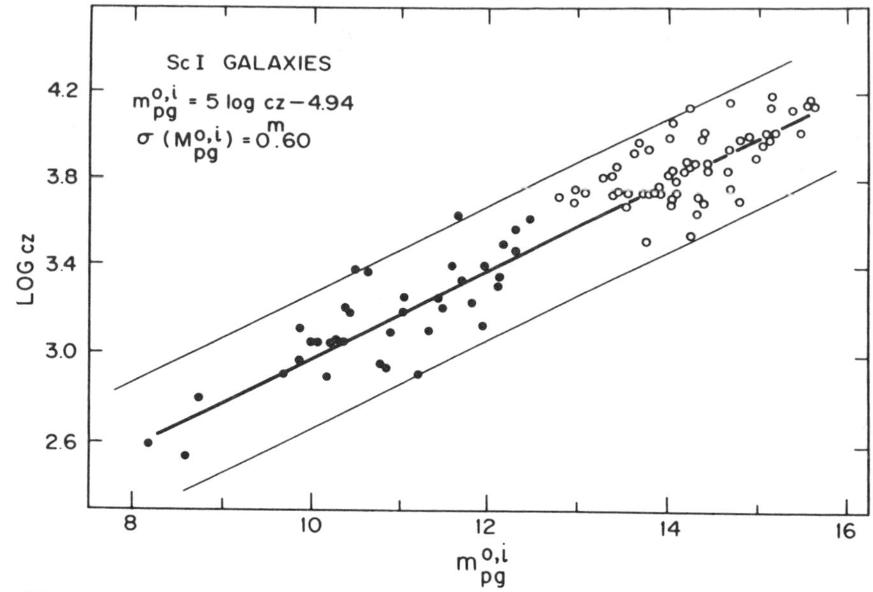

Fig. 2. The Hubble diagram for ScI galaxies in the general field showing the theoretical relation of slope 5 for a linear velocity-distance law as if no observational bias exists in the sample. Dark circles are ScI galaxies choosen from a bright magnitude-limited catalog with B_T brighter than ~13.0. The open circles are galaxies from a fainter catalog with a limit at B_T = 15.7. Using two catalogs in this way minimizes the Malmquist bias.

is surely an upper limit via this route.

We emphasize that we do not put as much credence in this method as with the remaining two, but find it impossible from these data to obtain H_0 near 100 km s^{-1} Mpc^{-1}.

2. The line width-absolute magnitude method of Öpik[34] was rediscovered by Tully and Fisher[35] and applied extensively by several groups to obtain $H_0 \simeq 100$. The most recent discussion by the short-distance-scale-group[36] quote $H_0 = 91 \pm 3$ km s^{-1} Mpc^{-1} following earlier infrared work by Mould, et. al.[37] Aaronson et. al.[38](with earlier references therein) and others who also obtained $H_0 \simeq 100$.

The method holds great promise but it is not yet mature. Different results are obtained according to the band pass of the observations. For example, Richter and Huchtmeier[39] obtained 45 ± 7 km s^{-1} Mpc^{-1} from their linewidth data in the blue. In an effort to reconcile the blue and infrared, we used the literature data quoted above, recalibrated via the local calibrators, and obtained H_0 = 55.[40] Kraan-Korteweg,[41] using the data of Bothum et. al. found a substantial Malmquist bias remaining in the corrected material which, after evaluating the effect, obtained $H_0 = 58 \pm 4$.

Until the method is understood more fully, it is not at a stage where either it or the data that go into it can yet be said to provide the value of H_0. It is however natural to put considerable weight on the analysis of Kraan-Korteweg (H_0~60), and of Richter and Huchtmeier[38] because of the care taken of the systematic errors for field galaxies and the extensive use of galaxy groups to calibrate the slope.

3. The route through brightest stars to supernovae of type I holds the most promise because (a) the absolute magnitudes of the brightest blue and red stars in the local calibrating galaxies are well defined,

with only a small intrinsic dispersion.[42-44] It is also suspected, following Kowal,[45] that SNeI have a well defined magnitude at maximum light. This is now known to be the case to at least $\sigma(M) = 0.3$ and probably better[46] using photometric data on light curves of only the highest quality. These authors also show that the 16 such SNeI now available follow the Hubble line of slope 5 in the m, log z diagram meaning that once $\langle M_B(\mathrm{SNeI})\rangle$ is known, H_0 follows immediately from the Hubble line through the data of $\langle M_B^{max}\rangle = (-19.73 \pm 0.24) + 5 \log(H_0/50)$.

We have followed this route to obtain H_0,[47] calibrating $\langle M_B\rangle_{SNI}$ in NGC4214 and IC4182 via their brightest red supergiants. The value $\langle M_B\rangle = -19.74 \pm 0.19$ obtained in this way requires $H_0 = 50 \pm 7$ km s^{-1} Mpc^{-1} using the observed Hubble diagram of the 16 SNeI.

A second method to calibrate $\langle M_B\rangle$ is to adopt the theoretical model of SNeI as a component of a binary with mass transfer, nudging one component over the Chandraseshar limit, producing an explosion. This model, calculated in detail by Sutherland and Wheeler[48] so as to produce the observed light-curve slope, requires $\langle M_{B_T}\rangle = -20.0$ leading to $H_0 = 44 \pm 5$ km s^{-1} Mpc^{-1}.

This method through SNeI seems to have the most promise, but needs to be more thoroughly applied to more fresh SNeI, which would have future proper photometry.

In the following we adopt $H_0 = 50 \pm 7$ as the global value of the Hubble constant, based on the present assessment of the evidence to date. We take this to be upper limit because all systematic errors of the nature of statistical bias always work in the direction that the true distances are larger than the inferred distances calculated when the biases are neglected. However, we also caution that much is unknown yet on basic questions of internal absorption in the relevant galaxies, and on systematic errors in distances to the local calibrators. The problems of zero-point of the Cepheids remain, and cry for a modern solution.

III. THE DENSITY PARAMETER Ω

A. Relationships

Of more fundamental importance than H_0 to the nature of the models is the density parameter Ω_0. Its value alone determines the sign of the spatial curvature in models where $\Lambda = 0$; it combines with Λ when $\Lambda \neq 0$ for the same purpose. A review of its remarkable properties is useful.

The two equations whose derivation, meaning, and solutions are known to all cosmologists, and which completely describe the homogeneous models are

$$\frac{\dot{R}^2}{R^2} + \frac{2\ddot{R}}{R} + \frac{8\pi G p}{c^2} = -\frac{kc^2}{R^2} + \Lambda c^2 \qquad (1)$$

$$\frac{\dot{R}^2}{R^2} - \frac{8\pi G\rho}{3} = -\frac{kc^2}{R^2} + \frac{\Lambda c^2}{3} \quad , \qquad (2)$$

where p is the total pressure (i.e. 2/3 the energy density of the random motions plus the radiation pressure, $aT_0^4/3$). These are derived from Einstein's equations of general relativity by specifying the g_{ij} components of the metric tensor that are demanded by the general homogeneous and isotropic line element.[49,50] The equations can also be derived heuristically from Newtonian mechanics[51] in a way that gives considerable insight into their meaning, once a repulsive acceleration of magnitude $(\Lambda c^2/3)R$ is postulated. An excellent review is given by McCrea.[52]

The necessary geometrical information can be obtained from Eqs. (1) and (2) with no need for their integration. Following Robertson[53] and Hoyle and Sandage[54] by introducing

$$H_0 \equiv \dot{R}_0/R_0 \tag{3}$$

$$q_0 \equiv - \ddot{R}_0/(R_0H_0^2) , \tag{4}$$

(where subscript zero means present values), subtracting Eq. (2) from Eq. (1) and considering only the present epoch (the equations are valid at any given time, and hence are valid at the present epoch), gives

$$\rho_0 = \frac{3H_0^2q_0}{4\pi G} - \frac{3p_0}{c^2} + \frac{\Lambda c^2}{4\pi G} . \tag{5}$$

It is easily shown[55] that $p_0c^2 = 0$ to high order. Hence, substituting the definition of

$$\Omega_0 \equiv \frac{\rho_0}{\rho_{crit}} = \frac{8\pi G\rho_0}{3H_0^2} , \tag{6}$$

into Eq. (5) gives

$$\Omega_0 = 2q_0 + \frac{2}{3}\frac{\Lambda c^2}{H_0^2} , \tag{7}$$

which we need as an auxiliary for the space-time curvature as follows. [Note from Eq. (7) that if $\Lambda = 0$, $\Omega_0 = 2q_0$ as is familiar.]

Substitution of Eq. (3) into Eq. (2) gives

$$\frac{kc^2}{R^2} = - H_0^2 + \frac{8\pi G\rho_0}{3} + \frac{\Lambda c^2}{3} , \tag{8}$$

for the curvature (k=1,0,-1), which, with Eq. (6) reduces to

$$\frac{kc^2}{R^2} = H_0^2\,(\Omega_0 - 1) + \frac{\Lambda c^2}{3} \quad \text{or} \tag{9}$$

$$= H_0^2\left(\Omega_0 + \frac{\Lambda c^2}{3H_0^2} - 1\right) , \tag{9'}$$

or using Eq. (7)

$$\frac{kc^2}{R^2} = H_0^2\left(\frac{3}{2}\,\Omega_0 - q_0 - 1\right) . \tag{10}$$

Equation (9) or (9') is the most useful. If $\Lambda = 0$, the condition for flatness (k=0) clearly is the standard (everybody knows) $\Omega_0 = 1$, or $q_0 = 1/2$. Closure (k=+1) requires $\Omega_0 > 1$, regardless of the value of H_0, justifying the first two sentences of this section.

We shall return to Eq. (9') in the penultimate section where we discuss the general conditions on Λ and the time scale parameter H_0T_0 (see Section IV) in order that $k/R^2 = 0$.

B. Three Methods to Determine Ω_0 from Observation

At the moment there are three, and we believe only three, reliable methods to determine Ω_0 observationally. These have been reviewed in summary in the ESO/CERN symposium volume;[16] the most promising (infall to the Virgo core) is discussed at length in the archive literature (cf. Tammann and Sandage 1985 with literature references).[27] We do not repeat the details here but merely mention the results.

1. <u>Counting the luminosity of visible matter</u> per unit volume and assigning an appropriate mass-to-luminosity (M/L) ratio is the most straightforward way to estimate a lower limit to the present mass density ρ_0 which, from the definition in Eq. (6), gives a lower limit to Ω_0. Counting luminosities of galaxies in the Revised Shapley Ames. Catalog[57] and correcting to the base luminosity level outside the local supercluster (i.e. for the Virgo-complex over density) gives blue luminosity densities of 5.6×10^7 and 2.2×10^7 solar blue luminosities per Mpc^{-3} for spiral and E + SO galaxies respectively. After correcting for internal absorption in spirals, the summed value is $L = 8.8 \times 10^7\ L_B \odot\ \text{Mpc}^{-3}$. If this luminosity has a mean mass-to-light ratio of M/L, the luminosity density transforms to

$$\rho_0 = 7.5(M/L) \times 10^{-33}\,\text{gm cm}^{-3} , \tag{11}$$

which, with $\rho_{crit} = 3H_0^2/8\pi G = 4.7 \times 10^{-30}(H_0/50)^2\ \text{gm cm}^{-3}$ gives

$$\Omega_0 \equiv \rho_0 / \rho_{crit} = 1.6(M/L)(H_0/50)^{-2} \times 10^{-3} \quad . \tag{12}$$

Reasonable values of M/L are controversial. If no invisible mass were present in galaxies or clusters of galaxies, M/L would be expected to be ~1 (in solar units) for spirals and ~3 for E galaxies. The observed rotation curves of spirals, the mass required to bind galaxy-pairs, and the mass required in the cores of the Abell and Zwicky-like clusters are all substantially higher than this (Faber and Gallager 1979 for a modern review).[58] Our best estimate is M/L = 25 for spirals and 325 for E galaxies in clusters (where they concentrate). We apply these values in the population ratio of the spirals to E galaxies that make up the sum in Eq. (12) to obtain <M/L> = 88 which, with $H_0 = 50$ gives

$$\Omega_0 \text{ (from binding of pairs and clusters)} \simeq 0.14 \ , \tag{13}$$

with an estimated error of a factor of ~2.

It is to be noted that invisible matter has been put into this calculation by at least a factor of ~10 larger than the visible matter. This is because M/L which is inferred only from the stellar content of galaxies is at most 10, and more likely M/L ≃ 5, which, with $H_0 = 50$ in Eq. (12) gives Ω_0 (visible matter) $\lesssim 0.02$. This leads to the well known result that if we demand Ω_0 ~1, we need 100 times the amount of dark matter as that in visible stars. As it is, we must postulate about 10 times more dark than light, strictly from the observation of bound pairs and clusters.

2. The dynamical perturbation of the Virgo-complex over-density on the local velocity field gives a confirmation of Eq. (13) to within a factor of ~2 providing only that any unseen matter in the Universe is distributed as the visible galaxies (i.e. is no more evenly spread). Said differently, the scale length for clustering must be the same for the dark and for the luminous matter for this perturbation method to work. Calculations of the effect of the over-density on our velocity relative to the Virgo core are now standard, following the pioneering Virgocentric model-building of Silk[59] and of Peebles.[60] All galaxies in concentric shells about the Virgo core (in an idealized case) suffer decelerations of their Hubble (free) expansion, the closer to the core the greater is the deceleration because the greater is the over-density mass within that shell. Depending on the exact mass distribution within the cluster core, here taken to go as r^{-2}, the deceleration of each shell at any distance from the core is calculated as if it were a small Friedmann universe.

This deceleration from the free Hubble flow appears as an "infall velocity" of the Local Group "toward" the Virgo core -- although it is actually the difference between the Hubble global flow and the decelerated (but still outward) velocity of the Local Group from Virgo. If the infall velocity v_{vc} can be measured, the effect of gravity on the expansion is found which, in fact, is what Ω_0 tells us -- the smaller is Ω_0, the smaller is the gravitational energy-density of the Universe relative to its kinetic energy-density.

Define the density contrast to be

$$\delta \equiv \frac{\rho_{cluster} - \rho_{backgound}}{\rho_{background}} \quad . \tag{14}$$

Linear analysis of a spherical mass over-density[61] shows that

$$\Omega_0 \approx \left[\frac{3v_{vc}}{(v_{vc}+v_0)\delta}\right]^{1.5} , \tag{15}$$

where v_0 is the observed mean velocity of the Virgo core.

From 1974 to the present, some 30 separate determinations of v_{vc} using a variety of methods have been made by about half as many different teams. There are two recent reviews; one by Davis and Peebles[62] and the other by Tammann and Sandage.[27] Although these differ in their adopted value of v_{vc} (450 km s^{-1} by DP; 220 by TS), the conclusion is the same; with $\delta = 2.8$[63,61] the value of $\Omega_0 \ll 1$. Using $v_0 = 967$ km s^{-1} gives $\Omega_0 = 0.20$ with the Davis Peebles infall velocity. We have been critical of their large infall value. If $v_{vc} = 220$ km s^{-1} which we adopt, then

$$\Omega_0(\text{infall to Virgo}) = 0.09 \pm 0.04, \tag{16}$$

or $q_0 = 0.045$ if $\Lambda = 0$.

The conclusion is that with v_{vc} as small as even the largest suggested value among the 30 determinations, the velocity perturbation due to Virgo requires gravity to be essentially absent (i.e. the $q_0 = 0$ limiting case) rather than so strong for the closure condition of $\Omega_0 = 1$ if $\Lambda = 0$.

We must again make the point that if large amounts of dark matter exist that is more widely spread than the galaxies, then the argument is not valid. A model with a longer clustering length for dark matter than for galaxies, by Hoffman and Salpeter,[64] shows this effect. But if such dark matter does exist with a different clustering scale, the agreement between Eq. (13) and (16) would be highly surprising. The case for no smoothed out dark matter becomes even stronger by noting the third coincidence with the nucleogenesis argument for ρ_0, and hence for Ω_0, discussed next.

3. The present value of the mean baryon density ρ_0 (baryon), calculated from nucleosynthesis arguments, given that the background radiation has a temperature of 2.7° K, is the third way to Ω_0. The prediction of the present expected abundances of D^2, He^3, He^4, and Li^7 under conditions that existed within the first few seconds of the hot early Universe[65-68] is the most brilliant early success of the detailed "big bang" model. The agreement of the predictions with the observed abundances is so good (see Audouze[69] for a review with the fundamental references) as to believe that the singularity did indeed occur.

If so, the theory shows that the present density of baryons alone is

$$\rho_0 = 4.7 \times 10^{-31}(T/2.7)^3 \text{ gm cm}^{-3} . \qquad (17)$$

With $\rho_{crit} = 4.7 \times 10^{-30}(H_0/50)^2$ gm cm^{-3}, we find

$$\Omega_0(\text{baryon}) \equiv \rho_0/\rho_{crit} = (0.10 \pm 0.06)(H_0/50)^{-2} , \qquad (18)$$

where the uncertainty of the value is from the estimate of the certainty of ρ_0(baryon) from nucleosynthesis theory.[68,69]

Note particularly that Eq. (17) does not contain H_0 which enters only in ρ_{crit} via Eq. (6). This permits a consistency check on our adopted value of H_0 and on the Ω_0 values from Eq. (13) and (16). Equation (18) is plotted in Fig. 3. The solid boundary lines are the upper and lower limits permitted by the ±0.06 quoted error. Any set of H_0, Ω_0 values within these boundary lines is a solution of Eq. (18). The striped region shows the restrictions imposed by the luminosity-density Eq.(13) and the infall Eq. (16).

The conclusion is that all three tests restrict the density parameter to be

$$0.08 \lesssim \Omega_0 \lesssim 0.15 , \qquad (19)$$

and the Hubble constant to be

$$25 \lesssim H_0 \lesssim 65 \text{ km s}^{-1}\text{Mpc} . \qquad (20)$$

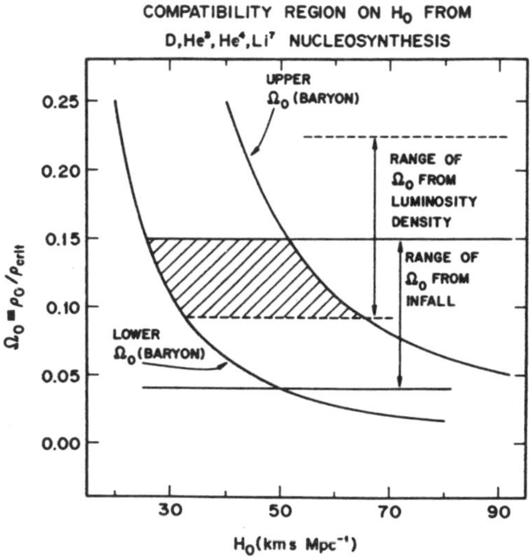

Fig. 3. The possible range of H_0 and Ω_0 by using Eq. (18) of the text derived from the nucleosynthesis of D, He^3, He^4, and Li^7 in the early Universe. Approximate limits on Ω_0 from Eqs. (13) and (16) from the luminosity density and the Virgo Cluster infall arguments are also shown. The adopted value of $\Omega_0 = 0.1$ is compatible with $H_0 = 50$.

Note that this "determination" of H_0 is independent of classical astronomical techniques of distance measurements. If valid, H_0 cannot be as large as 100 km s^{-1}Mpc^{-1}, in agreement with our conclusions from Section II.

One consequence of this low value of Ω_0, and hence the smallness of the Virgo infall velocity v_{vc}, is that the total mass of the Virgo complex inside the Local Group circle is <u>small.</u> A standard calculation with $v_{vc} = 220$ km s^{-1} and an over density of $\delta\rho/\rho = 2.8$ gives a total mass of $1.3 \times 10^{15}(H_0/50)$ solar masses (Eq. 32)[63] within the Local Group circle. The blue luminosity residing in all galaxies within the same volume is $L_B = 2.4 \times 10^{13}$ solar blue luminosities (data, corrected for internal absorption of the spirals following precepts in Yahil, Sandage, and Tammann).[70] Dividing the mass by the light gives the very low value of $(M/L)_B = 54$, $(H_0=50)$, in solar units, for the total matter averaged over the Virgo complex within the Local Group circle. This is much lower than $(M/L)_B \simeq 400$ required to bind the Virgo core (Zwicky[71], Smith[72], and up to modern times in Rivolo and Yahil[73]), or M/L between 240 and 480 needed in the inner 1.5° to bind the x-ray gas surrounding M87 into hydrostatic equilibrium.[74] The fact that M/L ≃ 400 in the inner Virgo core and only ~50 averaged over the 22 Mpc radius to the Local Group forces the conclusion that <M/L> for any aggregate does not increase as the aggregate size is increased, such as suggested by studies over scales of ~10 Mpc[75,61] but rather shows a sharp turnover near 10 Mpc. Such a turnover was suggested by Figure 2 of Davis et. al.,[61] but not by the drastic amount we require here. Because of this decrease of <M/L> for D ~ 10 Mpc, the hope to obtain $\Omega_0 \geq 1$ for closure by postulating the extremes of more widely distributed dark matter than in visible galaxies seems not possible.[76] However, as mentioned earlier and emphasized again in Section VII, $\Omega_0 \geq 1$ is not the general condition for closure (nor $\Omega_0 \equiv 1$ for flatness) in models where Λ is positive, hence the frenzy in the recent past to find as many observational possibilities for Ω_0 to be 1 seems to have been an unnecessary trip down a blocked avenue.

IV. THE AGE OF THE UNIVERSE

Even with H_0 and Ω_0 determined to this point, the model is not yet specified. Without knowledge of Λ we do not know the geometry (i.e. the Gauss-Riemann curvature value or even the sign of k) from Eq. 9 (or 9'), nor do we know if the expansion will continue forever or be decelerated, stopped, and eventually collapse.[77]

As is well known, only one more parameter -- the age T_0 -- is required to uniquely define the model. By uniquely we mean that if H_0, Ω_0, and T_0 were known precisely, then there is only one Freidman-Lemaître model that will fit. Our problem, of course, is that we shall "never" know these three parameters without a range of measuring error, and it is the assessment of this range which is a principle reason for this review.

There is now a large separate room within the castle of cosmology, called AGE DATING. The activities there are a bit different than those of Bishop Ussher, yet the problem is the same. Reviews of the modern methods via (1) age dating of the oldest stars and hence our Galaxy, and (2) age dating the chemical elements are many. An early overview of the results to ~1965[78] with emphasis on the stellar age dating via globular clusters in our Galaxy, compared with a modern review (cf. Audouze 1980 with its clear summary of the age of the chemical elements)[79] shows the progress in 15 years.

Details of the methods are well known and the basic literature is accessible; we quote here only the results.

a) Ages of the Galectic globular clusters are stabilizing near 16×10^9 years with uncertainities of $\sim\pm 3 \times 10^9$ years per cluster. All such clusters in the Galaxy are closely the same age.[80,81] Field subdwarfs of high velocity relative to the sun (and therefore with low rotational velocity about the Galactic center -- i.e. in highly eccentric orbits) are also known to be the same age, not older. Credence to the identification of these as the oldest stars comes from the fact that field subdwarfs and globular clusters have abundances of the elements heavier than He down by a factor of 100 or more compared with the sun, showing that they were formed at a very early epoch in the chemical enrichment of the interstellar medium from which modern, high metallicity stars formed later.

The age of globular clusters is not the age of the Universe but clearly forms a lower limit. The time between the "big bang" and the formation of the first stars in the Galaxy can be estimated, either from collapse models of galaxy formation, or from the direct observation of the upper redshift cutoff limit for quasars,[82] signalling the first galaxy formation in the Universe. Modern data[83] put this upper limit near $z = 4$, requiring the time between the big bang singularity and the formulation of the nuclei of galaxies (quasars are the dynamical run-away into a black hole of such nuclei) to be 2.3×10^9 years. This gestation time, added to the age of the globular clusters gives the age of the Universe as

$$T_0 = 18.3 \times 10^9 \text{ years} , \qquad (22)$$

to which we optimistically put an error of $\pm 3 \times 10^9$ years.

b) The age of the chemical elements that are made in stellar interiors (with charge $z > 7$) leads to a similar number. Thielemann, Metzinger, and Klapdor[84] from the radioactive actinides derive

$$T_0 = \left(20^{+2}_{-4}\right) \times 10^9 \text{ ys} ,$$

which is however model dependent on galactic evolution -- a general problem in all such determinations. Schramm, in a long series of studies with collaborators (cf. Schramm and Wasserburg 1970;[85] Symbalisty and Schramm 1981[86] with extensive references therein) favor $T_0 = (15\pm3) \times 10^9$ years. We take both values to be consistent with Eq. (22) and adopt its value, together with $H_0 = 50$ km s^{-1} Mpc^{-1}, to obtain the final number needed to specify the model as

$$H_0 T_0 = 0.94 \pm 0.15 \; (\sigma) . \qquad (23)$$

V. THE STANDARD MODEL ($H_0=50$, $\Omega_0 \simeq 0.1$, $\Lambda=0$) WORKS

A. Solution for Ω_0 Via the Equation of Time as a Test

Figure 4 is a reminder of the familiar Friedmann-Lemaître solutions to Eq. (1), together with the geometrical construction showing the relation of the numerical value of H_0^{-1} to the time T_0 from any point 0 (the present epoch) to the origin $R = 0$. The curve $\Omega_0 > 1$, turning downward after some time, applies only if $\Lambda = 0$ (cf. Eq. 9'); as mentioned several times before, Ω_0 need not be >1 for closure and collapse if $\Lambda \neq 0$. The $\Omega_0 = 0$ case (zero mass) is the non-decelerated straight line $R \sim t$. In this case $H_0 T_0 \equiv 1$ exactly, i.e. the tangent to the curve (here the straight line R~t) at the present epoch 0 extrapolated backward, intersects the time axis at $T_0 = 0$, $R_0 = 0$, giving the standard result that $H_0^{-1} = T_0$ (for $q_0=0$).

The general solution for the time as a function of Ω_0 and H_0^{-1}, if $\Lambda = 0$, is[55]

$$H_0T_0 = \frac{1}{1-\Omega_0} + \frac{\Omega_0}{2(\Omega_0-1)^{3/2}}\left[\frac{\pi}{2} + \sin^{-1}\left(\frac{\Omega_0-2}{\Omega_0}\right)\right], \tag{24}$$

for $\Omega_0 > 1$, i.e. the closed model (k=+1),

$$H_0T_0 = 2/3 \text{ for } k = 0, \text{ and} \tag{25}$$

$$H_0T_0 = \frac{1}{1-\Omega_0} - \frac{\Omega_0}{2(1-\Omega_0)^{3/2}} \ln\left[\frac{2-\Omega_0}{\Omega_0} + \frac{2(1-\Omega_0)^{1/2}}{\Omega_0}\right], \tag{26}$$

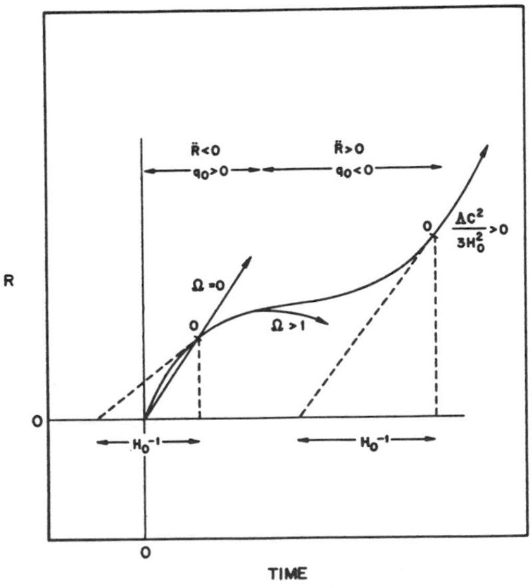

Fig. 4. Schematic reminder of the various solutions of the Friedmann-Lemaître Eq. (1). The construction for the relation between H_0^{-1} to the age T_0 (the time from "now" at point 0 to the origin at R=0) is shown by drawing the tangent to the R(t) curve at point 0 and extending it to the R = 0 axis. The empty Universe (Ω_0=0) is the straight line solution R ~ t, shown next to the Ω = 0 symbol.

for $\Omega_0 < 1$, i.e. the open model (k=-1).

It is because these equations contain only H_0, T_0, and Ω_0 that we have, several times above, said that the model is now entirely specified, i.e. because H_0T_0 is known observationally, Ω_0 can be calculated, and hence from Eq. (9') with $\Lambda = 0$ [i.e. Eq. (21)] the space-time curvature can be found. All aspects of the model, such as volumes enclosed within a given redshift value, the light travel time, angular diameters for a given redshift etc. follow then simply.

We call the $\Lambda = 0$ case the standard model. It works! Solving Eq. (26) using $H_0T_0 = 0.94$ gives

$$\Omega_0(\text{model}) \equiv 2q_0 = 0.05(+0.35,-0.05), \tag{27}$$

which agrees well enough with the observational determinations of Eqs. (13), (16), and (18). The errors are put from $\varepsilon(H_0T_0) = \pm 0.15$ by not permitting $\Omega_0 < 0$.

The observed redshift - "distance" relation has been attempted since 1956 as a test of any model (i.e. the magnitude-$\Delta\lambda/\lambda_0$ Hubble diagram). Well known difficulties of the luminosity-evolution of the standard candles in the light-travel time[87,88] complicate the test. However, correction for the predicted increase in brightness of E galaxies with increasing youth as we look backward in time by observing to ever larger redshifts, using the detailed models of Tinsley which are well approximated by the luminosity increase of

$$\Delta \text{ mag} = \ln[1/(1+\Delta\lambda/\lambda_0)] \ , \tag{28}$$

shows excellent agreement between the standard-model theory ($\Lambda=0$, $q_0=0$) and the redshift - apparent magnitude observations of 85 groups and clusters of galaxies over the redshift range of $0.003 \leq z < 0.75$. The fit is shown in Fig. 4 of the ESO/CERN report[16] and later in Fig. 7 here. Although necessary, the agreement is not sufficient to prove the standard model because the m,z diagram is degenerate in Λ, i.e. a high density $\Omega_0 = 1$ Universe with $\Lambda c^2/3 \simeq 2H_0^2$ has nearly the identical theoretical m,z line as the $q_0 = 0$, $\Lambda = 0$ model. For this reason we cannot say we have independent confirmation of the standard model from this once classic test.

B. Is $H_0 = 100$ Possible with the Standard Model?

We now inquire if any reasonable value of Ω_0 (given a lower limit to T_0 of say 10×10^9 years) is possible if $H_0 = 100$ km s^{-1}Mpc^{-1}. Solution of Eqs. (24), (25), and (26) with $H_0T_0 = 1.9$ (i.e. $H_0^{-1} = 9.8 \times 10^9$ years for $H_0 = 100$, $T_0 = 18.3 \times 10^9$ years) has no solution with $\Lambda = 0$. The only possible way to extend the age to 18.3×10^9 years is to introduce $\Lambda > 0$.[89]

Equations (24)-(26) are shown in Fig. 5, which is a concise summary of this section. Our solution of $\Omega_0 \simeq 0.08$, $H_0 = 50$, and $T_0 = 18 \times 10^9$ years is shown by the open circle. Boundary lines at the $\pm 3 \times 10^9$ years error limits are shown. Note explicitly that the largest value of H_0 permitted, given a lower age limit of $T_0 = 15 \times 10^9$ years, is $H_0 \leq 65$ km s^{-1}Mpc^{-1} for an empty ($\Omega_0=0$) case.

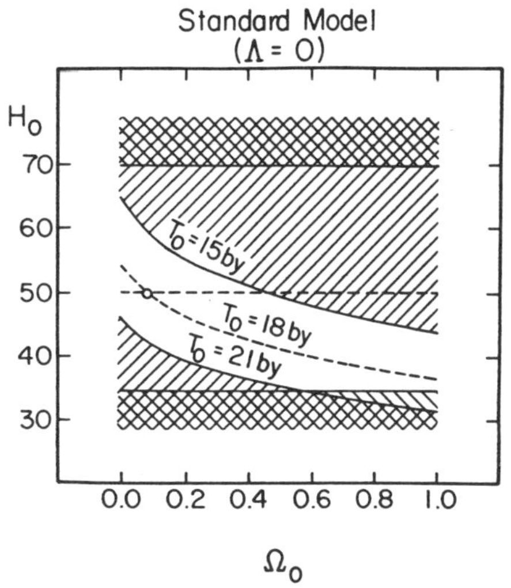

Fig. 5. Relation between Ω_0, H_0, and T_0 given by Eqs. (24)-(26). Our adopted solution from the direct observations, is the open circle.

We can finally ask for the conditions under which <u>flatness</u> can occur in the <u>standard</u> model. This requires $H_0 = 2/3\ T_0^{-1}$, $\Omega_0 = 1$. If $H_0 = 50$, then T_0 must be 13.1×10^9 years which is ~2σ lower than our preferred value and hence a possibility. Other combinations are, of course possible such as $H_0 = 44$ and $T_0 = 15 \times 10^9$ years which cannot be ruled out.

However, again, we can rule out $H_0 = 100$ on all accounts in this model because H_0^{-1} would equal 9.8×10^9 years; 2/3 H_0^{-1} for flatness would give $T_0 = 6.5 \times 10^9$ years which is out of the question. But, as concluded above, even with $\Omega_0 = 0$, the age $T_0 \equiv H_0^{-1} = 9.8 \times 10^9$ years is already too small. We conclude that $H_0 = 100$ km s^{-1}Mpc^{-1} cannot be fitted by the standard model.

What, then are the possibilities if we introduce a non-standard model with $\Lambda \neq 0$?

VI. THE RANGE OF POSSIBILITIES FOR MODELS WITH $\Lambda \neq 0$

A. The H_0T_0, Λ, Ω_0 Curves

We have seen that the standard model requires only three parameters (H_0, T_0, and Ω_0) to specify it completely. The addition of Λ, which is the coupling constant to the cosmic acceleration

$$\ddot{r} = \left(\frac{\Lambda c^2}{3}\right) r \equiv H_0^2 \left(\frac{\Lambda c^2}{3H_0^2}\right) r , \qquad (29)$$

produces families of models (found by varying the Ω_0, Λ pairs), where there was only one (the standard model) before. Most classical cosmological tests then become degenerate, i.e. non unique, in the sense that such correlations as the magnitude-redshift relation and the angular size-redshift relation, specified in the standard model by Ω_0 only, have a nearly identical form over the large redshift range of $0 < z \lesssim 1$ for several particular Λ, Ω_0 values. There is, then, no unique test for any model, only a consistancy check on some assumed Λ, Ω_0 pair.

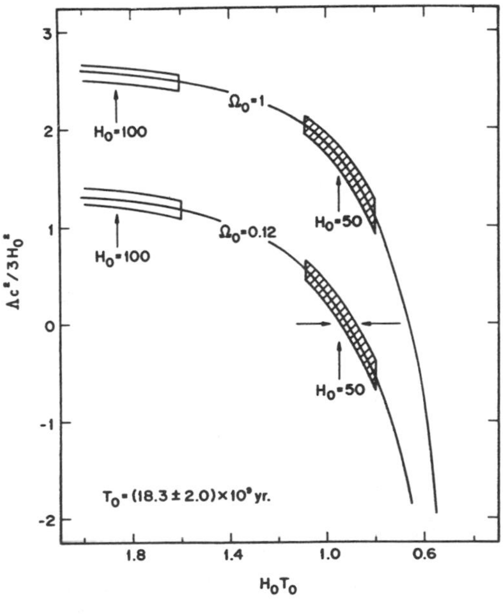

Fig. 6. The cosmological constant as a function of the time-scale parameter H_0T_0

Calculations of these models has been carried out in detail over a large range of the reasonable parameter space by Refsdal, Stabbel, and de Lange (hereafter RSL),[90] following the pioneer work of Solheim.[91] These integrations have been generally neglected in the past because observers have been reluctant to admit the $\Lambda \neq 0$ Worlds due to added complication in making definitive cosmological tests. But, in view of the flatness problem, the $\Lambda \neq 0$ models begin again to have charm.[92]

Straightforward interpolation in Table II of Stabell and Refsdal,[93] following Tomita and Hyashi,[94] shows that H_0T_0, Ω_0 and the dimensionless number $\Lambda c^2/3H_0^2$ are constrained to a particular $H_0T_0 = f(\Lambda c^2/3H_0^2)$ curve for a particular Ω_0 value. By varying Ω_0, a family of such curves is produced. Two curves of the family are shown in Fig. 6 for $\Omega_0 = 1$ and 0.12. The dimensionless time-scale number H_0T_0, still presumed to be known from the observations, is the dependent variable of the abscissa; the dimensionless $\Lambda c^2/3H_0^2$ is the ordinate. The solution (Section V) of $H_0 = 50$, $\Omega_0 \simeq 0.12$, $\Lambda = 0$ is indicated by three arrows on the $\Omega_0 = 0.12$ curve. The hatched area are 1σ error limits as if $\sigma(T_0) = 2 \times 10^9$ on a base of $T_0 = 18.3 \times 10^9$ years, a somewhat tighter error limit than reality. The point in Fig. 6 at $\Lambda = 0$, $H_0 = 50$, on the $\Omega_0 = 0.12$ curve is again merely the statement that the standard model works.

If one insists that $H_0 = 100$ km s^{-1}Mpc^{-1} (despite its rejection by Fig. 3 if all the matter in the Universe is baryonic), then Λ must be positive with a value in the range $1.2 \lesssim \Lambda c^2/3H_0^2 \lesssim 2.5$ for Ω_0 in the range between 0.1 and 1. This is the well-known result of Lemaître,[89] and follows from the construction shown in Fig. 4 of the tangent line to the R(t) curve extrapolated to R = 0, giving $H_0^{-1} < T_0$ for $\Lambda > 0$. Figure 6 shows by how much H_0^{-1} differs from T_0.

B. Possibilities of Determining Λ from Observation

A direct determination of Λ by measuring its effect on orbits within the solar system Eq. (29) is far beyond present technology. The added force is small. For example, if $\Lambda c^2/3H_0^2 = 2$ (a reasonable value according to Fig. 6), then $\ddot{r} = 2H_0^2 r$ by Eq. (29), which is $5 \times 10^{-36} r$ cm sec^{-2}. At the earth, the ratio of the sun's gravitational acceleration to this cosmic repulsion acceleration is 4×10^{21}. The effect of Λ on solar system dynamics and on the internal dynamics of the Galaxy are negligible.

The only other method known at the moment is to be able to measure q_0 independently of Ω_0 by using Eq. (7) to obtain

$$\Lambda c^2 = 3H_0^2 \left(\frac{\Omega_0}{2} - q_0\right) \quad . \qquad (7')$$

However, the methods to determine q_0 directly are not yet well understood.

<u>i. The volume test</u> is one such possibility in principle. The count-redshift relation differs between the $\Lambda = 0$ and $\Lambda \neq 0$ cases due to the effect of Λ on the comoving volume element caused by the coasting time for the $\Lambda \neq 0$ case[16,95] It is not yet known if the test has practical application.

<u>ii. The standard redshift-apparent magnitude test</u>, even if the effects of luminosity evolution and cluster cannibalism of first ranked cluster E galaxies were understood, is, as previously mentioned, degenerate. Different values for various Λ, q_0 pairs give closely the same predicted line in the Hubble

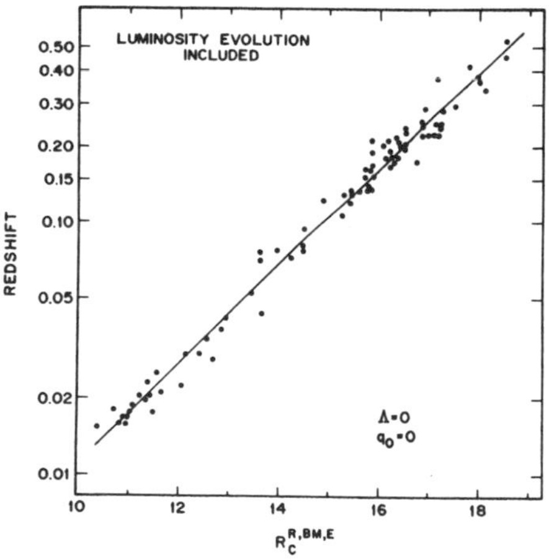

Fig. 7. The observed Hubble diagram, corrected for luminosity evolution of first ranked E galaxies in groups and clusters via Eq. (28). The abscissa is the observed (but corrected for technical effects) red ($\bar{\lambda} \simeq 6200$Å) magnitude. The line shown is from theory for a $\Lambda = 0$, $q_0 = 0$ model. The theoretical line for a non-standard model with $q_0 = -1$, $\Lambda c^2/3H_0^2 = 1.75$, $\Omega_0 = 1.5$ agrees with the drawn line to within 0.1 mag everywhere, hence the test is degenerate if redshifts only as large as ~0.6 are available.

redshift-magnitude diagram to redshifts of at least $\Delta\lambda/\lambda_0 \simeq 0.7$. This is illustrated in Fig. 7 which shows the largest sample of homogeneous data available on the R magnitude scale [corrected for the known technical effects of cluster richness, Bautz-Morgan cluster contrast, aperture effect, and luminosity evolution via Eq. (28)]. The fitted line is the theoretical z,R relation for $\lambda = 0$, $q_0 = 0$ (so labeled on the diagram), but the calculations of RSL show that an equally good fit to the observations can be obtained with $\Lambda c^2/3H_0^2 = 1.75$, $q_0 = -1$ which requires $\Omega_0 = 1.5$ from Eq. (7). The fits do become progressively less degenerate for larger redshifts, but until $\Delta\lambda/\lambda_0 \simeq 1$ the practical problem is as just stated. Nevertheless, what can be expected in the near future is a strong restriction on the size of the parameter space of the Λ, q_0 pairs that do satisfy the data, and this already constrains the problem. What we obviously need now is another relation, also involving Λ, q_0 pairs which, if different in principle from the Hubble diagram of Fig. 7, might permit a solution.

One might have thought that such a relation would be the angular diameter-redshift relation, calculation in closed form for the $\Lambda = 0$ case,[55] but calculated inextenso for all other cases by RSL. Again certain particular pairs of Λ, q_0 values give nearly the same curves in the $\Delta\lambda/\lambda_0$, θ redshift-angular diameter diagram. Now, however, we must be careful. If we were to use the angular diameter data for the same galaxies as have gone into Fig. 7 (and if the data were precise -- i.e. errorless), we must obtain identically the same Λ, q_0 pairs as by using Fig. 7 alone. This is because angular diameter, apparent magnitude, and redshift are related by

$$\text{surface brightness} \sim (1+z)^{-4} ,$$

independent of all model parameters. This means, in effect, that Fig. 7 contains no new information not already in the $\Delta\lambda/\lambda_0,\theta$ diagram, and again the case seems to be degenerate.

There may be other purely geometrical ways to obtain q_0, and hence Λ via Eq. (7'), but a search and an application is yet to be complete. At the moment we can only make consistency tests for pairs of Λ, q_0 values, as in Fig. 7.

VII. FLATNESS AND CLOSURE IN A LOW DENSITY UNIVERSE

By more extensive interpolation in Table II of Stabell and Refsdal,[93] Fig. 6 can be generalized to many values of Ω_0 (note that Ω_0 as used here is $2\sigma_0$ as used by Stabell and Refsdal) with the concomitant values of Λc^2 (which is λ in their notation) using Eq. (7'). The result is shown in Fig. 8, which is the most direct summary of most of this review.

The diagram is similar to Fig. 6, but now we have isolated the closed (k>0) from the open (k<0) domains, separated by a solid line marked k = 0 which is the flat space-time manifold obtained via Eq. (9') from the condition

$$\frac{\Lambda c^2}{3H_0^2} = 1 - \Omega_0 \quad . \tag{30}$$

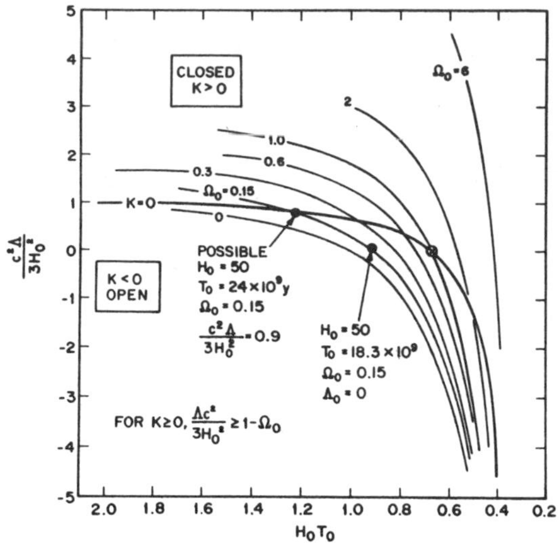

Fig. 8. Same as Fig. 6, but shown for more values of Ω_0, and also showing the regions of positive (k>0), negative (k<0), and zero curvature, as found from Eq. (9'). The flat (k=0) curve is of special interest for the inflationary Universe of Grand Unification.

Note again that flatness does not require $\Omega_0 = 1$ (nor closure $\Omega_0>1$) unless $\Lambda = 0$, and this special case (where of course $H_0T_0=2/3$) is shown by the crossed circle in the diagram.

Flatness is attractive because of the inflation scenario,[96–98] and we see that most of the data described in earlier sections fit quite nicely close to the k = 0 curve. Our acceptable solution with the standard model (Λ=0) is shown as a solid dot at $H_0 = 50$, $T_0 = 18.3$ by, $\Omega_0 = 0.08$, $\Lambda = 0$, but the model is open by Eq. (9).

The second closed circle on the k = 0 line gives quite acceptable observational parameters with $H_0T_0 = 1.2$, $\Omega_0 = 0.15$, and $\Lambda c^2/3H_0^2 = 0.9$. If $T_0 = 18.3 \times 10^9$ years, then $H_0 = 64$ km s^{-1} Mpc^{-1}, or if we believe in the direct observational determination of $H_0 = 50$, then $T_0 = 23 \times 10^9$ years -- all within the observable ranges.

The special conditions on the parameters along the k = 0 line can be seen more easily in Fig. 9 where the variables are displayed in an obvious way. The approximate upper and lower bounds on H_0 are put from Eq. (18) via Fig. 3, and the upper and lower bound on T_0 are from Eq. (22) using $\varepsilon(T_0) = \pm 3 \times 10^9$ years. Relaxation on these bounds on H_0 and T_0 will expand the open area in Fig. 9.

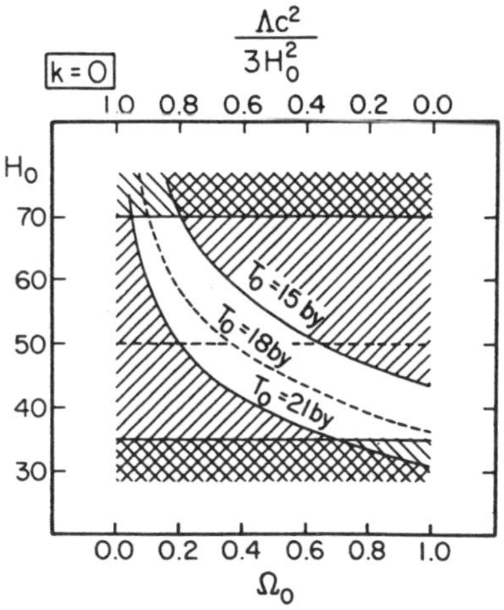

Fig. 9. Similar to Fig. 5 for the k = 0 line in Fig. 8. The Ω_0 values along the bottom correspond to the $\Lambda c^2/3H_0^2$ values along the top abscissa according to Eq. (30).

Peebles[92] has discussed this k = 0 flat case in detail and reaches the same general conclusions that $\Lambda c^2 \sim 2H_0^2$ and $\Omega_0 \simeq 0.3$. These conditions with Eq. (30) require flatness and do not contradict any of the other observational constraints (i.e. the Hubble diagram, and hence the angular size-redshift relation). At the moment this then is an acceptable observational model.

The next obvious step is to find an independent way to measure Λ observationally.

VIII. CONCLUSIONS

1) An open standard model ($\Lambda=0$) is possible with the observed parameters $H_0 = 50 \pm 15$ km $s^{-1}Mpc^{-2}$, $\Omega_0 = 0.1$ within a factor of two, $T_0 = 18 \pm 5 \times 10^9$ years where the quoted errors are larger than given in the text, but may be more realistic.

2) A flat (k=0) standard ($\Lambda=0$) model cannot be ruled out. The parameters would be $\Omega_0 = 1$, $H_0 = 44$ km $s^{-1}Mpc^{-1}$, and $T_0 = 15 \times 10^9$ years, i.e. $H_0T_0 = 2/3$. If this is the case, in order that $\Omega_0 = 1$ [in view of the restriction from D, He^3, He^4, Li^7 nucleosynthesis abundances that require Ω_0 (baryonic) $\simeq 0.1$] the remaining 90% of the dark matter needed for closure (i.e. $\Omega_0=1$) must be invisible, non-baryonic, and more widely distributed than the visible galaxies. Because the visible luminosity density with $(M/L)_B \simeq 10$ gives only $\Omega_0 \simeq 0.01$, this requires that 99% of the mass of the Universe be invisible. This review would then have said almost nothing about everything.

3) From the m, $\Delta\lambda/\lambda_0$ Hubble diagram, an $\Omega_0 = 1$, $\Lambda > 0$ model is possible with $H_0 = 50$, $T_0 = 18.3 \times 10^9$ years, and $\Lambda c^2/3H_0^2 \simeq 2$. The model is not particularly attractive because it is highly closed (k>0).

4) A flat (k=0) model with $\Omega_0 \simeq 0.1$, $H_0 = 50$, $T_0 = 23 \times 10^9$ years, and $\Lambda c^2/3H_0^2 \simeq 0.9$ is possible. The flatness is an important feature of the inflationary Universe of the Grand Unification Theory of elementary particles. If $\Omega_0 \simeq 0.1$, there is no reason to postulate any unseen mass greater than the factor of 10 already needed to bind galaxy pairs and the great clusters.

REFERENCES

[1]E. Hubble, Ap. J. **79**, 8 (1934).

[2]E. Hubble, Realm of the Nebulae, Yale University Press, New Haven, Chapter III (1936).

[3]H. Shapley, Har. Bull. **890**, 1 (1932).

[4]B.J. Bok, Har. Bull. **895**, 1 (1934).

[5]C.D. Shane, Vistas Astron. **2**, 1574 (1956).

[6]C.S. Frenk, S.D.M. White, and M. Davis, Ap. J. **271**, 417 (1983).

[7]A. Dekel, Ap. J. **284**, 445 (1984).

[8]J. Silk, in Large-Scale Structure of the Universe, Cosmology, and Fundamental Physics (Proc. CERN Symposium, Nov. 1983, CERN, Geneva), p. 225.

[9]J.H. Oort, in Large-Scale Structure of the Universe, Cosomology, and Fundamental Physics (Proc. CERN Symposium, Nov. 1983, CERN, Geneva), p. 209.

[10]G. de Vaucouleurs, Vistas Astron. **2**, 1584 (1956).

[11]S.A. Gregory and L.A. Thompson, Ap. J. **222**, 784 (1978).

[12]G.O. Abell, Ap. J. Suppl. **3**, 211 (1958).

[13]F. Zwicky, P. Wild, E. Herzog, M. Karpowicz, and C.T. Kowal, Catalog of Galaxies and Clusters of Galaxies, Vols. 1-6 (Calif. Inst. Tech., Pasadena), 1961-1966.

[14]D.T. Wilkinson, in Large-Scale Structure of the Universe, Cosmology, and Fundamental Physics (Proc. CERN Symposium, Nov. 1983, CERN, Geneva), p. 153.

[15]E. Lifshitz, J. Phys. (USSR) **10**, 116 (1946).

[16]A. Sandage and G.A. Tammann, in Large-Scale Structure of the Universe, Cosmology, and Fundamental Physics (Proc. CERN Symposium, Nov. 1983, CERN, Geneva) p. 127.
[17]A. Sandage, B. Binggeli, and G.A. Tammann, The Virgo Cluster (Proc. ESO Workshop held Sept. 1984), ed. B. Binggeli and M. Terenghi (ESO, Garching bei München).
[18]E. Hubble, Mon. N. R. Astron. Soc. **113**, 658 (1953).
[19]M.L. Humason, N.U. Mayall, and A. Sandage, A. J. **61**, 97 (1956).
[20]A. Sandage, Ap. J. **173**, 485 (1972).
[21]A. Sandage, Ap. J. **183**, 731 (1973).
[22]A. Sandage and E. Hardy, Ap. J. **183**, 743 (1973).
[23]J.E. Gunn and J.B. Oke, Ap. J. **195**, 255 (1975).
[24]A. Sandage, J. Kristian, and J.A. Westphal, Ap. J. **205**, 688 (1976).
[25]A. Bruzual and H. Spinrad, Ap. J. **220**, 1 (1978).
[26]S. Djorgovski and H. Spinrad, Ap. J. **251**, 417 (1981).
[27]G.A. Tammann and A. Sandage, Ap. J. 1985 July 1 issue.
[28]S. van den Bergh, Ap. J. **131**, 215 (1960).
[29]A. Sandage and G.A. Tammann, Ap. J. **197**, 265 (1975).
[30]G.A. Tammann, A. Yahil, and A. Sandage, Ap. J. **234**, 775 (1979).
[31]A. Sandage, B. Binggeli, and G.A. Tammann, A. J. **90**, 395 (1985).
[32]A. Sandage and G.A. Tammann, Ap. J. **194**, 223 (1974).
[33]A. Sandage, A. J. **88**, 1569 (1983).
[34]E. Opik, Ap. J. **55**, 406 (1922).
[35]B. Tully and J.R. Fisher, Astron. Astrophys. **54**, 661 (1977).
[36]G.D. Bothun, M. Aaronson, B. Schommer, H. Huchra, and J. Mould, Ap. J. **278**, 475 (1984).
[37]J. Mould, M. Aaronson, and J. Huchra, Ap. J. **238**, 458 (1980).
[38]M. Aaronson et. al., Ap. J. Suppl. **50**, 241 (1982).
[39]O.-G. Richter and W.K. Huchtmeier, Astron. Astrophys. **132**, 253 (1984).
[40]A. Sandage and G.A. Tammann, Nature **307**, 326 (1984).
[41]R.C. Kraan-Korteweg, in The Virgo Cluster, ESO Workshop held September 1984 (ESO), ed. B. Binggeli and M. Tarenghi.
[42]A. Sandage and G.A. Tammann, Ap. J. **191**, 603 (1974).
[43]A. Sandage, A. J. **89**, 630 (1984).
[44]A. Sandage and G. Carlson, A.J. submitted (1985).
[45]C.T. Kowal, A. J. **73**, 1021 (1968).
[46]G.A. Tammann and R. Cadonau, Ap. J. submitted (1985).
[47]A. Sandage and G.A. Tammann, Ap. J. **256**, 339 (1982).
[48]P.G. Sutherland and J.C. Wheeler, Ap. J. **280**, 282 (1984).
[49]H.P. Robertson, Proc. Nat. Acad. Sci. **15**, 822 (1929).
[50]A.G. Walker, Proc. London Math. Soc. **42**, 90 (1936).
[51]W.H. McCrea and E.A. Milne, Quart J. Math. (Oxford Ser.) **5**, 73 (1934).
[52]W.H. McCrea, Report on Progress in Physics (London: Physical Society), **16**, 321 (1953).
[53]H.P. Robertson, Pub. A.S.P. **67**, 82 (1955).
[54]F. Hoyle and A. Sandage, Pub. A.S.P. **68**, 301 (1956).
[55]A. Sandage, Ap. J. **133**, 355 (1961).
[56]A. Sandage and G.A. Tammann, in Supernovae (Cambridge Mass Workshop), Lecture Notes in Physics No. 224: ed. N. Bartel (1985), p. 1.
[57]A. Sandage and G.A. Tammann, A Revised Shapley Ames. Catalog, Carnegie Institution Publication No. 635 (1981).
[58]S.M. Faber and J.S. Gallagher, Ann. Rev. Astr. Astrophys. **17**, 135 (1979).
[59]J. Silk, Ap. J. **193**, 525 (1974).

[60]P.J.E. Peebles, Ap. J. **205**, 318 (1976).
[61]M. Davis, J. Tonry, J. Huchra, and D.W. Latham, Ap. J. Letters, **238**, L113 (1980).
[62]M. Davis and P.J.E. Peebles, Ann. Rev. Astr. Ap. **21**, 109 (1983).
[63]A. Yahil, A. Sandage and G.A. Tammann, in Physical Cosmology, eds. R. Balian, J. Audouze, and D.N. Schramm (Amsterdam: North-Holland), Les Houches 1979 School (Session 22), p. 127 (1980).
[64]G.L. Hoffman and E.E. Salpeter, Ap. J. **263**, 485 (1982).
[65]P.J.E. Peebles, Ap. J. **146**, 542 (1966).
[66]R.V. Wagoner, W.A. Fowler and F. Hoyle, Ap. J. **148**, 3 (1967).
[67]R.V. Wagoner, Ap. J. **179**, 343 (1973).
[68]J. Yang, M.S. Turner, G. Steigman, D.N. Schramm, and K.A. Olive, Ap. J. **281**, 493 (1984).
[69]J. Audouze, in Large-Scale Structure of the Universe, Cosmology, and Fundamental Physics: ESO/CERN Symposium ed. G. Setti and L. Van Hove: (Geneva), p 293 (1984).
[70]A. Yahil, A. Sandage, and G.A. Tammann, Ap. J. **242**, 448 (1980).
[71]F. Zwicky, Helv. Phys. Acta **6**, 110 (1933).
[72]S. Smith, Ap. J. **83**, 23 (1936).
[73]A.R. Rivolo and A. Yahil, Ap. J. **274**, 474 (1983).
[74]D. Fabricant and P. Gorenstein, Ap. J. **267**, 535 (1983).
[75]W.H. Press and M. Davis, Ap. J. **259**, 449 (1982).
[76]The argument is correct only if the unseen mass were postulated to be baryons because, from Eq. (18), Ω_0 (baryon) cannot be greater than ~0.2, making the infall method to Virgo consistent and hence presumably correct. Suppose, however, that the dark matter is not baryonic; then Ω_0 (baryon) from Eq. (18) is still correct, but Ω_0 (infall) from Eq. (16) is meaningless in the presence of more widely distributed matter, and we would then be left with the major curiosity of why Eqs. (13), (16), and (18) agree. This would seem a strong argument against the existence of non-baryonic matter.
[77]This, of course, is not true if $\Lambda = 0$ where the familiar dogma applies that

$$\frac{kc^2}{R^2} = H_0^2(\Omega_0 - 1) \equiv H_0^2(2q_0 - 1) \quad . \tag{21}$$

In this case, if $q_0 << 1/2$ ($\Omega_0 << 1$ as determined in the last section), then $k = -1$ (the geometry is open), and the Universe will escape from itself forever.
[78]A. Sandage, in Galaxies and the Universe, ed. L. Woltjer: Columbia University Press (New York), p. 75 (1968).
[79]J. Audouze, in Physical Cosmology, eds. R. Balian, J. Audouze, and D.N. Schramm (Amsterdam: North-Holland), Les Houches 1979 School (Session 22), p. 195 (1980).
[80]A. Sandage, Ap. J. **252**, 553 (1982).
[81]D.A. VandenBerg, Ap. J. Suppl. **51**, 29 (1983).
[82]A. Sandage, Ap. J. **178**, 25 (1972).
[83]P. Osmer, Ap. J. **253**, 28 (1982).
[84]F.-K. Thielemann, J. Metzinger, and H.V. Klapdor, Astron. Astrophys. **123**, 162 (1983).
[85]D.N. Schramm and G.J. Wasserburg, Ap. J. **162**, 57 (1970).

[86]E.M.D. Symbalisty and D.N. Schramm, Rep. Prog. Phys. **44**, 293 (1981).
[87]A. Sandage, Ap. J. **134**, 916 (1961).
[88]B.M. Tinseley, in Physical Cosmology, eds. R. Balian, J. Audouze, and D.N. Schramm (Amsterdam: North-Holland), Les Houches 1979 School (Session 22), p. 162 (1980).
[89]G. Lemaître, Ann. Soc. Sci. Bruxelles, **A53**, 51 (1933).
[90]S. Refsdal, R. Stabell, and F.-G. de Lang, Mem. Roy. Ast. Soc. **71**, 143 (1967).
[91]J.-E. Solheim, Mon. Not. R. A. S. **133**, 321 (1966).
[92]P.J.E. Peebles, Ap. J. **284**, 439 (1984).
[93]R. Stabell and S. Refsdal, Mon. Not. R. A. S. **132**, 379 (1966).
[94]K. Tomita and E. Hyashi, Prog. Theor. Phys., Osaka **30**, 691 (1963).
[95]Ya.B. Zel'dovich, private communication.
[96]A. Guth, Phys. Rev. **D23**, 347 (1981).
[97]A. Albrecht and P.J. Steinhardt, Phys. Rev. Lett. **48**, 1220 (1982).
[98]A.D. Linde, Phys. Lett. **114B**, 431 (1982).

On the Determination of Cosmological Parameters

John P. Huchra

I. INTRODUCTION

Advances in our understanding of Grand Unification are pointing once again to the connections between the very small and very large - particle physics and cosmology. The aim of this conference is to explore the interface between those two subjects. Unlike Dr. Tammann, who, in the preceding paper described a neat and compact cosmological model representing the most popular observational point of view, I'm here to present a somewhat more pessimistic view of the current state of observational cosmology. Rather than muddying the waters, I would like to show that there is still room for speculation. Ask not what particle physics can do for observational cosmology but rather what cosmology can do for Grand Unification!

Observational cosmology is often described as the search for two numbers. In reality, it is the search for two and one-half numbers - the expansion rate (Hubble constant), the deceleration parameter, and, with much trepidation, the cosmological constant. Following Einstein and not Eddington, we observational cosmologists would like very much to ignore this last parameter. These parameters are described in the simple Friedmann-Lemaitre models by:

Hubble's constant	$H_o = \dot{R}/R$
Deceleration parameter	$q_o = -R\ddot{R}/\dot{R}^2$
Cosmological constant	$\Lambda = 3H_o(\Omega/2 - q_o)$

where R is the "scale factor" of the universe at any given time, and Ω is the ratio of the mean mass density, ρ, to the mass density required to close the universe. Note that if the cosmological constant is zero (the fervent hopes of many!), then the deceleration parameter is just related to the cosmological mean mass density by:

$$2q_o = \Omega = 8\pi G\rho/3H_o^2 \cdot$$

The object of our game is to determine these parameters by observation - we can't play with the real universe in a laboratory so we have to do our best with passive techniques. We determine the Hubble constant by measuring both distances and redshifts of galaxies. We try to determine the deceleration parameter by measuring the distortion of the redshift-magnitude and redshift-diameter relations at high z (0.5 to 1.0). We determine Ω by measuring the luminosity density and mean mass-to-light ratio of galaxies and other material. And we resort to the cosmological constant if there is a measurable discrepancy between Ω and $2q_o$, or a measurable discrepancy between the age of the universe as determined from H_o and q_o and other (chemical, geological, stellar evolutionary) measures of age.

II. THE HUBBLE CONSTANT

The debate about the distance scale and thus the Hubble constant has been raging over several decades. Until recently there were only two major efforts: A. Sandage and G. Tammann,[27] who derive low values of the Hubble constant (50 km/s/Mpc - e.g. the previous paper), and G. de Vaucouleurs and his collaborators, who derive much higher values (100 km/s/Mpc)[8,9]. Other workers derive values spanning the whole range between the two and then some. The Supernovae model Baade-Wesselink values[19,4] go well below 50 (but neglect an important correction that we will discuss below), and Lynden-Bell's radio source expansion value[22] is well above 100.

Rather than champion any particular value (or try to detail their differences) I would like to describe one of the more recent determinations and discuss how it depends on other calibrations. In any review of this subject it is important to stress that the two basic experimental problems are: (1) accurate calibration of the local distance scale to allow the calibration of "extragalactic" standard candles, and (2) measurement of the Hubble flow on scales large enough to avoid problems with local inhomogeneities. A weaker alternative approach to this last could be very accurate measurement of the distortion of the Hubble flow caused by the Local "Supercluster".

It is also important to stress that "a measurement is a measurement". If the very best determination of the universal Hubble parameter and the deceleration parameter yield an age of the universe inconsistent with radioactive dating or stellar evolutionary theory then we will probably be forced to consider alternative, more complex cosmological models.

Over the last five years, Marc Aaronson, Jeremy Mould and I along with several groups of radio observers have been using the Infrared Tully-Fisher relation for the determination of relative distances to galaxies.[1,2,5,24]. This relation is based on a direct relation between the rotational velocities of spiral galaxies and their luminosities. More luminous galaxies are more massive and thus spin faster. When the luminosity is measured in the infrared (to lessen effects of galactic extinction, internal absorption in edge on galaxies, and differences in star formation rate that strongly affect optical luminosities) there appears to be little or no dependence on the morphological type of the galaxy and probably no dependence on local galaxy density (although this last point is still under debate). At present, this method is by far the most promising of all techniques for measurements of the relative distances of galaxies on large scales.

We have used these relative distances to galaxies in the local supercluster to measure the infall pattern or local distortion of the Hubble flow caused by the mass concentration of the Virgo cluster.[1] The pattern infall to Virgo at the local group is 250 ± 50 km/s; this value does not depend on the absolute value of the Hubble constant. It is also in excellent agreement with "infall" velocities derived by other means by other groups: 220 km/s by Sandage, Tammann, and Yahil[33] from deviations in the luminosity function of nearby galaxies, 300 km/s by de Vaucouleurs from an independent distance estimator, 250 km/s by Hoffman and Salpeter from detailed modeling of the velocity and velocity-dispersion field near Virgo. This infall velocity is important to consider in any local determination of the Hubble constant (e.g. supernovae in nearby galaxies) and has other cosmological implications that we will discuss later.

Armed with the infall velocity, we can easily use Tully-Fisher to derive the Hubble constant in the direction of the Virgo cluster. The cluster velocity relative to the local group is 1055 ± 40 km/s (velocities are trivial to measure).[16] Using Sandage and Tammann's old calibration[28] of the distances to M31 and M33, we derive an absolute calibration for the Infrared Tully-Fisher relation and find a distance to the Virgo cluster of 15.7 Mpc.[24] Thus

$$H_0(\text{Virgo}) = (1055 + 250)/15.7 = 83 \text{ km/s/Mpc},$$

voila! This result is also in fairly good agreement with the values of 85 to 90 km/s/Mpc that we find for distant clusters of galaxies[2] and distant Sc I galaxies.[5]

Sandage and Tammann have recently recalibrated distances to local group galaxies.[27] If we adopt their 1984 calibration rather than the earlier zero point, the above values change to about 80 km/s/Mpc. If we go a step further and adopt for our basic calibration of IR Tully-Fisher the new Sandage and Tammann distance to Virgo, 20 Mpc, the values drop to 65 km/s/Mpc (not 50). If we adopt de Vaucouleurs' Virgo distance of 11.5 Mpc, H_0 is 110 km/s/Mpc!

The most promising new local calibration technique is McAlary, MacLaren and Madore's measurements of galactic and extragalactic Cepheid variables in the Infrared.[23] Like the Infrared Tully-Fisher method for galaxies, this method gets around problems of dust absorption and metallicity difference. This technique is yielding distances to Local Group galaxies smaller than those of Sandage and Tammann. If we use the IR-Cepheid distances to nearby galaxies to calibrate IR Tully-Fisher scale, the Hubble constant is ≃ 90 km/s/Mpc.

High values for the Hubble constant produce ages for the universe that are inconsistent with measures of the ages of globular clusters and marginally consistent with radioactive dating and isotope ages. This is not yet a real problem - the path to globular cluster ages makes the distance scale problem simple by comparison. We can't yet produce solar models that predict its neutrino flux or even accurately describe its observed optical colors. When and if Hubble and cluster ages are found to disagree, we may have to resort to a cosmological constant. Until then ...

Bottom line: despite Tammann's optimism, I doubt that the Hubble constant is actually known to better than 25%. New techniques such as measurement of the Sunyaev-Zeldovich effect or use of Space Telescope to better the local calibration of the Cepheid scale and apply it directly to Virgo may reduce the uncertainty to 10% or less in the next decade.

III. DECELERATION PARAMETER

Classically, q_0 is directly measured by examining the redshift-magnitude relation (called the Hubble diagram) or the angular diameter-redshift relation. Again there have been two major measurement efforts, Sandage and Kristian and collaborators[21] and Gunn and Oke and collaborators.[13] Both attempt to use first ranked cluster galaxies as standard candles and yardsticks, and several students of Gunn have investigated the use of galaxy cluster diameters as yardsticks. Neither group has arrived at a definitive answer (not counting Dr. Tammann's backwards derivation from the Hubble time versus the ages of globular clusters!). These candles and yardsticks suffer from a major problem - evolution. The stellar population in a galaxy changes with time (stars evolve and the details of the stellar populations in galaxies are poorly known),[29,14] clusters of galaxies can and do appear in many dynamical states, and galaxies in clusters collide,[25] eat and are eaten.[26]

Other, less transparent, techniques rely on faint galaxy counts[30] or the QSO Hubble diagram. These approaches are even more plagued by the effects of evolution. The best that can be said at present about the deceleration parameter is that it is forcing us to learn about galaxy evolution and that the steady state theory is probably ruled out. The current best determinations are all consistent with $q_o = 0 \pm 1$, not very useful limits to either cosmologists or particle physicists. All is not lost yet - it will be possible to better determine the deceleration parameter from timing arguments applied to high redshift supernovae or by tests using gravitational lenses (e.g. Dr. E. Turner's paper).

IV. THE MEAN MASS DENSITY

Because of the open/closed universe question, the lack of information about q_o, and the strong prediction of the Inflationary Universe model (see Dr. A. Guth's papers) the mean mass density is of great interest to both particle physicists and cosmologists. It is also deceptively easy to derive.

The mean luminosity density in the universe is excruciatingly well known. The Revised Shapley-Ames catalog,[28] the Center for Astrophysics redshift survey,[7] the Kirshner, Oemler, Schechter and Shectman survey[20] and Ellis and collaborators faint galaxy counts[10] all give the same answer to within 10%:

$$L_{B(0)} \simeq 1.2 \times 10^8 \, L_\odot \, \mathrm{Mpc}^{-3} \, .$$

This is in the system of raw B(0)-Zwicky magnitudes[15] and H_o=100km/s/Mpc. (No one yet has tabulated more magnitude for bright galaxies than Zwicky and his collaborators and since that system is fairly linear and has known scatter it makes sense to use it until something better comes along). In this system, closure M/L is approximately 2200 in solar units. Remember, it does not matter for "cosmological" purposes what system of magnitudes and Hubble constant are used as long as they're used consistently. (Note: luminosity density and thus the closure M/L scale as H_o, thus the closure M/L is 1100 for H_o = 50 km/s/Mpc).

The mean mass-to-light ratio (M/L) is determined by a variety of dynamical techniques - rotation curves for spiral galaxies, velocity dispersions for elliptical galaxies, the virial and projected mass estimators for groups and clusters of galaxies. The estimators used for systems of galaxies almost always depend on the system being bound and dynamically relaxed (virialized).[3] An excellent review of techniques and early determinations was given by Faber and Gallagher.[11] Since then there have been several new results as a consequence of more detailed studies of dynamical systems.

New results on binary galaxies[32] have confirmed the earlier results of Turner,[31] but point out the extreme difficulty of selecting a sample of such objects. Construction of catalogs of groups of galaxies with full dynamical information has been aided tremendously by the production of complete galaxy redshift catalogs.[17] Our ability to rapidly measure radial velocities for galaxies now allows very detailed studies of cluster dynamics.

One fly has appeared in the ointment. Our more detailed studies of clusters have shown that very few of these beasts are the simple dynamical objects that we once thought they were.[16] Many have considerable substructure indicating that many systems of galaxies are dynamically young.[12] If clusters are bound but not relaxed, then the classical estimators like the virial theorem will overestimate their mass (a just bound cluster has its potential energy equal to its kinetic energy, not twice the kinetic energy). Many clusters are merely superpositions of groups which are not even bound![6] Projection effects are important - for example, the estimated mass of a system changes by a factor of two if the orbits are assumed to be radial instead of isotropic.

A good example of the projection problem is the globular cluster system surrounding M87. M87 is the dominant galaxy at the core of the Virgo cluster. Jean Brodie and I have recently measured velocities for 10 globular clusters around that galaxy in an attempt to measure its halo's mass. Unfortunately, because of the high surface brightness of the galaxy itself, we could only measure clusters outside its central region. In the determination of the halo mass we will assume that the distribution of orbital eccentricities of the globular cluster system is nearly isotropic. If the outer globulars are actually in more radial orbits we will underestimate the mass because we will underestimate the true velocity dispersion; conversely, if they are in more circular orbits we will overestimate the mass.

Almost all of the effects mentioned above work to decrease the masses that are measured for clusters.

Now we can also measure a M/L on a supercluster scale. The infall velocity to Virgo is related to the mean mass density through the overdensity of the local supercluster contained in a sphere centered on Virgo within the radius of the Local Group. We have measured the pattern infall fairly accurately - 250 km/s;[1] the supercluster overdensity as measured by the CfA redshift survey is about 2.2.[7] This implies an Ω of about 0.15 or a M/L of $\simeq$ 350. There has been some argument over whether the microwave background velocity towards Virgo (about 430 km/s) should be used instead of the pattern velocity. That would raise the M/L. However, we do not know that Virgo's mass is the source of all that acceleration, so the lower value must be preferred.

The best current results on M/L can be summarized as follows:

System	M/L	Scale
Galaxies, internal	$\simeq$ 10	in $\simeq$ 10 kpc
Galaxies, extended	$\simeq$ 50	in $\simeq$ 50 kpc
Galaxies, binary	$\simeq$ 100	in $\simeq$ 100 kpc
Groups of galaxies	$\simeq$ 300	in $\simeq$ 500 kpc
Clusters of galaxies	$\simeq$ 500	in $\simeq$ 1000 kpc
Local Supercluster	$\simeq$ 350	in $\simeq$ 20000 kpc

where M/L is again quoted in B(0)-Zwicky magnitudes in equivalent solar units. The M/L of systems of galaxies does increase with scale up to a few hundred kpc, ≃ the mean intergalaxy separation, but then levels off at values well below the closure value. M/L for systems of galaxies does not increase at large scales. (Note that M/L also scales as H_o except for the supercluster measure which scales as the Virgo distance). The ratio of the mean mass density to the closure density is only about one sixth. If this value is correct, and it would be very hard to squeeze a factor of 6 into the present measurements, the simple Inflationary cosmology is ruled out.

All of the observational techniques for measuring mass assume that the galaxy distribution traces the mass distribution, i.e. that we can determine the shape of the potential by the positions of galaxies. Those who wish to save $\Omega = 1$ cosmologies argue that this assumption is wrong. However, available tests such as the distribution of hot gas in clusters of galaxies generally argue in favor of this assumption. Anyway, I'm from New Jersey, which is a lot like Missouri... Given that, the best experimental results are that $\Omega = 0.15 \pm 0.1$ (where the error bars refer to the fraction of observational astrophysicists deviating in belief).

It is important to note a caveat for the theoretician or particle physicist who reads astronomical papers about M/L. The actual value depends on the magnitude system and color used to measure luminosity. For example, for de Vaucouleurs' B_T magnitudes, which integrate galaxies' luminosities to larger radii than B(0), $(M/L)_{BT} \simeq 0.7\ (M/L)_{B(0)}$. For visual total magnitudes (instead of blue) $(M/L)_{VT} \simeq 0.5\ (M/L)_{B(0)}$ because galaxies are redder than the Sun. The closure M/L scales exactly the same way so Ω is not affected if everything is calculated in the same system.

As an additional aside, galaxy population modelers, i.e. we dark halo folks, only have to explain a total integrated M/L of about 250 in the blue if the Hubble constant is ≃ 65 km/s/Mpc. Galaxy halos can still be made quite nicely of very cool K and M dwarf and subdwarf stars.

V. SUMMARY

My conservative view of the state of the observational determination of cosmological parameters is thus:

$$50 < H_o < 100 \text{ km/s/Mpc}$$

$$q_o \simeq \text{?} \quad \text{but probably close to 0}$$

$$0.05 < \Omega < 0.25$$

I think the Hubble constant will probably end up between 65 and 75 km/s/Mpc after all the debates about the local distance scale calibration are settled. The observers' universe would then be open and about 13 to 14 billion years old. We do not need to "invoke" the cosmological constant yet, but a good determination of q_o will settle the question.

Because of the brevity of this review, I have not prepared an extensive reference list, but have rather endeavored to mention the most recent papers in the field. A complete reference list would take up more than my allotted page limit. My apologies to people whose papers I have missed. I would like to thank M. Geller for a careful reading of parts of this manuscript. This work has been supported by the Smithsonian Institution and by NASA grant NAGW-201.

REFERENCES

[1]M. Aaronson, J. Huchra, J. Mould, P. Schechter, and R. B. Tully, Astrophys. J. **258**, 64 (1982).

[2]M. Aaronson, J. Mould, J. Huchra, W. Sullivan, R. Schommer, and G. Bothun, Astrophys. J. **239**, 12 (1980).

[3]J. Bahcall and S. Tremaine, Astrophys. J. **244**, 805 (1981).

[4]D. Branch, M.N.R.A.S. **186**, 609 (1979).

[5]G. Bothun, M. Aaronson, R. Schommer, J. Huchra, and J. Mould, Astrophys. J. **278**, 475 (1984).

[6]G. Bothun, T. Beers, M. Geller, and J. Huchra, Astrophys. J. **268**, 47 (1983).

[7]M. Davis and J. Huchra, Astrophys. J. **254**, 437 (1982).

[8]G. de Vaucouleurs, Astrophys. J. **268**, 451 (1983).

[9]G. de Vaucouleurs, Astrophys. J. **268**, 468 (1983).

[10]R. Ellis, in Early Evolution of the Universe and its Present Structure, eds. Abell and Chincarini, (Reidel:Dordrecht) p. 87 (1983).

[11]S. Faber and J. Gallagher, Ann. Rev. Astron. and Ap. **17**, 135 (1979).

[12]M. Geller and T. Beers, P.A.S.P. **94**, 562 (1982).

[13]J. Gunn and B. Oke, Astrophys. J. **195**, 255 (1975).

[14]J. Gunn, L. Stryker, and B. Tinsley, Astrophys. J. **249**, 48 (1981).

[15]J. Huchra, A. J. **81**, 952 (1976).

[16]J. Huchra, R. Davis, and D. Latham, in Groups and Clusters of Galaxies, Giuricin, Mezzetti, and Mardirossian, eds.

[17]J. Huchra and M. Geller, Astrophys. J. **257**, 423 (1982).

[18]S. Kent and J. Gunn, A. J. **87**, 945 (1982).

[19]R. Kirshner and J. Kwan, Astrophys. J. **193**, 27 (1974).

[20]R. Kirshner, G. Oemler, P. Schechter, and S. Shectman, A. J. **88**, 1285 (1983).

[21]J. Kristian, A. Sandage, and J. Westphal, Astrophys. J. **221**, 383 (1978).

[22]D. Lynden-Bell and W. Liller, M.N.R.A.S. **185**, 539 (1978); M.N.R.A.S. **185**, 539 (1978).

[23]C. McAlary and B. Madore, Astrophys. J. in press (1984).

[24]J. Mould, M. Aaronson, and J. Huchra, Astrophys. J. **238**, 458 (1980).

[25]J. Ostriker and S. Tremaine, Astrophys. J. **202**, L113 (1975).

[26]J. Ostriker and M. Hausman, Astrophys. J. **217**, L125 (1977).

[27]A. Sandage and G. Tammann, Nature **307**, 326 (1984).

[28]A. Sandage and G. Tammann, in Cosmology and Fundamental Physics, Vatican Study Week (1981).

[29]B. Tinsley, Astrophys. J. **151**, 547 (1968).

[30]B. Tinsley, Astrophys. J. **211**, 621 (1977).

[31]E. Turner, Astrophys. J. **208**, 304 (1976).

[32]S. White, J. Huchra, D. Latham, and M. Davis, M.N.R.A.S. **203**, 701 (1982).

[33]A. Yahil, A. Sandage and G. Tammann, Astrophys. J. **242**, 448 (1980).

Abundance of the Light Elements

B.E.J. Pagel

ABSTRACT

This is yet another review of the abundances of the light elements of cosmological interest, D, ^{3}He, ^{4}He and ^{7}Li. The results for D, ^{3}He and ^{7}Li constrain η in the standard Big Bang nucleosynthesis model to be between 3.5 and 6×10^{-10}, similar to the conclusions of Steigman[1] and other previous discussions. That model (with 3 neutrinos) then constrains the helium abundance to be within the limits $0.242 < Y_p < 0.253$, which is in quite good agreement with the value of 0.243 ± 0.010 found from extragalactic HII regions by Kunth and Sargent.[2] Some other astrophysical estimates suggest lower values of Y_p which would place the standard Big Bang nucleosynthesis model in difficulties, but the errors in these estimates are difficult to quantify at present.

I. INTRODUCTION

Gary Steigman has just indicated his view that standard Big Bang nucleosynthesis makes predictions that are in good agreement with astronomical observations. This view is quite correct within the uncertainties, but at the same time one has to admit that the results of astronomical observations involve ambiguities and difficulties which leave room for differing interpretations (cf. Audouze).[3,4]

What astronomers would like to do is to measure the abundances of elements in objects having either different ages or different degrees of effects of stellar evolution as measured, for example, by the abundance of oxygen or metals, and fit these to some relationship that can then be extrapolated to obtain pre-galactic abundances that can be checked against the Big-Bang predictions. Reference points that can be used for this purpose are (in order of increasing age) the present-day local interstellar medium (ISM), the solar system at the time of its origin (4.7×10^9 years ago) and extremely metal-deficient subdwarf stars and stars in globular clusters, both belonging to Population II and having essentially the age of the Galaxy of the order of 10^{10} years. Alternatively one may look at objects that are still young (or in a state of arrested development) in the sense that there has been little enrichment in heavy elements from carbon upwards attributable to stellar activity, particularly the compact HII galaxies in which helium is observed, and extrapolate their abundances to zero metallicity.[5] The assumption usually made for the particular case of helium is that there is a linear relation between helium mass-fraction Y and heavy-element mass fraction Z of the form

$$Y = Y_p + Z(dY/dZ) \ . \qquad (1)$$

Four basic problems arise in carrying out this program. To begin with, spectroscopic abundances are not very accurate. In most cases in astrophysics an error of ± 0.1 dex or 25 per cent can be considered as very good indeed. For helium in emission nebulae one can do considerably better than this if suitable precautions are taken,[6] perhaps about 5 per cent, but even this is not really good enough to make as critical a test of Big Bang predictions as one would like to have nowadays. Theoretical estimates, based on considerations of stellar structure, are subject to systematic errors which are hard to quantify. The second problem is that of obtaining a fair sample. Observations of D and ^3He in the ISM give results that vary over an unexpectedly wide range, like an order of magnitude, and so it is hard to choose values that one can wholeheartedly accept as being representative. This problem in turn may raise some doubts as to whether the Solar System was truly typical of the state of the local part of the Galaxy at the time of its formation. The third problem is that of fractionation, which makes it virtually impossible to make quantitative deductions from molecular spectroscopy and places a further question mark on any abundances measured in the Solar System. The fourth and final problem is that of extrapolation to pre-galactic conditions. In stars and planetary nebulae one has to consider the effects of self-enrichment and self-impoverishment, while in the ISM one has to consider the creation and destruction arising from stellar activity in the course of galactic chemical evolution. Theoretical models of these processes are inadequate or controversial, while purely empirical extrapolations lead to a multiplication of the inevitable observational errors.

Having expressed these reservations, I shall now look at the various relevant elements in turn, make some attempt to extrapolate them to pregalactic conditions and examine the possible window in parameter space for Big-Bang nucleosynthesis. In doing so I shall draw heavily on previous reviews of light element abundances by Pagel,[7] Yang et al.,[8] Steigman[1] and in the ESO Workshop on Primordial Helium,[9] all of which may be consulted for further details.

II. DEUTERIUM AND ^3He

Direct determinations of proto-solar deuterium have been made in the Solar System by molecular spectroscopy of the major planets.[10-15] The uncertainties are considerable, but they are not inconsistent with the value given in Table I, derived from solar winds, ancient and

TABLE I

D and ^3He

	ISM	pre-Solar	correction factor α
$10^5 \frac{D}{H}$	0.5:	2.3 ± 0.5	$1 < \alpha(<4)$*
$10^5 \frac{^3He}{H}$	4: (40 to <2)	1.3 ± 0.3	$(\alpha \simeq 1)$*
$10^5 \left(\frac{D+^3He}{H}\right)$	4.5:	3.6 ± 0.2	$1 < \alpha(<2.3)$*

*Values in brackets are based on a fit to standard Big Bang nucleosynthesis.

modern. Deuterium has also been observed in the ISM, both in molecular form (from which it is difficult to estimate a quantitative abundance) and in atomic form, since 1972. Deuterium abundances deduced from interstellar Lyman lines, observed mainly with the Copernicus satellite, range over an order of magnitude or more[16] which is more than was expected from mere errors in the determination, and various mechanisms were tried out in order to account for this.[17] More recently Vidal-Madjar et al.[18] discovered that the deuterium Ly-γ and Ly-δ lines towards ε Per varied with time,[19] suggesting that what was being observed was not interstellar deuterium at all, but a neutral hydrogen cloud moving out from the star at a velocity of 80 km s^{-1} mimicking the isotope shift. This conclusion receives support from observations of what are undoubtedly gas clouds moving at other velocities in ε Per and other stars and in some cases also varying with time. From the limited data so far available, Vidal-Madjar et al.[18] find a correlation between apparent deuterium abundance and stellar luminosity class, increasing from supergiants towards dwarfs. They suggest that the lowest value found, $D/H \simeq 5 \times 10^{-6}$, represents either the true ISM value or an upper limit thereto.

The resulting discrepancy with the solar system value has occasioned some surprise, because people had been brainwashed into believing that the surviving fraction of primordial deuterium today in the ISM, which is equal to the "virgin" fraction of interstellar gas that has not been cycled through stars, could not be less than about 1/2 according to models of galactic chemical evolution.[20] I pointed out that it could be considerably less than this, suggesting a lower limit of 0.2,[7] but I have to confess that even I did not go far enough, because perfectly good models now exist that make this fraction as little as 0.05[21] or possibly even less.[4] Such a large degree of astration of the ISM relative to the time of formation of the Solar System receives some support from one crumb of observational evidence, namely the virtual non-observation of interstellar ^{6}Li towards ζ Oph.[22] To summarise, we don't know the abundance of D in the ISM, and even if we did know it, we have no idea how much we would need to scale it up in order to extrapolate to pre-galactic conditions.

A direct approach to pre-galactic deuterium could be provided in principle by looking for deuterium lines in the Lyman-α "forest" of absorption lines in high red-shift QSOs, which appears to come from primordial hydrogen clouds of zero or very low metallicity.[23] The problem is that very few such clouds have a sufficiently high column density and probably only a fraction of these have sufficiently low velocity dispersion and freedom from blends. Searches reported to date have been dramatically unsuccessful,[24,25] with only an uninteresting upper limit of $D/H < 10^{-2}$ having been achieved.[24]

So much for interstellar deuterium. It is convenient at this point to mention the equally confusing results of the hunt for interstellar ^{3}He using the hyperfine λ3.46 cm transition of $^{3}He^{+}$ in HII regions. The latest work, by Rood, Bania and Wilson,[26] represents a heroic effort with unprecedented achievements in sensitivity and precision, and the results in six HII regions are even more spread out than the results for deuterium, ranging from $^{3}He/H = 40 \times 10^{-5}$ in W3, 12 kpc from the Galactic centre, to 4×10^{-5} in W43, 5 kpc from the galactic centre, to mere upper limits of 6×10^{-5} in Orion and only 2×10^{-5} in M17 and W49. One needs to be very brave to fit these results to any kind of abundance gradient or galactic chemical evolution model, but it looks as though the outer HII region W3 is showing effects of stellar enrichment, whereas the others are not inconsistent (bearing in mind all possible errors) with the sum of D + ^3He in the Solar System, to which I now turn.

The most precise results for D and ^3He in the Solar System seem to come from $^3He/^4He$ in various components of the solar wind, which we convert to $^3He/H$ assuming $^4He/H = 0.09$. The present-day solar wind, as observed in gas-rich meteorites, and foils on the Moon and in space, contains both proto-solar ^3He and ^3He resulting from the burning of protosolar deuterium. A third component could exist due to upward diffusion of ^3He resulting from hydrogen burning in the interior[27]; I shall neglect this for the present discussion, but the resulting possible complication needs to be borne in mind. Solar wind measurements[28-31] as summarised by Steigman[1] give (cf. Table I)

$$y_{23} \equiv \left(\frac{D}{H} + \frac{^3He}{H}\right)_{pre_\odot} = \left(\frac{^3He}{H}\right)_\odot = (3.6 \pm 0.2) \times 10^{-5} ,$$

which is also in good agreement with a spectroscopic determination of ^3He in a prominence.[32] The proto-solar deuterium abundance follows by subtraction of the ^3He in a more ancient solar wind released by stepwise heating of carbonaceous chondrites.[30,33] Various determinations from these authors and others[28,34,35] give (again following Steigman's summary)

$$y_2 \equiv \left(\frac{D}{H}\right)_{pre_\odot} = (2.3 \pm 0.5) \times 10^{-5}.$$

What constraints do these numbers impose on Big Bang nucleosynthesis? As we have seen, deuterium has to be divided by the protosolar virgin fraction f, which is unknown a priori but can be deduced from the Big Bang calculations if we fix on a particular value of η. In practice as we shall see later, η is rather tightly constrained to be between 3.5 and 7×10^{-10} (see Fig. 1) giving a theoretical primordal deuterium abundance

$$2.5 < 10^5\, y_{2p} < 7, \text{ or } f > 0.25 .$$

The upper limit on η comes from lithium and solar-system deuterium; the lower one comes from an argument introduced by Schramm,[36] Steigman,[37] Yang et al.[8] and Steigman,[1] using the sum of the abundances of D and ^3He. Following these authors we can write

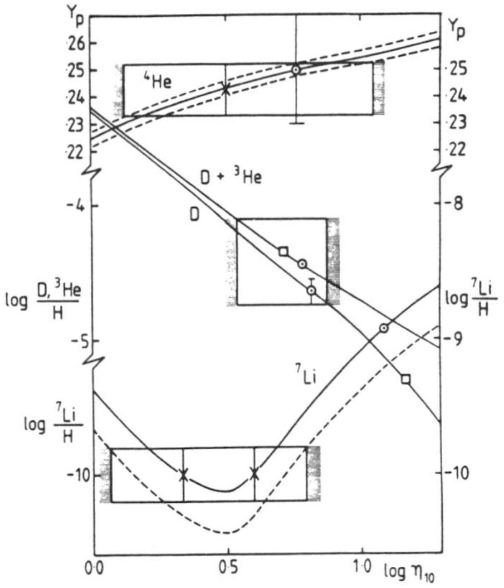

Figure 1: Primordial element abundances. Curves are theoretical values in the standard model with three neutrinos as a function of the baryon-photon ratio η after Steigman,[1] with broken-line curves indicating theoretical uncertainties. Crosses indicate primordial values deduced from observation, with boxes indicating the constraints that they impose on η within the framework of the standard model. Squares indicate abundances found in the local intersteller medium and Sun symbols indicate values in the Sun at the time of formation.

$$y_2 = f\, y_{2p} \tag{2}$$

and

$$y_3 \geq f\, y_{3p} + (1-f)g\, y_{23p} \tag{3}$$

where g is the fraction of ^{3}He which survives evolution in an average star and the inequality allows for any net synthesis of ^{3}He from H-burning followed by ejection into the ISM. Adding (2) and (3) we derive

$$y_{23p} \leq \frac{y_{23}}{f + g(1-f)} \tag{4}$$

g depends on stellar evolution and the initial mass function and according to estimates discussed by the above authors it may be somewhere between 1/4 and 1/2 or more.

Taking g > 1/4, f > 1/4, we find from equation (4)

$$\frac{y_{23p}}{y_{23}} \leq 2.3$$

$$y_{23p} \leq 8.7 \times 10^{-5} ,$$

whence $\eta \geq 3.5 \times 10^{-10}$ as we assumed at the outset. This agrees with the result obtained by Steigman using a slightly different argument, and we can calculate lower limits on η and f for any other value of g that may transpire from stellar evolutionary calculations; eg if g = 1/2 then $\eta > 4.6 \times 10^{10}$ and f > 0.45. (Thus the old estimate of a factor of 2 depletion may not be so wrong after all, at least as far as the Solar System is concerned.)

III. HELIUM

We now turn to helium, for which (in contrast to the other light elements) there is an enormous amount of data which has been reviewed at length in the references cited at the end of Section I. Some of the results are shown in Table II. The tension now rises, because it is clear from Fig. 1 that, while a primordial mass fraction Y_p of helium exceeding about 0.24 leads to a window of possible values of η in the framework of the standard Big Bang that is compatible with the constraints deduced from D + ^{3}He, a lower value would lead to a conflict such that this model of nucleosynthesis would have to be modified or abandoned unless the D and ^{3}He data have been misinterpreted. Many, indeed the majority of the values in Table II, are actually less than

TABLE II

Estimates of Y_p

Objects	Result	Method	Author	Reservations
Solar System	< 0.25 ± 0.02:	Sun int. and osc.	Gough[56]	Errors hard to quantify. ν-flux?
	< 0.20 ± 0.04:	Voyager, Jupiter	Gautier[15]	Model? Fractionation?
	< 0.28 ± 0.05	Prominence HeI	Heasley & Milkey[57]	Transfer model
Nearby B stars	< 0.28 ± 0.04	Abs. spectroscopy	Nissen and refs[58]	
h & χ Per	< Y(nrB) - 0.05 ± 0.02:	β-I(4026) ptmtry	Nissen[58]	Metal-line blanketing?
Hyades, NGC2301	< 0.2:	(c_o, β) for F stars	Nissen[58]	Rotation? CNO?
6 Subdwarfs	0.19 ± 0.05	ZAMS	Carney[38]	Convection, T_{eff}, parallax
Glob clusters	0.23:	RR and Δm(HB-ZAMS)	Caputo & Castellani[59]	Different methods discordant
	0.23 ± 0.02:	R=N(HB)/N(RG)	Buzzoni et al.[60]	Different methods discordant
	0.20 ± 0.03	R+N(HB)/N(RG)	Cole et al.[61]	Same method discordant
Galactic nebulae	0.22 ± 0.02	Planetaries	Peimbert[43]	Stellar and galactic enrichment
	0.22:	HII (radio)	Mezger & Wink[44]	He^0; gal enrichment
EG HII	0.233 ± 0.005	Irr. galaxies	Lequeux et al.[47]	He α icf
	0.216 ± .010	NGC 5471(M101)	Rayo et al.[51]	Icf = 1.00? dY/dZ = 3.8?
	0.24 ± ?	M101	Rosa[52]	Preliminary result
	0.243 ± 0.010	HII galaxies	Kunth & Sargent[2]	

0.24, so we need to decide whether this is a groundswell of accumulating evidence against the standard model or just an illustration of the systematic uncertainities in the results. At the same time, very few results fall below 0.20, giving a good reason to believe that the bulk of the helium has been synthesised in some universal process the results of which have an uncanny resemblance to the Standard Big Bang, especially when the observations of ^{7}Li are taken into account. Or, to speak more conservatively, it is a not unremarkable fact that these diverse methods, applied to a great variety of different objects, lead to results that are sufficiently similar to one another to suggest very strongly that there is such a thing as a primordial helium abundance that has a cosmological significance and that it is virtually certain that it lies between the limits $0.20 < Y_p < 0.25$.

There is not space to give a detailed review of the different methods here.[9] For the Sun one builds evolutionary models that give the solar luminosity and radius at the present age and reproduce the correct oscillation frequencies but not the low neutrino flux. Infra-red Voyager spectra and occultation observations of the atmosphere of Jupiter give an alternative estimate for the Solar System, assuming that no He-H_2 differentiation has taken place; this may or may not be true. Spectroscopic observations of solar prominences have a large uncertainty due to uncertainty in atomic level populations and radiative transfer effects. The upshot is that the helium abundance in the Solar System is not known very well.

For nearby stars of spectral type B, the situation is apparently somewhat better, because although the formal error is about the same, the mean value is in excellent agreement with the more accurate determinations for the Orion Nebula. However, the correction for galactic enrichment by stellar nucleosynthesis is probably substantial, in view of the results for extragalactic HII regions.[7] A substantial correction is also suggested by the results for B stars in open clusters further out from the Galactic centre such as h and χ Persei. Other results for open clusters like the Hyades, in the solar neighborhood, are based on Strömgren photometry of F stars, which measures the effective gravity; but owing to the possible complications of other effects like stellar rotation and CNO abundances, the systematic errors are totally unknown.

Old stars of Population II are of special interest, because their initial helium abundance should be very close to the pregalactic value. One effect that can be used to estimate it is the location of the unevolved main sequence in the Hertzsprung-Russell (HR) diagram, i.e. the relation between luminosity and effective temperature, and Carney[38] has used this fact to study six cool subdwarfs having distances from trigonometric parallax. The result depends quite sensitively on certain systematic corrections to the raw parallaxes which have brought it down by 0.05 compared to uncorrected parallaxes and there are uncertainties in the temperature scale and in the effect of convection on the theoretical main sequence. A much better determination, both for subdwarfs and for main-sequence stars in general, will be possible after completion of the mission of the astrometric satellite HIPPARCOS. Globular clusters have a number of properties that depend on initial helium abundance: the blue edge and width of the instability strip in the HR diagram occupied by RR Lyrae variables, the ratio R of relative lifetimes and numbers of horizontal-branch and red giant stars and the location of the horizontal branch in the HR diagram relative to the unevolved main sequence. Different methods, and even different people's application of the same method, lead to different results and the errors are again hard to quantify.

The "cleanest" determinations of helium abundance come from observations of emission lines in gaseous nebulae. Subject to certain precautions[6] such as allowing for underlying stellar absorption lines and for optical depth and collisional effects involving the metastable 2^3S level, the strengths of both helium I and hydrogen lines are given by straightforward recombination theory which is quite insensitive to electron temperature and density and is probably accurate to a few per cent or better. Their ratio, corrected for interstellar reddening, then gives the ratio of singly ionised helium to singly ionised hydrogen, weighted by the electron density along the line of sight.[39] Doubly ionised helium is rare in HII regions, and, when present, it is fairly easily observed and allowed for. So much for the good news. The bad news is that one now needs to apply an ionisation correction factor (icf) to allow for appreciable amounts of neutral helium in the HII region (i.e. the region where hydrogen is ionised), or in a few possible cases for the reverse, i.e. neutral hydrogen in the HeII region. There are basically two methods of estimating the icf. One is to observe certain other lines like [O^+], [O^{++}] and in some cases [S^+] and deduce the corresponding icf for helium by comparison with model HII regions, but the results of this are model-dependent. Another method, which is probably better, but can only be applied to apparently large objects like the Magellanic Clouds and Orion[40-42] is to study many HII regions in one irregular galaxy or many bits of one HII region where the abundances can reasonably be expected to be the same and plot He^+/H^+ against O^{++}/O; a good correlation is found in these cases, which can be extrapolated fairly well to the high ionisation limit which is assumed to correspond to icf = 1. This method was used in the classic investigations of the Magellanic Clouds by Peimbert and Torres-Peimbert[40,41] which first established that the helium abundance in low-metallicity HII regions is appreciably less than in the solar neighbourhood as typified by the Orion nebula.

In addition to HII regions, ionised by one or more hot young stars, there are planetary nebulae which are shells of gas expelled from old stars in an advanced stage of evolution. These have the advantage that in many cases the icf is probably close to 1.0, but the disadvantage that they contain helium manufactured by the star itself, as well as being enriched by galactic chemical evolution unless the planetary belongs to Population II. This leads to corresponding uncertainties in the primordial helium abundance deduced by Peimbert[43] from planetaries of different types. Galactic HII regions have been studied by Mezger and Wink[44] and by Shaver et al.[9] using combinations of optical and radio recombination lines. (The icf's are still more uncertain for radio than for optical data because of beam width effects and the lack of other lines to compare with.) Mezger and Wink[44] find a large-scale radial abundance gradient with helium decreasing outwards in the same sense as oxygen and metals, and extrapolate this using equation (1) to deduce Y_p somewhere near 0.22; but Shaver et al.[9] find no trace of a helium gradient and point out the uncertainty in the icf; also, galactic HII regions are too far from zero metallicity for a reliable extrapolation to be carried out in any case.

In most of the investigations discussed up to now, the primordial helium abundance is tied up with various purely astrophysical problems which are of great interest in themselves and which have often been considerably advanced by a by-product of the cosmological quest, which I have likened elsewhere to the quest for the holy grail.[45] But if we just want a number for Y_p, then these astrophysical problems get in the way. That is the basic reason for thinking that the best number for Y_p is to be derived from observations of extragalactic HII regions with low metallicity, of which three kinds are available having different luminosities but very similar physical conditions. They are the HII regions in nearby Irregular galaxies such as the Magellanic Clouds, the giant HII regions or complexes in the outer parts of late-type spirals like M101 and the very luminous HII galaxies or "isolated extragalactic HII regions" of which a few were first discovered as compact galaxies by Zwicky and many more have been discovered later, along with QSOs and Seyfert galaxies, in objective prism surveys for ultra-violet excess and emission-line objects. Some of these objects, such as I Zw18, have very low heavy-element abundances like 1/50 solar, so that the effects of enrichment by galactic chemical evolution should be small and relatively easy to estimate. Searle and Sargent,[46] in a historic investigation, showed that the helium in I Zw 18 and II Zw 40 was nearly normal, but Peimbert and Torres-Peimbert,[40] in another historic investigation, showed that it was a little bit less than in nearby HII regions like Orion.

Following the work of the Peimberts, Lequeux et al.[47] studied these two objects and a number of irregular galaxies (see Figure 1 of Ref. 7) and found $Y_p \simeq 0.23$ and $dY/dZ \simeq 3$. They argued that this value of dY/dZ was plausible on the basis of stellar evolutionary models with extensive mass loss, but Maeder[48,49] has argued that a value of 1 is more reasonable. Furthermore, the trend of Lequeux et al's[47] data is entirely accounted for by the icf's they adopted, which gives grounds for uneasiness.[50] Rayo et al.[51] deduced $Y_p = 0.216$ with a very small formal error, from a single HII complex, NGC5471 in M101, but their error is certainly much too low, because Rosa[52] showed that the apparent abundance varies considerably over the face of the complex and used an extrapolation procedure to deduce icf's for various HII regions of M101 which give $Y_p \simeq 0.24$. Ignoring a number of other investigations which give essentially no result because the errors are too large,[7] we come to the most convincing study of extragalactic HII regions, which is the one by Kunth and Sargent[2] giving the result on the bottom line of Table II. (I have adopted the result actually obtained from their maximum likelihood solution, rather than the straight mean recommended by them, mainly because the extra degree of freedom gives a much more realistic standard error, without altering the value significantly.)

This appears to be the best value we have and it can be seen from Fig. 1 that, while it formally lies outside the lower limit of η set by D + ^{3}He, the uncertainties are large enough to give a substantial window of possible values of η. If, however, more accurate observations come along in the future and establish an even slightly lower helium abundance like 0.23, or even 0.235, then the standard model will be in trouble.

Unfortunately, I see no prospect of improving on the extragalactic HII regions because of the icf problem, but there could be substantial advances in understanding globular clusters and planetary nebulae.

IV. LITHIUM 7

The remarkable discovery by Spite and Spite[53,54] of lithium lines in metal-deficient subdwarfs of Population II was somewhat unexpected because in the Sun, lithium has been virtually destroyed by convective mixing with deeper layers. The spectra are very clean and convincing and the abundance is remarkably constant for effective temperatures between 6500 and 5500 K, though at lower temperature there are the expected indications of convective depletion. The absence of ^{6}Li[54] and ^{9}Be[55] suggests that the lithium has not been produced by spallation, and the abundance obtained is in very impressive agreement with the standard Big Bang calculations, which can thus be credited with a highly successful prediction. The box shown in Fig. 1 takes account of the estimated errors in the determination and the factor 2 uncertainty in the prediction. The resulting constraints on η are in excellent agreement with those provided by the other light elements, with ^{7}Li providing (formally at least) the tightest upper limit on η, $\eta \leq 6 \times 10^{-10}$.

V. CONCLUSION

In the standard Big Bang model with three neutrinos, η is constrained to be between 3.5 and 6×10^{-10} by D + ^{3}He (in the Solar System) and ^{7}Li (in subdwarfs) respectively. This requires a primordial helium abundance by mass, $0.242 < Y_p < 0.253$, a result which is in quite good agreement with the range 0.243 ± 0.010 found from observations of extragalactic HII regions by Kunth and Sargent. Several astrophysical determinations of helium suggest lower values, but their errors are hard to quantify at present and so a more critical test of the model will have to await a better understanding of stellar evolution.

ACKNOWLEDGMENT

I thank Jean Audouze, Robert Carswell, Gary Steigman and Tom Wilson for their kindness in supplying results in advance of publication.

REFERENCES

[1]G. Steigman, in Challenges and Development in Nucleosynthesis, eds. W.D. Arnett and J.W. Truran, Yerkes Observatory, (1984).
[2]D. Kunth and W.L.W. Sargent, Ap. J. **273**, 81 (1983).
[3]J. Audouze, in ESO Workshop on Primordial Helium, eds. P. Shaver, D. Kunth and A. Kjär, Garching, p. 3, (1983).
[4]J. Audouze, in First ESO-CERN Symposium on Large Scale Structure of the Universe, Cosmology and Fundamental Physics, CERN, Geneva (1984).
[5]The word "metallicity" is used to denote the abundance of elements from carbon upwards.
[6]K. Davidson and T.D. Kinman, preprint, 1984.
[7]B.E.J. Pagel, Philos. Trans. R. S. London A., **307**, 19 (1982).
[8]J. Yang, M.S. Turner, G. Steigman, D.N. Schramm, and K.A. Olive, Astrophys. J. **281**, 493 (1984).
[9]P.A. Shaver, D. Kunth, and K. Kjär, (eds) ESO Workshop on Primordial Helium, Garching, (1983).
[10]J.T. Trauger, F.L. Roesler, N.P. Carleton, W.A. Traub, Ap. J. **184**, L137 (1973).
[11]J.T. Trauger, F.L. Roesler, and M.E. Mickelson, Bull. Am. Astron. Soc. **9**, 516 (1977).
[12]M. Combes and T. Encrenaz, Icarus **39**, 1 (1979).
[13]T. Encrenaz and M. Combes, Icarus **52**, 54 (1982).
[14]V. Kunde, R. Hanel, W. Maguire, D. Gautier, J.P. Baluteau, A. Marten, A. Chedin, N. Husson, and N. Scott, Ap. J. **263**, 443 (1982).
[15]D. Gautier, in ESO Workshop on Primordial Helium, eds. P.A. Shaver, D. Kunth and K. Kjär, Garching, p.139, (1982).
[16]C. Laurent, A. Vidal-Madjar, and D.G. York, Ap. J. **229**, 923 (1979).
[17]P. Bruston, J. Audouze, A. Vidal-Madjar, and C. Laurent, Ap. J. **243**, 161 (1981).
[18]A. Vidal-Madjar, C. Laurent, C. Gry, P. Bruston, R. Ferlet, and D.G. York, Astron. Astrophys. **120**, 58 (1983).
[19]C. Gry, C. Laurent, and A. Vidal-Madjar, Astron. Astrophys. **124**, 99 (1983).
[20]J. Audouze and B.M. Tinsley, Ap. J., **192**, 487 (1974).
[21]R. Güsten and P.G. Mezger, Vistas in Astron. **26**, 159 (1983).
[22]R. Ferlet and M. Dennefeld, in ESO Workshop on Prmordial Helium, p. 373, (1983).
[23]W.L.W. Sargent, P.J. Young, A. Boksenberg, and D. Tytler, Ap. J. Suppl. **42**, 41 (1980).
[24]D. Atwood, J.A. Baldwin, and R.F. Carswell, in 24th Liège Astrophysical Colloquium Quasars and Gravitational Lenses, p. 581, (1984).
[25]W.L.W. Sargent and A. Boksenberg, in 24th Liège Astrophysical Colloquium, Quasars and Gravitational Lenses (1984).
[26]R.T. Rood, T.M. Bania and T.L. Wilson, Ap. J. **280**, 629 (1984).
[27]E. Schatzman, in ESO Workshop on Primordial Helium, p. 137, (1983).
[28]P.M. Jeffrey and E. Anders, Geochim. Cosmochim. Acta **34**, 1175 (1970).
[29]D.C. Black, Nature Phys. Sci. **234**, 148 (1971).

[30]D.C. Black, Geochim. Cosmochim. Acta **36**, 347 (1972).
[31]J. Geiss and H. Reeves, Astron. Astrophys. **18**, 126 (1972).
[32]D.N.B. Hall, Ap. J. **197**, 509 (1975).
[33]E. Anders, D. Heymann, and E. Mazor, Geochim. Cosmochim. Acta **34**, 127 (1970).
[34]U. Frick and R.K. Moniot, Proc. 8th Lunar Science Conference, p. 229, (1977).
[35]P. Eberhardt, Proc. 9th Lunar Science Conference, p. 1027, (1978).
[36]D.N. Schramm, Philos. Trans. R. S. London, Ser. A. **307**, 43 (1982).
[37]G. Steigman, in ESO Workshop on Primordial Helium, p. 13, (1983).
[38]B. Carney, in ESO Workshop on Primordial Helium, p. 179, (1983).
[39]G. Stasinska, in ESO Workshop on Primordial Helium, p. 255 (1983).
[40]M. Peimbert and S. Torres-Peimbert, Ap. J. **193**, 327 (1974).
[41]M. Peimbert and S. Torres-Peimbert, Ap. J. **203**, 581 (1976).
[42]M. Peimbert and S. Torres-Peimbert, Mon. Not. R. Astron. Soc. **179**, 217 (1977).
[43]M. Peimbert, in ESO Workshop on Primordial Helium, p. 267, (1983).
[44]P.G. Mezger and J.E. Wink, in ESO Workshop on Primordial Helium, p. 281, (1983).
[45]B.E.J. Pagel, in ESO Workshop on Primordial Helium, p. 413, (1983a).
[46]L. Searle and W.L.W. Sargent, Ap. J. **173**, 25 (1972).
[47]J. Lequeux, M. Peimbert, J.F. Rayo, A. Serrano, and S. Torres-Peimbert, Astron. Astrophys. **80**, 155 (1979).
[48]A. Maeder, Astron. Astrophys. **101**, 385 (1981).
[49]A. Maeder, in ESO Workshop on Primordial Helium, p. 89, (1983).
[50]B.E.J. Pagel, ibid., p. 392, (1983b).
[51]J. Rayo, M. Peimbert, and S. Torres-Peimbert, Ap. J. **255**, 1 (1982).
[52]M. Rosa, in ESO Workshop on Primordial Helium, p. 317, (1983).
[53]M. Spite and F. Spite, Astron. Astrophys. **115**, 357 (1982).
[54]E. Maurice, F. Spite and M. Spite, in ESO Workshop on Primordial Helium, p. 361, (1983).
[55]P. Molaro and J.E. Beckman, this workshop, (1984).
[56]D. Gough, in ESO Workshop on Primordial Helium, p. 117, (1983).
[57]J.N. Heasley and R.W. Milkey, Ap. J. **221**, 677 (1978).
[58]P.E. Nissen, in ESO Workshop on Primordial Helium, p. 163, (1983).
[59]F. Caputo and V. Castellani, in ESO Workshop on Primordial Helium, p. 213, (1983).
[60]A. Buzzoni, F. Fusi Pecci, R. Buonanno, and C.E. Corsi, in ESO Workshop on Primordial Helium, p. 231, (1983).
[61]P.W. Cole, P. Demarque, and E.M. Green, in ESO Workshop on Primordial Helium, p. 235, (1983).

The ABC of Population III

B. J. Carr

ABSTRACT

We discuss the circumstances in which one would expect Population III stars to form and examine their cosmological consequences. Consideration of their production of light, metals, and remnants places a strong constraint on their formation epoch and mass spectrum. However, these constraints would still allow Population III stars to provide the dark matter in galactic halos, a helium abundance of around 25%, a detectable background of gravitational waves, a possible infrared background, and the generation of large-scale cosmological structure through explosions. All of these features could be achieved by VMOs in the range 10^2 to $10^5 M_{\odot}$. On the other hand, the Population III scenario would be most compatible with the standard Big Bang picture if the stars were SMOs with mass around $10^6 M_{\odot}$. Such stars could still provide the dark matter and detectable gravitational waves.

I. INTRODUCTION

Population III stars may be defined as the stars that formed before the Universe had any appreciable heavy element content. Since heavy elements can only be produced through stars, this definition requires the existence of at least a few such objects. The subject of this paper is the much more exciting and controversial possibility that most of the Universe may have been processed through them even before galaxies formed. One of the prime motives for this suggestion is the possibility that the dark matter in galactic halos and clusters may derive from Population III stars. This possibility has received rather scant attention at this conference, even though there is no compelling evidence for the alternative possibility, some type of elementary particle. However, the cosmological interest in Population III stars is not confined to the dark matter issue. A population of pregalactic stars would be expected to produce remnants, light, explosions, and nucleosynthesis products. As illustrated in the table below, each of these could have important cosmological consequences. Indeed, one might almost say that Population III stars have been invoked to explain everything of cosmological significance! Most people, of course, would regard that as far too ambitious a proposal. There are alternative explanations for most of the cosmological features that pregalactic stars are supposed to explain, and it is not clear that any observations unambiguously require their existence. A more conservative approach is to use the various effects indicated in the table to constrain the Population III star hypothesis.

The purpose of this paper is to summarize what conclusions such studies already permit. In Section II, I will indicate why one would expect pregalactic stars to form in some scenarios and in what respect they might differ from present epoch stars. In Section III, I will review the constraints that can be placed on their formation epoch and mass spectrum by considering the unstarred items in the table; this will be based on work done in collaboration with Bond and Arnett. In Section IV, I will focus on some consequences of Population III stars which have been the subject of recent developments (the starred items in the table). This will include a discussion of helium production and gravitational wave generation by Population III stars (based on further work by Bond and Arnett); the possible detection of an infrared background from such stars (based on work with McDowell and Sato);

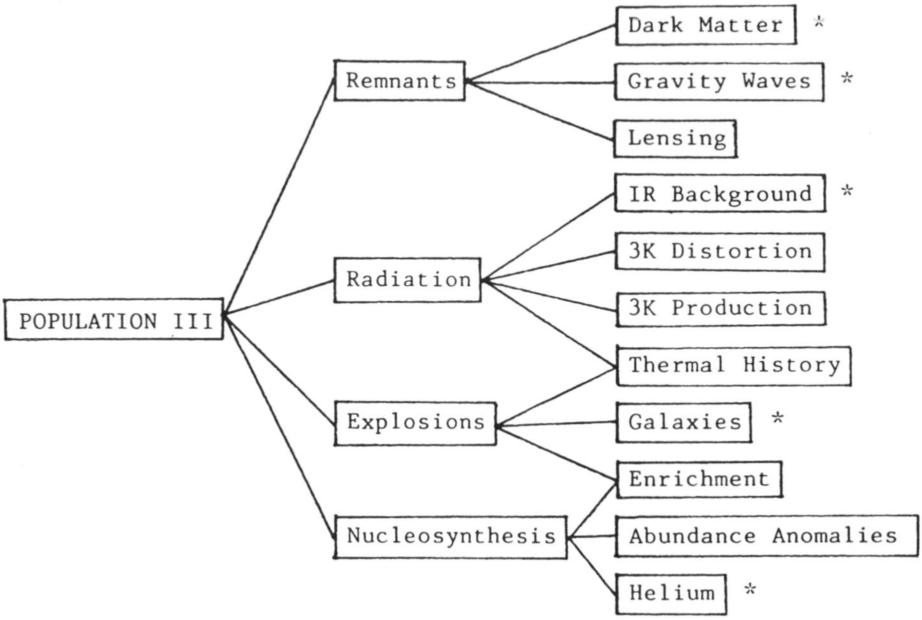

possible evidence for Population III remnants in our own halo (based on work by Lacey); and a discussion of the Hierarchical Explosion Scenario for the origin of cosmological structure (based on work with Ikeuchi). Some conclusions are drawn in Section V.

II. SCENARIOS FOR POPULATION III FORMATION

The existence of galaxies and clusters of galaxies implies that density fluctuations must have been present in the early Universe. The likelihood that these fluctuations could give rise to a population of pregalactic stars depends on what model one adopts for the nature of the fluctuations.

Isothermal fluctuations. In this case, the fluctuations are entirely in the baryons and one expects the smallest surviving scale after decoupling to be $M_{Jb} \simeq 10^6 \Omega_b^{-1/2} M_\odot$, corresponding to the baryon Jeans mass then.[1] (Here Ω_b is the baryon density in units of the critical density.) Larger structure would then build up through a process of hierarchical clustering.[2,3] The form of the galaxy correlation function [4,5] suggests that the fluctuations at decoupling must have had the form $\delta_{dec} = (M/M_1)^{-\beta}$ where $M_1 = 10^6 - 10^8 M_\odot$ and $\beta = 1/2 - 1/3$ (depending on the value of Ω_b). This implies that one would expect the first pregalactic clouds (of mass M_{Jb}) to bind shortly after decoupling ($z \approx 10^3$). These clouds would presumably fragment into stars, though the typical fragment mass and the efficiency of star formation is very uncertain. The lack of heavy elements[6-9] and substructure,[10] as well as the effects of the microwave background,[11] would tend to make the final fragments larger than at the present epoch and perhaps in the "VMO" range. Possibly there would be no fragmentation at all, in which case one could end up with a single "SMO".[12] On the other hand, enhanced molecular hydrogen formation[13] and associated thermal instabilities[14] could reduce the fragment mass to $0.01M_\odot$. The mass of the first stars is thus uncertain by 10^8 orders of magnitude! A hybrid scheme has been suggested in which each cloud forms a disc due to rotational effects; one then gets stars of mass $0.01M_\odot$ in the disc, together with a VMO which forms through relaxation at the center.[11]

Adiabatic fluctuations in a "cold" universe. If the fluctuations are contained in the total density and the Universe's mass is presently dominated by collisionless "cold" particles, like axions or photinos or primordial black holes, one expects bound clumps of these particles to form down to very small scales. Baryons would then fall into the potential wells, forming bound clouds, on baryon scales above $10^6 \Omega_b \Omega_a^{-3/2} M_\odot$ where Ω_a is the cold particle density.[1] These clouds could then form pregalactic stars just as in the isothermal scenario. In fact, the formation of the pregalactic clouds is even easier in this case because the cold particle fluctuations grow by an extra factor of $10\Omega_a$ between the time when the cold particles dominate the density and decoupling.

Adiabatic fluctuations in a "hot" universe. If the Universe's mass is dominated by baryons or collisionless "hot" particles, like neutrinos with non-zero rest mass, then adiabatic fluctuations are erased on subgalactic scales by photon diffusion (for $M < 10^{13} \Omega_b^{-5/4} M_\odot$) in the first case[15] or neutrino free-streaming (for $M < 10^{15} \Omega_\nu^{-2} M_\odot$) in the second.[16] Thus, the first objects to form are "pancakes" of cluster scale.[17] However, one still expects these pancakes to fragment into clumps of mass $10^8 M_\odot$ and these clumps might in turn fragment into stars before clustering into galaxies.[18] Even in this case, therefore, one might expect pregalactic stars to form, albeit at a relatively low redshift ($z<10$). The mass of the resulting stars is still very uncertain but at least two of the previous arguments for high mass fragments (lack of metallicity and substructure) would still pertain.

This discussion emphasizes that pregalactic stars could form in all of the most likely scenarios for the density fluctuations. However, it is not clear how much of the Universe would be expected to go into the stars. If they formed very efficiently, cosmological nucleosynthesis constraints on the baryon density ($\Omega_b<0.1$) would permit them to provide the dark matter in halos but not a critical density.[19] On the other hand, in the present epoch (e.g., in giant molecular clouds) star formation seems to be rather inefficient, so -- if one wants to argue that most of the Universe has been processed through Population III stars -- one needs to suppose that the conditions for star formation are very different at early times. One possibility is that the stars may be massive enough to generate a lot of explosive energy; in this case, the explosions may have amplified the fraction of the Universe going into stars,[20] as discussed in Section IV.

Assuming pregalactic stars do form, we can be reasonably certain how they will behave. Stars in the mass range above $10^5 M_\odot$ (Supermassive Objects - SMOs) are unstable to relativistic effects and collapse directly to black holes without burning their nuclear fuel if they have no metals.[21] (The situation is more complicated for Population I SMOs since these may explode in some mass range, especially if they have rotation.[22]) Stars in the mass range 4 - $100M_\odot$ (Massive Objects - MOs) explode, either at degenerate carbon ignition[23] for $M < 8M_\odot$ or at the Fe/Ni core phase[24] for $M > 8M_\odot$; in the latter case, one may get a neutron star or black hole remnant. Stars smaller than $4M_\odot$ (Low Mass Objects - LMOs) just leave white dwarf remnants, those smaller than $0.08M_\odot$ never even igniting hydrogen. The "missing" range of $10^2 - 10^5 M_\odot$ (Very Massive Objects - VMOs) had been rather neglected until a few years ago because there seemed to be little evidence that such stars actually exist. However, there are some VMO candidates at the present epoch[25,26] and they may have formed more prolifically at pregalactic epochs. This has motivated several groups[27-30] to study their properties in some detail. These properties may be summarized as follows:

(1) VMOs can be well modelled as n = 3 (radiation-dominated) polytropes with small radiative envelopes. They are unstable to nuclear-energized pulsations with consequent mass loss[31]; while the amount of loss is very uncertain, it would not necessarily be large enough to reduce the VMO to an MO.[32-36]

(2) VMOs have a luminosity close to the Eddington limit, $L_{ED} \simeq 10^{38}(M/M_\odot)\text{erg s}^{-1}$, and consequently a main-sequence lifetime $t_{MS} \simeq 3 \times 10^6$y, almost independent of their mass. Their surface temperature, $T_s \simeq 10^5$K, is also nearly mass-independent.

(3) Numerical calculations for a Population I star of 500 $M_\odot$ indicate that VMOs may eject their envelopes due to a super-Eddington luminosity during hydrogen-shell burning.[27] However, the mass range over which this occurs and whether or not it applies for Population III stars is not yet settled.[30]

(4) VMOs experience an instability during their oxygen core phase for a core mass $M_O > 40M_\odot$ because the temperature gets high enough to generate electron-positron pairs.[37] Sufficiently large cores collapse to black holes, while smaller ones explode.[38-41] Semi-analytical calculations,[27] as well as numerical results,[29] indicate that the critical dividing mass is $M_{OC} \simeq 100M_\odot$ to any accuracy of 10%. If the fraction of the mass which ends up in the oxygen core after hydrogen and helium burning is ϕ_O, this means that all stars larger than $M_C \simeq 10^2\phi_O^{-1}M_\odot \geq 200\ M_\odot$ end up collapsing to holes.

We will exploit all of these features in discussing the cosmological effects of Population III stars. The only qualitative feature which may depend on the amount of metallicity is point (3), even though the metallicity may play a crucial role in determining whether VMOs actually form.

III. COSMOLOGICAL CONSTRAINTS

Remnants. Only stars larger than M_C can generate black holes with high efficiency (i.e., with the fraction of the initial star mass which collapses ϕ_B being close to 1). Providing the Population III spectrum is such that most of the mass is in stars this large, the fraction of the Universe ending up in black hole remnants should be $f_B = f_* \phi_B$, where f_* is the fraction of the Universe going into the stars. Observationally, we require $0.9 \leq f_B \leq 0.99$ (the lower limit being required to explain the dark mass in halos[42] and the upper limit to ensure that enough visible material is left over). It might seem unlikely that f_B would be this high since we could not expect both ϕ_B and f_* to be so close to 1. In particular, ϕ_B must be less than 0.5 if envelope-ejection occurs at hydrogen-shell burning.[27] However, one could in principle boost f_B by invoking black hole accretion or many generations of stars. Dynamical constraints, associated with the puffing up of the disc in our own galaxy by traversing halo holes,[20,43,44] require that the mass of the hole which dominates the halo not exceed $10^6 M_\odot$. (Other dynamical limits, associated with the disruption of star clusters and dynamical friction effects[45,46] are somewhat weaker, as are accretion limits unless all the accretion-generated radiation is in the X-ray band.[47]) We therefore conclude that the mass of the halo holes must lie between $200 M_\odot$ and $10^6 M_\odot$ (i.e., they must derive from VMOs or low mass SMOs). The other possibility for explaining the dark matter is to suppose that the Population III stars are so small[42] that their mass-to-light ratio exceeds 100; this requires $M < 0.1 M_\odot$. In fact, infrared observations of other spiral galaxies would seem to require $M < 0.08 M_\odot$, excluding any hydrogen-burning stars.[48] Indirect arguments may also exclude "Jupiters": for it would be surprising if any fragmentation scenario could lead to fragments smaller than $0.004 M_\odot$ and, unless the stars have a very steep spectrum, this implies that there would still be too many light-producing stars above $0.1 M_\odot$.[49]

Lensing. Both high and low mass Population III remnants could produce interesting lens effects. If a galaxy is suitably positioned to image-double a distant quasar,[50] then it can be shown that there is also a high probability that an individual halo object will traverse one of the lines of sight, giving appreciable intensity fluctuations in one but not both images.[51] This effect would be observable for stars larger than $10^{-4} M_\odot$ but the timescale of the fluctuations, being of order $40(M/M_\odot)^{1/2}$yr, would only be detectable over a reasonable period for $M < 0.1 M_\odot$. The lensing effect associated with larger remnants is of a somewhat different nature, deriving from the fact that such objects would modify the ratio of the line to continuum output of the quasar[52]; the fluxes are affected differently because they derive from regions which act as extended and pointlike sources, respectively, unless M is very large. This already excludes a critical density of objects with $0.1 < M/M_\odot < 10^5$, though not necessarily the tenth critical density required for halos. SMOs with $M > 10^5 M_\odot$ are not excluded by this effect because the whole quasar nucleus would then act as a point-source; however, holes only slightly larger than this might be detected by direct image-doubling using VLBI or VLA.[50] Thus lensing could constrain the Population III mass spectrum over the entire range of relevance.

Background light. Since stars turn about 1% of their rest mass into radiation over their nuclear-burning time, the background light density they generate should be $\Omega_R \simeq 10^{-2}\Omega_* f_R (1+z_*)^{-1}$ where the densities are in units of the critical density, z_* is the redshift at which they burn most of their fuel, and f_R - the fraction of the radiation which goes into background light rather than heating effects - should be close to 1. Since the observed background density over all wavebands cannot exceed about 10^{-4} (with the possible exception of the far IR band, which is presently unobserved), this implies $\Omega_* < (z_*/100)$. Thus the stars must form before a redshift of 10 if $\Omega_* > 0.1$. This also requires that the stars be larger than about $10M_\odot$ in order to burn out by then; the precise value depends on rather uncertain cosmological parameters. Of course, these limits do not apply for stars with $M < 1M_\odot$ since these are still burning. For these, the background light constraint requires $\Omega_* < (M/0.4M_\odot)^{-3}$, which is somewhat weaker than the constraint associated with infrared observations of individual galaxies.[48] The background light limits would be more interesting if one knew that the starlight should presently be in a waveband where the density is less than 10^{-4}. If absorption by dust is unimportant, the light from VMOs should just have a redshifted black-body spectrum with temperature $10^5(1+z_*)^{-1}$K providing the Universe is sufficiently ionized that there is no cut-off shortward of the Lyman limit.[20,53-55] The spectrum should therefore peak around $300(1+z_*)$Å. This is in the UV or optical range (where observations require much smaller values of Ω_R than 10^{-4}) for $z_* < 20$, and it is in the observable IR range for $z_* < 80$. If there is a Lyman cut-off, the spectrum peaks at $1200(1+z_*)$Å and the conditions become $z_* < 5$ and $z_* < 20$, respectively. The limit on $\Omega_*(z_*)$ imposed by observations at 1400Å, 4100Å, 5300Å, 8700Å, 20000Å, and 24000Å is indicated in Fig. 1. One can already exclude a first generation of galactic stars producing the dark matter unless there is a Lyman cut-off; the latter possibility is rather unlikely since the stars themselves should produce high ionization.[20] This conclusion would be even firmer if the stars were smaller than VMOs.[55]

FIGURE 1

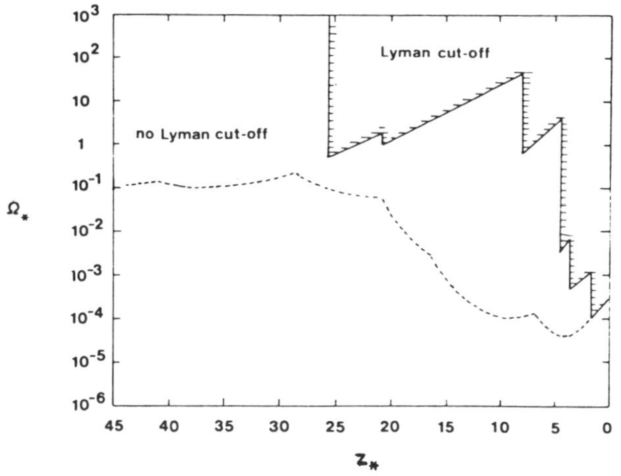

Microwave distortions. The previous discussion fails if absorption by dust occurs. For some range of values of z_* and the dust fraction χ, the starlight would be absorbed and re-radiated in the far IR. One can show[56] that it would presently reside in the waveband 300-900μ, with only a weak dependence on z_* and χ. This is presently unobservable, but could become so with future space experiments (e.g., SIRTF and LDR). For high values of z_* and χ (viz. $z_* > 100$ and $\chi > 10^{-5}$) the dust temperature would be held so close to the microwave background temperature that the re-radiated starlight would distort the microwave spectrum.[57-59] A few years ago, such a distortion was, in fact, reported: the best-fit black-body temperature around the peak was found to be 2.9K (significantly higher than the 2.7K found in the Rayleigh-Jeans region) and there also appeared to be a trough just shortward of the peak.[60] This could be explained if the 2.9K component was thermalized pregalactic starlight, with the trough reflecting the reduced absorption efficiency of grains in the 1-10μ region; the trough would be redshifted to the observed wavelength for $z_* \simeq 100$. This proposal would require that 25% of the 3K background energy come from stars, which in turn would require $\Omega_* \simeq 0.1$ and $M > 30M_\odot$. However, more recent data suggest that this effect may be going away.[61]

Generation of the 3K background. Even if the 3K background has an undistorted spectrum, it could still have been generated entirely by stars providing one can invoke a suitable thermalizing agent[62-64]; this may be feasible even at low redshifts if one adopts special types of elongated grains.[65-68] However, these models seem rather contrived: it is very difficult to ensure thermalization at both long and short wavelengths.[69] An alternative scheme is to assume that the black hole remnants of the stars generate the radiation through accretion[70]: energetic considerations then require that they produce it at a redshift $z \simeq 10^4 \Omega_B \eta$ (where η is the efficiency with which accreted material generates radiation energy) and this could precede decoupling if $\eta > 0.1$. In this case, free-free processes could thermalize the radiation if the gas were highly clumped.[70] The period in which the holes generate the radiation can be specified very precisely in this scenario: it has to be around 10^6y, corresponding to the collapse time of an SMO with $M \simeq 10^5 M_\odot$. Of course, any scheme which envisages the 3K background deriving from Population III stars or their remnants must also require that the early Universe was cold or tepid[71] (with the primordial photon-to-baryon ratio being much less than its present value of 10^9). The advantages originally claimed for the cold and tepid models were: (i) that the photon-to-baryon ratio could be explained naturally rather than being assigned arbitrarily as an initial condition of the Universe (a simple argument predicting $S \sim \alpha_G^{-1/4} \sim 10^9$, where α_G is the gravitational fine structure constant, if most of the Universe goes into stars[63]); and (ii) that the existence of a prolonged matter-dominated phase before decoupling would permit galactic-scale density fluctuations a longer period of growth. However, both these arguments have become less compelling in recent years. The grand unified theories[72-76] permit alternative explanations for the value of S by invoking baryon-non-conserving processes at 10^{-35}s; and the currently favored picture of galaxy formation requires that the early Universe was cold anyway, but in the sense of being dominated by non-relativistic particles for some period before decoupling rather than in the sense of having no radiation.[79,80] It is therefore more likely that the light from Population III stars will be found in the IR band (see Section IV).

Thermal history. The light generated by Population III stars could also have an important effect on the thermal history of the Universe. During its main-sequence phase, each star (or cluster of stars) would be surrounded by an HII region. The fraction of the Universe in such regions would progressively grow on account of both the increasing number of stars and the increasing size of the individual HII regions. For $z < 60\delta^{-1/3}$, where δ is the gas clumpiness, the recombination time within each HII region exceeds the lifetime of the stars, so the whole

Universe is ionized once the number of ionizing photons generated exceeds the number of atoms[81,82]; the fraction of the gas in stars is then just $f_* \simeq 3 \times 10^{-5}\delta$. For $z > 60\delta^{-1/3}$, one gets a fully developed HII region (whose structure is determined by the balance of ionizations and recombinations) and the situation is more complicated: one now needs a larger value of f_* to reionize the Universe but it is still small.[20] Such reionization could have important cosmological implications (e.g., in reducing anisotropies in the 3K background). After the HII regions have merged, the Universe would maintain a high degree of ionization, $1 - x \simeq 10^{-10} f_*^{-1} (T/10^4 K)^{-0.8}$, and the gas temperature T would tend to a value of $10^4 - 10^5$K until the stars cease burning.[20] If the stars finally explode, or if they leave black holes which heat the Universe through accretion,[78,83] the temperature may be boosted even higher.

Enrichment. One of the strongest constraints on the spectrum of Population III stars comes from the fact that stars in the mass range $4-100M_\odot$ should eventually explode, producing an appreciable heavy element yield.[20,84] For $100M_\odot<M<M_C$, one expects $Z_{ej} = 0.5$[27]; for $15M_\odot<M<100M_\odot$, $Z_{ej} = 0.5-(M/6.3M_\odot)^{-1}$;[85] and for $4M_\odot<M<8M_\odot$, $Z_{ej} \simeq 0.1$.[24] Since Population II stars are observed with metallicity as low as 10^{-5}, this implies that the pregalactic enrichment cannot exceed this amount unless Population II are themselves pregalactic.[86] In any case, it cannot exceed the lowest metallicity observed in Population I stars ($Z \simeq 10^{-3}$). The associated constraint on the Population III mass spectrum, assuming a gas density Ω_g=0.1, is shown in Fig. 2. If one wants to produce the dark matter without contravening this constraint, the most straightforward solution is to assume that the spectrum either begins above M_c or ends below $4M_\odot$. As mentioned in Section II, it is possible that the first stars were either very large or very small, so this solution is not unreasonable. Nevertheless, for some purposes it would be advantageous to have a slight pregalactic enrichment (e.g., to explain the G-dwarf problem[87,88] or the small grain abundance required to produce alleged distortions in the 3K spectrum[57-60] or the possible lack of a metallicity gradient in globular clusters[89]), so the question arises of whether it is likely that Population III stars could generate just a small enrichment (e.g., $Z \simeq 10^{-5}$). In principle, this could be achieved by a rather contrived choice of mass spectrum. A more natural solution would be to suppose that the first stars do indeed explode but that their formation is self-limited due either to reionization temporarily suspending star formation once f_* reaches 10^{-5} or due to the small metallicity shifting the characteristic star mass into the non-exploding range.

Abundance anomalies. Various chemical anomalies have been observed for metal-poor stars and these give clues as to the nature of the first stars.[90] For example, the high oxygen-to-iron ratio at low Z[91] suggests that the first stars were either MOs or VMOs (the latter producing a lot of oxygen since they explode in their oxygen burning phase)[92]; and the possible evidence for primary nitrogen[93,94] might also be explained by VMOs if the carbon and oxygen produced in helium-core burning could be convected through the hydrogen shell and there CNO-processed to nitrogen.[20,30,95] The other important nucleosynthetic consequence of Population III stars, the generation of helium, is discussed in Section IV.

The effects discussed above place interesting constraints on the density of stars with mass M which could have formed at a redshift z; these constraints are summarized in Fig. 2. The black hole constraint is most important for high M, the nucleosynthetic constraint for intermediate M, and the background light constraint for low M. Only the last constraint is dependent on z (cf. Fig. 1) and it is interesting in the intermediate mass range only for z<10. The figure indicates that the only mass ranges in which a large fraction of the Universe can be processed through Population III stars are $M < 0.1M_\odot$ and $M_C < M < 10^6 M_\odot$. It also restricts the form of the mass spectrum: for example, if the spectrum encompasses the exploding mass range, then the nucleosynthetic constraints require a spectral index which is either very large if the dark matter is in low mass stars or very shallow if it is in high mass stars.[20,84] Of course, the previous discussion shows that Population III stars could have had interesting effects even if they did not generate the dark matter; for example, only a small value of f_* would be needed to reionize the Universe or provide a burst of enrichment.

FIGURE 2

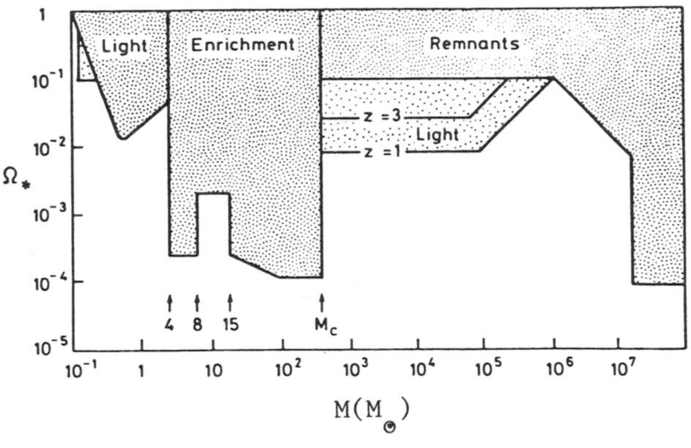

IV. RECENT DEVELOPMENTS

In this section we will consider some consequences of Population III stars which have been subject to recent attention. Whereas the considerations of the last section were primarily directed to placing constraints on the Population III hypothesis, our aim in what follows will be to stress what it might explain.

Helium Production by VMOs. Because the pulsational instability leads to mass-shedding of material convected from its core, a VMO is expected to return helium to the background medium during core-hydrogen burning.[96] The net yield depends sensitively on the mass loss fraction ϕ_L. If this is very high, the yield will be low because most of the mass will be lost before significant core burning occurs. However, for $\phi_L < (1-Y_i)/(2-Y_i)$, the mass loss is always slower than the shrinkage of the convective core and one can show that the fraction of mass returned as new helium is[97]

$$(\Delta Y) = (1-\tfrac{1}{2}Y_i)\,\phi_L^2 \leq 0.25\,\frac{(1-Y_i)^2}{(1-Y_i/2)} \quad . \qquad (1)$$

Here Y_i is the initial (primordial) helium abundance and the equality sign on the right applies only if ϕ_L has the critical value. This does not necessarily impose a constraint on the number of VMOs since (ΔY) is very small if ϕ_L is well below the critical value. However, there is some indication from numerical calculations that hydrogen-shell burning may produce a super-Eddington luminosity which completely ejects the stellar envelope.[27,30] If true, this means that the maximal helium production permitted by Eq. (1) is guaranteed. This would have profound cosmological implications. If $Y_i = 0.23$, corresponding to the conventional primordial value, $\Delta Y = 0.17$, so one would substantially overproduce helium if much of the Universe went into VMOs. In this case, the only remaining candidate for the dark matter would be SMOs in the mass range $10^5 - 10^6 M_\odot$. On the other hand, if $Y_i = 0$, corresponding to no primordial production, $\Delta Y = 0.25$, which is tantalizingly close to the standard primordial value. One can show that, if a fraction F of the Universe is processed through VMOs, then the resulting helium abundance is given by[97]

$$F = \int_0^Y \frac{2(2-Y)}{(1-Y)^2} \exp\left(\frac{2Y}{Y-1}\right) dY \tag{2}$$

For small Y, this just gives $F \approx 4Y$, so one naturally gets the sort of value required if F is close to 1. More precisely, Y lies between 0.20 and 0.25 for $0.8 < F < 0.9$.

This raises the question of whether the Population III VMOs invoked to produce the dark matter might not also generate the helium which is usually attributed to cosmological nucleosynthesis. Unlike stars smaller than M_C, which could not do so without overproducing heavy elements, stars larger than M_C may produce no appreciable yield since the heavy elements they generate collapse with the core to form a black hole. (A small amount of nitrogen may, in fact, be convected to the surface prior to core collapse; this is necessary since one needs CNO burning to drive envelope ejection.) Prior to accretion, the ratio of black hole density to surviving gas density is

$$\frac{\Omega_B}{\Omega_g} = \exp\left(\frac{2Y}{1-Y}\right) - 1 \tag{3}$$

and, for $Y \approx 0.25$, this is about 1.0. A reasonable amount of accretion would suffice to boost this enough to explain the dark matter problem.

One still has to find a way to suppress primordial helium production. One way to do this would be to suppose that the early Universe was cold. In this case, the amount of helium production is determined entirely by the neutrino degeneracy factor[98,99] and one could avoid any helium production at all for a lepton-to-baryon ratio exceeding 1.5. It must be stressed, however, that the added attraction of the hot Big Bang model is that it also predicts the observed abundances for the other light elements (deuterium, helium-3, and lithium).[100] While it is not impossible to conceive of astrophysical ways of generating these elements in a cold Universe,[101-104] the models are somewhat contrived. Furthermore, one still has to generate the 3K background in a cold Universe. Although stars or black hole accretion might achieve this (see Section III), the cold model clearly has a lot to answer for.

An infrared background from Population III stars. Recently the detection of a cosmological IR background in the waveband 2-5μ with a density $\Omega_{IR} \simeq 10^{-4}$ and a black-body spectrum with temperature 1500K has been reported.[105] Although this claim remains to be confirmed, the existence of an IR background is one of the main predictions of the pregalactic star scenario,[104] so it is obviously interesting to inquire what sort of stars would be required to explain the data. If we assume that the stars are VMOs with $M > M_C$ (in order to avoid overenrichment) and that their radiation is not modified by grain absorption (otherwise it would be at a much longer wavelength[56]), then the expected spectrum can be shown to be[106]

$$\Omega_R(\nu) \equiv \frac{4\pi\nu i(\nu)}{\rho_{crit}c^3} \simeq 6\times10^{-4}\left(\frac{f_b X_o}{0.6}\right)\left(\frac{\Omega_*}{1+z_*}\right)\left(\frac{x^4}{e^x-1}\right) \qquad (4)$$

where $i(\nu)$ is the intensity (in erg $cm^{-2}s^{-1}Hz^{-1}sr^{-1}$), $x = \hbar\nu(1+z_*)/kT_s$, and $f_b X_o$ is the fraction of the star's mass burnt to helium (normalized to the value appropriate for a VMO with $X_o = 0.75$). The quantity $\Omega_R(\nu)$ peaks at a present wavelength $\lambda_{max} = 0.04(1+z_*)\mu$, which lies in the observed waveband for $40 < z_* < 140$. A comparison with the data in Fig. 3 shows that no single curve passes through all the data points. If one excludes the 5μ point (which is the most dubious since it could be contaminated by interplanetary dust emission), the best fit curve (b) corresponds to $z_* = 75$, $\Omega_* = 1.6h^{-2}$ where $h = (H_o/50)$; if one excludes the 2μ point, the best fit (a) requires $z_* = 100$ and $\Omega_* = 2.5h^{-2}$.

FIGURE 3

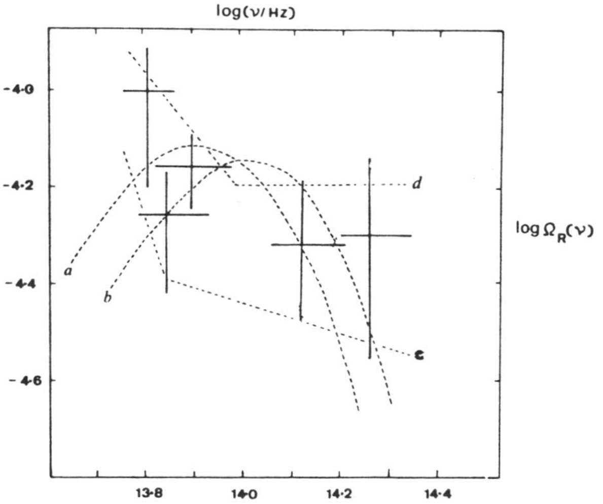

Both these values of Ω_* are embarrassingly high if one believes the cosmological nucleosynthesis constraint.[100] Ω_* can be reduced slightly if one assumes that the spectrum is cut off beyond the Lyman limit and that the stars form over a range of redshifts. In this case, the smeared out Lyman-α emission would dominate the spectrum and could, in principle, be made to fit any data by choosing a suitable dependence of star formation rate upon redshift (e.g., curves c and d). However, one still needs a high value for Ω_*.

Another possible explanation for the IR background, one that is more plausible energetically though less clean in its predictions, is to invoke accretion by the black hole remnants of the Population III stars. If the holes accrete gas with the mean cosmological density according to the Bondi formula[107] and generate black-body radiation with an efficiency η, then one can show that the present temperature and density of the background should be[106]

$$T_{bb} \simeq 2 \times 10^4[\eta\Omega_g h^2(1+z)^{-1}]^{1/4}T_4^{-3/8}K \tag{5}$$

$$\Omega_{IR} \simeq 8 \times 10^{-5}\eta M_6\Omega_B\Omega_g\Omega^{-1/2}(1+z)^{1/2}T_4^{-3/2}h \tag{6}$$

where T_4 is the background matter temperature in units of 10^4K (this should be close to 1 as a result of the heating effect of the holes), M_6 is the hole mass in units of $10^6M_\odot$, Ω_g is the gas density, and Ω is the total density. If the holes form accretion discs, the condition for black-body emission[77] is $\dot{M} < 10^{16}M^{7/8}$g s^{-1} and this implies that z must be less than a critical value

$$z_{bb} \simeq 17M_6^{-3/8}\Omega_g^{-1/3}h^{-2/3}T_4^{1/2} \ . \tag{7}$$

Most of the radiation density should derive from this redshift, so putting $z = z_{bb}$ in Eq. (5) gives a present characteristic temperature

$$T_{bb} \simeq 4 \times 10^3M_6^{0.1}\left(\frac{\eta}{0.1}\right)^{1/4}\left(\frac{\Omega_g}{0.1}\right)^{1/3} T_4^{-1/2}h^{2/3}K \ . \tag{8}$$

It is interesting that this is indeed in the IR band, with only a weak dependence on the parameters involved. Putting $z = z_{bb}$ in Eq. (6) shows that the holes can generate the required density for

$$M \simeq 4\times10^6\left(\frac{\eta}{0.1}\right)^{-1.2} \Omega_B^{-1.2}\Omega^{0.6}\Omega_g^{-1.0}h^{-3.2}T_4^{1.5}\left(\frac{\Omega_{IR}}{10^{-4}h^{-2}}\right)^{1.2} M_\odot \tag{9}$$

On the other hand, z_{bb} precedes the epoch of galaxy formation ($z \simeq 10$, say) only if $M < 10^8M_\odot$, which in turn [from Eq. (9)] requires $\Omega_B > 0.1$. Thus, in this scenario the holes have to be SMOs with mass in the range $10^6 - 10^8M_\odot$ and most of the radiation comes from a relatively recent epoch.

Dynamical effects of halo holes. It was mentioned in Section III that the most interesting constraint on the mass of any holes in our own halo is associated with their tendency to puff up the disc. A more detailed calculation of this effect suggests that halo holes could actually be responsible for the amount of disc-puffing which is observed.[43] The velocity dispersion of the disc stars in the radial, transverse, and vertical directions should evolve according to

$$\sigma_u(t)=(\sigma_{uo}^2+D_e t)^{1/2}, \quad \sigma_v(t)=\beta^{-1}\sigma_u(t), \quad \sigma_w(t)=(\sigma_{wo}^2+D_z t)^{1/2}, \qquad (10)$$

respectively, where

$$D_{e(z)} = 2\pi G^2 n_B M_B^2 \ln\left(\frac{V^3 T_{orb}}{GM_B}\right) F_{e(z)}(V/\sigma_B,\beta). \qquad (11)$$

Here σ_o is the initial velocity dispersion of the stars, σ_B is the velocity dispersion of the holes, V is the relative velocity of the holes and the stars (close to the circular velocity in the disc if the halo is assumed to have little rotation), M_B and n_B are the mass and number density of the holes, β is the ratio of the circular to epicyclic frequency, and F_e and F_z are determined in terms of the error function. The important features are: (1) that all three components of σ grow asymptotically as $t^{1/2}$, thus explaining the empirical relationship between the observed scale heights of stars and their age; and (ii) that the predicted ratio of the vertical and radial components σ_w/σ_u agrees with observation.[108] Attempts to explain the puffing up of the disc through heating by giant molecular clouds, for example, would seem to be less successful on both accounts: one then expects $\sigma \propto t^{1/4}$ and σ_w/σ_u is too small.[109] In order to normalize the $\sigma(t)$ relationship correctly, one needs $n_B M_B^2 \simeq 3 \times 10^4\ M_\odot^2\ pc^{-3}$. Combining this with the inferred halo density at our own galactocentric distance, implies a hole mass $M_B \simeq 2 \times 10^6 M_\odot$.

While one cannot definitely identify halo holes as the explanation for disc-heating (e.g., spiral density waves might also work[110]), it is interesting that the hole mass required can be specified so precisely. Note that, from the point of view of the preceding argument, one does not require that the halo object be a single hole; even a cluster of smaller holes would suffice. In this context, it may be relevant that M_B is close to the size of the first clouds one would expect to form in the "isothermal" or "cold" scenarios discussed in Section II.

Gravity waves from pregalactic holes. The formation of a population of pregalactic black holes would be expected to generate bursts of gravitational radiation. The characteristic period and density of the waves would be[111]

$$P_o \simeq 10(GM/c^3)(1+z_B) \simeq 10^{-4}(M/M_\odot)(1+z_B)\ s, \quad \Omega_g = \epsilon_g \Omega_B (1+z_B)^{-1} \qquad (12)$$

where z_B is the redshift at which the holes form and ϵ_g is the efficiency with which the collapsing matter generates gravity waves. One can show that the expected time between bursts (as seen today) would be less than their characteristic duration providing $\Omega_B > 10^{-2}\Omega^2 z_B^{-1}$. If the holes make up galactic halos, one would therefore expect the bursts to form a background of waves.[112] This background could be detectable by laser interferometry if M is below $10^3 M_\odot$ or by the Doppler tracking of interplanetary spacecraft if M is in the range $10^5 - 10^{10} M_\odot$. Thus the holes should be detectable unless M lies in the mass range $10^3 - 10^5 M_\odot$.

The prospects of detecting the gravitational radiation would be even better if the holes formed in binaries.[113] This is because two sorts of radiation would then be generated: (i) continuous waves as the binaries gradually spiral inwards due to quadrupole emission; and (ii) a final burst of waves when the components finally merge. The burst would have the same characteristics as that associated with isolated hole formation except it would be postponed to a lower redshift and ε_g would be larger ($\varepsilon_g \approx 0.08$) because of the larger asymmetry; from Eq. (12) both factors would increase Ω_g. The continuous waves would also be interesting since they would extend the spectrum to considerably longer periods, thus making the waves detectable by a wider variety of techniques. Over most wavebands, the spectrum of the waves would be dominated by the holes whose initial separation was such that they are coalescing at the present epoch. This corresponds to a separation $a_{crit} \approx 10^2 M_2^{3/4} R_\odot$ where $M_2 = M/10^2 M_\odot$. If the fraction of holes which form in binaries with this separation is f_{crit}, the spectrum should have the form:

$$P(d\Omega_g/dP) = 0.08\Omega_B f_{crit} \; (P/P_o)^{-2/3} \; (10^{-2}M_2 s<P<10^5 M_2^{5/8} s) \; . \qquad (13)$$

Providing $\Omega_B f_{crit}$ is not too small, this background should be detectable by laser interferometry for $M < 400M_\odot$ and by Doppler tracking of interplanetary spacecraft for $4 \times 10^4 M_\odot < M < 6 \times 10^{10} M_\odot$. One could also hope to observe the coalescences which are occurring in our own galactic halo at the present epoch. The average time between halo bursts and their expected amplitude would be $10M_2 f_{crit}^{-1} h^{-1}$y and $7 \times 10^{-17} M_2$, respectively. They would be detectable by lasers for $M < 10^4 M_\odot$ and by Doppler tracking for $3 \times 10^4 M_\odot < M < 6 \times 10^7 M_\odot$. Although the time between bursts would be uncomfortably long in the latter case, one could also detect bursts from the Virgo cluster every $4M_2 f_{crit}^{-1} h^{-1}$ days with somewhat improved sensitivity.

The crucial issue, of course, is whether one can expect black holes to form in binaries. Since at least 50% of O stars (the largest stars forming in the present epoch) appear to be in binaries,[114] it does not seem implausible that VMO binaries should also be abundant; indeed, if VMOs are born from protostars with the angular momentum characteristic of O stars, one would expect them to fragment into two before collapsing. While O stars appear to have a separation spectrum[115] which peaks at $a \approx 30R_\odot$, this could be raised by about a factor of 4 for VMOs due to mass loss effects. Therefore, a reasonably large fraction of the resulting black hole binaries could have the separation a_{crit}. In any case, the advent of gravitational wave astronomy should provide a crucial test of the binary black hole scenario, since laser detectors will be operating at the required sensitivity within the next few years.

The hierarchical explosion scenario. Stars in the mass range $4\text{-}200M_\odot$ should produce explosive energy[116] with an efficiency $\varepsilon = 10^{-5}\text{-}10^{-4}$; those in the range $200\text{-}10^5 M_\odot$ may explode with comparable efficiency providing the shell ejection mechanism discussed in Section II works.[27] This explosive release could have an important effect on the large-scale structure of the Universe.[20,117-122] One would expect the shock-wave generated by each exploding star (or cluster of stars) to sweep up a shell of gas. Under suitable circumstances, this shell could eventually fragment into more stars. If the new stars themselves explode, one could then initiate a bootstrap process in which the shells grow successively larger until they overlap. This mechanism has been proposed in three contexts: (i) as a means to boost the fraction of the Universe being processed through pregalactic stars[20]; (ii) as a means to generate many galaxies from a few seed galaxies[118-120]; and (iii) as a way of producing the giant voids and filaments, whose existence is indicated by observational data[123-125], from much smaller scale initial structure.[121]

This "Hierarchical Explosion Scheme" can be studied in some detail.[122] For $z > 10$, the dominant cooling mechanism is inverse Compton cooling off the 3K background radiation and, in this case, one can show that the fragment mass within the shell is likely to be in the exploding range (as required). The maximum amplification which can be achieved in the first step is

$$\xi_1 = (M_{shell}/M_1) \approx 4\times10^5(1+z)^{-1.7}\varepsilon_{-4}^{0.6}(M_1/10^6M_\odot)^{-0.4} \tag{14}$$

where M_1 is the seed mass. This assumes that the shell continues to sweep up material for a cosmological time; it may cease doing so before this in some circumstances. If one assumes that shells maintain their spherical symmetry, one can show that the amplification attained at the nth step is $\xi_n = \xi_1^{(0.6)^{n-1}}$; this decreases to 1 as $n \to \infty$. The shell mass thus evolves as

$$M_n = \xi_1^{2.5[1-(0.6)^{n-1}]}M_1 \tag{15}$$

and tends asymptotically to

$$M_\infty = \xi_1^{2.5}M_1 \approx 1 \times 10^{20}(1+z)^{-4.3}\varepsilon_{-4}^{1.5}M_\odot \tag{16}$$

providing the shells do not overlap first. Equation (16) specifies the maximum scale of structure that can be attained at a redshift z. By the end of the Compton era, M_∞ is already of order $10^{15}M_\odot$, independent of the original seed mass.

For $z < 10$, radiative cooling dominates and the situation becomes more complicated because the cooling time depends on the temperature, density, and metallicity within the shell. There is also a qualitatively important difference in that the fragment mass within the shell becomes comparable to that of a galaxy rather than an exploding star. Thus, the bootstrap may proceed for only one more step after $z = 10$, the extra amplification in the radiative era being[122]

$$\xi_R \approx 400\varepsilon_{-4}^{0.6}(1+z)^{-0.6}(M/10^{11}M_\odot)^{-0.4} \tag{17}$$

for negligible metallicity. One needs $\xi_R > 100$ to avoid overenrichment at the final step and this places an upper limit on M, the shell mass at the end of the Compton era. Equation (17) corresponds to a final shell size

$$R_{max} \approx 7\varepsilon_{-4}^{0.2}(1+z)^{-0.2}(\Omega h^2)^{-0.4}(M/10^{11}M_\odot)^{0.2}\mathrm{Mpc}\ . \tag{18}$$

This is comparable to the scale of the voids observed today providing M is of order $10^{15}M_\odot$; as we have seen, this is not impossible, even starting with seed objects as small as $10^6M_\odot$. An upper limit to R_{max} in all circumstances is $\sqrt{\varepsilon}ct_o \approx 60h^{-1}$ Mpc for $\varepsilon = 10^{-4}$.

Note that this scenario only works if the fraction of the Universe in the initial seeds (f_1) is so small that the shells do not merge too soon, so we must ask what determines f_1. One possibility is that $f_1 \approx 10^{-5}$ because of the self-limited ionization argument of Section III. However, this would only give a shell mass of galactic scale at merging if $M_1 \approx 10^6 M_\odot$. A more natural suggestion is that the value of f_1 is itself determined by the merging condition. For if the initial density fluctuations from which the seeds derive have a Gaussian distribution on the scale M_1, one would expect the fraction of the Universe in them to steadily increase until $f(z)\xi(z) \approx 1$. However, this still requires fine tuning if the final value of f is to be as small as required. A somewhat less exotic scheme for generating large-scale density fluctuations from pregalactic stars invokes pressure effects associated with their HII regions.[126,127]

V. CONCLUSION

The shaded regions in the table below indicate what sorts of Population III stars are required to explain the various cosmological effects alluded to in this paper. We have ordered these effects according to the extent to which they require a deviation from the standard Big Bang picture; only the last two require that one give up the cosmological nucleosynthesis scenario. The "IR background" refers

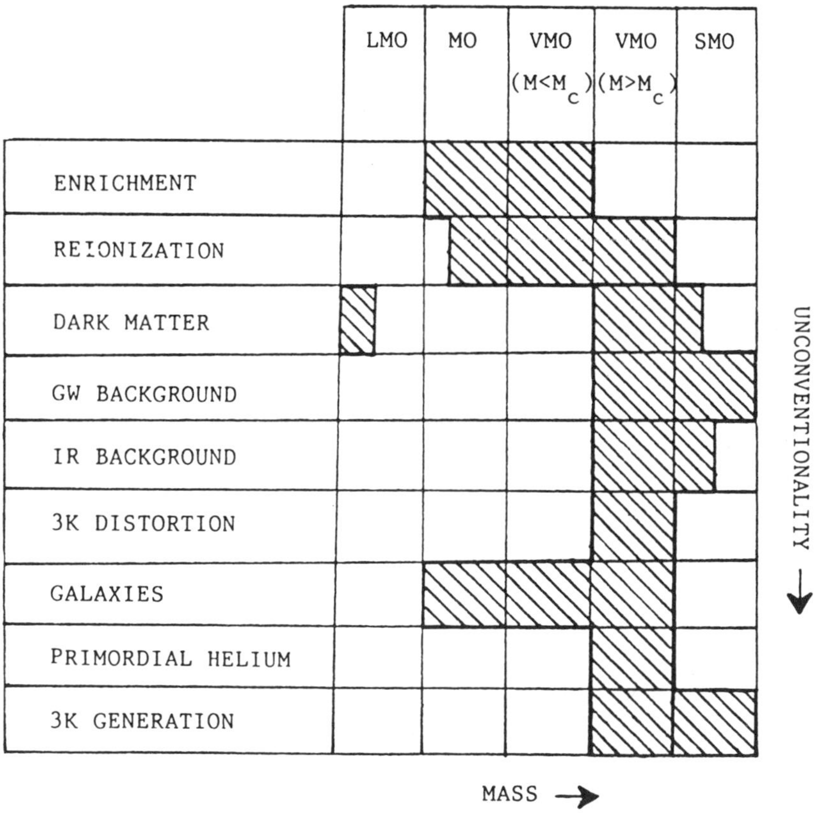

specifically to the Japanese measurement and the "GW background" is required to be detectable. The stars which solve most of the problems would seem to be the VMOs: indeed, the combination of exploding and collapsing VMOs would appear to explain everything. Therefore, if one is prepared to forego some of the attractions of the standard picture, VMOs would seem to be the Population III candidate with most explanatory power. On the other hand, if one wants to preserve the standard picture as much as possible, the most attractive Population III candidate would be SMOs of $10^6 M_\odot$: such objects could avoid making helium, light, enrichment, and explosions, while at the same time producing dark halos, disc heating, and detectable gravitational waves. Although one would no longer need some kind of elementary particle to explain galactic halos, such a particle might still serve to provide a critical density, should this be required for other reasons.

ACKNOWLEDGMENTS

The author thanks W. D. Arnett, J. R. Bond, S. Ikeuchi, J. McDowell, and H. Sato for the enjoyable collaboration which resulted in the papers on which most of this review is based. He is also grateful to the Institute for Theoretical Physics at Santa Barbara for hospitality received during this work. The research was supported in part by the National Science Foundation under Grant No. PHY77-27084, supplemented by funds from the National Aeronautics and Space Administration.

REFERENCES

[1]B. J. Carr and M. J. Rees, Mon. Not. R. Astron. Soc. **206**, 315 (1984).
[2]P. J. E. Peebles and R. H. Dicke, Astrophys. J. **154**, 891 (1968).
[3]S. D. M. White and M. J. Rees, Mon. Not. R. Astron. Soc. **183**, 341 (1978).
[4]P. J. E. Peebles, Astrophys. J. Lett. **189**, L51 (1974).
[5]S. M. Fall, Rev. Mod. Phys. **51**, 21 (1979).
[6]J. Silk, Astrophys. J. **211**, 638 (1977).
[7]R. Terlevich, Ph.D. thesis (Cambridge University, 1983).
[8]J. B. Hutchins, Astrophys. J. **205**, 103 (1976).
[9]T. Matsuda, H. Sato, and H. Takeda, Prog. Theor. Phys. **42**, 219 (1969).
[10]J. E. Tohline, Astrophys. J. **239**, 417 (1980).
[11]A. Kashlinsky and M. J. Rees, Mon. Not. R. Astron. Soc. **205**, 955 (1983).
[12]A. G. Doroshkevich, Ya. B. Zeldovich, and I. D. Novikov, Sov. Astron. AJ. **11**, 231 (1967).
[13]F. Palla, E. E. Salpeter, and S. W. Stahler, Astrophys. J. **271**, 632 (1984).
[14]J. Silk, Mon. Not. R. Astron. Soc. **205**, 705 (1983).
[15]J. Silk, Astrophys. J. **151**, 459 (1968).
[16]J. R. Bond, G. Efstathiou, and J. Silk, Phys. Rev. Lett. **45**, 1980 (1980).
[17]Ya. B. Zeldovich, Astron. Astrophys. **5**, 84 (1970).
[18]J. R. Bond and A.Szalay, Astrophys. J. **274**, 443 (1983).
[19]D. N. Schramm and G. Steigman, Astrophys. J. **243**, 1 (1981).
[20]B. J. Carr, J. R. Bond, and W. D. Arnett, Astrophys. J. **277**, 445 (1984).
[21]W. A. Fowler, Astrophys. J. **144**, 180 (1966).

[22]K. J. Fricke, Astrophys. J. **183**, 941 (1973).
[23]T. A. Weaver, G. B. Zimmerman, and S. E. Woosley, Astrophys. J. **225**, 1021 (1978).
[24]W. D. Arnett, Astrophys. Space Sci. **5**, 180 (1969).
[25]J. P. Cassinelli, J. S. Mathis, and B. D. Savage, Science **212**, 1497 (1981).
[26]C. D. Andriesse, B. D. Donn, and R. Viotti, Mon. Not. R. Astron. Soc. **185**, 771 (1978).
[27]J. R. Bond, W. D. Arnett, and B. J. Carr, Astrophys. J. **280**, 825 (1984).
[28]M. F. El Eid, K. L. Fricke, and W. W. Ober, Astron. Astrophys. **119**, 54 (1983).
[29]W. W. Ober, M. F. El Eid, and K. L. Fricke, Astron. Astrophys. **119**, 61 (1983).
[30]S. E. Woosley and T. A. Woosley, in Supernovae: A Survey of Current Research, eds. M. J. Rees and R. J. Stoneham (Reidel, Dordrecht, 1982).
[31]M. Schwarzschild and R.Harm, Astrophys. J. **243**, 1 (1959).
[32]R. Stothers and N. R. Simon, Astrophys. J. **160**, 1019 (1970).
[33]K. Ziebarth, Astrophys. J. **162**, 947 (1970).
[34]R. J.Talbot, Astrophys. J. **165**, 121 (1971).
[35]J. C. B. Papaloizou, Mon. Not. R. Astron. Soc. **162**, 169 (1973).
[36]I. Appenzeller, Astron. Astrophys. **9, 216 (1970).**
[37]W. A. Fowler and F. Hoyle, Astrophys. J. Supp. **9**, 201 (1964).
[38]Z. Barkat, C. Rakavy, and N. Sack, Phys. Rev. Lett. **18**, 379 (1967).
[39]G. S.Fralay, Astrophys. Space Sci. **2**, 96 (1968).
[40]W. D.Arnett, in Explosive Nucleosynthesis, eds. W. D. Arnett and D. N. Schramm (University of Texas Press, Austin, 1973).
[41]J. C. Wheeler, Astrophys. Space Sci. **50**, 125 (1977).
[42]S. M. Faber and J. S. Gallagher, Ann. Rev. Astron. Astrophys. **17**, 135 (1979).
[43]C. G. Lacey, in Formation and Evolution of Galaxies and Large Structures in the Universe, eds. J. Audouze and J. Tran Thanh Van (Reidel, Dordrecht, 1984).
[44]R. H. Miller, unpublished preprint (1982).
[45]B. J. Carr, Comm. Astrophys. **7**, 161 (1978).
[46]B. J. Carr and M. Sakellariadou, preprint (1985).
[47] B. J. Carr, Mon Not. R. Astron. Soc. **189**, 123 (1979).
[48]S. P. Boughn, P. R.Saulson, and M. Seldner, Astrophys. J. Lett. **250**, L15 (1981).
[49]D. J. Hegyi and K. A. Olive, Phys. Lett. B. **126**, 28 (1984).
[50]W. H. Press and J. E. Gunn, Astrophys. J. **185**, 397 (1973).
[51]J. R. Gott, Astrophys. J. **243**, 140 (1981).
[52]C. R. Canizares, Astrophys. J. **263**, 508 (1982).
[53]P. J. E. Peebles and R. B. Partridge, Astrophys. J. **200**, 527 (1975).
[54]J. R. Thorstensen and R. B. Partridge, Astrophys. J. **200**, 527 (1975).
[55]J. McDowell, preprint (1985).
[56]J. R. Bond, B. J. Carr, and C. Hogan, preprint (1985).
[57]M. Rowan-Robinson, J. Negroponte, and J. Silk, Nature **281**, 635 (1979).
[58]J. Negroponte, M. Rowan-Robinson, and J. Silk, Astrophys. J. **248**, 38 (1981).
[59]J. L. Puget and J. Heyvaerts, Astron. Astrophys. **83**, L10 (1980).
[60]D. P. Woody and P. L. Richards, Astrophys. J. **248**, 18 (1981).
[61]P. L. Richards, this volume (1985).
[62]D. Layzer and R. M. Hively, Astrophys. J. **179**, 361 (1973).

[63]M. J. Rees, Nature **275**, 35 (1978).
[64]E. L. Wright, Astrophys. J. **250**, 1 (1981).
[65]E. L. Wright, Astrophys. J. **255**, 401 (1982).
[66]N. C. Rana, Mon. Not. R. Astron. Soc. **197**, 1125 (1981).
[67]N. C. Wickramasinghe, M. G. Edmunds, S. M. Chitre, J. V. Narlikar, and S. Ramadurai, Astrophys. Space Sci. **35**, L9 (1975).
[68]H. Alfven and A. Mendis, Nature **266**, 699 (1977).
[69]S. M. Chandler and J. M. Scalo, preprint (1984).
[70]B. J. Carr, Mon. Not. R. Astron. Soc. **195**, 669 (1981).
[71]B. J. Carr and M. J. Rees, Astron. Astrophys. **61**, 705 (1977).
[72]A. D. Sakharov, JETP Lett. **5**, 24 (1967).
[73]D. Toussaint, S. Treiman, F. Wilczek, and A. Zee, Phys. Rev. D **19**, 1036 (1979).
[74]S. Weinberg, Phys. Rev. Lett. **42**, 850 (1979).
[75]D. V. Nanopoulos and S.Weinberg, Phys. Rev. D **20**, 2484 (1979).
[76]E. W. Kolb and M. S. Turner, Ann. Rev. Nuc. Part. Sci. **33**, 645 (1983).
[77]D. M. Eardley, A. P. Lightman, D. G. Payne, and S. L. Shapiro, Astrophys. J. **224**, 53 (1978).
[78]P. Meszaros, Astron. Astrophys. **38**, 5 (1975).
[79]M. Davis, G. Efstathiou, C. S. Frenk, and S. D. M. White, Astrophys. J., in press (1984).
[80]G. R. Blumenthal, S. M. Faber, J. R. Primack, and M. J. Rees, Nature **311**, 517 (1984).
[81]T. W. Hartquist and A. G. W. Cameron, Astrophys. Space Sci. **48**, 145 (1977).
[82]C. Hogan, Mon. Not. R. Astron. Soc. **188**, 781 (1979).
[83]B. J. Carr, Mon. Not. R. Astron. Soc. **194**, 639 (1981).
[84]P. W. Tarbet and M. Rowan-Robinson, Nature **298**, 711 (1982).
[85]T. A. Weaver and S. E. Woosley, Ann. New York Acad. Sci. **336**, 335 (1980).
[86]H. E. Bond, Astrophys. J. **248**, 606 (1981).
[87]J. W. Truran and A. G. W. Cameron, Astrophys. Space Sci. **14**, 179 (1971).
[88]J. E. Jones, preprint (1984).
[89]C. A. Pilachowski, G. D. Bothun, E. W. Olszewski, and A. Odell, Astrophys. J. **273**, 187 (1983).
[90]J. W. Truran, in Formation and Evolution of Galaxies and Large Structures in the Universe, eds. J. Audouze and J. Tran Thanh Van (Reidel, Dordrecht, 1984).
[91]C. Sneden, D. Lambert, and R. W. Whitaker, Astrophys. J. **234**, 964 (1979).
[92]C. Chiosi and F. Matteucci, in Formation and Evolution of Galaxies and Large Structures in the Universe, eds. J. Audouze and J. Tran Thanh Van (Reidel, Dordrecht, 1984).
[93]B. E. J. Pagel and M. G. Edmunds, Ann. Rev. Astron. Astrophys. **19**, 77 (1981).
[94]B. Barbuy, Astron. Astrophys. **123**, 1 (1983).
[95]J. Klapp, Astrophys. Space Sci. **93**, 313 (1983).
[96]R. J. Talbot and W. D. Arnett, Nature **229**, 250 (1971).
[97]J. R. Bond, B. J. Carr, and W. D. Arnett, Nature **304**, 514 (1983).
[98]M. Kaufman, Astrophys. J. **160**, 459 (1970).
[99]B. J. Carr, Astron. Astrophys. **56**, 377 (1977).
[100]J. Yang, D. N. Schramm, G. Steigman, and R. T. Rood, Astrophys. J. **277**, 697 (1979).

[101]R. L. Epstein, J. M. Lattimer, and D. N. Schramm, Nature **263**, 198 (1976).
[102]M. J. Rees, in Formation and Evolution of Galaxies and Large Structures in the Universe, eds. J. Audouze and J. Tran Thanh Van (Reidel, Dordrecht, 1984).
[103]J. Audouze and J. Silk, in Formation and Evolution of Galaxies and Large Structures in the Universe, eds. J. Audouze and J. Tran Thanh Van (Reidel, Dordrecht, 1984).
[104]S. Hayakawa, in Proc. Semin. Fundamental Studies: Present Attitude and Future Prospect (Colombo, Sri Lanka, 1982).
[105]T. Matsumoto, M. Akiba, and M. Murakami, in COSPAR Journal: Advances in Space Research, eds. G. F. Bignami and R. A. Sunyaev (Pergamon Press, Oxford, 1984).
[106]B. J. Carr, J. McDowell, and H. Sato, Nature **306**, 666 (1983).
[107]H. Bondi, Mon. Not. R. Astron. Soc. **112**, 195 (1952).
[108]R. Wielen, Astron. Astrophys. **60**, 263 (1977).
[109]C. Lacey, Mon. Not. R. Astron. Soc. **208**, 687 (1984).
[110]R. Carlberg, in Formation and Evolution of Galaxies and Large Structures in the Universe, eds. J. Audouze and J. Tran Thanh Van (Reidel, Dordrecht, 1984).
[111]B. Bertotti and B. J. Carr, Astrophys. J. **236**, 1000 (1980).
[112]K. S. Thorne, in Theoretical Principles in Astrophysics and Relativity, eds. N. R. Lebovitz, W. H. Reid, and P. O. Vandervoort (University of Chicago Press, 1978).
[113]J. R. Bond and B. J. Carr, Mon. Not. R. Astron. Soc. **207**, 585 (1984).
[114]C. D. Garmany, P. S. Conti, and P. Massey, Astrophys. J. **242**, 1063 (1980).
[115]V. Trimble and C. Cheung, in IAU Symposium No. 73, eds. P. Eggleton, S. Mitton, and J. Whelan (1976).
[116]J. Bookbinder, L. L. Cowie, J. H. Krolik, J. P. Ostriker, and M. J. Rees, Astrophys. J. **237**, 647 (1980).
[117]A. G.Doroshkevich, Ya. B. Zeldovich, and I. D. Novikov, Sov. Astron. AJ. **11**, 231 (1967).
[118]J. P. Ostriker and L. L. Cowie, Astrophys. J. Lett. **243**, L127 (1981).
[119]S. Ikeuchi, Pub. Astron. Soc. Japan **33**, 211 (1981).
[120]E. Bertschinger, Astrophys. J. **268**, 17 (1983).
[121]B. J. Carr and M. J. Rees, Mon. Not. R. Astron. Soc. **206**, 801 (1984).
[122]B. J. Carr and S. Ikeuchi, Mon. Not. R. Astron. Soc., **213**, 497 (1985).
[123]M. Davis, J. Huchra, D. W. Latham, and J. Tonry, Astrophys. J. **253**, 423 (1982).
[124]J. Einasto, M. Joeveer, and E. Saar, Mon. Not. R. Astron. Soc. **193**, 353 (1980).
[125]R. Kirshner, A. Oemler, P. Schechter, and S. Schectman, Astrophys. J. Lett. **248**, L57 (1981).
[126]C. Hogan, Mon. Not. R. Astron. Soc. **202**, 1101 (1983).
[127]C. Hogan and N. Kaiser, Astrophys. J. **274**, 7 (1984).

Nucleosynthetic Tests of Gravitation Theories

Clifford M. Will

I. INTRODUCTION

In recent years, the early universe has become a testing ground for new ideas in particle physics, quantum field theory, and quantum gravity. Unfortunately, these new ideas are interesting or relevant primarily in the very early universe, under conditions of temperature and curvature so extreme that precise, quantitative statements are difficult if not impossible to make. On the other hand, we think we understand enough about the universe in its "middle age" (~ minutes after the big bang) that we dare to risk being precise. For example, the standard model in general relativity is said to predict an abundance of primordial helium that is in agreement with that inferred from observations within errors that are sometimes quoted to be as small as ten per cent. Although this observation has been used to test particle-physics ideas by setting a limit on the number of neutrino species, it is useful to note that it can also be used to test a key ingredient of all early-universe studies, one that is normally taken for granted: the theory of gravitation. Many alternative theories of gravity, even those whose post-Newtonian limits are close to that of general relativity and thereby agree with solar-system tests, are different enough in their exact field equations that they may predict qualitatively different cosmological histories. Put differently, the deviations from flat space time in the solar system are $O(10^{-8})$, while during the era of nucleosynthesis, they are $O(10^{18})$, so nucleosynthetic tests are truly strong-field tests of gravitational theory.

In this paper we review the results of such tests. We discuss a general approach to constructing cosmological models in alternative theories of gravity (§ II), then describe a method of parameterizing the primordial helium abundance developed by Schramm and Wagoner[1] (§ III), and finally summarize the results for specific theories of gravitation (§ IV).

II. COSMOLOGICAL MODELS IN ALTERNATIVE THEORIES OF GRAVITY

We begin by making two important assumptions about the nature of the universe[2]:

(i) The Einstein Equivalence Principle (EEP): This principle restricts our attention to metric theories of gravity, which postulate that spacetime is endowed with a symmetric metric g in whose local Lorentz frames the non-gravitational laws of physics take on their special relativistic forms.

(ii) The Cosmological Principle: This principle amounts to the assumption that the universe is homogenous and isotropic on the largest scales. As a consequence, the line element of g may be written in the Robertson-Walker form

$$ds^2 = -dt^2 + a(t)^2[(1-kr^2)^{-1}\, dr^2 + r^2\, (d\theta^2 + \sin^2\theta d\phi^2)] \quad , \qquad (1)$$

where t is proper time as measured by an atomic clock, $a(t)$ is the expansion factor, and $k \in \{\pm 1, 0\}$ is a constant. During the era of nucleosynthesis, the value of k is irrelevant, and can be set to zero.

Because of these assumptions, the behavior of matter and non-gravitational fields can be related to the scale factor a(t) in a theory-independent manner. For example, the entropy per baryon is approximately constant throughout the expansion. During the radiation-dominated era, when the energy density is given by $\rho = fa_\rho T^4$, with a_ρ the radiation density constant, f a multiplicity factor set by the number of relativistic particles present, and T the temperature, T and a(t) are related by Ta(t) = constant. Thus the only input from the specific theory of gravity is the actual time dependence of a(t).

Most alternative theories of gravity to general relativity introduce auxiliary gravitational fields in addition to the metric, that mediate the manner in which matter generates spacetime curvature. These fields must first be expressed in a form compatible with the symmetries of the Robertson-Walker spacetime. These forms include

Scalar: $\phi(t)$,

Vector: $K_\mu dx^\mu = K(t)dt$,

Tensor: $B_{\mu\nu}dx^\mu dx^\nu = \omega_0(t)dt^2 + \omega_1(t)d\sigma^2$,

Flat-background metric (non dynamical):

$$\eta_{\mu\nu}dx^\mu dx^\nu = -\dot{\tau}(t)dt^2 + \tau(t)^2 d\sigma^2 \quad , \qquad [k = -1]$$

$$= -\dot{\tau}(t)dt^2 + d\sigma^2 \quad , \qquad [k = 0]$$

where $d\sigma^2$ is the spatial metric of the homogeneous hypersurface of the spacetime (quantity in square brackets in Eq. [1]). The matter variables have the form $\rho = \rho(t)$, $p = p(t)$, $u_\mu dx^\mu = -dt$, and so on. These variables are then substituted into the field equations of the chosen theory. It may be useful to set boundary conditions on the present values of some of the auxiliary fields, as these may be constrained by solar-system tests of post-Newtonian paramters or by limits on the rate of variation of G.[3] The equations can then be integrated backward to the era of nucleosynthesis to obtain the evolution of a(t) and $\rho(t)$.

III. PARAMETERIZED APPROACH TO HELIUM SYNTHESIS

Given a solution for a(t) and $\rho(t)$, it is useful to define the parameter

$$\xi(t) = \frac{\dot{a}/a}{(8\pi G\rho/3)^{1/2}} \quad . \tag{2}$$

In general relativity, $\xi \equiv 1$, expressing the fact that, with that value, the above equation is equivalent to one of the field equations at early times. In other theories, the relationship between ρ and $\dot{a}/a$ is generally more complicated than this, often involving the auxiliary gravitational fields. Nevertheless, the phenomenological function $\xi(t)$ defined above is useful because it turns out that the abundance of helium depends primarily on the value of ξ during the era of nucleosynthesis, provided it does not vary too rapidly.

The qualitative dependence of the helium abundance on ξ is easy to understand. At temperatures above $\sim 10^{10}$ K, neutrons and protons are maintained in equilibrium by weak interactions ($p + \bar{\nu}_e \rightleftarrows n + e^+$, for example) whose rates are strongly temperature (i.e., density) dependent. When the weak interaction rates drop below the expansion rate, the neutrons freeze out, in an abundance relative to that of protons given by the Boltzmann factor

$$n/p = e^{-Q/kT_f} = e^{-1.5/T_{f,10}} , \tag{3}$$

where Q is the n-p mass difference, and $T_{f,10}$ denotes the freeze-out temperature in units of 10^{10} K. The larger the value of ξ, the faster the expansion rate for a given temperature, the sooner the freeze-out, the higher the value of T_f, and consequently the larger the value of n/p at freeze-out. Following freeze-out, the neutrons decay freely (lifetime ~ 1000 s) until the temperature reaches 10^9 K, where deuterium is sufficiently stable that the chain of reactions leading to helium can occur. The larger the value of ξ, the sooner the temperature drops to 10^9 K, the larger the remaining fraction of neutrons. During nucleosynthesis, virtually all the neutrons are incorporated into helium, so the mass fraction X of helium depends only on the neutron-proton abundance ratio at the time t_N of nucleosynthesis, i.e.,

$$X(He^4) \simeq 2(n/p)(1+n/p)^{-1}\big|_{t_N} . \tag{4}$$

Thus $X(He^4)$ is an increasing function of the mean value $\bar{\xi}$ of $\xi(t)$ during nucleosynthesis. Detailed calculations by Schramm and Wagoner[1] lead to the specific dependence

$$X(He^4) \simeq 0.25 + 0.38 \log_{10}\bar{\xi} + 0.02 \log_{10} (\rho_0/10^{-30}\ gcm^{-3}) , \tag{5}$$

where ρ_0 is the present baryon density. Observational constraints on the primordial helium abundance,[4] $0.22<X_{obs}<0.25$, imply that $|\bar{\xi}-1|<0.2$ (we have assumed three neutrino species).

IV. TESTS OF ALTERNATIVE THEORIES OF GRAVITATION

For specific theories of gravity, there are sparse results for this test. No systematic study of cosmological models in alternative theories has been carried out, and of those analyses that have been performed within specific theories, few have addressed the helium abundance as a test. Thus the following is a brief list of those few results that are known.

General relativity: The standard big-bang model agrees well with the cosmic helium abundance ($\xi \simeq 1$).

Brans-Dicke theory: To agree with solar-system observations, the coupling constant ω must be constrained to be larger than 200.[3] Roughly speaking, when ω is large, all predictions of Brans-Dicke theory, including cosmological models, agree with those of general relativity to corrections of $O(\omega^{-1})$. Thus Brans-Dicke ($\omega>200$) theory satisfies this test.

Bekenstein's variable-mass theory (VMT)[5]: By contrast with Brans-Dicke theory, the VMT coupling constant ω is a function of the scalar field, and so the theory can have cosmological models that begin the expansion from a non-singular "bounce" at temperatures exceeding 10^{11} K (which presumably was preceded by a contraction phase). After numerical integration of the field equations for a variety of values of the curvature parameter k and the arbitrary constants r and q of the theory, Bekenstein and Meisels[6] reached the following conclusions: (i) Although the initial value of ω was quite small ($\sim -3/2$), its present value in most models exceeded 500, thus yielding close agreement with all experimental tests. (ii) The observable gravitational constant G could decrease in some models by as many as 40 orders of magnitude between the initial moment and the present, thereby accounting for the "large number" puzzle that $G m_p^2/hc \simeq 10^{-38}$, where m_p is the proton mass. Because of the large variation in G, this ratio would be initially near unity. Nevertheless, the predicted present value of $\dot{G}/G$ was well below the experimental upper limits. However, the predicted synthesis of helium is compatible with the observed abundance only in models in which G varies by no more than a factor of 300 over all time.[7] Thus any hope of accounting for the large numbers puzzle within VMT was dashed.

Rosen's bimetric theories: An early bimetric theory by Rosen was strongly ruled out by observations of the orbital period change in the binary pulsar.[8] A more recent version[9] has a background de Sitter metric whose "Einstein tensor" plays the role of a cosmological term in the Einstein equation for the physical metric. Up to corrections of order [size of local system/Hubble radius)^2, its predictions for the solar system and the binary pulsar agree with those of general relativity. Only closed cosmological models that begin from a non-singular bounce at the Planck scale have been analyzed,[9] and these have not been meshed completely with current observational constraints (such as the Hubble constant and the mean matter density). As presented in Ref. 9, the models give a value of $\bar{\xi} \sim 1.1$, so a more detailed calculation could yield a definitive test.

Some of these issues have also been addressed in the cosmological models of Rastall's theory,[10] and Canuto's theory.[11] Although the results presented here are very sketchy, they illustrate that, for some theories of gravity, early universe cosmology may provide do-or-die tests.

This work was supported in part by the National Science Foundation [PHY 83-13545].

REFERENCES

1. D. N. Schramm and R. V. Wagoner, Ann. Rev. Nucl. Sci. **27**, 37 (1977).
2. For details, see C. M. Will, Theory and Experiment in Gravitational Physics (Cambridge University Press, Cambridge, 1981), § 13.1.
3. For a review, see C. M. Will, Phys. Reports, **113**, 345 (1984); or Ref. 2 §§7-9.
4. B. E. J. Pagel, this volume; J. Yang, M. S. Turner, G. Steigman, D. N. Schramm, and K. A. Olive, Astrophys. J., **281**, 493 (1984).
5. J. D. Bekenstein, Phys. Rev. D **15**, 1458 (1977).
6. J. D. Bekenstein and A. Meisels, Astrophys. J. **237**, 342 (1980).
7. A. Meisels, Astrophys. J. **252**, 403 (1982).
8. Ref. 2, § 12.3.
9. N. Rosen, Found. Phys. **10**, 673 (1980).
10. P. Rastall, Astrophys. J. **220**, 745 (1978); **244**, 1 (1981).
11. V. Canuto and S. H. Hsieh, Astrophys. J. Suppl. **41**, 243 (1979); T. Rothman and R. Matzner, Astrophys. J. **257**, 450 (1982).

Looking for Cosmological Neutrinos Through Resonant Absorption of Cosmic Ray Neutrinos

Thomas J. Weiler

ABSTRACT

The proposal to search for relic neutrinos through their absorption effect on the neutrino cosmic ray flux is discussed. It is shown that within the context of big bang cosmology, detection of relic neutrinos with this method appears very unlikely but not impossible.

I. INTRODUCTION

Quite analogous to the 2.7°K photon background permeating the universe, the existence of a 1.9°K neutrino gas is predicted by big bang cosmology.[1] In this talk we discuss a possible indirect measurement of the neutrino gas. Whereas the properties of the photon gas were determined when the photon interaction rate fell below the universe's expansion rate some 10^5 years after the initial singularity, the properties of the neutrino gas were determined at a much earlier decoupling time of one second. Thus the proposed measurement tests first second cosmology.

If the cosmological neutrinos are nonrelativistic ($m \gg kT$), then sizeable absorption dips are predicted[2] in the energy spectrum of neutrino cosmic rays originating from a large z (cosmological red-shift) extragalactic source. The absorption mechanism is ν-$\bar{\nu}$ annihilation into the recently discovered Z boson ($M_Z \sim 90$ GeV) of electroweak theory.[3] The resonant absorption energy for each neutrino mass eigenstate is prescribed, $E_R \sim M_Z^2/2m(1+z)$, or $E_R \sim 4 \times 10^{21}$ $eV^2/m(1+z)$. In models unifying the strong and electroweak interactions, neutrino masses in the eV range arise quite naturally.[4] Thus a dip energy of $10^{20\pm1}$ eV may be expected for neutrinos eminating from a large z source.

Suggested sources of ultraenergy neutrinos include quasars, active galactic nuclei, pulsars, supernovae and accreting black holes. The emission mechanism is presumed to be the collision of hadrons, accelerated by gravitational infall, shock waves, and/or by enormous magnetic fields, yielding pions with decay products $\nu_\mu \bar{\nu}_\mu e \nu_e$. Could 10^{20} eV neutrinos be produced? Cosmic rays have been detected with energies a little beyond 10^{20} eV. If these particles received a major portion of their energy in the act of emission from their source, then neutrino cosmic rays should exist at similar energies, and perhaps even at higher energies because the neutrinos escape with their ultraenergies intact, while all other quanta endure strong absorption and energy degradation in the high density environment of the source.

II. SINGLE SOURCE ABSORPTION

As a result of neutrino annihilation on the Z-resonance of the standard electroweak model, the transmission probability for the neutrino cosmic ray flux is[2]

$$\exp\left\{ -K\, \tilde{E}^{-3/2}(\Omega_0+\tilde{E}(1-\Omega_0))^{-1/2} \right\}, \qquad (1)$$

for $(1+z)^{-1} < \tilde{E} < 1$, and unity elsewhere. Ω_0 is the present ratio of the universe's energy density to the closure density, $\tilde{E} \equiv 2mE_0/M_Z^2$ is the incident neutrino (antineutrino) energy at earth scaled by $M_Z^2/2m$, and $K = 2\pi\sqrt{2}G_F\eta_\nu H_0^{-1} = 0.021\ h^{-1}\ (\eta_\nu/53\ cm^{-3})$, with G_F the Fermi constant, H_0 the present value of the Hubble parameter, and η_ν the present density of relic antineutrinos (neutrinos). h is the present Hubble parameter in units of 100 km/s/Mpc. $53\ cm^{-3} \times (T_\gamma/2.7°K)^3$ is the neutrino density per flavor predicted by cosmology,[1] under the assumption of zero chemical potential for the neutrino-antineutrino gas. T_γ is the present photon blackbody temperature. If the assumption of zero chemical potential is dropped, then either the neutrino or antineutrino density must exceed 53/cc, and densities up to that value saturating the energy density of the universe, viz. $\eta_\nu \leq 0.90 \times 10^4 \times h^2\Omega_0/(m/eV)cm^{-3}$, cannot be ruled out.

As shown in the solid line of Fig. 1, significant absorption may be anticipated for neutrinos originating at a $z \gtrsim 2$ source. E.g., a $z \sim 3$ source implies 30% absorption at the low energy end of the dip.

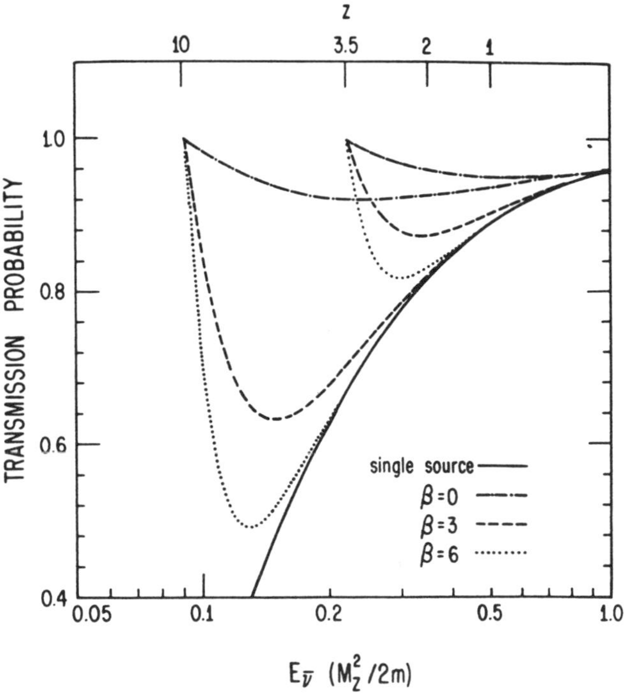

Figure 1: Transmission probability for neutrino flux. Assumed values of (h^{-1}, Ω_0, η_ν, T_γ, α) are (2, 1, 53/cc, 2.7°K, 0).

Unfortunately, even if there exist high z sources of 10^{20} eV neutrinos, it is unlikely that a single source spectrum can be measured at the energies here required. Consider as a measurement criterion, detection of 10^3 events per year in the 10^{20} eV energy region by a 100% efficient Fly's Eye type detector. At 10^{20} eV the effective fiducial volume is 400 km^3 of earth.[5] Since the ultrahigh energy neutrino-nucleon cross section is approximately $10^{-33} cm^2$ at 10^{20} eV, an incident neutrino flux of $F \sim 10^{-13}/cm^2/s$ is required. The distance to a highly red shifted source is a large fraction of the Hubble radius, so the power generating this minimum flux is $P \sim F \cdot E_\nu \cdot 4\pi H_0^{-2} \sim 10^{52}$ erg/s, or $10^5\ M_\odot/yr$. This may be constrasted with a 10^{44} to 10^{49} erg/s optical emission from quasars. With a DUMAND type experiment (1 km^3 of water detector), the flux and power requirement are similar.

III. SUM OVER SOURCES

In order to enhance the flux of 10^{20} eV neutrinos incident on earth, a sum over sources may be carried out. Taking the energy density of the universe to be $\sim 10^{-29}$ g/cc, the minimum flux for Fly's Eye or DUMAND detection of the dip results if 10^{-7} of this energy exists as 10^{20} eV neutrinos. This is possible, and we proceed to discuss the absorption of the neutrino flux integrated over sources.

Let $\mathscr{L}$ be the source luminosity (#ν/physical volume/time/energy) as measured at the sources. For simplicity we will assume $\mathscr{L}(E,z)$ factorizes in E and z, with the energy dependence in the region of absorption ($\tilde{E}\sim 1$), given by a power law spectrum. Factoring out the trivial $(1+z)^3$ behavior resulting from cosmic expansion, we have

$$\mathscr{L}(E,z) = AE^{-\alpha}(1+z)^3 f(z) = AE_0^{-\alpha}(1+z)^{3-\alpha} f(z) \ .$$

f(z) is the function, as yet unspecified, containing the evolutionary history of the source luminosity per comoving volume. We may normalize it to f(0) = 1. Changes in source spectra with time, and the creation or destruction of sources, are characterized by the deviation of f(z) from unity.

Assuming a Robertson-Walker metric, zero cosmological constant and a matter dominated universe, one finds[6] for the transmission probability

$$1 - D(\tilde{E},\alpha,\Omega_0)\left\{1 - \exp \frac{-K}{\tilde{E}^{3/2}\sqrt{\tilde{E}(1-\Omega_0)+\Omega_0}}\right\}, \tag{2}$$

for $\tilde{E} < 1$, unity for $\tilde{E} \geq 1$. We have

$$D = \int_{1/\tilde{E}-1}^{\infty} dz\, I(z) \Big/ \int_0^{\infty} dz\, I(z), \text{ and } I(z) = (1-z)^{-(1+\alpha)}(1+\Omega_0 z)^{-1/2} f(z).$$

This result states that the absorption of the integrated neutrino flux is just the single source expression (cf. expression (1)) weighted by a dilution factor D whose value lies in the interval [0,1]. The more dramatic the source evolution is, the more peaked is f(z) toward larger z, and the greater is the magnitude of D. It is clear from setting D to unity in expression (2) that the single source transmission probability is a lower bound for multiple source transmission.

To analyze the degree of dilution expected from the sum over source redshifts, we must specify the source history, f(z). Lacking better direction, we shall model neutrino source evolution after quasar evolution. A power law parameterization of f(z) is commonly used in studies of quasar evolution:

$$f(z) = (1+z)^{\beta}\, \Theta(z_c - z) \ .$$

The theta function defines the critical time, z_c, at which the sources began emitting neutrinos, and $\beta \neq 0$ characterizes source evolution.

We may obtain simple analytic results for the D factor by inserting into I(z) the value 0 or 1 for Ω_0. These values probably bracket the true value and are therefore interesting. The resulting dilution factor is

$$D(\tilde{E},z_c,\Gamma) = \frac{1 - [\tilde{E}(1+z_c)]^{-\Gamma}}{1 - (1+z_c)^{-\Gamma}},$$

with $\Gamma = \beta - \alpha - (0,1/2)$ for $\Omega_0 = (0,1)$. Predictably, D increases with increasing Γ or z_c.

IV. RESULTS AND CONCLUSIONS

To obtain quantitative results, α, β and z_c must be guessed. If the analogy with quasar optical emissions is maintained, then some observational results become relevant. Quasar counts as a function of z show definite evolutionary effects,[7] at least for $z \lesssim 2$, corresponding to a mean $\beta \sim 6$. Furthermore, the number of contributing sources could be large. For example, statistical studies[7,8] imply a quasar number of several million for $z \lesssim 3$. The apparent quasar density peaks around $z \sim 2$ (perhaps a selection effect), and vanishes abruptly above $z \sim 3.5$. This cut-off at $z \sim 3.5$ may be interpreted as signifying the age of quasar production, or alternatively, as signaling the end of a radio, optical and x-ray dense area. Thus if neutrinos are to endure red-shifts beyond 3.5, their source may be quasars under the latter hypothesis, but must be something other under the former hypothesis.

Transmission probabilities for the integrated neutrino flux are shown in Fig. 1, with z_c values set to 3.5 and 10. (For nonzero α, assign the given values of β to $\beta - \alpha$ instead.) The dependence of the transmission upon values of (h^{-1}, Ω_0, η_ν, T_γ) is given in Ref. 2. We infer that with the standard value for the relic density, significant neutrino absorption can occur only if there exist very red-shifted sources ($z>2$) and considerable evolution (large β).

In conclusion, the requirements for detection of massive cosmological neutrinos are: extreme source red-shift and evolution, extreme neutrino flux, and extreme neutrino energy. Although each is unlikely, none can be ruled out. Neutrino astronomy will some day settle the issue.

This work is supported in part by the Department of Energy contract DE-AT03-81ER-40029.

REFERENCES

[1]See, e.g., S. Weinberg, Gravitation and Cosmology, Wiley, New York (1972).

[2]T. Weiler, Phys. Rev. Lett. **49**, 234 (1982).

[3]G. Arnison et al., Phys. Lett. **126B**, 398 (1983); P. Bagnaia et al., Phys. Lett. **129B**, 130 (1983).

[4]P. Ramond, Proc. of the First Workshop on Grand Unification, eds., P. Frampton, S. Glashow, A. Yildiz, U.N.H. (1980); E. Witten, ibid.

[5]P. Sokolsky, Proc. Utah Cosmic Ray Workshop (1983).

[6]T. Weiler, Ap. J. (October, 1984).
[7]M. Schmidt, Ap. J. **151**, 393 (1968); **162**, 371 (1970).
[8]M. Schmidt and R.F. Green, Ap. J. **269**, 352 (1983).

Are Baryonic Galactic Halos Possible?

K. A. Olive and D. J. Hegyi

There is little doubt from the rotation curves[1] of spiral galaxies that galactic halos must contain large amounts of dark matter. In this contribution, we will review our arguments[2-4] which indicate that it is very unlikely that galactic halos contain substantial amounts of baryonic matter. While we would like to be able to present a single argument which would rule out baryonic matter, at the present time we are only able to present a collection of arguments each of which argues against one form of baryonic matter. These include: 1) snowballs; 2) gas; 3) low mass stars and Jupiters; 4) high mass stars; and 5) high metalicity objects such as rocks or dust. Black holes,[5,6] which do not have a well defined baryon number, are also a possible candidate for halo matter. We shall briefly discuss black holes.

We will use a $10^{12} M_\odot$ halo extending out to 100 kpc as our standard halo in the discussions which follow. The average halo density is then $\rho_H = 1.6 \times 10^{-26}$ g cm^{-3}. Also, we will assume a mass-to-light ratio consistent with that deduced from binary galaxies so that[7] $(M/L)_H > 70\ h_o$ where $h_o = H_o/(100$ km Mpc^{-1} s$^{-1})$ and H_o is the present value of the Hubble parameter. We take $1/2 < h_o < 1$. This value of M/L corresponds to a fraction of critical density, $\Omega_H = (M/L)_H/(M/L)_C > 0.05$.

We will now briefly present our arguments against the forms of baryonic matter listed above beginning with snowballs. A snowball refers to a condensation of hydrogen which is electrostatically bound. At typical halo velocities, $v \sim 250$ km s^{-1}, the kinetic energy per hydrogen atom is ~300 eV, implying that any collisions between snowballs would vaporize them. Therefore, snowballs, if they exist, must have radii $R_s > 4$ cm. However, the existence of snowballs today, requires that they be in equilibrium with the microwave background at $T \sim 2.7$°K. In order not to have totally evaporated by today, $R_s > 10^{14}$ cm. Such an object is not electrostatically bound but is gravitationally bound. Snowballs quickly evaporate and cannot constitute the dark matter in galactic halos.

We shall now discuss the possibility that the halo is gaseous by separately considering the problems associated with a hot and cold gaseous halo. If the halo gas is in hydrostatic equilibrium, it is hot; its temperature is $T = 1.3 \times 10^6$ °K. The assumption that the gas is in equilibrium is a good one since the collapse timescale for the gas is $\sim 5 \times 10^8$ yrs, much less than the age of the galaxy. One possibility for cold gas is to break up the halo into small clouds. We shall return to this possibility below.

Hot gas is ruled out by the observed upper limits on the X-ray background. To derive the amount of radiation emitted by a hot halo, one must consider both the clumpiness and density of the halo. The clumpiness is defined by

$$C = \langle \rho^2 \rangle / \langle \rho \rangle^2 \tag{1}$$

where ρ is the average density of the universe. The quantity $C\Omega$ can be expressed as[8]

$$C\Omega = \rho_H / \rho_c \tag{2}$$

where $\rho_c = 1.88 \times 10^{-29} h_o^2 g\ cm^{-3}$ is the critical density. Silk[9] has converted the observed X-ray limits into a limit or $C\Omega^2 h_o^3$. With $C\Omega \sim 900\ h_o^{-2}$ and $\Omega \geq \Omega_H = 0.05$ for $h_o > 1/2$, we have $C\Omega^2 h_o^3 \geq 22$. Silk's constraint at $T \sim 10^6$°K is $C\Omega^2 h_o^3 \leq 3$. For higher temperatures the constraint becomes more severe.

A possible way around the X-ray constraint might be to break up the halo into many cooler gas clouds.[10] Depending on the number of clouds, N, we can find a temperature and a clumpiness to satisfy the constraints. For N < 50, however, there is no acceptable temperature; the clouds are too hot. For N < 500, the cooling timescales are still long enough so that cooling and hence star formation does not occur. Although, for N = 500, the cloud temperature is cooler, and the X-ray limits are relaxed by a factor of ~28 so that for this case $C\Omega^2 h_o^3 \leq 84$, the clumpiness is greatly enhanced so that under these conditions one expects $C\Omega^2 h_o^3 > 5 \times 10^4$. For N > 500, some star formation takes place and the halo would become visible. However, it is possible for just enough stars to form to keep the clouds cool and prevent them from further collapse, and at the same time for these stars to be hidden in the clouds. We feel that this is a highly contrived situation since it must hold true in the vast majority of gas clouds.

We now move to low mass stars and Jupiters. In order to make any statement regarding the possibility of Jupiters, it is necessary for us to make some assumptions. We shall assume an initial mass function of the form

$$\phi(m) \propto m^{-(1+x)} \tag{3}$$

where ϕ is the number of stars formed per unit volume per unit mass. We will also assume that the slope of the mass function, x, extends below $m_o = 0.08\ M_\odot$, which is the lower cutoff for nuclear burning. The minimum mass for Jupiters will be the minimum mass taken from fragmentation calculations,[11] $m_{min} = 0.004\ M_\odot$.

Using the initial mass function ϕ, we can compute a mass-to-light ratio for the halo. The total mass density is given by

$$\rho_m = \int_{m_{min}}^{m_G} m\phi(m)dm \tag{4}$$

where $m_G = 0.75\ M_\odot$ is the mass of a giant. The luminosity density is given by

$$\rho_L = \int_{m_o}^{m_G} L(m)\phi(m)dm + L_G \tag{5}$$

where $L \propto m^D$ is the luminosity of a star of mass m and L_G is the total luminosity density due to giants. Using the data of Tinsley[12] and Tinsley and Gunn,[13] we have computed L_G and M/L for the halo.

We have compared these calculations to observational data discussed by Hegyi,[14] and Boughn, Saulson and Seldner[15] for the halo if NGC 4565. These measurements were made in two different spectral bands (I and K). The observations indicate that $M/L|_I > 60$ and $M/L|_K > 38$ in solar units. The limits on M/L correspond to a lower limit on the slope of the initial mass function, $x > 1.7$. This is somewhat higher than the preferred value[13,16] of $x < 1$ found from observational determinations.

The limit $x > 1.7$, was found using $m_{min} = 0.004$. For smaller m_{min}, one can reduce the limit on x. For $x = 1.35$, the calculated value of M/L is consistent with the observations for $m_{min} \lesssim 10^{-4}\ M_{\odot}$, a factor of 40 below the calculated value. For $x < 1$, no value of m_{min} brings the calculation into agreement with the observations. Earlier, Dekel and Shaham[17] used a similar discussion to explore the possibility of Jupiters and low mass stars in the halo of NGC 4565. However, they did not introduce a lower cutoff on the mass of planets.

An obvious escape from the above limit, would be to scrap the initial mass function for low mass stars and replace it with a delta function peaked at low masses. One possibility for forming very low mass objects was suggested by Kashlinsky and Rees[18] in which massive (10^6-$10^8\ M_{\odot}$) objects formed thin disks and then fragmented down to Jupiters. However, we do not understand how this scenario is consistent with the Toomre stability condition. One can show[3] that these disks would be unstable to form somewhat massive objects, ~10 $M_{\odot}$, rather than Jupiters. To form $10^{-3}\ M_{\odot}$ objects out of a $10^5\ M_{\odot}$ disk, one needs the ratio of the thickness of the disk to the disk radius to be $\varepsilon \sim 10^{-4}$. Under these circumstances one can show that the Jeans mass is $M_J > \varepsilon M_D \sim 10\ M_{\odot}$, where M_D is the mass of the disk.

We now consider stars with masses larger than the sun's and will use metal abundance arguments to exclude stars in this mass range. Since these stars eject mass, if we assume that at most 1% of the stars ejecta is in the form of heavy elements, then $\Omega_{metals} = r(0.01)\Omega_H$ where

r is the fraction of the initial mass ejected. Taking $r > 0.3$ is probably conservative so that $\Omega_{metals} > 1.5 \times 10^{-4}$. The minimum metalicity for the disks of spiral galaxies will then be

$$Z > \Omega_{metals}\varepsilon/\Omega_D$$

where $\Omega_D < 0.014$ is found from[7] $(M/L)_D < 20\ h_0$, and ε is the mass fraction of the halo which mixes with the disk. Because of the existence of very low metalicity stars[19] ($Z \sim 10^{-5}$) we require $Z_{min} < 10^{-5}$ or $\varepsilon \lesssim 10^{-3}$. We would expect over the age of the galaxy that $\varepsilon \sim 1$. Hence the production of metals poses a severe constraint on the various candidates for the halo material.

Black holes are the final possibility which we consider, though they do not have a well defined baryon number. Black holes of mass ~$10M_{\odot}$ run into the same difficulty as high mass stars because of their metal production. The obvious way around the metals constraint is for the black holes to accrete all of their ejected material. In these Proceedings, Carr[5] has discussed other constraints on black holes in different mass ranges. If black holes do make up the dark matter and significant amounts of the dark matter resides in the disk[20] then black hole accretion in the disk leads to other difficulties.[21]

We conclude therefore that it is very difficult to hide baryons in galactic halos. If pushed to extremes, we could ask what is the upper limit on the fraction of the halo mass in baryons. For an initial mass function with slope $x = 1$ and $m_{min} = 0.004M_{\odot}$, we could tolerate an $\Omega_B = 0.005$ in the form of Jupiters. An additional 0.007 could be in the form of hot gas and perhaps 0.005 could be contained in cold gas which has not yet gone through star formation. Thus we have $\Omega_B < 0.017$, or about 35% of the halo mass could be in baryons.

REFERENCES

[1]J. S. Gallagher, these proceedings.
[2]D. J. Hegyi and K. A. Olive, Phys. Lett. **126B,** 28 (1983).
[3]D. J. Hegyi, in the Proceedings of the IInd Moriond Astrophysics Meeting, ed. J. Audouze and J. Tran Thanh Van, 149 (D. Reidel, 1984).
[4]D. J. Hegyi and K. A. Olive, submitted to Astrophys. J. (1984).
[5]B. J. Carr, these proceedings.
[6]B. J. Carr, J. R. Bond, and W. D. Arnett, Astrophys. J. **277,** 445 (1984).
[7]S. M. Faber and J. S. Gallagher, Ann. Rev. Astron. Astroph. **17,** 135 (1979).
[8]G. B. Field, Ann. Rev. Astron. Astroph. **10,** 227 (1972).
[9]J. Silk, Ann. Rev. Astron. Astroph. **11,** 269 (1973).
[10]J. Tarter and J. Silk, Q. Jl. R. Ast. Soc. **15,** 122 (1974).
[11]F. Palla, E. E. Salpeter and S. W. Stahler, Astrophys. J. **271,** 632 (1983). See also: C. Low and D. Lynden-Bell, Mon. Not. R. Astron. Soc. **176,** 367 (1976); M. J. Rees, Mon. Not. R. Astron. Soc. **176,** 483 (1976); J. Silk, Astrophys. J. **256,** 514 (1982).
[12]B. M. Tinsley, Astrophys. J. **203,** 63 (1976).
[13]B. M. Tinsley and J. E. Gunn, Astrophys. J. **203,** 52 (1976).
[14]D. J. Hegyi in Cosmology and Particles, ed. J. Audouze, P. Crane, T. Gaisser, D. Hegyi, and J. Tran Thanh Van, 321 (Frontieres, Oreux, 1981).
[15]S. P. Boughn, P. R. Saulson and M. Seldnor, Astrophys. J. **250,** L15 (1981).
[16]B. M. Tinsley, Astrophys. J. **222,** 14 (1978).
[17]A. Dekel and J. Shaham, Astron. Astroph. **74,** 186 (1979).
[18]A. Kashlinsky and M. J. Rees, Mon. Not. R. Astron. Soc. **205,** 955 (1983).
[19]H. E. Bond, Astrophys. J. **248,** 606 (1981); J. G. Hills, Astrophys. J. **258,** L67 (1982).
[20]J. Bahcall, Astrophys. J. **276,** 169 (1984).
[21]D. Hegyi, E. W. Kolb and K. A. Olive, Astrophys. J. (submitted) (1984).

EXAM: An Experiment to Search for Antimatter in Distant Clusters of Galaxies

S. P. Ahlen

It is often claimed that the absence of antimatter in the universe is evidence in favor of Grand Unified Theories (GUT's) of particle physics. This is due to the three requirements initially enumerated by Sakharov, for the evolution of a matter-antimatter symmetric universe to an asymmetric universe: 1) Existence of baryon number non-conserving processes such as predicted by GUT's, 2) Existence of CP violating processes in the hot early universe, 3) Deviations from thermal equilibrium in the early universe. However, before this argument can be accepted, one must examine the evidence against antimatter in the universe. This has been done by Steigman[1] and it is concluded that direct experimental observations rule out mixing of matter and antimatter on scales smaller than and including clusters of galaxies. However, no direct evidence rules out the possibility that 50% of clusters of galaxies consist entirely of antimatter. It seems that the strongest "evidence" against antimatter comes from the difficulties of constructing consistent, symmetric cosmological models. Nevertheless, models of symmetric cosmologies do exist[2] and Stecker and collaborators have argued[3] that cosmic γ-rays and low energy cosmic ray antiprotons provide direct evidence in support of a symmetric universe.

Clearly resolution of the issue requires experimental input. Ahlen and co-workers have proposed a scheme[4,5] whereby cosmic rays can be probed for antimatter at a sensitivity (~ 1 part in 10^6 - 10^7) well below the expected fraction of extragalactic cosmic rays in the earth's neighborhood, even when allowance is made for modulation by a galactic wind (~ 10^{-5}). The proposed scheme relies on response characteristics of Cerenkov counters, plastic scintillators and CR-39 track etch detectors. The Cerenkov counters and track detectors have been shown to respond only to the magnitude of projectile charge while the scintillators have been shown to respond differently to negatively charged nuclei than to positively charged nuclei due to higher order terms in the Mott cross section. Thus, a detector can be made with very large area for the same weight as previous antimatter searches involving calorimeters or magnets which had small collecting power. The experiment is referred to as EXAM (for Extragalactic AntiMatter) and is currently being constructed at Indiana, Michigan and Berkeley. A negative result would provide the most convincing evidence to date for baryon-non-conserving processes. However, observation of just one unambiguous heavy antinucleus (EXAM is sensitive from silicon to nickel) will prove the existence of at least one antistar and probably an anti-cluster of galaxies.

REFERENCES

[1]G. Steigman, Ann. Rev. Astron. Astrophys. **14**, 339 (1976).
[2]R. W. Brown and F. W. Stecker, Phys. Rev. Lett. **43**, 315 (1979).
[3]F. W. Stecker and A. W. Wolfendale, Nature, in press (1984).
[4]S. P. Ahlen, P. B. Price, M. H. Salamon, G. Tarle, Ap. J. **260**, 20 (1982).
[5]S. P. Ahlen, P. B. Price, M. H. Salamon, G. Tarle, Nucl. Instr. Meth. **197**, 485 (1982).

PART II

MICROWAVE BACKGROUND RADIATION

The big bang theory emerged as the standard cosmological model in 1965, with the discovery by Penzias and Wilson of the 3K microwave background radiation. The microwave background gives us a clear picture of the Universe when the photons last scattered, at a red shift of about 1000, long before the formation of any known astrophysical object. The blackbody nature of the spectrum is strong evidence that the Universe was once hot and dense enough to ionize matter. Spatial fluctuations in the microwave background (or more precisely thus far, the lack of fluctuations) give us an indication of the perturbations in the mass density of the Universe at the time of recombination. In the gravitational instability theory of galaxy formation, these perturbations will grow to form galaxies and other large-scale structures in the Universe. In this chapter the observational status and theoretical interpretation of the microwave background is reviewed. Peterson, Richards, and Bonomo discuss the most recent data on the spectrum, and the evidence that it is indeed blackbody. Three papers by Wilkinson, Uson, and Partridge present recent limits on anisotropies in the background radiation. Silk presents a theoretical overview of the significance of the microwave background in the formation of galaxies and large-scale structure, and connects the subject of this chapter with that of the next. Next Bond and Efstathiou discuss the importance of the cosmic background anisotropies for particle physics and cosmology, by using them to constrain the properties of dark matter. Finally, Abbott and Traschen present theoretical papers on the origin and evolution of anisotropies.

Our knowledge of the microwave background bears on many of the new cosmological problems discussed in this volume. In the next chapter, the matter fluctuations responsible for galaxy formation, and the distribution of dark matter, are both shown to be constrained by the near isotropy of the microwave background. A variety of candidates for the dark matter is reviewed in Chapter VII. In Chapter IV, Abbott shows how the inflationary theories provide, by causal processes, a very uniform Universe, in accord with the observed isotropy of the microwave background; nonetheless, it also predicts small fluctuations which should be manifest in the background at some level.

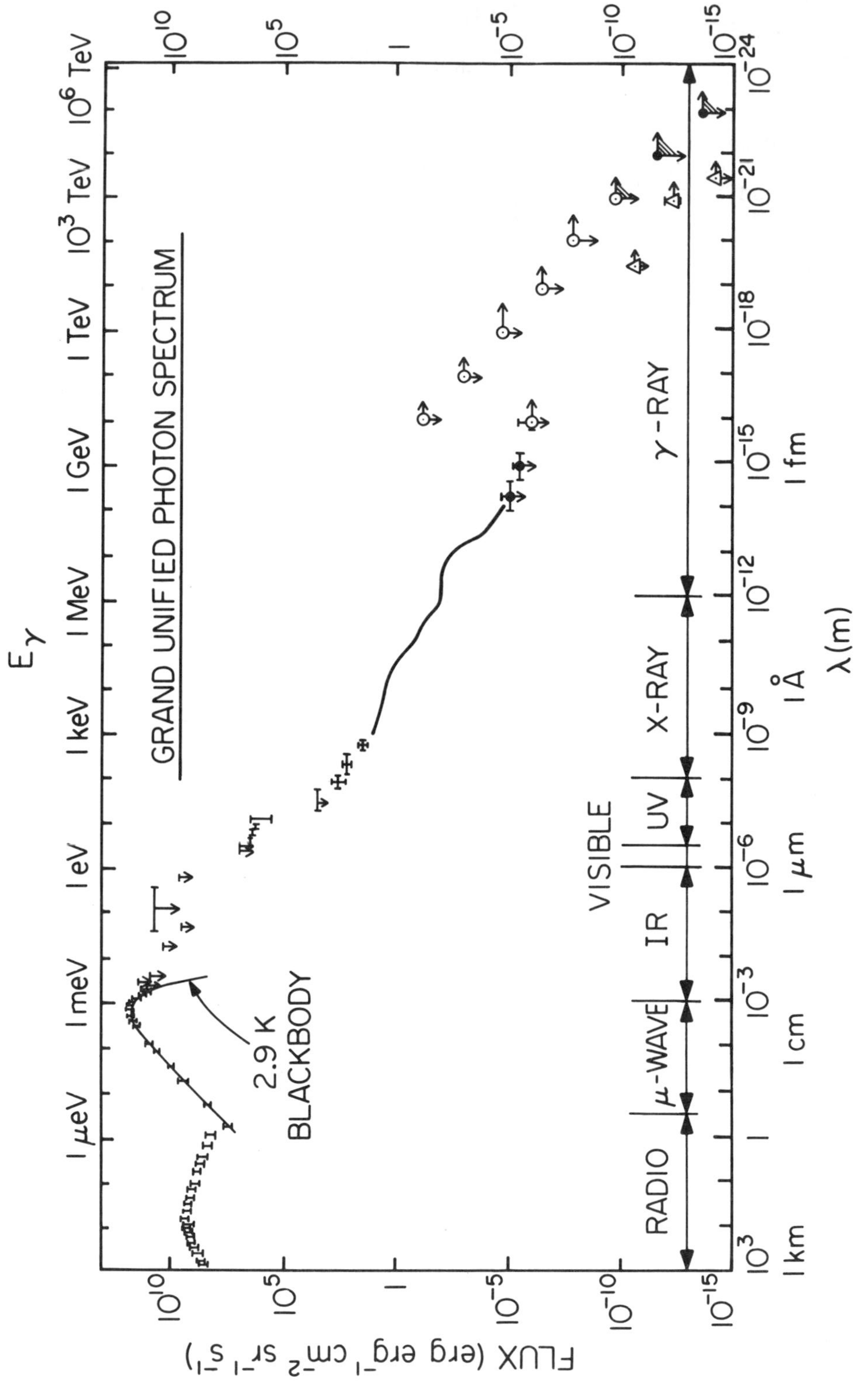
GRAND UNIFIED PHOTON SPECTRUM
E_γ
1 μeV
1 meV
1 eV
1 keV
1 MeV
1 GeV
1 TeV
10^3 TeV
10^6 TeV
2.9 K
BLACKBODY
RADIO
μ-WAVE
IR
VISIBLE
UV
X-RAY
γ-RAY
FLUX (erg erg^-1 cm^-2 sr^-1 s^-1)
λ(m)
1 km
1 cm
1 μm
1 Å
1 fm

New Measurements of the Spectrum of the Cosmic Microwave Background

J.B. Peterson, P.L. Richards, J.L. Bonomo, and T. Timusk

Accurate measurements of the spectrum of the cosmic microwave background (CMB) can provide useful tests of cosmological theories. The data set existing in 1982 has been summarized on a number of occasions[1-3] and is shown in Fig. 1. To first approximation the CMB is

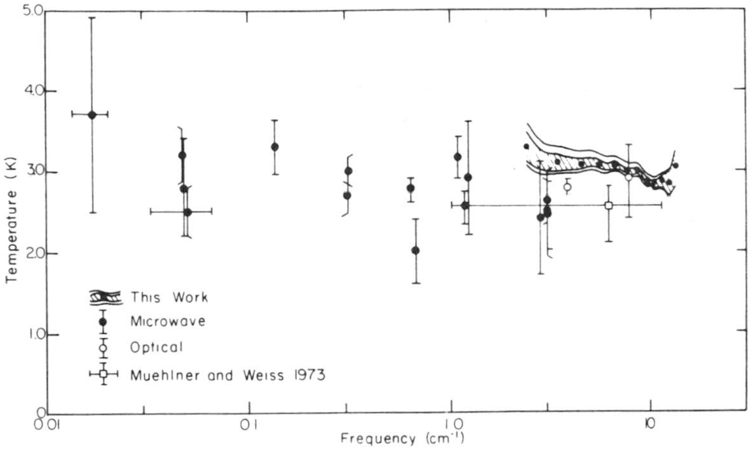

Figure 1: Measurements of the CMB temperature available before 1982.

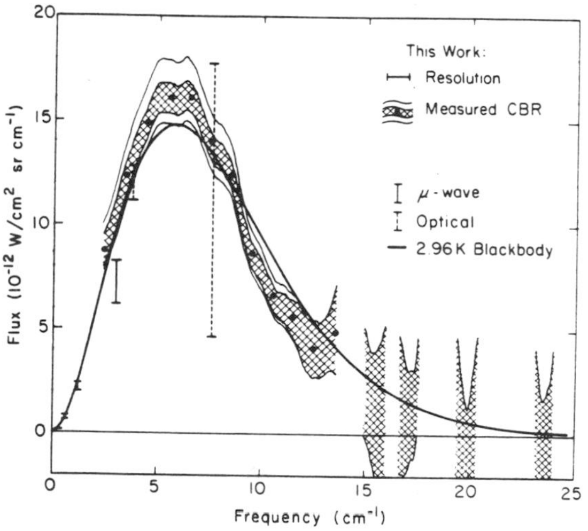

Figure 2: CMB spectral flux measured by Woody and Richards compared with microwave and optical results available at that time.

characterized by a single temperature and thus has a blackbody spectrum over the frequency range from 0.02 to 24 cm^{-1}. The error limits given for these experiments are dominated by systematic errors and are often very subjective. Consequently, it is not clear how to analyze the data set in a valid way. The general impression, however, is of a scatter in the high frequency data that is somewhat larger than would be expected from the given error limits.

The infrared experiment of Woody and Richards[4] produced data over a significant spectral range as is shown in Fig. 2. When these data are tested for agreement with a best-fit blackbody curve, the fit is rather poor. If we make the doubtful assumption that the errors are statistical, there is only a 0.3 percent chance that the data are consistent with a 2.90 K blackbody. There is an excess of flux near the peak.

Deviations from the Planck curve are expected at some level, and their observation is of the highest importance for the refinement of cosmological models. Compton scattering of the CMB by hot electrons, radiation damping of turbulence, and the annihilation of residual anti-matter are some of the mechanisms which could lead to deviations from a blackbody spectrum.[5] The net result of these mechanisms is to scatter low-energy photons to higher energy and hence to shift the peak in the spectrum to higher frequencies. These models do not fit the data as well as a simple Planck curve.[4]

Models of the CMB which do not involve establishing complete thermodynamic equilibrium,or which utilize frequency-dependent emission processes, are relatively unconstrained and can be made to fit the 1982 data set. Rowan-Robinson, Negroponte and Silk[6] have carried out calculations using the red-shifted dust features from a pre-galactic generation of stars to increase the CMB in the 3 to 8 cm^{-1} frequency range relative to that on either side of this range. They obtained a satisfactory fit to at least one plausible weighted average of the observations by this theory.

The fact that a fit can be obtained should not be interpreted as theoretical evidence for the suggested deviation. The question of the existence of a deviation must be answered experimentally.

NEW MICROWAVE AND OPTICAL EXPERIMENTS

An international collaboration involving groups from Berkeley and Haverford in the U.S.A. and Bolonga, Milano, and Padova in Italy began in 1982 to carry out new microwave measurements of the CMB from White Mountain, California. The results announced thus far are generally consistent with previous microwave measurements and the error estimates are relatively small.[7] These results are plotted in Fig. 3. A group at

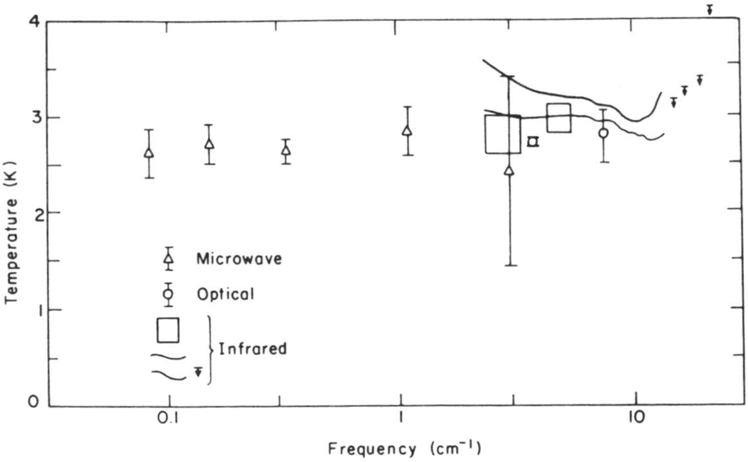

Figure 3: Recent microwave, optical and infrared measurements of the CMB temperature. The boxes, whose width represents the spectral band, are preliminary results of the new Berkeley experiment described here.

U.C.L.A. has used measurements of the relative strengths of optical transitions[8] to deduce new values for the excitation temperature of cyanogen in the molecular could ζ Oph. Their results, shown in Fig. 3, give upper limits for the temperature of the CMB at two frequencies.

NEW BERKELEY INFRARED EXPERIMENT

We have designed a new apparatus to measure the spectrum of the CMB in the frequency range from 3 to 10 cm^{-1}. This range includes the peak of the measured spectrum and more than 80 percent of flux in the CMB. The new experiment is based on the balloon gondola of Woody and Richards. To avoid the possibility of repeating undetected systematic errors in that experiment, however, the new experiment has been changed in most important respects.

Although there is no specific reason to expect that the spectroscopy of the Woody-Richards experiment introduced significant errors, the Fourier transform spectrometer used by them has been replaced by a filter wheel with five band-pass filters. The fabrication of these filters and the measured filter bands have been described elsewhere.[9] Two of the filter bands lie below the peak of the CMB spectrum, one is at the peak and two lie above it. Although the change to filter spectroscopy permits us to do a relatively independent experiment, it introduces several potential problems. Resonant filters of the type used are subject to leakage at higher order resonant frequencies. Stray radiation from hot sources within the apparatus is somewhat more likely to be chopped in a filter spectrometer than to be interference modulated in a Fourier spectrometer. Great care has been exercised in the design of the new experiment to avoid these difficulties.

Emission from the atmosphere is greatly reduced but not eliminated by measuring from balloon altitudes. In the Woody-Richards experiment, a 512 point spectrum of the sky emission was measured and compared with a theoretical calculation based on a simple atmospheric model and experimental parameters. The fit of the model spectrum to the measured sky spectrum was excellent in the frequency range beyond 12 cm^{-1} where they could be accurately compared. Below 12 cm^{-1} the atmospheric emission predicted from the model was a small fraction of the measured flux. Therefore, the model spectrum almost certainly provided an adequate correction for atmospheric emission in the frequency range where deviations from the blackbody curve were seen.

In the new experiment a detailed sky spectrum will not be available. The filter bands have been selected so as to minimize atmospheric emission. The amount of this emission has been estimated from the model to be negligible in the two lowest frequency bands and to be comparable to the CMB at the highest frequency. The contribution of the atmospheric emission to the measurement is evaluated by measuring the flux in each filter band as a function of the angle of the apparatus from the zenith.

One interesting feature of the Woody-Richards experiment is the fact that arbitrarily reducing the overall calibration factor by 30 percent brings the data into excellent agreement with a blackbody curve at the temperature of 2.77 K given by plausible averages of the microwave data.[4] If this change is made, then all of the data in Fig. 1 are consistent with a single blackbody temperature. Although no reason for such a large error in calibration factor has been found, the suspicion has arisen that the calibration factor might have been different in flight from the value measured in the laboratory.[10]

Great care must be taken in the calibration of any new infrared measurement of the spectrum of the CMB. The ideal location for the calibrator is outside of the apparatus and filling the entire beam. Although a low temperature calibrator in this location is possible on satellite experiments, it is very difficult to provide in the balloon environment. Radiation from warm objects must be prevented from reflecting or scattering from the calibrator surface and entering the throughput of the photometer. The ambient pressure air must be prevented from freezing on the cold calibrator surface. If a window is used for this purpose, then reflections from it cause additional problems.

These considerations have led us to employ a variable temperature blackbody inside of our cryostat which can be used to calibrate the instrument during the flight. In principle this calibrator removes the effects of unwanted signals which arise from locations between the calibrator and the detector. Care must be used to avoid signals which could arise between the calibrator and the sky.

EXPERIMENTAL APPARATUS

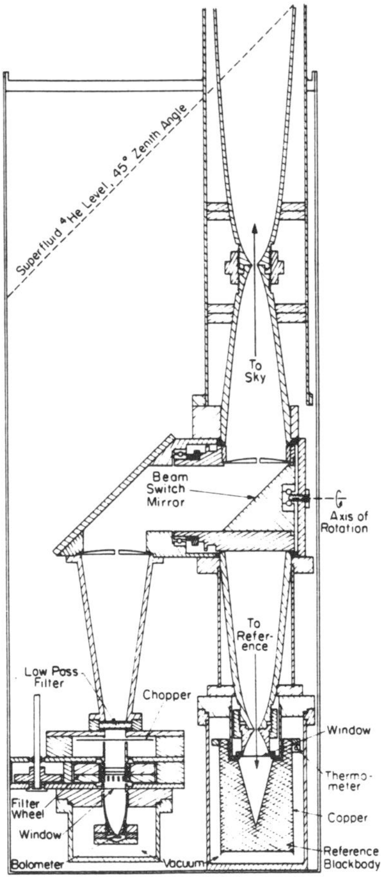

Figure 4: Apparatus used for the new Berkeley balloon experiment.

The liquid helium cooled spectrophotometer used by Woody and Richards has been redesigned to include the features discussed in the preceding section. The entrance to the apparatus is surrounded by a large earthshine shield. The beam enters the top-looking cryostat through a removable window and an apodizing horn. The beam on the sky is defined by a long Winston[11] light concentrator. These portions of the apparatus were used previously and are described fully elsewhere.[4] Fig. 4 shows the lower portion of the apparatus which is enclosed in a copper box that is kept filled with superfluid liquid helium throughout the flight. A combination of a shortened Winston concentrator and a lens is used to collimate the beam arriving from the sky. This collimated beam is then shifted to the left by reflection from two diagonal mirrors. The beam is then refocused by a lens-and-cone combination onto a glass low-pass filter, an oscillating chopper, a filter wheel with five band-pass filters, and a ^{3}He-temperature composite bolometric detector. The filters[9] and detector[12] have been described fully elsewhere. The chopper and the filters, along with their drive mechanisms, are immersed in superfluid helium and are carefully baffled with absorber to minimize the possibility that radiation from the warm upper parts of the cryostat can be chopped and reach the detector.

A portion of the apparatus including the lens and the first diagonal mirror rotates by 180° around a horizontal axis so that the detector can look either up to the sky, as is shown in Fig. 4, or down into a reference blackbody. The throughput to the calibrator is determined by the lens and a back-to-back pair of Winston concentrators. The dimensions of these concentrators are matched as closely as possible to the pair of back-to-back concentrators leading to the sky. The lowest concentrator has been truncated so as to permit the installation of a conical black absorbing surface.

The calibration system also includes two designs of variable temperature laboratory calibrator which can be inserted into the upper Winston concentrator. The purpose of these laboratory calibrators is to measure any asymmetry in the throughput or transmission between the sky and the internal cold calibrator.

CALIBRATION

The first step in the calibration of our apparatus is carried out in the laboratory by measuring the detector signal as a function of the temperature of the internal flight calibrator and of the two laboratory calibrators which are placed in the throat of the Winston cone antenna. These two laboratory calibrators provide the primary calibration standard. Comparisons between the signals from them confirm that their design is adequate. A full description of the design and testing of these calibrators will be published elsewhere. The flight calibrator is used as a transfer standard. Signals from it differ by 2 percent from those coming from the laboratory calibrators. This up-down asymmetry is ascribed to the machining tolerances of the apertures which define the throughput up and down, and is assumed to be the same during flight as in the laboratory.

As a test of our understanding of the performance of the apparatus, we have plotted in Fig. 5 the measured signal as a function of the signal calculated from the blackbody temperature, the filter function, and the Planck spectrum. The data fit a straight line to ~1%. This experiment can be interpreted as the most accurate test ever made of the Planck function.

The intercept in Fig. 5 is very close to zero, indicating very little leakage of warm radiation. Some previous measurements of the CMB have had large values of this intercept.[13] The danger then exists that the leakage radiation causing the intercept will be different in flight from the value determined in the laboratory.

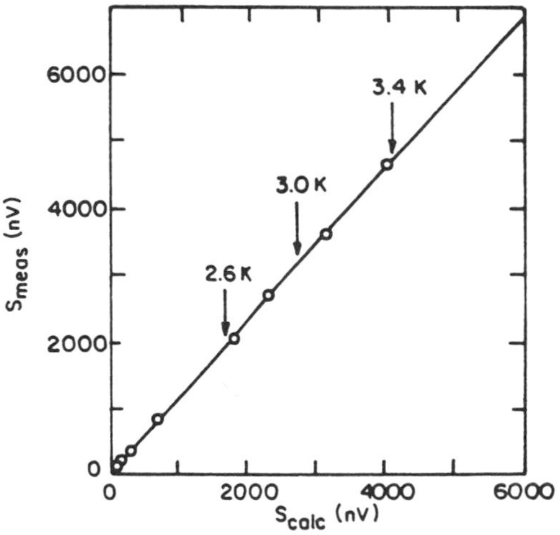

Figure 5: Comparison of the measured signal from the blackbody calibrator operated at various temperatures with the signal computed from the temperature, the filter transmittance and the Planck law.

Except for the measurement of the small up-down asymmetry mentioned above, the calibration of our instrument is carried out in flight. Signals from the sky are compared to signals from the flight calibrator which is maintained at 3.1 K. Tests are made for spurious radiation from the upper part of the antenna by heating it in flight. Tests for earthshine are made by measuring the sky signal as a function of zenith angle close to the horizon and comparing it with the sec θ dependence expected from a horizontally stratified atmosphere.

NEW INFRARED MEASUREMENT

Our apparatus was successfully flown from the National Scientific Balloon Launch Facility in Palestine, Texas on November 16, 1983. The data were telemetered to the ground and recorded on magnetic tape. The gondola was recovered undamaged.

The first step in data analysis was the removal of glitches caused by cosmic rays hitting the bolometer. The noise in the deglitched data was comparable to that seen in the laboratory. The calibration was very stable in time.

Data for each frequency channel were plotted as a function of air mass. Such data were measured both at fixed zenith angle as the balloon approached float altitude, and also by varying the zenith angle while the balloon was at (approximately) constant altitude. Because of extreme uncertainty about the anticipated duration of the flight, the scenario used was not optimum. The dominant error in these data occurred as a result of changes in the atmospheric ozone during the flight. These changes can be modeled and corrected to some extent. A better procedure, which is permitted by the high sensitivity of the apparatus, is to make all measurements for one channel in a short time.

At present, the focus of the data analysis is on estimating the importance of systematic errors. These estimates are nearly complete and will produce a set of new measurements of the spectrum of the CMB which will be more accurate than the Woody-Richards result.[4] Preliminary results of this flight are available for the two lowest frequency bands centered at 3 and 5 cm^{-1}. The measured temperatures of 2.80 ± 0.2 and 2.95 ± 0.15 K are shown as boxes in Fig. 3. It is anticipated that results for all five bands will be available during 1984. A reflight of the apparatus in late 1984 should lead to further improvements in the data.

ACKNOWLEDGMENTS

This work was begun with support from the Director, Office of Energy Research, Office of Basic Energy Sciences, Materials Sciences Division of the U.S. Department of Energy under Contract No. DE-AC03-76SF00098, and has been completed with support from the Office of Space Sciences of the National Aeronautics and Space Administration, and by the Natural Sciences and Engineering Research Council of Canada.

REFERENCES

[1]P.L. Richards, Phys. Scripta **21**, 610 (1980).
[2]P.L. Richards and D.P. Woody, IAU Symposium 92, Objects at High Redshifts, G. Abell and J. Peebles, eds. (Dordrecht, 1980) p. 283.
[3]P.L. Richards, Phil. Trans. R. Soc. London **A307**, 77 (1982).
[4]D.P. Woody and P.L. Richards, Astrophys. J. **248**, 18 (1981).
[5]Ya.B. Zel'dovich, A.F. Illarionov, and R.A. Sunyaev, Zh. Eksp. Teor. Fiz. **62**, 1217 (1972) [English translation in Sov. Phys.-JETP **35**, 643 (1972)].
[6]M. Rowan-Robinson, J. Negroponte, and J. Silk, Nature **281**, 635 (1979).
[7]G. Smoot et al., Phys. Rev. Lett. **51**, 1099 (1983), and private communication.
[8]D.M. Meyer and M. Jura, Astrophys. J. Lett. **276**, L1-L3 (1984).
[9]T. Timusk and P.L. Richards, Appl. Opt. **20**, 1355 (1980).
[10]R. Weiss, A. Rev. Astr. Astrophys. **18**, 489 (1980).
[11]R. Winston, J. Opt. Soc. Am. **60**, 245 (1970).
[12]N.S. Nishioka, P.L. Richards, and D.P. Woody, Appl. Opt. **17**, 1562 (1978).
[13]H.P. Gush, preprint.

Anisotropies in the 2.7 K Cosmic Radiation

David T. Wilkinson

INTRODUCTION

The first precision searches for large-scale[1] and small-scale[2] anisotropies in the 2.7 K background radiation are now 15 years old. Both were motivated by theoretical estimates of interesting effects (motional dipole and primordial perturbations) that should be present at the level of $\Delta T/T \sim 10^{-3}$. It was another 10 years before the large-scale (dipole) effect was clearly seen,[3] and the tantalizing small-scale anisotropy still eludes the most careful observations with our best radio telescopes. The early observations gained orders of magnitude in our knowledge of anisotropies; current programs, involving a few person-years of effort, are content with a factor of two improvement in the precision of upper limits. Fortunately, the fundamental importance of the phenomena, and incessant badgering by theorists, keep the observers interested in this difficult problem.

In the past, major experimental advances have been made by introducing new instrumentation, such as maser amplifiers and cooled mixers. But the point has now been reached where experimental precision is being limited by effects other than detector noise and integration time; ground and atmospheric radiation, Galactic radiation (radio and sub-millimeter) and point source confusion are causing serious problems in several of the current best measurements.

Other speakers[4,5] at this conference will review the status of theoretical work on sources and consequences of anisotropy in the 2.7 K radiation. Here, let me just note a few general features of the Standard cosmological model which are being tested by anisotropy measurements. Figure 1 shows the recent (for this audience) cosmological epochs probed by background radiation anisotropies. The key event is decoupling at $z \sim 10^3$ (T~4000 K) after which the thermal radiation streams to us, carrying the imprints of that surface. We know remarkably little about the important events following decoupling. In the Standard model, small- and large-scale mass structure forms at this time, but the dominant processes and

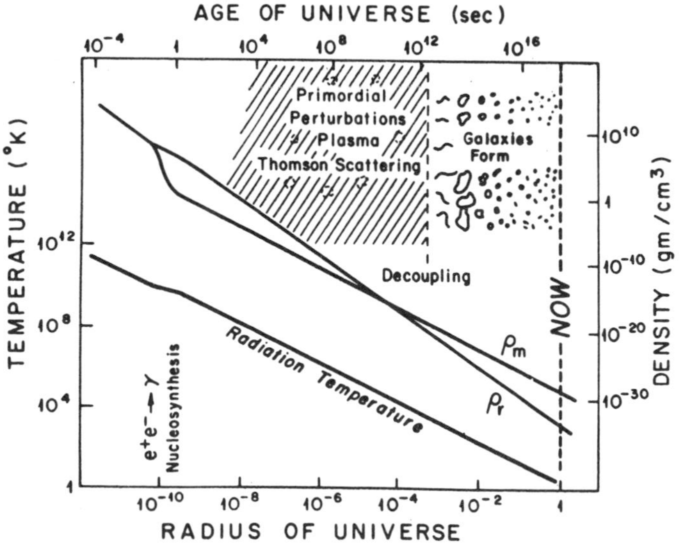

Figure 1: The end game for a Standard cosmological model. Most effects of interest to anisotropy measurements occur very early (Planck, GUTS or inflation epochs) or at decoupling. Agreement between calculations and observations of cosmic abundances of light elements implies tight specification of conditions at the nucleosynthesis epoch. So far, anisotropy measurements only impose constraints on the cosmological model.

the sequence of events are subjects of active study and lively debate. A question of major importance for anisotropy studies is: Was the matter reionized (by stars) and thus able to again scatter the thermal photons? This process could alter the original anisotropy, but in predictable ways. The detection and measurement of anisotropy will tell us much about the period from $z = 10^3$ to now and, hopefully, they will also reveal the all-important spectrum of primordial perturbations which are probably the seeds of large-scale structure, and may originate in very early (particle physics) epochs.

The absence of intrisic anisotropy in the 2.7 K radiation on large angular scales (> 10°) argues that the early universe was remarkably homogeneous and isotropic, even in regions not causally connected in the Standard model (without inflation). Apparently, Einstein's Cosmological Principle is well satisfied by our universe. Finally, the accurately measured "dipole" effect measures the Galaxy's peculiar velocity (with respect to the comoving frame) which has important consequences for studies of cluster dynamics and the local Hubble flow.

To the great relief of experts in the field (and the conference organizers) I will now discuss the current observational situation. Anisotropy searches fall naturally into three classes based on angular scale. The phenomena expected and the experimental techniques and problems are quite different, depending on the size and separation of the antenna beams.

SMALL-SCALE ANISOTROPY

The smallest angular scale expected for anisotropies emerging from the decoupling process is about 3 arcminutes. Estimates of the fluctuations on this scale predict anisotropy at $\Delta T \sim 3 \times 10^{-5}$ K for currently interesting models.[4,5] These numbers define the observational problems 1) Relatively large antenna are needed, making balloon and satellite observations more difficult. So far, only ground-based measurements have been made at small angular scales. With the best receivers (T_{sys}~30 K), the minimum integration time to average down detector noise is a few hours per spot. Even if the receiving system integrates down to $\Delta T \times 10$ μK (few do), one expects real observations to be limited by atmospheric inhomogeneity and variable ground pickup at higher levels. These sources typically contribute a few degrees of radiation temperature to the received antenna power. So, observing strategies must be found that cancel atmospheric and ground effects to an accuracy of better than one part in 10^5, over several hours of tracking a spot across the sky. Uniform atmospheric conditions and a stable, reproducible telescope are essential.

The observing wavelength is constrained on the long wavelength side by point source confusion; currently, large corrections are needed for observations with $\lambda > 3$ cm.[6] At wavelengths shorter than 3 mm, atmospheric fluctuations and receiver noise are the limiting problems. Broadband bolometric systems have the sensitivity required for these measurements,[7] but even from mountaintops the atmosphere is a problem.[8]

Observations of small-scale anisotropy in the 2.7 K radiation are summarized in Figure 2. At the smallest angular scales the Very Large Array is being used by two groups.[9,10] The limits shown (solid triangles) are the first results and can probably be improved, now that the first order technical problems have been isolated. The lowest limit at any angular scale is currently $\Delta T/T \leq 2.1 \times 10^{-5}$ at 95% confidence.[11]

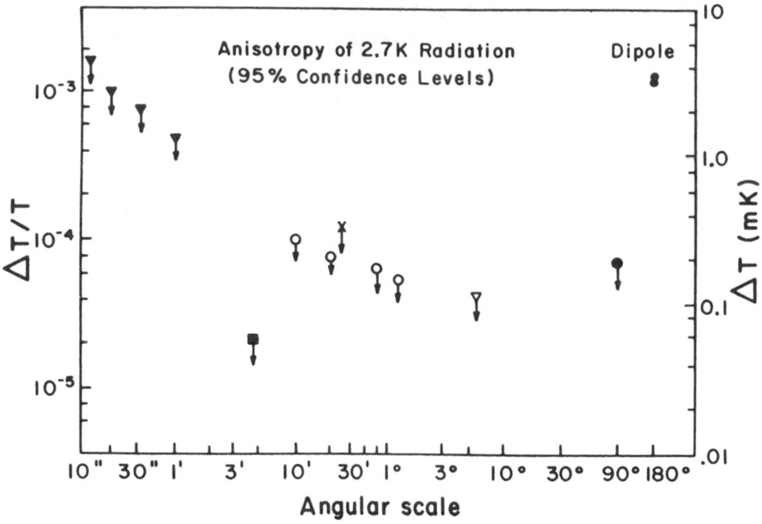

Figure 2: Summary of results giving the tightest constraints on anisotropy in the 2.7 K radiation, as of December, 1984. Discussion and references for each point are given in the text.

The angular scale is 4.5 arcminutes, obtained by beam chopping with the 140' NRAO telescope into a 50 K maser receiving system. At this meeting Uson[12] discusses the measurement in more detail and Silk[4] and Bond[5] report interesting consequences for Standard models with cold dark matter.

At angular scales from 10' to 1° the observations are more hampered by gradients in atmospheric and ground radiation. The best limits have been set by Parijskij, et al.[13] with RATAN-600 telescope in the Caucasus. The values in Figure 2 are those given by Lasenby and Davies[14] based on an improved statistical analysis of the data. (The original values have been increased by about a factor of 2.5.) The RATAN-600 group have reported[15] new results: $\Delta T/T < 10^{-5}$ (1 σ level) on angular scales from 4.5' to 9', and $\Delta T/T < 3 \times 10^{-5}$ (1 σ level) at a scale of 1°. These results are not shown in Fig. 2 because no details of the observations or analysis are yet available. The observing wavelength was 7.6 cm, so substantial corrections had to be made for source confusion. In this painstaking business, observational results should be subjected to detailed scrutiny before being accepted and used. (And we all know that theorists only look at graphs and tables.)

The point at 25' is of particular interest because it was obtained with a broadband bolometric detector, from a high altitude site.[7] Since current limits are straining the noise performance of the best hetrodyne systems, broadband detectors may be required for major advances in the search for small-scale anisotropy.

INTERMEDIATE-SCALE ANISOTROPY

Causally connected regions of the decoupling surface ($z \sim 10^3$) are a few degrees in size. Therefore, this is an interesting angular scale for anisotropy measurements. Also, the Sachs-Wolfe process operates on this scale. Here the 2.7 K photons suffer different gravitational redshift from statistical variations in the mass clumping.[16,17]

Atmospheric inhomogeneities are a serious problem for ground-based measurements, few of which have been attempted. The best limit is at 6°, as shown in Figure 2. This result was obtained by bolometric detectors at balloon altitudes.[18] Thermal emission from patchy Galactic dust is the main problem for these observations. Radio and infrared sky maps can be used to estimate signals due to dust,[19] but this procedure has limited accuracy. Observations with filters chosen to discriminate against dust emission, and others chosen to measure dust signals, are now being used to address this problem.

LARGE-SCALE ANISOTROPY

The prediction by Peebles of a dipole signal due to the sun's peculiar velocity motivated the early searches for large-scale anisotropy.[1,20] The effect is now clearly seen in long[3,21,22] and short[23,24] wavelength regions and the spectrum of the dipole is that expected for Doppler shifted thermal radiation.[25]

To date, balloon borne radiometers at 3 mm and 1.2 cm wavelength[21,22] have given the most accurate measurements of the dipole effect. Statistical errors are smaller for the 1.2 cm observation (30 K system noise), but systematic errors due to Galactic radiation are smaller at 3 mm. Each radiometer was successfully flown three times, once in the southern hemisphere, so sky coverage exceeded 80%. Instrument noise was low enough that the dipole effect could easily be seen on real-time chart records. After removing Galactic contamination (small at λ=3 mm) the following dipole distributions are found:

T(mK)	Declination(°)	Right Ascension (hr)	Group
3.46 ± 0.17	-6.0 ± 1.4	11.3 ± 0.1	Berkeley[26]
3.18 ± 0.17	-8.0 ± 0.7	11.18 ± 00.5	Princeton[22]

Nearly all of the uncertainty in T is due to inadequate calibration accuracy (5%). Uncertainty in the direction depends mainly on statistical errors which are 1 to 2%. The agreement between these two measurements of the dipole effect is surprisingly good.

An important result of this work is that the Galaxy's (or Local Group's) velocity is measured with respect to the 2.7 K radiation -- presumably the comoving frame. The magnitude of this velocity is 600 km/sec, which is large compared to typical random velocities of galaxies. The explanation is that we are probably bound to the Virgo cluster -- a conclusion supported by the dynamics deduced from surveys of galaxy redshifts. In coordinates centered on Virgo our velocity is

$$(v_r, v_\theta, v_z) = (410 \pm 25,\ 297 \pm 25,\ -308 \pm 25) \text{ km/sec.}$$

The large non-radial components suggest that the Virgo cluster may not be the only mass concentration affecting the peculiar velocity of the

Galaxy. Comparison of our velocity with respect to the 2.7 K radiation to the velocity measured against galaxies at various distances, can be used to study dynamics on galaxy clustering scales.[27]

Two of the Princeton balloon flights were made at times of the year when the earth's orbital motion added to, and opposed, the dipole effect. This alignment maximized sensitivity to the earth's motion. The data from these two flights detected the earth's velocity through the 2.7 K radiation at the 7σ level.[22]

As for the smaller angular scales, no intrinsic anisotropy has been found in the 2.7 K radiation on scales between 10° and 180°. Hopes of a quadrupole detection[28,25] faded when not confirmed by more sensitive measurements.[21,22] Upper limits of $\Delta T/T < 10^{-4}$ argue for a remarkably homogeneous isotropic universe on large scales. The long-standing causality problem raised by large-scale isotropy is a key feature of our universe. The Standard Big Bang model might be rescued by inflation or possibly by yet unknown quantum gravity effects in the Planck era. Unlike the situation at small angular scales, the cosmological models have not yet yielded specific large-scale predictions to be tested.

FUTURE PROSPECTS

The lack of detections of anisotropy and the increasing difficulty of the observations is not encouraging to experimenters. Large improvements in sensitivity due to new instrumentation or techniques are unlikely, because systematic and astrophysical limitations are being encountered on all angular scales and at all wavelengths.

At small angular scales where the standard model is being closely tested, factors of two are important. Perhaps a few such advances are possible. For ground-based observations lower system noise and weather-dependent telescope scheduling are the main hope. (Green Bank engineers have recently achieved a 60% reduction in minimum noise for the system used to get the lowest point in Fig. 2.) Owens Valley Observatory is planning a major commitment of their maser-based system to the small-scale anisotropy problem.

Another approach[29] is being considered by several groups. One could capitalize on the much greater sensitivity of bolometers by building a narrow-beam radiometer and flying it on a pointed balloon platform. Bandpass filters must be carefully designed to avoid (and measure) Galactic dust emission. Eventually the Large Deployable Reflector may provide sufficient aperture and shielding to do the experiment in orbit, away from atmospheric problems.

The first step toward doing large-scale observations in orbit have already been taken. (Small antennas make large-scale experiments the natural first choice.) NASA has made a major commitment to the Cosmic Background Explorer[30] satellite. It will carry radiometers at three wavelengths (9 mm, 6 mm and 3 mm) to measure large-scale anisotropy. (COBE will also make high precision measurements of the 2.7 K spectrum, and measure the infrared background flux from 1 to 300 μm.) A 1987 launch is scheduled.

The Russian Prognos-9 satellite in a highly eccentric orbit is carrying an 8 mm radiometer designed to measure large-scale anisotropy. Reports[31] of early results are very encouraging. The dipole has been seen with good statistical accuracy and the upper limit on a quadrupole effect is already at the level shown in Fig. 2. The systematic limitations due to Galactic radiation and the earth and moon in the (rather large) antenna sidelobes are not yet known. The group plans to follow this success with flights using shorter wavelength receivers of higher sensitivity. The satellite's orbit and orientation is nearly ideal for measuring large-scale anisotropy.

REFERENCES

[1]R.B. Partridge and D.T. Wilkinson, Phys. Rev. Lett. **18**, 557 (1967).
[2]E.K. Conklin and R.N. Bracewell, Nature **216** 777 (1967).
[3]G. Smoot, M. Gorenstein, and R. Muller, Phys. Rev. Lett. **39**, 14, 898 (1977).
[4]N. Vittorio and J. Silk, these proceedings, 1984.
[5]J.R. Bond and G. Efstathiou, these proceedings, 1984.
[6] L. Danese, G. DeZotti, and N. Mandolesi, Early Evolution of the Universe and Its Present Structure, (Eds. Abell and Chincarini) I.A.U. Symposium No. 104, pp. 131-133, 1983.
[7]N. Caderni, V. DeCosmo, R. Fabbri, B. Melchiorri, F. Melchiorri, and V. Natale, Phys. Rev. **D16**, 2424 (1977).
[8]S.S. Meyer, A.D. Jeffries, and R. Weiss, Ap. J. Lett. **271** L1, (1983).
[9]E.B. Fomalont, K.I. Kellerman, and J.V. Wall, Ap. J. Lett. **277**, L23 (1984).
[10]J.E. Knoke, R.B. Partridge, M.I. Ratner, and I.I. Shapiro, Ap. J. **284**, 479 (1984).
[11]J.M. Uson and D.T. Wilkinson, Nature (in press 1984).
[12]J.M. Uson, these proceedings 1984. J.M. Uson and D.T. Wilkinson, Ap. J. Lett. **277**, L1 (1984), Ap. J. **283**, 471 (1984); Phys. Rev. Lett. **49**, 1463 (1982).
[13]Yu. N. Parijskij, Z.E. Petrov, and L.N. Cherkov, Sov. Astron. Lett. **3**, 263 (1977).
[14]A.N. Lasenby and R.D. Davies, Mon. Not. R. Astron. Soc. **203**, 1137 (1983). Also see this paper for discussion of early results of a new program at λ = 6 cm on angular scales 10 to 60 arcminutes.
[15]A.B. Berlin, E.V. Bulaenko, V.V. Vitkovsky, V.K. Kononov, Yu. N. Parijskij, and Z.E. Petrov, Early Evolution of the Universe and Its Present Structure (Eds. Abell and Chincarini) I.A.U. Symposium No. 104, pp. 121-124, (1983).
[16]M.L. Wilson and J. Silk, Ap. J. **243**, 14 (1981).
[17]P.J.E. Peebles, Ap. J. Lett. **243**, L119 (1981).
[18]F. Melchiorri, B.O. Melchiorri, C. Ceccarelli, and L. Pietraner, Ap. J. Lett. **250**, L1 (1981).
[19]C. Ceccarelli, G. Dall'Oglio, B. Melchiorri, F. Melchiorri, and L. Pietranera, Ap. J. **260**, 484 (1982).
[20]E.K. Conklin, Nature **222**, 971 (1969). P. Henry, Nature **231**, 516 (1971). Both detected the dipole effect with large uncertainties.
[21]P.M. Lubin, G.L. Epstein, and G.F. Smoot, Phys. Rev. Lett. **50**, 616 (1983).
[22]D.J. Fixsen, E.S. Cheng, and D.T. Wilkinson, Phys. Rev. Lett. **50**, 620 (1983).
[23]R. Weiss, Ann. Rev. Astron. Astrophys. **18**, 489 (1980).
[24]C. Ceccarelli, G. Dall'Oglio, P. de Bernardis, S. Masi, B. Melchiorri, F. Melchiorri, G. Moreno, and L. Pietranera, The Birth of the Universe, 7th Rencontre de Moriond (Eds. Audouze and Tran Thanh Van) p. 175, (1982).
[25]S.P. Boughn, E.S. Cheng, and D.T. Wilkinson, Ap. J. Lett. **243**, L113 (1981).

[26]P.M. Lubin and T.V. Neto, 1984 (preprint). Data from a southern hemisphere balloon flight are combined with the data used in Ref. 21 where the result was 3.63 ± 0.15 mK, -4 ± 2°, and 11.3 ± 0.1^h.
[27]For a review see: M. Davis and P.J.E. Peebles, Ann. Rev. Astron. Astrophys. **21**, 109 (1983).
[28]R. Fabbri, I. Guidi, F. Melchiorri, and V. Natale, Phys. Rev. Lett. **44**, 1563; **45**, 401(E) (1980).
[29]M. Dragovan, these proceedings (1984).
[30]J. Mather and T. Kelsall, Physica Scripta **21**, 671 (1980).
[31]I.A. Strukov and D.P. Skulacher, Sov. Astron. Lett. **39**, 898 (1984).

Small-Scale Structure in the Microwave Background

Juan M. Uson

The cosmic microwave background is a remnant of the early Universe.[1] As such, it may carry imprints of initial conditions or important cosmological processes. Observations of the small scale structure of this radiation provide a direct observational probe of the power spectrum and characteristic lengths and masses of the initial density fluctuations[2], although this picture of the early Universe could be blurred by subsequent scattering of the microwave photons by ionized matter at a much later epoch.[3] The expected anisotropy in the microwave background can be derived from the observed structure of the distribution of matter[4]; although, so far, ways have been found to lower theoretical estimates of the "necessary" anisotropy of the microwave background whenever they have conflicted with the observations.

A number of observers have published upper limits on small scale variations in the temperature of the microwave background at various angular scales and frequencies.[5] No small scale anisotropy has been detected. The data presented here imply an upper limit of $\Delta T_{rms}/T < 2.4 \times 10^{-5}$ at the 95% confidence level on fluctuations at an angular scale of 4.5 arc-minutes.

We used the N.R.A.O. 140-foot telescope at Green Bank, West Virginia, operating at a frequency of 19.5 GHz.[6] The maser receiver was mounted at the Cassegrain focus and nutation of the secondary subreflector was used to switch the telescope beam between two nearby positions on the sky. A data point, recorded every 0.6 sec, is a measure of the difference (δT_A) between the antenna temperature of the radiation collected from the two beam positions on the sky, plus the difference of ground and atmospheric radiation into the two beams. Some recent improvements have lowered the theoretical rms fluctuations in δT_A to about 8 mK for each data point. One then expects that integrating over a time t_{int} will lower this rms fluctuations by $t_{int}^{-1/2}$; but this theoretical limit is in principle only a worthwhile goal, which is normally very hard to achieve.

In high sensitivity radiometry, an important factor is the receiver power spectrum near the switch frequency. As we started working on this experiment (1980), there were about a dozen lines in the power spectrum below 5 Hz, most of them variable in frequency and amplitude, on top of a strong "1/f" component; so that for integrations of several minutes the actual rms fluctuations were as large as twenty times the theoretical value. Radiometry at the millikelvin level was impossible. The current power spectrum is clean and stable, thanks to the efforts of the Green Bank staff (see Ref. 6, figs. 1a and 1b). During the last observing run (February 25 to March 5, 1984), the measured noise was as low as only a factor of 1.2 above the theoretical level for two hours of integration. I would like to stress that most of the reasons that make the actual performance of such a system worse than the ideal limit, cannot be predicted in advance. The observing technique is very important, and the best choice is most likely different for each system as it depends on where the small systematic effects originate. Estimates of achievable sensitivity based simply on system noise temperature, receiver bandwidth and integration time are misleading, and results reported without a careful discussion of statistical and systematic effects should be regarded with skepticism.

When we started this experiment we did not know how we would do it. The first few observing schemes that we tried were either not possible due to physical limitations imposed by the telescope or simply did not work because of systematic effects. After trying several procedures we found that ground pickup was the dominant systematic effect, being strongly dependent on telescope position and only slowly varying with time. The observing procedure was as follows. (See Ref. 6 for a more detailed description.) Each of twelve Fields was tracked for two hours each day. To minimize telescope motion with respect to the ground and atmosphere, the twelve Fields were chosen as close to the north pole as the safe operation of the 140-foot telescope allows, Dec = 86°51' with RA's = 1^h30^m, 3^h30^m,....23^h30^m (epoch 1950). The motion of the telescope as it follows a Field is then slow and the systematic variation in δT with hour angle is only about 55 mK/hr.

To further reduce the effects of variable ground pickup we used a double subtraction technique. To each of the twelve Fields we associate two reference fields on opposite sides and separated 4.5 arcminutes from the corresponding field. The observations then consist of comparing the Field with one of the references by beamswitching at a rate of 3.33 Hz. The Field is alternatively placed in the "+" beam (ON scans) and the "-" beam (OFF scans) which provides for each Field two sets of data:

$$\delta T_{ON} = T_{FIELD} - T_{REF2} \text{ and } \delta T_{OFF} = T_{REF1} - T_{FIELD} \quad , \qquad (1)$$

where REF1 and REF2 denote respectively the westerly and easterly reference positions.

The quantity measured for each Field,

$$\Delta T_{FIELD} = (\delta T_{ON} - \delta T_{OFF})/2 = T_{FIELD} - (T_{REF1} + T_{REF2})/2 \quad , \qquad (2)$$

compares each Field to the average of the two associated reference patches. Careful timing of the OFF-ON sequence allows both scans to be taken with the telescope going through the same range of positions with respect to the ground, thus allowing the variable offset due to variable ground pickup to be subtracted out. (The OFF and ON scans last 4 min 48 sec each).

The data presented here were taken during 1982 December 27 to 31, 1983 May 16 to 20 and 1984 February 25 to March 5. These runs gave a total of 174 hours of useful data. After correcting the data for the beam efficiency of the antenna and converting them to thermodynamic temperatures we compute the mean values ΔT^i_{FIELD} for the observations of each Field on each day (i). The corresponding standard deviations, σ^i_{FIELD}, are calculated from the scatter of the individual 10 minute OFF-ON measurements and are an estimate of the statistical uncertainty in the mean of two hours of data. A typical value for data taken with clear sky is σ^i_{FIELD} = 0.25 mK whereas one expects 0.14 mK from system noise alone. The excess noise at this stage is due to the 1/f component in the maser noise spectrum and in the variable components of the ground and atmospheric contributions to δT_{ON} and δT_{OFF}.

The final values, $\overline{\Delta T}_{FIELD}$, for each one of the Fields were obtained by weighting by $(\sigma^i_{FIELD})^2$ each one of the two hour observations for a given Field. The results are plotted in Figure 1 and show no statistically significant signal in any of the Fields. The errors associated to each point correspond to the standard deviations estimated from the measurement statistics in the usual way. As the data will be used to decide what level of a sky signal is excluded by these measurements, it is crucial to estimate the errors in Figure 1 accurately. The scatter in the values of ΔT^i_{FIELD} for the different days that a given Field was observed could be larger than expected from the values σ^i_{FIELD}; we are asking for exceptional cancelation of any systematic effects. In principle, one could perform standard Chi-square tests for each one of the Fields. Any such test would compare the scatter of the different two-hour observations of a given Field with respect to the final value ΔT_{FIELD}, with the corresponding values of σ^i_{FIELD}. Let me emphasize that such a test does not depend on whether there is a true signal coming from the sky, as the scatter is computed with respect to the "local" mean $\overline{\Delta T}_{FIELD}$; it is truly an "instrumental" Chi-square test. Although we do not have enough observations on each of the Fields to give twelve strong tests (one per Field), we can add up all the Chi-square distributions to perform a good test. Figure 2 shows a histogram of the weighted residuals $(\Delta T^i_{FIELD} - \overline{\Delta T}_{FIELD})/\sigma^i_{FIELD}$ for the 87 two-hour observations. The histogram is reasonably gaussian with a width that is consistent with what we would expect if all the scatter in the values of ΔT^i_{FIELD} with respect to their corresponding average value $\overline{\Delta T}_{FIELD}$ were due to fluctuations contained in σ^i_{FIELD}. Indeed, the corresponding value of Chi-square is of 78 with 75 degrees of freedom. Therefore, we conclude that it is legitimate to estimate the standard deviations $\bar{\sigma}_{FIELD}$ in Figure 1 in the standard way, as $\bar{\sigma}_{FIELD} = [\sum_i(\sigma^i_{FIELD})^{-2}]^{-1/2}$. The variation of σ_{FIELD} from Field to Field is mainly due to the different number of days over which each Field is observed. The weighted mean of the points in Fig. 1 is $\Delta T_{AVE} = (26 \pm 39)\mu K$.

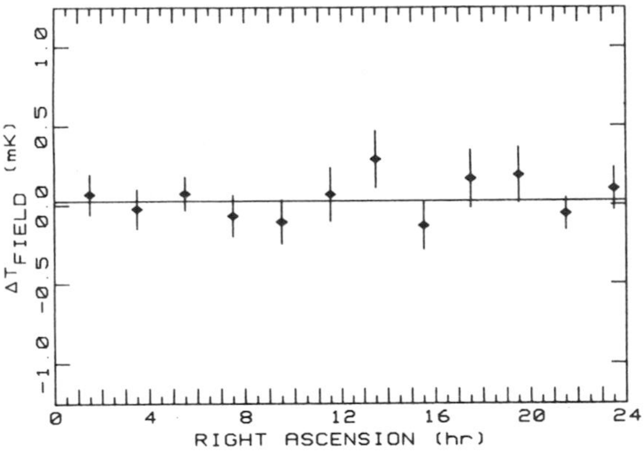

Figure 1

It follows from visual examination of Fig. 1 that none of the points differs significantly from the mean, so it is clear that, conforming to the unfortunate fate of this kind of measurements, no anisotropy has been detected. One then has to consider what level of anisotropy is excluded by these data. An underlying signal in the sky is expected to be gaussian, characterized by its standard deviation σ_{SKY}. Assuming there is no correlation between the signal arriving to a given Field and the ones arriving at the associated reference beams, the rms residual in ΔT_{FIELD} due to a true sky signal is (from Equation 2)

$$\sigma^2_{rms} = (4\,\sigma^2_{SKY} + \sigma^2_{SKY} + \sigma^2_{SKY})/4 = 1.5\,\sigma^2_{SKY} \quad . \qquad (3)$$

In this case, the OFF-ON procedure is a probe of σ_{SKY} with $\sqrt{1.5}$ efficiency. This "efficiency" depends on the assumed spatial correlation of the underlying sky signal and has to be computed for each given model.

The question is then what limits do the data place on σ_{SKY} from the fact that we only see the scatter shown in Figure 1. The Neyman-Pearson lemma prescribes the optimal statistical estimator for testing the hypothesis that $\sigma_{SKY} \neq 0$. The optimal statistic turns out to be essentially a weighted Chi-square test[6], which can be used to find out which values of σ_{SKY} are ruled out at a certain confidence level. The data in Figure 1 give

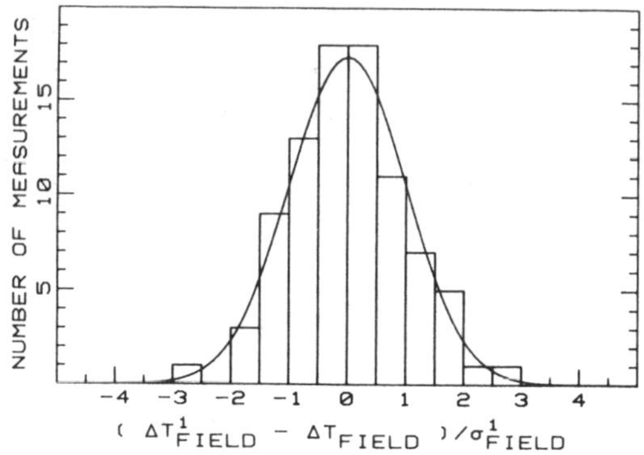

Figure 2

$$\sigma_{rms} < 7.8 \times 10^{-5} \text{ K at the 95\% confidence level} \quad . \qquad (4)$$

Using (3) and T = 2.7 K gives

$$\sigma_{SKY}/T < 2.4 \times 10^{-5} \text{ at the 95\% confidence level} \quad . \qquad (5)$$

Limits on anisotropy given by different experiments are normally plotted together in order to compare them. This raises the question of which angular scale corresponds to a given measurement. The answer is not unique and depends on the model that one assumes for the underlying fluctuations. Most experiments are sensitive over a range of angular scales. The combination of telescope beamwidth and observing procedure determine an "instrumental transfer function".

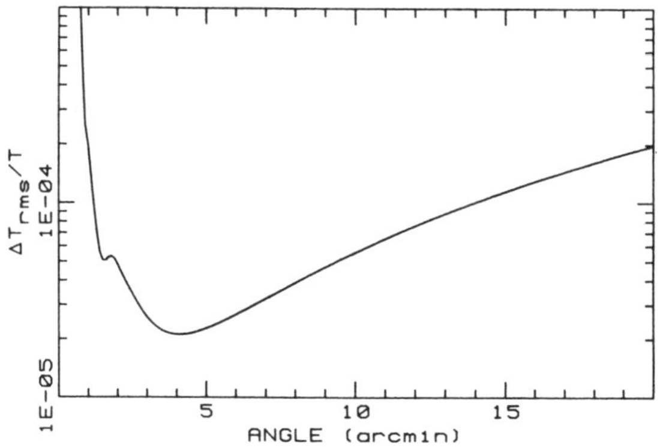

Figure 3

Figure 3 shows the limits that our measurement implies on a given scale, assuming that all the anisotropy is the result of a superposition of (spatially) monochromatic sine waves with random orientations and phases. The strongest limit corresponds to a scale (half-wavelength) of 4 arcminutes and is

$$\Delta T_{rms}/T < 2.1 \times 10^{-5} \text{ (at the 95\% confidence level)} \quad . \qquad (6)$$

This limit is more stringent than the one given in (5) because for this kind of fluctuations the gaussian beam pattern couples more efficiently to the "signal" than for the case considered before, which can be understood as patches with a constant value of ΔT that fill the beam and are uncorrelated over the scale of the beam-throw. For the models in Figure 3, anisotropy is limited to less than 5×10^{-5} for half-wavelengths between 1.5 arcminutes and 10 arcminutes.

A physical interpretation of the results depends on the power spectrum assumed for the underlying fluctuations. A number of authors have computed the expected signal that our measurement should yield for various cosmological models and initial conditions.[4,8] Comparing our results with their predictions rules out all possible models of adiabatic fluctuations dominated by baryonic matter. Models dominated by massive neutrinos require $\Omega \sim 1$, although they suffer other problems.[9] Models dominated by cold dark matter require that $\Omega h > 0.2$ (Ω is the density parameter and h is Hubble's constant in units of 100 km sec^{-1} Mpc^{-1}). Isothermal models, which seem momentarily out of fashion, require $\Omega > 0.1$.

The observations described were made in collaboration with Dave Wilkinson. We are indebted to Chuck Brockway and Rick Fisher who are responsible for the excellent performance of the telescope, receiver and data acquisition system. During these four years of collaboration, they have made continuous improvements to the equipment, always beyond any previous realistic expectations. We are grateful to everybody on the Green Bank staff for helping to solve our problems at any time - day or (more often!) night.

This work is supported in part by the National Science Foundation.

REFERENCES

[1]R. H. Dicke, P. J. E. Peebles, P. G. Roll and D. T. Wilkinson, Ap. J. **142**, 414 (1965).

[2]P. J. E. Peebles, Lectures in Applied Mathematics **8**, Relativity Theory and Astrophysics, Ed: J. Ehlers (1967), p. 274.

[3]G. Dautcourt, M.N.R.A.S. **144**, 255 (1969).

[4]P. J. E. Peebles and J. T. Yu, Ap. J. **162**, 815 (1970); R. A. Sunyaev and Ya. B. Zeldovich, Astrophys. and Sp. Science **6**, 358 (1970).

[5]A. N. Lasenby and R. D. Davies, M.N.R.A.S. **203**, 1137 (1983) present a detailed discussion on their experiment with a review on previous results; see also R. B. Partridge in The Origin and Evolution of Galaxies, VIIth course of the International School of Cosmology and Gravitation (Eds. B. J. T. Jones and J. E. Jones, 1983), p. 121.

[6]J. M. Uson and D. T. Wilkinson, Ap. J. (August 15, 1984, in press) presents a detailed discussion of the experimental technique that we use.

[7]P. E. Boynton in Objects at High Redshift, I.A.U. Symposium 92 (Eds. G. O. Abell and P. J. E. Peebles, 1980), p. 293.

[8]See the contributions of J. R. Bond and J. Silk to this volume; also J. R. Bond and G. Efstathiou (1984 preprint) and N. Vittorio and J. Silk, Ap. J. Lett. (1984, in press) and references therein.

[9]See S. D. M. White, this volume.

Very Small Scale Anisotropies of the Microwave Background

R. B. Partridge

The successes of the inflationary model and of new theories in particle physics in explaining some of the difficult puzzles of Big Bang cosmology have been noted here by other participants. Particle physics and inflation have combined to offer explanations, or retrodictions, of the large-scale isotropy in the Universe, of its approximate flatness, and of the baryon to photon ratio.

Of more interest to most of us in the field are the (as yet untested) predictions of the new inflationary cosmology and Grand Unified Theories.

Among these predictions are proton decay, the existence of non-baryonic matter (to explain the very small or zero spatial curvature of the Universe), and a set of predictions about the properties of density perturbations in the material content of the Universe. It is this last set of predictions, ones we may be able to test with astronomical observations, which will concern us here. New theories of the very early Universe provide an realistic source for small amplitude perturbations in density--quantum fluctuations. Further, these theories permit us to calculate, at least roughly, the relevant amplitudes $\Delta\rho/\rho$ at any given epoch in the history of the Universe. Thus, unlike earlier versions of the Big Bang model, the new inflationary models provide an explanation for density perturbations rather than building them in as initial conditions. GUTs also predict that density perturbations will be adiabatic, rather than isothermal; that is that they will have a constant baryon to photon ratio. Finally, inflation suggests a scale-invariant spectrum for the perturbations, which leads to a mass spectrum of the following sort: $\Delta\rho/\rho \propto M^{-2/3}$.

The importance of these new theories in cosmology makes any attempt to confirm such predictions worthwhile--even if the attempt involves some messy astrophysical details.

I. FLUCTUATIONS IN THE MICROWAVE BACKGROUND

One potential check on these predictions is a measurement of the amplitude of density perturbations at a particular epoch early in the Universe. That epoch corresponds to the surface of last scattering of the microwave photons, which is conventionally taken to lie at a redshift of $z \sim 1000$. If density perturbations are present, temperature fluctuations will be introduced into the cosmic microwave background.[1]

Detailed calculations have been made of the amplitude and spectrum of density perturbations required to produce the structure we see in the Universe today. These suggest $\Delta\rho/\rho \sim 10^{-3}$ at $z = 1000$.[2] Since $\Delta\rho/\rho \ll 1$, the perturbations are still in the linear regime. Hence observations of the density perturbations at this early epoch may give a better picture of the actual distribution of all mass in the Universe than observations of the present distribution of luminous matter only, since the latter has clearly been affected by non-linear processes.

Calculations of $\Delta\rho/\rho$ from the other direction, based on the growth of quantum fluctuations, are not yet in as polished a state. On the other hand, the absence of very large amplitude perturbations at the surface of last scattering has already been used to rule out some theories of the early Universe.

In very crude qualitative form, the models predict a spectrum of temperature fluctuations in the microwave background which rises sharply to a critical angle Θ_c, and then levels off. It is worth noting explicitly that the predicted amplitude and spectrum of temperature fluctuations in the microwave background is exceedingly model dependent. That dependence, of course, enhances the ability of observations to decide among the various theories. For instance, models containing massive neutrinos or other -inos make quite different predictions about the spectrum of microwave background fluctuations than models in which the density of the Universe is dominated by baryonic matter.[2]

II. SEARCHES FOR FLUCTUATIONS IN THE MICROWAVE BACKGROUND

We turn now to the observational evidence. Unfortunately, all we have to date are a set of upper limits at various angular scales.

The first of these is a series of observations made on angular scales of a few arcminutes, generally using conventional radio telescopes. The most recent and exciting of such measurements is the one to be reported here by Juan Uson. The limits set by Uson and David Wilkinson[3] are the most sensitive available to date. Further, the angular scale they investigate is close to Θ_c, the scale at which the fluctuation spectrum is expected to be a maximum. Their work suggests that the amplitude of density perturbations at the surface of last scattering was $<10^{-4}$, at least if pure baryonic, adiabatic, density perturbations were responsible for producing the fluctuations.

Another angular scale, investigated less fully, is a few degrees of arc; such an angular scale is of particular interest because larger regions were not causally connected at the time of the last scattering of the microwave photons. Thus, absent inflation, there is no way in which regions a degree or so in size could be uniform in their properties, including density or temperature. More simply put, in standard Big Bang models we have no reason not to expect very substantial anisotropy on scales larger than a degree or so; in the inflationary models, on the other hand, uniformity on such angular scales has a natural explanation. The best limits on angular scales of a few degrees are those set by the Italian group of Francesco Melchiorri and his colleagues:[4] $\Delta T/T \leq 3.5 \times 10^{-5}$ on an angular scale of 6°.

Finally, there is the angular scale of 0.1' - 1.0', spanning the range of angular scales corresponding to perturbations of galactic mass. In conventional models, $\Delta T/T$ on arcsecond scales should be smaller than on arcminute scales, but less conventional models[5] (including those with a secondary scattering epoch at $z \ll 1000$) predict larger values.

III. UPPER LIMITS ON $\Delta T/T$ AT $6'' < \Theta < 60''$

A microwave search for fluctuations on arcsecond scales requires the use of aperture synthesis, since the angular resolution of filled-aperture telescopes is typically $\geq 1'$. The use of arrays of telescopes, such as the Very Large Array, allows one to improve the angular resolution, since that is determined by the size of the entire array. Advantages of using aperture synthesis are: increased resolution, freedom from some systematic errors present in filled aperture observations[6] and, perhaps of most general interest, the fact that aperture synthesis provides a two-dimensional map of a portion of the surface of last scattering. Given the exotic morphology suggested for density perturbations--including pancakes, cusps, strings and perhaps even asparagus--the advantage of a two-dimensional map rather than a one-dimenional slice is clear. There are, however, disadvantages to using aperture synthesis. Other sorts of systematic errors may arise. More important, aperture synthesis observations are inherently inefficient compared to filled aperture observations. Only a small fraction of the photons arriving at the array are actually collected by the constituent telescopes of the array. Nevertheless, the large number of telescopes in the VLA make it the best suited instrument for the detection of fluctuations in the microwave background on arcsec scales.

Two groups have attempted such observations.[7,8] In our work, we attempted to separate fluctuations in the microwave background from instrumental noise as follows. First, we constructed a map of the sky covering an area substantially larger than the solid angle of the beam of the individual telescopes of the array. thus the system had a very low response to fluctuations in the sky at the edge of our map, and a much higher response to sky fluctuations at the center of the map. On the other hand, the instrumental noise in an aperture synthesis map is essentially constant across the map. The radial dependence of the response to sky fluctuations permitted us to extract them from the instrumental noise; in effect we took the quadrature difference of the noise at the center and at the edge of our maps. Formalont et al[7] used a somewhat related technique.

Both groups encountered a fundamental instrumental problem, more-or-less circular rings concentric with the phase center of the map. Those interested in the gory details of aperture synthesis may consult references 7 or 8. Here, it suffices to note that such effects were probably present in the maps analyzed by both groups. Hence our upper limits on sky fluctuation are conservative, since at least some of that apparent sky fluctuation was in fact instrumental noise which was largest at the phase center of the maps. Once this problem was identified, the solution to it also became clear--it is possible electronically to shift the phase center of the map away from the actual center of the region being studied.

As part of another observing program,[9] we used just this procedure in the summer of 1983. We were studying a cluster of galaxies, so that the region of the sky we examined was not randomly chosen. On the other hand, shifting the instrumental noise from the center of the map permitted somewhat higher sensitivity than had been achieved before. This new work is reported in column 4 of table 1, which summarizes the results to date. We must emphasize that the center of our map coincided with a cluster of galaxies which is known to contain some radio sources. Thus our upper limits on fluctuations in the microwave background are again conservative in the sense that at least some of the fluctuation was surely due to sources in the cluster. Even if we remove from the map the obvious sources in the cluster, fainter radio sources may still have been present, and may be "contaminating" the limits set out in column 4 of the table. As an estimate of the minimum contribution of radio sources to the sky fluctuation, we may take a value of $\Delta T/T \sim 1.2 \times 10^{-4}$ at $\Theta = 60''$, derived from the work of Danese, De Zotti and Mandolesi.[10] Note that our recent observations are within a factor of 2 of detecting the predicted sky fluctuations introduced by weak radio sources alone.

We have underway more careful observations in a region of sky free of known radio sources and with a longer integrating time than used by us in 1983, which may be able to reduce the upper limits of the table by a further factor of 3 at $18'' < \theta < 60''$.

IV. CONCLUSIONS

The conclusion of all this work, which will be reemphasized in the following presentation by Dr. Uson,[6] is that no fluctuations in the cosmic microwave background have been reliably detected. The upper limit on such fluctuations is at or below one part in 10^4 over a range of angular scales from a minute of arc to roughly 6° and is being pushed to smaller angular scales as well. These observations place constraints on the amplitude of density perturbations at a particular, early, epoch in the history of the Universe. It is already the case that these limits in turn have an impact on theories of elementary particles. Some GUT models for the early Universe predicted much larger density perturbations, and have had to be discarded because they violate the observational material. And that is a reasonable note on which to end--how interesting it is that relatively straightforward radio astronomical observations may in fact provide one way of checking the most fundamental theories of microscopic physics.

Angular	$\Delta T/T \times 10^3$		
Scale	Ref.7	Ref.8	New Results
6"	--	3.2	--
12"	--	1.7	--
18"	1.0	1.2	0.5
30"	0.8	--	0.4
60"	0.5	--	0.25

In the table are listed 2σ upper limits on the amplitude of temperature fluctuations, $\Delta T/T$. All the observations were made at a wavelength of 6 cm using the VLA. The results of Fomalont et al[7] and of new work reported here were made with the array in its densest configuration, and hence with the lowest angular resolution. The work of Knoke et al[8] was made with the array in a somewhat larger configuration, and hence provides enhanced angular resolution at the cost of decreased sensitivity.

REFERENCES

[1]J. Silk, Astrophys. Journal , **151**, 459; (1968). R. A. Sunyaev, and Ya. B. Zel'dovich, Astrophys, and Space Sci., **7**, 3; (1970). For a summary of the observations see papers by Uson and by Wilkinson in this volume and an older summary by Partridge, Physica Scripta, **21**, 624 (1980).

[2]J. Silk, these proceedings.

[3]J. M. Uson, and D. T. Wilkinson, Astrophys. Journal (Letters), **277**, L1.(1984).

[4]R. Fabbri, B. Melchiorii, F. Melchiorii, V. Natale, N. Caderni and K. Shivanandan Phys. Rev., **D21**, 2095, (1980).

[5]C. J. Hogan, Monthly Notices Roy. Astron. Soc. **192**, 891; Astrophys Journal, **252**, 418; Astrophys. Journal (Letters), **256**, L33, (1982).

[6]J. Uson, these proceedings.

[7]E. B. Formalont, K. I. Kellerman, and J. V. Wall Astrophys. Journal (Letters), **277**, L23, (1984).

[8]J. E. Knoke, R. Partridge, and M. I. Ratner, Astrophy. Journal **284** 479,(1984).

[9]N. Mandolesi, R. B. Partridge, and R. Perley, submitted to Astrophys. Journal, (1984).

[10]L. Danese, G. De Zotti, and N. Mandolesi, Astron. and Astrophys., **121**, 114, (1983).

The Cosmic Microwave Background Radiation: Implications for Galaxy Formation

Joseph Silk

ABSTRACT

The isotropy of the cosmic microwave background sets strong constraints on the cosmological density parameter and on the nature of the primordial density fluctuations that give rise to the observed large scale structure. If these fluctuations were scale-invariant, there can be a significant dipole component at the present epoch. Both this and the predicted angular anisotropy are lower in a universe dominated by cold particles, such as axions or photinos, than in a baryon or massive neutrino-dominated universe once the residual large-scale fluctuations are normalized to the variance in the observed galaxy distribution. Anti-inflationary scenarios with density parameter $\Omega \lesssim 0.4(h/0.5)^{-1}$, where h is the Hubble constant in units of 100 kms^{-1} Mpc^{-1}, lead to excessive fine-scale radiation anisotropy in a cold dark matter-dominated universe. This result already suggests that the underlying dark matter is more weakly correlated on large scales than the luminous galaxy distribution, as must be the case for an inflationary cosmological model. Consideration of the spectral distortions induced in the microwave background radiation by activity associated with early stages of galaxy formation suggests that one cannot exclude the possibility that the universe underwent considerable reheating at epochs as early as a redshift of 1000. Astrophysical sources for heating the intergalactic medium at an epoch much earlier than a redshift of 10 are only available in the context of a theory of fluctuation origin that allows primeval, and possibly rare, seeds with masses of order 10^6 solar masses to have gone non-linear sufficiently early.

I. INTRODUCTION

The cosmic microwave background is a unique tool for probing the early universe. It contains information about conditions at a redshift far in excess of that of any known object. Angular anisotropies are produced by density fluctuations that are still in the linear regime at a redshift of about 1000, corresponding to the epoch of last scattering of the radiation. Even if reionization can occur, primordial anisotropies on angular scales above several degrees persist, and new fine-scale anisotropy is regenerated at a diminished level. The blackbody nature of the spectrum of the cosmic microwave background radiation arose, according to the standard model, at a far earlier epoch in the history of the Universe, when the density and temperature were high enough to thermalize the radiation.

The large angular scale anisotropy of the cosmic microwave background radiation directly reflects the seed inhomogeneities from which structure evolved. Unique signatures are imposed on the large-scale anisotropy by different theories of galaxy origin. With adequate sky coverage of the multipole structure of the radiation background, one could hope to distinguish between the hierarchical clustering and pancake fragmentation theories of galaxy formation, unfold the role of massive neutrinos or photinos in modifying the primordial fluctuation spectrum, and even measure the curvature radius of the Universe. All of these enigmas are encoded in the large-scale anisotropy: deciphering it promises to be one of cosmology's greatest and most rewarding challenges. The low order multipoles of the background radiation distribution on the sky describe causally disconnected regions of the Universe, and the radiation anisotropy probes the origin both of the large-scale structure of the Universe and of the galaxies themselves. For it is simply not credible that galaxies could have formed and left no trace of the initial fluctuation spectrum on larger scales. Moreover, there is increasing evidence from deep redshift surveys of structure on scales as large as ~100 Mpc.

Similarly, distortions are expected to arise at some level from the blackbody spectrum, simply because the universe that we observe is inhomogeneous. Sources and sinks of radiation are present now and were present in the remote past, and should result in deviations, from a perfect blackbody spectrum. It is only by searching for such deviations that we can ever hope to understand the processes by which the universe evolved to reach its present state. Inflationary scenarios have provided tentative resolutions of several outstanding cosmological puzzles. Of noteworthy interest for the present discussion are the predictions that the universe is flat and that the primordial fluctuation spectrum is scale-invariant. Armed with these predictions, precise estimates of the angular anisotropy of the cosmic background radiation can be made, the principle uncertainty remaining in the type of particle that dominates the cosmological mass density. In Section II, I discuss the primordial fluctuation spectrum, for different choices of dark matter. Then in Sections III and IV, I review the large and small-scale background anisotropy measurements. Finally, Section V is devoted to a discussion of spectral distortions of the cosmic microwave background radiation.

II. THE PRIMORDIAL FLUCTUATION SPECTRUM

Non-baryonic dark matter not only provides an attractive means of accounting for the high mass-to-light ratio in the outer regions of galaxies and in galaxy groups and clusters,[1] and offers a means of combining the high density Ω=1 universe favored by inflationary arguments with primordial nucleosynthesis of deuterium,[2] but it leads to a rather striking prediction for the large-scale distribution of matter. Structure on large scales is enhanced relative to smaller scales. The dividing line is determined by the horizon scale when the universe is first dominated by non-relativistic particles. Larger scale structure can subsequently grow after entering the horizon, thereby enhancing the density fluctuations amplitude above that of the initial curvature fluctuations. On the other hand, smaller scale structure entered the horizon when the universe was radiation-dominated, and further growth was inhibited. If the pressure of the collisionless relics is negligible compared to that of the radiation, this critical comoving scale is $L_{eq} = 12(\Omega h^2)^{-1}$ Mpc, and will be effectively imprinted on the residual density fluctuation spectrum prior to galaxy formation.

A more detailed discussion requires one to distinguish three types of collisionless relics, determined by the epoch at which they decouple from thermal equilibrium.[3] Massive neutrinos thermally decouple at ~1 MeV, become non-relativistic at $kT \sim m_\nu c^2$, and now have a velocity dispersion $\sim 6(m_\nu/30\ \mathrm{eV})^{-1}\ (1+z)\ \mathrm{km\ s^{-1}}$. More exotic particles decoupled earlier, and can be much colder by now. Consequently, they can cluster on smaller scales. In the most extreme case, weakly interacting massive particles can decouple when already non-relativistic: these particles are always cold, and hence cluster on all scales. Examples of weakly interacting particles which decouple relativistically are gravitinos, photinos, and left-handed neutrinos. Particles which decouple when non-relativistic include very massive (> 1 MeV) lepton: these could be photinos, right-handed neutrinos, or monopoles. Axions, or bose condensates, are effectively a species of particles that have always been non-relativistic. The number density of relativistic decouplers is determined by the number g of interacting species present at decoupling. The earlier decoupling occurs, the more species there are present to share the entropy, and hence the lower the number per baryon of the exotic species destined to eventually dominate the universe. This domination occurs provided the particle is long-lived and has sufficient mass. For neutrinos, $g = 43/4$, while for particles which decouple prior to the quark-hadron phase transition at 200 MeV, $g \approx 100$.

Particles which decouple relativistically are subject to strong damping on increasing scales as the universe expands until the particles became non-relativistic. The minimum surviving scale of fluctuations corresponds to the horizon size at the epoch z_{nr} when the particle velocity is eventually becoming non-relativistic. This is because the comoving distance over which a collisionless particle can stream freely increases until this epoch, and subsequently decreases. Fluctuations are erased as a consequence of the free streaming. For massive neutrinos, the minimum surviving comoving scale of fluctuations is $L_\nu = 41(m_\nu/30\ \text{eV})^{-1}$ Mpc, equivalent to a mass of $M_\nu \sim 1.8\ m_{pl}^3 m_\nu^{-2} \sim 10^{15}(m_\nu/30\ \text{eV})^{-2}\ M_\odot$. Particles which decoupled earlier, such as gravitinos or photinos, are more massive (for given Ω) and the surviving scale is reduced to $M_x \sim 0.06\ m_{p\ell}\ m_x^{-2}\ (g/100)^{-4/3}$; typically, $m_x \sim 1$ keV for $\Omega \approx 1$. Now for neutrinos, the horizon size at matter-radiation equality $\lesssim 2\ L_\nu$; hence the residual fluctuation spectrum is determined primarily by the damping cut-off and the initial shape over scales in excess of the horizon mass M_{eq} at z_{eq}, these latter scales undergoing uninterrupted growth. However, for the particles which decoupled earlier, $L_x \ll ct_{eq}$; for $g \sim 100$ and $m_x \sim 1$ keV, the damping scale is about one-tenth of the horizon scale at t_{eq}. This means that over the range $M_{eq} \approx 4 \times 10^{15}\ M_\odot > M > M_x \approx 10^{11}\ M_\odot$, the initial fluctuation spectrum flattens from $|(\delta\rho/\rho)_k|^2 \propto k^n$ to k^{n-4}, growth being suppressed during the radiation-dominated phase on sub-horizon scales. In the extreme case of cold particles, or non-relativistic decouplers, damping is unimportant above stellar mass-scales (since decoupling occurred at $kT \gg 1$ MeV when the horizon contained $\ll 1\ M_\odot$). Hence the $|(\delta\rho/\rho)_k|^2 \propto k^{n-4}$ scaling applies on scales well below M_{eq}.

In fact, numerical integration of the fluctuation growth equations shows that while the damping cut-off is fairly abrupt,[3] the transition at M_{eq} is gradual over several decades in mass-scale for the case of cold particles.[4] The asymptotic k^{n-4} slope is not attained until a comoving scale of $\lesssim 0.1$ Mpc. Between 1 and 30 Mpc, the slope is

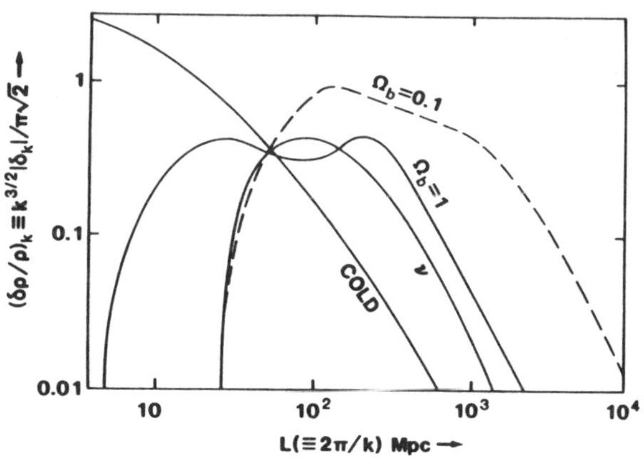

Figure 1: Density fluctuation spectra. Density fluctuation amplitude, defined to be power per logarithmic interval of wavelength, plotted versus wavelength for a universe dominated by baryons (Ω_b=1 and 0.1), massive neutrinos (Ω=1) and cold relics (Ω=1). The Hubble constant has been set equal to 50 km s^{-1} Mpc^{-1} (h=1/2). From Silk.[5]

approximately $k^{n-3/2}$, yielding a nearly white-noise-like residual spectrum if n = 1 initially over this limited range of comoving scales that is of importance for galaxy clustering. Only above ~30 Mpc (or the horizon size at t_{eq}) does the residual spectrum approach the primeval slope, k^n. Examples of the residual fluctuations spectra for baryonic models, for massive neutrinos, and for cold relics are shown in Fig. 1. These describe the fluctuation spectra $k^{3/2}(\delta\rho/\rho)_k$, equivalent to the power in density fluctuations per logrithmic decade of wavelength, for the dark matter (for massive neutrino and cold relic models) and for the case of a baryon-dominated universe.

III. LARGE-SCALE ANISOTROPY OF THE MICROWAVE BACKGROUND

It is convenient to divide our discussion of the large-scale anisotropy into the implications of the dipole anisotropy, which is associated with our peculiar velocity, and the quadrupole and low order multipoles of the radiation anisotropy.

A. Dipole Anisotropy

The peculiar motion induced by the local non-linearity associated with the Virgo Supercluster amounts to

$$\delta v \sim L(G\,\delta\rho)^{1/2} , \qquad (1)$$

where $\delta\rho$ is the density excess within scale L. In the linear regime, for L >> 10 Mpc, we have

$$\delta v \sim L(\delta\rho/\rho)^{1/2}(G\rho)^{1/2} . \qquad (2)$$

Now the large-scale limit of a scale-invariant spectrum is $\delta\rho/\rho \propto L^{-2}$, so that linear theory predicts that the peculiar velocity should be dominated by contributions from scales which are just going non-linear. The linear theory prediction for our infall velocity to the Local Supercluster is

$$v_{sc} = 1/3\ H_0 R\ \Omega^{0.6}\ (\delta M/M) , \qquad (3)$$

where $\delta M/M$ is the mass excess within a radius R. Adopting $\delta M/M \approx 2$ yields

$$v_{sc} \approx 670\ \Omega^{0.6}\ \text{km s}^{-1}\ . \qquad (4)$$

The measured infall velocity from analysis of the non-linear velocity field of Local Supercluster galaxies is $v_{sc} \approx 250(\pm 50)$ km s^{-1}. Comparison of the predirection and the observational result yields an estimate of the local value of Ω over scales $\lesssim 10\ h^{-1}$ Mpc. This approach has been reviewed by Davis and Peebles[6] and by Sandage and Tammann.[7]

In addition to the local contribution, distant irregularities in the linear regime still contribute to our peculiar velocity according to (2). To evaluate the role of large-scale fluctuations, one can try to take the vector difference of our velocity relative to the CBR with respect to our motion relative to Virgo. This has been done by Sandage and Tammann,[7] who deduce a net motion of the entire Virgo Supercluster of ~450 km s^{-1} relative to the CBR. A key assumption in this is the neglect of any substantial transverse component of our motion towards Virgo.

An independent approach, due to Hart and Davies,[8] utilizes observations of galaxies spread over a shell at a distance ~$25h^{-1}$ Mpc. Bypassing the Virgo cluster, these authors measure the motion of our galaxy relative to this distant shell. Subtraction of the dipole anisotropy motion yields the result that the shell of galaxies has a velocity of only 130 (±70) km s^{-1} with respect to the cosmic background radiation. Hart and Davies[8] use the Fisher-Tully correlations to determine galaxy distances.

These results can be compared with the predicted one-point velocity correlation obtained from linear theory. Convolving the perturbed velocity with a Gaussian window function, Kaiser[9] obtained

$$\langle v_p^2 \rangle^{1/2} = 1400\ \left(\frac{1+z_{gf}}{4}\right) \left(\frac{0.5}{h}\right)\ \text{km/s}\ ,$$

in a neutrino-dominated universe. With non-linearity occuring at redshift $z_{gf} = 4$, the predicted velocity as measured at a randomly chosen point only falls below that claimed by Sandage and Tammann[7] about five percent of the time. This would suffice to eliminate the neutrino-dominated universe if the lower Hart-Davies result for $\langle v_p^2 \rangle^{1/2}$ is eventually confirmed. Occurrence of first non-linearity at a later epoch reduces the scale that is presently going non-linear and hence tends to suppress the peculiar velocity.

That linear fluctuations are probably contributing significantly to our motion relative to the CBR is inferred from the fact that the apex of the dipole motion is about 45° away from the Virgo cluster, and well away from any other substantial concentration of galaxies. A large-scale shearing motion would be necessary if the dipole anisotropy is entirely of local origin. This would be rather probable in a neutrino-dominated universe, where large-scale coherent motions develop. However in a cold particle-dominated universe, large-scale motions will tend to be suppressed.

In order to predict the peculiar velocity, the fluctuation spectrum must be normalized appropriately. This can be done by evaluating the rms mass fluctuations, and requiring them to be unity at present on a scale of $8h^{-1}$ Mpc, where $\delta M/M$ is observed to be unity in the galaxy distribution.[10] Predictions of $\langle v_p^2 \rangle$ as a function of the smoothing scale are shown in Fig. 2, with 95 percent confidence levels indicated by the

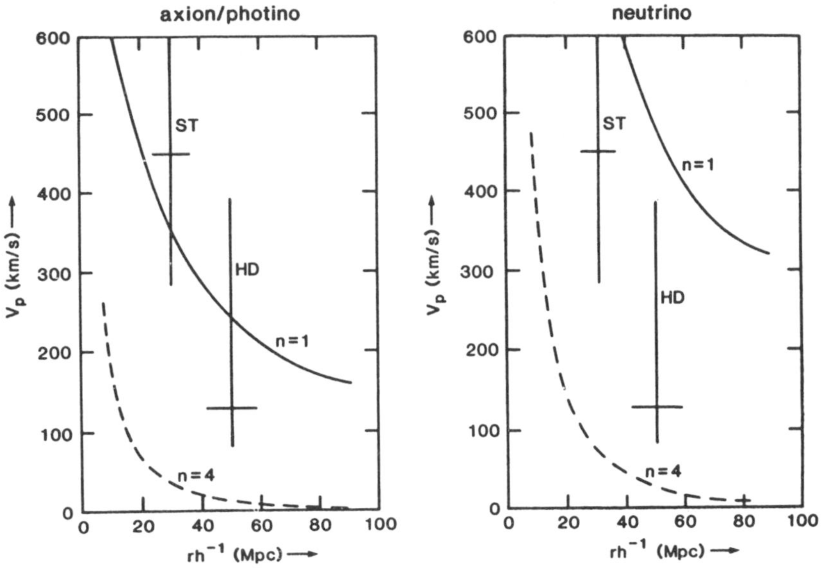

Figure 2: Peculiar velocity due to large-scale density fluctuations. The rms streaming motion induced by linear fluctuations on a distant shell of matter is plotted versus the shell radius for scale invariant (n=1) and causal (n=4) primordial fluctuations spectra in the cold particle and neutrino-dominated flat universe models. Comparison is shown with the results of Sandage and Tamman (ST) on the motion of the Virgo Supercluster relative to the cosmic frame, and with the result of Hart and Davies (HD) for the motion of a more distant shell of galaxies (see text). The vertical error bars give the 95 percent confidence limit for the measured velocity to coincide with the predicted value. From Vittorio and Silk[11] (in preparation, 1984).

vertical bars for a cosmological model with $\Omega = 1$. The predicted velocity decreases as $\Omega^{1/2}$. Evidently, if $\Omega = 1$ and normalization is made to the observed galaxy distribution, there is an appreciable contribution to the dipole anisotropy that can plausibly account for the difference with the Virgo infall velocity vector. It would be of considerable interest to improve the observational determination of $\langle v_p^2 \rangle^{1/2}$ especially with respect to scales outside the Virgo supercluster.

B. Quadrupole Anisotropy

The large-scale radiation anisotropy is especially simple to evaluate. It arises principally from the Sachs-Wolfe[12] effect, due to large-scale gravitational potential fluctuations $\delta\psi$ at the last scattering surface and around the observer. This contribution arises because photons are gravitationally redshifted as they climb out of a fluctuation on the last scattering surface and acquire a slight blueshift as they fall into the potential well associated with local fluctuations. This results in a temperature anisotropy of order

$$\delta T/T \sim \delta\psi/c^2 \sim \begin{cases} [(\delta\rho/\rho)(L/ct)^2]_{t_0} & (L<ct_0) \\ [\delta\rho/\rho]_{t_0} & (L>ct_0) \end{cases}$$

where L is a comoving scale and t_0 denotes the present epoch. The quadrupole component of the anisotropy directly measures $\delta\rho/\rho$ at horizon crossing: it can be related, within a factor of order unity, to the gauge-invariant quantities defined by Bardeen.[13] A more precise definition of the predicted quadrupole anisotropy may be obtained by expanding the calculated radiation intensity in spherical harmonics.

Early reports of detection of a quadrupole anisotropy have not been confirmed. The upper limit on the rms (95 percent confidence) amplitude of the quadrupole anisotropy is 7×10^{-5}. The predicted quadrupole anisotropy for scale-invariant neutrino or cold matter fluctuations is below this limit: for neutrinos, $Q = 3 \times 10^{-5}$, and for cold dark matter (Ω=1), we expect $Q = 1 \times 10^{-5}$. A baryon-dominated model, compatible with primordial nucleosynthesis, with $\Omega = 0.1$, however leads to excessive quadrupole anisotropy unless the large-scale power in the primordial fluctuation spectrum is reduced. The large coherence length associated with baryon damping guarantees that the large-scale anisotropy will be important.

There are two possibilities for suppressing the quadrupole anisotropy. One is to increase the power law-index of the initial power spectrum to $n > 2$. This results in a divergence of the curvature fluctuations, $\delta\varepsilon$, on small scales. For example, with $\delta\varepsilon < 10^{-4}$ on scale $\sim 10^4$ Mpc and $\delta\varepsilon \propto L^{(1-n)/2}$, we infer that non-linearity occurs at $L \lesssim 1$ Mpc if $n = 3$. We would need to require that $\delta\varepsilon < 0.1$ to avoid problems of non-linear mode-mode coupling that could drive premature heating and collapse. Evidently power on scales below $L \sim 10$ Mpc must be suppressed below the extrapolated power-law spectrum if $n \approx 3$. There seems to be no natural way of introducing such a scale into the initial conditions.

A second loophole for a baryon-dominated universe involves the resurrection of isothermal fluctuations. The inflationary argument for a scale invariant (n=1) initial spectrum no longer necessarily applies, since the associated curvature fluctuation are negligible in the radiation-dominated era. On scales larger than the horizon size L_{eq} at equal matter-radiation densities, there are two possibilities for the isothermal mode. If defined to be strictly orthogonal to the adiabatic mode, then there is no associated curvature fluctuation. To arrange this, there has to be a small negative radiation density fluctuation along with the positive matter fluctuation. The principal consequence arises at the decoupling epoch, when the observable radiation anisotropy is partially suppressed. Temperature fluctuations are still produced as velocity fields develop on the last scattering surface. However fluctuations must enter well within the horizon before potential irregularities develop and produce a Sachs-Wolfe effect. An alternative definition is that matter irregularities are superimposed on a uniform radiation field, in which case there is appreciable anisotropy on large angular scales due to the large-scale curvature and potential fluctuations.[14] Nevertheless, whichever definition of isothermal fluctuations is considered appropriate, the large-scale anisotropy does end up being considerably reduced relative to that of the adiabatic baryon fluctuation model because of the persistence of entropy fluctuations on scales below the baryon damping length. Viable isothermal models are possible in a baryon-dominated universe provided that a primordial spectrum is adopted in which small-scale power is important.[15]

IV. FINE-SCALE ANISOTROPY

The most sensitive observational constraints on the anisotropy of the microwave background radiation arise at small angular scales. Here we have to take considerable care in doing the computation, because of smearing associated with the finite thickness of the last scattering surface. At last scattering, the horizon subtends an angle $(\Omega_0/z)^{1/2} \sim 2\ \Omega_0^{1/2}$ degrees at $z \approx 1000$. The thickness of the last scattering surface may be defined by the half-thickness of the effective visibility function, $e^{-\tau}\ d\tau/dz$, where the optical depth back to epoch t is

$$\tau = \int_t^{t_0} n_e\ \sigma\ c\ dt\ .$$

The visibility function peaks at $z \approx 1055$, and has a half-thickness $\Delta z \approx 150$, corresponding to the duration of the transition from a fully ionized to a neutral baryonic component of the universe. The angular scale subtended by fluctuations which are smeared out by the finite thickness of the last scattering surface is $\Delta\sigma \approx 7'.6\ \Omega_0^{1/2}$, corresponding to a linear scale of ~10 Mpc (if $\Omega_0 = 1$).

If recombination preceeds on schedule, as in the standard uniform model, primordial fluctuations in $\delta T/T$ should survive on scales in excess of a few arc-minutes. In order of magnitude, the small angular scale contributions to $\delta T/T$ can be decomposed into an adiabatic mode

$$\delta T/T = (f/3)(\delta\rho/\rho)_{LS},$$

where detailed computations of the transition between complete and very low ionization yield $f \approx 0.1$, and a velocity mode

$$\delta T/T \sim \int (v/c)e^{-\tau}d\tau \sim (\delta\rho/\rho)_{LS}(L/ct)_{LS},$$

which arises even for isothermal initial conditions and is due to the potential motions gravitationally induced by fluctuations on the last scattering surface. The dominant contribution on large angular scales arises from gravitational potential fluctuations up to the scale of our present horizon: the range $L_{eq} << L \lesssim ct_0$ is important, since $\delta\rho/\rho \sim L^{-2}$ in the asymptotic regime. Similarly, the velocity-induced perturbations contribute over comoving length scales of order L_{eq}, and the adiabatic mode contribution is important over an angular scale of order $\Delta\sigma$.

Even if the universe becomes reionized at an early epoch, since the maximum baryon density is unlikely to exceed $\Omega_b/\Omega \sim 0.1$, the latest redshift at which the universe remains opaque is large, amounting to

$$1 + z \sim 70(\Omega/0.2)^{1/3}(\Omega_b/0.02)^{-2/3} .$$

This means that primordial fluctuations of scale larger than the horizon at this epoch would not be erased by any causal scattering process. The corresponding angular scale is about 10°: large angular scale anisotropy probes the primordial fluctuations that must have arisen early in the history of the Big Bang.

However reionization at such an early epoch is unnatural in any theory of galaxy formation from adiabatic fluctuations. Non-linearity generally occurs too late, when the density of the universe is already too low for ionization of the intergalatic medium by early generations of massive stars to have any effect on smearing temperature fluctuations in the microwave background. Because of this, the fine-scale anisotropy limits, which are lower than the large-scale limits, pose the most stringent constraints on the primordial fluctuations spectra. In fact, a baryon-dominated universe is inconsistent with the Uson-Wilkinson[16] upper limit ($\delta T/T < 3 \times 10^{-5}$ at 4'.5) even if a power-law power spectrum $|\delta_k|^2 \propto k^4$ is adopted: this is the minimal spectrum expected on large scales if any causal process operates on small scales to yield the seed fluctuations which form galaxies.

Consider next the case of a universe dominated by weakly interacting particles. Contributions from different Fourier components for an experiment with an effective resolution of 4'.5 and 10° are shown in Fig. 3, and compared with the contributions to the dipole and

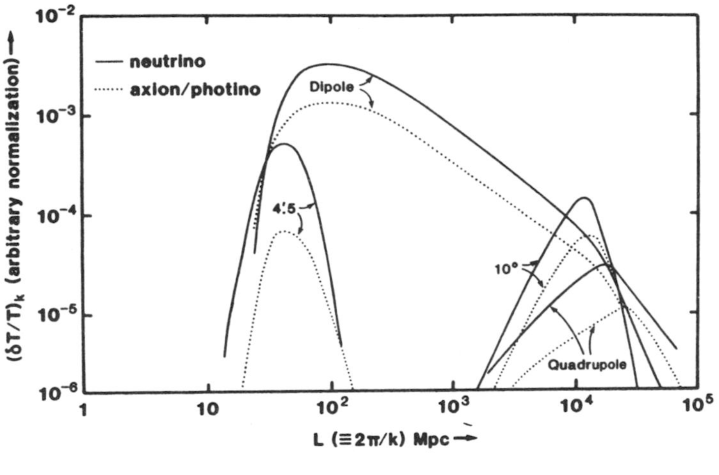

Figure 3: Contribution of different wavelengths to δT/T, measured with 4'.5 resolution (appropriate to the Uson and Wilkinson[16] experiment), with 10° resolution (appropriate to the Melchiorri, et al.[17] experiment, and to the dipole and quadrupole anisotropy). Predictions are shown for neutrino and axion/photino models of a flat universe with a primordial scale-invariant fluctuation spectrum. From Vittorio and Silk[11] (in preparation).

quadrupole anisotropies. The presence of small-scale structure in the cold particle distribution on scales below the normalization scale means that fluctuations in a neutrino universe exceed those in a cold particle-dominated universe. However when Ω_0 and the Hubble constant h are reduced, the predicted fluctuations increase. This leads to an important constraint in a low Ω cold particle-dominated universe, which otherwise yields in many respects a satisfactory model for the large-scale structure of the universe. Numerical simulations find that the galaxy correlation function slope and amplitude can be explained in such a model, with a reasonable epoch inferred for the formation of galaxies. Moreover, all astronomical determination of Ω_0 by dynamical studies of galaxy clustering indiate that $\Omega_0 \approx 0.2(\pm 0.1)$.

Fluctuation growth is sensitive to choice of Ω_0, because in a low Ω_0 universe, the effects of gravity eventually become unimportant, the expansion rate being dominated by the curvature term, at a redshift below $\sim\Omega_0^{-1} - 1$. Hence since we hope to observe temperature fluctuations that were generated at a definite epoch, that of last scattering, we may expect their amplitude to be enhanced by a factor of order Ω_0^{-1}. An additional effect arises because the cold particle fluctuation spectrum changes its slope in the vicinity of comoving scale $L_{eq} \sim 12/\Omega_0 h^2$, and is moreover normalized at a scale $\sim 8h^{-1}$ Mpc, where the fluctuations in the number counts of galaxies have unit variance. Normalization to the flatter region of the spectrum further enhances $\delta T/T$. A final complication is that a scale of proper length L (at Ω_0 z >> 1) subtends an angle $\sigma = (\Omega_0 H_0/2c)L$. Hence lowering Ω_0 shifts larger scales into a given angle of observation σ: this effect tends to reduce $\delta T/T$ on scales large enough that smearing due to the finite extent of the last scattering surface is unimportant.

These effects have been incorporated into numerical calculations of $\delta T/T$ in a cold matter-dominated universe.[18,11] Predictions are shown in Fig. 4 of $\delta T/T$ for a beam switching experiment that alternately subtracts two nearby beam positions in the sky from one central position. The experimental upper limit refers to the 4'.5 beam throw of the Uson-Wilkinson[16] experiment. Evidently, values of Ω_0 as low as 0.2 yield excessive anisotropy in the case of a scale-invariant initial fluctuation spectrum.

The inferred limits can be translated into constraints in the $\Omega - H_0$ plane. There is an additional important constraint, namely that we obtain a reasonably long age for the universe. Dating of the oldest globular clusters yields an age of 17(±2) billion years.[19] Even conservatively requiring that the universe be at least 13 billion years old leads (Fig. 5) to an allowed regime in the (Ω_0-H_0) plane which is barely compatible with the inflationary $\Omega_0 = 1$ universe. Models with $\Omega_0 < 0.4$ (0.5/h) can already be ruled out.

Reionization cannot occur sufficiently early to smooth out the fluctuations even though there is small-scale power in the initial fluctuation spectrum. This is because of the flatness of the spectrum on sub-galactic scales. With the adopted normalization, even rare 2σ fluctuations do not go non-linear and collapse prior to a redshift of 30.

One underlying assumption deserves to be questioned. Normalization of the linear theory predictions to the observed variance in galaxy number counts assumes that the underlaying dark matter is well correlated with the galaxy distribution. If Ω is indeed unity, it is very doubtful that this is the case. Failure to observe $\Omega = 1$ on the

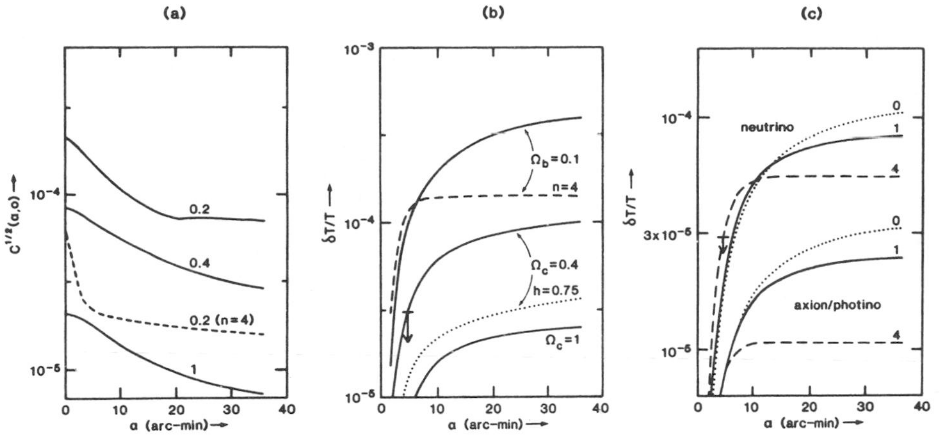

Figure 4: Predictions of δT/T. (a) The angular correlation function measured by an antenna with perfect resolution is shown for cold matter-dominated cosmological models with $\Omega = 0.2$, 0.4 and 1 and a scale-invariant spectrum (n=1). Also shown are predictions for minimal fluctuations (n=4) in the $\Omega = 0.2$ model: all models have Hubble constant h = 0.5. (b) The rms temperature fluctuations measured in a triple beam switching experiment with a beam of 1'.5 are shown for various low density models. Predictions are given for a baryon-dominated universe (Ω_b=0.1, n=1 or 4, h=0.5), and for cold matter-dominated models with n=1, and two values of Ω: 0.4 (h=1 and 0.75) and $\Omega = 1$. The experimental upper limit (95 percent confidence) is also indicated.[16] (c) The rms temperature fluctuations measured in a triple beam switching experiment with a beam of 1'.5 are shown for flat cosmological models dominated either by massive neutrinos or cold matter, with h taken to be 0.5 and either a scale-invariant (n=1) or causal (n=4) primordial spectrum. The experimental upper limit is again indicated. From Vittorio and Silk.[11]

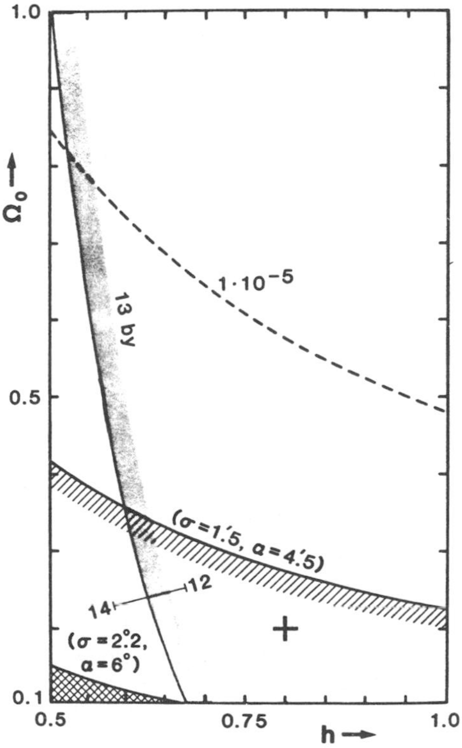

Figure 5: Constraints on Ω_0 and h from δT/T. The upper limits on δT/T at 4'.5 due to Uson and Wilkinson[16] and at 6° due to Melchiorri et al.[17] are expressed as constraint contours in the Ω_0-h plane. The excluded region (below the hatched line) results in excessive δT/T in a cold-particle dominated universe provided the large-scale mass distribution follows that of the galaxies. The dashed line indicates the improvement potentially available were the current upper limits obtained in the Uson and Wilkinson[16] experiment lowered to $\delta T/T < 1 \times 10^{-5}$. The light hatched line shows the constraints on Ω_0 and h if the universe is older than the oldest globular cluster stars, taken to have formed 13 billion years ago. The bar shows the Ω_0-h limit for ages of 12 and 14 billion years (with zero cosmological constant). Also, the cross is inferred from N body simulations, assuming matter and light are correlated on large scales.[20] From Vittorio and Silk.[11]

scale, say, of our Local Supercluster implies that the cold matter is more uniformly distributed than the galaxies over this region. If matter and light indeed are not directly correlated, a possible normalization scheme might involve the dipole anisotropy, as advocated by Abbott and Wise.[21] After all, this is the only measured radiation anisotropy. Part of it probably is due to fluctuations beyond the Virgo cluster that are possibly still in the linear regime. This would clearly introduce considerable uncertainty into our estimates of the fine-scale anisotropy. Inspection of the peculiar velocity predictions of Fig. 2 suggests that, if we normalize by requiring our velocity to be within 2σ of the Hart-Davies measurement, the predicted fine-scale anisotropy should be reduced by a factor of 3 or 4. This may be considered to yield the minimal anisotropy expected in a cold particle dominated universe.

Schemes of biased galaxy formation in a cold-particle-dominated universe have recently been proposed that also require a similar suppression factor for the underlying dark matter correlations[22,23] and enable the inflationary universe ($\Omega_0 = 1$) to be reconciled with the observed universe.[24] Biasing schemes, wherein galaxies form only from the rare 2σ or 3σ correlations which are more correlated than the 1σ correlations, also promise to explain other aspects of the large-scale distribution of galaxies, including the low peculiar motions of galaxies and the luminosity and multiplicity functions.[25]

V. SPECTRAL DISTORTIONS

The cosmic microwave background radiation has a blackbody spectrum, to a precision of better than 20 percent.[26] Tentative evidence for deviations from a thermal spectrum have not been confirmed.[27] This provides us with a powerful tool for probing the early universe, since thermalization occurred at $t < 10^7$ s.[28] This has been of interest in constraining out-of-equilibrium particle decays, such as the gravitino.[29,30] Of relevance for the present discussion, however, are the distortions induced after the decoupling epoch, when galaxy formation has already commenced.

Two distortion mechanisms have been studied in considerable detail. One involves massive pregalactic stars shrouded by intergalactic dust grains[31,32] and plausible grain emissivities were found to yield distortions near the K spectral peak that mimicked those originally reported by Woody and Richards,[26] namely a slight excess at frequencies below that of the peak, and a deficiency in the Wien tail.[33] These models produced distortions of up to 25 percent of the energy density in the cosmic background, but astrophycially plausible models of spherical grains are incapable of thermalizing photons of wavelength longward of about 10 cm. Wright[34] found that highly elongated grains could evade this restriction, and with a modest abundance of heavy elements locked up in the grains, showed that the entire 3 K background could be thermalized at a late epoch.

For such a model to be plausible, one needs to demonstrate that pregalactic objects could have been sufficiently numerous and luminous to meet the energetic requirements. This is where such a scheme becomes even more highly contrived: Bond[35] et al. find that the initial mass function (i.m.f.) of pregalactic massive stars must be carefully adjusted to avoid over-producing heavy elements by nucleosynthesis. The i.m.f. must be abruptly cut-off below a few hundreds of solar masses, yet at the same time, enough heavy elements must be synthesized to produce the thermalizing dust grains. While production of deviations from the 3 K spectrum seems a not unreasonable consequence of pregalactic massive objects, thermalization of the entire microwave background seems most unlikely.

A second mechanism for producing distortions involves Compton scattering of the 3 K photons by a hot intervening medium. Originally discussed by Zel'dovich and Sunyaev,[36] comparisons of the distorted background spectrum express the distortion in terms of the Comptonization parameter y, which describes the energy transfer per scattering, integrated over look-back time, and is defined by

$$y = \int^{t_0} k(T_e - T_r)(m_e c^2)^{-1} n_e \sigma_T c \, dt .$$

Current observations are consistent with the upper limit $y < 0.05$.[37,38] The frequency variation of the distortion, expressed as a brightness temperature relative to the undistorted Rayleigh-Jeans brightness temperature (defined by $T_{RJ} = \lambda^2 I/2k$ at wavelength $\lambda >> 1$ mm), is

$$\Delta T/T_{RJ} = y \left\{ \frac{h\nu/kT}{\tanh(h\nu/2kT)} - 2 \right\} .$$

The characteristic shape of the Comptonization distortion is to enhance the Wien region and suppress the Rayleigh-Jeans region near the peak, since the photon number is conserved by Compton scattering. For a region of uniform hot gas of dimension L, the maximum amplitude of the "cooling" longward of the 3 K peak is

$$\Delta T/T_{RJ} = -4 \frac{kTe}{m_e c^2} \sigma_T n_e L .$$

This effect has been extensively sought along lines of sight towards rich clusters of galaxies, known from x-ray observations to contain intergalactic gas at $T \gtrsim 10^8$ K, and several detections have been tentatively reported.[39]

However the lack of any definite distortions to the 3 K background along arbitrary lines of sight means that one can only use the upper limit on y as a constraint on possible thermal histories of the intergalactic medium. This leaves one with a large degree of latitude, even when other constraints are incorporated. These include the diffuse x-ray and optical backgrounds, and the lack of distortions to the cosmic microwave background in the long wavelength Rayleigh-Jeans spectral region. Here the constraint is not as good as at higher frequencies, because galactic non-thermal emission dominates the background longward of 100 cm wavelength. An important contributor to low frequency distortion would be free-free emission from a relatively cool, ionized intergalactic medium at $T \sim 10^4$ K, if the mean density is high enough. This can occur if the medium is reionized sufficiently early.

Some examples of the distortions that are possible are shown in Fig. 6. If we adopt an IGM corresponding to a value of Ω_b in the range 0.01 to 0.1, as is inferred from the requirements of primordial nucleosynthesis, then interesting distortions only arise, due to the uniform IGM, at a redshift of 10 or larger. The heating requirements are modest, if expressed as a fraction of the mass density of the universe, but of course require some ad hoc hypothesis invoking rare primordial fluctuations to supplement any adiabatic fluctuation scenario in which first non-linearity occurs at a late epoch. Distortions of the

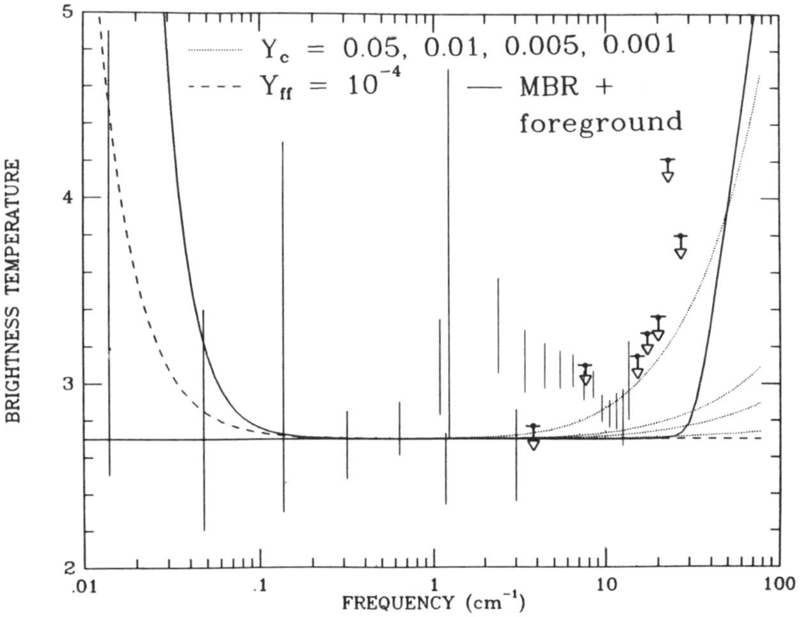

Figure 6: Comparison of possible distortions with the observations of the brightness temperature of the diffuse MBR. The vertical lines give the ± 1 σ limits for various experiments as quoted in Weiss.[40] The upper limits give the +1 σ values where no lower limits are quoted. Included also are the upper limits of Meyer and Jura.[41] The dashed line shows the spectrum with the maximum acceptable free-free distortion as defined by the parameter y (see text) ≡ Y_c. The parameter Y_{ff} measures the amount of low frequency distortion and can be written.[29]

$$Y_{ff} = 10^{-6}\ \Omega_b^2 h^3 \int \chi^2\ (T_r/T_e)^{3/2} (T_e/T_r - 1)\ dz$$

where χ is the fractional ionization and T_e is the electron temperature at redshift z. The solid line gives a model of the distorted MBR including a contribution from high latitude galactic foreground emission as estimated by Weiss.[40] Note that the two lowest frequency radio measurements have had a modeled galactic foreground subtracted out.[42] From Stebbins and Silk.[29]

3 K spectrum lie within acceptable bounds even if the universe never recombined at $z \sim 1000$.[29] Isothermal fluctuation scenarios allow the possibility of gravitational collapse as soon as the matter fluctuations are sufficiently neutral to overcome Compton drag. Continuing ionization and heating could be self-limiting, as once the universe is reionized, Compton drag inhibits further collapse until $z \lesssim 100$.[43]

It may be depressing to some that unless we accept certain precepts, such as that of a scale-invariant primordial fluctuation spectrum, as matters of faith, it becomes exceedingly difficult to constrain the early universe between the epoch $z = 10$ and $z = 1000$. This is, after all, the pregalactic era, the time when gravitational stability reigns relatively freely, and when the initial conditions that determine the characteristic features of the galaxies around us are acquired. There is little hope of ever finding a discrete source at such high redshift: it is only by studying the microwave background radiation that we may hope to learn more about this cosmological desert. Observers are to be encouraged to strive even harder to search for deviations from a blackbody spectrum and to discover angular anisotropies over all angular scales. Only when such blemishes on the cosmic backdrop are found will we have full confidence in our theory of the evolution of structure in the context of the Big Bang theory.

ACKNOWLEDGMENTS

I am indebted to my collaborator Nicola Vittorio for many discussions of the microwave background anisotropy and to my student Albert Stebbins for his insights into the spectral distortions of the background radiation. I have also had helpful discussions with several other colleagues during the Santa Barbara Workshop on The Large-Scale Structure of the Universe, including J.R. Bond, M. Davis, C. Frenk, N. Kaiser, and S.D.M. White.

REFERENCES

[1] G. Blumenthal, S. Faber, J. Primack and M. Rees, Nature **311**, 517 (1984).
[2] S. Yang, M.S. Turner, D.N. Schramm and K.A. Olive, Ap. J. **281**, 493 (1984).
[3] J.R. Bond and A.S. Szalay, Ap. J. **276**, 443 (1983).
[4] P.J.E. Peebles, Ap. J. **263**, L1 (1982).
[5] J. Silk, 14th GIFT Seminar "Relativistic Astrophysics and Cosmology," ed. X. Fustero, E. Verdaguer (World Scientific: Singapore), p. 249 (1984).
[6] M. Davis and P.J.E. Peebles, Ann. Rev. Astr. Ap. **21**, 109 (1983).
[7] A. Sandage and G. Tammann, Proc. First ESO-CERN Conference on Elementary Particles and Large-Scale Structure of the Universe, (in press), (1984).
[8] L. Hart and R.D. Davies, Nature **297**, 191 (1982).
[9] N. Kaiser, Ap. J. Letters **273**, L17 (1983).
[10] M. Davis and P.J.E. Peebles, Ap. J. **267**, 465 (1983).
[11] N. Vittorio and J. Silk, Ap. J. Letters (in press), (1984).
[12] R.K. Sachs and A. Wolfe, Ap. J. **147**, 73 (1967).
[13] J.M. Bardeen, Phys. Rev. **D22**, 1882 (1980).
[14] M.W. Wilson and J. Silk, Ap. J. **243**, 14 (1981).

[15]M. Wilson, Ap. J. **273**, 2, (1983).
[16]J.M. Uson and D.T. Wilkinson, these proceedings, (1984).
[17]F. Melchiorri, B. Melchiorri, C. Ceccarelli and L. Pietranera, Ap. J. **250**, L1 (1981).
[18]J.R. Bond and G. Efstathiou, Ap. J. Letters (in press), (1984).
[19]I. Iben and A. Renzini, Reports on Progress in Physics (in press), (1984).
[20]S. White, these proceedings, (1984).
[21]L. Abbott and M. Wise, Ap. J. Letters **282**, L47 (1984).
[22]J.M. Bardeen, these proceedings, (1984).
[23]N. Kaiser, these proceedings, (1984).
[24]M. Davis, G. Efstathiou, C.S. Frenk and S.D.M. White, Ap. J. (submitted), (1984).
[25]R. Schaeffer and J. Silk, preprint, (1984).
[26] D.L. Woody and P.L. Richards, Ap. J. **248**, 18 (1981).
[27]P. Richards, these proceedings, (1984).
[28]L. Danese and G. De Zotti, Astr. Ap. **107**, 39 (1982).
[29]A. Stebbins and J. Silk, Ap. J. (submitted), (1984).
[30]J. Ellis, J.E. Kim and D.V. Nanopoulos, CERN preprint TH 38439, (1984).
[31]M. Rowan-Robinson, J. Negroponte and J. Silk, Nature **281**, 635 (1979).
[32]J. Puget and J. Heyvaerts, Astr. and Ap. **83**, L10 (1980).
[33]J. Negroponte, M. Rowan-Robinson and J. Silk, Ap. J. **248**, 38 (1981).
[34]E.L. Wright, Ap. J. **255**, 401 (1982).
[35]J.R. Bond, D. Arnett and B. Carr, in press, (1984).
[36]Ya.B. Zel'dovich and R.A. Sunyaev, Astrophys. and Sp. Sci. **4**, 301 (1969).
[37]A.F. Illarionov and R.A. Sunyaev, Astron. Zh. **51**, 698, 1162 (Sov. Astron. **18**, 413 691 [1975]).
[38]G.B. Field and S.C. Perrenod, Ap. J. **215**, 717 (1977).
[39]M. Birkinshaw and S.F. Gull, Mon. Not. R. Astron. Soc. **206**, 359 (1984).
[40]R. Weiss, Ann. Revs. Astr. Ap. **18**, 489 (1980).
[41]D. Meyer and M. Jura, Ap. J. Letters **276**, L1 (1984).
[42]T. Howell and J. Shakeshaft, Nature **216**, 753 (1967).
[43]C.J. Hogan, Mon. Not. R. Astron. Soc. **188**, 781 (1979).

Constraints on Dark Matter From Cosmic Background Anisotropies

J. R. Bond and G. Efstathiou

ABSTRACT

We discuss the major stages in the linear evolution of the statistical ensemble of adiabatic fluctuations in radiation, baryons and dark matter. If we assume that the distribution of light emitters, i.e. of galaxies, follows the distribution of mass, i.e. of dark matter, then universes dominated by massive collisionless relics of the Big Bang must have $\Omega > 0.2h^{-4/3}$ to avoid exceeding the current observational limits on small-scale anisotropies in the microwave background. However, values of $\Omega \sim 0.2$ are indicated by dynamical studies of galaxy clustering. We conclude that universes dominated by cold dark matter in which light traces mass are probably not viable models.

I. INTRODUCTION

Most cosmological observations of our past light cone tell us about nonlinear structure emitting, absorbing or scattering radiation at relatively small redshifts. An important exception is observation of anisotropies in the cosmic background radiation, which gives information about the universe when density fluctuations were linear. We have calculated the linear evolution of the coupled equations for metric perturbations and for the transport of photons (including polarization), of baryons, of massless neutrinos and of dark matter, whether hot (e.g. massive neutrinos) or cold (e.g. photinos, gravitinos, axions, right-handed neutrinos, monopoles). We find that anisotropies on scales smaller than the size of the horizon at the epoch of photon decoupling, which subtends an angle $\sim 2°$, can be used to constrain models of the universe with $\Omega < 1$ dominated by dark matter. To probe a given small angular scale θ in Friedmann-Robertson-Walker geometry is equivalent to probing a fixed comoving separation along the past light cone, $d \sim 100(\Omega h)^{-1}(\theta/1°)$Mpc, providing the redshift exceeds ~ 10. Thus, typical small-angle anisotropy experiments provide information on the sources of radiation fluctuations for a given length scale; i.e., on the electron density and velocity fields, which generate radiation fluctuations via anisotropies in Thomson scattering, and also on the gravitational metric perturbations (Sachs-Wolfe effect). The new upper limit on the anisotropy at 4.5' reported by Uson and Wilkinson in these proceedings can be used to strongly constrain the amplitude of density fluctuations around recombination, and hence on the level to which they can have grown by the present epoch. Details of these results are reported in Bond and Efstathiou[1] (BE). Here, we focus on the physical issues involved in such calculations.

II. INITIAL CONDITIONS FOR THE PERTURBATION ENSEMBLE

Processes occurring in the early universe are assumed to determine the initial conditions. We adopt adiabatic scalar fluctuations characterized by a gauge invariant gravitational potential perturbation[2], $\phi_H(x,\tau)$, where x is comoving space and τ is conformal time. ϕ_H is taken to be a random variable distributed as a Gaussian random process. When ϕ_H is expanded in spatial eigenmodes (exp(ik.x) for Einstein-deSitter Universes) labelled by a comoving wavenumber k and characterized by a complex amplitude $\phi_H(k,\tau)$, this is equivalent to all k-modes being statistically independent, each distributed with random phases for $\theta_k = \arg(\phi_H(k))$, and with a Rayleigh distribution in $|\phi_H(k)|$:

$$P(|\phi_H(k)|,\theta_k)d|\phi_H(k)|d\theta_k$$
$$\sim \exp[-|\phi_H(k)|^2/(2<|\phi_H(k)|^2>)]\ |\phi_H(k)|\ d|\phi_H(k)|\ d\theta_k/2\pi\ . \qquad (1)$$

The ensemble average is assumed to be of the Harrison-Zeldovich constant curvature spectral form:

$$k^3<|\phi_H(k)|^2>\ dk/k \sim dk/k. \qquad (2)$$

At early times, when waves of a given wavenumber are outside of the horizon ($k\tau << 1$), the density fluctuations for all of the species of particles present are proportional to $\phi_H(k,\tau)/k^2$ and each of the density fields is statistically distributed according to Eq. (1). The joint distribution of all of the density fields and the metric consists of Eq. (1) multiplied by a series of delta functions expressing the relation of each density to $\phi_H(k)$, the one independent random variable. Fluctuation generation as a result of quantum processes arising during inflation apparently predicts such a form. On the other hand, our universe is only one ensemble member. Nonetheless, if a spatial ergodic theorem is assumed to hold, whereby averages over spatial volumes are equated to ensemble averages, Eq. (1) will describe the initial probability density.

III. LINEAR EVOLUTION OF THE ENSEMBLE

In the linear regime of evolution, Eq. (1) continues to be valid, though the ensemble averages evolve. In particular, if the linear evolution equation for the mode amplitude is schematically $\mathbf{L}\Delta(k,\tau) = 0$, where $\mathbf{L}$ is a second order ordinary differential matrix operator and $\mathbf{\Delta}$ is a vector which includes all of the various perturbed components of densities, velocities, distribution functions and metric potentials, then the equation for P is simply $(\mathbf{L}\Delta(k,\tau))P(\mathbf{\Delta}(k),\tau) = 0$. The solution, $P(\mathbf{\Delta}(k,\tau),\tau) = P(\mathbf{\Delta}(k,0),0)$, preserves the Gaussian form, Eq. (1), though the variances evolve. Hence, we need only solve for each k-mode once, irrespective of phase, which factors out; i.e., the phase is conserved during linear evolution, apart from sign changes due, for example, to sound wave oscillations. In the nonlinear regime of evolution, mode-mode coupling complicates the development of P. In N-body studies, this regime is attacked by evolving enough realizations from the ensemble until averages converge, where the averages are a combination of limited spatial and restricted ensemble expectations. We are fortunate that such procedures are unnecessary for the CBR fluctuation problem, for the extra complications of hydrodynamic and radiative transfer in nonlinear random media would make such studies extremely complex.

IV. MODE DEVELOPMENT IN THE LINEAR REGIME

We choose a specific gauge, the synchronous one[2], to solve for the k-mode evolution of the gravitational field and the several species of matter present. In Fig. 1, we show how the fractional overdensities evolve through various regimes: τ^2 growth occurs outside of the horizon; oscillations in the tightly coupled photon-baryon gas occur once inside the horizon, but before recombination; cold dark matter fluctuations always continue growing due to the Jeans instability, but the growth is suppressed due to dominance of the energy density by relativistic particles prior to $a_{eq} = 4 \times 10^{-5}$ (Meszaros effect), where $a = 1$ now; Silk damping occurs due to the viscous coupling of photons and baryons, and is most pronounced in the intermediate regime between optically thick and thin; the freeing of the baryons after recombination allows their perturbations to catch up to the level of the cold dark matter fluctuations; finally, the Sachs-Wolfe effect, due to the gravitational redshift, is evident in the offset of δ_γ from zero at late times. For small wavenumbers ($k < k_{Heq} = \Omega h^2(5.2\text{Mpc})^{-1}$), there is no Meszaros effect. Massive neutrino fluctuations suffer strong damping for $k > k_{\nu m} = \Omega_\nu h^2(2.0\text{Mpc})^{-1}$ due to phase mixing; and massless neutrino fluctuations quickly decay in amplitude once they enter the horizon. (See Bond and Szalay[3].) When many such evolutionary histories for different k's are added together, we get fluctuation spectra for dark matter, baryons and radiation.

FIGURE 1

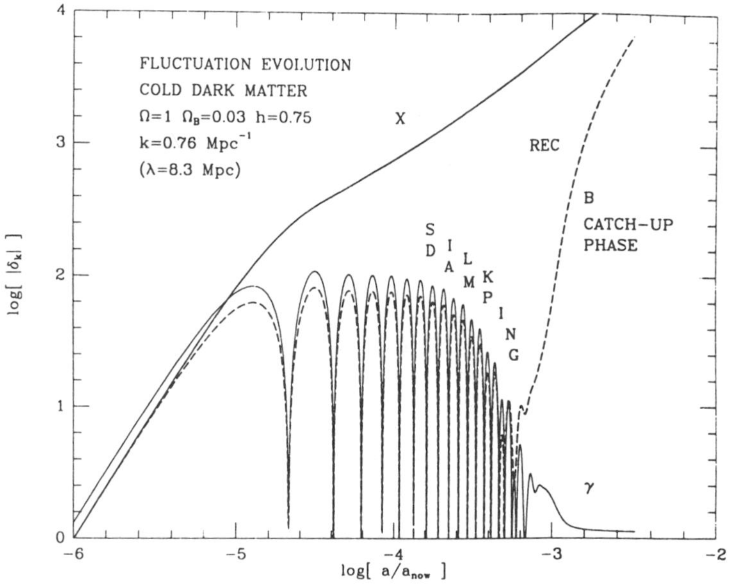

V. TRANSFER FUNCTIONS AND FLUCTUATION SPECTRA

The growth of cold dark matter (X) fluctuations prior to recombination followed by the baryon catch-up phase implies that Silk damping leaves little imprint on the final transfer functions for X's and for baryons if $\Omega_B \ll \Omega$. In this case, the transfer function - final fluctuation amplitude over initial, normalized to one at $k = 0$ (see Fig. 2) - for the baryons equals that for the X's provided k is smaller than the Jeans wavenumber of the gas just after recombination, $k_{BJ} = (\Omega h^2)^{1/2}(1.1\ \mathrm{kpc})^{-1}$; and T_k is a universal function of k/k_{Heq}, as is illustrated in Fig. 2. As Ω_B approaches Ω, this form ceases to hold, and small oscillations superimposed upon the net decline appear in T_k for both X's and the gas: Silk damping does leave an imprint in the $\Omega_B = \Omega_X = 0.1$ case. If $\Omega_B = \Omega$, T_k oscillates about zero, and Silk damping is severe for wavenumbers in excess of $k_S \sim 3(\Omega h^2)^{5/6}\ \mathrm{Mpc}^{-1}$.

FIGURE 2

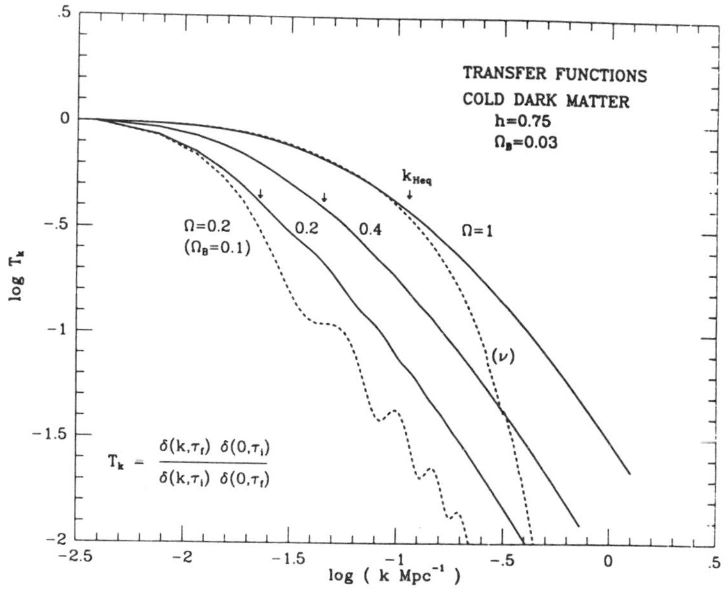

The catch up phase does leave an imprint in the final CBR fluctuations however, arising primarily from Thomson scattering off the rapidly moving electrons as they fall into the X-particle potential wells. The fractional temperature difference of CBR photons arriving from two directions separated by angle θ is predicted to be Gaussian-distributed for our ensemble, with variance $2^{1/2}(C(0) - C(\theta))$, where $C(\theta)$ is the correlation function for temperature fluctuations at the present epoch τ_0:

$$C(\theta,\tau_0) = \langle\Delta_T(\hat{q},x,\tau_0)\Delta_T(\hat{q}',x,\tau_0)\rangle/16$$
$$= \int \frac{Vk^2dk}{2\pi^2}\ \frac{1}{16}\sum_\ell (2\ell+1)\,|\Delta_{T\ell}(k,\tau_0)|^2 P_\ell(\theta)\ . \qquad (3)$$

Here, $\cos\theta = \hat{q}\cdot\hat{q}'$ defines the angle θ between the two photon momenta directions, $\hat{q}$ and $\hat{q}'$, Δ_T is the relative radiation intensity fluctuation, and $\Delta_{T\ell}$ are its moments with respect to Legendre polynomials in the angular variable $\hat{q}\cdot\hat{k}$. We find $C(\theta)$ is well fit by the formula $C(\theta)/C(0) = (1+(\theta/\theta_c)^2/(2\beta))^{-\beta}$, where β is a parameter ~ 1, and the coherence angle, $\theta_c = (-C''(0)/C(0))^{-1/2}$, ranges from ~4' to ~10', corresponding to the comoving distances d_c ~ 20 - 60 Mpc, the former appropriate to $\Omega = 1$, $h = 0.75$ and the latter to $\Omega = 0.2$, $h = 0.5$. This range coincides with the comoving distance over which the universe passes from 90% ionized to 99% recombined; e.g., for the model of Fig. 1 this transition is shown by the position and width of 'REC'.

The coherence angle θ_c is a damping angle for the radiation, arising as a result of interference of positive and negative contributions to the Thompson scattering source terms for those fluctuations whose wavelengths are smaller than the fuzziness of the last scattering surface. This is evident in Fig. 3, where we plot the radiation spectra, $C''(0,k) = dC''(\theta=0)/d\ln k$, which shows the contribution of different wavevectors to the small angle anisotropy integral $C(0) - C(\theta)$. If we let $k_c = 2\pi/d_c$, then for the $\Omega = 1$ and $\Omega = 0.2$ models we get $k_c = 0.3$ and 0.1 Mpc^{-1} respectively. These are just the wavenumbers at which the dramatic damping onsets. The shapes of the gas density and velocity spectra are compared with the $\Omega = 1$ radiation fluctuation spectrum; these illustrate that the velocities gained during catch-up before recombination is complete play a significant role in determining $\Delta T/T$ on small scales. The spectrum appropriate to measurements on scales >> θ_c, $C(0,k)$ for the total intensity and $C_P(0,k)$ for the net linear polarization, are shown for the $\Omega = 0.2$ model; these two spectra are the only ones properly normalized in Fig. 3, though the relative spacing of the C'' curves does indicate the normalization difference between the high and low Ω models.

FIGURE 3

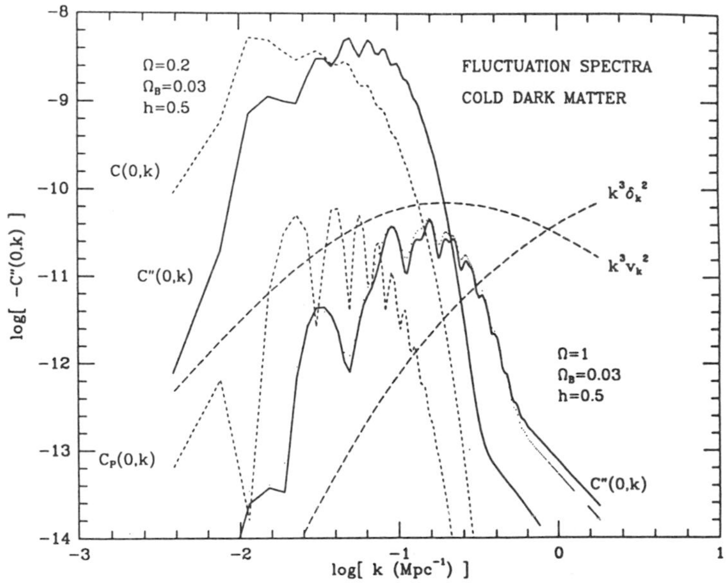

VI. NORMALIZATION OF THE AMPLITUDE

The overall amplitude of the density fluctuations is usually fixed by normalizing to the statistical distribution of galaxies in some way. This one number then fixes the amplitude of the predicted radiation anisotropy. We equate $<(\Delta M/M)(x_o)\delta(x=0)>$ determined from our transfer functions, where $\Delta M/M$ is the volume-averaged overdensity in a sphere centered at $x = 0$ of radius x_o, to observational estimates from the CfA redshift survey[4] to obtain this normalization. (Note that this quantity is just the volume average of the density-density correlation function, ξ; i.e., $3J_3(x_o)/x_o^3$.) Peebles and Groth[5] have shown that $<(\Delta M/M)(x_o)\delta(x=0)>$ grows as in linear theory provided $\xi(x_o) << 1$, even if there is nonlinearity on smaller scales within the sphere. Accordingly, we can relate our linear calculations to the observations even though much of the universe may be nonlinear. We choose $x_o = 10h^{-1}$Mpc, where $<(\Delta M/M)\delta(0)> = 0.27$, so as to be as close to linearity as possible subject to data constraints. The normalization amplitude is relatively insensitive to variations in x_o over the range 7.5 - $20h^{-1}$Mpc, especially for low Ω models.

Forcing the equality of $<(\Delta M/M)\delta(0)>$ for the mass to that determined for galaxies is the 'light tracing mass' assumption. If the gas collapses with the dark matter on galactic scales, and galaxies always form in such places, then this assumption would be valid. Since it is difficult for the gas to heat enough ($> 10^6$K) to avoid collapse on galactic scales, the light tracing mass hypothesis might be valid in cold dark matter models, at least in the simplest versions. In neutrino dominated universes, it seems unlikely that light traces mass[6]; instead, we normalize by the nonlinear redshift, defined by $\xi(0,z_{nl}) = 1$, when ~ 10% of the gas would have shocked in pancake collapses.

VII. COMPARISON WITH OBSERVATION AND CONSTRAINTS

Uson and Wilkinson[7,8] (UW) have looked at 12 uncorrelated places in the sky, employing a 'second derivative' procedure in each field, beam switching through an angle $\theta = 4.5'$ plus wagging the beams by $4.5'$. For such an experimental setup, our theoretical prediction is that the net observed CBR temperature anisotropies would be Gaussian-distributed with variance $2(C_F(0) - C_F(\theta)) - (C_F(0) - C_F(2\theta))/2$. (Due to the finite beam pattern F of radio telescopes, the radiation correlation function must be appropriately convolved with F. See BE). UW give a 95% confidence upper limit of $(2.9 \times 10^{-5})^2$ for this variance.

In ν-dominated models, $\Omega = 1$ is preferred over lower values[6]. Nonetheless, the $\Delta T/T$ constraints can be interesting. For example, we require $z_{nl} < 3.5$ for $h = 0.5$ and $\Omega_B = 0.03$ in order to satisfy the UW limit.

For the cold particle models of Fig. 3, we predict dispersions of 0.28×10^{-5} and 7.2×10^{-5} for the $\Omega = 1$ and 0.2 universes respectively. The former is safely under the UW limit, the latter greatly exceeds it. More generally, we find that cold dark matter universes with $\Omega \leq 0.2h^{-4/3}$ violate the limit if $\Omega_B = 0.03$. In BE, we show that it is very unlikely that reionization of the universe can occur early enough to damp the small scale anisotropy. With the assumption that light traces mass, observations of the peculiar velocities between galaxy pairs imply[4,9] $\Omega = 0.14 \times 2^{\pm 1}$. Also, the N-body studies[10] discussed by White in this volume show that, with the light tracing mass assumption, an Ω = 0.2 cold dark matter model with h ~ 0.5 - 0.8 can match many of the features of the observed clustering pattern, whereas models with Ω = 1 cannot. Thus, though cold dark matter models with low values of Ω may seem promising, they are ruled out by CBR limits.

To avoid this conclusion, we must assume that mass is less strongly clustered than galaxies, thereby lowering the normalization amplitude and the predicted anisotropy. Not only would the low Ω models then be viable, but it may be possible to reconcile the Ω = 1 models with the observed clustering properties and peculiar velocities, as discussed by Bardeen, Kaiser and White in this volume.

We thank A. Lasenby, B. Partridge, J. Uson and D. Wilkinson for discussions regarding their various CBR experiments. We also acknowledge stimulating discussions with A. S. Szalay. N. Vittorio and J. Silk[11] have done calculations similar to ours and come to similar conclusions. This work was supported in part by the UK SERC and by the NSF under Grant No. PH77-27084, supplemented by funds from NASA.

REFERENCES

[1] J. R. Bond, and G. Efstathiou, Astrophys. J. (Lett.), in press (1984).
[2] J. M. Bardeen, Phys. Rev. D **22**, 1882 (1980).
[3] J. R. Bond, and A. S. Szalay, Astrophys. J. **227**, 443 (1983).
[4] M. Davis, and P. J. E. Peebles, Astrophys. J. **267**, 465 (1983).
[5] P. J. E. Peebles, and E. J. Groth, Astr. & Astrophys. **53**, 131 (1976).
[6] J. R. Bond, J. Centrella, A. S. Szalay, and J. Wilson, Mon. Not. R. Astron. Soc., **210**, 515 (1984).
[7] J. M. Uson, and D. T. Wilkinson, Astrophys. J. (Lett.) **227**, L1 (1984).
[8] J. M. Uson, and D. T. Wilkinson, Proceedings of the Inner Space/Outer Space Workshop, Fermilab (1984).
[9] A. J. Bean, G. Efstathiou, R. S. Ellis, B. A. Peterson, and T. Shanks, Mon. Not. R. Astron. Soc., **205**, 605 (1983).
[10] M. Davis, G. Efstathiou, C. S. Frenk, and S. D. M. White, preprint (1984).
[11] N. Vittorio, and J. Silk, preprint (1984); also J. Silk, this proceedings.

Inflation and the Isotropy of the Microwave Background

L. F. Abbott

ABSTRACT

I discuss the constraints imposed on inflationary cosmologies by the isotropy of the microwave background radiation.

By solving the horizon problem, inflationary cosmology[1] has made it possible to understand the large-scale homogeneity and isotropy of the universe. However, even with inflation, it is non-trivial to account for the extreme isotropy observed in the microwave background radiation. This is because quantum fluctuations appearing during the inflationary period expand into long-wavelength perturbations which cause microwave anisotropies. In this talk I discuss the constraints imposed on inflationary models by present limits on the quadrupole anisotropy and by measurements of the dipole anisotropy. This work was done in collaboration with Mark Wise.

At first, I would like to be as general as possible and consider inflation in a model-independent way.[2] To be inflationary a cosmology must have an epoch during which two things happen. First, the ratio of the matter density term $(8\pi G/3)\rho$ to the curvature term $1/R^2$ in the Friedmann equation

$$\left(\frac{\dot{R}}{R}\right)^2 = \frac{8\pi G}{3}\rho \pm \frac{1}{R^2} \tag{1}$$

(where R is the Robertson-Walker scale factor) must become very large. This solves the flatness problem. From Eq. (1) we see immediately that this implies that $\dot{R}$ must increase with time or equivalently $\ddot{R} > 0$ during inflation. Second, to solve the horizon problem the ratio of the distance between fixed coordinate points, which grows like R, to the horizon size, which is proportional to $R/\dot{R}$, must increase with time during inflation. This ratio goes like $R(\dot{R}/R) = \dot{R}$ so this condition is also met if $\ddot{R} > 0$.

That fact that fixed coordinate lengths get pushed outside the horizon in any inflationary cosmology has an important consequence. Short-wavelength quantum fluctuations will get red-shifted out of the horizon during inflation. There they remain until much later, during the radiation or matter dominated eras, they re-enter the horizon as long-wavelength perturbations. These perturbations introduce anisotropy into the microwave background through the Sachs-Wolfe effect.[3] The procedure for calculating the anisotropy induced in this manner is as follows. First, we compute the quantum-mechanical two-point function in the inflating universe and relate this to a statistical ensemble of classical field fluctuations.[4] We then let these fluctuations evolve with time and evaluate the resulting microwave anisotropy. In this talk I will present the results of our analyses without going into computational details.[5]

To keep things as general as possible for the time being, I will only assume that the inflationary model contains gravity as one of its interactions. In this case the fluctuations of interest are gravitational waves. The large-scale anisotropy induced by gravitational waves is dominated by physical wavenumbers of order the present Hubble constant, H_0. We have shown[2] that the amplitude of these waves in any generalized inflationary cosmology is of order $H_{H.C.}/m_{PL}$ where $H_{H.C.}$ is the value of the Hubble constant at the time when a wave of present physical wavenumber H_0 crossed the horizon during the inflationary era. These waves produce a $\delta T/T$ of order their amplitude, $H_{H.C.}/m_{PL}$. Since observations indicate a large-scale isotropy of less than 10^{-4} we find

$$H_{H.C.} < 10^{-4}\ m_{PL} \quad . \tag{2}$$

This result applies to any model and it indicates that at least some part of the inflation must occur at scales well below the Planck mass where quantum gravitational effects are small.

If we know R as a function of time during the inflationary period we can be more precise and use the present limit on the quadrupole anisotropy along with energy conservation to constrain the reheating temperature after inflation.[2] The results for $R \sim t^p$ with $p > 1$ are shown in Fig. 1. Note that if we believe that baryons can only be produced at temperatures above 10^{15} GeV, for example, Fig. 1 implies $p > 6$. For large p, power-law inflation becomes essentially equivalent to exponential inflation so the asymptotical limit of the curve in Fig. 1 gives the bound[6]

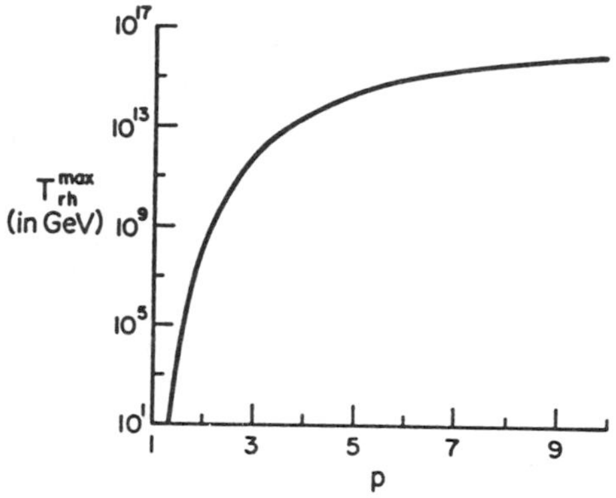

The maximum reheating temperature as a function of p for power law inflation.

$$T_{reheat} < 10^{17}\ \text{GeV} \tag{3}$$

for standard exponential inflation.

I will now consider a specific inflationary scenario, the so-called new inflationary cosmology.[7] In this model there is an additional source of perturbations, the scalar field responsible for inducing the exponential inflation. Fluctuations in this field result in a scale-invariant spectrum of energy-density perturbations.[4] Perturbations of coordinate wavenumber k can be characterized by Bardeen's gauge-invariant variable[8]

$$\varepsilon = \frac{\delta\rho}{\rho} + 3\left(\frac{\rho+p}{\rho}\right)\left(\frac{\dot{R}}{k}\right)(v-B) \tag{4}$$

where ρ and p are the unperturbed energy density and pressure and δρ, v and B are the amplitudes of fluctuations in the energy density, velocity and off-diagonal metric respectively. For wavelengths which enter the horizon during the matter-dominated era we define ε_H as the value of ε when $(\dot{R}/k) = 1$.

From the inflationary prediction for ε we can compute the induced anisotropy in the microwave background. If we expand the microwave background temperature in multipoles,

$$\frac{\delta T_0}{T_0} = \sum_{\ell m} a_{\ell m} Y_{\ell m} \tag{5}$$

and define

$$a_\ell^2 = \sum_{m=-\ell}^{\ell} |a_{\ell m}|^2 , \tag{6}$$

then the result for the expectation value of a_ℓ^2 is[9]

$$<a_\ell^2> = \frac{2\pi^2 \varepsilon_H^2 \ (2\ell+1)}{\ell \ (\ell+1)} \tag{7}$$

Also we can show that at 90% confidence level we expect that a_2^2 will be greater than $0.3 < a_2^2 >$. Comparing this with the present limit[11]

$$a_2^2 < 5.2 \times 10^{-8}$$

gives the 90% confidence level bound

$$\varepsilon_H < 1.0 \times 10^{-4} \ .$$

A more stringent bound on ε_H can be obtained from the observed dipole anisotropy. Wavelengths with $k < k_{max}$, which entered the horizon during the matter-dominated era, give a contribution to the dipole moment

$$< a_1^2 > = 6\pi^2 \ \varepsilon_H^2 \ k_{max}^2 \ t_0^2 \ . \tag{8}$$

Unlike the higher multipoles, the dipole moment is sensitive to short-wavelength fluctuations. This is because the dipole moment is dominated by the peculiar velocity of the observer. However, the observed dipole moment can still be used to obtain a bound on ε_H without having to consider modes which have gone nonlinear. If we assume that the short-wavelength modes that have gone nonlinear are uncorrelated with the long-wavelength modes that are evolving linearly then the contributions of both of these sets of modes to $<a_1^2>$ is positive. The full probability distribution for the dipole moment is a convolution of probability distributions for the linear and nonlinear contributions. With a Gaussian distribution for the linear modes it is straightforward to show that the nonlinear modes can only make the full probability distribution broader than that of the linear modes alone. We thus obtain a bound on ε_H by demanding that the computable long-wavelength contribution to the dipole anisotropy is no larger than the observed dipole anisotropy[11]

$$a_1^2 = 5.5 \times 10^{-6} \ . \tag{9}$$

If we cutoff the linear analysis at a wavelength of 60 Mpc for a Hubble constant of 100 km/sec Mpc we obtain the 90% confidence level bound

$$\varepsilon_H < 3.4 \times 10^{-6} . \tag{10}$$

In keeping with the prediction of inflation that the total energy density is critical, we have assumed for the above result that the universe is presently dominated by some uncoupled form of dark matter. The bound of Eq. (10) is valid for any form of cold dark matter. For neutrinos (hot dark matter) our bound is modified somewhat due to free streaming. Using the results of Bond and Szalay[12] we find that

$$\varepsilon_H < 7 \times 10^{-6} \tag{11}$$

for neutrinos using a 60 Mpc cutoff. Eqs. (10) and (11) impose a severe constraint on inflationary models.

We have seen that present limits on the quadrupoole anisotropy of the microwave background radiation constrain the scale of any sort of inflation to lie well below the Planck mass and severely restrict the reheating temperature for models with power-law inflation. In standard inflationary models the small value of the observed dipole anisotropy requires that the amplitude of energy-density perturbations be constrained to quite small values.

REFERENCES

[1] A. Guth, Phys. Rev. **D23**, 347 (1981).

[2] L. Abbott and M.Wise, Nucl. Phys. **B244**, 541 (1984).

[3] R. Sachs and A. Wolfe, Astrophys. J. **147**, 73 (1967).

[4] A. Guth and S.-Y. Pi, Phys. Rev. Lett. **49**, 1110 (1982); J. Bardeen, P. Steinhardt, and M. Turner, Phys. Rev. D **28**, 679 (1983); A. Starobinskii, Phys. Lett. **117B**, 175 (1982); S. Hawking, Phys. Lett. **115B**, 295 (1982).

[5] For details see L. Abbott and M. Wise, Ref. 9 and 10, or L. Abbott, R. Schaefer and M. Wise, in preparation.

[6] A. Starobinskii, JETP Lett. **30**, 683 (1979); V. Rubakov, M. Sazhin and A. Veryaskin, Phys. Lett. **115B**, 189 (1982); R. Fabbri and M. Pollock, Phys. Lett. **125B**, 445 (1983).

[7] A. Linde, Phys. Lett. **108B**, 389 (1982); A. Albrecht and P. Steinhardt, Phys. Rev. Lett. **48**, 1220 (1982).

[8] J. Bardeen, Phys. Rev. **D22**, 1882 (1980).

[9] P. J. E. Peebles, Astrophys. J. **263**, L1 (1982); L. Abbott and M. Wise, Phys. Lett. **135B**, 279 (1984).

[10] L. Abbott and M. Wise, Astrophys. J. **282**, L47 (1984); see also N. Kaiser, Astrophys. J. **273**, L17 (1983); J. Silk and M. Wilson, Astrophys. J. **243**, 14 and **244**, L37 (1981).

[11] P. Lubin, G. Epstein and G. Smoot, Phys. Rev. Lett. **50**, 616 (1983); D. Fixen, E. Cheng and D. Wilkinson, Phys. Rev. Lett. **50**, 620 (1983).

[12] J. Bond and S. Szalay, Astrophys. J. **274**, 443 (1983).

Integral Constraints on Stress Energy Perturbations in General Relativity: Applications to the Sachs-Wolfe Effect[1]

J. Traschen

In special relativity, energy and momentum conservation impose constraints on the density perturbations which are allowed. For example, if a perturbation $\delta\rho$ is localized in space, then Gauss' law tells us that the monopole and dipole moments of $\delta\rho$ vanish. In general relativity, on the other hand, energy and momentum are not in general conserved. However, we have found that in some (general relativistic) spacetimes, there is a generalized Gauss' law for stress energy perturbations δT^{μ}_{ν}. In particular, in the Robertson-Walker universes these constraints must be satisfied.

For example, in the flat Robertson Walker Universe suppose there is a perturbation which is localized in space and $\delta T^{o}_{k} = 0$. Then the general relativistic constraints reduce to the special relativistic statements that the monopole and dipole moments of $\delta\rho$ vanish. Any perturbation which is created by causal processes with local initial conditions will remain local. In special and general relativity, locality of $\delta\rho$ implies that the boundary term in Gauss' Law is zero.

Next, we apply the constraints to the Sachs-Wolfe effect. Sachs and Wolfe[2] computed the contribution to the anisotropy in the microwave background due to perturbations of the photon path. If $A(x)$ is the gravitational potential for $(\delta\rho/\rho)$ in a Poisson equation, then $(\delta T/T) = A(x_E)$, where x_E is the emission event. Suppose $\delta\rho$ is a local perturbation. Then A can be solved for in a multipole expansion. The constraints tell us that the monopole and dipole moments vanish, meaning that the magnitude of $(\delta T/T)$ is less than what one would find forgetting the constraints. Finally consider a scenario with a bunch of uncorrelated separate perturbations, each which is causal (and hence local). One finds $\langle\delta T_1 \delta T_2\rangle^{1/2} \propto \sin\theta^{-3/2}$, and $\langle(T_1 - T_2)^2\rangle^{1/2}$ is independent of θ, using the constraints. This should be compared to Peebles' calculation[3] in which the constraints are not imposed. He finds $\langle(T_1 - T_2)^2\rangle^{1/2} \propto \sin\theta^{1/2}$, and with a magnitude that is larger. Indeed, his dominant term is proportional to the monopole moment of $\delta\rho$.

REFERENCES

[1]J. Traschen, Phys. Rev. D**29**, 1563 (1984).
[2]R. K. Sachs and A. M. Wolfe, Astrophys. J. **147**, 73 (1967).
[3]P. J. E. Peebles, Astrophys. J. Lett. **243**, L119 (1981).

PART III

ORIGIN AND EVOLUTION OF LARGE-SCALE STRUCTURE

An explanation for the origin and evolution of the large-scale structure in the Universe is one of the fundamental goals of cosmology. The influx of new ideas from particle physics has opened up new possibilities in theories for the formation of galaxies and other large-scale structure. Theories of inflation offer the possibility of predicting the amplitude and spectrum of primordial perturbations necessary for galaxy formation. Massive neutrinos, axions, gravitinos, photinos, cosmic strings, and other candidates from particle physics could play an important role in galaxy formation and account for dark matter in the Universe.

This chapter is devoted to the description of the present large-scale structure in the Universe, and theories of how it developed. The observational view of structure in the Universe is reviewed by Davis, and brief reports are given by Batuski and Burns, Vishniac, and Shaver. The question of dark matter in galaxies is reviewed by Gallagher, and Bahcall discusses the possibility of dark matter in the local solar neighborhood. Various models to explain the existence of dark matter are presented by Bardeen, Stecker, Bludman and Hoffman, and Sumi and Smarr. The development of computers have allowed numerical simulations of gravitational clustering. White reviews the subject, and short papers on recent results are given by Occhionero, Melott, and Dekel. A new wrinkle in the theory of galaxy formation is the introduction of a threshold to bias the formation of structure toward the most pronounced peaks in the matter distribution. This idea is presented by Kaiser and applied to the formation of QSO's by Szalay. The possibility that structure formation may include physics from outside the traditional view of gravitational clustering is discussed in a review on cosmic strings by Vilenkin, and in a paper by Hogan on formation of structure.

The input for formation of structure includes the nature of the dark matter and of the initial perturbations. The amplitude and spectrum of the initial perturbations are constrained by the microwave background radiation as discussed in the previous chapter, and can, in principle, be predicted by inflation, as discussed in Chapter IV. Different possibilities for the dark matter can be found on a tour of the particle physics zoo, which includes stops in Chapter V on monopoles, Chapter VI on supersymmetry, and Chapter VII on cosmological constraints on particle properties.

Drawing by Angela Gonzales

An Observational View of Large Scale Structure

Marc Davis

ABSTRACT

A summary of recent observations of galaxy clustering is presented, including a brief review of redshift maps and galaxy clustering statistics. The current major uncertainty is the nature of clustering on scales in excess of $10h^{-1}$Mpc. How should one reconcile the small correlation length measured for galaxy-galaxy clustering with the large correlation length associated with the clustering of Abell clusters? Simple arguments are presented that argue the underlying mass fluctuations are most likely associated with a clustering scale no larger than that of individual galaxies. I also discuss the acceleration of the local group from the comoving frame of the Universe, and its connection to the microwave dipole anisotropy. A final topic for consideration is the existence of large voids and clusters, and whether they are consistent with Gaussian initial conditions. The extreme size and depth of the Bootes void, if real, does present a puzzle. Finally I discuss future directions for observational study of large scale structure.

I. REDSHIFT SURVEYS

The observational study of large scale structure in the Universe has continued steadily through the past decade. Prior to the completion of large redshift surveys, the statistical studies were confined to the analysis of catalogs of positions of galaxies on the sky, such as the Zwicky catalog,[1] the Shane-Wirtanen catalog,[2] and the Jagellonian survey.[3-5] The Shane-Wirtanen catalog, completed in the early 1950's, lists counts of galaxies in cells 10' by 10' in size over the entire sky observable from Lick Observatory. It contains more than 10^6 galaxies, and is still unsurpassed for delineating structure on the largest scales. Another venerable catalog that has seen continuous extensive analysis is the Abell catalog of rich clusters of galaxies, completed in 1958.[6]

Within the past five years, a number of redshift catalogs have been published and most attention has shifted to three dimensional analysis afforded by the redshift for each galaxy. Existing redshift surveys have either been wide angle studies to modest depth[7-9] or deeper surveys within a limited solid angle.[10-16] Both of these approaches are useful and should continue, although it is important to obtain data with very well defined selection criteria, so that meaningful statistical analyses can be performed on the resulting catalogs. Future work will hopefully push the full sky surveys to deeper depths; it is quite misleading to survey only regions selected for an interesting cluster as this can lead to false impressions of the overall clustering patterns.

The general nature of the galaxy clustering is quite remarkable, and is seen in all redshift surveys. The galaxy clustering is extremely strong on small scales; it has a loose, filamentary appearance that can stretch into long supercluster chains with a length in excess of $50h^{-1}$Mpc (h is Hubble's constant in units of 100km/sec/Mpc). With the strong clustering, there must also be large regions of low density, known as voids. These voids are indeed quite deficient in luminous galaxies, but we really don't know if they are as deficient in baryons. A recent review of superclustering is given by Oort.[17]

As an example of redshift survey results, consider Figures 1 through 4, derived from the largest existing survey, the Center for Astrophysics redshift survey, which incudes all galaxies in the northern sky with galactic latitude b $\geq$ 40 and apparent magnitude m $\lesssim$ 14.5.[9] Figure 1 shows the CfA northern catalog projected onto the sky, after the catalog has been trimmed of galaxies not sufficiently luminous to be visible to a redshift of 4000km/sec. The catalog is also limited to galaxies with redshift v < 10000km/sec, although there exist few galaxies beyond this limit with m < 14.5. The effect of this semi-volume limitation of the catalog is to de-emphasize the strong foreground Virgo cluster and associated supercluster at a redshift of 1000 km/sec. In Figure 1 it is in fact difficult to locate the Virgo cluster center (12.4^h, $+6°$). The dense cluster at the north galactic pole is the Coma cluster, and the extensions of galaxy clustering to the east and west are parts of its associated supercluster. The clustering is evident through the loose, irregular concentrations and voids.

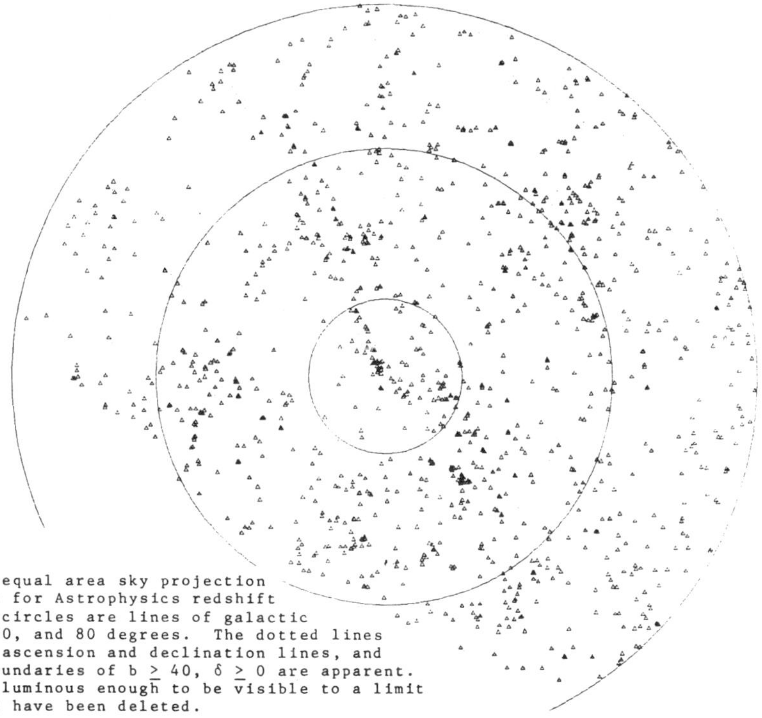

Figure 1: An equal area sky projection of the Center for Astrophysics redshift survey. The circles are lines of galactic latitude 40,60, and 80 degrees. The dotted lines denote right ascension and declination lines, and the survey boundaries of b $\geq$ 40, $\delta \geq$ 0 are apparent. Galaxies not luminous enough to be visible to a limit of 4000km/sec have been deleted.

An alternative vantage point is provided in Figure 2, which is a projection of the trimmed catalog onto a Cartesian plane. The distance to each galaxy is derived using the Hubble law, v = Hℓ. The earth is at the origin, and the north galactic pole is in the z direction. Each point is again a galaxy, but now the area of each point is a measure of the reciprocal of the probability of selection at that distance. Each point with a large area statistically represents multiple galaxies, most

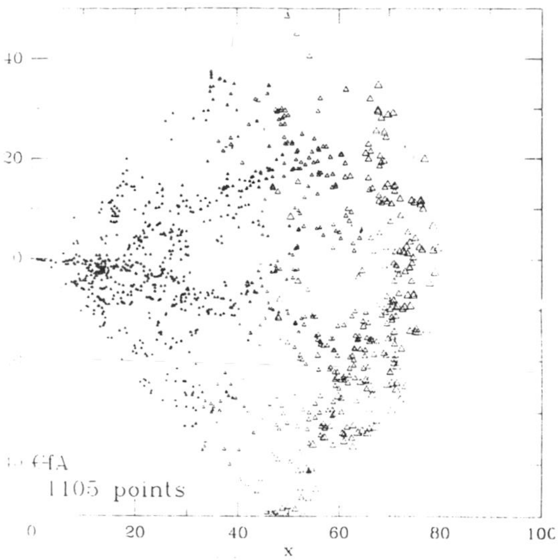

Figure 2: A cartesian projection of the same data as in Fig. 1. We sit at the origin of the plot. The axes are labeled in Megaparsecs, and the galactic pole is oriented in the x direction. The Virgo supercluster is apparent as the high density region in the foreground. Larger symbols at larger distances reflect the selection effects in this magnitude limited catalog.

of which were too faint to be included within the survey limit. The entire foreground is seen to be higher than average density; this is the local Virgo supercluster, which is an amalgamation of numerous loose groupings roughly centered around the Virgo cluster. The entire middle region is of lower density than the mean, and contains several unconcentrated clusters and conspicuous voids. The most distant part of the catalog contains a few large clusters and a very large void. Much of the clustering has a random, incoherent appearance, as should be expected since the survey probes well beyond the length scale of individual clusters.

Figures 3 and 4 are plots in redshift space, in which velocity is plotted versus right ascension for selected windows of declination.[9] Figure 3 shows the data from the CfA catalog in the window $0 < \delta < 10$ degrees, and Figure 4 is for the window $20 < \delta < 30$. The Virgo cluster center is prominent in the foreground of Figure 3, and the Coma cluster and the cluster Abell 1367 are prominent in Figure 4. Note how these 3 rich clusters are elongated in the redshift direction. This, of course, is caused by peculiar velocities within these clusters, which in Coma have an RMS value of nearly 1000km/sec in the line of sight. These slices are best for demonstrating the elongated, loose structure that defines the large scale clustering. Redshift distortion effects are not especially prominent except in the rich clusters. In Figure 4 it is clear that the Coma and A1367 clusters are simply regions of highest local density within an extended supercluster. The void to the east of the Coma-A1367 supercluster is the largest void within the CfA survey; it is a region of diameter $30h^{-1}$Mpc in which the galaxy density is no more than 25 percent of the mean.

Other redshift surveys have detailed larger voids and richer complexes of superclustering, but the CfA data is fairly representative. This sample is perhaps large enough to represent a fair sample volume of the Universe. We shall return to the discussion of large voids and clusters in Section V. More detailed discussions of the clustering are given in Refs. 9 and 17.

II. CORRELATION STATISTICS

The standard lowest order statistic of galaxy clustering is the two point correlation function, which has been extensively applied to galaxy catalogs in two and three dimensions. The definition of the correlation function $\xi(r)$ is given by the incremental probability dP of finding a galaxy in a volume element dV a distance r from a randomly chosen galaxy

$$dP = n(1 + \xi(r))dV \tag{1}$$

where n is the number density of galaxies in the sample. In two-dimensional catalogs, only angular separations are available for measuring the correlations, but in three dimensions it is possible to study the galaxy correlations by binning the pair separations either in redshift separation s, or in projected separations r_p. The resulting correlations are well fit by a power law model $\xi(r) = (r/r_o)^{-\gamma}$ with $\gamma = 1.8$, the same as the slope derived from the larger, two-dimensional catalogs. Details are given by Davis and Peebles.[18]

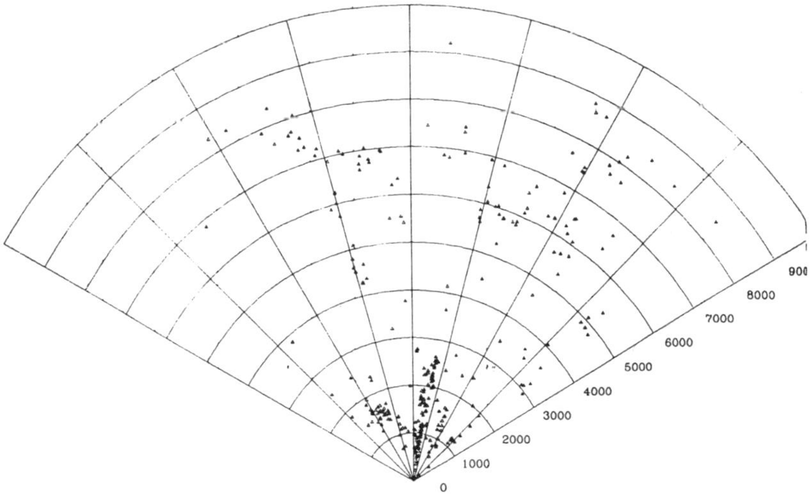

Figure 3: Right ascension is plotted versus observed velocity for galaxies in the range 0 < δ < 10 degrees. The radial elongation of the Virgo cluster in the foreground is conspicuous.

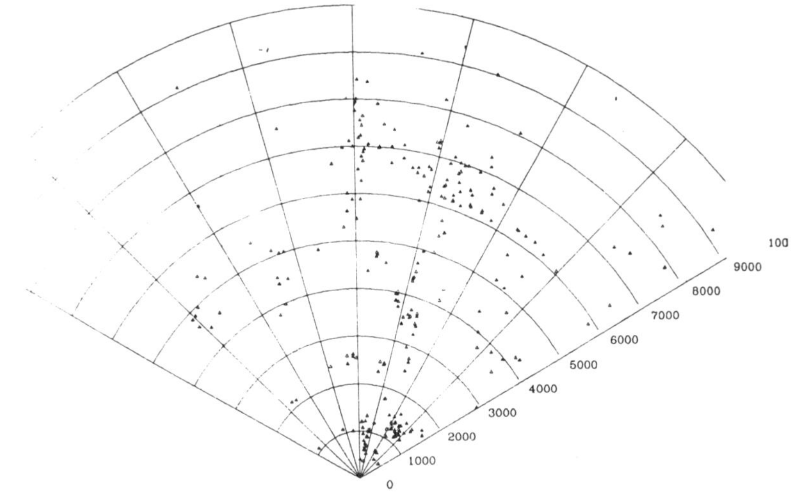

Figure 4: Right ascension is plotted versus observed velocity for galaxies in the range 20 < δ < 30 degrees. The Coma cluster and A1367 are the two concentrations at 6000 - 7000km/sec and both clusters are substantially elongated in the radial direction. Note the large void to the left of these rich clusters.

Apparently power law correlations are quite consistent with the filamentary, loose patterns of clustering observed about us. Furthermore, the correlation length scale r_o is $5h^{-1}$Mpc in spite of the existence of structure 5-10 times larger. The correlations simply count neighbors in spherically symmetric shells about random galaxies, and are a poor indicator of linear structures for scales where $\xi(r) < 1$.

For separations larger than $10h^{-1}$Mpc the correlations become quite uncertain because the background density is uncertain at the 10 percent level. Within the CfA survey and the AAT redshift survey,[9,15] the correlations are reported to be negative on large scale, although there is disagreement on the length scale at which $\xi(r) < 0$. This result has also been reported from careful analysis of several angular correlation studies,[19] but it is fair to say that all evidence for anticlustering is extremely weak. In the CfA survey a one sigma upward adjustment (10 percent) of the mean number density of galaxies will cause the anticlustering to completely disappear. The statistical significance of the anticlustering in the other surveys is even weaker. The clustering on large scales is extremely important because linear perturbation theory should adequately describe the evolution of large scale fluctuations, and hence the post-recombination spectrum of perturbations is in principle directly observable. If the inflationary models are correct, the initial spectrum of perturbations should be of the Harrison-Zeldovich constant curvature form, and there should be anticlustering on some scale. I shall discuss below the question of the reality of anticlustering on scales $r > 20h^{-1}$Mpc.

Within redshift surveys it is possible to measure statistically the relative peculiar velocity of pairs by measuring the radial distortion of the redshift space maps of clusters. The observed two point correlation function of two variables, $\xi = \xi(r_p,\Pi)$ where r_p is the projected separation of a pair, and Π is the line of sight velocity difference of a pair. The stretching of contours of equal amplitude of ξ in the radial direction is a measure of the peculiar velocity difference of the pairs. The distribution function of pair differences has a longer tail than in a Gaussian broadening model; an exponential model is a good fit to the broadening function of the CfA data, whereas the AAT data is best fit by a functional form intermediate to a Gaussian and exponential. The measured RMS line of sight velocity difference of galaxy pairs $\sigma(r)$ depends only slightly on the projected separation of the pairs; from the CfA data $\sigma(r)$ can be represented as

$$\sigma(r) = 340 \pm 40\ (r_p/1h^{-1}\mathrm{Mpc})^{0.13\pm0.04}\mathrm{km/sec} \qquad (2)$$

over a range $10h^{-1}\mathrm{kpc} < r_p < 3h^{-1}$Mpc. Results from the AAT survey are in substantial agreement, but their best fit velocity is independent of projected separation. The velocity field measured in all redshift surveys is very small, and is a powerful discriminant for models of the evolution of large scale structure.

The low amplitude of the velocity field argues for an open universe, if galaxies trace the matter distribution. The behavior of $\sigma(r)$ is consistent with a stable hierarchy of clustering of constant mass to light ratio,[4] in which one would expect

$$\sigma(r)^2 \propto r^{2-\gamma} \qquad (3)$$

and leads to[18] $\Omega = 0.2\exp(\pm\ 0.4)$. An Einstein-de Sitter Universe, Ω =

1, is possible only if galaxies are not fair tracers of the mass distribution. Whether this is possible or reasonable depends on the details of galaxy formation and on the nature of the dark matter and is clearly a subject for further investigation.

Another method of measuring large scale clustering is to study the correlation properties of selected samples of Abell clusters. These are rich clusters that have a mean intercluster spacing of $\approx 50h^{-1}$Mpc, versus an interparticle spacing of $5h^{-1}$Mpc for bright galaxies. Neta Bahcall will discuss the clustering properties of Abell clusters in detail in her presentation. Correlation analysis was first applied to the Abell catalog by Hauser and Peebles,[20] and more recently with complete redshift data for the nearest 100 Abell clusters by Bahcall and Soneira.[21] The results in both of these analyses are in agreement; the Abell clusters appear to have a power law correlation function with slope roughly 2, but with a correlation length of $25h^{-1}$Mpc, five times that of the galaxy clustering. This fact has been known for a decade and presents a challenge of interpretation. Which length scale, if either, traces the underlying mass fluctuations? What do these two disparate length scales tell us about the initial fluctuation spectrum?

Very recently Nick Kaiser[22] suggested that the enhanced amplitude of the Abell clusters was a natural effect of the statistics of rare events in Gaussian noise. In Gaussian noise the apparent clustering amplitude of rare, high sigma peaks is enhanced over that of the overall mass fluctuations. Furthermore, the higher and rarer the peaks, the more strongly enhanced is the clustering. The observational data show a similar trend.[21] Nick will discuss this mechanism in some detail in his talk later this week. In this model, the Abell clusters do not trace the mass fluctuations, but only amplify with a positive factor $A > 1$ the large scale matter correlations. The Abell clustering is quite strongly positive to scales larger than $50h^{-1}$Mpc, and is detected to be positive up to length $100h^{-1}$Mpc, whereas the galaxy clustering seems to go negative on much smaller scale, as discussed above. If Kaiser's amplification mechanism is the correct explanation for the Abell clustering anomaly, then in fact the galaxy correlations cannot be negative on these large scales, and one must simply ignore the reports of weak anticlustering. Apparently the rich clusters are a better probe of clustering on these large scales.

A further important result concerns the length scale at which ξ does become negative. If the Universe is dominated by cold dark matter such as axions, photinos, or primordial black holes, and if the primordial power spectrum of perturbations was the constant curvature type, then the linear spectrum of perturbations is specified within a normalization factor by the standard cosmological parameters.[23] In linear theory, the correlations of the matter fluctuations are expected to be negative for length $R > 18(\Omega h^2)^{-1}$Mpc. It is therefore of considerable interest to measure ξ to the largest possible scales. Bahcall and Soneira[21] report positive correlations on a length of $100h^{-1}$Mpc, which would imply $\Omega h < 0.18$. Their sample, however, contains only four independent spheres of $100h^{-1}$Mpc radius, and there are several selection biases that could contaminate their results. Shectman[23a] has recently reported correlations for smaller clusters drawn from the Shane-Wirtanen catalog. He finds $\xi < 0$ for scales $r > 30h^{-1}$Mpc in some subsamples, although the scatter is large. It is clearly important to improve the existing database to properly address this question.

III. THE CORRELATION LENGTH OF THE MASS FLUCTUATIONS

In the above section we described two clustering lengths which differ by a factor of five. There is a simple test one can apply to determine which of these length scales is the fairer tracer of the underlying matter fluctuations. This is the cosmic energy equation which relates the RMS peculiar velocity v_p of random particles to the energy stored in gravitational fluctuations. The exact differential equation can be adequately approximated as the simple algebraic expression[4,5]

$$v_p^2 \approx (3/2)\Omega H_0^2 \int_{r_\ell}^{\infty} \xi(r) r dr \tag{4}$$

If we use a truncated power law model,

$$\xi(r) = (r/r_0)^{-1.8}, \quad r < 4r_0$$

$$\xi(r) = 0, \quad r > 4r_0 \tag{6}$$

$$r_\ell = 0.5h^{-1}\mathrm{Mpc}$$

then we find

$$v_p \approx 2.3\Omega^{1/2}(H_0 r_0) \tag{7}$$

Not surprisingly the peculiar velocity is of order the Hubble velocity across the correlation length. Inserting $r_0 = 25h^{-1}$Mpc, as suggested by the correlations of rich clusters, we would expect $v_p = 6400\Omega^{1/2}$km/sec. This can be compared to the departure of the Milky Way from the comoving frame of the Universe, as measured by the dipole anisotropy of the microwave radiation, $v_{MW} = 600$km/sec, with an uncertainty of roughly 20 percent. Assuming the one dimensional velocity distribution function to be gaussian, the probability that a random point would have a three dimensional velocity as low as ours is

$$P(v \leq 600) = 2 \times 10^{-4}, \quad \Omega = 1$$

$$P(v \leq 600) = 7 \times 10^{-3}, \quad \Omega = 0.1$$

Such low probability would imply that the Milky Way is in a special place and is anti-Copernican. In numerical simulations of large scale clustering the distribution function of peculiar velocities has a Gaussian core, and a slightly extended exponential tail, which will not seriously change the above probability calculations.

If we repeat the calculation using $r_0 = 5h^{-1}$Mpc, as suggested by the galaxy correlations, we expect $v_p = 1130\Omega^{1/2}$km/sec, leading to cumulative probabilities

$$P(v \leq 600) = 0.036, \quad \Omega = 1$$

$$P(v \leq 600) = 0.58, \quad \Omega = 0.1$$

again demonstrating the consistency of the $\Omega = 0.1$ model if the galaxies trace the matter distribution. If we insist that inflation is correct and $\Omega = 1$, then we should consider the case of still smaller clustering lengths, e.g., $r_o = 2.5h^{-1}$Mpc. In this event, we get a lower velocity $v_p = 635\Omega^{1/2}$km/sec and a cumulative probability

$$P(v \leq 600) = 0.17, \quad \Omega = 1.$$

Thus we again reach the conclusion that an $\Omega = 1$ universe is likely only if the galaxies are significantly more clustered than the matter. Suggestions that the Abell clustering length best traces the matter distributions are extremely difficult to reconcile with the observed peculiar velocities for any reasonable value of Ω.

IV. THE GENERATION OF PECULIAR VELOCITIES

The local inhomogeneity of space is well documented by existing redshift surveys. If we assume the galaxies to be a tracer of the mass distribution when averaged over a suitable volume, we can use the observed spatial distribution of galaxies to compute the peculiar acceleration **g** of our local group of galaxies from the comoving frame of the Universe. To do this properly requires a uniformly sampled redshift survey over the entire sky. There exists at present good survey data for the Northern galactic sky, but the Southern sky has been surveyed to only half the depth of the North. Relatively little survey work has been done for galaxies at low galactic latitude, so only the component of **g** directed toward the galactic pole can be reliably computed.

The expression for the local acceleration is simply

$$\mathbf{g}(0) = G\rho \int_0^\infty \delta(r)\mathbf{r}/r^3 d^3r = G\rho \sum_{\text{shells}} \delta_{\text{shell}}\theta_{\text{shell}} \tag{8}$$

where δ is the fluctuation from the mean density at location r and θ_{shell} is the direction of acceleration by the shell. The integral converts to a sum over equally weighted spherical shells, and convergence at large r occurs only if δ averaged over angles on distant shells approaches zero. With the CfA survey, we can sum the shells to a distance of $80h^{-1}$Mpc in the North, and to 40^{-1}Mpc in the south, using the Shapley Ames catalog. We presume that regions outside the volume of the summation do not contribute to **g**. Details of this procedure are given by Davis and Huchra.[24]

Since we are interested in large scale effects, the fluctuations should obey linear perturbation theory, which leads to substantial simplication. In linear theory, assuming the growing mode dominates the present density fluctuations, the induced peculiar velocity, $\mathbf{v}_v$ is proportional to **g**.[1]

$$\mathbf{v}_v = 2\Omega^{.6}\mathbf{g}/3H\Omega. \tag{9}$$

The results for the summation are given in Figure 5, where we plot the galactic polar component of the cumulative induced velocity for $\Omega = 1$ as a function of the shell distance in the north and south. Note that as the summation begins in the north, the gravitational acceleration grows rapidly; this is the effect of the Virgo supercluster, of which the local group is an outlying member. Beyond the Virgo region, the cumulative velocity drops slightly because this low density region has a repulsive effect on us. Similarly the low density region in the foreground of the Southern sky repels us toward the North.

At greater distance the large fluctuation of the Coma supercluster has relatively little effect on our acceleration because its attraction is mostly compensated by the repulsion of the large void at the same distance. It seems as though the cumulative peculiar velocity has converged, although a deeper survey, particularly in the South, would be a helpful check. The basic conclusion of the peculiar gravity calculation is that the Virgo supercluster is clearly the major large scale gravitational perturbation to the local group of galaxies, and that the combined gravitational influence of all the other well surveyed clusters and voids is relatively small.

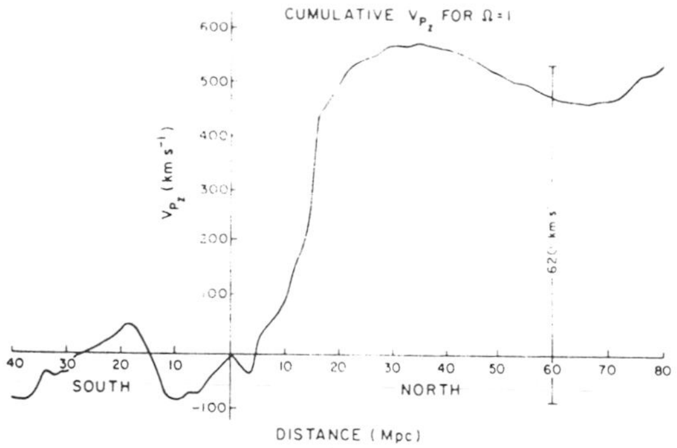

Figure 5: The expected cumulative acceleration of the local group, expressed as the velocity induced in linear theory, is plotted as a function of radius. Note that the curve quickly levels off at $30h^{-1}$Mpc, just beyond the far side of the Virgo cluster.

The component of the predicted peculiar velocity directed toward the North galactic pole is, as indicated in Figure 5, $620\Omega^{.6}$km/sec. This would be a powerful measure of Ω if we knew what observed velocity to compare this prediction. The component of the microwave dipole anisotropy directed toward the north galactic pole is 400km/sec, and if our entire peculiar velocity is due to the above computed peculiar acceleration, we derive $\Omega = 0.5$. This, however, is a risky step, because at least part of the peculiar velocity of the local group could be induced by non-linear perturbations, or perhaps even by non-gravitational processes. Furthermore if the spectrum of initial perturbations places sufficient power on very large scales, then the entire sphere of radius $60h^{-1}$Mpc surrounding us could be coherently moving toward some distant perturber, so that local shear flows would account for only part of the microwave dipole anisotropy. Although we do not have good information on the density of galaxies in the galactic plane, it does seem that the Virgo cluster will dominate the local gravity anomaly. On the other hand, the vector of the microwave anisotropy is directed 40 degrees away from the Virgo cluster center, suggesting either that the Virgocentric velocity flow has begun to pancake, or that some other more distant fluctuation has induced a substantial component of the dipole anisotropy. There have been suggestions[25] that a major fraction of the dipole anisotropy is induced by the Hydra-Centaurus cluster, located at redshift 3000km/sec, and close to the galactic plane in the southern sky. The mass anomaly associated with this cluster complex would need to be some twenty times that of the entire Virgo supercluster, which is quite extreme.

To address the question of the expected contribution by very large scale matter flows, we should compare the microwave anisotropy measurements to studies of the local anisotropy of the Hubble flow. There are a variety of astronomical methods used to measure distances independently of redshifts, which can be combined to map the distortion of the local velocity field from a purely radial Hubble flow. At present the results are quite preliminary, and again only the component of velocity directed toward the galactic pole can be measured with any precision. A review of the literature is given by Davis and Peebles[26]; the observational situation is somewhat confusing. The best value of the local shear field that measures the infall of the local group toward the Virgo cluster, is in the range 200 - 300km/sec. The shear field relative to a frame of galaxies at redshift ≈ 5000km/sec is somewhat higher, ≈ 400km/sec, which agrees with this component of the microwave anisotropy. There have been reports that the shear velocity of the local group relative to galaxies on shells at distance ≥ 4000km/sec is fully consistent with the microwave anisotropy vector.[27] The data will gradually improve for this type of study, and hopefully future results will be more definitive.

The theoretical expectation for the velocity field depends on the nature of the dark matter. Shown in Figure 6 is a plot of the contribution to the variance of the velocity field by fluctuations of various wavelengths, assuming linear perturbation theory and a spectrum of perturbations consistent with a universe dominated by cold dark matter, a current favorite. Linear theory should be appropriate for large scale velocity fields and in any event, the effects of nonlinearity on small scale do not substantially change the argument. The total variance of the microwave dipole velocity is given by the area under the curve. Within the cold dark matter models, the only length scale adjustment permitted is that given by the scaling of the product Ωh^2. As shown the peak wavelength for the contribution to the velocity field is $12.6(\Omega h^2)^{-1}$Mpc, so that for any reasonable value of cosmological parameters, most of the contribution to the velocity field is expected to be induced by fluctuations with size less than $100h^{-1}$Mpc, less than the size of existing redshift surveys. Therefore we should expect eventual convergence of the microwave anisotropy with the local studies of perturbations from the Hubble flow. It is quite unlikely that the microwave dipole anisotropy arises from fluctuations the size of our present horizon.

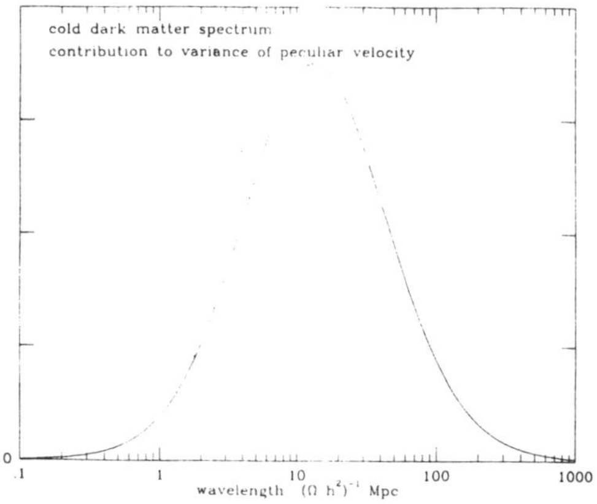

Figure 6: The contribution of fluctuations of various wavelengths to the peculiar velocity dispersion of particles in a universe dominated by cold dark matter. The velocity dispersion is proportional to the area under the curve. Note how most of the power arises from wavelengths within a factor of three of the peak.

V. EVIDENCE FOR NON-GAUSSIAN INITIAL CONDITIONS?

The most natural initial condition for the early Universe is random phases for the initial perturbations. Random phase fluctuations would be generated via quantum processes during an inflationary epoch.[28] Non-random phase fluctuations could be generated at relatively late epochs if cosmic strings exist and form closed loops.[29] The superposition of a large number of random phase waves will lead to a gaussian density field, but the density field will not retain its gaussian distribution once the fluctuations grow to non-linear amplitude. The question is, are there objects observed in the universe that are difficult to reconcile with Gaussian, random-phase, initial fluctuations? To address this we should consider large scale fluctuations, which should obey linear theory and have nearly gaussian statistics. I shall discuss the two most extreme cases reported in the literature. The first is the void in Bootes,[30] and the second is the Serpens Virgo cluster.[31-32]

The Bootes void is a region of diameter $60h^{-1}$Mpc at a mean distance $150h^{-1}$Mpc that was discovered by accident. It contains no luminous galaxies, whereas 18 objects were expected in the redshift surveys that have studied this region. It is estimated that the 90 percent confidence upper limit to the density of this void is 25 percent the mean value. Present studies constrain only the density of luminous galaxies, and set no constraint on the density contrast of low surface brightness objects, or on uncondensed material. Based on the measured behavior of the correlation function, the RMS fluctuation of counts of galaxies within a sphere of radius R is

$$\delta\rho/\rho_{rms} \approx (R/8h^{-1})^{-1} \approx 0.25 \qquad (10)$$

on the scale of the Bootes void. Thus the Bootes void is a 3σ downward fluctuation. The volume of the void is approximately 10 percent of the volume of space that has been surveyed via complete redshift surveys, although the void has only been sampled in thin needles. An examination of N-body results of clustering in the universe shows that for radii $R > 4r_o$, the fluctuations of counts of objects in random cells is fairly well described by a gaussian distribution; this statement is true whether the counts are measured in real space or redshift space.[33] Voids as large and as deep as the Bootes void are not found in numerical simulations. If a complete redshift survey of the Bootes region fails to find any galaxies in the void, there will be three (or more) possible explanations:

(1) The extrapolation of Eq. (10) is wrong because we have underestimated $\xi(r)$ on large scales.

(2) The initial fluctuations do not have random phase; the large void did not develop from a chance superposition of random waves.

(3) Galaxies are very poor mass tracers; the void is not as empty as it appears.

At the present time, the data is not compelling, but perhaps is telling us that there is more power on large scales than expected from the extrapolation of smaller scale correlations.

On the other extreme is the Serpens-Virgo cloud, conspicuous as a 5 degree patch on the Shane-Wirtanen maps that has twice the mean surface density of galaxies.[32] Given its large distance of $300h^{-1}$Mpc, this cluster is an upward density fluctuation of 20 times the mean in a sphere of diameter $20h^{-1}$Mpc and is the richest cluster in the catalog.[30,32] As a comparison, the Virgo cluster is three times the mean density, averaged in a similar volume. Since the RMS fluctuations in a volume this size are close to 1, they are expected to show a large departure from gaussian statistics. The Serpens-Virgo cloud appears to be a 20σ fluctuation today; it is a non-linear fluctuation, whose density contrast is growing faster than lower density fluctuations that are still linear. If the Serpens-Virgo cloud were uniformly dispersed into a sphere having the mean density of the Universe, its radius would be $27h^{-1}$Mpc, which represents approximately one part in 1000 of the volume surveyed by the Shane-Wirtanen catalog. In a spherical model, Serpens-Virgo was a 4.4σ upward initial fluctuation on the scale of $27h^{-1}$Mpc, and it is consistent that only one such fluctuation is present in the Shane-Wirtanen volume. This cluster is probably consistent with random phase initial fluctuations, although just barely.[34]

VI. PROJECTS FOR THE FUTURE

There has been considerable recent progress in understanding the large scale structure of the Universe, both from an observational and theoretical viewpoint. Some of the outstanding observational problems were discussed above, and in the remaining section I would like to put forth a "wish list" of observational projects to pursue in the next decade. All of them are possible, but none will be easy.

First, it is time to generate new full sky catalogs of faint galaxies using automated procedures, either by digitizing existing plate material or using new plate material that would be better calibrated. Much effort will be needed to compare diameters and magnitudes of galaxies in overlapping plates to control large scale sensitivity gradients. From this project one would hope to measure the positons, diameters, and magnitudes of 10^{7-8} galaxies, and thus to generate better, more controlled versions of the Shane-Wirtanen catalog and the Abell catalog. This is the best hope for getting improved data of clustering on scales in excess of $20h^{-1}$Mpc, the region where our knowledge is weakest.

A second project is intended to improve our understanding of the evolution of correlations. This is a subject I didn't discuss, partly because we know so little. The evolution of $\xi(r,t)$ is a sensitive probe of the large scale structure; the various models of cosmogony predict rather different behavior for $\xi(r,t)$. For example, in cold dark matter models in which galaxies trace the matter, the correlations are expected to evolve in redshift quite rapidly. For cold dark matter models in which galaxies do not trace the matter ξ is expected to be almost constant in time. In a scale free hierarchical model the shape of ξ should be time invariant, but if a characteristic length scale is present in the initial conditions, $\xi(r)$ will steepen with time. There are a few observational projects nearing completion in which angular correlations are being studied for very faint galaxies, with median redshift $z \approx 0.3$[35] At these large distances most observed pairs are chance superpositions, and without redshifts for each object, the correlations are very small. Very deep redshift surveys containing samples of a thousand redshifts $z > 0.5$ in a small region of the sky would give very useful information. With the advent of multiple object fiber spectroscopy, such a project is not beyond the realm of possibility on a 10 meter class telescope. Redshift surveys can also be done via absorption line studies of distant QSO's. More work along the lines discussed by Peter Shaver this week would be very useful for constraining the evolution of ξ.

A third project, which can be done without heroic effort, is to continue wide angle redshift surveys. One could either survey all galaxies to a limiting magnitude or diameter, or observe only a random subset from a complete galaxy list. The calculation of peculiar gravity **g** discussed above would be much improved if the redshift surveys covered the southern sky to a depth comparable to the northern sky. A further dramatic improvement on the technique would occur if we could somehow generate a galaxy list at low galactic latitude. Hopefully the recently completed IRAS mission will lead to a complete list of spiral galaxies at low latitude whose redshifts could be measured via 21 cm techniques. Until that time we really cannot adequately judge the significance of the fact that the microwave anisotropy does not point at the Virgo cluster.

A final project will require use of the Space Telescope, and is potentially the most significant observational approach for the next decade. Cosmologists generally assume that galaxies are in some sense a fair tracer of the underlying mass distribution, but this is only an assumption. We cannot be very confident that the voids seen in the redshift surveys are really empty. Perhaps galaxy formation went to completion only in the most favorable of circumstances and only the original 2σ density fluctuations ever became galaxies. This would be more likely to occur in an incipient cluster rather than in an incipient void and so the voids would be more depopulated of galaxies than of baryons. Later during this conference Jim Bardeen will discuss physical processes that might lead to galaxy suppression in low density regions. We have tried a simple recipe along these lines for galaxy formation in our numerical simulations of cold dark matter dominated universes; we find it gives a rather good match to the observed clustering.[33] If the voids contain low surface brightness galaxies or slowly cooling clouds, they should be observable as absorption line systems in the spectra of distant QSO's. Many Lyα clouds with absorption redshifts $z > 2$ are observed in high redshift QSO's, and it is clear that these clouds do not cluster like the galaxy distribution.[36] The physical nature of these clouds is uncertain; are they gravitationally bound or are they externally pressure confined by a hot intergalactic medium? The advantage of the Space Telescope is that it will permit us to study the Lyα clouds at low redshift, $z < 0.05$, a region for which we have good maps of the three dimensional galaxy distribution. Will there be an abundance of absorption clouds or will they all have evaporated by the present epoch? Based on the studies at $z \approx 2$, we would expect only one or two clouds with $z_{abs} \lesssim 0.05$ per QSO; therefore many QSO's must be examined before we can address the statistical questions. Will the clouds be correlated or anticorrelated with the galaxy distribution? If the galaxies and the clouds anti-correlate, then the galaxies must represent a biased tracer of the mass distribution, and very likely the voids contain sufficient matter to close the Universe.

ACKNOWLEDGEMENTS

This paper was written at the Institute for Theoretical Physics, Santa Barbara, during a workshop on the Evolution of Structure in the Universe. I thank the director and his staff for extending gracious hospitality and providing a convivial research environment. This work was partially supported by NSF grants AST81-18557 and PHY77-27084 (supplemented by funds from the National Aeronautics and Space Administration).

REFERENCES

[1]F. Zwicky, P. Wild, E. Herzog, M. Karpowicz, and C. T. Kowal, Catalogue of Galaxies and Clusters of Galaxies, Vols. 1-6, (Pasadena, Calif. Inst. of Tech.), (1961-1968).
[2]D. Shane and C. A. Wirtanen, Publ. Lick Obs. **22**, 1 (1967).
[3]K. Rudnicki, T. Z. Dworak, P. Flin, B. Baranowski, and A. Sendrakowski, Acta Cosmologica **1**, 7 (1973).
[4]P. J. E. Peebles, in The Large Scale Structure of the Universe, Princeton Press (1980).
[5]M. Fall, Rev. Mod. Phys. **51**, 21 (1979).
[6]G. Abell, Ap. J. Suppl. **3**, 211 (1958).
[7]J. R. Fisher and R. B. Tully, Ap. J. Supp. **47**, 139 (1981).
[9]M. Davis, J. Huchra, D. W. Latham, J. Tonry, Ap. J. **253**, 423 (1982); J. Huchra, M. Davis, D. W. Latham and J. Tonry, Ap. J. Suppl. **53**, 89 (1983).
[10]R. P. Kirshner, A. Oemler, P. L. Schechter, Ap. J. **83**, 1549 (1978).
[11]S. A. Gregory and L. A. Thompson, Ap. J. **222**, 784 (1978).
[12]S. A. Gregory, L. A. Thompson, and W. G. Tifft, Ap. J. **243**, 411 (1981).
[13]M. Tarenghi, W. G. Tifft, G. Chincarini, H. J. Rood, and L. A. Thompson, Ap. J. **234**, 793 (1979).
[14]M. Tarenghi, G. Chincarini, H. J. Rood, and L. A. Thompson, Ap. J. **235**, 724 (1980).
[15]R. Giovanelli, M. P. Haynes, and G. L. Chincarini, in press (1984).
[16]A. J. Bean, G. Efstathiou, R. S. Ellis, B. A. Peterson, and T. Shanks, M.N.R.A.S. **205**, 605 (1983); T. Shanks, A. J. Bean, G. Efstathiou, R. S. Ellis, and Peterson, Ap. J. **274**, 529 (1983).
[17]J. H. Oort, Ann. Rev. of Astr. and Astrophys. **21**, 373 (1983).
[18]M. Davis, and P. J. E. Peebles, Ap. J. **267**, 465 (1983).
[19]P. C. Hewett, M.N.R.A.S. **201**, 867 (1982).
[20]M. G. Hauser, and P. J. E. Peebles, Ap. J. **185**, 757 (1973).
[21]N. A. Bahcall and R. M. Soneira, Ap. J. Lett. **258**, L17 (1982); N. A. Bahcall and R. M. Soneira, Ap. J. **277**, 27 (1984).
[22]N. Kaiser, Ap. J., in press (1984).
[23]P. J. E. Peebles, Ap. J. **277**, 470 (1984); G. R. Blumenthal and J. R. Primack, Fourth Workshop on Grand Unification, eds. H. A. Weldon, P. Langacker, and P. J. Steinhardt, (Birkhauser, Boston), p. 256 (1983); J. R. Bond and G. Efstathiou, Ap. J. Lett., submitted (1984).
[23a]S. A. Shectman, preprint (1984).
[24]M. Davis and J. Huchra, Ap. J. **254**, 425 (1982).
[25]A. Sandage and G. A. Tammann, in High Energy Physics and Cosmology, CERN Conference Proceedings, November 1983.
[26]M. Davis and P. J. E. Peebles, Ann. Rev. of Astr. and Astrophys. **21**, 109 (1983).

[27]L. Hart and R. D. Davies, Nature **297**, 191 (1982); G. deVaucouleurs and W. L. Peters, submitted to the Ap. J. (1984).
[28]J. M. Bardeen, P. S. Steinhardt, and M. S. Turner, Phys. Rev. D **28**, 679 (1983).
[29]N. Turok and D. Schramm, preprint (1983).
[30]P. J. E. Peebles, Ap. J. **274**, 1 (1983); P. J. E. Peebles Science **224**, 1385 (1984).
[31]R. P. Kirshner, A. Oemler, P. L. Schechter, and S. A. Shectman, Ap. J. Lett. **248**, L57 (1981); Early Evolution of the Universe and Its Present Structure, IAU Proceeding, eds. G. O. Abell and G. Chincarini, 197 (1983).
[32]D. Shane, in Stars and Stellar Systems, **9**, Galaxies and the Universe, eds. A. Sandage, M. Sandage, J. Kristian, **647** (1975).
[33]M. Davis, G. Efstathiou, C. Frenk, S. White, preprint (1984). See White's contribution in this volume.
[34]N. Kaiser and M. Davis, in preparation.
[35]D. Koo and A. Szalay, Ap. J. in press (1984); T. Tyson, private communication (1984).
[36]W. L. W. Sargent, P. J. Young, A. Boksenberg, and D. Tytler, Ap. J. Supp. **42**, 41 (1980).

Observations of Candidate Superclusters of Galaxies in Pisces-Cetus and Sextans-Leo

David J. Batuski and Jack O. Burns

We have developed a catalog of candidate superclusters[1] by applying the percolation technique to the entire Abell catalog of rich clusters of galaxies. Our catalog contains 363 supercluster candidates, two of which are very large (28 member clusters in Sextans-Leo and 36 in Pisces-Cetus) and nearby ($z<0.08$). Since both of these candidates contained clusters with only a single measured redshift or with only estimated redshifts, many redshift measurements were needed to reliably determine the spatial distribution of member clusters.

In September 1983, we obtained redshifts for nine Abell clusters and five poor Zwicky clusters in and near Pisces-Cetus, using the Intensified Image Dissector Scanner on the 2.1-m. telescope at Kitt Peak. Figure 1 shows the projected 3-D arrangement of the clusters in this region, with our data included. There appears to be much structure having 100-200 h^{-1} Mpc ($h=H_0/(75$ km/sec/mpc)) scale size, with large, very empty regions alternating with dense clumpings of clusters. In particular, there is an apparent single filament of clusters with a total length of over 310 h^{-1} Mpc, which includes the much-studied Perseus-Pisces supercluster. Structures of this size, if they were the rule throughout the universe, would not support the hierarchical model of superclustering. Such structures would also severely constrain the pancake model, requiring that a hot particle (e.g. a massive neutrino) have a mass $\lesssim$ 4eV in order to produce the damping of smaller size density perturbations[2].

In March 1984, we observed the Sextans-Leo region, measuring redshifts of 14 Abell clusters and six Zwicky clusters. Figure 2 shows the distribution of all the Abell clusters in this region and the six poor Zwicky clusters. While there are some suggestions of structure in the region, nothing of very large scale is obvious.

Is the structure seen in Figure 1 a statistical aberration, with most of space looking more like Figure 2? The Bootes void[3] and the void/supercluster structure pointed out by Bahcall and Soneira[4] are other examples of Abell cluster arrangements with scale size ~200 h^{-1}. Thus there are now several cases of reported very large scale structure, but these come from a very limited volume of space ($z<0.1$), and many more cluster redshifts are needed to ensure a representative sample. More galaxy redshift surveys are also necessary to confirm whether or not Abell clusters are good tracers of the overall distribution of luminous matter.

REFERENCES

[1]D. J. Batuski and J. O. Burns, submitted to MNRAS.
[2]D. J. Batuski and J. O. Burns, submitted to Nature.
[3]R. P. Kirshner, et al., Astrophys. Lett. **248**, L57 (1981).
[4]N. A. Bahcall and R. M. Soneira, Astrophys. J. **262**, 419 (1982).

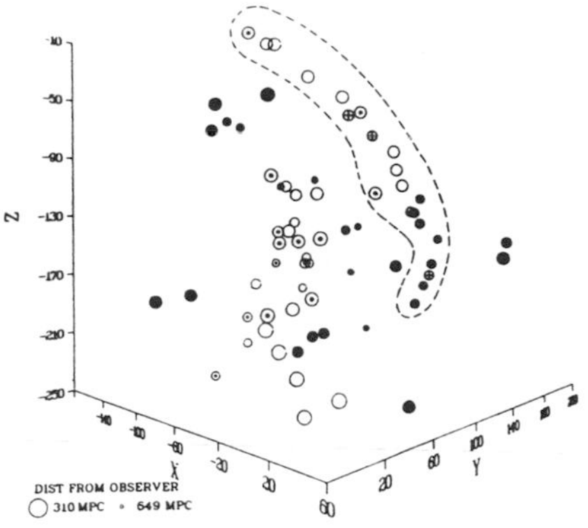

Figure 1: Plot of projection of the three-dimensional arrangement of clusters in the Pisces-Cetus supercluster candidate and neighboring clusters. Units for the axes are 1 h^{-1} megaparsecs. Earth is at X=Y=Z=0, Z-axis is along north galactic pole and X-axis is along l=b=0°, so large empty regions in upper third of cube are result of obscuration. Each symbol represents one cluster and is sized to show distance from the observer's "position" in space, 480 h^{-1} Mpc from center of cube. Symbols: ● - non-member of Pisces-Cetus, ⊙ - rich (R>0) Abell Cluster member, O - poor (R=0) Abell cluster member, ⊕ - poor Zwicky cluster. The filamentary arrangement within the dashed contour is over 310 h^{-1} Mpc in length.

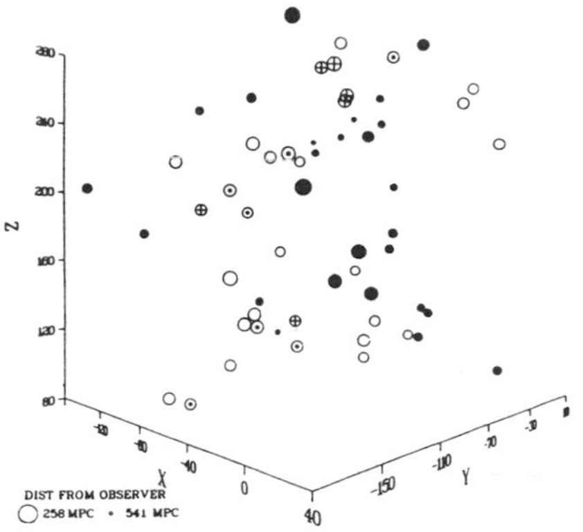

Figure 2: Projection plot of Sextans-Leo region with same legend as Figure 1. This region is 200 h^{-1} Mpc on a side and is drawn from perspective of observer 400 h^{-1} Mpc from center. Obscuration affects only small portion of region along lower left front edge of cube.

The Shape of Large Scale Structure in the Universe

Ethan Vishniac

ABSTRACT

We propose a statistic designed to measure the degree of filamentary structure in a two dimensional map as a function of scale. We point out some advantages of this statistic over previously proposed measures of filamentarity. We use our statistic to analyze the Shane-Wirtanen galaxy counts, and briefly discuss the results.

The compilation of galaxy counts from the Lick survey into a map of the distribution of galaxies on the sky, down to a magnitude of about 18.5, constitutes our major source of information about the nature of very large scale clustering of galaxies in the universe.[1,2] One of the more striking features of this map is that the clustering of the galaxies appears to form a "frothy" pattern on the sky. In fact, there are many places where the eye can pick out strings of high density regions extending large distances across the sky.

If these features are real then they contain a powerful clue to the origin of large scale structure in the universe. In fact, they have often been used as an argument in favor of theories in which the large scale structure of the universe forms before, or at the same time as, individual galaxies. Examples of such models are the "pancake" theory advanced by Zel'dovich and his collaborators[3] and more recent variations on this theory involving some species of massive neutrino.[4] In contrast, models that rely on the hierarchical formation of structure through gravitational forces, e.g., the "isothermal" model of Peebles and collaborators,[5] tend to produce roughly spherical structures with no sign of large scale filaments. The galaxy formation models involving "cold" particles[6,7,8] may be intermediate in this respect. Unfortunately, the appearance of such features in the Shane-Wirtanen counts is not conclusive proof that galaxies actually cluster in such patterns. The Lick survey is actually a mosaic, consisting of 1242 6 degree by 6 degree plates. In spite of work done on calibrating the plates by comparing regions of overlap, it seems necessary to regard any large scale features with suspicion.

Clearly, there is a need for a means of statistically analyzing this structure so that we can gain some insight into the nature of the filaments, and their significance. The statistical tools which have been used with the most success to analyze galaxy clustering are the n-particle correlation functions.[5] Briefly stated, the two particle correlation function at a given separation r is the fractional enhancement in the probability of finding a galaxy at a distance r from another galaxy. This can be generalized to m particles in a straightforward manner, although the mathematical labor involved in actually determining the mth correlation function increases exponentially. Unfortunately the lower order correlation functions do not constitute an adequate description of the filamentary structure.[9] Theoretically this could be solved by extending the hierarchy of known correlation functions. In practice this is unreasonable, both because the significance of the higher order correlations is unclear (and the mathematical labor prodigious), and because the mth correlation function is weighted according to the local density to the mth power, and therefore tells us primarily about the cores of rich clusters. Although the entire correlation hierarchy does contain the relevant information it does so in an indigestible way. It is necessary to invent some alternate tools to study filaments.

What are the criteria for devising an appropriate statistic? First, it should be sensitive to the shape of clustering, rather than its strength. Second, it should be weighted by the local density to the first power. Third, it should be well calibrated, in the sense that its value for distributions whose statistical properties are well known should be calculable. Fourth, it should be predictable for a map which is a superposition of maps of known statistical properties. This last point is important because it will enable us to compare galaxy surveys of varying depths to decide whether the perceived structure is intrinsic or due to intervening absorption or selection effects.

We are now ready to define a statistic which is a measure of filamentarity as a function of scale, and which meets the above criteria. Consider the neighborhood of some galaxy, where the neighborhood is defined by a radius R. If there are N neighboring galaxies, then we define the moments of the distribution as

$$M^{ij\ldots k} \quad \frac{1}{N} \sum_{s=1}^{N} x^{i}(s) x^{j}(s) \ldots x^{k}(s) \tag{1}$$

We will construct our statistic from coordinate invariant combinations of the matrices M^{i} and M^{ij}. The higher order moments are increasingly dependent on the positions of the galaxies closest to the window edges and are therefore unsuitable. The general form of the statistic is

$$S = \frac{C_0 M^{ij} M^{ij} + C_1 (M^{ii})^2 + C_2 M^{ij} M^{i} M^{j} + C_3 M^{i} M^{i} M^{j} M^{j}}{(M^{ii})^2} \tag{2}$$

where the implicit summation rule is followed for repeated indices. The denominator is chosen to minimize the effect of clustering within the window.

We then constrain the coefficients C_i by the following choices. If the window contains a straight line of uniform density which passes through the window center then we require that the signal strength S be equal to one. If the window contains a uniform distribution with a uniform gradient of arbitrary size then we require S = 0. Finally, if the window contains only a circle of uniform density and size r, displaced from the center but contained entirely within the window, then we require S = 0. We obtain

$$S = \frac{2M^{ij}(M^{ij} - M^{i}M^{j}) - M^{ii}(M^{jj} - M^{j}M^{j})}{(M^{ii})^2} \tag{3}$$

The statistic defined in Eq. (3) has a calculable value for a random distribution of points, or even an isotropic realization of a given hierarchy of correlation functions. Calibrating its value to a given cosmological model is a more involved process that can only be satisfactorily done by a direct comparison between various n-body calculations and actual data.

It is useful to note that since widely separated regions are (probably) uncorrelated, we expect to see a crude scaling law of the form

$$\frac{\langle S\rangle}{\langle S_P\rangle} = F(D_{eff}\theta) \tag{4}$$

apply to statistics derived from galaxy samples of different effective depths. In Eq. (4), the symbol $\langle S\rangle$ denotes the value of the statistic averaged over all the galaxies in the uncorrected data. The symbol $\langle S_P\rangle$ denotes the average of the Poisson expectation values for a sample with the same number of galaxies within the same number of windows. D_{eff} is the effective depth of the sample and θ is the angular size of the window.

Figure 1 shows some preliminary results due to applying the statistic given in Eq. (3) to the Shane-Wirtanen counts. The ordinate of $\langle S\rangle/\langle S_P\rangle-1$ gives the excess amount of filamentarity on a given scale. The horizontal axis shows the physical size of the window diameter scaled to the mean depth of the sample. This is taken to be 220 h^{-1}Mpc. Two sets of points are shown. One set shows the results for plates whose center coordinate has a galactic latitude between 40 and 55 degrees. The other set is for plates with a galactic latitude between 70 and 90 degrees.

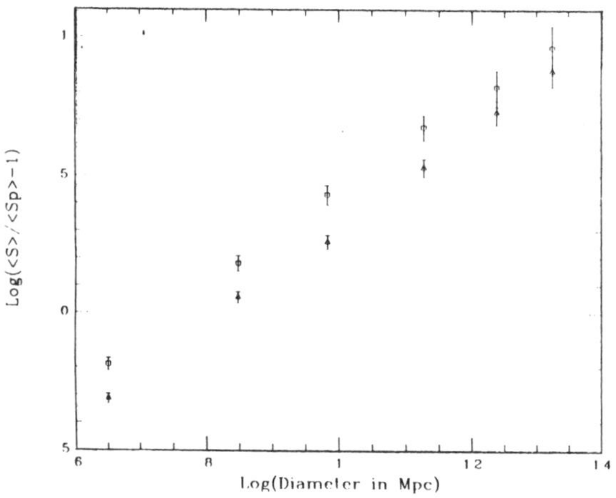

Figure 1: The fractional excess in the average signal $\langle S\rangle$, above what would be expected for Poisson statistics, is shown for two data sets taken from the Shane-Wirtanen counts, for a variety of sampling radii. The points denoted by Δ are taken from plates with $40° < b < 55°$. The points denoted by $\square$ have $70° < b < 90°$. The error bars are 1σ errors determined from the variance of $\langle S\rangle$ from separate plates.

Several points are worth noting. First, the two sets of points trace out roughly parallel lines in accordance with what we expect from Eq. (4). The average horizontal displacement is a factor of 1.21. The displacement expected from assuming the real depth of the sample to vary as the galaxy number density to the one third is 1.12, which is significantly less. Whether or not this constitutes a serious discrepancy is unclear. The sample depth which this statistic is sensitive to is not necessarily exactly the same as the effective depth determined from number counts.

Second, if the density perturbation spectrum has the form $|(\delta\rho/\rho)(\underset{\sim}{K})|^2 \propto |\underset{\sim}{K}|^n$ where $n>0$, for scales greater than some wavelength L, then we would expect to see the function $\langle S\rangle/\langle S_P\rangle-1$ bend over when the window diameter is comparable to L. There is no indication in Fig. 1 of any curvature to the fitted line. In view of the reported bend in the correlation function between 20 h^{-1} and 40 h^{-1}Mpc,[10] it would be interesting to try this on a data set with a larger transverse size.

Third, although <S>/<S_p>-1 is increasing as a function of window diameter, <S> itself is decreasing. The amplitude of the signal is more or less consistent with galaxies lying on sheets of size 10-20 h^{-1}Mpc. It is inconsistent with all galaxies being contained in filaments of that size.

Finally, this work has been applied[11] to a series of deep galaxy counts made by Tyson and Jarvis (1978).[12] The results are consistent with the Shane-Wirtanen counts, although the transverse scales are less than 10 h^{-1}Mpc. To be more precise, the slope recovered from the analysis of the deep plates matches the slope obtained from the Lick sample. The amplitude scales slightly faster with depth than Eq. (3) would indicate. This shows that, at least on such scales, the impression of filamentary structure is not due to variable galactic absorption, but is an intrinsic property of the galaxy clustering.

REFERENCES

[1]C. D. Shane and C. A. Wirtanen, Publ. Lick Obs. **22**, part 1 (1967).

[2]M. Seldner and P.J.E. Peebles, Ap. J. **215**, 703 (1976).

[3]For a review see Ya. B. Zeldovich, in The Large Scale Structure of the Universe, eds. M. S. Longair and J. Einasto (Dordrecht: Reidel) p. 409 (1978).

[4]For example: A. G. Doroshkevich, M. Yu. Khlopov, R. A. Sunyaev, A. S. Szalay and Ya. B. Zeldovich, in the Tenth Texas Symposium on Relativistic Astrophysics, eds. R. Ranaty and Frank C. Jones, Annals of the New York Academy of Sciences **375**, 32 (1981); J. R. Bond, G. Efstathiou and J. Silk, Phys. Rev. Lett. **45**, 1980 (1980).

[5]P.J.E. Peebles, The Large Scale Structure of the Universe, (Princeton University Press 1980).

[6]J. R. Bond, A. S. Szalay and M. S. Turner, Phys. Rev. Lett. **48**, 1636 (1982).

[7]P.J.E. Peebles, Ap. J. **263**, L1 (1982).

[8]J. R. Primack and G. R. Blumenthal, in Formation and Evolution of Galaxies and Large Structures in the Universe, eds. J. Audouze and J. Tran Thinh Van (Dordrecht- Reidel) p. 163 (1983).

[9]R. M. Soneira and P.J.E. Peebles, A.J. **83**, 845 (1978).

[10]M. Davis and P.J.E. Peebles, Ap. J. **267**, 465 (1983).

[11]E. T. Vishniac and J. A. Tyson, in preparation (1985).

[12]J. A. Tyson and J. F. Jarvis, Ap. J. **230**, L153 (1978).

22

Clustering at High Redshifts

P.A. Shaver

ABSTRACT

Evidence for clustering of and with high-redshift QSOs is discussed. QSOs of different redshifts show no clustering, but QSOs of similar redshifts appear to be clustered on a scale comparable to that of galaxies at the present epoch. In addition, spectroscopic studies of close pairs of QSOs indicate that QSOs are surrounded by a relatively high density of absorbing matter, possibly clusters of galaxies.

CLUSTERING OF QSOs

A. QSOs of Different Redshifts

The clustering of QSOs of different redshifts can be studied by examining the number of QSO pairs as a function of angular separation. Figure 1 shows that the shape of this distribution depends on the extent to which the QSOs are clustered.

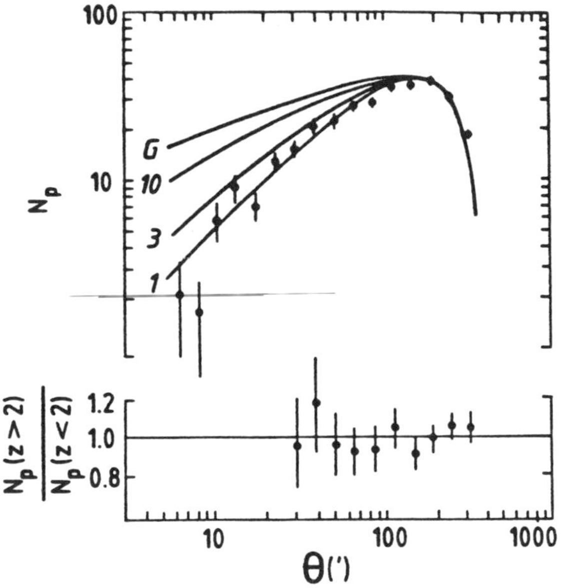

Figure 1: Differential frequency distribution of QSO pairs with $\Delta z > 0.05$ as a function of angular separation. The data points are from objective prism surveys listed in Shaver.[1] Curves 1, 3, and 10 represent Monte Carlo simulations, for no clustering (curve 1), and for cluster/field contrast ratios of 3 and 10. Curve G represents galaxy clustering from the Zwicky catalog. All have been normalized to the peak of the distribution.

The data points in Fig. 1 are consistent with a random distribution of QSOs on the sky -- no clustering is evident. Figure 2 shows similar distributions for three different samples of QSOs.

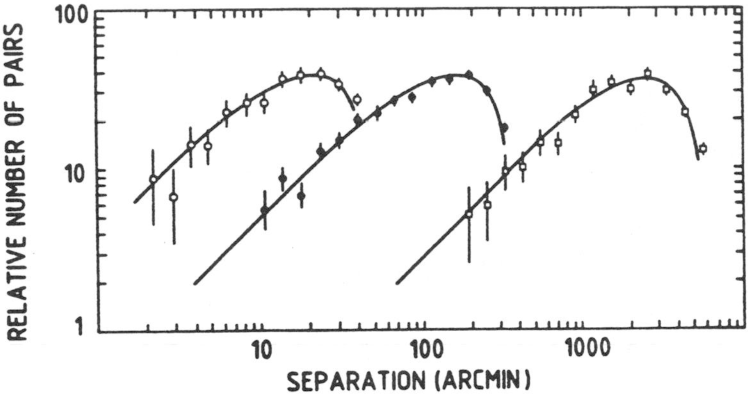

Figure 2: Differential frequency distribution of QSO pairs with $\Delta z > 0.05$ as a function of angular separation. The data points are from grism surveys (O), objective prism surveys (●), and a UVX survey (□).[1] The curves are Monte Carlo simulations of random distributions.

In all of them, the distributions of QSOs appears to be random. As these three distributions overlap, we may conclude that the sky distribution of QSOs of different redshifts is random on all scales from arcminutes to tens of degrees.

The results given here and in (b) below are elaborated in Shaver[1]; the data are from the Véron QSO catalog.[2] Other recent studies also find no compelling evidence for any widespread physical clustering amongst QSOs of different redshifts.[3-7] This is consistent with a cosmological interpretation of QSO redshifts, for which the projection on the sky of QSOs of arbitrary redshifts is expected to be random. The close pairs, groupings, and alignments of QSOs of different redshifts which have sometimes been taken as evidence for non-cosmological redshifts are apparently no more than chance coincidences.

B. QSOs of Similar Redshifts

The physical clustering of QSOs should show up amongst those QSO pairs which have small separations and redshift differences. By contrast, the vast majority of QSO pairs with large separations and/or redshift differences should exhibit no clustering (as shown above). The clustering of QSOs can therefore be estimated by <u>comparing</u> these two groups. Both should be subject to the same selection effects, which should then cancel out in such a comparison. Such a study can therefore be done using large QSO catalogs, in spite of their heterogeneous character. Previous searches for the physical clustering of QSOs were based on small, homogeneous samples, and the only evidence found were some groupings of QSOs in position and redshift which may conceivably be physical entities (e.g. Oort et al., 1981, and references therein).[8]

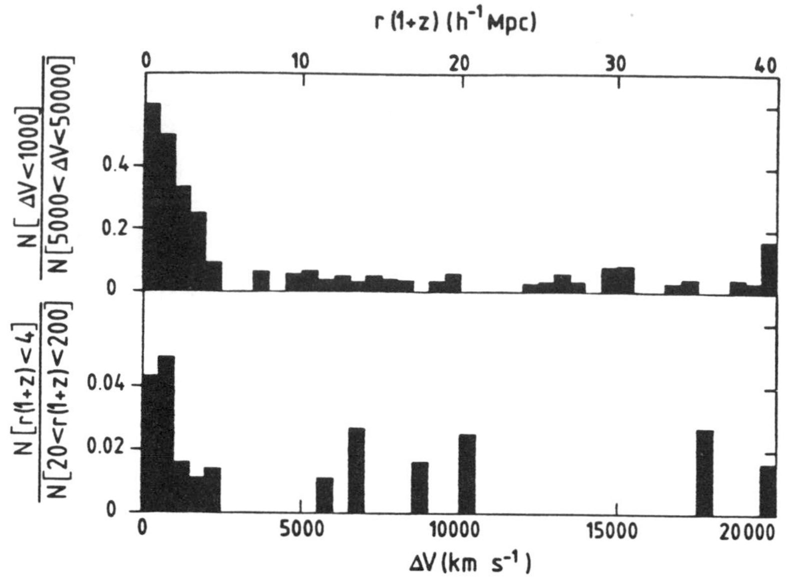

Figure 3: Top: ratio of the number of QSO pairs with small redshift difference ($\Delta V < 1000$ km s^{-1}) to the number with large redshift difference, against projected linear separation. Bottom: Ratio of the number of QSO pairs with small projected linear separation ($r(1+z) < 4$ h^{-1} Mpc) to the number with large separation, against velocity difference.

Figure 3 shows that there is indeed an excess of QSO pairs of small projected separation (< 5 h^{-1} Mpc in comoving coordinates, with $q_0 = 0$) and small velocity difference (< 2500 km s^{-1}). Extrapolating from larger separations and velocity differences, 2.4 such pairs would be expected if there were no clustering, whereas 12 were found. Evidently, the probability is high that QSOs which are close to each other on the sky have similar redshifts. Two of these pairs may be gravitationally lensed QSOs, which are irrelevant for clustering, but the others have angular separations which are too great for plausible gravitational lenses. And there is one selection effect which gives rise to some uncertainty: the possible inclusion of close pairs which were published preferentially because they also have similar redshifts. This would affect 3 of the remaining pairs at most,[1] so the number of unbiased close pairs of distinct QSOs probably lies between 7 and 10, still well in excess of the expected 2.4.

It can be seen in Fig. 4 that the QSO-QSO correlation function, generated from the data in Fig. 3, appears to be comparable with the correlation function for galaxies at the present epoch. Any corrections for the selection effects mentioned above are well within the error bars shown. We may tentatively conclude that the degree of clustering at z ~2 was not radically different from that at present, but more definitive conclusions clearly await larger and deeper QSO samples.

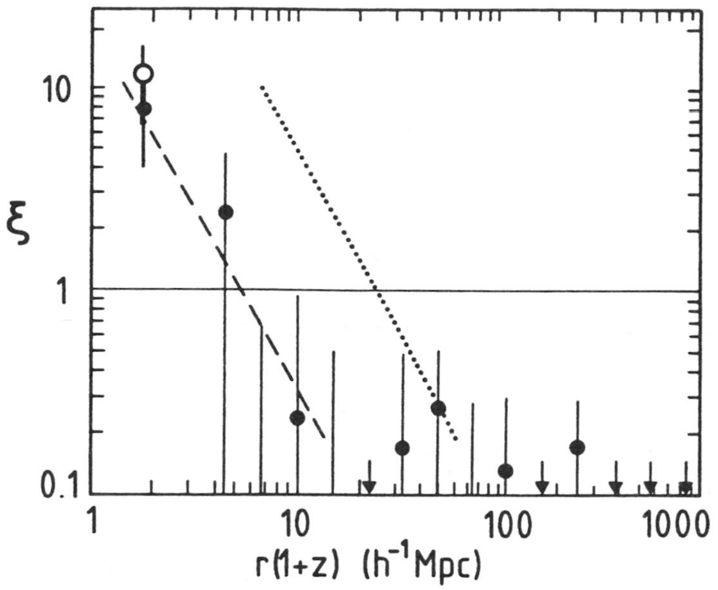

Figure 4: (Projected) spatial correlation function from the data in Fig. 3. The open circle includes two possible gravitational lens QSOs. The dashed and dotted lines represent the galaxy-galaxy and cluster-cluster spatial correlation functions respectively.[9] The mean difference between true and projected scale of ~15% has not been applied here.

ABSORPTION ASSOCIATED WITH QSOs

Further evidence for clustering at high redshift comes from absorption-line observations of pairs of QSOs of very different redshifts which happen to be located close to each other on the sky. Narrow CIV absorption systems have been found in the spectrum of the background QSO at the redshift of the foreground QSO (within 2000 km s^{-1}, for angular separations < 5') in 4 cases out of 8.[10] Examples are shown in Fig. 5. QSOs are evidently located in regions of relatively high matter density. Where the absorption-emission redshift difference is small (~100-200 km s^{-1}) the absorption may conceivably arise in large gaseous halos around the foreground QSOs, but in other cases it is most likely due to clusters of galaxies associated with the foreground QSOs. When adequate statistics have been built up, such absorption studies will provide an independent means of examining the clustering at high redshifts.

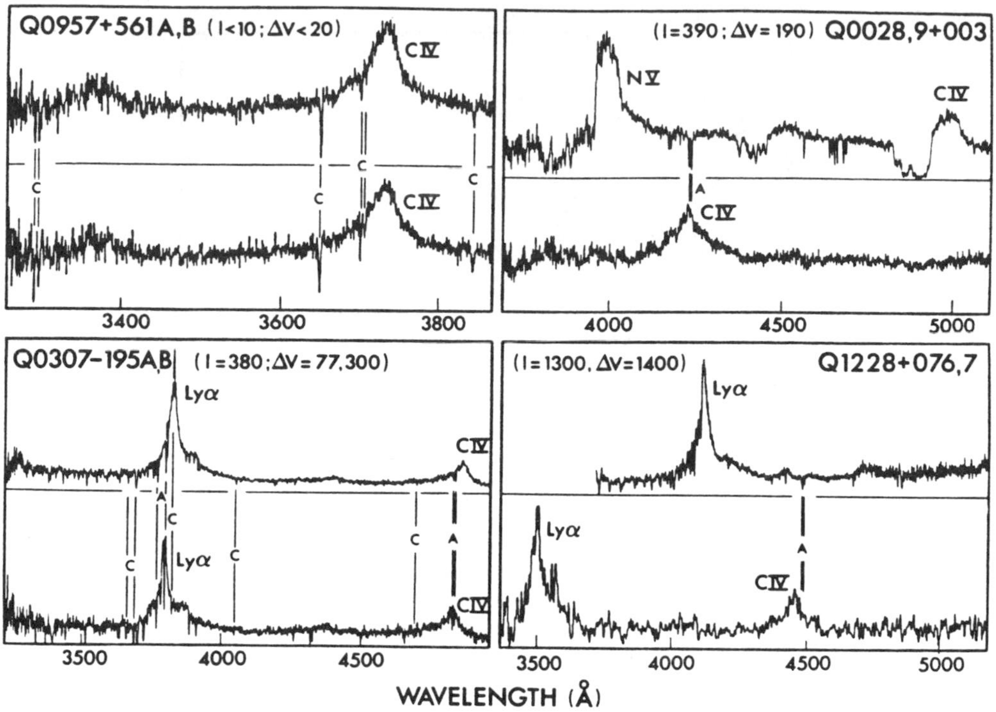

Figure 5: Spectra of four QSO pairs, showing common (C) and associated (A) absorption. Also indicated are the projected separation (ℓ, in h^{-1} kpc) and the velocity difference (ΔV, in km s^{-1}).

REFERENCES

[1]P.A. Shaver, Astron. Astrophys. (in press).

[2]M.P. Véron and P. Véron, ESO Scientific Report No. 1, (1984).

[3]P.S. Osmer, Astrophys. J. **247**, 762 (1981).

[4]A. Webster, Mon. Not. R. Astron. Soc. **199**, 683 (1982).

[5]Y. Chu and X. Zhu, Astrophys. J. **267**, 4 (1983).

[6]T. Shanks, R. Fong, M.R. Green, R.G. Clowes, and A. Savage, Mon. Not. R. Astron. Soc. **203**, 181 (1983).

[7]B.J. Boyle, R. Fong, and T. Shanks, Quasars and Gravitational Lenses, Proc. 24th Liège International Astrophysical Colloquium, p. 368.

[8]J.H. Oort, H. Arp, and H. de Ruiter, Astron. Astrophys. **95**, 7 (1981).

[9]N.A. Bahcall and R.M. Soneira, Astrophys. J. **270**, 20 (1983).

[10]P.A. Shaver and J.G. Robertson, Mem. Soc. Astron. Ital. **54**, 665 (1984).

Masses of Galaxies

John S. Gallagher, III

I. INTRODUCTION

It is both a remarkable and fundamental property of the present Universe that the majority of 'visible' matter (i.e., matter emitting photons) is concentrated into gravitationally bound systems which we know as galaxies. While galaxies have been recognized since the pioneering work of E. Hubble in the 1920s,[1] our basic model of galaxies as spatially well separated 'island universes' has undergone significant changes during the past decade. These have arisen from the realization that the optically detectable components of galaxies (consisting mainly of thermally radiating stars) may contain only a minor fraction of the total mass. This conflict between mass whose existence can be identified from measurements of electromagnetic radiation and masses derived from dynamical arguments is sometimes known as the 'missing mass', or more appropriately 'missing light', problem in galaxies.[2] Until this problem is resolved, we will not be able to develop a full understanding of the structure and evolution of galaxies, or properly estimate the total contribution of galaxies to the cosmological mass density.

II. ROTATION CURVES, LIGHT DISTRIBUTIONS, AND GALACTIC MASSES

In common spiral-type galaxies (such as the Milky Way) much of the luminous matter is located in highly flattened, rotationally supported galactic disks. Velocity fields for matter within the disks can be measured along the line of sight from Doppler shifts of known spectral features; e.g., stellar absorption lines, emission lines from ionized gas, or the hyperfine transition at $\lambda 21.1$ cm from low density interstellar HI gas. Because galaxies are both small in angular size and are quite faint relative to natural and detector noise sources, velocity field measurements until the late 1960s were largely limited to the bright inner regions of spiral galaxies. These measurements showed matter in disks to be moving in nearly circular orbits and the derived rotation laws appeared to be consistent with the mass following the observed distribution of stars, although the uncertainties were recognized to be large.[3]

To illustrate these expectations, we consider a particularly structurally simple class of galaxies, i.e., those in which the galactic disk contains virtually all of the stars and gas (see Figure 1). Studies of the light distributions in common spiral galaxies show that the projected radial fall off in intensity roughly follows an exponential law for galactic disks; i.e., $I(r) = I(0) \exp(-\alpha r)$ where α^{-1} is a scale length of typically 2-5 kpc (1 kpc = 3.1×10^{21} cm; h = 0.5 is used throughout this paper).[4] We then see that the total optical radiation power received from the disk rapidly converges, $L_\infty = 2\pi \int_0^\infty I(r)\, dr$, and $L_{Holm} \approx 0.95\, L_\infty$, where the Holmberg radius is a standard definition of galactic optical size defined by the radius where $I \sim 5\ L_\odot$ pc^{-2} in blue light (about 1% of the dark night sky). We conclude that a sphere with a Holmberg radius contains most of the stars (as well as known gas) and therefore presumably most of the visible mass in a galaxy. Thus the rotation velocities should begin to decline beyond the Holmberg radius in a disk galaxy consisting only of visible matter.

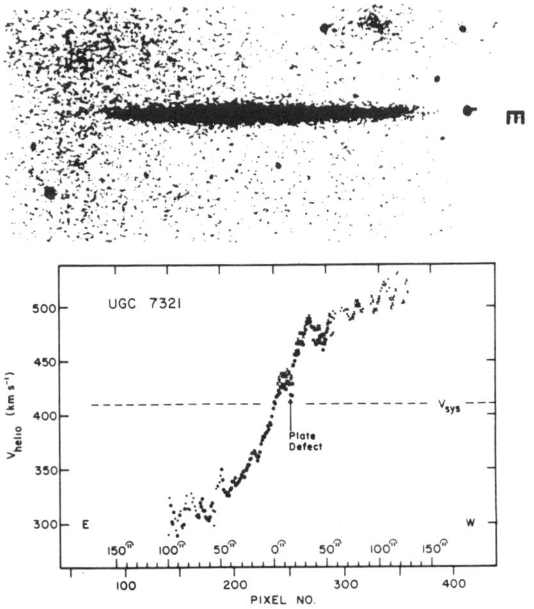

Figure 1: The upper panel shows a blue light image of a disk galaxy seen almost exactly edge-on, and below at approximately the same linear scale is a radial velocity curve derived from Doppler shifts of the Hα Balmer emission line arising from ionized gas within the galaxy's disk. While the total light has converged by the edge of the optical image, these observations show the rotation velocities of disk material are continuing to increase, implying that an optically invisible mass component is present. These data are from a study by Goad and Roberts.[23]

For a 2-D exponential disk model of infinite extent, Freeman derived a rotation curve which depends only on α^{-1} and the central projected mass density μ_0[5]. The calibration of the model in terms of observable quantities is obtained by assuming that the local mass density is proportional to the intensity at some optical wavelength λ, i.e., $\mu_0 = (M/L)_\lambda I(0)$. This procedure is especially useful as empirically I(0) is roughly constant in spiral galaxy disks. We can make this calibration in two ways: (1) Assume that the distribution function which describes the range of stellar masses in the immediate vicinity of the Sun is a universal property of galactic disks. This is a necessary assumption since the types of stars which produce most of the light at any wavelength contain only a small fraction of the mass. Thus the low mass stars only can be detected in the solar neighborhood, where individual stars are observed, and faint stars are not overwhelmed by their rare but luminous counterparts. After a correction for mass in the form of detectable interstellar gas, we deduce the $(M/L)_\lambda$ for identified mass in the neighborhood of the Sun, and can extend this result to other galaxies via stellar population modelling and suitable corrections for gaseous matter. (2) Dynamical analyses of stellar velocities perpendicular to the disk of the Milky Way in which the Sun resides yield an estimate for the total surface density of the Galactic disk near the Sun.[6] From properties of the Milky Way, we use this total mass estimate (which is about twice the identified disk mass value) to find μ_0 for the Milky Way, which we take to have a typical galactic disk. The resulting model rotation curves from procedures (1) and (2) are shown in Figure 2.

By the early 1970s important advances in experimental techniques led to new kinematic observations which conflicted with predictions based on simple models of the visible disks of galaxies. High sensitivity optical detectors allowed rotation curves derived from slit spectra to be extended to near the Holmberg radii of spiral galaxies.[7] Radio facilities underwent similar improvements in sensitivity and aperture synthesis techniques further provided the spatial resolution to map velocity fields in HI gas to beyond the optical limits of many spirals.[8] Both of these approaches yielded flat rotation curves in the majority of galaxies; i.e., the circular rotation velocity was found to

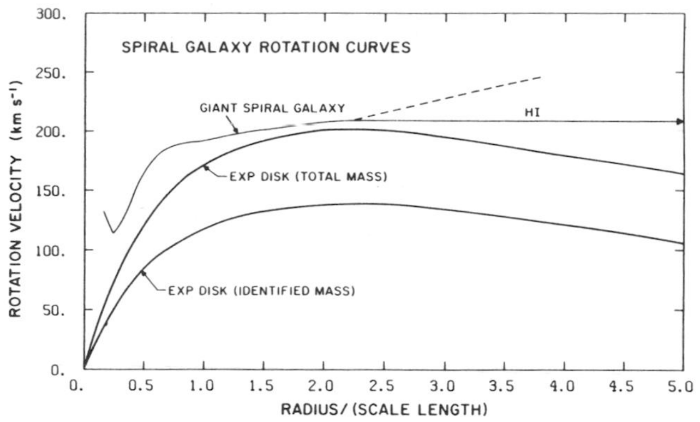

Figure 2: A galactic rotation curve for a typical spiral galaxy derived from optical and radio observations is compared with expectations for an infinite exponential disk model. Radii are in units of disk density e-folding scale lengths, and the Holmberg radius typically corresponds to about 5 scale lengths in giant spirals. A disk model based only on mass identified by emitted light in the solar neighborhood for its normalization falls far short of the observations, while a model normalized to a dynamically determined local Galactic disk density provides a reasonable fit to observations at smaller radii. The divergence between the models and observed flat rotation curves at large radii provides prime evidence for an extended, invisible mass component in many spiral galaxies.

be constant or even slightly rising out to and beyond the Holmberg radii of typical spiral galaxies[9] (e.g., Figures 1 and 2). As shown in Figure 2, these results seriously differ from expectations derived from 'visible' matter, and thus imply the existence of an unseen mass component in spiral galaxies.

Using rotation velocities, we can derive a minimum mass within radius r. Assuming Newtonian gravity, $M(r) = \varepsilon(r)V^2(r)rG^{-1}$ where $0.6 \leq \varepsilon(r) \leq 1$ is a correction for flattening of the galactic mass distribution ranging from a thin disk to a sphere.[10] Also note that for non-spherical, finite models, M(r) is in general sensitive to the total mass distribution, and not just the interior mass. Flat rotation curves with $V^2(r) \approx$ constant thus imply isothermal sphere-like diverging mass distributions with $M(r) \propto r$, in marked contrast to the rapid radial convergence of the light from stars and diffuse gas.[11] A representative spiral galaxy might have a Holmberg radius of 25 kpc (7.7×10^{22} cm) and a flat rotation curve at 225 km s^{-1}, implying a minimum dynamical mass of $3\times10^{11}\ \varepsilon(r)\ M_\odot$.

The lack of consistency between the apparent distributions of total visible mass has led astronomers to consider three major possibilities: (1) Spiral galaxies contain substantial amounts of non-luminous matter, (2) spirals are not in a state of dynamical equilibrium or the models are in other ways incomplete, or (3) Newtonian gravity does not work on the scale of galaxies.[12] Most astronomers have adopted the first alternative with some worries about the second point.

III. MAPPING MASS DISTRIBUTIONS IN GALAXIES

The discovery of flat rotation curves fostered a broad reexamination of the properties of spiral galaxies which might bear on the dark matter problem; i.e., how much dark matter is there, is it all in one dynamical component, and how is it spatially distributed? A brief summary of some of the major areas receiving attention follows:

1. An increased emphasis has been placed on the determination of velocity fields in the outer regions of galaxies, particularly via HI measurements with radio synthesis telescopes. These observations show the outer HI often has complex properties, e.g., the outer HI disk may be strongly warped rather than lying in a plane and the

overall spatial distribution of gas may be quite asymmetrical. As a result, there are difficulties in obtaining highly accurate rotation curves at large radii.[13]

2. Improved studies have been made of the low level distribution of light from galaxies. These are designed to check for faint, extended stellar components that might previously have been missed. No unusual features have been found, and increasingly strict limits have been placed on optical/IR luminosity from the dark mass component.[14]

3. A test of the Milky Way galaxy mass distribution has been obtained by utilizing globular star clusters and small satellite galaxies as probes of the Galactic tidal field. These results appear to be consistent with the type of model derived from the flat rotation curves; i.e., a spherical isothermal mass distribution containing perhaps 10^{12} M_o extends well beyond the optical limits of the Galaxy.[15]

4. It had long been hoped that binary galaxy samples could provide a statistical sampling of the large-scale gravitational potentials of galaxies in the 30-100 kpc range where HI is often no longer detectable. A variety of problems have hobbled this approach. Initially there were difficulties in selecting samples relatively free from contaminating physically unrelated binary galaxies, which are near each other only when seen in projection, and in obtaining sufficient precision in measuring projected velocity differences. While these difficulties have been largely surmounted, a more fundamental obstacle remains. As pointed out by White et al. and Lynden-Bell et al.,[16] the mean mass derived from a sample of binary galaxies depends sensitively on orbital eccentricities while the observed parameters of binary samples are insensitive to orbital eccentricities. Thus intrinsic uncertainties of factors of 4 or more remain in typical masses assigned to binary galaxy samples. Furthermore, the miniumum masses that are derived for binary systems are on the order of those found from rotation curve studies, and thus do not provide a clean test for properties of extended mass distributions. Representative results are shown in Figure 3.

5. Gravitational lensing characteristics of individual galaxies have the potential to reveal mass distributions on large scales, but initial results are not yet conclusive.[17]

6. More attention has been paid to producing realistic comparisons between observations and model predictions. This involves inclusion of the visible multiple structural components of galaxies (e.g., stellar bulges and halos) and taking account of the leveling of rotation curves at the edges of sharply truncated stellar disks in making models. The improved models generally require an extended dark matter component to fit spiral galaxy rotation curves.[18]

7. Theoretical studies have been undertaken to examine the stability of pure disk galaxies. Ostriker and Peebles pointed out that self-gravitating, axisymmetric cold stellar disks are unstable against formation of linear, centered density enhancements known as

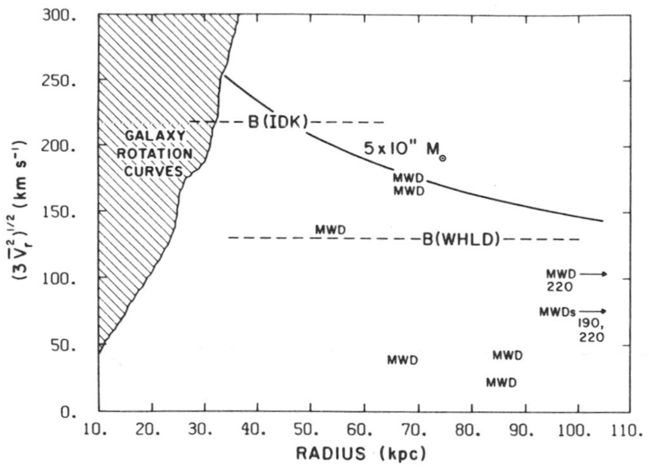

Figure 3: A schematic summary of binary galaxy data is shown for two major samples of binary systems (IDK = Karachentsev; WHLD = White et al.) and for dwarf galaxy companions of the Milky Way (MWD, Lynden-Bell et al.).[16] Since we are located near the center of the spatial distribution of MWDs, Doppler shift velocity measurements are most sensitive to radial orbital motions. Also in this case the distances of the MWDs from the center of mass are relatively well known. A minimum mass of $\sim 2\times10^{11}$ M_o is obtained from MWDs for our Milky Way system by assuming the radial orbital components dominate. External binary galaxy samples are seen in projection, and thus true mean radial separations are model dependent and minimum masses are obtained for near circular orbits. A Keplarian velocity curve for a 5×10^{11} M_o point mass is also illustrated as is the area covered by observable disk rotation velocities. Because of the large but poorly known corrections for orbital eccentricities, the binary galaxy data do not yield precise information on galaxy halo masses.

bars. Yet many galaxies appear to have all their visible matter in dynamically cold disks, but do not have bars. Since the bar instability is eliminated if the disk is embedded in a massive, dynamically hot halo, this line of reasoning suggests the presence of invisible massive envelopes around some disk galaxies,[19] although barred spiral galaxies are also common.

8. Similarly the questions of the origins and persistence of warps in outer disks of spirals remain unsolved. Warps are most readily understood if the overall mass distributions of even galaxies which optically appear to be pure thin disks are spheroidal or triaxial in form.[20]

From these kinds of considerations, a 1980s consensus standard model has emerged for typical luminous spiral galaxies which attempts to interpret flat rotation curves in a simple way. Stars and detectable gas are concentrated in the disk with an additional spheroidal stellar component being present in most instances. The visible mass extends to a radius of 20-30 kpc and amounts to $\sim 10^{11}$ M_o. These tips of galactic mass distributions are embedded within non-radiating spheroidal 'dark envelopes' or 'massive halos' which are tidally limited by nearby galaxies at radii of ~100 kpc and contain $\sim 10^{12}$ M_o. The blue M/L for such a system is ~100. This working model has the advantage of naturally fitting most of the observations, but it is important to emphasize that due to the circumstantial nature of much of the evidence, it has not yet rigorously been proven correct.

A number of significant questions thus remain regarding the validity and universality of the standard dark halo model: (1) Do all galaxies of similar luminous mass content contain the same relative mass fractions of dark material? Among luminous galaxies, for example, it is not clear whether galaxies consisting only of stellar spheroids (the ellipticals) contain dark mass, and some spirals seem to be particularly deficient in visible mass content. (2) In disk galaxies there is good evidence for undetected matter within the dynamically cold disks, and thus the association of dark matter primarily with dynamically hot, extended spheroids can be questioned. (3) There is as yet no agreement on the presence of dark matter in low luminosity, low mass dwarf galaxies. This problem is made difficult by the low internal velocities of dwarf galaxies, by the small amounts of visible matter which they contain, and often by the tendency for dwarfs to be rich in gas that in molecular form can be effectively a mundane form of dark matter. Preliminary reports concerning detections of dark mass components in dwarfs are thus subject to a variety of uncertainties, and can only be considered as suggestive.[21]

IV. COSMOLOGICAL IMPLICATIONS

1. Even if we adopt the maximal dark halo model by assigning a blue M/L = 100 to all galaxies, the mass bound in galaxies is insufficient for closure of the Universe. Following the arguments presented in this symposium by J. Huchra, we see that by using a mean M/L to convert the observed blue luminosity density of galaxies to a mean mass density, the Ω_0 associated with matter in galaxies is $\lesssim 0.1$. We also recognize, however, that visible galaxies may only represent peaks of much larger density fluctuations, and thus the true Ω_0 may considerably exceed 0.1.

2. The very existence of large amounts of dark matter within galaxies, of course, is cosmologically interesting, and implies that we still have an incomplete inventory of even our local region of the Universe. There are good indications that some of the dark matter is 'cold' since it is found in galactic disks and perhaps in dwarf galaxies. On the other hand, indirect arguments for massive dark halos imply that much of the mass could be 'warm' or even 'hot'. The form of the dark mass in galaxies is unknown, but as G. Steigman has emphasized, it could all be baryonic. Baryonic material, perhaps in the form of very low mass, non-luminous stars, is an especially appealing candidate for cold dark matter near the Sun and elsewhere in galactic disks, where luminous and dark matter appear to be spatially coextensive. There are no compelling reasons why the dark halos should be made of baryons, and several indirect reasons why they should not.[22] At this stage all we can say is that non-baryonic, massive dark envelopes could exist in galaxies. Future studies of the stellar content and dynamics of galactic halos, and particularly the stellar halo of the Milky Way galaxy, which we can probe in 3 spatial dimensions and eventually in both tangential and radial velocities, thus will be of key importance in discovering the nature and significance of galactic dark matter.

I would like to thank Nigel Sharp for discussions which clarified several points regarding binary galaxy dynamics, Jean Goad for help in preparing Figure 1, and Garth Illingworth for comments on masses of elliptical galaxies. Many of the issues raised here have been the subjects of stimulating discussion at the Aspen Astrophysical Workshops over the past several years, and support of the Workshops by NASA and the Aspen Center for Physics is gratefully acknowledged.

REFERENCES

[1]See E. Hubble, Realm of the Nebulae (New Haven, Yale University Press, 1936).

[2]Recent reviews of masses of galaxies are given by P. C. van der Kruit and R. J. Allen, Ann. Rev. Astron. Astrophys. **16**, 103 (1978); S. M. Faber and J. S. Gallagher, Ann. Rev. Astron. Astrophys. **17**, 135

(1979), and in several papers in IAU Symposium 100 Internal Kinematics and Dynamics of Galaxies, ed. E. Athanassaula, (Dordrecht, Holland, D. Reidel, 1983).

[3]For example, see the summaries by E. M. Burbidge and G. R. Burbidge, in Galaxies and the Universe, eds. A. Sandage, M. Sandage, and J. Kristian (Chicago, University of Chicago Press, 1975), p. 81, and by M. S. Roberts in the same volume, p. 309.

[4]G. de Vaucouleurs, in Photometry, Kinematics and Dynamics of Galaxies, ed. D. S. Evans (Austin, University of Texas press, 1979), p. 1, J. Kormendy, in ESO Workshop on Two Dimensional Photometry, eds. P. Crane and K. Kjar (Geneva, Switzerland, ESO, 1980), p. 191 and references therein.

[5]K. C. Freeman, Astrophys. J. **160**, 811 (1970).

[6]This is known as the Oort limit, J. H. Oort, Bull. Astron. Soc. Netherlands **6**, 284, (1932) and Bull. Astron. Soc. Netherlands **15**, 45 (1960). Most recently this problem has been reanalyzed in detail by J. Bahcall, see this volume and 1984 preprint.

[7]Much of the optical work on global rotation curves has been carried out by V. Rubin and collaborators; e.g., see V. Rubin, W. K. Ford, Jr., and N. Thonnard, Astrophys. J. Lett. **225**, L107 (1978); V. Rubin et al., Astrophys. J. **238**, 1171 (1980); V. Rubin et al., Astrophys. J. **261**, 439.

[8]An important example is the study of extended HI in the nearby galaxy by M31 M. S. Roberts and R. N. Whitehurst, Astrophys. J. **201**, 327 (1975).

[9]The forms of rotation curves measured at large galactic radii was the subject of heated debate in the mid-1970s, see IAU Symposium 77 Structure and Properties of Nearby Galaxies, eds. E. M. Berkhuijsen and R. Wielebinski (Dordrecht, Holland, D. Reidel, 1978). However, the optical rotation curves by V. Rubin et al. (see above references) and HI maps and velocity fields from the Westerbork Radio Synthesis Telescope by, e.g., A. Bosma, Astron. J. **86**, 1721 and ibid., 1825 (1981) served to confirm flat rotation curves as a feature of many spiral galaxies.

[10]This point is summarized by J. Lequeux, Astron. Astrophys. **125**, 394 (1983).

[11]Constant rotation velocity disks were theoretically explored as resulting from collapse of rotating proto-galactic clouds by L. Mestel, Mon. Notices Roy. Ast. Soc. **126**, 553 (1963). The mass distributions in disk-dominated galaxies are empirically determined by D. Burstein, V. C. Rubin, N. Thonnard, and W. K. Ford, Jr., Astrophys. J., **253**, 330 (1982), and the conflict between dynamically deduced mass distributions and visible mass is critically reviewed by P. C. van der Kruit, Proc. Australian Astron. Soc. **5**, 136 (1983).

[12]A modification to Newtonian dynamics in galaxies is suggested by M. Milgrom, Astrophys. J. **270**, p. 365, ibid., p. 389 (1983).

[13]Observational descriptions of HI warps are given by D. H. Rogstad, I. A. Lockhart, and M.C.H. Wright, Astrophys. J. **193**, 309 (1974) and A. Bosma, Astron. J. **86**, 1791 (1981). Difficulties in deriving rotation curves from HI data are discussed by A. Bosma, Astron. J. **86**, 1825 (1981) and by R. Sancisi and R. J. Allen, Astron. Astrophys. **74**, 73 (1979).

[14]Examples of observational efforts to describe the structure of galaxies at low surface brightness levels include J. S. Gallagher and H. S. Hudson, Astrophys. J. **209**, 389 (1976); D. J. Hegyi and G. L.

Gerber, Astrophys. J. Lett. **218**, L7 (1977); E. B. Jensen and T. H. Thuan, Astrophys. J. Suppl. **50**, 421 (1982); R. G. Hohlfeld and N. Krumm, Astrophys. J. L **244**, 476 (1981), and a series of papers by P. C. van der Kruit and L. Searle, e.g., Astron. Astrophys. **110**, 79 (1982).

[15]F.D.A. Hartwick and W.L.W. Sargent, Astrophys. J. **221**, 512 (1978); C. A. Frenk and S.D.M. White, Mon. Not. R. Ast. Soc. **193**, 295 (1980); K. A. Innanen, W. E. Harris, and R. F. Webbink, Astron. J. **88**, 338 (1983).

[16]S.D.M. White, J. Huchra, D. Latham, and M. Davis, Mon. Not. R. Ast. Soc. **203**, 701 (1983); D. Lynden-Bell, R. D. Cannon and P. J. Godwin, Mon. Not. R. Ast. Soc. **204**, 87P (1983). For other recent discussions of the binary galaxy problems, see I. D. Karachentsev, Astrofizika **17**, 675 (1981); C.P. Blackman and G. A. van Moorsel, Mon. Not. R. Ast. Soc. **208**, 91 (1984); E. M. Sadler and N. A. Sharp, Astron. Astrophys. **133**, 216 (1984).

[17]See paper by E. Turner. J. A. Tyson, F. Valdes, J. F. Jarvis, and A. P. Mills, Astrophys. J. Lett. **281**, L59 (1984) suggest that the absence of lens-induced distortions in background galaxies which potentially could be gravitationally lensed by nearer galaxies is inconsistent with the standard massive halo model.

[18]A good example of model fitting is in J. H. Bahcall, Astrophys. J. **267**, 52 (1983); see also S. Casertano, Mon. Not. R. Ast. Soc. **203**, 735 (1983).

[19]J. P. Ostriker and P.J.E. Peebles, Astrophys. J. **186**, 467 (1983). A more recent investigation of the problem is given by G. Efstathiou, G. Lake, and J. Negroponte, Mon. Not. R. Ast. Soc. **199**, 1069 (1982), although the problem of bar formation as an instability in galactic disks is still not fully understood.

[20]For example, J. Binney, Mon. Not. R. Ast. Soc. **183**, 779 (1978); A. Toomre, in _IAU Symposium 100 Internal Kinematics and Dynamics of Galaxies_, ed. E. Athanassoula (Dordrecht, Holland, D. Reidel, 1983), p. 177.

[21]S. M. Faber and D.N.C. Lin, Astrophys. J. Lett. **266**, L17 (1983) and D.N.C. Lin and S.M. Faber, _ibid._, L21 argue for dark matter in small spheroidal galaxies, and M. Aaronson, Astrophys. J. Lett. **266**, L11 (1983) provides some supporting observations. More recent unpublished studies by P. Seitzer and J. A. Frogel, as well as a different approach by J. G. Cohen, Astrophys. J. Lett. **270**, L41 (1983), suggests much less mass. J. S. Gallagher and D. A. Hunter, Ann. Rev. Astron. Astrophys. 22, in press (1984) discuss problems in determining masses for small, gas-rich galaxies.

[22]See S.D.M. White and M.J. Rees, Mon. Not. R. Astron. Soc. **183**, 341 (1978); G. Efstathious and J. Silk, Fund Cosmic Rays. **9**, 2 (1983).

[23]J. W. Goad and M. S. Roberts, Astrophys. J. **250**, 79 (1981).

Missing Matter in the Vicinity of the Sun

John N. Bahcall

ABSTRACT

The Poisson and Vlasow equations are solved numerically for realistic Galaxy models which include multiple disk components (between 14 and 28), a Population II spheroid, and an unseen massive halo. The total amount of matter in the vicinity of the Sun is determined by comparing the observed distributions of tracer stars, samples of F dwarfs and of K giants, with the predictions of the Galaxy models. Results are obtained for a number of different assumed distributions of th unseen disk mass. The major uncertainties, observational and theoretical, are estimated. For all the observed samples, typical models imply that about half of the mass in the solar vicinity must be in the form of unobserved matter. The volume density of unobserved material near the Sun is about 0.1 $M_{\odot} pc^{-3}$; the corresponding column density is about 30 $M_{\odot} pc^{-2}$. This so far unseen material must be in a disk with an exponential scale height of less than 0.7 kpc.

The problem of determining the total volume density and the total column density of matter at the solar position is a classical one in astronomy. Some of the early investigations of this problem by Oort[1,2] led to one of the first astronomical suggestions of a large "missing mass." The method of weighing the matter in the local neighborhood can be summarized as follows. The density distribution of a set of tracer stars (e.g., K giants or F dwarfs) is measured perpendicular to the galactic plane (typically out to distances of order a few hundred parsecs). Theoretical models are then computed for the expected distribution of tracer stars in different gravitational potentials (mass distributions). The amount of matter that is actually present in the Galaxy is determined by comparing the observed and computed distributions.

The availability of modern computers has made possible significant advances in the theoretical analysis of this problem at the same time that better observational samples of tracer stars have been obtained. I have taken advantage of these developments to improve and strengthen the determinations of the total amount of matter in the solar vicinity, using more realistic Galaxy models and more accurate theoretical solutions. I have solved numerically the combined Poisson-Vlasow equation for the gravitational potential of Galaxy models consisting of realistically large numbers of individual isothermal disk components in the presence of a massive unseen halo. The calculations were carried out with different assumptions about the unseen matter and compared with the observed number densities of F dwarfs and K giants versus height above the plane.

The principal result of the present work is that about half of the matter in the vicinity of the Sun is in the form of unseen disk material which has a scale height of less than 0.7 kpc. The unseen material that is inferred from galaxy rotation curves at large galactocentric distances and from applying the virial theorem to groups and clusters of galaxies may not be the same as the unobserved disk matter discussed in the present paper.

Table I summarizes the Galaxy models for the observed mass components that were derived--using data from many sources--by Bahcall and Soneira[3] and by Hill, Hilditch, and Barnes,[4] hereafter referred to

Table I
The Galaxy Model for Observed Components[a]

Component (1)	B & S Mass Fraction (A_i) (2)	$<v_z^2>^{1/2}$ ($km\ s^{-1}$) (3)	HHB Mass Fraction (A_i) (4)
Main Sequence Stars			
$M_V < 2.5\ mag$	0.021	4	0.038
$2.5\ mag \le M_V \le 3.2\ mag$	0.015	8	0.019
$3.2\ mag \le M_V \le 4.2\ mag$	0.031	11	0.033
$4.2\ mag \le M_V \le 5.1\ mag$	0.035	21	0.034
$5.1\ mag \le M_V \le 5.7\ mag$	0.025	20	0.023
$5.7\ mag \le M_V \le 6.8\ mag$	0.037	17	0.036
	0.0358	8	
	0.0626	13	
$M_V \ge 6.8\ mag$	0.0536	15	0.0262
	0.0626	20	
	0.0834	24	
Subgiants and Giants	0.016	~20	
White dwarfs	0.052	21	0.185
Atomic H and He			0.287
and	0.469	4	
Molecular H and dust			0.083
Spheroid	0.001	~100 km/sec	
Total	0.0958 $M_\odot pc^{-3}$		0.108 $M_\odot pc^{-3}$

[a] Disk luminosity functions and velocity dispersions from Wielen.[10]

as the B&S and the HHB Galaxy models. The models contain many observed disk components (typically 14) whose characteristics are determined by local measurements, a Population II spheroid inferred from faint star counts, different models for the unobserved disk components, and an unseen massive halo whose normalization is fixed by the solar rotation velocity. The mass fractions are defined in terms of the total observed mass density (in stars, gas, and dust), i.e.,

$$A_i = \frac{\rho_i(0)}{\rho_{obs}(0)} \qquad (1)$$

Previous studies have been limited mainly to simplified Galaxy models with one or, at most, a few disk components and no spherical component.

The previous solutions were also limited either by what was tractable analytically or by assuming a numerical form for the total matter density that was independent of the potential. I have computed numerical models with many different sets of input data and several assumptions about how the unseen material is distributed. I estimate the major uncertainties in the determination of the distribution of unseen matter by comparing an extensive collection of theoretical models with the available data.

The basic equation used is the combined Poisson-Vlasow equation for the potential. This equation describes how the gravitational potential at a given height above the plane can be calculated from the mass densities and velocity dispersions that are specified in the plane of the disk for any number of isothermal disk components--some observed as stars, dust, or gas and some unobserved--plus a halo mass density (constant, to first approximation, with height above the plane). The dimensionless form of the combined Poisson-Vlasow equation is

$$\frac{d^2\phi}{dx^2} = 2\left[\sum_{i=1}^{N_{obs}} A_i\, e^{-\alpha_i\phi} + \sum_{j=1}^{N_{unobs}} B_j e^{-\beta_j\phi} + \varepsilon\right] , \tag{2}$$

with $\phi(0) = (d\phi/dx)_0 = 0$. The gravitational potential has been divided by the square of a velocity dispersion which is taken here to be $(10\ \mathrm{km\ s^{-1}})^2$ for numerical convenience. The quantities $\alpha_i = [(10\ \mathrm{kms^{-1}})^2/\langle v_z^2\rangle_i]$, with a similar definition for the unobserved β_j. The height z above the plane is taken to be $z = z_0 x$, where the unit of length is $z_0 = [(10\ \mathrm{kms^{-1}})^2/2\pi G\rho_{obs}(0)]^{1/2}$. The quantity N_{obs} is the total number of observed mass components. The unobserved mass fractions B_j are defined, by analogy with Eq. (1), as the ratio of the mass density in component j to the total observed mass density. Finally ε is defined as the ratio of $\rho^{eff}_{halo}(0)$ to $\rho_{obs}(0)$. The effective halo mass density is equal to the total halo mass density for a constant rotation curve but is slightly different if the rotation curve is not exactly flat.

Table II gives the ratio of unobserved to observed mass density for twenty-eight detailed models (see Bahcall[5] for a description of these models) that fit the observed distribution of K giants. The models represent numerical solutions of the combined Poisson-Vlasow equation for different input parameters, as well as for several assumptions about the distribution of the unobserved disk material. There are separate columns referring to the observed K-giant samples of Oort[2] and to the Upgren[6] K giant density distributions. For both the volume and the column density, the typical best-fit model has, for the Oort densities, about equal amounts of unobserved and observed material. For the Upgren densities, the typical best-fit model has about 40% more unobserved than observed matter. These averages are only illustrative since at most one of the models considered for the distribution of unseen matter can be correct. Similar results are obtained by comparing theoretical models to the observed sample of F dwarfs.[7]

I conclude that a typical best-fit model implies that about half of the disk material at the solar position has not yet been observed. This conclusion is in qualitative agreement with the previous major studies,[1,2,4,8,9] although I find a larger ratio of unobserved to observed matter than in some of the earlier analyses. The present investigation established more firmly and specifically the existence of

Table II
Ratio of Unobserved to Observed Disk Material

Row[a]	$\frac{\rho_{unobs}(0)}{\rho_{obs.}(0)}$	$\frac{\sigma_{unobs}}{\sigma_{obs}}$	$\frac{\rho_{unobs}(0)}{\rho_{obs}(0)}$	$\frac{\sigma_{unobs}}{\sigma_{obs}}$
(1)	(2)	(3)	(4)	(5)
	Oort Densities		Upgren Densities	
1	1.1	1.1	1.5	1.6
2	1.6	1.6	2.1	2.1
3	0.6	0.6	1.0	1.0
4	0.9	0.9	1.3	1.4
5	1.3	1.3	1.8	1.8
6	1.3	1.1	1.8	1.6
7	0.6	1.2	0.8	1.6
8	0.7	0.7	1.0	1.0
9	0.4	0.7	0.5	1.0
10	2.4	0.5	2.6	0.5
11	1.5	0.3	2.2	0.5
12	0.6	2.5	0.7	3.2
13	1.1	1.1	1.5	1.6
14	1.5	1.5	2.0	2.0
Average	1.1	1.1	1.5	1.5

[a]The models used here are described in the corresponding rows of Tables 5 and 6 of Bahcall.[5]

unobserved disk material. The added confidence in the results arises because: 1) more realistic Galaxy models are used; 2) the Poisson and Vlasow equations are solved self-consistently; 3) improved (and more homogeneous) observational data is utilized; and 4) many theoretical models are compared with the observations in order to estimate the uncertainties.

If the missing material is in the form of stars that are not massive enough to burn hydrogen ($M < 0.1\ M_{\odot}$), then the nearest such brown dwarf is probably less than a parsec away and has a proper motion of more than an arcsec per year. Brown dwarfs of the required number density might be detected in future dedicated large area surveys for very red, high proper motion objects. If the unseen material has a typical mass like that of Jupiter, the nearest such object would be about 0.2 pc from the Sun, moving with a proper motion of order 5 arcseconds per year. Such remarkable objects might be discoverable with IRAS.

The unseen material must be mostly in a disk form, i.e., be dissipational. If all of the material were in a relatively round halo, then the rotation velocity at the solar position would have to be as large as 500 km s^{-1}. For a given local volume density of unseen mass, the total amount of mass required in a round halo is larger than the amount of mass needed in a disk by about the ratio of the galactocentric distance of the Sun to the disk scale height, i.e., by more than an order of magnitude. The largest scale height of the unseen disk material that is consistent with the solar rotation velocity is 0.7 kpc (see row 12 of Tables 5 and 6 of Bahcall[5]).

The largest source of uncertainty in the Oort limit is the unknown form of the distribution of unseen matter (see the last row of Table 9 of Bahcall[5]). In the future, it should be possible to constrain sharply the distribution of unseen matter by requiring consistency with observations of several carefully selected samples of tracer stars with different scale heights.

This work was supported in part by NASA contract number NAS8-32902 and by NSF contract number PHY-8217352.

REFERENCES

[1]J. H. Oort, Bull. Astr. Inst. Netherlands **6**, 249 (1932).
[2]J. H. Oort, Bull. Astr. Inst. Netherlands **15**, 45 (1960).
[3]J. N. Bahcall and R. M. Soneira, Ap. J. Suppl. **44**, 73-100 (1980).
[4]G. Hill, R. W. Hilditch, and J. V. Barnes, M.N.R.A.S. **186**, 813 (1979).
[5]J. N. Bahcall, Ap. J., December 15.
[6]A. R. Upgren, A. J. **67**, 37 (1962).
[7]J. N. Bahcall, Ap. J. **276**, 169 (1984a).
[8]R. Woolley and J. M. Stewart, M.N.R.A.S. **136**, 329 (1967).
[9]C. T. Lacarrieu, Astron. and Astrophys. **14**, 95 (1971).
[10]R. Wielen, Highlights of Astronomy **Vol.** 3, 395, ed. G. Contopoulos, (Dordrecht: D. Reidel).

Galaxy Formation in an Ω=1 Cold Dark Matter Universe

James M. Bardeen

ABSTRACT

A model for galaxy formation is proposed which assumes that bright galaxies form where the primordial density fluctuations exceed a high threshold. Most of the mass in the universe is uncondensed or associated with low surface brightness galaxies. I discuss physical mechanisms and predictions for the galaxy-galaxy correlation function.

I. INTRODUCTION

A cold dark matter dominated inflationary universe is theoretically attractive for its predictive power. Barring extraordinary fine tuning of the present vacuum energy density (cosmological constant), the cosmological density parameter Ω should be very close to one.[1] A Harrison-Zel'dovich scale independent spectrum of primordial density fluctuations[2] arises naturally,[3] with only the overall amplitude as a free parameter. The processing of the spectrum due to subsequent dynamical evolution is simple to calculate in the linear regime, and introduces no additional free parameters beyond the overall scaling associated with the value adopted for the Hubble constant.

However, an Ω=1 cold dark matter universe is apparently inconsistent with dynamical estimates of Ω from virial tests applied to clusters of galaxies,[4] with infall models for the local supercluster,[5] and with estimates based on peculiar velocity correlations from redshift surveys.[6] All these tests give values for Ω of at most about 0.2 to 0.4, making the conventional assumption that galaxies trace the overall mass density on large scales in the universe.

In the cold dark matter scenario there does not seem to be any mechanism which would separate baryons and the dark matter on large scales. However, galaxies may well not form with the same efficiency everywhere, particularly the luminous, high-surface brightness galaxies that are counted in the redshift surveys.

Primordial density perturbations which arise from quantum fluctuations during an inflationary epoch are expected to obey, to a good approximation, Gaussian statistics as long as their amplitude is small. Kaiser[7] has shown this implies enhanced clustering of rare objects which form where the linear density perturbation field, smoothed on an appropriate mass scale, exceeds a threshold ν times the root mean square fluctuation. While originally applied to explain how the correlation function for Abell clusters[8] can be several times larger in amplitude than the correlation function for galaxies, the same mechanism may enhance the correlations of galaxies relative to the correlations of the matter density fluctuations.

Some of the consequences of a threshold for galaxy formation are discussed elsewhere in these proceedings by Kaiser.[9] Here I will focus on specific results for the two-point galaxy correlation function in a cold dark matter dominated universe and on possible physical mechanisms for producing a threshold.

II. THE DENSITY PERTURBATIONS

The evolution of density perturbations in a cold dark matter dominated universe was originally calculated by Peebles,[10] with subsequent refinements by Blumenthal and Primack[11] and by the author.[12] The results presented here are for "adiabatic" perturbations, which have an initially unperturbed ratio of cold particle number density to photon number density.

In considering the formation of structure on various scales it is convenient to introduce a smoothed density perturbation field $\delta\rho_s$. Denote the smoothing scale in comoving cosmological coordinates by r_s and define

$$\delta\rho_s(\vec{x},r_s)/\rho = \int d^3r\delta\rho(\vec{x}+\vec{r})\exp(-r^2/r_s^2)/\int d^3r\rho \ \exp(-r^2/r_s^2). \tag{1}$$

The Fourier power spectrum of $\delta\rho_s$ differs from that of $\delta\rho$ by a factor $\exp(-k^2r_s^2/2)$. Let $\sigma_o(r_s)$ be the root mean square value of $\delta\rho_s/\rho$. One can also introduce a smoothed mass correlation function ξ_s defined by

$$\xi_s(r,r_s) = \langle(\delta\rho_s(\vec{x},r_s)/\rho)(\delta\rho_s(\vec{x} + \vec{r},r_s)/\rho)\rangle. \tag{2}$$

The matter dynamics is linear on a given scale as long as $\sigma_o(r_s) \ll 1$. In this regime the amplitude of the perturbations grows in time proportional to the cosmological scale factor.

The shape of the cold matter fluctuation spectrum as represented by the Fourier power spectrum or by $\sigma_o(r_s)$ is independent of the value of the Hubble constant if r is scaled relative to the particle horizon at matter domination, for which $r \approx 1$, say. I define r so that, with a microwave background temperature of $2.7^\circ K$, three species of massless neutrinos, and $\Omega=1$ the mass inside a sphere of coordinate radius r is

$$M = 1.86 \times 10^{16} \ h^{-4} \ r^3 \ M_\odot, \tag{3}$$

and a proper distance R in the present universe is

$$R = 25.1 \ h^{-2}r\text{Mpc}. \tag{4}$$

The Hubble parameter h is defined by H_o = 100 h km/s/Mpc. Age estimates for globular clusters[13] suggest that $h \leq 0.5$. Then primordial nucleosynthesis predicts that the fraction of the closure density in baryons, Ω_b, is 0.1 or a bit larger.[14]

The range of galactic masses, say $10^{10}M_\odot$ to $10^{12}M_\odot$, corresponds to r in the range $.0025 < r < 0.1$ for $h \approx 0.4$. The galaxy-galaxy correlation function is observed[6] to be one at the correlation length R_o = $5h^{-1}$Mpc, or

$$r_o = 0.2h. \tag{5}$$

III. A THRESHOLD FOR GALAXY FORMATION?

What physical mechanisms can set a threshold for galaxy formation? To shut off galaxy formation completely requires heating the intergalactic medium to a temperature above $10^{6\circ}K$. While this may eventually happen as the result of galactic winds and explosions and virialization in clusters of galaxies, the heating would be most effective in regions of high galaxy density. The inhomogeneity of the heating would seem likely to counteract strongly any statistical enhancement of galaxy clustering.

A more promising approach is to look for mechanisms which would make star formation relatively inefficient, perhaps by favoring the formation of massive stars whose explosions would make further star formation difficult and might expel much of the gas from the galaxy. Gas would still fall into a potential well produced by the dark matter, but would not produce a high surface brightness galaxy which would be counted in redshift surveys.

There is a general decrease in cooling efficiency of gas as the universe expands, since most relevant cooling rates go as the square of the density. Whether this alone can set a sufficiently sharp threshold for galaxy formation is not clear, and should be explored further.

Quasars seem to turn on rather abruptly at a redshift $z \approx 3$; perhaps a sudden increase in the overall flux of UV radiation at this time could prevent efficient formation of low mass stars by putting the gas on a higher adiabat.[15] But a substantial UV flux could well be produced by young galaxies or massive stars at earlier times.[16]

Inverse Compton cooling of electrons scattering off the microwave background radiation is a universal cooling agent for ionized gas. At a redshift $z > 6$ (for $h \approx 0.4$) the IC cooling time is less than the dynamical time. Competition between photo-ionization heating, compressional heating, and IC cooling could conceivably set a sharp threshold at a somewhat higher redshift for efficient formation of low mass stars from gas falling into a protogalaxy.

The last possibility seems most promising, but none are particularly compelling at this point. I would like to think of the threshold as controlling the formation of the spheroid component of a galaxy. The disk of a spiral is presumably associated with inefficient star formation in gas accreting after the collapse of the spheroid.

IV. STATISTICAL PROPERTIES

Now adopt the hypothesis that a typical spiral galaxy of the sort dominating the redshift surveys will form where the linear density fluctuation, smoothed on a scale r_s corresponding to a mass in baryons the order of the final spheroid mass, is a local maximum exceeding a threshold ν times $\sigma_o(r_s)$. The statistics of these maxima should correspond to the statistics of galaxies on scales for which nonlinear dynamical evolution is unimportant. The value of ν is not a free parameter, since the density of maxima must correspond to the observed density of galaxies.

The density of galaxies in the CfA survey[5] is roughly $.01h^3\text{Mpc}^{-3}$, which translates into a coordinate number density

$$n_g = 160h^{-3} = 2500(h=0.4). \tag{6}$$

According to Doroshkevich,[17] the number density of maxima of the smoothed density field is given with reasonable accuracy for $\nu > 2$ by

$$n_m(r_s,\nu)=(4\pi^2 r_s^3)^{-1}(r_s^3\sigma_1\sigma_2/\sigma_o^2)(\nu^2-1)\exp(-\nu^2/2). \tag{7}$$

The factor $(r_s^3\sigma_1\sigma_2/\sigma_0^2)$ depends only on the shape, not the amplitude, of the density fluctuation spectrum and is a slowly varying function of r_s. The quantities σ_1 and σ_2 are the root mean square values of the first and second spatial derivatives of the smoothed density.

Politzer and Wise[18] have derived an analytic expression for the two-point correlation function of points above threshold in the limit

$$\xi_{PW}(r) = \exp[\nu^2\xi_s(r)/\xi_s(0)] - 1 \qquad (8)$$

(note that $\xi_s(0) = \sigma_0^2$). The two-point correlation function of density maxima above threshold should be well approximated by Eq. (8) when ν is very large and overestimated when ν is not so large. Equation (8) can be considered a crude estimate of the two-point galaxy correlation function. Note that ξ_{PW} at any r is independent of the overall amplitude of the density perturbations. Equation (8) applies to the correlations of protogalaxies as they are sprinkled on the cosmological background, before their gravitational interaction can move them appreciably on the scale r.

Numerical evaluation of Eqs. (7) and (8), for the "adiabatic" mode with $n_m = n_g$ and h = 0.4, gives the results shown in Table I for ν and for ξ_{PW} at the observed galaxy correlation length r_o. Since $\xi_{PW}(r_o)$ turns out to be insensitive to the only free parameter, r_s, the agreement with $\xi_g(r_o) = 1$ is encouraging, particularly since nonlinear dynamic effects[8] are likely to increase the value of the correlation function a bit. Larger values of n_m/n_g or h decrease $\xi_{PW}(r_o)$, to less than 0.5 if $n_m/n_g > 2$ or h > 0.5.

TABLE I. Predictions for the adiabatic cold particle spectrum with $n_m = n_g$ and h = 0.4.

				ν/σ_o	
r_s	$M_{GW}/M_\odot$	ν	$\xi_{PW}(r_o)$	$z_c=3$	$z_c=8$
.00316	3.05×10^{10}	3.80	.60	2.14	.95
.00412	7.24×10^{10}	3.54	.67	1.85	.82
.00562	1.72×10^{11}	3.25	.73	1.56	.69
.00750	4.07×10^{11}	2.92	.76	1.26	.56
.01000	9.66×10^{11}	2.49	.72	.92	.41

The exponential in Eq. (8) rapidly makes the correlation function much steeper than the observed one as r decreases below r_o. There are good reasons to suppose the Politzer-Wise expression is inaccurate in this regime, but getting a correlation function at small r compatible with observations promises to be a severe test of the threshold hypothesis. The tendency toward a strong statistical enhancement of spatial correlations at small separations is in the right direction to make observed velocity correlations compatible with $\Omega = 1$.

The linear amplification factor of Kaiser,[9] ν/σ_o, where σ_o is the root mean square linear density fluctuation extrapolated to the present, can be calculated if the redshift z_c at which threshold fluctuations collapse is specified. As defined by a spherical top hat model $1.69(1+z_c) = \nu\sigma_o$, or

$$\nu/\sigma_o = \nu^2/[1.69(1+z_c)]. \tag{9}$$

Values are given in Table I for $z_c = 3$ (quasars?) and for $z_c = 8$ (inverse Compton cooling?).

V. DISCUSSION

We have seen that it is possible to replace the assumption that galaxies trace the mass on large scales in the universe by a simple alternative hypothesis which in some ways is even more tightly constrained than the conventional one. Strong statistical enhancement of galaxy clustering on small scales and, as discussed by Kaiser,[9] in rare bound objects on larger scales, seems likely to be associated with only modest statistical enhancement of the correlation function on scales larger than the correlation length. Dynamical estimates of Ω are possible precisely in those regions where the galaxy density is enhanced, which biases the empirical estimates of Ω toward values less than the true global value. Numerical simulations[19] based on identifying galaxies in this way tentatively seem to fit the observations reasonably well, consistent with $\Omega = 1$.

Unfortunately, the physical mechanism which would set a threshold for galaxy formation remains obscure, mainly because the theory of star formation is not well enough developed to predict with confidence what would happen at some high redshift.

ACKNOWLEDGMENTS

I am indebted to all the participants in the Large Scale Structure Program at the ITP for many stimulating discussions. I would particularly like to acknowledge helpful comments from M. Davis, N. Kaiser, M. Rees, J. Silk, and S. White. This material is based upon research supported in part by the National Science Foundation under Grant No. PHY77-27084, supplemented by funds from the National Aeronautics and Space Administration.

REFERENCES

[1]A. Guth, Phys. Rev. D **23**, 347 (1981).
[2]E. R. Harrison, Phys. Rev. D **1**, 2726 (1970); Ya. B. Zel'dovich, Mon. Not. R. Astr. Soc. **160**, 1P (1972).
[3]A. A. Starobinskii, Phys. Lett. **117B**, 175 (1982); A. Guth and S.-Y. Pi, Phys. Rev. Lett. **49**, 1110 (1982); J. M. Bardeen, P. S. Steinhardt, and M. S. Turner, Phys. Rev. D **28**, 679 (1983).
[4]S. M. Faber and J. S. Gallagher, Ann. Rev. Astr. Astrophys. **17**, 135 (1979).
[5]For example, M. Davis and J. Huchra, Astrophys. J. **254**, 437 (1982).
[6]M. Davis and P. J. E. Peebles, Astrophys. J. **267**, 465 (1983).
[7]N. Kaiser, Astrophys. J., to be published.

[8]N. A. Bahcall and R. M. Soneira, Astrophys. J. **270**, 20 (1983).
[9]N. Kaiser, in these proceedings.
[10]P. J. E. Peebles, Astrophys. J. **263**, L1 (1982).
[11]J. R. Primack and G. R. Blumenthal, Clusters and Groups of Galaxies, eds. F. Mardirossian, G. Giuricin, and M. Mezzetti (Reidel, Dordrecht, Holland), in press.
[12]J. M. Bardeen, in preparation.
[13]K. Janes and P. Demarque, Astrophys. J. **264**, 206 (1983).
[14]J. Yang, M. S. Turner, G. Steigman, D. N. Schramm, and K. A. Olive, Astrophys. J., in press.
[15]J. Silk, private communication.
[16]B. J. Carr, J. R. Bond, and W. D. Arnett, Astrophys. J. **277**, 445 (1984).
[17]A. G. Doroshkevich, Astrophys. **6**, 320 (1970).
[18]H. D. Politzer and M. B. Wise, preprint.
[19]S. D. M. White, in these proceedings.

Note added in proof: Recently J. R. Bond, N. Kaiser, A. Szalay, and the author have derived a more accurate expression for the number density of peaks above a threshold ν. When used in place of equation (7), together with a numerical evaluation of the correlation function in place of equation (8), the results for $\xi(r_0)$ are about a factor of two less than the values given in Table I.

26

Axions, Neutrinos and Strings: the Formation of Structure in an SO(10) Universe

F.W. Stecker

I will report on work with Qaisar Shafi where we consider a class of grand unified theories in which cosmologically significant axion and neutrino energy densities arise naturally. To obtain large scale structure we consider (1) an inflationary scenario, (2) inflation followed by string production, and (3) a non-inflationary scenario with density fluctuations caused solely by strings. We show that inflation may be compatible with the recent observational indications that $\Omega < 1$ on the scale of superclusters, particularly if strings are present.

Axions with a cosmologically significant energy density provide an important component in the mechanism for generating structure in the universe on scales up to 10^{15} $M_{\odot}$.[1,2] An SO(10) GUT framework which leads to the production of cosmologically significant axions has been given.[3]

As an example of a grand unified theory which gives $\Omega_a \simeq \Omega_\nu$, consider an SO(10) model[3] where both the Pecci-Quinn[4] U(1) symmetry and the local B-L symmetry are broken at a scale of order 10^{12} GeV. The value of the intermediate scale is not put in by hand, but is determined from the renormalization group equations of the gauge couplings. From the results of Ref. (5), it follows the $\Omega_a \simeq 0.1$-1.

The breaking of B-L at scale f_a, caused by a $\underset{\sim}{126}$-plet of Higgs fields, induces a Majorana mass term for the right-handed neutrino ν_{Ri} of order $h_i f_a$, where h_i denotes the Yukawa coupling of the ith generation. The breaking of $SU(2) \times U(1)$ to $U(1)_{em}$ is achieved by a Higgs $\underset{\sim}{10}$ plet and gives rise to Dirac mass terms $m_{\nu i}^{(D)}$ linking the left and right-handed neutrinos. Moreover, it can be shown that an effective Majorana mass term for the left-handed neutrino ν_{Li}, of order $c_i \simeq h_i(\lambda_1/\lambda_2) \langle\phi_{10}\rangle^2/f_a$ is also induced.[6] Here λ_1 denotes the quartic higgs coupling between the $\underset{\sim}{126}$ and the $\underset{\sim}{10}$, λ_2 is the quartic self-coupling of $\underset{\sim}{126}$, and $\langle\phi_{10}\rangle$ is the vacuum expectation value of the $\underset{\sim}{10}$. With $f_a \simeq 10^{12}$ GeV, λ_1/λ_2 of order unity, and $h_i \sim O(g^2)$ (where g denotes the SO(10) gauge coupling), c_i is in the electron volt range. Diagonalization of the neutrino mass matrix (neglecting, for simplicity, mixings between generations) yields the eigenvalues $(m_{\nu_i})_{heavy} \simeq h_i f_a$, $(m_{\nu_i})_{light} \simeq c_i - (m_{\nu i}^{(D)})^2/(m_{\nu i})_{heavy}$.

Due to the presence of the c_i term, the light neutrino of each generation can have a mass in the electron volt range. The second term involving the Dirac masses can be made small so that the masses of the different neutrino flavors can be almost degenerate, providing a possible explanation for the lack of observed neutrino oscillations.[6] It is this possibility which will be of particular interest to us here.

We now discuss the implications of significant axion and neutrino energy densities for the evolution of structure in the universe. Two mechanisms for producing density fluctuations in the early universe have been extensively discussed, viz., inflation[7] and strings.[8] Recently, it was pointed out[9] that one could obtain another scenario in which inflation is followed by string production. The inflationary phase is associated with the transition from SO(10) to $SU(3) \times SU(2)_L \times SU(2)_R \times U(1)_{B-L}$. It can be implemented by generalizing the arguments of Ref. (10) where the SU(5) model is discussed. The breaking of B-L and the U(1) symmetry can occur during, or at the end of the inflationary era. The spectrum of density fluctuations produced in this scenario is scale invariant.

According to recent observations,[11] the value for Ω obtained on scales up to $\sim 10^{15}$ $M_\odot$ is $\simeq 0.2 \pm 0.1$, considerably less than unity, the value predicted by an inflationary cosmology. As a reasonable upper limit for Ω_{sc} of superclusters,[12] we may take $\Omega_{sc} \lesssim 0.5$. Therefore, since axions and baryons cluster on scales smaller than rich clusters and superclusters,[1] their contribution to Ω must be $\lesssim 0.5$. The balance of the total Ω in the universe must therefore be in the mass density of the neutrino component if we are to have Ω=1 as predicted by inflationary cosmology.

We must therefore require that the neutrinos be light enough so that they will not cluster on scales below $\sim 10^{16}$ $M_\odot$. In order to arrange this, especially since the neutrino Jeans mass drops significantly between the redshift z_{nr} when the neutrinos become nonrelativistic and the present time, we invoke neutrino phase space limits using the arguments of Tremaine and Gunn[13] in reverse to get an upper limit on m_ν. These authors find that for neutrinos to be able to cluster on the scale of rich clusters, their mass must be greater than $\sim 4\ h_{50}^{-1/2}$ eV (where h_{50} is the Hubble constant in units of 50 $\mathrm{kms^{-1}Mpc^{-1}}$). We require a mass less than this limit to prevent clustering of the neutrino component. The neutrino contribution to Ω is $\Omega_\nu = 4.56 \times 10^{-2}\ m_\nu(\mathrm{eV}) N_f\ h_{50}^{-2} T_{2.8}^3$, where N_f is the number of neutrino flavors of approximately equal mass and $T_{2.8}$ is the present temperature of the cosmic blackbody radiation in units of 2.8 K. We require Ω_ν to be $\gtrsim 0.5$ so that the total Ω = 1. Thus, one needs at least three flavors of neutrinos, each of approximately 3-4 eV. As discussed above, this situation is readily obtained in the SO(10) model (see above). (If the efficiency of neutrino clustering is low, m_ν could be somewhat larger.)

The maximum neutrino Jeans mass for three neutrinos of roughly equal mass is[14] $M^*_{J\nu} = 2.7 \times 10^{18}\ [m_\nu(\mathrm{eV})]^{-2} M_\odot$ which, for N_f = 3 and $m_\nu \simeq 3.6$ eV gives $M^*_{J\nu} \simeq 2 \times 10^{16}\ M_\odot$. The corresponding spatial scale at present for pancaking structure would be $\sim$150 Mpc. This scale may correspond to the "superpancaking" scale[15] for clustering of superclusters.[16] Structure on this scale would correspond to density perturbations $\delta \equiv \delta\rho/\rho$ just becoming nonlinear (δ = 0.5-1) at the present time (z = 0).

The spectrum of linear perturbations in a universe dominated by axions and neutrinos is readily estimated by adopting the arguments previously given for a baryon-neutrino universe.[17] It is convenient to define $\xi = \Omega_a/(\Omega_a+\Omega_\nu)$ such that $\xi \leq 1/2$ (we assume, for simplicity, that $\Omega_b \ll \Omega_a,\ \Omega_\nu$).

For $z < z_{eq} \simeq 0.93 \times 10^4\ (1-\xi)^{-1}\ \Omega_\nu h_{50}^2\ T_{2.8}^{-4}$ the neutrino Jeans mass decreases as $(1+z)^{3/2}$. Neutrino perturbations on scales below $M^*_{J\nu}$ are erased at $z \simeq z_{eq}$. The axion perturbations, however, grow like $\delta_\alpha \propto t^\alpha \propto (1+z)^{-3\alpha/2}$ where $\alpha = (\sqrt{1+24\xi}\ -1)/6$. Thus,

$$\delta_a\ (z) \simeq \delta_a\ (z_{eq})\ \left(\frac{1+z_{eq}}{1+z}\right)^{3\alpha/2} \quad . \tag{1}$$

This continues until $z \simeq z_M$ when the neutrino Jeans mass becomes $\simeq M$,

$$(1+z_M) \sim \left(\frac{M}{M^*_{J\nu}}\right)^{2/3} (1+z_{eq}) \ . \tag{2}$$

For $z < z_M$ the overall density fluctuation $\delta\rho/\rho \propto t^{2/3} \propto (1+z)^{-1}$. Thus,

$$\frac{\delta\rho}{\rho}\,(z<z_M) \simeq \xi\ \delta_a(z_M)\left(\frac{1+z_M}{1+z}\right) \simeq \xi\ \delta_a(z_{eq})\left(\frac{1+z_{eq}}{1+z}\right)\left(\frac{M}{M^*_{J\nu}}\right)^{(2/3-\alpha)} . \tag{3}$$

As a rough approximation, $\delta_a(z_{eq}) \approx$ constant when $M < M^*_{J\nu}$ for a scale invariant initial spectrum. This gives

$$\frac{\delta\rho}{\rho} \propto M^{(2/3-\alpha)} \qquad (M < M^*_{J\nu}) \ , \tag{4}$$

which is an increasing function of M since $\alpha < 2/3$. For $M > M^*_{J\nu}$, the neutrino perturbations are not damped and $\delta\rho/\rho \propto M^{-2/3}$.

From this discussion we conclude that even in the most optimistic case where $\xi = 1/2$, $\alpha \simeq 0.43$, so that the scales between the present neutrino Jeans mass and $M^*_{J\nu}$ cannot collapse before $M^*_{J\nu}$. We thus run into the timing problems which are becoming well known for the neutrino pancaking scenario. In particular, it is hard to envision the development of quasars[18] and substructure[19] with such a model.

The presence of strings, which provide an additional source of density fluctuations, can eliminate the above difficulty (see also the discussion below). Assume that topologically stable strings, with mass per unit length characterized by a superheavy (GUT) scale, appear at or near the end of the inflationary phase.[10] (This is readily achieved in the present case either by appending a new spontaneously broken global U(1) symmetry to the SO(10) model or using an E_6 model. It can also be obtained naturally in a Kaluza-Klein model (Wetterich, private communication)). The strings can intercommute forming closed loops[20] which produce axion density perturbations $\delta_a(z_{eq}) \propto M^{-1/3}$ below the Jeans mass scale. It then follows that

$$\frac{\delta\rho}{\rho} \propto M^{(1/3-\alpha)} \qquad (M < M^*_{J\nu}) \ , \tag{5}$$

as compared with Eq. (4).

For $\xi = 1/2$, $\alpha = 0.43$ and $\delta\rho/\rho \propto M^{-0.1}$. Therefore, if $\delta\rho/\rho \sim O(1)$ on scales $\sim 10^{16} - 10^{17}\ M_\odot$ at $z = 0$ as suggested by Dekel,[15] scales $\sim 10^{10}\ M_\odot$ go non-linear at $z \simeq 4$, corresponding to the epoch of quasar formation. Thus, in the presence of axions and neutrinos, an inflationary scenario supplemented by strings offers a better prospect of explaining the observed large scale structure in the universe than one without strings. It should also be noted that growth of axion perturbations during the radiation era[21] will have the effect of increasing α. This effect may be enough to make the spectrum from Eq. (4) flat without strings and allowing $\Omega_a < 0.5$ to be acceptable.

Suppose we dispense with inflation and assume that the density fluctuations are produced solely by strings. In this case, since Ω need not be unity, ξ can be greater than 1/2 and α can be >0.434. (Of course, we need have only one ν flavor in the eV mass range to get Dekel's[15] scale.) In particular for $\Omega_a >> \Omega_\nu$, $\alpha = 2/3$. A natural extension of SO(10) which gives the desired strings is provided by E_6 symmetry breaking to SO(10) at a scale $\eta \simeq 10^{16}$ GeV. The energy per unit length of the strings formed is $\mu \simeq \eta^2 \simeq 10^{32}$ GeV^2. Then at z = 0

$$\frac{\delta\rho}{\rho}(M^*_{J\nu}) \sim 30\ G\mu\ (1+z_{eq}) \sim O(1)\ . \qquad (6)$$

Thus, neutrino perturbations are on the verge of becoming non-linear at the "superpancake" scale, in agreement with observations.[15,16]

We are grateful to Dr. Alexander Vilenkin for many helpful discussions.

REFERENCES

[1]F.W. Stecker and Q. Shafi, Phys. Rev. Lett. **50**, 928 (1983).

[2]M.S. Turner, F. Wilczek and A. Zee, Phys. Lett. **125B**, 35 (1983); M. Axendides, R. Brandenberger and M. Turner, Phys. Lett. **126B**, 178 (1983); M. Fukugita and M. Yoshimura, Phys. Lett. **127B**, 181 (1983).

[3]R. Holman, G. Lazarides and Q. Shafi, Phys. Rev. **D27**, 995 (1983).

[4]R.D. Peccei and H. Quinn, Phys. Rev. Lett. **38**, 1440 (1977).

[5]J. Preskill, M.B. Wise, and F. Wilczek, Phys. Lett. **120B**, 127 (1983); L.F. Abbott and P. Sikivie, ibid. pg. 133; M. Dine and W. Fischler, ibid. pg. 137.

[6]C. Wetterich, Nucl. Phys. **B187**, 343 (1981) and references therein; F.W. Stecker in Electroweak Interactions (Proc. 21st International Winterschool on Theoretical Physics, Schladming, Austria) ed. H. Mitter, Springer-Verlag, Vienna, 307 (1983).

[7]S.W. Hawking, Phys. Lett. **115B**, 295 (1982); A.H. Guth and S.Y. Pi, Phys. Rev. Lett. **49**, 1110 (1982); A. Starobinsky, Phys. Lett. **117B**, 175 (1982); J. Bardeen, P.J. Steinhardt and M.S. Turner, Phys. Rev. **D28**, 679 (1983).

[8]Ya.B. Zeldovich, Mon. Not. R. Astron. Soc. **192**, 663 (1980); A. Vilenkin, Phys. Rev. Lett. **46**, 1169, 1496 (E) (1981) and Phys. Rev. **D24**, 2082 (1981).

[9]Q. Shafi and A. Vilenkin, Bartol preprint BA-84-11, to appear in Phys. Rev. D.

[10]Q. Shafi and A. Vilenkin, Phys. Rev. Lett. **54**, 691 (1984).

[11]M. Davis and J. Huchra, Astrophys. J. **254**, 437 (1982); J.P. Huchra, Highlights in Astronomy **6**, 749 (1983); M. Davis, J. Huchra and D. Latham in Early Evolution of the Universe and Its Present Sturcture ed. G.O. Abell and G. Chincarini, Reidel Pub. Co., Dordrecht, p. 167 (1983); J. Bean, et al., ibid., p. 175; R.J. Harms, et al., ibid. p. 285.

[12]M. Davis and P.J.E. Peebles, Ann. Rev. Astron. Astrophys. **21**, 109 (1983).
[13]S. Tremaine and J.E. Gunn, Phys. Rev. Lett. **42**, 407 (1979).
[14]J.R. Bond and A.S. Szalay, Proceedings Neutrino 81 International Conference (ed. R.J. Cence, E. Ma and A. Roberts, Univ. of Hawaii) **1**, 59 (1981).
[15]A. Dekel, Astrophys. J., in press; also preprint.
[16]N.A. Bahcall and R.M. Soniera, Astrophys. J. **277**, 27 (1983).
[17]J.R. Bond, G. Efstathiou and J. Silk, Phys. Rev. Lett. **45**, 1980 (1980); A.G. Doroshkevich, Ya.B. Zeldovich, R.A. Syunyaev and M. Yu, Sov. Astron. Lett. **6**, 252 (1981).
[18]M. Davis, J. Huchra, D.W. Latham and J. Tonry, Astrophys. J. **253**, 423 (1982); S.D.M. White, C.S. Frenk and M. Davis, Astrophys. J. **274**, L1 (1983); P.J.E. Peebles, Astrophys. J. **274**, 1 (1983); N. Kaiser, Astrophys. J. **273**, L17 (1983).
[19]I.M. Gioia, et al., Astrophys. J. **255**, L17 (1983); J.P. Huchra and M.J. Geller, Astrophys. J. **264**, 356 (1982); G.D. Bothum, M.J. Geller, T.C. Beers in Early Evolution of the Universe and Its Present Structure, ibid., p. 231.
[20]A. Vilenkin and Q. Shafi, Phys. Rev. Lett. **51**, 1716 (1983).
[21]P.J.E. Peebles, Astrophys. J. Lett. **263**, L1 (1982).

The Universe Cannot be Closed by Unclustered Dark Matter

S.A. Bludman and Y. Hoffman

ABSTRACT

Any matter unclustered on the supercluster scale must be "superhot" and became non-relativistic only very recently. Such an unclustered collisionless background would dominate the radiation universe at nucleosynthesis and would later dampen the growth by gravitational instability of structure in the clustered component. If structure evolved by gravitational instability, the universe cannot be dominated by unclustered collisionless, non-relativistic dark matter. Thus, if $\Omega_{c\ell} = 0.35 \pm 0.15$ and there is no present cosmological constant, the universe must be open, rather than flat.

Dynamical studies presently agree that, if dark matter clusters along with luminous galaxies, then $\Omega_0 \equiv \rho/\rho_{cr} < 1$[1,2,3] and the universe must be open. These studies apply the virial and cosmic virial theorems to clusters of galaxies assumed to be in dynamical equilibrium or to the Virgocentric flow and use the observed luminous structure as a tracer of the total matter density. These dynamical observations do not measure any dark matter that is not clustered about the luminous galaxies. The global mean density can therefore be inferred from the dynamical observations only by assuming some correlation between dark and luminous matter, a correlation deriving from the complex and poorly understood processes of galaxy formation.

Observations show that the mass to light ratio increases with scale to an asymptotic value corresponding to $\Omega_0 = 0.1 - 0.3$. This increase of M/L to an asymptotic value suggests not only the presence of dark matter but also suggests that, on the supercluster scale, it is clustered as the luminous matter.[4,5] The universe can be flat ($\Omega_0 = 1$), as preferred by some aesthetic considerations and by the inflationary model, only if the majority of the dark matter, instead of clustering as the galaxies, is either smoothly distributed or anticorrelated to the luminous matter.

We assume the present universe to be dominated by luminous and by dark non-relativistic matter with negligible vacuum energy density (cosmological constant) and radiation energy density. We show that unclustered dark matter (x), presently smooth at least on the scale of superclusters and voids, cannot dominate the clustered component $\Omega_{c\ell} = 0.35 \pm 0.15$ observed.

Any fluctuations in this free-streaming x-component must be damped over scales smaller than $R_{c\ell} = 30\ h^{-1}$ Mpc $= 9 \times 10^7\ h^{-1}\ \ell yr$, the scale over which superclusters and voids are observed. This means that such unclustered x-matter must have mean speed $v_x > R_{c\ell}/T_0$, where

$T_0 \leq 2/3\ H_0^{-1} = 6\ h^{-1}$ Gyr is the present age of our matter-dominated

universe. (The Hubble constant $H_0 \equiv 100\ h = (50\text{-}100)$ kms^{-1} Mpc^{-1}). Thus any x-matter has present speed $v_x > c/66\ R_{30}$ where $R_{30} \equiv R_{c\ell}h/30$ Mpc. This free-streaming speed is much faster than any non-relativistic dark matter previously considered and implies that the x-matter was relativistic as recently as red-shift $1 + z \sim 66\ R_{30}^{-1}$. There are then three arguments against x-matter.

(1) Using the classification of dark matter according to when it decoupled from thermal equilibrium, the unclustered x-matter must be "superhot". Hot dark matter such as neutrinos of mass $m_\nu \simeq 30$ eV would be much slower, $v_\nu = 2 \times 10^{-5} c\, m^{-1}_{\nu/30\text{ev}}$, and therefore would cluster on the supercluster scale.

If the x-matter were ever in thermal equilibrium when it was relativistic, its number density relative to microwave background photons would have been frozen in the ratio $n_x/n_\gamma = 2\, g_x/g_*$ where g_x and g_* are the effective number of degrees of freedom of the x-particles and of all relativistic particles in thermal equilibrium when x decoupled. Depending on when this happened, g_* lies between 10.75 and 161 for any standard grand unification theory. Since x became non-relativistic when $m_x v_x/T_x = m_\nu v_\nu/T_\nu$, its mass would be $m_x < (0.02 \text{ eV})(100/g_*)^{1/3} R_{30}$ and at present[6]

$$\Omega_x = m_x n_x/\rho_{cr} = (2 \times 10^{-5}) \left(\frac{100}{g_*}\right)^{4/3} \left(\frac{g_x}{1.5}\right) \frac{R_{30}}{h^2} << 1 .$$

If x-matter were to be appreciable, $\Omega_x \sim 0.65$, it cannot ever have been in thermal equilibrium! This seems very hard for superhot matter.

(2) Because the x and photon energy densities are now in the ratio $\rho_{xo}/\rho_{\gamma o} = \Omega_x h^2 (4\times 10^4)$, and increase as $(1+z)^3$ and $(1+z)^4$ respectively back to redshift 66 R^{-1}_{30}, their energy density ratio before then is frozen at $\rho_x/\rho_\gamma \sim \Omega_x h^2$ 600 R_{30}. Although the x matter now dominates the clustered mater density by only a factor ~2, in the radiation universe, allowing $h \geq 0.5$, it dominates photons by a factor ≥ 150! At nucleosynthesis, this would lead to an expansion rate at least twelve times faster and a decoupling temperature at least five times higher than in the standard scenario, so that the cosmological helium abundance would need to be far in excess of what is observed. This objection rules out an x-dominated universe, unless the x-matter were created after nucleosynthesis.

(3) On the largest scales, the growth of structure in the clustered component must be dominated by gravitational instability. Now perturbations in the clustered component can only start growing very late, when matter dominates at $z < 66\, R^{-1}_{30}$, and the growth thereafter is dampened by the smooth x-component. In the linear regime, the primordial perturbations in the clustered and x-component are coupled by[7]

$$\left[\frac{d^2}{dt^2} + \frac{2\dot{a}}{a}\frac{d}{dt} - 4\pi G\bar{\rho}\, \omega_{c\ell}\right] \delta_{c\ell} = 4\pi G\bar{\rho} \left(\frac{\Omega_x}{\Omega}\right) \delta_x ,$$

where $\Omega_o \equiv \Omega_x + \Omega_{c\ell}$, $\omega_{c\ell} \equiv \Omega_{c\ell}/\Omega$ and the cosmological expansion factor $\underline{a}$ is governed by $\dot{a}/a = 2/3t$ in a matter-dominated universe. On scales below the free-streaming scale, $\delta_x = 0$ and $\delta_{c\ell}$ evolves via

$$\left[\frac{d^2}{dt^2} + \frac{4}{3t}\frac{d}{dt} - \frac{2}{3}\frac{\omega_{c\ell}}{t^2}\right] \delta_{c\ell} = 0 .$$

For the growth mode $\delta_{c\ell} \sim t^{\alpha_+}$, where

$$\alpha_+ = \frac{-1 + \sqrt{1 + 24\,\omega_{cl}}}{6} .$$

If there were no unclustered component $\omega_{c\ell} = 1$ and $\alpha_+ = 2/3$, but if $\Omega_o = 1$, $\omega_{c\ell} = 0.35 \pm 0.15$, $\alpha_+ = 0.3 \pm 0.1$. With the growth rate suppressed by the free-streaming component, the perturbation $\delta_{c\ell}$ can only grow by a factor 7 since red-shift 66 $R_{c\ell}^{-1}$. To reach the non-linear regime would require initial amplitude $\delta_{c\ell} > 0.1$ far above the limit 3×10^{-4} set (in the adiabatic scenario) by the small-scale anisotropy of the background radiation.

Thus nucleosynthesis and the growth of structure show that free-streaming dark matter cannot dominate the universe. We cannot rule out a universe dominated by dark matter clustered away from the matter in galaxies and hot gas, i.e. selection or negative feedback effects that prevent galaxies or hot gas from forming about the clustered dark matter. We also have not considered a present cosmological constant[8,9] or radiation density[9] finely tuned to $1 - \Omega_o$. Barring these possibilities, the universe must be open rather than flat.

REFERENCES

[1]M. Davis and P.J.E. Peebles, Ann. Rev. Astron. Astrophys. **21**, 109 (1983).
[2]M. Davis and P.J.E. Peebles, Ap. J. **257**, 465 (1983).
[3]J. Huchra, this meeting.
[4]Y. Hoffman, J. Shaham and G. Shaviv, Ap. J. **262**, 314 (1982).
[5]The observations allow a constant ratio of dynamical to (baryonic) luminous matter for all galaxy systems (Blumenthal, Faber, Primack, and Rees, Nature (submitted), (1984)).
[6]J.R. Bond and A. Szalay, Ap. J. **274**, 443 (1983).
[7]I. Wasserman, Ap. J. **248**, 1 (1981).
[8]P.J.E. Peebles, preprint (1984).
[9]M.S. Turner, G. Steigman and L.M. Krauss, Phys. Rev. Lett. **52**, 2090 (1984).

Are the Inner Cores of Galaxy Clusters Dominated by Dark Matter?

Dean Sumi and Larry Smarr

One of the basic assumptions used in the study of the dark matter in clusters of galaxies (essentially any nonluminous mass source: massive neutrinos, axions, blackholes, bricks) is that this matter has a distribution similar to the galaxies--galaxies are "good" tracers of the dark matter. Unfortunately, when using galaxies, determination of the distribution of the dark matter in the cores of the cluster is very uncertain.[1] Furthermore, the assumption that galaxies are good tracers will be violated at small enough scales where galaxies begin to interact.[2] Therefore, new types of tracers are necessary. Stars from the halos of centrally located galaxies and hot intracluster gas are ideal for these scales. Kinematic/hydrostatic studies[3,4] using these tracers are consistent with a substantial concentration of dark matter in the cores of some clusters. For example, M87 is at the kinematic center of the Virgo cluster. It is undistinguished optically yet it possesses an extension halo of ~10^7 K intracluster gas.[5] If this gas is in hydrostatic equilibrium, the central region containing M87 and the halo must have mass a factor of 10 greater than the mass accounted for by stars in M87 and by the gas in the halo.[4] We argue that this central concentration or "Black Pit" is an extension of the cluster's dark matter and not a halo as one finds around many spiral galaxies.[6] Galaxy halos, probably dissipationless structures, are easily disrupted and stripped from a galaxy during cluster virialization.[7]

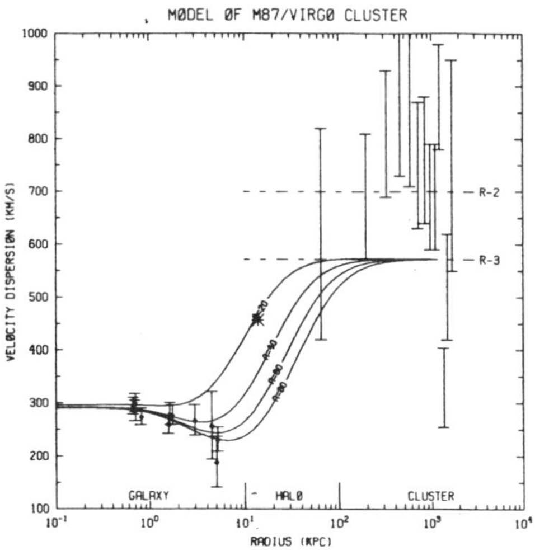

Figure 1 - Models of the velocity dispersion in a Black Pit Potential. Stellar[3], and galaxy[9] velocity dispersion of M87 and the Virgo cluster are plotted versus radius (Virgo cluster distance = 15 Mpc). Model stellar dispersions for Black Pit Models of 20, 40, 60, and 80 kpc are represented by solid lines. Model galaxy dispersions with density distributions of the form r^{-2} and r^{-3} are represented by dashed lines. Also indicated (*) is the approximate result from globular cluster observations[10].

A Black Pit can be characterized roughly by a two component model, one for a central galaxy and the other for the dark matter.[8] The dark matter component lengthscale, A, ranges from ~200 kpc, the usually assumed value, down to 10 to 20 kpc, on the order of a galaxy scale. For the Virgo cluster and M87 we are able to fit a Black Pit model using a combination of the observed intracluster gas[4], stellar[3] and galaxy[9] velocity and spatial distributions. In Fig. 1, the velocity dispersion of M87 is calculated using a r^{-3} density distribution (a King Model) and the cluster velocity distribution using an isothermal model with r^{-3} and r^{-2} density distributions. The corresponding mass distribution is shown in Figure 2. The critical region between the stellar and galaxy regions is filled using masses derived from the distribution of the hot intracluster medium (Fig. 2). Globular cluster data in this critical region is now just being obtained[10], and is tentatively identified in both figures. The best fit roughly occurs at dark matter length scale of 60 kpc as compared with the commonly quoted scale of ~250 kpc. More complete details will be published elsewhere. This research was partially funded by the National Science Foundation grant number PHY 83-08826.

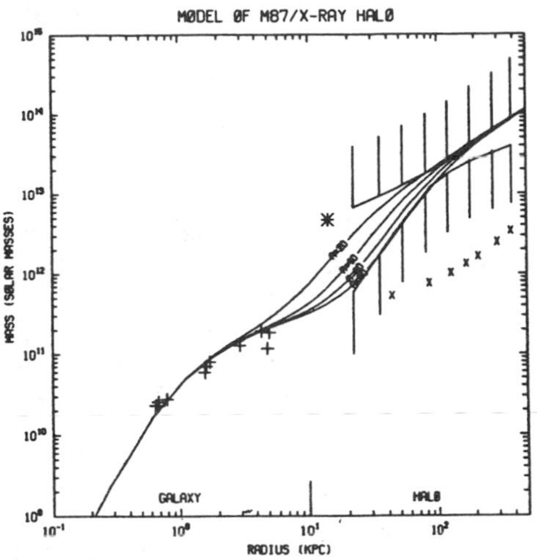

Figure 2 - Models of the mass distribution in a Black Pit Model. Masses inferred from the stellar velocity dispersion[3] are indicated by crosses (+). The total mass in X-ray gas is indicated by X's. The mass range excluded by X-ray observations[4] is indicated by the hatched regions between 20 and 400 kpc. Solid lines represent the mass distribution for a Black Pit Model of core radius 20, 40, 60, and 80 kpc. Again indicated (*) is the approximate result from globular cluster observations[10].

REFERENCES

[1]G. Chincarini, X-ray Astronomy, edited by R. Giacconi, G. G. Setti, D. Reidel, **197** (1979).

[2]F. Żwicky, Morphological Astronomy, Springer-Verlag, (1957).

[3]W. L. Sargent, W. L. Young, A. Boksenberg, K. Shortridge, C. R. Lynds, F. D. A. Hartwick, Astrophys. J. **221**, 731 (1978).

[4]D. Fabricant, P. Gorenstein, Astrophys. J. **267**, 535 (1983).

[5]W. Forman, C. Schwarz, C. Jones, W. Liller, A. C. Fabian, Astrophys. J. **234**, L27 (1979).

[6]V. C. Rubin, W. K. Ford Jr., N. Thonnard, Astrophys. J. **225**, L107 (1978).

[7]D. O. Richstone, Astrophys. J. **200**, 535 (1975).

[8]R. D. Blandford, L. L. Smarr, Univ. of Illinois Astronomy Department Preprint IAP 82-39, (1982).

[9]I. D. Karachentsev, Astrophysics **18**, 279 (1983).

[10]J. Huchra, these proceedings (1985).

29

The Evolution of Large-Scale Structure

Simon D.M. White

I. INTRODUCTION

The crossfertilisation between particle physics and cosmology has introduced a number of new elements into the old problem of understanding the large-scale structure of the Universe. At the simplest level particle physics has come up with a number of elementary particle candidates for the dark matter which appears to dominate the gravitational dynamics of galaxy clusters and of the Universe as a whole. The first of these, and still the most plausible from the point of view of particle physics, was a massive neutrino, but many other possibilities have been suggested. The major requirement on any such candidate is that the product of its mass, m, and its present abundance, n, both of which can in principle be derived from particle physics and from the theory of the Hot Big Bang, should equal the density of the Universe inferred from cosmological parameters:

$$nm = 2 \times 10^{-29} \Omega h^2 \text{ g cm}^{-3} , \tag{1}$$

where h is Hubble's constant in units of 100km s^{-1} and Ω is the mean density divided by that of an Einstein-de Sitter universe with the same Hubble constant. Although Ω can be constrained by dynamical studies of large systems and by observations of the global geometry of space-time, such measurements at present suggest only that the global value of Ω lies in the range 0.1 - 2. Particle physics has had important input here; it has strengthened the theoretical prejudice that Ω should be very close to unity by providing a specific model for the early universe which leads naturally to this result. Inflation similarly strengthens the prejudice in favor of a random phase, constant curvature primordial fluctuation spectrum, since structure of this kind can be produced by quantum fluctuations during the inflationary epoch.

More detailed discussion of the above ideas can be found in other contributions to this volume[1]; the present article is concerned solely with the problem of the growth of structure in the kind of universe which they imply. A great virtue of this general framework is that once a candidate "dominant particle" (for example, the massive neutrino) has been chosen, the initial conditions for the nonlinear formation of structure are determined apart from a normalisation constant. The form of the fluctuation distribution at early times can be calculated from linear theory, and for Ω = 1 its characteristic scale is uncertain only to the extent that the present value of h is poorly known. Its normalisation can be chosen to reproduce the observed level of clustering as well as possible. Thus, in contrast to most alternative pictures for the origin of structure, there is some hope that these theories are well enough specified to be falsifiable. The problem of extrapolating them from the linear regime to the present is formidable. Although direct simulation provides a reliable method for following the nonlinear evolution of the gravitationally dominant, but invisible component, further hypotheses about galaxy formation are required to relate the mass distribution to the observed galaxy distribution. Our poor understanding of galaxy formation introduces considerable uncertainty into this final step and is the major obstacle to using large-scale structure as a definitive test. If only we could see the **mass** distribution...

In Section II below I give a brief introduction to the linear phases of the evolution of structure and I display the linear predictions for the mass fluctuation distribution in a universe dominated by collisionless elementary particles. Section III then describes how N-body techniques can be used to follow the nonlinear growth of structure from such initial conditions, and discusses their technical limitations. The two final sections analyse the growth of structure in universes dominated by stable massive neutrinos, and by "cold" dark matter respectively. In each case I show how the interpretation of the mass distribution predicted by direct simulation is complicated by our ignorance of how galaxies form. Nevertheless, the predictions for a universe dominated by massive neutrinos seem fatally at variance with observation. Cold dark matter **may** be acceptable in a flat universe if galaxy formation occurs only at high peaks of the mass distribution. Most of the material presented in the last three sections comes from an ongoing collaboration with Marc Davis, George Efstathiou and Carlos Frenk.

II. OVERVIEW OF THE EARLY EVOLUTION OF STRUCTURE

The isotropy of the microwave background provides strong evidence that the Universe was very nearly uniform on large scales when the primordial plasma recombined (at a redshift $z \sim 10^3$). In addition we intuitively expect gravitational effects to drive the Universe towards states of increasing inhomogeneity. It is thus natural to describe the early evolution of structure in terms of the linear gravitational instability of an expanding Friedmann universe. At early times curvature can be neglected, and the evolution of the background model is determined by the energy densities of the matter and radiation fields. Within any particle physics model these energy densities can, in principle, be calculated as a function of redshift by following their evolution from the very early epochs when they were in thermal equilibrium. The linear instability problem then reduces to that of studying the growth of plane wave perturbations of this expanding multi-fluid system. The major components that need to be followed in the present context are photons, baryons and electrons, massless neutrinos, and the massive particles which provide the dark matter. The photons and the charged particles were strongly coupled until recombination, but the neutrinos and the dark matter candidates decoupled at much earlier epochs and can be considered collisionless.

Growing mode perturbations in an expanding universe can be divided into two types; those which are associated with a fluctuation in the local curvature of space-time and those which are not. This separation is simplified if the former perturbations are taken to involve no fluctuation in the relative energy densities of the various components; they are then termed adiabatic. The latter involve relative energy density fluctuations with zero sum and are termed isothermal.[2] It is usually assumed that adiabatic modes are likely to dominate the later evolution of structure on large scales, but this assumption can only be justified within particular models for the origin of fluctuations. Fluctuation spectra with significant power on large scales ($n < 4$ in the notation of Eq. 3 below) can be produced only if the relevant scales lay within the causal horizon at the generation epoch. This appears to require the kind of horizon evolution found in inflationary models. Quantum effects in such models lead to adiabatic fluctuations if the dominant perturbations are those associated with the field which drives inflation; however, they can also produce isothermal perturbations if the dominant perturbations are instead associated with an uncoupled free field that only acquired a mass at a much later epoch (for example, the axion field).[3]

The process which generated fluctuations determined not only the kind of mode emerging from the early universe, but also the relative power in waves of different scale. The density fluctuation field is usually specified in terms of $\delta(\underline{r}) = (\rho(\underline{r})/\bar{\rho}) - 1$, or of its Fourier transform $\delta_{\underline{k}}$. The fluctuation field at late times (from which nonlinear structures form) may be thought of as,

$$\delta_{k}(t) = T(k,t)(\delta_{k})_{p} , \tag{2}$$

where $(\delta_{k})_{p}$ is the field generated initially, and $T(k,t)$ is a transfer function describing its subsequent evolution. Because it seems unnatural that the generation process should have a characteristic scale in the range of interest here, the primordial power spectrum is usually taken to be a power law.

$$|\delta_{k}|^{2} \propto k^{n} . \tag{3}$$

If the Fourier components of the fluctuation field are assumed to have random phase, its statistical properties are completely specified by Eq. 3. This random phase assumption seems natural and is almost always employed, but it can be rigorously justified only for specific models of the origin and evolution of fluctuations. Cosmic strings offer an example of a fluctuation generation mechanism in which nonlinear effects may lead to strong phase correlations.[4] Fluctuations generated by quantum effects during inflation have random phase, however, and their power spectrum is of the form (3) with the index taking the value $n = 1$.[3] This is the Harrison-Zel'dovich "constant curvature spectrum" and has the property that the rms gravitational potential depth of fluctuations is independent of their scale. Larger values of n lead to small-scale divergence and require truncation of the spectrum to avoid black hole production on scales where $\delta\phi \sim c^{2}$. Smaller values of n lead to universe models that become highly inhomogeneous on large scales at late times. The value $n = 4$ corresponds to the large-scale fluctuation spectrum generated by purely local rearrangement of matter (as might occur below the horizon scale at a phase transition), while $n = 0$ is the white noise spectrum corresponding to a Poisson distribution of mass points and appears to have no particular cosmological significance.

The transfer function $T(k,t)$ can be calculated precisely for any chosen particle mix, and for both adiabatic and isothermal fluctuations. Reliable results are now available for all cases of interest here.[5] When the wavelength of a given perturbation exceeds the particle horizon, causal effects cannot act across it, and it evolves with a constant fluctuation in curvature. The associated density fluctuation may be thought of as amplifying in proportion to the square of the expansion factor during the radiation-dominated era, and in direct proportion to it thereafter. Once the horizon has grown to encompass the perturbation, pressure and diffusive effects can alter its evolution. Perturbations in a pressure-free ("dust") fluid continue to grow like the expansion factor if they enter the horizon when the dust dominates the energy density. On the other hand, if they enter the horizon when the universe is still radiation-dominated, their growth is arrested because the dominant photon-baryon fluid begins to execute acoustic oscillations at constant amplitude. This slowing of growth is known as the Meszaros effect, and, in the absence of other processes, it results in a bending of the power spectrum to an effective slope of $n - 4$ for wavenumbers above the characteristic value,

$$k_{eq} = 2\pi/(13(\Omega h^2)^{-1}\mathrm{Mpc}) \, , \tag{4}$$

corresponding to the present size of the diameter of the horizon at the epoch of equal density of matter and radiation ($z_{eq}=2.5\times10^4\Omega h^2$). In addition photon-baryon sound waves are damped exponentially by viscous effects on scales which are short enough for photons to diffuse out of their peaks before recombination. This is known as Silk damping, and leads to a characteristic scale,

$$k_s = 2\pi/(3.2(\Omega_b h^2)^{-1/2}(\Omega h^2)^{-1/4}\mathrm{Mpc}), \tag{5}$$

where Ω_b is the density parameter for the baryonic component.[6]

A collisionless particle of mass m, which was once in thermal equilibrium, becomes nonrelativistic when the radiation temperature is,

$$T_c = (g_*/3.9)^{1/3}mc^2/k_B \, , \tag{6}$$

where k_B is Boltzmann's constant and g_* is the total number of effective degrees of freedom of the relativistic species present when the particle decoupled; g_* is about 11 for neutrinos, and about 100 for more weakly interacting particles such as photinos.[7] Fluctuations in the collisionless particles which enter the horizon while $T > T_c$ are wiped out by relativistic streaming of the particles away from the peaks. This leads to a characteristic damping scale,

$$k_d = 2\pi/(6g_*^{-1/3}(30\mathrm{eV}/m)\mathrm{Mpc}) \, , \tag{7}$$

corresponding to the present diameter of the horizon at T_c. For sufficiently massive particles, or for particles like axions which formed out of thermal equilibrium in states of low momentum, this damping scale is smaller than any scale of interest for the formation of large-scale structure. Because of their small thermal velocities such dark matter candidates are generically termed cold dark matter. Weakly interacting particles with $g_* \sim 100$ and $m \sim 1$ keV might provide warm dark matter, but they are currently out of fashion; the damping scale of Eq. (7) is comparable to that of a bright galaxy in this case. Abundant particles such as neutrinos are the best candidates for hot dark matter; Eq. (7) then predicts the erasing of structure on all scales below those of large galaxy clusters.

Figure 1 shows power spectra at late times for adiabatic, constant curvature initial fluctuations imposed on a universe now dominated by collisionless particles. I have plotted $k^3|\delta_k|^2$ in this diagram, because this quantity is closely related to the amplitude of density fluctuations averaged over regions of size k^{-1}, and thus gives an indication of the order in which objects of different scale fragment out of the general expansion. For cold dark matter the curve has no peak, and the only scale corresponds to k_{eq} and characterises the gradual transition between the asymptotic regimes $n = 1$ and $n = -3$. For warm dark matter the spectrum is truncated exponentially above k_d, while for hot dark matter the truncation scale has lengthened until it almost coincides with k_{eq}. Together with the random phase assumption, these power spectra determine the statistical properties of the mass distribution until the quite recent epochs when nonlinear structures first form. For hot dark matter these first structures are expected to be large "pancakes" which must fragment to form galaxies, whereas for warm dark matter the pancakes themselves will be galaxy sized. For cold dark matter a whole hierarchy of structures is expected to grow rapidly from star cluster scales up to those of galaxy clusters. The detailed investigation of these nonlinear phases requires direct numerical simulation, and I turn now to a discussion of simulation methods.

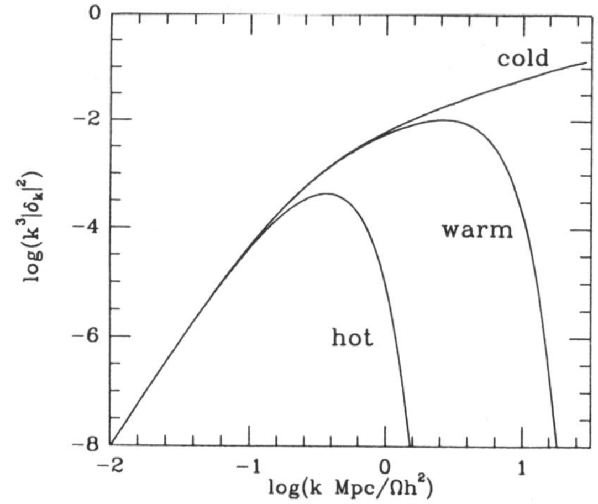

Figure 1: The power per decade as a function of spatial frequency for the density fluctuations expected in a universe dominated by collisionless elementary particles. These are the linear power spectra at late times in a universe which initially had the adiabatic, constant curvature fluctuations predicted by inflationary models. The three cases shown are differentiated by the magnitude of the random velocities of the particles. These results are taken from work by Bond & Szalay and Bond & Efstathiou (Ref. 5).

III. N-BODY METHODS

If the Universe is indeed dominated by collisionless particles, the evolution of the large-scale mass distribution should, to a good approximation, be driven by gravity alone. It is then natural to follow the nonlinear phases of this evolution by direct simulation. The governing equations are those of Vlasov and Poisson; however, since the general problem has no symmetries, the distribution function in the Vlasov equation is a function of all six phase space coordinates and of time, and, as a result, direct integration on a grid is not possible.

Instead the most practical method of solution is to follow the trajectories of a representative sample of phase space elements. The gravitational field in which these elements move can be estimated in some appropriate way from their positions. One is thus led to N-body methods in which each simulation particle can represent as many as 10^{80} of the elementary particles of the physical model. It is clearly crucial to ensure that this severe discretisation of the problem does not invalidate any of the conclusions drawn from simulations.

In order to get a good representation of a smooth density distribution with N discrete particles, it is desirable to make N as large as possible. Grid schemes which compute forces by applying transform methods to Poisson's equation are much faster for large N than direct summation methods, and are thus generally to be preferred. In addition the use of Fourier methods allows periodic boundary conditions to be applied at the edge of the computational volume. This seems the most natural way to model part of an effectively infinite universe with a finite calculation, and it works very efficiently if the computational volume is allowed to expand with the background universe so that the dominant motion can be taken out. The use of a density and potential grid in PM (particle-mesh) codes of this kind leads to interparticle forces with are anisotropic, and are weaker than the desired inverse square law on scales of order the mesh size. This property imposes a resolution limit which expands with the grid. It can be overcome, at the expense of a substantial reduction in speed, by direct evaluation of the forces between close pairs of particles. This technique, known as P^3M (particle-particle/particle-mesh) appears necessary to get reliable results in most situations where structure builds up from small scales. Detailed discussion of the application of N-body methods of this type to the problem of large-scale structure can be found in Ref. 8.

There are two main criteria for judging the merits of cosmological simulations. The initial conditions of the model should correspond as closely as possible to the predictions of linear theory. In addition the code should follow evolution from these initial conditions without undue distortion from artificial effects such as the softening of the forces, the discreteness of the particles, or the finite size of the region simulated. To set up initial conditions one needs to choose particle positions and velocities so that the deviation from a uniform, uniformly expanding distribution is a good representation of growing mode fluctuations with random phase and the desired initial power spectrum. The most effective way to do this appears to be to start with the particles on a comoving lattice, and then assign them displacements and velocities obtained from a particular realisation of δ_k using Zel'dovich's general Lagrangian formulation of linear theory.[9] The longest wavelength mode that can be imposed on the distribution corresponds, of course, to the side, L, of the computational volume, while the shortest mode that can be represented corresponds to the Nyquist frequency of the **particle** grid $2N^{-1/3}L$. For 32^3 particles this gives a dynamic range in length of 16 in the initial conditions, corresponding to a dynamic range of 4000 in mass. As a simulation evolves, the reliable dynamic range in mass cannot change. However, clustering increases the useful dynamic range in length, because bound systems collapse and cease to expand with the computational volume. As a result, at late times the major resolution limit usually comes from the softening of the forces, rather than from discreteness effects. This is the reason why P^3M codes give better results than codes without a short-range force correction, despite the fact that they cannot handle as many particles.

Figure 2 illustrates some of the points just discussed by showing the power spectra of a variety of simulations.

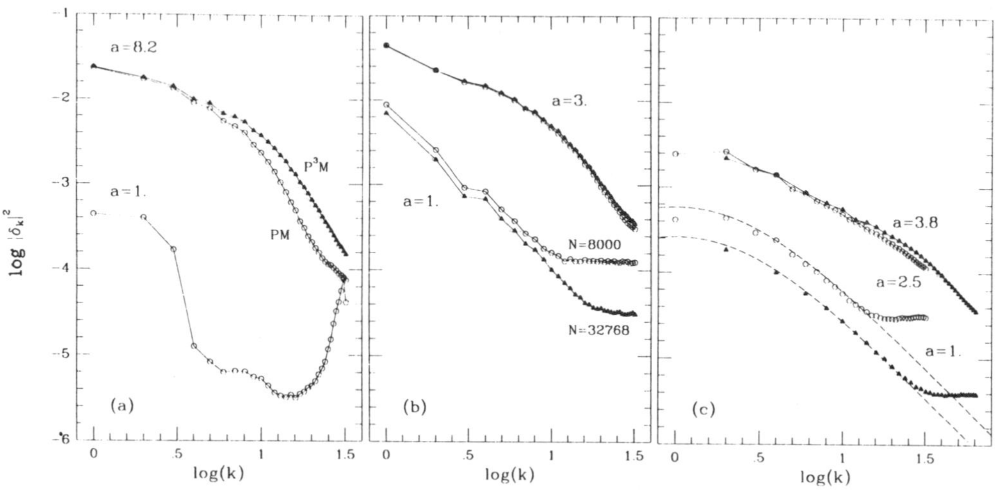

Figure 2: Power spectra illustrating the tests of N-body methods discussed in that text. In (a) the same initial conditions were evolved using codes with high and low resolution force calculations. Excessive force softening results in inadequate modeling of small-scale (high frequency) structures. In (b) the same initial fluctuation distribution was modeled using different numbers of particles. The expansion factor at the later time is 2.7 rather than 3 for the 8000-particle model in order to compensate for its larger initial fluctuation amplitude. In (c) Two ensembles of simulations are compared which model the theoretical spectrum shown as a dashed curve over regimes differing by a factor of two in scale. The difference in the later times compensates for the different initial fluctuation amplitudes.

Figure 2a shows the result of using two different codes to follow evolution from the same initial conditions; a 32^3 cloud-in-cell PM code with one particle per cell, and a P^3M code with a length resolution about 8 times better. The initial particle grid in these models was perturbed with a flat power spectrum, truncated so that there was no power in waves with k-vectors outside a cube of side twice the fundamental frequency. The power spectra shown were computed on a 64^3 grid. The initial conditions are seen to have very little power on frequencies between $k = 3$ and $k = 32$, the fundamental frequency of the particle grid. When this distribution is evolved it produces large pancakes, filaments and clusters (see Fig. 5 of Ref. 8) which correspond in the power spectra to a nonlinear transfer of power from low to high wavenumber. Very strong phase correlations are present at late times, so that although the power spectrum contains power on a wide range of frequencies, the density distribution has a very pronounced scale. The substantial power deficit at high frequency in the PM simulation results from its lack of force resolution, and clearly makes it unsuitable for studying structure on any but the largest scales ($k<6$).

Figure 2b shows the result of trying to set up the same realisation of a near power law initial power spectrum using $N = 32768$ and $N = 8000$. Apart from the small normalisation difference, the particle distributions clearly have the same spectrum at frequencies below the particle Nyquist frequency (k = 16 and k = 10 for these two cases); at higher frequencies the power flattens off at the white noise level ($|\delta_k|^2 = 1/N$). The later output times in these two P^3M models differ by the amount required for linear theory to predict that they should have the same clustering amplitude. This expectation is extremely well fulfilled, showing that discreteness has had almost no effect on the final power spectrum; further comparisons are given in Ref. 8. For this kind of initial condition the relatively large initial difference at high frequency is rapidly swamped by power generated from lower frequencies by nonlinear effects. Apparently even 8000 particles are sufficient to study the mass distribution in these models over a range of 32 in scale. Figure 2c shows a test of the effects of the finite size of the computational volume. It compares the power spectra of two ensembles of P^3M models designed to follow the evolution of universes dominated by cold dark matter (see Section V below). The initial conditions were set up so that the side of the computational box corresponded to length scales differing by a factor of two. The later times plotted were chosen so that linear theory would predict fluctuations of the same amplitude in the two sets of models; in the absence of any artificial effects due to the finite box size, the power spectra should therefore coincide. At large k differences are visible which are due to the mesh size of the grid used to calculate the power spectra, but at small k the agreement is very good. Thus "edge" effects have not influenced the evolution of the Fourier components accessible to the small box calculation.

Many tests of the kind just described are necessary to delineate the range of validity of cosmological N-body simulations. A more complete description of such tests can be found in Ref. 8. Nevertheless, it is clear that reliable results for the nonlinear evolution of the mass distribution can be obtained over a moderate range of scales. I now go on to describe the results of using these methods to study the evolution of the mass distribution in universes dominated by massive neutrinos and by cold dark matter.

IV. NEUTRINO-DOMINATED UNIVERSES

Massive neutrinos were the first weakly interacting particle candidate for the dark matter, and they are still a priori the most plausible, since we do at least know that neutrinos exist. The most important feature of the fluctuation spectrum in a neutrino-dominated universe is the free streaming cutoff which eliminates all small-scale structure. The characteristic scale usually quoted for this cutoff is not that in Eq. (7) but rather,

$$\lambda_c = 41(m_\nu/30\text{eV})^{-1}\text{Mpc} = 13(\Omega h^2)^{-1}\text{Mpc}, \tag{8}$$

where Eq. (1) has been used to relate the mass of the dominant neutrino to cosmological parameters (see, for example, the article of Bond and Szalay[5]). λ_c is the wavelength at which the transfer function T(k,t) has fallen to 10% of its value for small k; as Fig. 1 shows there is no significant power at shorter wavelengths. As pointed out long ago by

Zel'dovich and his colleagues, a distribution of this kind has a large coherence length, and the initial nonlinear phases of its evolution will involve the formation of caustics (or "pancakes") with a characteristic scale given by Eq. (8).[9,10] A cell-like structure of such caustics is expected to form and then evolve into a network of filaments which eventually collapse into dense clumps. This general picture has been confirmed through a series of one-, two- and three-dimensional simulations of evolution from distributions with a large coherence length.[11] In most of this work the initial conditions were only schematically related to the linear predictions for a neutrino-dominated universe. However, White, Frenk and Davis modeled their initial conditions directly on the linear calculations of Bond and Szalay, and so were able to eliminate any ambiguity in how the results should be scaled for comparison with the real Universe. We also ran an ensemble of realisations for each power spectrum we considered, and so were able to get a realistic estimate of the statistical uncertainities in our results; such uncertainties can be quite large when the characteristic scale of structure approaches the size of the simulation. Figure 3 shows a series of projections of one of the models of White, Frenk and Davis. These models used a PM scheme to follow the motion of 32,768 particles with a 64^3 density and potential grid; the neutrino fluctuation spectrum of Fig. 1 was imposed on an Einstein-de Sitter universe, setting the side of the box equal to $5\lambda_c$ and the rms initial relative density fluctuation at 22%. The three left-hand panels of the figure show projections of all the points after the model has expanded by factors of 4.0, 6.1 and 10; the general expansion has been taken out by rescaling. Over this period the model goes from a relatively uniform state to one in which much of the mass is concentrated in condensed clusters. It is important to realise, however, that these pictures represent the **neutrino** distribution. In this kind of universe galaxies can only form in regions which have undergone local collapse; the spaces between the collapsed objects in the left-hand panels are thus expected to be devoid of galaxies. We have tried to identify those particles which **could** represent galaxies by following the evolution of each cell of the initial particle grid, and noting the time when its volume goes to zero. When this occurs, we tag the corresponding particle as a galaxy candidate. One percent of the particles are tagged as "galaxies" by an expansion factor of 2.9; this seems a reasonable time to identify with the onset of significant galaxy formation. Later times can then be identified by their expansion factor, $1 + z_{GF}$, since this epoch; the panels of Fig. 3 correspond to $z_{GF} = 0.4$, 1.1 and 2.5. The right-hand panels show the evolution of the projected distribution of "galaxies", and contain 5, 23 and 52% of the particles at the three times shown. The formation of sheets, filaments and clusters is much clearer in the "galaxy" pictures than in the "neutrino" pictures. Notice that clusters begin to form very soon after the first collapse of structure, and that any cell-like structure is a transient state that lasts only for a relatively short time.

In the absence of a cosmological constant, $h < 0.54$ is required to get the age of a flat universe to exceed 12Gyr. This appears to be a very conservative lower bound on the ages of globular star clusters. The side of the box in Fig. 3 must thus exceed 200 Mpc, and it is clear that the model predicts strong inhomogeneity on very large scales. This may be quantified by calculating the autocorrelation function, $\xi(r)$, of the three-dimensional point distributions in the models.

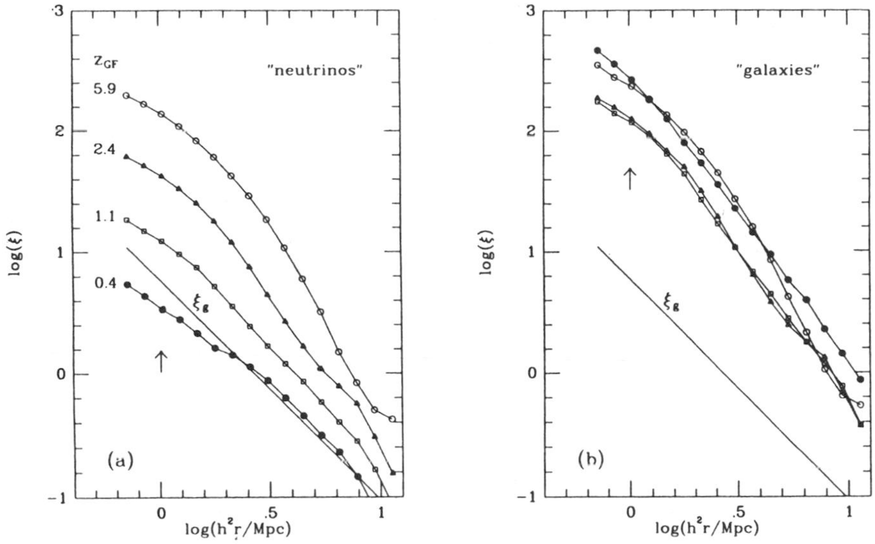

Figure 4: Autocorrelation functions of the particle distribution in an ensemble of Einstein-de Sitter, neutrino-dominated models. Results for all the particles are shown in (a), while results for the particles tagged as "galaxies" are shown in (b). The different curves are labelled by the expansion factor, 1 + z, since the epoch when the first collapsed structures formed. The straight lines in these plots are the power law model for the observed galaxy correlation function scaled assuming h = 0.54. Arrows denote the mesh size of the potential grid.

In Fig. 4 I give the results found for "neutrinos" and for "galaxies" in an ensemble of 4 models similar to the one illustrated in Fig. 3. The growth of structure produces a steady amplification and a steepening of $\xi(r)$ for the "neutrino". For $z_{GF} > 1$, the neutrino function is significantly steeper than the $\xi \propto r^{-1.8}$ observed for real galaxies. The correlation function for "galaxies" in the simulations is also rather steep and evolves very little with time; this reflects the fact that the "galaxy" selection algorithm picks out nonlinear structure determined by the form of the **linear** neutrino distribution. I have plotted the power law model of the observed galaxy distribution ($\xi \propto (r/5h^{-1}\mathrm{Mpc})^{-1.8}$) in Fig. 4 assuming h = 0.54; this scaling produces the best possible agreement between theory and observation for $t_0 > 12$ Gyr and $\Omega \leq 1$. (The Ω scaling of Eq. (8) results in even poorer agreement for open universes than for the flat models shown in the figure). The clustering of real galaxies is clearly quite incompatible with that of "galaxies" in the model, and it is also significantly weaker than that of the "neutrinos" for any acceptable redshift of galaxy formation.

The scale discrepancy just described can be illustrated graphically by imagining oneself to be an observer situated at some point in a properly scaled neutrino model, and choosing particles in the same way as observers have chosen galaxies for inclusion in complete redshift surveys of nearby space. The observations can then be compared directly with the simulated catalog in any desired plane. The upper left-hand picture in Fig. 5 shows an equal area projection of the distribution on

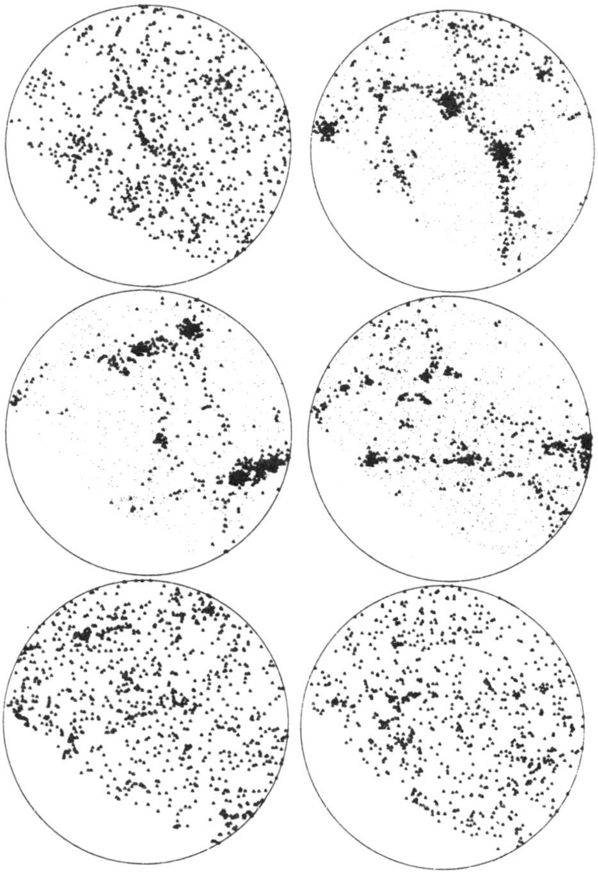

Figure 5: Equal area projections of the galaxy distribution on the northern sky and in artificial galaxy catalogs made from simulations. The outer circular boundary corresponds to Galactic latitude +40, while the inner dashed boundary corresponds to declination 0. The upper left-hand picture is the real data of the CfA survey. The bottom two pictures are made from simulations of open universes dominated by cold dark matter. The other three pictures were made from simulations of a neutrino-dominated universe; the triangles represent "galaxies", while the dots are points in whose neighborhood the matter has not yet collapsed. These models assume a flat universe with $h = 0.54$ in which galaxy formation began at $z = 2.5$; the cold dark matter models assume $h = 1.1$ in order to match their 2-point correlation function to that observed for galaxies.

the sky of part of the Center for Astrophysics redshift survey.[12] The galaxies shown are a complete sample subject to the selection criteria: galactic latitude, $b > 40°$, declination, $\delta > 0°$, apparent magnitude $m_z \le 14.5$, radial velocity, $v_r < 10{,}000$ km s^{-1} and absolute magnitude bright enough that the galaxy would still pass the apparent magnitude limit if its distance were increased by a factor $(v_r/4{,}000 \text{ km s}^{-1})^{-1}$. This sample is thus volume limited to 4,000 km s^{-1} and magnitude limited from 4,000 to 10,000 km s^{-1}. Figure 5 compares it to artificial catalogs over the same area of sky and with the same mean distribution in depth; three of these were made from neutrino-dominated models with $\Omega = 1$, $z_{GF} = 2.5$ and $h = 0.54$, and two were made from the cold dark matter models discussed in Section V below. In the neutrino models "galaxies" are shown as open triangles, whereas other points are shown as dots. It is important to remember that the triangles mark those points which could, in principle, represent real galaxies; in a real neutrino-dominated universe only a biased subset of them might succeed in making visible objects. On the other hand, none of the dots can represent a real galaxy, so it is significant that there are much larger empty areas in the neutrino models than in the real data.

Although Figs. 4 and 5 suggest that the scale discrepancy between the observed galaxy distribution and the predictions of a neutrino-dominated universe is too great for the latter to be acceptable, galaxy formation is a sufficiently complex and poorly understood process that these arguments cannot be accepted without reservation. It is conceivable that various feedback mechanisms might suppress galaxy formation in dense regions to such an extent that, in contrast to our crude model, observable objects are actually more weakly clustered than the neutrinos. Arguments of this kind, although rather implausible, are difficult to dismiss entirely. It therefore seems worthwhile to try to use the mass distribution as a direct constraint on neutrino models, instead of using it to infer the galaxy distribution. By the last time shown in Fig. 3 (z_{GF} = 2.5), much of the mass appears to be in dense clumps, and it turns out that the predicted properties of these clumps are so extreme that they cannot be identified with any known population of objects.[13] Table 1 summarises the properties of clumps in our ensemble of four Ω = 1 models at z_{GF} = 2.5 and 6. (We

TABLE I. Properties of ν Clusters

z_{GF}	2.5	6	scaling	observed clusters
Mean cluster spacing (Mpc)	65	70	f^{-1}	50 h^{-1} [c]
Fraction of total mass in clusters[b]	0.39	0.61	-	-
Cluster mass (10^{15} $M_\odot$)[a,b]	26	54	f^{-2}	1.5 h^{-1} [d]
Maximum circular velocity (km s^{-1})[a]	3739	5227	$f^{-1/2}$	1780[d]
Maximum temperature (keV)[a]	45	88	f^{-1}	7[e]
Total x-luminosity (10^{45} erg s^{-1})[a,b]	8	30	$g^2f^{-3/2}$	0.5 h^{-2} [e]
Number expected in HEAO-1 survey	496	1950	$g^3f^{9/4}$	≤ 7[f]

Notes: a) These values are the fifth most extreme for clusters in the ensemble of four simulations.

b) Masses and luminosities are measured within 11 f^{-1} Mpc.

c) Appropriate to Abell clusters with richness R $\geq$ 1.[14]

d) For the Coma cluster within 3.5 h^{-1} Mpc.[15]

e) For the Perseus cluster.[16]

f) Unidentified sources in the HEAO-1 survey.[17]

consider only clumps which contain more than $8\times10^{14}f^{-2}$ $M_\odot$ within a radius of $11f^{-1}$ Mpc). The scaling parameters in this table, $f = h^2/0.3$ and g, defined as ten times the baryon fraction in the clumps, are both expected to be less than or of order one. The models predicted about half the entire mass of the universe to be in big clumps at these times. The masses and binding energies of the clumps are too large for them to be identified with real galaxy clusters. However, even if they succeed in suppressing all galaxy formation, they will provide a highly visible (and unobserved) population of x-ray sources, unless the fraction of baryons in their central regions is much smaller than the ~ 10% required to get cooling and fragmentation in a neutrino pancake.[18]

While the above arguments against the standard neutrino picture appear very strong, they are not conclusive. However, they do force one to abandon most of its attractive features. For any acceptable epoch of galaxy formation, the dominant structures in the neutrino distribution are dense clumps rather than sheets or filaments. Although these clumps contain most of the mass of the universe, they must somehow be rendered invisible. The galaxy formation process is required to be strongly biased against dense regions; this is exactly the opposite of the simplest and most plausible expectation. In addition our observed velocity with respect to the microwave background appears implausibly low, unless galaxy formation also singles out regions of low peculiar velocity.[19] All these difficulties arise because of the very large scale implied by Eq. (8). Although this scale is comparable to that of certain observed structures, a neutrino-dominated universe requires density contrasts on these scales far in excess of observation if galaxies are to form before a redshift of 2. The coherence length of Eq. (8) can be reduced by introducing a cosmological constant, or by allowing the universe to be dominated by an unstable particle at some stage of its evolution.[20] However, these possibilities are sufficiently unattractive, and introduce enough extra difficulties, that it seems best to look elsewhere for a resolution of the nature of the dark matter; at least, that is, until experiment forces us to accept the existence of a massive neutrino.

V. COLD DARK MATTER

If the large coherence length of a neutrino-dominated universe makes it incompatible with observation, one is forced to much more exotic candidates in order to maintain that the dark matter is made up of weakly interacting elementary particles. In the present context the exact nature of these particles is of little consequence, and I now discuss the evolution of the mass distribution in a universe dominated by axions, photinos, gravitinos or any other kind of cold dark matter. Once again I assume the adiabatic constant curvature initial fluctuation spectrum predicted by inflationary models. As shown in Fig. 1, this leads to a fluctuation spectrum at late times with no damping cutoff, but with the power per decade increasing only very slowly on the smallest scales. Such a distribution is expected to evolve by hierarchical aggregation into larger and larger clumps, the characteristic clump mass increasing very rapidly at early times.[21] This situation is difficult to simulate because nonlinear behavior occurs simultaneously on a wide range of scales. Quite soon after the smallest scales accessible to a model have collapsed, structures comparable in size to the simulation volume begin to go nonlinear; once this has occurred the model is no longer a good representation of a subregion of a larger universe.

Figure 6 shows the evolution of a P^3M simulation of an $\Omega = 1$ universe filled with cold dark matter (one of the models presented in Ref. 22). The initial perturbations were set in such a way that the side of the computational volume corresponds to $32.5h^{-2}$Mpc, implying a mass of $3;\times 10^{11}h^{-4}M_{\odot}$ for each particle in the model. This should be compared with the mass per observed bright galaxy which is ~ $3\times10^{13}h^{-1}M_\odot$ if our Universe is flat. The simulation cannot therefore represent aggregates smaller than the halo of a bright galaxy. The structure in this simulation clearly evolves rapidly from small to large scale in a more or less hierarchical fashion. It is interesting to compare these point plots with similar diagrams for simulations evolved from Poisson initial conditions.[23] A much wider range of scales is apparent in the clustering in the present models. Filamentary structures, superclusters of clumps, and large low density regions are all in evidence in these plots; quantitative study is, of course, necessary to decide if these are truly analogous to observed structures. The model of Fig. 6 is one of five similar models which make up the smaller scale ensemble for which power spectra are plotted in Fig. 2c. Both diagrams show quite clearly the considerable large-scale power in the initial conditions of these models.

Figure 7a shows the evolution of the autocorrelation function of

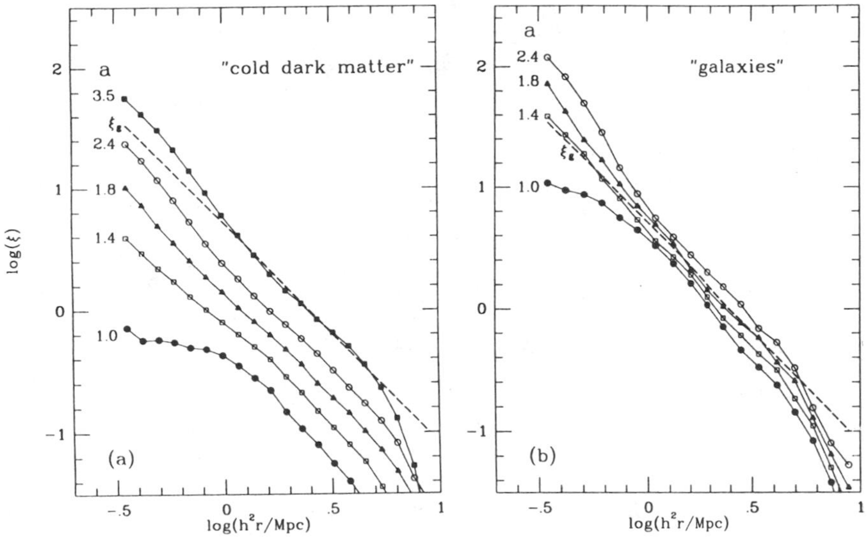

Figure 7: Autocorrelation functions for an ensemble of Einstein-de Sitter models of a universe dominated by cold dark matter. Results for all the points are shown in (a), while (b) gives results for these particles identified as "galaxies" by our algorithm for modeling biassed galaxy formation. The curves are labelled by expansion factor since the start of the calculation. The dashed straight lines give the power law model for the observed galaxy correlation function scaled assuming h = 0.5. The softening length of the interparticle force is too small to appear in this diagram.

the same ensemble of five cold dark matter models. Even in the initial conditions ξ exceeds unity on small scales and has a relatively steep slope on intermediate scales. The nonlinear evolution of clustering steepens ξ and flattens the fluctuation power spectrum (see Fig. 2). As a result there is only one time when the mass autocorrelation function of the models has approximately the same slope as the observed autocorrelation function of galaxies. However, at this time (which occurs after expansion by a factor of only 1.8 from the initial conditions) the amplitude of the model function is smaller than that observed unless $h \leq 0.22$. Such values of the Hubble constant are unacceptably low. After expansion by a factor of 3.5 the model function agrees with observation in the region $\xi \sim 1$ for $h = 0.5$, but by this time it is much too steep on small scales. There is thus no time when the mass distribution of these Einstein-de Sitter models is a good match to the observed galaxy distribution.

An even more serious difficulty with these models is found when particle motions with respect to the expanding background are compared with observations of galaxies. The scale of clustering in the models may be characterised by the separation, r_0, defined by $\xi(r_0) = 1$. If the expansion velocity across r_0 is set to its observed value of 500 km s^{-1}, the rms relative peculiar velocity of close pairs of particles is found to be of order 1800 km s^{-1}, whereas observation suggests that the corresponding number for pairs of galaxies is 700 km s^{-1} or less.[24] This is a manifestation of the well known fact that the mass-to-light ratios inferred for groups and clusters of galaxies from their internal velocity dispersions are too low to close the universe by a factor of 5 - 10; if $\Omega = 1$, then we already know that the galaxy distribution cannot trace the mass distribution on the scale of groups and clusters. A possible way out of this difficulty is clearly to consider cold dark matter universes with $\Omega < 1$. We have run an ensemble of such open models. We find that if $\Omega = 0.2$ at present, the mass autocorrelation function can match the observed galaxy function fairly well for values of h near 1; in addition the inferred relative peculiar velocites on small scales drop by about a factor of two and so become closer to those observed. Simulated catalogs from two of these models are shown at the bottom of Fig. 5; they clearly resemble the observations much more closely than the neutrino models. Nevertheless, we have found that the peculiar velocity distribution in open models does not agree in detail with observation, and those models which are in rough agreement turn out to predict excessive fluctuations in the microwave background on small angular scale.[25]

If we accept that Ω is significantly less than one, we cast doubt on the whole inflationary model for the early evolution of the Universe. Since, in addition, open models suffer from other difficulties just discussed, it is natural to see whether agreement with observation can be obtained in a flat universe for any plausible relaxation of the assumption that galaxies trace the mass distribution on large scales. In the kind of situation under discussion, galaxies may be expected to form when gas has time to collect, cool and collapse within the preexisting potential wells provided by clumps of cold dark matter.[21] We have investigated the possibility that galaxies form only in the neighborhood of high peaks in a suitably smoothed version of the linear fluctuation distribution. It is easy to think of reasons why galaxy formation should occur more easily around an unusually high peak (which turns into an unusually dense "halo" of dark matter); in certain circumstances such effects might be strong enough to lead to an effective threshold for galaxy formation. We modeled this situation by smoothing the linear density field of our initial conditions, identifying all peaks higher than ν times the rms fluctuation, and calling the particle initially nearest each peak a "galaxy". The properties of the "galaxy" distribution can then be followed as a simulation evolves. As discussed by Kaiser and Bardeen elsewhere in this volume, a prescription of this kind picks out a biased subset of the particles which is more strongly clustered than the distribution as a whole.

Figure 7b shows the evolution of ξ for a population of "galaxies" obtained by smoothing lightly the 64^3 density grid of our initial conditions and then identifying peaks above a threshold, $\nu = 2.5$. These choices led to us finding about 1000 "galaxies" in each simulation volume $((32.5h^{-2}Mpc)^3)$; this corresponds roughly to the number density of bright galaxies for h = 0.5. The distribution of "galaxies" shown in the bottom right-hand frame of Fig. 6 corresponds to the mass distribution shown at bottom left. The enhanced clustering of the galaxies is visually apparent in these pictures and is displayed quantitatively in the correlation functions of Fig. 7. The correlation function of the "galaxies" evolves much more slowly than that of the mass distribution. This is because it is due, in large part, to a pattern "painted onto" the initial mass distribution by the statistical properties of the linear density fluctuation field. At the initial time ξ for the "galaxies" exceeds 10 on small scales, and it is considerably steeper than the mass autocorrelation function. By an expansion factor of 1.4 the "galaxy" correlation function is already slightly steeper than the observed galaxy function and its amplitude is correct for h = 0.44, slightly less than the value h = 0.5 used to plot the power law model of the observations in Fig. 7, but large enough to be acceptable. (Remember that in a flat universe $h \leq 0.54$ is required to get an age in excess of 12 Gyr.) If at this time the expansion velocity across the "galaxy" correlation length is set equal to 500 km s^{-1}, the velocity corresponding to the mass correlation length is only 200 km s^{-1}. Thus peculiar velocities in these biased formation models are inferred to be about a factor of 2.5 lower than in models where galaxies are asumed to be unbiased tracers of the mass distribution; this almost entirely eliminates the discrepancy with observation.

While the model for galaxy formation just elaborated is quite plausible, it has, as yet, no detailed physical basis. Nevertheless, it illustrates several important points. Although the galaxy distribution in a universe dominated by cold dark matter is expected to be closely linked to the underlying mass distribution, we have no reason to expect galaxies to be a **fair** sample of the mass. In fact the shape of the linear power spectrum predicted for such a universe suggests that galaxies and galaxy clusters may form at close enough epochs for galaxy formation to be significantly influenced by structure on much larger scale. Our model indicates that large biases may result from this, and that they may be of the right kind and of about the right size to reconcile observations of the galaxy distribution with an Einstein-de Sitter universe filled with cold dark matter.

ACKNOWLEGEMENTS

The original research reported in this article was carried out in collaboration with Marc Davis, George Efstathiou and Carlos Frenk, and I am grateful for their permission to reproduce results from a number of our joint papers. Much of the more recent work was carried out at the Institute for Theoretical Physics, Santa Barbara, during its program on the evolution of structure in the universe. I thank the director and his staff for their warm hospitality during our stay. Various stages of this work were supported by NSF grants AST-8114715, AST-8118557, PHY77-27084 and by DOE contract AT03-82E1240069.

REFERENCES

[1]The articles by J.S. Gallagher and A. Guth are particularly relevant.

[2]More complete discussions of fluctuation growth are given by P.J.E. Peebles, The Large-Scale Structure of the Universe, Princeton, Univ. Press (1980), and J.M. Bardeen, Phys. Rev. **D22**, 1882 (1980).

[3]A. Guth and S.-Y. Pi, Phys. Rev. Lett. **49**, 1110 (1982); S. Hawking, Phys. Lett. **115B**, 295 (1982); A. Starobinskii, Phys. Lett. **117B**, 175 (1982); J.M. Bardeen, P.J. Steinhardt and M.S. Turner, Phys. Rev. **D28**, 679 (1983). A review is given by R. Brandenberger, Rev. Mod. Phys., in press (1984). J.M. Bardeen explained to me how isothermal fluctuations can be dominant in certain models of an axion-dominated universe.

[4]A. Vilenkin, this volume.

[5]P.J.E. Peebles, Astrophys. J. **258**, 415 (1982); J.R. Bond and A.S. Szalay, Astrophys. J. **174**, 443 (1983); P.J.E. Peebles, Astrophys. J. **263**, L1 (1982); J.R. Bond and G. Efstathiou, Astrophys. J. Lett., in press (1984); G.R. Blumenthal and J.R. Primack, Fourth Workshop on Grand Unification, (eds. H.A. Weldon, P. Langacker and P.J. Steinhardt), Birkhauser Boston, P. 256, (1983).

[6]J.I. Silk, Astrophys. J. **151**, 459 (1968); M.R. Wilson, Ph.D. thesis, Univ. of Calif., Berkeley (1981).

[7]J.R. Bond, A.S. Szalay and M.S. Turner, Phys. Rev. Lett. **48**, 1636 (1982).

[8]G. Efstathiou, M. Davis, C.S. Frenk and S.D.M. White, Astrophys. J. Suppl., in press (1985).

[9]Ya.B. Zel'dovich, Astron. Astrophys. **5**, 84 (1970).

[10]A.G. Doroshkevich, M.Yu. Khlopov, R.A. Sunyaev, A.S. Szalay and Ya.B. Zel'dovich, Ann. NY. Acad. Sci. **375**, 32 (1980).

[11]A.G. Doroshkevich, E.V. Kotok, I.D. Noikov, A.N. Polyudov, S.F. Shandarin, and Yu.S. Sigov, Mon. Not. R. Astro. Soc. **192**, 321 (1980); A. Melott, Phys. Rev. Lett. **48**, 894 (1982); A. Melott, Mon. Not. R. Astr. Sco. **202**, 593 (1983); A.A. Klypin and S.F. Shandarin, Mon. Not. R. Astr. Soc. **204**, 891 (1983); C.S. Frenk, S.D.M. White and M. Davis, Astrophys. J. **271**, 417 (1983); J. Centrella and A. Melott, Nature, **305**, 196 (1983); S.D.M. White, C.S. Frenk and M. Davis, Astrophys. J. **274**, L1 (1983).

[12]M. Davis, J. Huchra, D. Latham and J. Tonry, Astrophys. J. **253**, 423 (1982).

[13]S.D.M. White, M. Davis and C.S. Frenk, Mon. Not. R. Astro. Soc., **209**, 27P (1984).

[14]N.A. Bahcall and R.M. Soneira, Astrophys. J. **270**, 20 (1983).

[15]S.M. Kent and J.E. Gunn, Astron. J. **87**, 945 (1982).

[16]J.D. McKee, R.F. Mushotzky, E.A. Boldt, S.S. Holt, G.E. Marshall, S.H. Pravdo and P.J. Serlemitsos, Astrophys. J. **242**, 843 (1980).

[17]G. Piccinotti, R.F. Mushotzky, E.A. Boldt, S.S. Holt, G.E. Marshall, P.J. Serlemitsos and R.A. Shafer, Astrophys. J. **253**, 485 (1982).

[18]P.R. Shapiro, C. Struck-Marcell and A. Melott, Astrophys. J. **275**, 413 (1983); J.R. Bond, J. Centrella, A.S. Szalay and J.R. Wilson, Mon. Not R. Astr. Soc., **210** 515 (1984).

[19]N. Kaiser, Astrophys. J. **273**, L17 (1983).

[20]M. Davis, M. Lecar, C. Pryor and E. Witten, Astrophys. J. **250**, 423 (1981); P. Hut and S.D.M. White, Nature, **310** 637 (1984).
[21]S.D.M. White and M.J. Rees, Mon. Not. R. Astr. Soc. **183**, 341 (1978); G.R. Blumenthal, S.M. Faber, J.R. Primack and M.J. Rees, Nature, **311** 517 (1984).
[22]M. Davis, G. Efstathiou, C.S. Frenk and S.D.M. White, Astrophys. J., in press (1985).
[23]G. Efstathiou and J.W. Eastwood, Mon. Not. R. Asto. Soc. **194**, 503 (1981); G. Efstathiou and J. Barnes, Formation and Evolution of Galaxies and Large Structures in the Universe (eds. J. Audouze and J. Tran Thanh Van), Reidel, P. 361 (1983).
[24]M. Davis and P.J.E. Peebles, Astrophys. J. **267**, 465 (1983).
[25]J.R. Bond and G. Efstathiou, Astrophys. J. Lett., in press (1984); N. Vittorio and J. Silk, Astrophys. J. Lett., in press (1984). See also the contributions by J.R. Bond and J. Silk in this volume.

Cold Dark Matter in Universes with Critical and Low Density

S. Achilli, F. Occhionero, and R. Scaramella

Dark (i.e. collisionless and non-baryonic) matter (DM) is now widely believed to constitute a significant fraction of the cosmic density. DM is for instance necessary, although not automatically sufficient as we shall see, to grow condensations in the matter-dominated era without conflicting with the observed smoothness of the microwave background (MWB).

FIG.1

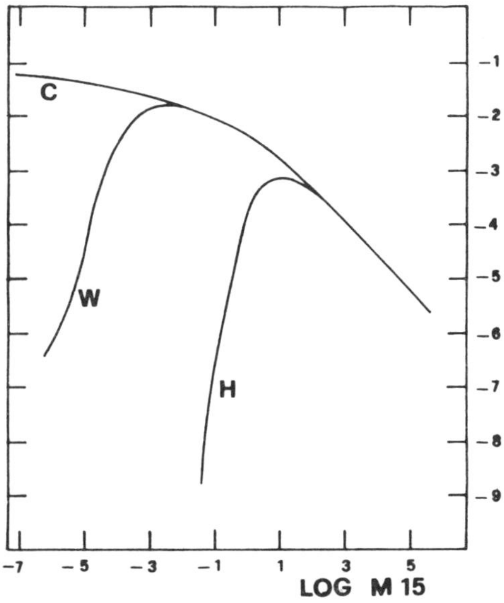

In Fig. 1 we plot vs. mass[1] the power spectrum $k^{(3+n)/2}\ T(k)$ (with primordial spectral index n=1) for 1) the hot case (H; e.g. 30 eV neutrinos; free-streaming mass=$M_J=10^{15}\ M_\odot$), 2) the warm case (W; e.g. 1 KeV gravitinos; $M_J=10^{12}\ M_\odot$), and 3) the cold case (C; e.g. photinos or axions; M_J vanishingly small). In agreement with a vast literature, in 1) the degeneracy of M_J and M_{HEQ}, the mass which enters the horizon at the radiation-matter equivalence, singles out the supercluster size and forces the top-down pancake scenario;[2] in 2) the degeneracy is removed and a whole mass range between the galactic and the supercluster scale opens up;[3] finally in 3) there is power on much smaller scales with very interesting perspectives[4] perhaps reviving the bottom-up scenario. Indeed CDM may be preferred over its competitors[5] because it does not have the shortcomings of the very late entrance into non-linearity at the supercluster scale[6] and of the too broad correlation length[7] and also solves the problem of the missing mass in dwarf ellipticals.[8] Incidentally in 3) we have found good agreement with Peebles' analytical fit.[7]

The above computations refer to $\omega_0 = \Omega_0 h^2 = 1$ $(1/2 \leq h \leq 1)$. However, although inflation demands very strongly that $\Omega_0 = 1$, we can not ignore the observational request[9] that, at least for clumped matter, $\Omega_0 < 1$. Work is in progress in order to investigate whether CDM can give the observed structure even if the Universe is open or in presence of a light diffused component. By the numerical integration of the canonical equations, we study the linear evolution of adiabatic perturbations in a Universe composed of radiation, baryonic matter and CDM (treating the first two components initially as a single relativistic fluid, then as two fluids coupled by Thomson scattering with instantaneous decoupling at $z = 10^3$; a fourth diffuse component is easily added to our scheme).

We explore a wide mass range in two cases: a) $\omega_0 = 1$, $z_{EQ} = 40{,}000$, $\varepsilon = \rho_{BAR}/\rho_{TOT} = 0.01$ and b) $\omega_0 = 0.1$, $z_{EQ} = 2{,}500$ (three massless neutrinos being added to the radiative component), $\varepsilon = 0.5$.

From the direct inspection of the evolution of the density contrasts, one can see already that b) is marginally in conflict with the MWB: indeed the relevant growth of the baryon component between the moment each mass enters the horizon and the present does not exceed 1,000. In Fig. 2 we give the power spectrum for CDM (n = 1, z = 10) for case b).

FIG. 2

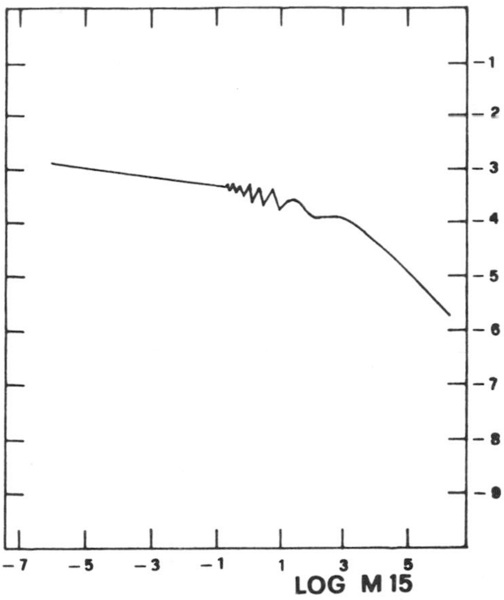

REFERENCES

[1]R. Scaramella, Dissertation, Univ. of Rome (1984).
[2]P.J.E. Peebles, Ap. J. **248**, 885 (1981); J.R. Bond and A.S. Szalay, Ap. J. **274**, 443 (1984).
[3]G.R. Blumenthal, H. Pagels and J.R. Primack, Nature **299**, 37 (1982); J.R. Bond, A.S. Szalay and M.S. Turner, Phys. Rev. Lett. **48**, 1636 (1982).
[4]P.J.E. Peebles, Ap. J. **277**, 470 (1984).
[5]G.R. Blumenthal, S.M. Faber, J.R. Primack and M.J. Rees, preprint (1984).
[6]C. Frenk, S.D.M. White and M. Davis, Ap. J. **271**, 417 (1983).
[7]P.J.E. Peebles, Ap. J. Lett. **263**, L1 (1982).
[8]D.N.C. Lin and S.M. Faber, Ap. J. Lett. **266**, L21 (1983).
[9]G.A. Tammann, this volume (1984).

Some Comments on Gravitational Clustering Simulations of "Ino" Dominated Universes

Adrian L. Melott

ABSTRACT

Large-scale numerical simulations of universes dominated by collisionless elementary particles possess a number of ambiguities: (1) identification of the "present", (2) identification of the onset of galaxy formation, and (3) the relationship between simulation points and galaxies. It is possible to modify the interpretation of simulations in such a way as to make various alternatives more closely resemble the observed universe. Some statistical measures to favor so-called "hot" particles, while others favor "cold" particles. It seems that at this time neither can be ruled out.

The hypothesis that the universe, and systems within it such as clusters and galaxies, are dominated by collisionless relics of a hot Big Bang (which we shall call "inos") is an extremely elegant hypothesis which has attracted a great deal of attention in recent years. It is amenable to exploration by numerical simulation, because it is possible to pose well-defined initial conditions, and evolve them in a well-defined way, i.e., the operation of gravity in comoving coordinates.

Inflationary models lead to the prediction of closure density for the Universe, and n = 1 where the power spectrum of density fluctuations is

$$\langle |\delta_k|^2 \rangle \propto k^n \tag{1}$$

(see A. Guth, this volume). Furthermore, in ino-dominated models, there is a characteristic length scale associated with the horizon at the time the universe becomes dominated by nonrelativistic inos, which is of order tens of megaparsecs in standard hot Big Bang models, below which the spectrum is modulated.

Thus we have a reasonably well-posed problem: various sets of initial conditions can be evolved with known laws of physics and the results compared to observations of the large-scale structure of the Universe. The three main uncertainties in this enterprise, as we shall see, concern the multiplicity of theoretical models, the difficulty of numerical modeling of the complexities of ordinary matter (as opposed to inos), and the present primitive state of our knowledge of the distribution of matter in the Universe.

Pure power-law spectra, lacking the feature at supercluster scales, were explored starting about ten years ago.[1,2] However, such spectra are unlikely to arise in ino-dominated models; they also give a distribution lacking the interesting (some say filamentary) structure on large scales which we associate with superclusters.

The first numerical simulations appropriate to a neutrino-dominated universe were undertaken before the experimental suggestions of nonzero neutrino rest mass.[3] These simulations showed the characteristic cell structure of two-dimensional neutrino models. The cell structure arises as a consequence of the truncated power spectrum, which reflects the neutrino damping discussed by S. White and A. Szalay in this volume. It is interesting to note that the contours of average density divide space into many unconnected regions here, while in three dimensions they divide it into only two regions.

One early objection to neutrino models, which has been resolved by numerical simulation, concerns the velocity dispersion of the particles surrounding the large voids. This velocity dispersion is much smaller than the product of the void diameter and the Hubble constant. In addition, it has a multistream character,[3] in which there exists a population of low velocity particles[4,5] which can fragment to form interesting substructure[6,7] (see Fig. 1).

Another non-problem concerns the two-point correlation of the matter in these simulations. It is not the case, as had been widely supposed, that there are large features in the two-point function at the characteristic scale of the voids.[8]

A number of groups have done three-dimensional simulations of neutrino models.[9-11] These works generally confirm conclusions taken from lower dimensionality. The amplitude of the two-point correlation function is a bit high, about a factor of two higher than the two-point correlation of galaxies. This has led to the claim that neutrino models can be ruled out.[12]

This conclusion depends on the identification of the two-point correlation of neutrinos and galaxies, which is certainly not the case, as one system is collision-less, the other subject to collisions and gas cooling as galaxies are born. The explosion of pregalactic objects may inhibit galaxy formation in dense regions, and galaxy mergers may lower the number density of galaxies in clusters.

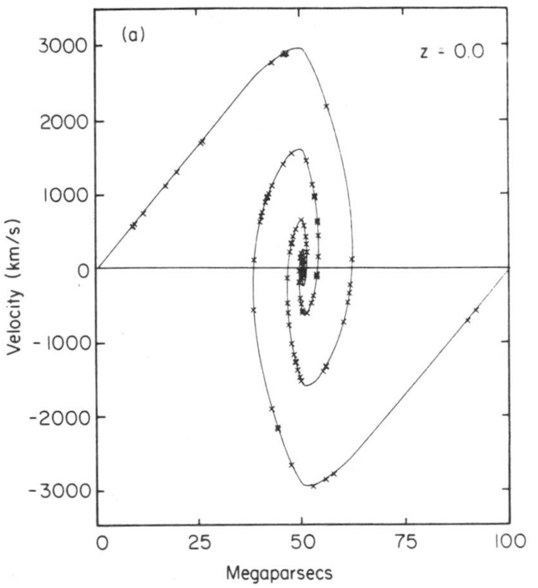

Figure 1: The phase space of a one-dimensional cross-section of a sheetlike pancake collapse.[5] Note the condensation with low velocity-dispersion and high phase-space density near the center, the site of galaxy formation in these models. Such particles may be captured to form galactic halos.

It is possible to avoid this problem even without introducing the caveat of inequality between the two autocorrelations. If the spectral index n is taken to be 3 or 4, then this problem is greatly reduced.[12,13] Such a steep power spectrum diverges at small scales without a cutoff, since metric fluctions $\delta\phi \sim R^{(1-n)/2}$. There exists at least one scenario[14] which produces a very steep spectrum, n = 3 from quantum fluctuations. A period of deSitter-like growth may carry the horizon to a scale which will not re-enter the horizon until well after nucleosynthesis, but early enough to form galaxies.

It is also important to identify carefully the onset of galaxy formation in neutrino models, since the velocity dispersion of the system and the amplitude of the two-point correlation grow rapidly as a function of time. One criterion which sounds conservative (S. White, this volume) is to identify the onset of galaxy formation with a time when 1% of the cells in the simulation have collapsed, and assign this a redshift ~3.5. This results in the high clustering amplitude discussed earlier. Let us examine this more closely.

The choice of a redshift approaching 4 is primarily motivated by the observation of QSOs at such redshift, and their apparent association with galaxies. However, the space density of QSOs at such redshift is a strong function of the assumed evolutionary model. In a model in which strong luminosity evolution is assumed,[15,16] the space density can be very low--about one QSO per supercluster. We would therefore identify the onset of QSO formation as the moment when a similar fraction, about 1 part in 3×10^4 of the simulation volume has collapsed.

Although the distribution of density is not precisely gaussian in pancake models, we can estimate from such a distribution that there will be an expansion factor of order 1.5 between this moment and the one when 1% of the cells have collapsed. It should be noted that in order to verify this expectation in a numerical simulation[13] it is necessary to begin in the linear regime, since at density peaks, second order errors are of the same amplitude as one's approximation method. Many simulations[9,10,12] have begun late enough that this identification of peaks could be a real problem, since the approximation scheme used to set up initial conditions does not explicitly include gravity!

What is the appropriate fraction of matter to associate with the onset of galaxy formation? It is important to remember that estimates of the ages of low metallicity Population II stars are only relevant to about ten percent of the stellar mass of the galaxy; these are not the nearly zero metallicity objects which must have formed at the time of pancaking. Furthermore, the mass associated with the visible parts of galaxies constitutes only about one percent of the critical density and only about ten percent of that allowed by nucleosynthesis constraints (G. Steigman, this volume.) The ages of objects are a weak function of the time of collapse; in particular we find that the collapse of 1% of the system takes place 80% of the proper time back to the Big Bang, at a redshift of 2, if we make the aforementioned normalization to the QSO onset. Therefore, most pregalactic material would have collapsed between redshifts of about 2 and 1. Most material collapsing later would be unable to cool.[17,18] In this picture, primordial galaxies would continually accrete matter, with a roughly constant star formation rate. Their early phases would be expected to be rather blue, but not necessarily highly luminous.

The combination of a steep primordial power spectrum, galaxy formation at a redshift ~2, and QSO luminosity evolution results in a self-consistent model in which clustering amplitudes and velocity dispersions are reasonable.

So-called cold dark matter presents a different challenge. The primordial power spectrum is bent rather than cut off, because small scale fluctuations have damped growth in the radiation-dominated phase, rather than being destroyed by free-streaming.

In the first simulation of this picture,[19] it was found that interesting large-scale coherence arises on the scale of the bend in the spectrum. That is, unlike pure power-law simulations, but similarly to neutrino models, a pattern of voids and filaments emerged. However, the voids would not be completely empty of galaxies as in a neutrino model.

They would rather have a low-space density of visible matter. It might be rendered invisible. (J. Bardeen, this volume.)

Simulations of cold matter have the opposite problem of neutrinos: the amplitude of two-point correlation is too low, implying $\Omega h \sim 0.3$. It is easy to correct this, however; if we calculate the two-point correlation of only a small fraction of points--the density peaks--we find that the correlation amplitude is magnified. It is possible to adjust the normalization of Ωh arbitrarily by this procedure; if we make the threshold high enough, $\Omega = 1$ becomes possible given current values $(1/2)<h<1$.

But this is intended to illustrate the extent to which interpretation is arbitrary, not to suggest a particular conclusion. We could force a threshold by demanding, for example, $\Omega = 1$ and $h = (1/2)$ to fit theoretical prejudice, but will not.

It seems at this time that both hot and cold collisionless particles are viable candidates for the "missing mass". They both have problems, but at the level of uncertainty in the observations and interpretation of simulations, both seem allowed.

It is important to mention a statistic which may help discriminate between hot and cold models. The integral of the two-point function ξ is J_3

$$J_3(R) = \int_0^R \xi(r)r^2dr \qquad (2)$$

Previous work has shown the importance of anticorrelation in pancake models.[8,11] Although anticorrelation exists in cold particle simulations, its amplitude is typically too weak to be observable. This difference shows up dramatically in J_3:J_3 becomes negative at an $R\sim\lambda_c$ where λ_c is the neutrino free-streaming length in pancake simulations, just as predicted in an approximation scheme.[20] This does not happen at any radius for which ξ has been determined from galaxy counts in cold particle models.

It is found that this behavior is not a strong function of time in the models, nor is it affected qualitatively by the "tagging" procedure described above. It happens at large enough radii not to be affected by local collisional processes. The data generally exhibit the behavior found in pancake models,[21-23] although they are subject to systematic errors.

No firm conclusions seem possible at this time. Hot particles give values of J_3 in better accord with the (uncertain) observations, but demand rather recent galaxy formation. Cold particles are less stringently constrained. They do not, however, give true voids, but rather low density regions in which galaxies should exist, but have not been found. Also, we must appeal to unknown energetic processes to ionize the remaining baryons so that they neither form galaxies nor show up in absorption as neutral hydrogen.

Neither model is perfect, but both seem worth continued investigation.

One area which may prove interesting is to relate the topology of the observed matter distribution to that of various initial conditions. In a neutrino dominated model, for example, we can plot isodensity contours at the initial conditions and compare the type of structure to the observed distribution of matter. Models with a truncated power spectrum show a continuous surface which divides space into only two regions (Fig. 2). This is very different, for example, from the behavior of white noise initial conditions. It is hoped that further exploration of this[24] will lead to tests based on topology.

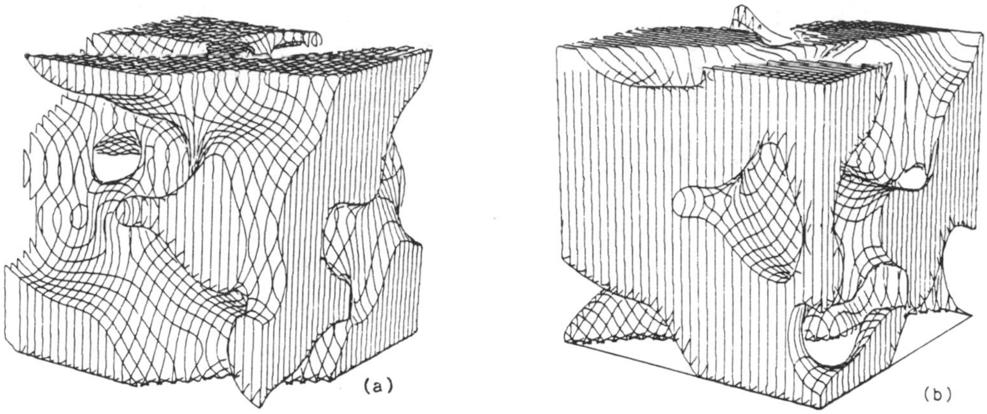

Figure 2: An isodensity surface divides space into (a) overdense and (b) underdense regions in a neutrino-dominated model.

REFERENCES

[1]S. J. Aarseth, J. R. Gott, and E. L. Turner, Astrophys. J. **228**, 644 (1979).

[2]G. Efstathiou and J. W. Eastwood, Mon. Not. Roy. Astron. Soc. **194**, 503 (1981).

[3]A. G. Doroshkevich, E. B. Kotok, I. D. Novikov, A. N. Polyudov, S. F. Shandarin, and Yu. S. Sigov, Mon. Not. Roy. Astron. Soc. **192**, 321 (1980).

[4]A. Melott, Phys. Rev. Lett. **48**, 894 (1982).

[5]A. Melott, Astrophys. J. **264**, 59 (1983).

[6]A. Melott, Nature (London) **296**, 721 (1982).

[7]J. R. Bond, A. Szalay, and S. White, Nature (London) **301**, 584 (1983).

[8]A. Melott, Astrophys. J. Lett. **273**, L21 (1983).

[9]A. Klypin and S. Shandarin, Mon. Not. Roy. Astron. Soc. **204**, 891 (1983).

[10]C. Frenk, S. White, and M. Davis, Astrophys. J. **271**, 417 (1983).

[11]J. Centrella and A. Melott, Nature (London) **305**, 196 (1983).

[12]S. White, C. Frenk, and M. Davis, Astrophys. J. Lett. **274**, L1 (1983).

[13]A. Melott, Astrophys. J., to be published.

[14]D. A. Kompaneets, V. N. Lukash, and I. D. Novikov, Astron. Zh. **59**, 424 (1982), [Sov. Astron. **26**, 259 (1982)].

[15]G. Chanan, Astrophys. J. **252**, 32 (1982).

[16]M. F. Khodyachikh, Astron. Zh. **60**, 840 (1983), [Sov. Astron. **27**, 487 (1983)].

[17]P. Shapiro, C. Struck-Marcell, and A. Melott, Astrophys. J. **275**, 413 (1983).

[18]J. R. Bond, J. Centrella, A. Szalay, and J. R. Wilson, Mon. Not. Roy. Astron. Soc., in press (1984).
[19]A. Melott, J. Einasto, E. Saar, I. Suisalu, A. Klypin, and S. Shandarin, Phys. Rev. Lett. **51**, 935 (1983).
[20]R. Schaeffer and J. Silk, Astron. Astrophys. **130**, 131 (1984).
[21]M. Davis and P. J. E. Peebles, Astrophys. J. **267**, 465 (1983).
[22]A. R. Rivolo, Ph. D. dissertation, State University of New York at Stony Brook (1981).
[23]E. Sadler and N. Sharp, Astrophys. J. (to be published) (1984).
[24]J. R. Gott and A. Melott, in preparation (1984).

Superpancakes and the Nature of Dark Matter

Avishai Dekel

The correlation function of rich clusters of galaxies, $\xi_c(r)$, is observed to be in excess of the galaxy correlation function, $\xi_g(r)$; $\xi_c = 1$ at $r_c \simeq 25h^{-1}$Mpc for Abell ($R \geq 1$) clusters as analyzed by Bahcall and Soneira[1], and at $r_c \simeq 18h^{-1}$Mpc for less rich clusters as measured by Shectman,[2] while $\xi_g = 1$ at $r_g \simeq 5h^{-1}$Mpc.[3] In order to find out whether this effect is reproduced in standard clustering scenarios, Barnes, Dekel, Efstathiou, and Frenk[4] have applied cluster finding algorithms to cosmological N-body simulations, comparing ξ_c for clusters of a certain richness to ξ_g as given by the individual bodies when the slope of ξ_g is $\gamma \simeq 1.8$. Figure 1 summarizes the results. For a Poisson noise initially (A), ξ_c and ξ_g practically coincide. In pancake scenarios (C; also Ref. 5), ξ_c is steeper than ξ_g, but $r_c/r_g \lesssim 2$, with no richness dependence. In the case (B) of a very flat spectrum, $|\delta_k|^2 \propto k^n$ with $n = -2$, there is a significant growth of ξ_c with richness, but at a rate which seems weaker than the observed rate; while the richest clusters analyzed in model B are comparable to Shectman's, the excess detected in B is smaller, and a naive extrapolation to Abell-like clusters indicates $r_c/r_g \sim 2$ only. At an earlier time, when ξ_g is too flat with $\gamma \simeq 1.4$, the excess in model B comes close to Shectman's, but the extrapolation to Abell clusters still indicates $r_c/r_g \sim 3$ only. Also, model B does not represent a realistic cosmological scenario, having too much power on large scales. Thus, it seems that the simple scenarios do not fully reproduce the observed effect.

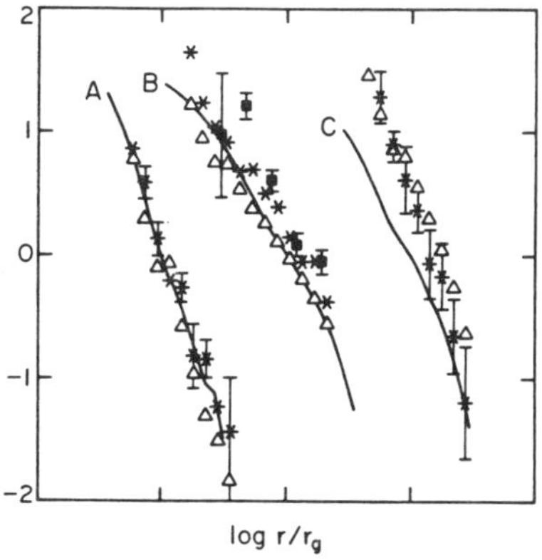

Figure 1: $\xi_g(r)$ (solid curves) and $\xi_c(r)$ (symbols) in N-body models A, B, C.[4] The symbols Δ, *, □ are in order of growing richness. Separation between large ticks is one decade.

A trend with richness, as in B, can be qualitatively understood in terms of statistics of rare events in a linear Gaussian noise. Kaiser[6] showed that ξ for mass in high density regions, above a given threshold, increases with the height of the threshold, indicating a trend with richness for ξ_c. Such an idealized model, however, cannot make realistic quantitative predictions, especially in pancake scenarios, mainly because initial density peaks are not uniquely related to Abell-like clusters; many peaks end up in pancakes or filaments, so that some degree of local isotropy is required to guarantee an Abell-like cluster, and nonlinear effects may also be important. Actually, the deviations from a Gaussian distribution are significant already in the range where ξ_c is best observed even though $\xi_g < 1$ there. Also, it is difficult to calculate in this model a ξ_c which is not weighted by the mass of each cluster. In $n = -2$ models,[7] r_c derived from linear density peaks was found to overestimate the one derived for realistic clusters by a factor of 2 or more, indicating that the statistical amplification of rare events cannot explain the whole observed effect even in such models.

A hint may come from indications that clusters tend to populate elongated or flattened structures[8] of $\lambda_c \sim 100 - 150h^{-1}$Mpc, perhaps to be called "superpancakes" in analogy with "pancakes" of galaxies[9] of $\lambda_g \sim 20 - 30h^{-1}$Mpc. ξ near unity is strongly affected by the dimensionality of the distribution of points[10]; for structures of thickness d much smaller than their length λ, $1 + \xi(r) \simeq [r/(\lambda/2)]^{-\gamma}$ in the range $d < r < \lambda$, with $\gamma = 1$ or 2 for oblate or prolate systems respectively. Such pancakes originate from a spectrum of adiabatic perturbations if a coherence length[11] is present at a scale λ. $\xi(r)$ measured in cosmological N-body simulations that started from such a spectrum[10,12] exhibit this behavior, with $\xi = 0$ typically at $\simeq 0.5\lambda$ and $\xi = 1$ at $\simeq 0.2\lambda$. It has been argued[10] that λ_g pancaking can indeed be responsible for the observed[3] excess of ξ_g in the range 2 - $10h^{-1}$Mpc.

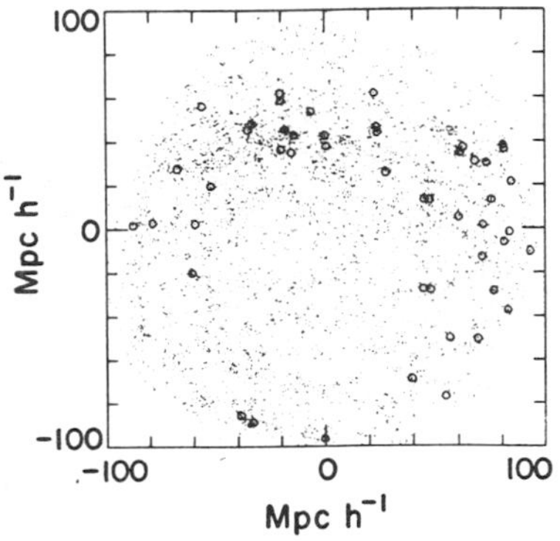

Figure 2: A projected distribution of galaxies (dots) and rich clusters (circles) in a kinematic model with two coherence lengths for pancakes and superpancakes at $\lambda_g = 20$ and $\lambda_c = 100h^{-1}$Mpc respectively.[7]

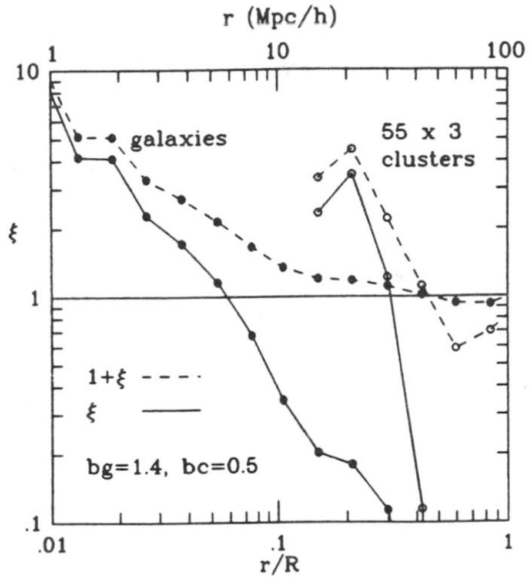

Figure 3: $\xi_g(r)$ and $\xi_c(r)$ averaged over three kinematic models.[5]

An analogous geometrical effect may produce the observed excess of ξ_c on larger scales: an additional coherence length at $\lambda_c \simeq 5\lambda_g$ would produce $\sim 100 - 150h^{-1}$Mpc superpancakes of small amplitude, inducing preferential cluster formation along them, yielding $r_c \simeq 5r_g$. The present "linear" rms amplitude of perturbations on the pancake scale should be $\delta(\lambda_g) \simeq 1.5$ in order to fit the appropriate slope of ξ_g.[8,10] The amplitude on the superpancake scale should be large enough to affect the formation of clusters, but small enough not to affect ξ_c too much; $\delta(\lambda_c) \sim 0.5$ is appropriate. This process is demonstrated in Figure 2 by a kinematic N-body simulation[7] in which the points represent galaxies and clusters are identified at peaks where a local 3D collapse is anticipated, *i.e.*, where the lowest eigenvalue of the deformation tensor[11] is above some critical positive value chosen to give the appropriate number density for $R \geq 1$ Abell clusters. Figure 3 shows the correlation functions from these simulations, with $r_c/r_g \simeq 5$ as required. The bias towards cluster formation in superpancakes is enhanced if the universe is open, as marginally bound lumps end up as clusters while marginally unbound lumps freeze out in the expanding universe.[13]

I can see two reasonable possibilities for the cosmological origin of the feature at λ_c. If the universe is dominated by baryons (b), the photon-baryon Jeans length prior to decoupling produces a coherence length[14] at $\lambda_J \simeq 25 - 50(\Omega h^2)^{-1}$Mpc, provided that $\Omega h^2 > 0.03$. To match λ_c it requires an open universe with $\Omega h^2 \sim 0.1$, in marginal agreement with nucleosynthesis constraints.[15] The pancake scale, λ_g, is provided in this case by Silk damping of adiabatic perturbations,[16] at $\lambda_s \simeq 3(\Omega h^2)^{-5/4}$Mpc. The small-angle isotropy of the Microwave Background[17] then requires reionization by early formation of "pop-III"[18] due to a perturbations component that has not been damped out on small scales. Such a component seems to be required anyway to make galaxies before pancakes,[10,19] even if the universe is dominated by massive neutrinos.[20]

Massive neutrinos (ν) provide a free-streaming damping scale[21] at $\lambda_\nu \simeq 15(\Omega_\nu h^2)^{-1}$Mpc. If $\Omega_\nu = 1$, then $\lambda_\nu \sim \lambda_g$, but the λ_J feature is smeared out by the neutrino perturbations. If, however, $\Omega_\nu \sim 0.1$, then $\lambda_\nu \sim \lambda_c$, and the Silk feature still survives at $\lambda_s \sim \lambda_g$. Hence, $\Omega_\nu \sim \Omega_b \sim 0.1$ is acceptable. "Cold" particles (c) do not produce the required features,[21] although they give rise to small scale filamentary structure[22] due to the very flat portion of the spectrum. They may naturally provide the small scale perturbations that are responsible for galaxy formation but they should not dominate the mass. The combinations $\Omega_c \sim \Omega_b \sim 0.1$, and even $\Omega_c \sim \Omega_\nu \sim \Omega_b \sim 0.1$, would allow the required features.

In any case, the model requires an open universe, giving up the apparent simplicity of an Einstein-de Sitter universe that is favored by Inflation. It is encouraging, though, that a simple baryonic universe ("inner space") can naturally explain the features indicated by its large scale structure ("outer space").

REFERENCES

[1]N. Bahcall, this volume (1984), and references therein.

[2]S. Shectman, preprint (1984).

[3]M. Davis and P. J. E. Peebles, Astrophys. J. **267**, 465 (1983).

[4]J. Barnes, A. Dekel, G. Efstathiou, and C. S. Frenk, Astrophys. J. (1985).

[5]S. F. Shandarin and A. A. Klypin, preprint (Moscow-65) (1984).

[6]N. Kaiser, this volume (1984); H. D. Politzer and M. B. Wise, preprint (1984).

[7]A. Dekel, Astrophys. J. **284**, 445 (1984).

[8]R. Giovanelli and M. P. Haynes, A. J. **87**, 1355 (1983); R. Ciardullo, H. Ford, F. Bartko, and R. Harms, preprint (STSI-7) (1984); D. J. Batuski, this volume (1984).

[9]R. B. Tully, Astrophys. J. **257**, 389 (1982); E. Vishniac, this volume (1984).

[10]A. Dekel, and S. J. Aarseth, Astrophys. J. **183**, 1 (1984).

[11]Ya. B. Zeldovich, Astron. Astroph. **5**, 84 (1970).

[12]A. A. Klypin and S. F. Shandarin, M.N.R.A.S. **204**, 891 (1983); C. S. Frenk, S. D. M. White, and M. Davis, Astrophys. J. **271**, 417 (1983).

[13]A. Dekel, Astron. Astroph. **101**, 79 (1981).

[14]J. R. Gott, III and M. J. Rees, Astron. Astroph. **45**, 365 (1975); P. J. E. Peebles, in The Large Scale Structure of the Universe (Princeton University Press), 92 (1980); C. J. Hogan and N. Kaiser, Astrophys. J. **274**, 7 (1983).
[15]J. Yang, M. Turner, D. Schramm, G. Steigman, and K. Olive, Astrophys. J. **281**, 000 (1984); B. E. J. Pagel, this volume (1984).
[16]J. Silk, Astrophys. J. **151**, 459 (1968).
[17]J. M. Uson and D. T. Wilkinson, this volume (1984).
[18]B. Carr, this volume (1984).
[19]A. Dekel, Astrophys. J. **164**, 373 (1983); A. Dekel, in the Eighth John Hopkins Workshop on Current Problems in Particle Theory, eds. G. Domokos and S. Kovesi-Domokos (World Scientific Publishing, 1984).
[20]S. D. M. White, C. S. Frenk, and M. Davis, Astrophys. J. (Lett.) **274**, L1 (1983).
[21]J. R. Bond and A. S. Szalay, Astrophys. J. **274**, 443 (1983).
[22]J. R. Primack and G. R. Blumenthal, in Formation and Evolution of Galaxies and Large Scale Structure in the Universe, eds. J. Audouze and J. Tran Thanh Van (Reidel), p. 163; P. J. E. Peebles, Astrophys. J. (Lett.) **263**, L1 (1982).
[23]S. D. M. White, this volume (1984).

Statistics of Density Maxima and the Large-Scale Matter Distribution

N. Kaiser

ABSTRACT

High peaks in Gaussian noise display enhanced clustering. The enhancement takes two forms: On large scales one obtains a **linear** amplification of the correlation function which is independent of scale. On smaller scales, but larger than the mass scale of the peaks themselves, a **non-linear** (exponential) enhancement of the number density of high peaks in overdense regions arises. The large-scale correlations of Abell's rich clusters can be understood as a manifestation of this phenomenon. If the formation of bright galaxies favours the high overdensity peaks then the number of galaxies (per unit mass) in clusters and groups may be considerably enhanced. Consequences of these ideas for the density parameter and the large-scale matter distribution are discussed.

INTRODUCTION

Currently popular theories of high energy physics and the early universe make specific predictions for the value of the density parameter and for the spectrum of primordial fluctuations. It is therefore of great interest to study the present density fluctuations and the peculiar velocity field in order to test these ideas. Unfortunately it is impossible to measure the mass distribution directly; we can only map out the pattern of the luminous material. It is necessary then to have some model for the way the spatial distribution of luminous material reflects that of the density. The simplest and most commonly adopted assumption is that light fairly samples the mass. However, observations of the clustering properties of various classes of objects over a wide range of scales cast considerable doubt on the validity of this assumption. On scales $\simeq 5$ - $50h^{-1}$Mpc (here h is the Hubble parameter in units of $100\text{kms}^{-1}\text{Mpc}^{-1}$) rich clusters of galaxies have correlation strength which is one order of magnitude or more greater than that of galaxies. There is also a strong increase in correlation strength with richness (or mass) of clusters.[1] On smaller scales $\lesssim 5h^{-1}$Mpc a dependence of galaxy correlation strength on morphological type is observed.[2] There may also be a conflict between the present clustering strength of galaxies on scales $\simeq 0.5$ - $1h^{-1}$Mpc and the lack of clustering of Lyα clouds.[3] The assumption that any one particular class of objects (bright galaxies say) is fairly sampling the mass distribution then seems to be rather arbitrary and so conclusions about the spectrum of primordial fluctuations (which are best constrained by the large-scale clustering) and virial estimates of Ω (which derive from studies of clustering at small-scales) should be treated with caution.

Recently[4] it has been shown that the clustering properties of Abell clusters can be understood as being due to a statistical effect whereby the rare high peaks of a Gaussian density field have a correlation function which has the same form (at least on large-scales) as the underlying density field but with an amplitude which can be considerably enhanced. The connection with rich clusters is that one can argue that these objects will form only at positions where the primordial overdensity was sufficiently high.

In this paper I shall briefly review the mathematical results and the application to Abell clusters. I shall then discuss the consequences of a similar threshold for the formation of galaxies, though in this case it is difficult to estimate how strong this effect should be. The aim is to see whether this statistical effect can plausibly give rise to strong segregation of luminosity from mass and to examine the observational consequences.

ENHANCEMENT OF CLUSTERING OF PEAKS IN GAUSSIAN NOISE

The idea we shall exploit below is that the sites of formation of objects such as galaxies and clusters of galaxies may be, at least approximately, identified with the high peaks of the primordial density field when smoothed on a suitable mass scale. The problem is therefore to give a statistical description of the spatial distribution of these peaks in terms of the 2-point correlation function $\xi(r)$ of the underlying density field $\delta(x)$ (since δ is assumed to be a Gaussian process it is completely described by its 2-point correlation function). A simpler calculation, which illustrates the expected behaviour for peaks, is to consider the non-linear (Heaviside) function $\delta^*(x) = H(\delta(x)-\nu\sigma)$ which is non-zero only in those regions of space in which the density field lies above some threshold $\nu\sigma$ (where σ is the rms fluctuation). The exact result for the 2-point function was given by Kaiser[4] who showed that, on large-scales, where the clustering of these regions is not too large, ξ^*, the 2-point function of δ^*, is

$$\xi^*(r) \simeq \left[\frac{\exp(-\nu^2/2)}{\int_\nu^\infty \exp(-y^2/2)\,dy}\right]^2 \xi(r)/\xi(0) . \quad (1)$$

This formula shows that, while ξ^* has the same form as ξ, the higher peaks have enhanced correlation function. For the case of very high peaks ($\nu >> 1$) Politzer and Wise[5] showed that this result could be extended into the regime where $\xi^* \geq 1$. They found

$$1 + \xi^*(r) \simeq \exp\left(\frac{\nu^2}{\sigma^2}\,\xi(r)\right) \quad (2)$$

and they also gave an analogous expression for the higher order correlation functions to arbitrary order N.

The complete set of correlation functions gives a complete statistical description of the non-linear and non-Gaussian function δ^* in terms of ξ, which is valid when $\nu << 1$ and on scales such that $\xi(r) << 1$. Provided one is only interested in these large-scales it turns out that one can find an alternative expression which directly relates the large-scale fluctuations in $\delta^*(x)$ to those of the underlying perturbation $\delta(x)$. In fact one can show[6] that the top-hats of the Heaviside function δ^* can be considered to be a fair sample from an **effective density field** ρ which is a different non-linear function of δ:

$$\rho(x) = \exp\left(\frac{\nu}{\sigma}\,\delta_s(x)\right) \quad (3)$$

where δ_s is obtained from δ by smoothing on a scale l_s which is intermediate between the coherence length of $\delta(x)$ and the larger scales of interest. One can verify that the N-point functions of ρ are identical, on the relevant scales, to those of δ^* which were obtained by Politzer and Wise;[5] the dependence on the (completely artificial) smoothing length l_s is only in the normalisation of ρ.

The operation of locating density peaks must be thought of as being performed at a time when the underlying fluctuations are linear ($\xi << 1$). However, the effective density field may have **non-linear** fluctuations when the magnitude of the argument of the exponential in Eq. (3) is not much less than unity. Provided the true density fluctuations on the scales of interest remain linear the correlation function of the peaks remains constant in time and the formulae for the correlation functions are valid. In some cases of interest it may still be legitimate to equate the sites of formation with high peaks in the linear density field but Eqs. (1), (2) may be invalid because larger structures may have gone non-linear. In this case one can use Eq. (3) to calculate the enhancement of number of objects per mass which exists prior to the collapse of the larger-scale structure. These objects will then follow the collapse of the matter and their spatial distribution at later times can thereby be obtained.

RICH CLUSTERS

For the case of rich clusters it is not unreasonable to argue that such objects may, at least approximately, be associated with high overdensity peaks. This is because clusters, by definition, are very massive and condensations of this mass are rare. A model for locating rich clusters is described in Ref. 4: First a linear smoothing filter (or radius R) is applied to the primordial density field in order to eliminate peaks where lower mass objects are destined to form and then peaks exceeding the threshold ν are located. For any class of objects (e.g., Abell richness = 1 etc.) the two parameters R and ν are determined by the requirement that the number density of clusters agree with that observed and that the clusters should have collapsed recently. Both R and ν increase with decreasing number density of objects and, since σ decreases with increasing R, one expects the correlation function amplification factor $A \equiv \xi^*/\xi \simeq \nu^2/\sigma^2$ to be a strongly increasing function of richness (or equivalently with intercluster separation). This is in good accord with the clustering behaviour of Abell clusters (Ref. 1 and references therein) and the weaker clustering of the groups identified by Shectman,[7] which are more numerous than Abell $R \geq 1$ clusters, provides further support for the enhancement mechanism described here.

In order to obtain these results it was necessary to assume that the primordial fluctuations on the scale of rich clusters had a Gaussian distribution and the observed correlations are clearly in support of this hypothesis. The inferred density correlations at large separations are discussed below.

GALAXIES

If galaxies are identified with moderately high peaks of the primordial density field, when suitably filtered, then they too would give a biased measure of the mass distribution. On large scales the correlation function for galaxies ξ_g would be a constant multiple of the density correlation function ξ. On scales somewhat larger than galaxies one can expect a non-linear enhancement of the galaxy density in overdense regions. It is very interesting to ask then whether this segregation could have been important for clusters and groups of galaxies as, if so, this would enable one to reconcile the low values of $\Omega \simeq 0.2$ obtained from virial studies[8] of such objects with ones prejudice that $\Omega = 1$. Let us assume then (the physical basis for this assumption is discussed by Bardeen[9]) that galaxies are identified with peaks in excess of $\nu\sigma_g$ (here σ_g is the rms fluctuation of the filtered density field at some high redshift Z_i when all fluctuations of interest are still in the linear regime) and explore the consequences.

In order to explain the virial results one requires a non-linear enhancement of number density of galaxies by a factor ~5 in proto-objects with overdensity sufficient to have collapsed by the present. If the proto-object has overdensity δ_i at Z_i then by Eq. 3 one requires $\nu\delta_i/\sigma_g \sim \log(5) = 1.6$. According to the dissipationless spherical collapse model $\nu\sigma_g = 1.68(1+Z_f)/(1+Z_i)$, where Z_f is the redshift at which the galaxy collapses, and $\delta_i = 1.68/(1+Z_i)$ so if

$$\nu^2 \simeq 1.6(1+Z_f) \qquad (4)$$

then objects which collapsed recently would have a large enhancement of L/M (or more precisely N_g/M where N_g is the number of galaxies in the object).

The large-scale galaxy correlation function is $\xi_g \sim \nu^2/\sigma_g^2\xi \sim (1.6/1.68)^2 \times (1+Z_i)^2\xi$ at Z_i. However, the density correlations ξ will have subsequently grown by a factor $(1+Z_i)^2$ so one would expect only a small enhancement of correlations at large separations (i.e., where $\xi \ll 1$) at the present. The estimate given here is approximate so one should not rule out the possibility of moderate enhancement of ξ_g.

The present discussion is limited because no explanation for the threshold has been given nor have competing effects been considered. The point I wish to make is that if there is a threshold for the formation of bright galaxies and Eq. 4 is satisfied (and this implies a fairly flat spectrum of fluctuations on scales intermediate between galaxies and clusters or a large value for ν) then this statistical enhancement provides a strong segregating influence for recently collapsed systems which are just those objects naturally chosen for virial analysis. A consequence of this hypothesis is that one would expect higher N_g/M in denser objects. However, the analysis is only valid for systems much more massive than individual galaxies so the predicted enhancement of N_g/M will become unreliable for small groups.

DISCUSSION

By the simple expedient of identifying the sites of formation of luminous objects with the peaks of the primordial density field one can understand the observational fact that different classes of objects have different clustering properties. While the mathematical model discussed here is highly idealised its key features are not physically unreasonable. In particular the sharp threshold can be justified by considerations of the non-linear evolution of density perturbations; two perturbations with only slightly different initial overdensities give rise to objects with very different overdensities around the time when the regions collapse and freeze out of the expansion. Thus the probability of an overdense region of rich cluster mass having been included in Abell's catalogue is essentially a step-function of the initial overdensity. Similarly, if the fate of a galactic mass perturbation is determined by the nature of the gas at the time of collapse, which may be rapidly evolving at that epoch, a sharp threshold may arise. The approach adopted here presupposes a hierarchical picture and would not be applicable in a scenario in which the objects form by fragmentation of large pancakes.

While the applicability of this effect to galaxies is rather uncertain, the possibility of strong segregation of luminous galaxies from mass obtained for those objects one would naturally choose for virial analysis is a very attractive feature for those who find the idea that $\Omega \neq 1$ unpalatable. The hypothesis (of associating galaxies with the primordial peaks) can best be tested by observations of galaxy clustering (or of absorption systems associated with galaxies) at high redshift. One predicts strong and unevolving clustering with a characteristic form for the higher order correlation functions, e.g., the 3-point function ζ contains a term proportional to ξ^3 (Ref. 5).

The application to rich clusters is on firmer ground since the formation of these objects, involving essentially only the gravitational force, is much better understood. Cluster finding algorithms applied to N-body simulations[11] give results which are in reasonable accord with the predictions of the simple model.[4] The observed relationship between the correlation functions of different classes of clusters provides support for the statistical enhancement mechanism and for the hypothesis that the primordial fluctuations had a Gaussian distribution.

The picture of the present distribution of matter which emerges is that, on small scales $\lesssim 5h^{-1}$Mpc, the slope of the correlation function may be considerably shallower than that of the galaxies while on large-scales the galaxy data should not be strongly biased. Unfortunately, the galaxy data become unreliable on scales not much larger than $5h^{-1}$Mpc. On large scale the strong enhancement of ξ for rich clusters make these the preferred indicator of the form (and the sign) of the density correlation function. The choice of an ideal class of clusters for probing the large-scale structure involves a compromise between the enhancement of signal and the enhancement of noise due to the small number of very rich clusters. The rich cluster data are consistent with $\xi \propto r^{-2}$ or slightly shallower out to very large separations[1] though Klypin and Kopylov[10] reported evidence for anticorrelation at $r \gtrsim 50h^{-1}$Mpc. Since $\xi = (hr/5\text{Mpc})^{-2}$ implies logarithmically diverging peculiar velocities of $\sim\Omega(500\text{kms}^{-1}\text{Mpc}^{-1})^2$ contribution to the velocity variance per e-folding in length scale. For $\Omega = 1$ these velocities are strongly constrained by the observed microwave dipole anisotropy. The prospect is good that we may be able to see anticorrelation at large r for a suitably chosen class of clusters and thereby determine the linear power spectrum of fluctuations over the range in which the transition to positive index n occurs (as it must in order to avoid violating the limits to the large scale microwave anisotropy). In addition, since although the density fluctuations are small the predicted peculiar velocities are considerable, one should also, by carefully mapping the peculiar velocity field, be able to confirm the role of gravity in the evolution of the large-scale structure.

ACKNOWLEDGMENTS

This material is based upon research supported in part by the National Science Foundation under Grant No. PHY77-27084, supplemented by funds from the National Aeronautics and Space Administration, and by Grant No. AST82-13345.

REFERENCES

[1]N.A. Bahcall and R.M. Soneira, Ap. J. **270**, 20 (1983).
[2]M. Davis and M.J. Geller, Ap. J. **208**, 13 (1976).
[3]W.L.W. Sargent, P.J. Young, A. Boksenberg, and D. Tytler, Ap. J. Suppl. **42**, 41 (1980).
[4]N. Kaiser, to appear in the Ap. J. (1984).
[5]D.H. Politzer and M.B. Wise, preprint (1984).
[6]N. Kaiser, in preparation (1984).
[7]S. Shectman, preprint (1984).
[8]S.M. Faber and J.S. Gallagher, Ann. Rev. Astr. Astrophys. **17**, 135 (1979).
[9]J. Bardeen, these proceedings.
[10]A.A. Klypin and A.I. Kopylov, Sov. Astron. Lett. **9**, 41 (1983).
[11]J. Barnes, A. Dekel, G. Efstathiou, and C. Frenk, preprint (1984).

Gaussian Galaxy Formation and the Density of QSO's

Alexander S. Szalay

ABSTRACT

If galaxy formation is due to a random Gaussian process, then the comoving density of galaxies and quasars is expected to approach $\log \rho \approx -\gamma\,(1+z)^2$ at the high redshift limit. It is suggested, that the absence of high redshift quasars may be due to this effect. The above expression agrees with recent data on QSO densities remarkably well.

For several years the general belief was that the density of quasars is increasing towards higher redshifts, until a sharp cutoff at $z \sim 3.5$ is reached[1]. This was based upon observations of bright quasars. Recently fainter objects[2], high redshift radio sources[3] or very high redshift QSO's[4] all seem to suggest that this may not be the case: the comoving density is actually gradually decreasing, only the quasars at high redshifts may be brighter. The combined effect of these two phenomena on the bright end of the luminosity function can give the illusion of the rise and sharp cutoff of the density with increasing redshift. A schematic model illustrating the behaviour of the luminosity function (based upon Koo's paper) is shown on Fig. 1.

If quasars are active galactic nuclei, AGN's, their density and formation is related to that of the parent galaxies. Here we discuss the high redshift limiting behaviour of the galaxy and quasar densities in as general terms as possible, and show that a gradual falloff in the comoving density is expected both for the quasars and the galaxies.

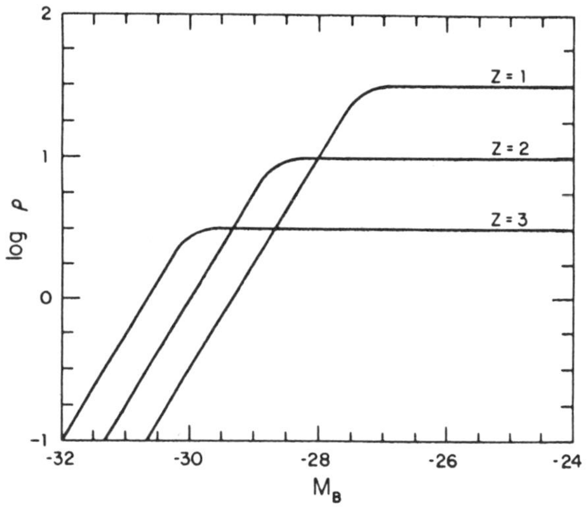

FIGURE 1

For a given power spectrum the local overdensity is distributed as a Gaussian, with dispersion $\sigma = d\rho/\rho$, if the phases of the different Fourier components are random, i.e. they are independent. This assumption of random phases is a very general principle, true for practically all mechanisms capable of generating small fluctuations.

As galaxies originate from these primordial density fluctuations we can derive some general properties of the galaxy formation process by knowing the initial statistical distribution, because even though today the galaxy density is nonlinear, its evolution has followed the initial density pattern. As it will be shown below, these properties we consider here are based upon the Gaussian nature of the fluctuations only, and are fairly insensitive to the specific details of the galaxy and QSO formation, or the density parameter Ω.

Through the rest of the paper we assume, that we have a fluctuation spectrum, which represents the distribution of galaxies, and above: the contributions from smaller scales are filtered out, there is no substructure below galactic scale. We consider the three most likely possibilities of the many how galaxy formation can be related to the initial fluctuations: formation at points of high density, at peaks, and at Zeldovich 'pancakes'.

The probability, that the overdensity at a given point is $\delta = \nu\sigma$ is given by

$$P(\nu)d\nu = A\ e^{-\nu^2/2}\ d\nu \tag{A}$$

Galaxies may be formed at peaks of the overdensity. The probability of finding a peak of height ν is slightly different, given by[5]

$$n_{pk}(\nu)\ d\nu\ dV = 0.01\ R_*^{-3}\ g(\nu)\ e^{-\nu^2/2}\ d\nu\ dV \tag{B}$$

where $g(\nu) \rightarrow \nu^3$ for high ν, and R_* is a characteristic length scale of about 1 Mpc.

If galaxies are forming in 'pancakes', the parameter relevant for galaxies is λ_{11}, the largest eigenvalue of the deformation tensor of the comoving perturbations[6]. For high values λ_{11} is also distributed as a Gaussian, with $\nu = \lambda_{11}/\sigma_\lambda$

$$P(\nu)d\nu = A\ \nu^2\ e^{-\nu^2/2}\ d\nu \tag{C}$$

The tail of $P(\nu)$ for high ν is dominated by the Gaussian term in all the above cases, so we can represent either of the three cases by considering a distribution of the form

$$P(\nu)d\nu = A\ \nu^\beta\ e^{-\nu^2/2}\ d\nu$$

where only the value of β is different from case to case. If the formation of a galaxy is tied to a threshold $f_{cr} = \nu_{cr}\ \sigma_f$ in the parameter f, where f can be either δ, the height of the peak, or λ_{11}, then the threshold corresponds to a $\nu_{cr} = f_{cr}/\sigma_f$. Since for all the above cases in an $\Omega = 1$ universe $\sigma_f = \sigma_0/(1+z)$, to each ν_{cr} there is a corresponding redshift, related by

$$\nu_{cr} = (f_{cr}/\sigma_0)\ (1+z) = \nu_0\ (1+z)$$

Thus higher ν levels form galaxies at higher redshifts. Combining this with the distribution of ν we obtain the number of galaxies forming at a given z:

$$dn_g/dz = -A\ (1+z)^\beta\ e^{-\gamma(1+z)^2}$$

with $\gamma = \nu_0^2/2$. As long as we are in the high tail of the Gaussian, the integrated equation will be dominated by the source term:

$$n_g = B\ (1+z)^{\beta-1}\ e^{-\gamma(1+z)^2}$$

Until now there are no observations of high redshift galaxies. Only quasars were found in substantial numbers at redshifts above 1. There is increasing evidence, that quasars are active galactic nuclei. There are even suggestions, that QSO's form from interacting galaxies[7].

We will show below, that in this case the above simple form of the comoving density evolution is expected to hold for quasars as well, relatively independent of the details of the formation process or the QSO lifetimes.

Let us attempt to model the formation of QSO's with a simple model. In the rate equation for quasar formation the source term is related to the density of galaxies. It is either proportional to n_g (spontaneous) or to n_g^2 (interacting).

$$\frac{dn_q}{dz} = -\left(\frac{n_q}{t_q}\right)\left(\frac{dt}{dz}\right) + \begin{cases} a\ n_g & \text{(spontaneous)} \\ b\ n_g^2 & \text{(interacting)} \end{cases}$$

In either case the dominant Gaussian behaviour is not affected, only the constant in the exponent. Also, due to the extreme growth of the source term all effects of a QSO lifetime t_q are negligible. The $(1+z)^2$ in the exponent is virtually indestructable. Integrating this equation in the high z limit gives

$$n_q = \begin{cases} (1+z)^{\beta-2}\ e^{-\gamma(1+z)^2} & \text{(spont)} \\ (1+z)^{2\beta-3} e^{-2\gamma(1+z)^2} & \text{(int)} \end{cases}$$

We can avoid the actual normalization, which is hard to determine, by taking the logarithmic derivative with respect to $(1+z)^2$

$$\frac{d \log n_q}{d(1+z)^2} = \begin{cases} -\gamma + (\beta-2)/2(1+z)^2 & \text{(spont)} \\ -2\gamma + (2\beta-3)/2(1+z)^2 & \text{(int)} \end{cases}$$

At higher redshifts ($z > 2$) we can neglect the second terms, so the slope approximates $-\gamma$ or -2γ respectively. Note, that the prediction is that the logarithm of the comoving quasar density changes with $(1+z)^2$!

There is one effect, which has been neglected so far. If the quasars are due to a massive black hole in the center of the present galaxy, then it takes about 0.1 - 1 Gy to build up that black hole. This may influence the above equation, introducing a 'retardation'. This will be considered in a later paper.[8]

Two data sets were found, which can be used to compare to our predictions. One is given by Condon (1985)[3], and the other one by Koo (1985)[4]. Condon considered the luminosity function of faint, high redshift radio sources. He found that the luminosity function is very well described by a constant shape, with simultaneous density and luminosity evolution. His fit to the data can be used to estimate the amount of relative density evolution in the range $1 < z < 4$, where the actual data were taken. His models extend below that range, but there they are weakly locked to the data.

The Koo and Kron sample of faint quasars is on the field SA57. They preselected quasar candidates on the basis of multicolor photometry, then took actual spectra. A fit to the luminosity function was made by using a constant shape with separate density and luminosity evolution. Some of the bins contained only 1 or 2 quasars, so the actual error bars are quite large, on the other hand they find 1 quasar, where the Schmidt and Green model predicted about 100! The first Koo and Kron data point represents actually the density of Seyfert galaxies, which may be a factor of 2 too high. Altogether, both data sets show a remarkably similar decreasing tendency. For comparison, the Koo and Kron data were originally presented for $\Omega = 0$. We scaled them to $\Omega = 1$, in order to compare directly to Condon. We have plotted log ρ, the comoving density against $(1+z)^2$. Since the absolute densities are not used, the two data sets have been shifted on top of each other. A crude fit to the slope gives $\gamma = 0.15$. Were both f_{cr} and the value of this slope known, this could in principle provide a normalization of the density fluctuations in the universe, independent of the fact whether the light does trace the mass, or not. This value of γ corresponds to the respective values

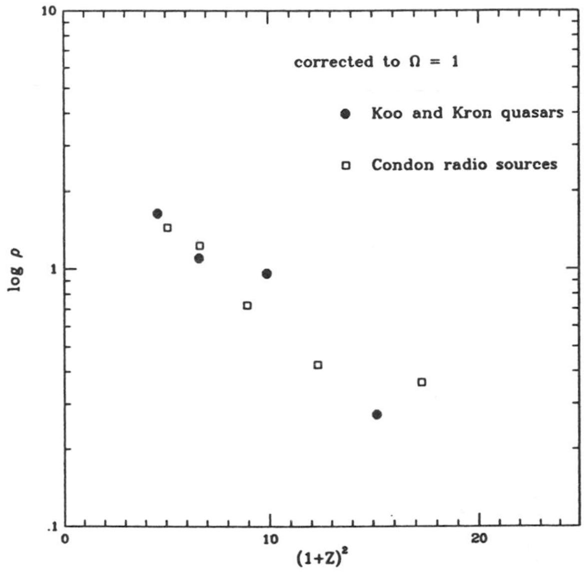

$$\nu_o = \begin{cases} 0.53 \text{ (spont)} \\ 0.37 \text{ (int)} \end{cases}$$

The value of f_{cr}, the projected linear overdensity in case (A) and (B) is usually taken to be about 1.6 - 2 for nonlinear collapse to take place, for pancakes (C) $f_{cr} = 1$. On the other hand for pancakes $\delta\rho/\rho = \sqrt{5}\sigma_f$. Taking all of this into account, one can estimate $\delta\rho/\rho$ today for the different models. This is shown in the following table.

		A,B	C
$\delta\rho/\rho$	sp	3.10	4.08
	int	4.39	5.77
f_{cr}		1.70	1.00
σ_f/σ_o		1.00	0.45

Table 1 - The value of $\sigma_o = \delta\rho/\rho$ for various models (A, B and C), (sp, int) is shown for a density evolution with $\gamma = 0.15$. The threshold f_{cr} and σ_f/σ_o is indicated for the different models. Note, that for (A,B) σ_o is measured on galactic scales. The σ_o values are upper limits, effects of a delay in QSO formation would cause σ_o decreasing.

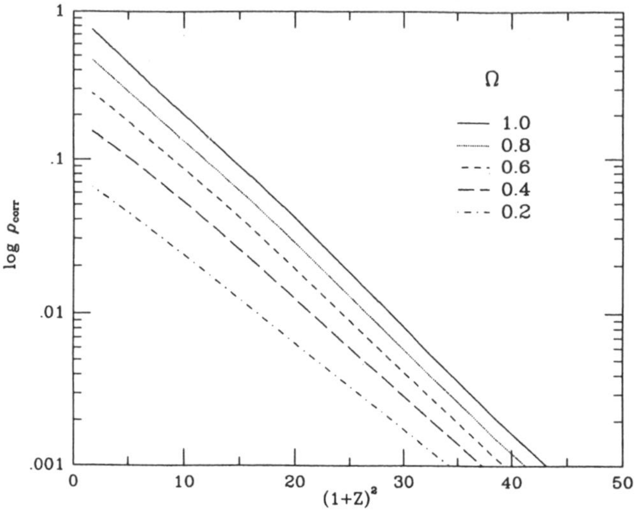

FIGURE 3

Another uncertainty is in the value of Ω. All the above arguments were using the value Ω = 1. If this is not the case, two effects will occur: the growth rate of fluctuations changes, but so does the comoving volume element. Not knowing the real value of Ω, astronomers usually scale to either Ω = 0 or to Ω = 1, as we have done. For the case we are interested in, the two changes will have a marginal combined effect only, assuming that Ω > 0.2. The comoving density corrected to Ω = 1 for different real values of Ω is shown for model (A,sp) as a function of $(1+z)^2$ on Figure 3. One can see, that the deviations from a linear relation are minimal, even though the slope is slightly changing. The present quality of the comoving density data does not enable us to compare finer details of the predictions. Such data will soon be available (Boyle etal.)[9], settling many uncertainties.

REFERENCES

[1]M. Schmidt, and R. F. Green, Ap. J. **269**, 357 (1983).

[2]D. C. Koo, Proc. 24th Liege Astrophysical Colloquium 'Quasars and Gravitational Lenses' (1983) p. 240; D. C. Koo, STScI preprint 48 (1985).

[3]D. W. Weedman, Ap. J. Suppl. **57**, 523 (1985); F. F. Condon, Ap. J. **287**, 461 (1984).

[4]C. Hazard, and R. McMahon, Nature **314**, 238 (1985).

[5]J. R. Bardeen, J. R. Bond, N. Kaiser, and A. S. Szalay, submitted to Ap. J. (1985).

[6]A. G. Doroshkevich, Astrofizika **6**, 581 (1970); A. G. Doroshkevich, and S. F. Shandarin, Sov. Astron. **22**, 653 (1978).

[7]M. Gaskell, Univ. Texas preprint (1985).

[8]A. Cavaliere, and A. S. Szalay, submitted to Ap. J. (1985).

[9]B. Boyle, et al., to be published.

String Review

Alexander Vilenkin

I. INTRODUCTION

Phase transitions in the early universe can give rise to topologically stable defects of various kinds. The basic types of defects are monopoles, vacuum domain walls and strings.[1] Hybrid topological animals - monopoles connected by strings[2] and walls bounded by strings[3,4] - can also be formed. Monopoles and domain walls are disastrous for cosmological models and should be avoided. The hybrid structures are harmless, but not very interesting: they rapidly break into pieces and decay practically without trace.[5] Strings, on the other hand, cause no harm but can lead to very interesting cosmological consequences. In this talk I would like to review the evolution of strings and their cosmological implications.

II. FORMATION OF STRINGS

The simplest model leading to the formation of strings is the Abelian Higgs model,

$$L = D_\mu\phi^+ D^\mu\phi - \frac{1}{4} F_{\mu\nu}F^{\mu\nu} - \frac{1}{4} \lambda(\phi^+\phi - \eta^2)^2. \tag{1}$$

Here, ϕ is a complex scalar field coupled to the Abelian gauge field A_μ, $D_\mu = \partial_\mu + ieA_\mu$ and $F_{\mu\nu} = \partial_\mu A_\nu - \partial_\nu A_\mu$. Below the critical temperature, $T_c \sim \eta$, the Higgs field develops a vacuum expectation value, $\langle\phi\rangle = \eta e^{i\theta}$. The magnitude of $\langle\phi\rangle$ is fixed by the model, but the phase θ is arbitrary and can be different in different regions of space. One can introduce the correlation length ξ such that the values of θ are uncorrelated for points separated by a distance greater than ξ. In the early universe, causality requires that ξ cannot exceed the causal horizon: $\xi < t$. (I assume that the phase transition is not of inflationary type. For the model (1), the phase transition is second-order and[1] ξ is $\sim \lambda^{-1/2}\eta^{-1} \ll t$). Chaotic spatial variation of the phase leads to the formation of strings. Whenever θ changes by 2π around a closed loop, thin tubes of the false vacuum ($\langle\phi\rangle = 0$) must be caught somewhere inside the loop. This is due to the fact that such a loop cannot be contracted to a point without a discontinuous change of phase. In the Higgs model of Eq. (1) it is possible to assign a direction to the strings. It can be taken as the direction of the magnetic gauge field trapped inside the string. An important property of topologically stable strings is that they cannot have free ends: the strings can be either infinite or closed.

In the general case, let G and H be the symmetry groups before and after the phase transition, respectively. Strings are formed if the first homotropy group $\pi_1(G/H)$ is nontrivial, or equivalently, if the manifold of degenerate vacuum states after the symmetry breaking is not simply connected.[1] For the Higgs model (1), $G/H = U(1)$ and $\pi_1(G/H) = Z$. If G is a simple group, then $\pi_1(G/H) = \pi_0(H)$, and so strings are formed if the new symmetry group H has disconnected components.[6] A fairly common example of this sort is when the group H is a product of a continuous group and Z_2, e.g. SO(10) → SU(5) × Z_2. In this case the strings have no direction and any two parallel strings can be annihilated by one another. This analysis applies to local, as well as global symmetry breakings. The simplest example of a model with global strings[4] is the Goldstone model, which is given by Eq. (1) with A_μ set equal to zero. Unlike monopoles, strings are not mandatory for grand unified theories (GUTs), but they are not at all uncommon. Examples of realistic GUTs with strings are given in Refs. 6, 7, and 8.

The width of the strings is microscopically small; for the model (1) it is $\delta \sim \lambda^{-1/2}\eta^{-1}$. The energy per unit length of the strings is $\mu \sim \rho_v\delta^2$, where $\rho_v \sim \lambda\eta^4$ is the false vacuum energy density. This gives

$$\mu \sim \eta^2 \ . \tag{2}$$

For global strings, the gradient terms in the energy density are not compensated by the gauge field, and μ acquires a large logarithmic factor,[4]

$$\mu \sim 2\pi\eta^2 \ln(R/\delta) \ . \tag{3}$$

Here the cutoff distance R is the typical distance between the strings.

At the time of formation the strings are expected to have the shape of brownian trajectories with a persistence length ~ ξ and typical distance between the neighboring string segments also ~ ξ. T. Vachaspati and I did a computer simulation of the string formation.[9] We randomly assigned a phase θ at the vertices of a cubic lattice and assumed that θ varies smoothly between the vertices. For simplicity we allowed the phase at the vertices to take only three values: $\theta = 0$, $2\pi/3$, and $4\pi/3$. The size of the cubic cell is identified with the correlation length, ξ. A string passes through the face of a cubic cell if the Higgs field rotates through 2π when traced around the face. It can be checked that the construction is such that all strings are either closed or end at the boundaries of the lattice. Using this prescription, we found all strings in the lattice and studied the statistical properties of the system. As expected, we found that long strings are brownian, so that the length of string between two points separated by a distance $R \gg \xi$ is $\ell \sim R^2/\xi$. Results of the simulation for various sizes of the lattice indicate that a large fraction (~80%) of the total string length is due to infinite strings. The remaining strings are closed loops with a scale-invariant distribution,

$$dn \sim R^{-4} dR \ . \tag{4}$$

Here, R is the typical size of the loop, which can be defined as the diameter of the smallest sphere enclosing the loop, and dn is the number of loops with sizes from R to R+dR per unit volume.

III. EFFECTS OF EXPANSION

Once it is formed, how will the system of strings evolve? Tension in convoluted strings tends to develop accelerations inversely proportional to the local curvature radius. In the very early stages the motion of strings is damped by friction due to their interaction with surrounding matter. Friction becomes negligible at time t_* which is roughly given by[1,10] $t_* \sim m_p^3/\eta^4$, where m_p is the Planck mass. (For superheavy strings with $\eta \sim 10^{16}$GeV, $t_* \sim 10^{-31}$s). Thereafter there is little damping and the strings reach relativistic speeds.

The action for a string is given by the area of the 2-dimensional world sheet it sweeps in space-time,[11]

$$S = -\mu \int [-g^{(2)}]^{1/2} d^2x \,. \tag{5}$$

From this action one can derive the equation of motion for a string in an expanding universe and try to solve it analytically or numerically. The case tractable analytically is that of small perturbations on a straight string.[12] In flat space-time, perturbations of wavelength λ and amplitude $A << \lambda$ result in oscillations of constant amplitude and period λ. In an expanding universe λ grows like the scale factor, $\lambda \propto a(t)$. Perturbations of wavelength greater than the horizon ($\lambda > t$) have not started oscillating yet. The amplitude of such perturbations also grows like $a(t)$, so that the whole string is just stretched by the expansion, its shape remaining unchanged. For $\lambda << t$ the amplitude of the oscillations remains constant, the curvature radius $R \sim \lambda^2/A$ grows like $a^2(t)$, and thus the perturbations are smoothed out. Numerical results indicate[13] that strongly convoluted strings behave in a similar manner. They are conformally stretched by the expansion on scales greater than the horizon and are straightened out on scales smaller than the horizon, so that the persistence length of long strings at time t is $\xi \sim t$.

For a brownian string, the length of string between two points separated by a distance R is $\ell \sim R^2/\xi$. If $R >> t$, then R grows like the scale factor, while $\xi \sim t$. In a radiation-dominated universe $a(t) \propto t^{1/2}$ and the length of string remains roughly constant. This means that the mass of strings per co-moving volume remains constant, and thus $\rho_s/\rho_\gamma \propto t^{1/2}$, where ρ_s and ρ_γ are the energy densities of strings and of radiation respectively. In this case the universe would rapidly become string-dominated. This, however, is not our final conclusion, since there are important physical effects that still have to be taken into account. These are discussed in the following two sections.

IV. CLOSED LOOP FORMATION

Intersecting strings can intercommute (or "change partners").[1] Pairs of strings intercommuting at two points can form closed loops. Loops can also be formed by self-intersection of individual strings. These processes are important, since, as it will be shown in the next section, loops eventually dissipate their energy and save the universe from string domination.

Let ν be the typical number of segments of infinite strings (or very large closed loops) per horizon volume, t^3. Then the density due to infinite strings is $\rho_{inf} \sim \nu\mu t^{-2}$ and each segment has about ν intersections per Hubble time ($\sim t$). The total number of intersections per time interval dt is $\sim \nu^2 t^{-1} dt$ and the rate of loop formation per unit volume is

$$dn/dt \sim p\nu^2 t^{-4} \,. \tag{6}$$

Here, p is the probability of loop formation per intersection. The typical curvature radius of strings at time t is ~ t, and we expect that loops formed by their intercommuting will have size ~ t. Then, by energy conservation, we must have

$$\frac{d\rho_{inf}}{dt} \sim -\mu t \frac{dn}{dt}, \tag{7}$$

which, together with Eq. (6) gives $\nu \sim p^{-1}$. The intercommuting probability of strings is unknown, and one's best guess is probably that $p \sim 1$. (The case $p \ll 1$ has been discussed in Ref. 14.) Then there are no more than a few segments of infinite strings per horizon volume at any time, and Eq. (6) tells us that about one loop of size ~ t is formed per horizon volume per Hubble time.

Intercommuting of strings can also destroy the existing closed loops. If a long segment of string intercommutes at one point with a closed loop, the loop gets absorbed into the segment. However, the probability for a loop smaller than the horizon to be hit by a string decreases with time like t^{-1}, and we expect a large fraction of loops to survive.

V. FATE OF CLOSED LOOPS

Closed loops much smaller than the horizon are not affected by the expansion, and one can use the flat space equations of motion[11]:

$$\ddot{\vec{r}} = \vec{r}''; \quad \dot{\vec{r}} \cdot \vec{r}' = 0; \quad \dot{\vec{r}}^2 + \vec{r}'^2 = 1. \tag{8}$$

Here, σ is a parameter on the loop, $\dot{\vec{r}} = \partial\vec{r}/\partial t$, $\vec{r}' = \partial\vec{r}/\partial\sigma$, and the loop trajectory is represented as $\vec{r}(\sigma,t)$. The parameter σ has the dimension of length and changes from 0 to M/μ, where M is the mass of the loop. The general solution of (8) is

$$\vec{r}(\sigma,t) = \frac{1}{2}[\vec{a}(\sigma-t) + \vec{b}(\sigma+t)], \tag{9}$$

where $\vec{a}$ and $\vec{b}$ are arbitrary periodic functions, with period M/μ, satisfying

$$\vec{a}'^2 = \vec{b}'^2 = 1. \tag{10}$$

It is easily seen from (9) that the motion of the loop is periodic with period $T = M/2\mu$.

What is the fate of the closed loops? One possibility is that they break into smaller and smaller loops as a result of multiple self-intersections, and eventually decay into elementary particles. Here, an important result has been obtained by Kibble and Turok[15] who have demonstrated the existence of a large class of loop trajectories which never self-intersect. Turok[16] has analyzed a representative class of loop trajectories and has shown that a large portion of the parameter space corresponds to non-self-intersecting loops. Even if a particular loop does self-intersect, chances are that at least one of the two resulting loops will be of non-intersecting variety. These results indicate that a substantial fraction of loops is not destroyed by intercommutings.

For such loops, the dominant energy loss mechanism is the

gravitational radiation.[17] Loops of size R have a typical frequency $\omega \sim R^{-1}$, and one can estimate the radiation rate using the quadrupole formula,

$$dM/dt \sim -GM^2R^4\omega^6 \sim -G\mu^2. \quad (11)$$

Here, $M \sim \mu R$ is the mass of the loop. The lifetime of the loop is:

$$\tau \sim M/|\dot{M}| \sim R/G\mu. \quad (12)$$

The motion of loops is relativistic, and it can be shown[16] that at one moment during the period certain points of the loop actually reach the velocity of light. This indicates that the quadrupole formula is quantitatively incorrect.

Combined analytic and numerical analysis[19] for the class of loops found by Kibble and Turok gives

$$dM/dt \sim -\gamma G\mu^2, \quad (13)$$

where γ depends on the particular loop trajectory, but is typically ~ 100.

It should be noted that Kibble and Turok's results[15,16] do not apply directly to global strings, in which case there is a long range interaction between different parts of the loop. There is no reason to expect that the motion of such loops is periodic, and they can be subject to rapid decay via multiple self-intersections.

VI. DENSITY OF LOOPS

Loops of size $\sim R$ are formed at $t \sim R$; at that time their number density is $\sim R^{-3}$. At later times, the density of such loops is

$$n_R(t) \sim [a(R)/a(t)]^3R^{-3}, \quad (14)$$

where $a(t)$ is the scale factor. For the discussion in the following sections, it will be useful to be more precise in normalization of $n_R(t)$. We shall assume that R stands for the <u>length</u> of the loop, so that its mass is $M = \mu R$, and introduce in (14) a <u>numerical</u> coefficient, β, which can be determined by a computer simulation of the string evolution. During the radiation era, $a(t) \propto t^{1/2}$ and

$$n_R(t) = \beta(tR)^{-3/2}. \quad (15)$$

It follows from Eq. (13) that the loops surviving at time t have (initial) lengths greater than $\gamma G\mu t$. The dominant contribution to the mass density of strings is given by the smallest loops with $\ell \sim \gamma G\mu t$:

$$\rho_s/\rho_\gamma \sim 30\beta\gamma^{-1/2}(G\mu)^{1/2} \sim 3\beta(G\mu)^{1/2}, \quad (16)$$

where $\rho_\gamma \sim (30Gt^2)^{-1}$ is the radiation density.

Using Eq. (2) we can write

$$G\mu \sim (\eta/m_p)^2, \quad (17)$$

where m_p is the Planck mass. For $\eta \ll m_p$ we have $\rho_s/\rho_\gamma \ll 1$, and the strings never dominate the universe.

We now turn to the discussion of various cosmological consequences of strings.

VII. GALAXY FORMATION

Density fluctuations produced by strings can lead to formation of galaxies and clusters of galaxies.[20,17] On scales greater than the horizon, the density fluctuations due to strings are balanced by the corresponding variations of the radiation density; so at any time t the fluctuations can only be produced on scales smaller than t. If one assumes[20] that closed loops rapidly decay as a result of multiple self-intersections (which may be the case for global strings), then the density of strings is $\rho_s \sim \mu t/t^3 \sim \mu t^{-2}$. The density fluctuations produced on each scale at horizon crossing are of the order

$$(\delta\rho/\rho)_{hor} \sim 30G\mu. \tag{18}$$

(Of course, there may be an additional numerical factor in this equation.) This is the well-known scale-invariant (Zel'dovich) spectrum. The following evolution of density fluctuations depends on the composition of the dark matter. A distinctive feature of the string scenario is the formation of planar wakes behind relativistically moving strings.[21] These wakes can help to explain the formation of large-scale structure in a universe dominated by weakly interacting cold particles (e.g. axions).

An alternative scenario[17] assumes that galaxies and clusters condense around oscillating closed loops, while the loops gradually decay by gravitational radiation. Loops of size R introduce a density fluctuation on a co-moving scale $\ell \sim n_R^{-1/3}$. Using Eq. (14), one can find[7] the spectrum of density fluctuations produced by loops for various assumptions about the dark matter. On scales coming inside the horizon during the matter-dominated era, the magnitude of fluctuations produced by loops is comparable to that for the open strings, Eq. (18). Requiring $(\delta\rho/\rho)_{hor} \sim 3 \times 10^{-5}$ we get $G\mu \sim 10^{-6}$. This value of $G\mu$ corresponds to $\eta \sim 10^{16}$ GeV for local strings and $\eta \sim 10^{15}$ GeV for global strings. For a detailed discussion of the spectra of fluctuations I refer the reader to Ref. 7 (see also Ref. 16). Here, I would like to mention some of the new features introduced by the presence of long-lived loops.

In the standard model of galaxy formation with a primordial spectrum of adiabatic fluctuations, all density perturbations are damped below the Silk mass ($M \sim 10^{14} M_\odot$) or neutrino free-streaming mass ($M \sim 4 \times 10^{14} M_\odot$) in the baryon- and neutrino-dominated cases, respectively. The same is true if the density fluctuations are due to inflation. In the string scenario the fluctuations are preserved on smaller scales.[7] Baryons and neutrinos erase their own fluctuations, but when the Jeans mass drops to the corresponding scale, they pick up the perturbations produced by surviving loops. This allows the possibility of early formation of some galaxies and quasars.

An important feature of the string scenario is that the fluctuations are not in the form of waves with random phases. Closed loops have sizes much smaller than those of the galaxies and clusters condensing around them and produce large density perturbations in their immediate vicinity. This can result in accretion of matter onto the loops and formation of massive compact objects, which can be identified with quasars and active galactic nuclei[21,22]. On larger scales, rare supergiant loops could produce localized regions of density contrast much greater than one might expect from gaussian fluctuations (such regions may have actually been observed).[23] Occasional splitting of oscillating loops can also be used to explain the observed cluster-cluster correlations.[14,24]

VIII. OBSERVATIONAL EFFECTS OF STRINGS

If the strings are really there, how can we observe them? Nongravitational interactions of strings with surrounding matter are negligible,[1,10] and thus the strings can only be detected through their gravitational effects. One possibility is to look for the stochastic gravitational wave background produced by oscillating loops. Loops of size R radiate at frequencies $\omega \sim R^{-1}$; after the waves are emitted, their frequency is redshifted like $\omega \propto (1+z)$. Presently surviving loops have extremely low frequencies, and the flux they produce is well below the observational capabilities.[16,25] Waves which now have periods less than 1000 years correspond to loops that formed and decayed during the radiation era.

A convenient measure of the intensity of the radiation is

$$\Omega_g(\omega) \sim \frac{\omega}{\rho_c} \frac{d\rho_g}{d\omega} . \tag{19}$$

Here ρ_g is the energy density of the gravitational waves and ρ_c is the critical density. $\Omega_g(\omega)$ gives the energy density in units of ρ_c per logarithmic frequency interval. Using Eq. (16) and noticing that gravitational waves redshift in the same way as radiation, one finds[26]

$$\Omega_g\ (\omega) \sim 3\beta\ (G\mu)^{1/2}\ \Omega_\gamma, \tag{20}$$

where $\Omega_\gamma \rho_c$ is the radiation density, $\Omega_\gamma = 2 \times 10^{-5}\ h^{-2}$. With $G\mu \sim 10^{-6}$, Eq. (20) gives $\Omega_g(\omega) \sim 10^{-7} h^{-2} \beta$. Recent observations of the millisecond pulsar imply[27] that $\Omega_g \lesssim 10^{-5}$ for gravitational waves with periods ~ few months. However, the accuracy grows rapidly with the time of observation, and $\Omega_g \sim 10^{-7}$ will become detectable within several years.[28]

Kaiser and Stebbins[29] have pointed out that strings should leave a characteristic signature on the microwave background: the background temperature should have steplike discontinuities on curves on the sky. Strings can also produce double images of quasars and galaxies.[30-32] This is discussed in the next section.

IX. STRINGS AS GRAVITATIONAL LENSES

The gravitational field produced by a straight string is rather different from that of a linear distribution of ordinary matter. The energy-momentum tensor of a string lying along the z-axis is[30]

$$T_\mu^\nu = \mu\ \mathrm{diag}\ (1,0,0,1)\ \delta(x)\delta(y) \tag{21}$$

The pressure is negative and equal to the energy density in the direction parallel to the string and is equal to zero in the perpendicular directions. Note that T_μ^ν is invariant under Lorentz boosts in the z-direction, as it should be, since the string is described by a z- and t-independent solution of Lorentz-invariant field equations. The solution of Einstein's equations with T_μ^ν from Eq. (21) is[30,32]

$$ds^2 = dt^2 - dz^2 - dr^2 - (1-4G\mu)^2 r^2 d\phi^2 \tag{22}$$

A coordinate transformation $\phi' = (1-4G\mu)\phi$ brings the metric to a locally Minkowskian form, but now the angle ϕ' changes from 0 to $(1-4G\mu)2\pi$. Thus, Eq. (22) describes a "conical space", that is, a flat space with a wedge of angular size $8\pi G\mu$ taken out and the two faces of the wedge identified.

Stationary massive objects located near a straight string do not experience any gravitational attraction. This unexpected result can be understood in the following way. For weak gravitational fields and low velocities, Einstein's equations can be reduced to

$$\nabla^2 \Phi = 4\pi G(\rho + P_1 + P_2 + P_3), \tag{23}$$

where Φ is the gravitational potential and P_i are eigenvalues of pressure. In our case, $P_1 = P_2 = 0$, $P_3 = -\rho$, and thus $\vec{\nabla}\Phi = 0$. It should be emphasized that these results apply only to portions of strings which can be regarded as straight. A closed loop of size R produces a regular Schwarzshild field at distances >>R.

Light propagating in the plane perpendicular to the string is deflected by the angle

$$\delta = 4\pi G\mu, \tag{24}$$

and we can see double images of objects located behind the string.[30] It is not impossible that some of the known double quasars are due to strings. If d and ℓ are, respectively, the distances from the observer and from the quasar to the string, then the angular separation between the images is

$$\delta\phi = 2\ell(d + \ell)^{-1}\delta \tag{25}$$

With $G\mu \sim 10^{-6}$, δ is ~ few arc seconds. The maximum value of $G\mu$ still consistent with observations is $G\mu \sim 10^{-5}$, which gives $\delta \sim 1$ arc min.

A double image is obtained if the quasar is within the angle $\sim\delta$ from the string. An estimate of the probability for a quasar to be lensed by a string gives[31] $p \sim \delta \ell n \delta^{-1}$ (or $p \sim \delta$ if closed loops have short lifetimes). With $\sim 3 \times 10^3$ known quasars and for $G\mu \sim 10^{-5} - 10^{-6}$ we do not expect more than a few double quasars due to strings. Galaxies are much more numerous than quasars, and one can expect to see lines of double galaxies along the strings.[31,32]

X. PERSON-STRING INTERACTION

Now I turn to the burning question of what happens if a string passes through a person. This problem was addressed by Jim Peebles, and here I report the calculations he did on a napkin.[33]

In the frame of reference of the string, the person is moving towards the string with a velocity $v \sim 1$. After she/he passes the string, the velocities of parts of the body on opposite sides of the string are deflected towards one another by an angle $\sim\delta$. In other words, the head and the feet start moving towards one another with a velocity $\delta v \sim 5$ km/s (for $G\mu \sim 10^{-6}$). This is unhealthy, and you should avoid these things if at all possible. (Who said that cosmology does not address human needs?).

XI. CONCLUDING REMARKS

Strings can give rise to a rich assortment of astrophysical phenomena, and experimental techniques are now approaching the level where some of the predictions of the string scenario can be directly tested. At this time, the experimental bounds from the millisecond pulsar[28] and from the background anisotropy measurements[29] are about one order of magnitude off the target. The accuracy of the millisecond pulsar measurements grows like the fourth power of the observation time, and there is an exciting possibility that gravitational signals from strings will make their appearance within the next few years.

On the other hand, theoretical predictions of the string scenario can now be made only in the form of rough order-of-magnitude estimates. Here I would like to mention some problems for the future, which I think are important for sharpening the predictions of the model and for our general understanding of the evolution of strings.

(i) Intercommuting probability, p. Strings are described by classical solutions of the field equations, and their intercommuting probably occurs at the classical level as well. Then the intercommuting process is totally deterministic, and the word "probability" refers to averaging over collision angles and relative velocities of the strings. p can well be model-dependent, and its calculation will probably involve a numerical solution of nonlinear field equations.

(ii) Computer simulation of the evolution of strings for various values of p. In particular, this will help to determine the rate of closed loop formation. One can use the results of Ref. 9 as initial conditions.

(iii) Evolution of global strings. In particular, the existence of non-self-intersecting loop trajectories has not been demonstrated.

(iv) Strings and inflation. To have cosmologically interesting strings in an inflationary scenario, one has to arrange for strings to be formed after (or near the end of) inflation. This is not easy, since the required energy scale of strings is rather high. One "realistic" model has been suggested[8] in which global strings form by the end of the inflationary phase transition. It would be interesting to see if local strings can coexist with inflation.

REFERENCES

[1]T. W. B. Kibble, J. Phys. **A9**, 1387 (1976); Phys. Rep. **67**, 183 (1980).
[2]A. Vilenkin, Nucl. Phys. **B196**, 240 (1982).
[3]T. W. B. Kibble, G. Lazarides and Q. Shafi, Phys. Rev. D**26**, 435 (1982).
[4]A. Vilenkin and A. E. Everett, Phys. Rev. Lett. **48**, 1867 (1982).
[5]See, however, F. Stecker and Q. Shafi, Phys. Rev. Lett. **50**, 928 (1983).

[4]A. Vilenkin and A. E. Everett, Phys. Rev. Lett. **48**, 1867 (1982).
[5]See, however, F. Stecker and Q. Shafi, Phys. Rev. Lett. **50**, 928 (1983).
[6]T. W. B. Kibble, G. Lazarides and Q. Shafi, Phys. Lett. **113B**, 237 (1982); D. Olive and N. Turok, Phys. Lett. **117B**, 193 (1982).
[7]A. Vilenkin and Q. Shafi, Phys. Rev. Lett. **51**, 1716 (1983).
[8]Q. Shafi and A. Vilenkin, Phys. Rev. D**29**, 1870 (1984).
[9]T. Vachaspati and A. Vilenkin, Phys. Rev. D **30**, 2036 (1984).
[10]A. E. Everett, Phys. Rev. D**24**, 858 (1981).
[11]P. Goddard et. al., Nucl. Phys. **B56**, 109 (1973).
[12]A. Vilenkin, Phys. Rev. D**24**, 2082 (1981).
[13]N. Turok and P. Bhattacharjee, Phys. Rev. D **29**, 1557 (1984).
[14]J. Preskill and M. Wise, to be published.
[15]T. W. B. Kibble and N. Turok, Phys. Lett. **116B**, 141 (1982).
[16]N. Turok, UC at Santa Barbara Preprint TH-3 (1984).
[17]A. Vilenkin, Phys. Rev. Lett. **46**, 1169, 1496(E) (1981).
[18]It can be shown that non-gravitational radiation from strings is negligible. See T. Vachaspati, A. E. Everett and A. Vilenkin, Phys. Rev. D **30**, 2046 (1984).
[19]T. Vachaspati and A. Vilenkin, Harvard preprint HUTP-84/A065.
[20]Y. B. Zel'dovich, Mon. Not. R. Astr. Soc. **192**, 663 (1980).
[21]J. Silk and A. Vilenkin, Phys. Rev. Lett. **53**, 1700 (1984).
[22]C. J. Hogan, Caltech Preprint, 1984.
[23]P. J. E. Peebles, private communication.
[24]N. Turok and D. Schramm, University of Chicago Preprint (1983).
[25]E. Witten, Phys. Rev. D **30**, 272 (1984).
[26]A. Vilenkin, Phys. Lett. **107B**, 47 (1981).
[27]J. Taylor, private communication.
[28]C. J. Hogan and M. J. Rees, Nature **311**, 109 (1984).
[29]N. Kaiser and A. Stebbins, Nature **310**, 391 (1984).
[30]A. Vilenkin, Phys. Rev. D**23**, 852 (1981).
[31]A. Vilenkin, Ap. J. Lett. **282**, L51 (1984).
[32]J. R. Gott, Princeton University Preprint (1984).
[33]P. J. E. Peebles, private communication.

Neogeny and Paleogeny

Craig J. Hogan

ABSTRACT

A didactic dichotomy is drawn between conventional "paleogenic" scenarios for the origin of large scale structure, which assume that the binding energy fluctuations in galaxy clustering date from the very early universe, and "neogenic" scenarios in which nongravitational mechanisms at late times dominate the evolution of structure on these scales and obliterate any traces of the primordial fluctuations. Observational strategies for distinguishing these possibilities are briefly discussed.

The most prevalent theoretical approach to the origin of large scale structure since the pioneering work of Lifshitz has been to follow the evolution of preexisting small-amplitude density perturbations. Linear theory tells us that the binding energy of such perturbations is constant as the universe expands (see e.g. Bardeen),[1] so if we identify galaxy clusters as the end result of this evolution then their binding energy is preexisting, and is not accounted for by the theory (in fact in this sense it is misleading to regard the initial perturbations as "small" because the perturbations in energy, if not density, are quite large). Another approach, which has historically been relatively neglected, is to treat the structure as being primarily the result of nongravitational perturbations at late times to a universe which is initially smooth on these scales. Various scenarios of this type have been proposed through the years (e.g. Doroshkevich, Zeldovich, and Novikov),[2] and some recent directions for these scenarios are briefly reviewed below. For brevity I refer to these two possibilities as "paleogeny" and "neogeny" respectively. In this paper I draw attention to these alternatives because the physics of the initial conditions has become a subject of serious study and it is important to ask which aspects of the present-day universe are likely to be direct relics of processes at early times.

This is a natural dichotomy to draw because in fact there are two distinct "windows" of cosmic time during which it is possible to alter the Friedmann metric on large scales. The first occurs before baryogenesis, when inflation or some other mechanism can introduce enormous entropy per comoving volume without excessively diluting the baryon number of the universe. As Guth[3] showed, this makes it possible in principle for very large comoving scales to have been in causal contact and offers the opportunity to alter the metric. We are reasonably sure that the large-scale uniformity of the universe originated during this pre-baryogenesis window. It is possible in principle that this era also created significant perturbations; specific examples of this occur in various calculations of perturbations produced in inflationary universes (see e.g. Gibbons, Hawking, and Siklos).[4] However, it is also possible that negligible perturbations were introduced at this epoch. In that case there is a second window for altering the pure Friedmannian evolution, namely when a given scale enters the horizon in the classical sense and ordinary non-gravitational forces can come into play and alter the pure gravitational equations of motion. Other forces enter, for example, before recombination in standard paleogenic models, when radiation pressure, photon diffusion, and other processes alter the (previously purely gravitational) linear evolution of perturbations after they enter the horizon. However, in these models nongravitational forces only enter into damping terms; they do not generate binding energy, they can only dissipate it. (In linear theory, damping terms dissipate binding energy; unlike the situation in good old nonlinear Newtonian gravity, where dissipation increases binding energy.)

In the neogenic models, these nongravitational forces are not merely responsible for modifying preexisting perturbations, but are held responsible for actually generating them. This obviously requires the release of some form of free energy. At early times, free energy is available in the form of latent heat in a supercooled phase transition. At the late times under consideration, after the scales in question ($\approx 10^{15} M_\odot$) come within the horizon, the natural sources of free energy are astrophysical ones, such as stars or quasars.

It is a surprising fact that if one does rough estimates of the possible effects of such sources, one finds that they are just marginally capable in principle of producing the observed structure. The coincidence is interesting because it would be fortuitous in a paleogenic scenario. In other words, a priori one would either expect primordial fluctuations to be obliterated by the release of stellar free energy, or to completely overwhelm the effect of such energy, i.e. to generate structure at such a large scale of mass and binding energy that the stars would not have a commensurate effect. Since the latter is not the case, economy dictates that one investigate the former possibility, and suppose that all of the observed structure is "neogenic" in origin.

There are several ways of expressing the coincidences. I will cite just a few examples. (1) We can measure the binding energy density contained in the large scale structure of the universe by measuring the first moment of the galaxy autocorrelation function (see e.g. Fall).[5] A typical estimate of the fraction of closure density contained in gravitational binding energy of galaxies is $\Omega_{grav} \approx 5 \times 10^{-6} \Omega^2$.

Apparently unconnected with this is the observed extragalactic energy density of starlight from luminous galaxies, which is of the order of $\Omega_{starlight} h^2 \approx 10^{-6}$. Thus even in the present universe there is almost

enough free energy available to produce the large scale structure in an expansion time. (2) The largest mass scale over which nongravitational forces would operate, the radiation Jeans mass $M \approx 10^{16} M_\odot (\Omega h^2)^{-2}$, is commensurate with the scale of clustering $\approx 10^{15} M_\odot \Omega h^{-1}$; this is probably the largest scale on which a neogenic model would be possible. (3) The timescale on which this structure was created coincides with the astronomical timescales (e.g. the "Salpeter time" $e^4/Gm_p m_e^2 c^3$) for free energy release. (4) Energy injected by supernovae is known to be important in the dynamical evolution of gaseous components of galaxies today, and simple extrapolation to gaseous protogalaxies indicates that the observed distribution of stars within galaxies was probably strongly affected by such events.

It is useful to divide neogenic scenarios up according to which force is used to generate large scale structure.

(1) Hydrodynamic scenarios invoke the thermal energy of gas heated by supernovae to create pressure gradients and large-scale motions. Recent work includes that of Bertschinger[6], Carr and Ikeuchi[7], and Vishniac[8]; see also the review by Ostriker.[9]

The most direct observational constraint on using these scenarios appears to come from the microwave anisotropy they generate via the Sunyaev-Zeldovich effect. Pregalactic gas must be very hot in order to generate nonlinear structures which today have a Hubble velocity of order 1000 km/sec; and it must be distributed inhomogeneously in order to generate inhomogeneity. These two constraints can be quantified, and one can estimate the variation of total thermal energy contained in radio telescope beams on arc minute scales, and hence the anisotropy in Sunyaev-Zeldovich spectral distortion. Simplified models[10] seem to indicate that such schemes can only generate structure (without excessive anisotropy) on ≈10Mpc scales if the structure reflected in the galaxy distribution is not a good indication of structure in the pregalactic gas distribution. In this case I would classify it as a type 2 scenario:

(2) Another possibility is that galaxy clustering is due primarily to large-scale correlated spatial variation in the efficiency with which baryonic material is converted into luminous galaxies. This correlation could in principle be neogenic or paleogenic in origin. As an example, suppose one could arrange to vary the Jeans mass in protogalactic gas over large scales by suitably varying the temperature. It might turn out that a large fraction of baryonic material within apparent "voids" collapsed into invisible low-surface-brightness galaxies, or never collapsed at all and now forms an ionized intergalactic plasma. These scenarios tend to be difficult to constrain because so little is known about what protogalactic properties determine the observable properties of galaxies. Schemes of this type have been investigated recently by Bardeen[11] and Rees.[12] Since this type of model automatically segregates mass and light, it is naturally attractive from the point of view of solving the observational "Ω"-problem.[15]

(3) The third type of neogenic scenario invokes the radiation pressure of the microwave background to generate large-scale perturbations.[13,14] These models may be regarded as modest generalizations of standard paleogenic models because they follow the time-worn path of linear theory. The only difference is in the addition of a "source" term in the equations of radiation transport, by which nongravitational free energy can actively produce fluctuations and binding energy. We envision a population of pregalactic quasars or similar objects at redshifts of order several hundred when ionized matter and microwave radiation were still coupled by Thomson scattering. The effect at late times is very similar to the effect of primordial isothermal fluctuations, except for differences in the microwave anisotropy. The amplitude and angular spectrum of anisotropy can be calculated quite precisely in these models. The amplitude is sensitive to the redshift at which the sources introduce their perturbation. For realistic sources, the peak amplitude is of order greater than $\delta T/T \gtrsim 3 \times 10^{-5}$. The angular spectrum is insensitive to any parameters of the model, although it depends weakly on Ω and Ω_g, the gas density. We find that the largest variation in temperature should be found at angular scales of order a few degrees, and would urge more observational programs to concentrate on this (relatively neglected) angular scale.

The most direct way to verify the importance of paleogenic fluctuations on large scales is by examining the angular spectrum of the microwave anisotropy. For a "constant-curvature" spectrum $\delta T/T$ should be roughly constant (to within calculable sampling errors) on angles larger than a few degrees. No neogenic sources of anisotropy can mimic this spectrum, so its observation would imply directly the existence of primordial fluctuations with significant amplitude.

Observations such as these are clearly needed to sort out whether the phenomenon of galaxy clustering -- especially, the origin of large-scale binding energy -- will ultimately be explained by astronomers or by physicists. Physicists obviously hope that a paleogenic explanation is the correct one, because in that case the observable phenomena such as galaxy clustering and microwave anisotropy carry information about the very early universe, and probably about physics at very high energies. Astronomers, I suspect, will remain skeptical; as a group they would probably be upset if it turned out that the most spectacular aggregations of matter in the universe, such as the great clusters of thousands of galaxies, are bound because of some random quantum fluctuations in a Higgs field or perturbations produced by vacuum strings.

This work was supported by a Bantrell Fellowship, and by NSF grant 82-14126.

REFERENCES

[1]J. M. Bardeen, Phys. Rev. D**22**, 1182 (1980).
[2]A. G. Doroshkevich, Ya. B. Zeldovich, and I. D. Novikov, Sov. Astr. A. J. **11**, 231 (1967).
[3]A. Guth, Phys. Rev. D**23**, 347 (1981).
[4]G. W. Gibbons, S. W. Hawking, and S. T. C. Siklos, The Very Early Universe, (Cambridge University Press), 1983.
[5]S. M. Fall, Rev. Mod. Phys. **51**, 21 (1979).
[6]E. Bertschinger, Astrophys. J. **268**, 17 (1983).
[7]B. J. Carr and S. Ikeuchi, preprint (1984).
[8]E. T. Vishniac, Astrophys. J. **274**, 152 (1983).
[9]J. P. Ostriker and L. L. Cowie, Astrophys. J. Lett. **243**, L127 (1981). J. P. Ostriker, in Astrophysical Cosmology, eds. H. A. Bruck, G. V. Coyne, and M. S. Longair (Rome:Specola Vaticana, 1982).
[10]C. J. Hogan, Astrophys. J. Lett. **284**, L1 1984.
[11]J. M. Bardeen, these proceedings (1984).
[12]M. J. Rees, private communication (1984).
[13]C. J. Hogan, M.N.R.A.S. **202**, 1101 (1983).
[14]C. J. Hogan and N. Kaiser, Astrophys. J. **274**, 7 (1983).
[15]N. Kaiser, these proceedings (1984).

Galactic Cycles and their Relationship to Life on Earth

Arne P. Olson

Episodic events in the geologic time scale for evolution of life on earth are identified relative to galactic "structure", i.e. variations in composition and density with latitude. Potentially cyclic behavior arises from the ~700 My period of time for the solar system to pass through all existing structure. The orbital period[1] of 237 ±21 My and the spiral arm "pattern speed" if ~2/3 earth's orbital speed,[2] produce this fundamental galactic cycle. Higher harmonics due to galactic arms, shocks, magnetic fields, or non-uniformly distributed molecular clouds or condensed objects may have meaningful causal connection to transition events on earth marking major geologic epochs.[3-7]

Major impacts of cometary bodies on earth could have, by chance or by other means, clustered in time at those transitions and thereby provided the necessary additional environmental stress. But if not by chance, then what mechanism would be responsible? The answer is postulated to be of galactic origin: the galactic shock. Our galaxy, The Milky Way, is a spiral. A very recent picture of its large-scale structure is provided by Blitz, Fich, and Kulkarni.[8] They identify four known arms (Carina, Orion, Perseus, and Cygnus), and they postulate the existence of a fifth unnamed arm on the opposite side of the galaxy, about 180° from the Perseus arm. Shock waves are believed to be responsible for formation of galactic arms.

The density wave theory of galactic spirals[9,10] predicts a large-scale shock in the interstellar gas along spiral arms. It is believed to be a triggering mechanism for star formation, via sudden compression.

At the center of the galaxy, some form of "engine" exists which controls (or at least influences) its evolution. It has been suggested that a massive black hole could be there. The important thing, in terms of galactic structure, is to try to understand processes happening near the galactic center. Is matter falling into or being ejected from the galactic center? What overall dynamic mass flows are there? Can shock waves be initiated by "sputtering" or intermittent turning on and off of these flows? Additional details of how galactic structure may arise naturally from fusion processes is given in a companion paper.[11] Arms are proposed to be emitted from the galactic center as a consequence of fusion processing of biaxial infalls of "dark matter".

Geophysical and biological evidence was presented[12-16] which is very suggestive of cyclic patterns of galactic origin having an effect on the earth and/or the sun. If it truly is cyclical, then the geologic record for cratering, extinctions, life proliferation, ore bodies, etc. can be better understood as to how the solar system evolved, and as to potential dangers ahead. The inferences drawn support the conclusion that the pattern speed of the galaxy is above 2/3 the rotation speed. Finally, the geologic scale of time should be redefined in correspondence to galactic cycles. The Cenozoic will then extend to ~+120 My (the far side of the Perseus arm), and the Mesozoic corresponding to the gap to the "unnamed" arm of Blitz. The Paleozoic would be better split into Upper and Lower, using the start of the Silurian at 435 My ago as the interface. The Lower Paleozoic would extend to the Pre-Cambrian-Cambrian interface, and the Ediacarian Era would take up the remainder in the inter-arm gap to the Perseus arm. Ediacarian time extends back to ~670 My ago, at which time soft-bodied fauna first appeared.[16] From a biological point of view, the Milky Way has (or had!) five arms. Furthermore, two major boundaries per cycle occur every 340 My, with "now" ±2 My being the time to begin a major cycle.

REFERENCES

[1]G. DeVancouleurs, Astrophys. J. **268**, 451 (1983).

[2]B. J. Bok, Sci. Am. **244** (3), 92 (1981).

[3]K. J. Hsu, et al., Science **216**, 249 (1982).

[4]D. McLaren, New Scientist, pp.588-593 (1983).

[5]L. W. Alvarez, W. Alvarez, F. Asaro, H. V. Michel, Science **208**, 1095 (1980).

[6]A. Mazaud, C. Laj, L. de Seze, and K. L. Verosub, Nature (London) **304**, 328 (1983).

[7]B. P. Glass and B. C. Heezen, Sci. Am. **217** (1), 32 (1967).

[8]L. Blitz, M. Fich, and S. Kulkarni, Science **220**, 1233 (1983).

[9]C. Yuan and C. Y. Wang, Astrophys. J. **252**, 508 (1982).

[10]Shu, et al., Astrophys. J. **172**, 557 (1982).

[11]A. P. Olson, "A Three-Dimensional Model for Fusion Processes", Inner Space/Outer Space Conference, Fermi National Accelerator Laboratory, May 1984.

[12]M. Morris and L. J. Rickard, Ann. Rev. Astron. Astrophys. **20**, 517 (1982).

[13]P. J. Cook and J. H. Shergold, Nature (London) **308**, 231 (1984).

[14]K. E. Apt, et al., in Natural Fission Reactors (IAEA, Vienna, 1978), p. 677.

[15]L. Blitz, Sci. Am. **246** (4), 84 (1982).

[16]P. Cloud and M. F. Glaessner, Science **217** (4562), 783 (1982).

PART IV

INFLATION

Foremost at the interface of particle physics and cosmology is the theory of inflation. A big-bang Universe that undergoes inflation at some time in its history can lead to a Universe that closely resembles our own, i.e., spatially flat, age much greater than the Planck time, homogeneous and isotropic to a high accuracy but with perturbations present at some level, and devoid of magnetic monopoles. However, a completely acceptable particle physics model to implement the inflationary Universe scenario has not yet been found. Several models have come tantalizingly close but have failed some crucial test in the particle physics or in the cosmology. These near misses have encouraged people to search even harder for the particle physics framework for inflation.

This chapter provides a good representation of the state of inflation. Guth introduces the subject with a long review article explaining why the inflationary scenario is so attractive. He also reviews the basic features any particle physics model implementing inflation should have. At present, the most promising models for inflation are those with either global or local supersymmetry. Steinhardt reviews recent progress in supersymmetric inflation, and papers by Srednicki and Holman present details of attempts to make an inflationary Universe with supersymmetric models. One of the most attractive features of inflation is the possibility of generating small density perturbations for galaxy formation. Brandenberger discusses inflation and fluctuations. The inflationary Universe scenario involves calculations in curved space-time, phase transitions, and quantum mechanics. These subjects are covered by Wald, Guth and Pi, Sato and Kodama, and Gott. Finally, Seckel presents a non-standard inflationary scenario. Other non-standard scenarios are to be found in the chapter on extra dimensions.

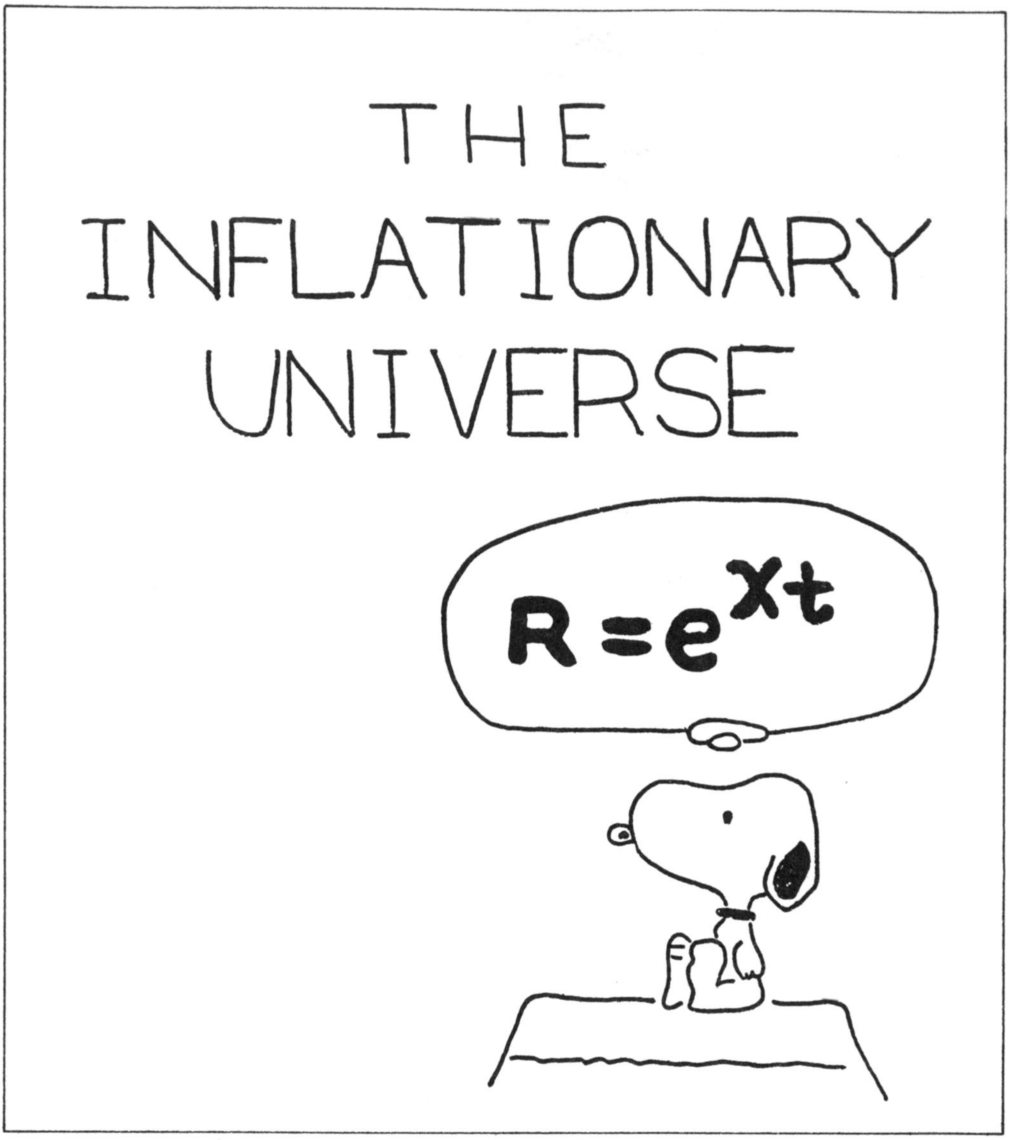

(Drawing by Alan Guth with a salute to Charles Schultz)

The New Inflationary Universe, 1984

Alan H. Guth

I. INTRODUCTION

Over the past five years, the interface between particle physics and cosmology has grown to become what the organizers of this conference have aptly called the Inner Space/Outer Space Connection. From the point of view of the particle theorists who have been involved in this work, the "connection" is strongly motivated by the developments in particle physics over the last decade or so.

During this period, experimental particle physics has tended to confirm the notion that the standard SU(3) × SU(2) × U(1) model is correct, and that it in fact accounts for essentially all the physics that we have seen. In hopes of discovering new physical laws, many particle theorists have turned to speculations on what happens beyond the standard model, and much of this speculation has centered on grand unified theories (GUTs).[1] Grand unified theories explain the quantization of charge and also make a very good prediction for $\sin^2\theta_W$ (θ_W = Weinberg angle), giving the idea a certain amount of plausibility. But the most dramatic predictions of grand unified theories occur only at the extraordinary energy scale of 10^{14} GeV. By the standards of your local power company, this is not an extraordinary amount of energy--it is roughly what it takes to light a one-hundred-watt light bulb for about a minute. However, the idea of having that much energy on a single elementary particle is genuinely extraordinary. If we were to try to build a 10^{14} GeV accelerator with present technology, we could in principle do it, more or less. It would be a linear accelerator with a length of about one light-year. Since such an accelerator is unlikely to be funded, we must turn to other means to see the 10^{14} GeV consequences of grand unified theories. According to standard cosmology, the universe had a temperature with $kT = 10^{14}$ GeV at about 10^{-35} sec after the big bang, and thus the universe itself becomes the best laboratory for studying this type of physics.

The new inflationary universe is a scenario in which the hot matter of the early universe supercools by many orders of magnitude below the critical temperature of a phase transition predicted by grand unified theories; in the process the universe expands exponentially by many orders of magnitude[2]--hence the name "inflationary". The word "new" refers to a modification of my original proposal which was suggested independently by Linde[3] and by Albrecht and Steinhardt.[4] They suggested a new mechanism by which the phase transition could take place, solving some crucial problems[2,5,6] which were created by my original proposal. The inflationary model is very attractive because it can solve several very fundamental cosmological problems--which I will talk about later. If correct, it would also mean that grand unified theory mechanisms are responsible for the production of essentially all the matter, energy, and entropy in the observed universe.

In today's talk I will review the new inflationary universe scenario,[7] placing particular emphasis on the basic features and motivations. I will also summarize some of the recent developments, and I will try to give a simple explanation for the generation of density fluctuations.

Before going on, let me specify the units I will be using. Since I am a particle theorist, I will set $\hbar = c = k = 1$, and I will take the GeV as the fundamental unit for everything. Since I am not a general relativist, I will not set Newton's constant G equal to one. Instead I will set $G \equiv 1/M_P^2$, where $M_P = 1.22 \times 10^{19}$ GeV is the Planck mass. Note that 1 GeV = 1.16×10^{13} °K = 1.78×10^{-24} gm, and that 1 $\mathrm{GeV}^{-1} = 1.97 \times 10^{-14}$ cm = 6.58×10^{-25} sec. I have been attempting to communicate with astronomers, and I have learned that 1 megaparsec (Mpc) is their way of denoting 1.56×10^{38} GeV^{-1}.

II. THE STANDARD SCENARIO

Now let me discuss what may be called the standard scenario of the very early universe.[9] The universe is assumed at the outset to be homogeneous and isotropic, and hence can be described in comoving coordinates by a Robertson-Walker metric. Whether it is open, closed, or flat, the curvature is negligible at very early times. Thus, the metric takes the simple form

$$ds^2 = -dt^2 + R^2(t)d\vec{x}^2. \tag{2.1}$$

The expansion of the universe is described by the scale factor R(t), which is governed by the Einstein field equation

$$\left(\frac{\dot{R}}{R}\right)^2 = \frac{8\pi}{3} G\rho, \tag{2.2}$$

where ρ is the energy density, and the dot denotes differentiation with respect to the time t. The above equation would have an additional term if the curvature were not negligible. The quantity $H \equiv \dot{R}/R$ is the Hubble "constant" (which of course varies as the universe evolves). The mass density is dominated by the thermal radiation of effectively massless (i.e., $M \ll T$) particles at temperature T:

$$\rho = cT^4. \tag{2.3}$$

The constant c depends on the number of effectively massless particle species; for the minimal SU(5) grand unified theory[10] at the highest temperatures, c is about 50. The expansion is taken to be adiabatic, which implies (as long as the number of effectively massless particle species remains unchanged)

$$RT = \text{constant}. \tag{2.4}$$

Putting these relations together, one finds

$$T^2 = \left(\frac{3}{32\pi c}\right)^{1/2} \frac{M_P}{t} \approx \frac{M_P}{40t} \tag{2.5}$$

and

$$R \sim \sqrt{t}. \tag{2.6}$$

To develop some feeling for the relevant numbers, consider $T = 10^{14}$ GeV, which is approximately the critical temperature T_c of the GUT phase transition. The equations then give $t \approx 1.9 \times 10^{-35}$ sec, and $\rho \approx 1.2 \times 10^{75}$ gm-cm^{-3}. (For comparison, recall that in these units the density of an atomic nucleus is about 10^{15}.) Objects which are separated by 10^{10} light years today were separated by only about 6 cm at this early time.

Finally, there must be a phase transition (or possibly several) when the temperature falls to T_c. In the standard scenario it is assumed that this phase transition occurs quickly, without any significant supercooling. The effects on the evolution of $R(t)$ are then negligible.

III. PROBLEMS OF THE STANDARD SCENARIO

The standard scenario suffers from four problems, which I will now describe. It is the existence of these problems which motivates the inflationary scenario. The first is the monopole problem--that is, the standard scenario leads to a tremendous overproduction[11] of magnetic monopoles. To see how this comes about, recall that the monopoles are in fact topologically stable knots in the Higgs field expectation value. If the Higgs field has a correlation length ξ, then one would expect a density of monopoles given roughly by the Kibble relation[13]

$$n_M \approx 1/\xi^3. \tag{3.1}$$

When the universe cools below the critical temperature $T_c \approx 10^{14}$ GeV of the GUT phase transition, it becomes thermodynamically favorable for the Higgs field to align uniformly over large distances. However, it takes time for these correlations to be established. Note that the horizon length, defined as the distance which a light pulse could have travelled since the time of the initial singularity, is given in the radiation-dominated era by $\ell_H = 2t$, where t is the age of the universe. Thus, causality alone implies[14] that

$$\xi \lesssim 2t. \tag{3.2}$$

Eqs. (3.1) and (3.2) can be used to obtain an approximate lower bound on n_M immediately after the phase transition. One can also calculate the entropy density s, and one finds

$$n_M/s \gtrsim 10^{-13}. \tag{3.3}$$

At these densities the rate of monopole-antimonopole annihilation can be shown to be insignificant.[11,15] Entropy is essentially conserved in this scenario, so the ratio n_M/s should be about the same today. If the mass of each monopole is about 10^{16} GeV, then the bound (3.3) implies that today

$$\Omega \equiv \rho/\rho_c \gtrsim 3 \times 10^{11}, \tag{3.4}$$

where $\rho_c = 3H^2/8\pi G$ is the critical mass density (which gives a precisely flat ($k = 0$) universe). Such a huge value of Ω is clearly impossible. It would imply, for example, that the current age of the universe would be $\lesssim$ 30,000 years. (I am told that an age of this order is considered acceptable in some circles--but I have checked with my friends, and they are all confident that such an age can be ruled out.) Thus, some mechanism must be found to suppress the production of magnetic monopoles.

The second problem of the standard scenario is known as the horizon problem; it was first pointed out by Rindler[16] in 1956. The observational basis for this problem is the uniformity of the cosmic background radiation, which is known to be isotropic in temperature to at least one part in 10^4. This fact is particularly difficult to understand when one considers the existence of the horizon length, the maximum distance that light could have travelled since the beginning of time. Consider two microwave antennae pointed in opposite directions. Each is receiving radiation which is believed to have been emitted (or "decoupled") when the hot plasma of the early universe converted to neutral atoms--when T was about 4000 °K at $t \approx 10^5$ yr. At the time of emission, these two sources were separated from each other by over 90 horizon lengths.[17] The problem is to understand how two regions over 90 horizon lengths apart came to be at the same temperature at the same time. Within the standard scenario this large scale homogeneity cannot be explained; rather, it must be assumed as an initial condition.

The third problem of the standard scenario was pointed out by Dicke and Peebles[18] in 1979, and is known as the flatness problem. The basis of the problem is the fact that today the ratio between the actual mass density and the critical mass density (i.e., $\Omega \equiv \rho/\rho_c$) is conservatively known to lie in the range

$$0.01 < \Omega < 10. \tag{3.5}$$

No one is surprised by how narrowly this range brackets $\Omega = 1$. However, in the context of the standard model, the value $\Omega = 1$ is an unstable equilibrium point. Thus, to be near $\Omega = 1$ today, the universe must have been very near to $\Omega = 1$ in the past. When T was 1 MeV (at $t \approx 1$ sec), Ω had to equal one to an accuracy of one part in 10^{15}. At the time of the GUT phase transition, when T was about 10^{14} GeV, Ω had to be equal to one to within one part in 10^{49}! In the standard model this precise fine-tuning must be assumed (without explanation) to be a property of the initial conditions.

The fourth problem is the difficulty of understanding the formation of structure in the universe. While the universe appears homogeneous when one averages over lengths of a few times 10^8 light years, on smaller scales there is a complicated clustering of matter into galaxies, clusters of galaxies, superclusters of clusters, etc. In order to account for the evolution of this structure, the standard scenario relies on the assumption of an ad hoc spectrum of inhomogeneities in the initial conditions. The fact that this spectrum is unexplained is a drawback in itself, but the situation becomes even more puzzling when the starting time is taken as early as $t \approx 10^{-35}$ sec. Gravitational instabilities[19] cause the inhomogeneities $\delta\rho/\rho$ to grow linearly with t, as long as the wavelength of the inhomogeneity exceeds the Hubble length $H^{-1} = 2t$. Thus, an early starting time requires peculiarly small initial inhomogeneities.

For example, a galactic scale inhomogeneity has a wavelength which implies linear growth until $t \approx 10^9$ sec; at that time one requires $\delta\rho/\rho \approx 10^{-4}$ to account for subsequent galactic evolution. Thus, at

$t = 10^{-35}$ sec one needs $\delta\rho/\rho \approx 10^{-48}$. For comparison, one might consider an ordinary hot gas in thermal equilibrium. Such a gas would have $\delta\rho/\rho \approx 1/\sqrt{N} \approx 10^{-39}$, where $N \approx 10^{78}$ is the number of particles (mostly radiation of effectively massless particles) which make up the protogalaxy. Thus, the standard scenario requires the assumption that the matter in the universe began in a peculiar state of extraordinary but not quite perfect uniformity. The necessity for this assumption is called the smoothness problem, and it was pointed out by Dicke and Peebles in the same paper[18] as the flatness problem.

IV. THE NEW INFLATIONARY UNIVERSE

The (original) inflationary universe scenario[2] was developed to solve the first three of the problems discussed in the previous section. The scenario contained the basic ingredients necessary to eliminate these problems, but unfortunately the scenario also contained one fatal flaw: the all-important phase transition occurred by a mechanism which produced gross inhomogeneities. This flaw is completely avoided in a variation known as the new inflationary universe, developed independently by Linde[3] and by Albrecht and Steinhardt.[4] This new model not only avoids the flaw of the older model, but it also provides a possible solution to all four of the problems described above.

In order for the new inflationary scenario to occur, the underlying particle theory must contain a scalar field ϕ which has the following properties:[20]

1) The potential energy function $V(\phi)$ must have a minimum at a value of ϕ not equal to zero.
2) $V(\phi)$ must be very flat in the vicinity of $\phi \approx 0$. The value $\phi = 0$ is usually assumed to be a local maximum of $V(\phi)$.
3) At high temperature T, the thermal equilibrium value of ϕ (i.e., the minimum of the finite temperature effective potential) should lie at $\phi = 0$.

A potential energy function of this general form is shown in Figure 1. The scalar field ϕ which drives the inflation was originally taken to be the Higgs field, but in Section V I will explain why it now appears that ϕ must be a gauge singlet.

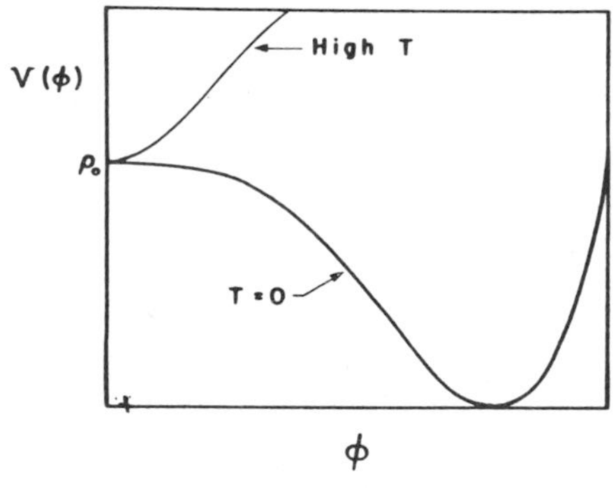

Figure 1: The form of the potential energy function required for the new inflationary universe model.

Thus, at high temperatures one expects the scalar field to have a mean value around zero. As the universe expands and cools, the thermal excitations disappear, and the scalar field finds itself in a state of essentially zero temperature, with $\phi \approx 0$. This state is called the false vacuum, and its peculiar properties are the driving force behind the inflationary model.

The false vacuum is clearly unstable, as ϕ will not remain forever at a local maximum of $V(\phi)$. However, if $V(\phi)$ is sufficiently flat, then the time which it takes for ϕ to move away from $\phi = 0$ can be very long compared to the time scale for the evolution of the early universe. Thus, for these purposes the false vacuum can be considered metastable.

Since the false vacuum has $\phi = 0$ and no other excitations, the energy density has a fixed value which is determined by the effective potential $V(\phi)$ (see Figure 1). For a typical grand unified theory, this value is given roughly by

$$\rho_0 \approx (10^{14}\ \mathrm{GeV})^4 \approx 10^{74}\ \mathrm{gm\text{-}cm}^{-3}. \tag{4.1}$$

The pressure p of the false vacuum is completely determined by the fact that the energy density has a fixed value ρ_0. To see this, think of an imaginary piston which is filled with false vacuum and surrounded by ordinary true vacuum, as illustrated in Figure 2. The true vacuum has zero energy density and zero pressure.[21] Suppose now that the piston is pulled out so that the volume of the chamber increases by ΔV. The energy of the system then increases by $\rho_0 \Delta V$, and therefore the agent which moved the piston must have done precisely this amount of work. Since the work done is $-p\Delta V$, it follows that

$$p = -\rho_0. \tag{4.2}$$

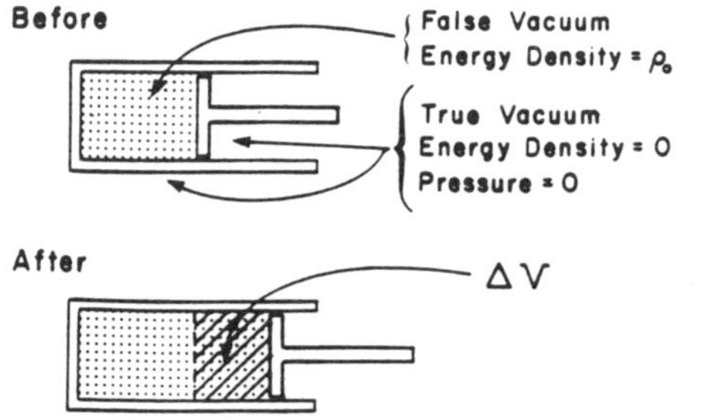

Figure 2: A thought experiment for the calculation of the pressure of the false vacuum.

The above result can also be demonstrated by noting that the false vacuum is a Lorentz-invariant state, and therefore must have a Lorentz-invariant energy-momentum tensor. The form is then restricted to

$$T_{\mu\nu} = -\rho_0 g_{\mu\nu}, \tag{4.3}$$

from which Eq. (4.2) follows.

When the energy-momentum tensor for the false vacuum is inserted into the Einstein field equations, one finds that its contribution is identical in form to that of a positive cosmological constant. To see the effect, recall the Robertson-Walker equation of motion

$$\ddot{R} = -\frac{4\pi}{3} G(\rho + 3p)R. \tag{4.4}$$

This equation says that both the pressure and the energy density contribute to the retardation of the universal expansion. For the false vacuum, however, $\rho + 3p < 0$, and it follows that $\ddot{R}$ is positive--the false vacuum creates a gravitational repulsion! It is this repulsion which will drive the colossal expansion of the inflationary scenario.

We can now go through the new inflationary scenario step by step. The starting point of a cosmological scenario is, unfortunately, still somewhat a matter of taste and philosophical prejudice. Some physicists find it plausible to assume that the universe began in some highly symmetrical state. Many others, however, consider it more likely that the universe began in a highly chaotic state, since the number of chaotic configurations is presumably much larger. One advantage of the inflationary scenario, from my point of view, is that it appears to allow a wide variety of starting configurations. It requires only that the initial universe is hot ($T > 10^{14}$ GeV) in at least some places, and that at least some of these regions are expanding rapidly enough so that they will cool to T_c before gravitational effects reverse the expansion.

In these hot regions, thermal equilibrium would imply $\langle\phi\rangle = 0$, where $\langle\phi\rangle$ denotes the expectation value of ϕ. (Actually, though, the universe has not had time at this point to thermalize.[23] Thus, I need to assume that there are some regions of high energy with $\langle\phi\rangle \approx 0$, and that some of these regions lose energy with ϕ being trapped in the false vacuum.) Such a region will cool to T_c, and nucleation rate calculations[24] indicate that it will continue to supercool well below T_c. The energy density ρ will approach ρ_0, the energy density of the false vacuum. To see what happens next, it is easiest to begin by assuming that the region is homogeneous, isotropic, and flat. (Later I will describe what happens when this assumption is dropped.) The region can then be described by the metric of Eq. (2.1), and the equation of motion (2.2) reduces to

$$\left(\frac{\dot{R}}{R}\right)^2 = \frac{8\pi}{3} G\rho_0. \tag{4.5}$$

The solution is given by

$$R(t) = e^{\chi t}, \tag{4.6}$$

where

$$\chi = \sqrt{\frac{8\pi}{3} G\rho_0}. \tag{4.7}$$

This exponential expansion is of course the hallmark of the inflationary model. (For our parameters, $\chi \approx 10^{10}$ GeV, and $\chi^{-1} \approx 10^{-34}$ sec.) Such a space is called a de Sitter space.[25]

Now let us consider what would happen if the initial region were not homogeneous and isotropic. In that case, one must examine the behavior of perturbations about the metric (2.1). These perturbations seem to be governed by a "cosmological no-hair theorem", which states that whenever the energy-momentum tensor is given by Eq. (4.3), then any locally measurable perturbation about the de Sitter metric is damped exponentially on the time scale of χ^{-1}. Any initial particle density is diluted to negligibility, and any initial distortion of the metric is stretched (i.e., redshifted) until is it no longer locally detectable. The theorem has been demonstrated[26] in the context of linearized perturbation theory, and it is conjectured to hold even for large perturbations.[27,28] Its validity in the nonperturbative regime has also been verified in certain exactly soluble models.[29] Thus, a smooth de Sitter metric arises naturally, without any need to fine-tune the initial conditions.

(The above paragraphs described the new inflationary universe with a hot beginning, but there are certainly other possibilities. Vilenkin[30] and Linde[31] have investigated speculative but attractive scenarios in which the universe tunnels from absolutely nothing,[32] entering directly into a de Sitter phase. In a similar spirit Hartle and Hawking[33] have proposed a unique wave function for the universe, incorporating dynamics which leads to an inflationary era. Linde[34] has also proposed the idea of chaotic inflation, in which inflation is driven by a scalar field which is initially chaotic but far from thermal equilibrium.)

As the space continues to supercool and exponentially expand, the energy density is fixed at ρ_0. Thus, the total energy (i.e., all energy other than gravitational) is increasing! If the inflationary model is right, this false vacuum energy is the source of essentially all the matter, energy, and entropy in the observed universe. (This seems to violate our naive notions of energy conservation, but we must remember that the gravitational field can exchange energy with the matter fields. A conserved total energy (matter plus gravitational) could be defined if the space were asymptotically Minkowskian, but there exists no energy-conservation law which holds for general spacetimes.)

After the region has undergone exponential expansion for some time, the phase transition must eventually take place. The scalar field is in an unstable configuration, perched at the top of the hill of the potential energy diagram (Figure 1). It will undergo fluctuations due to thermal and/or quantum effects. Some fluctuations begin to grow, and at some point these fluctuations become large enough so that their subsequent evolution can be described by the classical equations of motion.[35,36] I will use the term "coherence region" to denote a region within which the scalar field is approximately uniform. The coherence regions are irregular in shape,[4,35] and their initial size is typically of order χ^{-1}. Note that χ^{-1} is only about 10^{-11} proton diameters; the entire observed universe will evolve from a region of this size or smaller.

The Higgs field ϕ then "rolls" down the potential of Figure 1, obeying the classical equations of motion:

$$\ddot{\phi} + 3\,\frac{\dot{R}}{R}\,\dot{\phi} = -\,\frac{\partial V}{\partial \phi}\,. \tag{4.8}$$

If the initial fluctuation is small, then the flatness of the potential for $\phi \approx 0$ will imply that the rolling begins very slowly. Note that the second term on the left-hand-side of Eq. (4.8) is a damping term, helping to further slow down the speed of rolling. As long as $\phi \approx 0$, the energy density ρ remains about equal to ρ_0, and the exponential expansion continues. The expansion occurs on a time scale χ^{-1} which is short compared to the time scale of the rolling.

For the scenario to work, it is necessary for the length scale of homogeneity to be stretched from χ^{-1} to at least about 10 cm before the scalar field ϕ rolls off the plateau in Figure 1. This corresponds to an expansion factor of about 10^{25}, which requires about 58 time constants (χ^{-1}) of expansion. The expected duration of the expansion depends on the precise shape of the scalar field potential,[37,35] and models have been constructed[38] which yield much more than the minimally required amount of inflation.

When the ϕ field reaches the steep part of the potential, it falls quickly to the bottom and oscillates about the minimum. The time scale of this motion is a typical GUT time of $(10^{14}\ \mathrm{GeV})^{-1}$, which is very fast compared to the expansion rate. The scalar field oscillations are then quickly damped by the couplings to the other fields, and the energy is rapidly thermalized.[39] (The scalar field oscillations correspond to a coherent state of scalar particles; the damping is simply the decay into other species.) The release of this energy (which is just the latent heat of the phase transition) reheats the region back to a temperature of order 10^{14} GeV.

From here on the standard scenario ensues, including the production of a net baryon number. The length scale of homogeneity increases to $\gtrsim 10^{10}$ light-years by the time T falls to 2.7°K.

V. SOLUTIONS TO THE COSMOLOGICAL PROBLEMS

Let me now explain how the four problems of the standard cosmological scenario discussed in Section III are avoided in the inflationary scenario. First, let us consider the monopole problem. Recall that in the standard scenario, the tremendous excess of monopoles was produced by the disorder in the Higgs field (i.e., by the Kibble mechanism). In some versions of the new inflationary model, the Higgs field and the scalar field ϕ which drives the inflation are one and the same. In such cases the Higgs field is approximately uniform throughout the coherence region, which has been stretched by inflation to be much larger than the observed universe--the Kibble mechanism thereby produces less than one monopole in the observed universe. (Some monopoles would still be produced by thermal fluctuations after reheating, but this number would be negligible in the SU(5) model[40,12] and presumably in most other models as well. Monopole production by fluctuations in the Higgs field as it begins to roll down the potential energy hill can be significant[41] if particle physics is governed by the minimal SU(5) model, but in more realistic inflationary models one would expect these monopoles to be enormously diluted by the subsequent exponential expansion.) In many recent versions of the inflationary model, ϕ and the Higgs field are distinct. In such cases one must arrange for the Higgs field to acquire its nonzero expectation value either before or during the inflationary era, so that the monopole density is diluted by the inflation.

The horizon problem is clearly avoided in this scenario, since the entire observed universe evolves from a single coherence region. This region had a size of order χ^{-1} at the time when the fluctuation began to grow classically. The region was causally connected, and the scalar field is expected to have been homogeneous on this length scale. The exponential expansion causes this very small region of homogeneity to grow to be large enough to encompass the observed universe.

The flatness problem is avoided by the dynamics of the exponential expansion of the coherence region. As ϕ begins to roll very slowly down the potential, the evolution of the metric is governed by the energy density ρ_0. Assuming that the coherence region (or a small piece of it) can be approximated locally by a Robertson-Walker metric, then the scale factor evolves according to the standard equation:

$$\left(\frac{\dot{R}}{R}\right)^2 = \frac{8\pi}{3}G\rho_0 - \frac{k}{R^2} , \tag{5.1}$$

where $k = +1$, -1, or 0 depending on whether the region approximates a closed, open, or flat universe, respectively. (There could also be perturbations, but the cosmological no-hair theorem guarantees that they would die out quickly.) In this language, the flatness problem is the problem of understanding why the k/R^2 term on the right-hand-side is so extraordinarily small. But as the coherence region expands exponentially, the energy density ρ remains very nearly constant at ρ_0, while the k/R^2 term is suppressed by at least a factor of 10^{50}. This provides a "natural" explanation of why the value of the k/R^2 term immediately after the phase transition is smaller than that of the other terms by a factor of 10^{49} or more.

Except for a very narrow range of parameters, this suppression of the curvature term will vastly exceed that required by present observations. This leads to the prediction that the k/R^2 term of Eq. (5.1) should remain totally negligible until the present era, and even far into the future. In the absence of a cosmological constant, this implies that the value of Ω today is expected to be equal to one with a high degree of accuracy. However, if the cosmological constant Λ is nonzero,[42] the prediction is that

$$\Omega + \frac{\Lambda}{3H^2} = 1, \tag{5.2}$$

where $\Omega \equiv 8\pi G\rho_{matter}/3H^2$.

Finally, I come to the smoothness problem. Although this problem is only partially solved, I still consider it a major success of the new inflationary scenario. The problem was not even considered when the scenario was formulated, so it is impressive that the scenario offers a possible solution.

The evolution of density fluctuations in the new inflationary universe was discussed actively by a number of physicists[43] at the Nuffield Workshop on the Very Early Universe, held in the summer of 1982 at Cambridge University. The results can be described in a number of different ways, but I will summarize the approach developed by Pi and me.[44] As already discussed, the inflationary process smoothes out any inhomogeneities which may have been present in the initial conditions of the universe, leaving only zero-point quantum fluctuations. The problem, then, is to understand how these zero-point quantum fluctuations give rise to inhomogeneities in the mass density of the universe.

To understand this, consider again the time period during which the scalar field ϕ is just beginning to roll down the hill of the potential energy diagram. In Section IV I described the scalar field as being spatially homogeneous, but we are now interested in looking at the corrections to this approximation. The field can be decomposed as

$$\phi(\vec{x},t) = \phi_0(t) + \delta\phi(\vec{x},t), \tag{5.3}$$

where $\phi_0(t)$ is a classical solution to the equations of motion, and

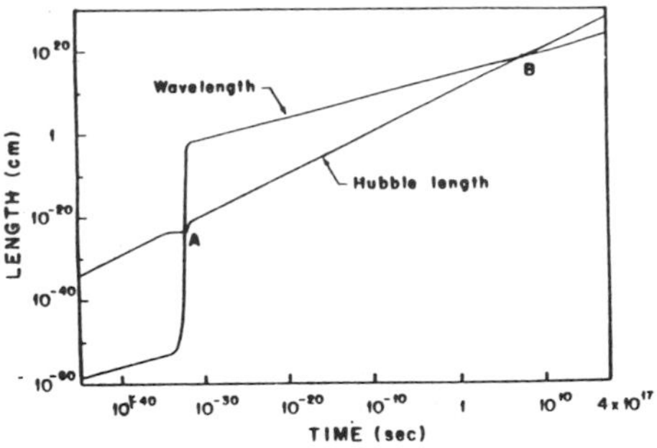

Figure 3: Wavelengths and the Hubble length as a function of time. The first curve shows the physical wavelength λ of a fluctuation with a typical galactic length scale. The second curve shows the evolution of the Hubble length H^{-1}. In standard cosmology these curves would cross only once, but in the inflationary model they cross at two points, denoted on the graph as A and B. (These curves show an expansion factor of about 10^{50} during the inflationary era, but the actual expansion factor depends sensitively on the form of the scalar field potential.)

$\delta\phi(\vec{x},t)$ represents the quantum fluctuations about this classical background.[45] Suppose that $\delta\phi(\vec{x},t)$ is Fourier transformed with respect to the comoving coordinate $\vec{x}$, so that we can talk about contributions with a definite coordinate wavelength λ_{coord}. Figure 3 illustrates how the physical wavelength $\lambda = R(t)\lambda_{coord}$ varies with time. The coordinate wavelength shown corresponds to the scale of a typical galaxy. For comparison, the value of the Hubble length H^{-1} is shown on the same graph. Note that for a fixed coordinate wavelength, the corresponding physical wavelength λ is equal to the Hubble length at two different times, denoted in the figure by A and B.

At time A, during the inflationary era, the wavelength $\lambda = H^{-1} = \chi^{-1}$. At this stage one can estimate the zero-point fluctuations in $\delta\phi$ by using quantum field theory. Since the potential energy curve is very flat for the regions of interest, it is reasonable to approximate the dynamics by that of a free quantum field theory. One then finds that[46]

$$(\delta\phi)_A \approx \chi. \tag{5.4}$$

The fluctuations can be viewed as a time delay (or advance) in the rolling of the scalar field, given by

$$\delta\tau = \frac{\delta\phi}{\dot{\phi}_0(t_A)}. \tag{5.5}$$

This time delay means that the energy density of the false vacuum will be released at slightly different times in different places, giving rise to inhomogeneities in the mass density of the universe. The effect can be calculated using linearized general relativity, but the result can also be guessed by dimensional analysis:

$$\left(\frac{\delta\rho}{\rho}\right)_B \approx \chi\delta\tau. \tag{5.6}$$

Combining with Eqs.(5.5), (5.4), and (4.7), one has

$$\left(\frac{\delta\rho}{\rho}\right)_B \approx \frac{\chi^2}{\dot{\phi}_0(t_A)} \approx \frac{G\rho_0}{\dot{\phi}_0(t_A)}. \tag{5.7}$$

The shape of the spectrum is fairly insensitive to the details of the scalar field potential, and can be extracted immediately from the above equation. Since $\phi_0(t)$ and its time derivative are slowly varying functions of t at the relevant times (i.e., it is a slow-rollover), one has

$$\left(\frac{\delta\rho}{\rho}\right)_B \approx \text{const} \quad (\text{independent of } \lambda). \tag{5.8}$$

This is the well-known Harrison-Zeldovich[47] scale-invariant spectrum, often advocated as a plausible spectrum to account for the observed properties of the universe.

The magnitude of $\delta\rho/\rho$, on the other hand, is very sensitive to the details of the scalar field potential. Figure 4 shows two possible shapes for the potential, illustrating the different possibilities. Curve (a) shows the potential which was originally proposed[3,4] for the new inflationary universe model: the Coleman-E. Weinberg potential[48] for the minimal SU(5) grand unified theory.[10] This potential has been found[43] to give fluctuations $(\delta\rho/\rho)_B \approx 10^2$, which is about six orders of magnitude too large. However, fluctuations can be suppressed if the potential looks more like curve (b). This potential has a smaller value of ρ_0, but more importantly it has a steadier slope. The potential curve (a) gives rise to large fluctuations because the scalar field evolves very slowly at first, accelerating rapidly near the end of its rollover. Since t_A is an early time, $\phi_0(t_A)$ is very small. The potential curve (b), however, leads to a rather steady value of $\dot{\phi}_0(t)$ (recall the viscous damping term in the evolution Eq. (4.8)), which can then be rather large.

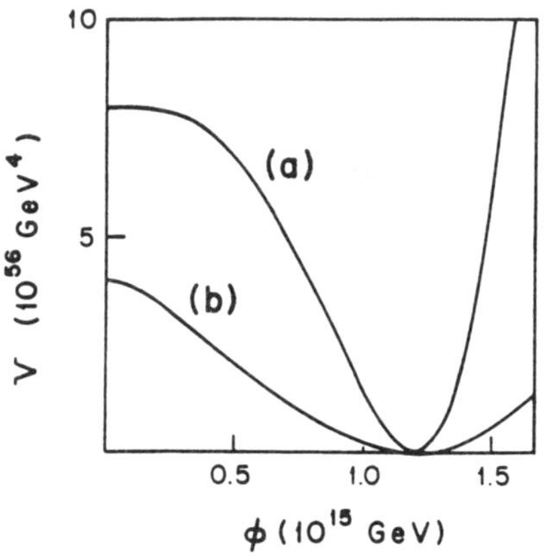

Figure 4: Forms of the scalar field potential energy function. The Coleman-Weinberg potential for the minimal SU(5) model is shown as curve (a)--it leads to density fluctuations which are far too large. The potential energy function described by curve (b) produces much smaller density fluctuations.

Over the past year there has been much effort[38] in constructing particle theories which give rise to the desired density fluctuations. Since gauge couplings tend to generate effective potentials that resemble curve (a) of Figure 4, interest has focussed on models in which ϕ is a gauge singlet, with scalar couplings which are unusually small $(O(10^{-10}-10^{-12}))$. A variety of workable models has been constructed, but it is certainly fair to say that none of them has yet been generally accepted as a correct description of the laws of physics. At least we have a demonstration that models of this type exist. The issue of density fluctuations and particle-physics model-building remains an active area of research.

VI. CONCLUSION

In conclusion, I want to say that the basic idea of inflation--the idea that the universe went through a period during which it expanded exponentially while trapped in a false vacuum--appears to me to be probably correct. It is a very simple and natural idea in the context of spontaneously broken gauge theories, and it seems to solve some very fundamental cosmological problems. On the other hand, we clearly do not yet have the details straight. In order to understand the density fluctuations, we must at the same time understand the details of particle physics at GUT energy scales. We are presumably some distance from that goal. Thus, I expect that the "Inner Space/Outer Space Connection" will remain an exciting area of research for some time to come.

I want to acknowledge the support of the U.S. Department of Energy (DOE) under contract DE-AC02-76ER03069, the National Aeronautics and Space Administration (NASA) under grant NAGW-553, and an Alfred P. Sloan Fellowship.

REFERENCES

[1]H. Georgi and S. L. Glashow, Phys. Rev. Lett. **32**, 438 (1974); H. Georgi, H. R. Quinn, and S. Weinberg, Phys. Rev. Lett. **33**, 451 (1974). For a review, see P. Langacker, Phys. Rep. **72C**, 185 (1981).

[2]A. H. Guth, Phys. Rev. **D23**, 347 (1981).

[3]A. D. Linde, Phys. Lett. **108B**, 389 (1982).

[4]A. Albrecht and P. J. Steinhardt, Phys. Rev. Lett. **48**, 1220 (1982).

[5]A. H. Guth and E. J. Weinberg, Nucl. Phys. **B212**, 321 (1983).

[6]S. W. Hawking, I. G. Moss, and J. M. Stewart, Phys. Rev. **D26**, 2681 (1982).

[7]For the reader who would like a more detailed review of this subject, let me point out that Ref. 8 contains reviews by both A. D. Linde and by me.

[8]Proceedings of the Nuffield Workshop on The Very Early Universe, eds. G. W. Gibbons, S. W. Hawking, and S.T.C. Siklos (Cambridge University Press, Cambridge, England, 1983).

[9]For a general background in cosmology, see S. Weinberg, Gravitation and Cosmology (Wiley, New York, 1972). At a less technical level, see J. Silk, The Big Bang (W. H. Freeman and Co., San Francisco, 1980) or S. Weinberg, The First Three Minutes (Bantam Books, New York, 1977).

[10]H. Georgi and S. L. Glashow, Ref. 1.

[11]J. P. Preskill, Phys. Rev. Lett. **43**, 1365 (1979); Ya. B. Zeldovich and M. Y. Khlopov, Phys. Lett. **79B**, 239 (1978). For a more detailed review of this topic, see Ref. 12 and the article by G. Lazarides in the same volume.

[12]A. H. Guth, in Magnetic Monopoles (Proceedings of a NATO Advanced Study Institute on Magnetic Monopoles, Wingspread, Wisconsin, 1982), eds. R. A. Carrigan and W. P. Trower (Plenum, New York, 1983).

[13]T.W.B. Kibble, J. Phys. **A9**, 1387 (1976).

[14]J. P. Preskill (private communication); A. H. Guth and S.-H. Tye, Phys. Rev. Lett. **44**, 631, 963(E) (1980); M. B. Einhorn, D. L. Stein and D. Toussaint, Phys. Rev. **D21**, 3295 (1980).

[15]T. Goldman, E. W. Kolb, and Toussaint, Phys. Rev. **D23**, 867 (1981).

[16]W. Rindler, Mon. Not. R. Astron. Soc. **116**, 663 (1956). See also S. Weinberg, Gravitation and Cosmology (Ref. 9), pp. 489-490 and 525-526.

[17]A. H. Guth, in Asymptotic Realms of Physics: Essays in Honor of Francis E. Low, eds. A. H. Guth, K. Huang, and R. L. Jaffe (MIT Press, Cambridge, Massachusetts, 1983).

[18]R. H. Dicke and P.J.E. Peebles, in General Relativity: An Einstein Centenary Survey, eds. S. W. Hawking and W. Israel (Cambridge University Press, Cambridge, England, 1979).

[19]See, for example, D. W. Olson, Phys. Rev. **D14**, 327 (1976). Other relevant references include J. M. Bardeen, Phys. Rev. **D22**, 1882 (1980); W. H. Press and E. T. Vishniac, Ap. J. **239**, 1 (1980).

[20]The properties required of the scalar field are described in more detail by P. J. Steinhardt and M. S. Turner, "A Prescription for Successful New Inflation," Fermilab preprint FERMILAB-PUB-84-019-AST (1984).

[21]The statement that the energy density of the vacuum is zero is equivalent to the statement that the cosmological constant (of the Einstein field equations) is zero. It is known (Ref. 22) that the energy density of the vacuum is extremely small ($\lesssim (3 \times 10^{-3} \text{eV})^4$), but there is no generally accepted theoretical explanation of this remarkable fact.

[22]C. W. Misner, K. S. Thorne, and J. A. Wheeler, Gravitation (W. H. Freeman and Co., San Francisco, 1973), p. 411.

[23]G. Steigman, in Unification of the Fundamental Particle Interactions II (Proceedings of the Europhysics Study Conference, Erice, Italy, Oct. 6-14, 1981), eds. J. Ellis and S. Ferrara (Plenum, New York, 1983).

[24]M. Sher, Phys. Rev. **D24**, 1699 (1981).

[25]The properties of a de Sitter space are well described in S. W. Hawking and G.F.R. Ellis, The Large Scale Structure of Space-Time (Cambridge University Press, Cambridge, England, 1973).

[26]J. A. Frieman and C. M. Will, Ap. J. **259**, 437 (1982); J. D. Barrow, in Ref. 8; W. Boucher and G. W. Gibbons, in Ref. 8; P. Ginsparg and M. J. Perry, Nucl. Phys. **B222**, 245 (1983).

[27]G. W. Gibbons and S. W. Hawking, Phys. Rev. **D15**, 2738 (1977).

[28]S. W. Hawking and I. G. Moss, Phys. Lett. **110B**, 35 (1982).

[29]R. M. Wald, Phys. Rev. **D28**, 2118 (1983).

[30]A. Vilenkin, Phys. Lett. **117B**, 25 (1982); Phys. Rev. **D27**, 2848 (1983); "The Quantum Origin of the Universe," Tufts preprint TUTP-84-5 (1984).

[31]A. D. Linde, Nuovo Cim. Lett. **39**, 401 (1984); ZhETF **87**, 369 (1984).

[32]For an early reference suggesting this general idea, see E. P. Tryon, Nature **246**, 396 (1973).

[33]J. B. Hartle and S. W. Hawking, Phys. Rev. **D28**, 2960 (1983); see also I. G. Moss and W. A. Wright, Phys. Rev. **D29**, 1067 (1984); S. W. Hawking and J. C. Luttrell, "The Isotropy of the Universe," Cambridge Univ. preprint DAMTP-84-0479 (1984) and Phys. Lett. **143B**, 83 (1984).

[34]A. D. Linde, Pis'ma Zh. Eksp. Teor. Fiz. **38**, 149 (1983) [JETP Lett. **38**, 176 (1983)]; Phys. Lett. **129B**, 177 (1983); A. B. Goncharov and A. D. Linde, Phys. Lett. **139B**, 27 (1984).

[35]A. H. Guth and S.-Y. Pi, "The Behavior of the Higgs Field in the New Inflationary Universe," to be published in this volume. A more detailed manuscript is in preparation.

[36]R. H. Brandenberger, "Quantum Fluctuations as the Source of Classical Gravitational Perturbations in Inflationary Universe," Santa Barbara (ITP) preprint (1984) and Nucl. Phys. **B245**, 328 (1984).

[37]A. Vilenkin and L. H. Ford, Phys. Rev. **D26**, 1231 (1982); A. Vilenkin, Nucl. Phys. **B226**, 504 (1983); A. D. Linde, Phys. Lett. **116B**, 335 (1982).

[38]A. Albrecht, S. Dimopoulos, W. Fischler, E. W. Kolb, S. Raby, and P. J. Steinhardt, Nucl. Phys. **B229**, 528 (1983); J. Ellis, D. V. Nanopoulos, K. A. Olive, and K. Tamvakis, Nucl. Phys. **B221**, 524 (1983); D. V. Nanopoulos, K. A. Olive, M. Srednicki, and K. Tamvakis, Phys. Lett. **123B**, 41 (1983); D. V. Nanopoulos, K. A. Olive, and M. Srednicki, Phys. Lett. **127B**, 30 (1983); G. B. Gelmini, D. V. Nanopoulos, and K. A. Olive, Phys. Lett. **131B**, 53 (1983); D. V. Nanopoulos and M. Srednicki, Phys. Lett. **133B**, 287 (1983); B. A. Ovrut and P. J. Steinhardt, Phys. Lett. **133B**, 161 (1983), Phys. Rev. Lett. **53**, 732 (1984), Phys. Lett. **147B**, 263 (1984), and Rockefeller preprint RU84/B/88 (1984); B. A. Ovrut and S. Raby, Phys. Lett. **134B**, 51 (1984); Q. Shafi and A. Vilenkin, Phys. Rev. Lett. **52**, 691 (1984); P. Q. Hung, Phys. Rev. **D30**, 1637 (1984) and U. Virginia preprint 84-0006 (1984); S. Gupta and H. R. Quinn, Phys. Rev. **D29**, 2791 (1984) and SLAC preprint 3269 (1983); R. Holman, P. Ramond, and G. G. Ross, Phys. Lett. **137B**, 343 (1984); S.-Y. Pi, Phys. Rev. Lett. **52**, 1725 (1984); Q. Shafi and F. W. Stecker, Phys. Rev. Lett. **53**, 1292 (1984) and Bartol preprint BA-84-17 (1984); (See also the contributions in this volume by R. Holman, M. Srednicki, F. W. Stecker, and P. J. Steinhardt.)
[39]A. Albrecht, P. J. Steinhardt, M. S. Turner, and F. Wilczek, Phys. Rev. Lett. **48**, 1437 (1982); L. F. Abbott, E. Farhi, and M. B. Wise, Phys. Lett. **117B**, 29 (1982); A. D. Dolgov and A. D. Linde, Phys. Lett. **116B**, 329 (1982).
[40]G. Lazarides, Q. Shafi, and W. P. Trower, Phys. Rev. Lett. **49**, 1756 (1982).
[41]I. G. Moss, Phys. Lett. **128B**, 385 (1983) and Nucl. Phys. **B238**, 436 (1984).
[42]P.J.E. Peebles, "Tests of Cosmological Models Constrained by Inflation," Princeton University preprint (1983); P. H. Frampton and G. B. Lipton, "Age and Flatness of Inflationary Universe with Cosmological Constant," North Carolina Univ. preprint IFP-203-UNC (1983).
[43]A. A. Starobinsky, Phys. Lett. **117B**, 175 (1982); A. H. Guth and S.-Y. Pi, Phys. Rev. Lett. **49**, 1110 (1982); S. W. Hawking, Phys. Lett. **115B**, 295 (1982); J. M. Bardeen, P. J. Steinhardt, and M. S. Turner, Phys. Rev. **D28**, 679 (1983). See also the contributions of these authors in Ref. 8.
[44]A. H. Guth and S.-Y. Pi, Ref. 43.
[45]For a more detailed discussion of these ideas, see Ref. 35.
[46]For a precise definition of $\delta\phi$, see Ref. 44.
[47]E. R. Harrison, Phys. Rev. **D1**, 2726 (1970); Ya. B. Zeldovich, Mon. Not. R. Astr. Soc. **160**, 1P (1972).
[48]S. Coleman and E. J. Weinberg, Phys. Rev. **D7**, 1888 (1973).

Supersymmetric Inflation: Recent Progress

Burt A. Ovrut and Paul J. Steinhardt

ABSTRACT

Recent progress in obtaining a successful inflationary universe scenario based on locally supersymmetric field theories is reviewed.

INTRODUCTION

The new inflationary universe scenario[1-3] is, in principle, a simple and powerful approach to resolving a large number of fundamental cosmological problems. However, in order for the scenario to be considered a complete theory, one critical question remains to be answered: What is the physics responsible for the phase transition that triggers the exponential expansion (inflation) of the universe? One possibility that we and several other groups have been pursuing is that the physics responsible for the phase transition involves (local) supersymmetry. The goal of this paper is to review the present status of "Supersymmetric Inflation", particularly emphasizing some very exciting results that we have recently obtained.

Supersymmetric inflation is worth considering for several reasons: Firstly, inflation triggered by the spontaneous breaking in grand unified theories (GUTs) cannot be successful. GUT inflation leads to an unacceptably large amplitude for energy density fluctuations following the inflationary epoch.[4] Inflation automatically leads to a scale invariant spectrum of energy density fluctuations. A density fluctuation amplitude ($\delta\rho/\rho$) is required that is large enough to account for galaxy formation but small enough to be consistent with the observed isotropy of the cosmic microwave background ($\lesssim 10^{-4}$). In order to obtain an acceptably small amplitude, the order parameter fields for the phase transition (e.g. the Higgs field in a GUT)--henceforth referred to as inflation fields -- must be very weakly coupled to themselves and other fields.[4] The GUT gauge coupling that couples Higgs fields to gauge mesons is ten orders of magnitude too big.[4,5] (Recently, the suggestion has been made that "extra" scalar gauge singlet fields be added to a GUT that couple very weakly to themselves and other GUT fields, and which lead to an independent phase transition that can trigger inflation.[6] This suggestion is consistent with all constraints, but seems to us to be somewhat ad hoc, since the extra fields have nothing to do with ordinary GUTs.)

Supersymmetric Inflation is certainly worthy of consideration by those (such as ourselves) who believe that supersymmetry may play an important role in high energy physics. Even though supersymmetry and inflation appear to be quite independent of one another, they both depend critically on the properties of elementary particles at very high temperatures (energies). A careful examination should be made to determine whether or not supersymmetry is ultimately compatible with inflationary cosmology.

By far the most significant reason for considering supersymmetric inflation is that locally supersymmetric theories appear, at least naively, to produce the "slow rollover" phase transitions necessary for inflation with little or no finetuning of parameters. The basic reason is that all couplings in a supersymmetric model are scaled by some power of (μ/M), where μ is the supersymmetry scale and $M = M_P/\sqrt{8\pi} = 2.4 \times 10^{18}$ GeV for M_P = Planck mass; the power of (μ/M) varies from term to term in a prescribed way. Inflation requires all the couplings of the scalar fields to be weak and sets constraints on the ratios of various couplings. In a supersymmetric model, (nearly) all these constraints can be met by setting just one parameter - the supersymmetric scale, μ. All dimensionless parameters can have values of O(1).

The appearance that supersymmetry and inflation can be combined easily and naturally has caused many groups, including ourselves, to investigate Supersymmetric Inflation. However, as we shall point out in this review, appearances can be deceiving. A little more than a year ago, we made a more systematic analysis of all the constraints required for inflationary cosmology and showed that all Supersymmetric Inflation models previously proposed fail to satisfy those constraints.[7]

In fact, except for some extreme possibilities, we further showed that only a new class of special supersymmetric models appears capable of satisfying all the constraints for inflationary cosmology.[7] This conclusion appears depressing, since some may conclude that inflation and supersymmetry really are incompatible. However, we have recently shown that our new class of supersymmetric models has a number of properties that make it quite attractive. We can demonstrate a subclass of such models that: (1) satisfies all the constraints of inflationary cosmology; (2) spontaneously breaks supersymmetry;[7] (3) sets a supersymmetry breaking scale that is consistent with the mass hierarchy;[8] (4) produces baryon asymmetry consistent with astrophysical observations;[9] and (5) avoids the overproduction of entropy caused by the domination of the energy density by gravitinos and/or coherent field oscillations (a problem which occurs in nearly all supersymmetry models).[9] Such models encourage us to hope that Supersymmetric Inflation may yet be the best approach to achieving inflationary cosmology, as well as resolving various problems in particle phenomenology.

The review is divided into three sections. In Section I we discuss how the constraints of inflationary cosmology rule out nearly all but a special class of supersymmetric models. All models in this class spontaneously break supersymmetry and, thereby, set a mass hierarchy. In Section II we show how to maintain all the constraints for inflationary cosmology and obtain the correct mass hierarchy. In Section III, we discuss how the models in Section II can produce baryon asymmetry and can avoid the overproduction of entropy from gravitinos and coherent field oscillations.

I. WHAT SUPERSYMMETRY MODELS CAN MEET THE CONSTRAINTS OF INFLATIONARY COSMOLOGY?[7]

The efforts at Supersymmetric Inflation prior to our analysis can be divided into two categories. In the first category, only a special subclass of supersymmetric potentials was studied - the O'Raifeartaigh models.[10] This class of supersymmetric models was studied for several reasons. First, O'Raifeartaigh models[11] appeared, at the time, to be attractive models for phenomenological purposes; examples of such models are geometric (or "inverse") hierarchy models[12] and Polonyi potentials.[13] Second, the models automatically provide phase transitions in which the effective potential is rather flat without any fine-tuning of parameters. Let ϕ be the expectation value of the "inflation field", the field which serves as the order parameter for the phase transition. Let μ be the fundamental mass scale in the theory. Then for ϕ greater than μ the globally supersymmetric effective potential decreases logarithmically with increasing ϕ for an O'Raifeartaigh model without any fine-tuning of parameters. Thus, the potential becomes flatter and flatter with increasing ϕ, which is ideal for inflation. In the investigation of inflationary O'Raifeartaigh potentials, it was presumed that for large ϕ (near the Planck scale) gravitational effects would cause the potential to steepen, forming a minimum near $\phi = M_p$ which could be adjusted to have zero cosmological constant. As ϕ evolves along the logarithmic portion of the potential, the universe inflates and then ϕ oscillates around the minimum of the potential.

As originally advertised, the model was described as inflating but not reheating. The problem is that the inflation field, ϕ, is necessarily greatly decoupled from ordinary matter and radiation in the O'Raifeartaigh models.[10] (The coupling is suppressed by may powers of $1/M_p$). Describing the model as not reheating is a bit of an exaggeration, though, since as long as there is some non-zero coupling there will eventually be some time when the energy density once again becomes dominated by ordinary matter and radiation. For $\mu \simeq 0\ (10^{12}$ GeV) as appears in the Dimopoulos-Raby geometric hierarchy model,[12] the reheating temperature is of 0(10 MeV). This is much too low to generate a baryon asymmetry after the transition.[10] (Even in this very depressing case the reheating temperature is sufficient to recover all the other successful predictions of the hot big bang model, which involve events at temperatures less than 1 MeV or so.) The reheating temperature is very sensitive to μ; by raising μ to 10^{16} GeV or so the reheating temperature can be raised to 10^{10} GeV, marginally sufficient for baryon asymmetry generation. At the time this approach to Supersymmetric Inflation was proposed, though, raising μ to this large value was not pheonomenologically attractive. Therefore, the O'Raifeartaigh approach to inflation lay on shaky foundations even before we began our analysis.

A second approach to supersymmetric inflation has been to introduce a new superfield into the theory called an "inflaton" whose sole purpose is to play the role of the inflation field.[14] In the superpotential, the inflaton couples only to itself so that, in the absence of gravity, the effective potential for the inflaton depends only on its self-couplings. The inflaton plays no role in ordinary phenomenology. The self-couplings in the superpotential can be adjusted term by term so that the effective potential has the correct form for a successful inflationary model (or so it was thought). Reheating is possible after the inflation because gravity induces some coupling between the inflaton field and the superpotential for ordinary particles. This approach is not "pretty", since an extra field must be introduced with specially adjusted couplings, but it could at least serve as an example to show that all the constraints for successful inflationary cosmology can be met.

Surprisingly, our subsequent investigation has shown that neither of these previous approaches to Supersymmetric Inflation is workable (as posed).[7] The fact that the first approach does not work may not seem surprising since it already seemed that reheating is problematic for the O'Raifeartaigh models. The fact that the inflation approach is unworkable is very surprising, since it appears that one can introduce through the superpotential even an infinite number of adjustable parameters, if necessary, to obtain a successful inflationary model.

The key feature of our investigation of locally supersymmetric potentials is that all the constraints necessary for a successful inflationary cosmology have been systematically and simultaneously applied to a wide class of supersymmetric models. In the past, a few of the necessary constraints were considered and it was assumed that the others could be satisfied by adjusting extra parameters. However, if there is one lesson that we have surely learned, it is that supersymmetry is very subtle. Constraints that appear to have nothing to do with one another can be subtlely and non-linearly related. In many cases where it was known that two constraints could be satisfied individually, we discovered that they could not be satisfied simultaneously by the same choice of parameters.

Consider a completely general superpotential, $W(\phi_i)$ (where i is an index that labels different chiral superfields). The prescription for the zero-temperature, locally supersymmetric scalar potential is given by:[15]

$$V = e^{\sum_i |\phi_i|^2/M^2} \left\{ \sum_i |D_{\phi_i} W^2| - \frac{3}{M^2} |W^2| \right\} \tag{1.1}$$

where

$$D_{\phi_i} W = \frac{\partial W}{\partial \phi_i} + \frac{\phi_i^+}{M^2} W \tag{1.2}$$

is the i-th Kahler derivative. We consider here only N=1 local supersymmetry with a flat Kahler manifold. The Kahler derivative is a very significant physical quantity since its expectation value is zero for a state that is supersymmetric and non-zero for a state that is supersymmetry breaking. It can be thought of as the order parameter for supersymmetry.

We consider superpotentials which include even an infinite number of terms (and adjustable parameters) in our investigation. If the theory were a globally supersymmetric model, a W with terms more than cubic in the fields would generate non-renormalizable terms in the potential and would be considered unacceptable. However, in a locally supersymmetric potential one always obtains an infinite number of non-renormalizable terms for any non-trivial W because of the exponential pre-factor in Eq. (1.1). We cannot see any reason not to consider terms of higher order than cubic. The non-renormalizable terms can be interpreted as being due to the couplings of matter to gravity; N=1 local supersymmetry models are usually considered effective low energy theories in which the Planck scale is introduced as a cutoff.

When gravity is turned off or the Planck (cutoff) scale is taken to infinity, all non-renormalizable terms should then disappear. Such is the case for the theories we will consider.

(To convince skeptics that studying W's with infinite numbers of terms is not ludicrous, consider the locally supersymmetric potential generated according to Eq. (1.1) from a W which has at most cubic terms, i.e., which generates a renormalizable potential in global supersymmetry. Now shift all the fields by some finite amount; if the theory was acceptable before, it should remain acceptable after such a trivial change. Note that the exponent in the exponential prefactor in

Eq. (1.1) now includes linear and constant terms in the shifted field. The new potential can be reorganized so that the exponent is now only quadratic in the shifted field, just as in Eq. (1.1), but there is a new superpotential W'. The new W' absorbs the terms generated by shifting the fields and it contains an infinite number of terms. In other words, one can show that a potential generated from a W with a finite number of terms can be equivalent to a potential generated from a W with an infinite number of terms. There is nothing sacred about a W with a finite number of terms in local supersymmetry.)

At high temperatures, the potential is modified to:

$$V \to V + \Delta V \tag{1.3}$$

were V is the T=0 potential; roughly,[16]

$$\Delta V = \frac{\pi^2}{45} k^4 T^4 + \frac{1}{24} k^2 T^2 \mathrm{Tr}\left[\frac{\partial^2 V}{\partial\phi_i \partial\phi_j{}^*}\right] + \ldots, \tag{1.4}$$

where k = Boltzmann's constant. A more accurate expression has been recently obtained by Olive and Srednicki.[17] We will be interested in the second, ϕ-dependent term.

To begin our systematic analysis, let us first consider W's which are functions of a single chiral superfield, ϕ. The chiral superfield contains other components, but we will only be interested in the scalar components and the scalar potential implied by W. Because the scalar field is very weakly coupled to itself and all other components, the corrections to the tree-level couplings due to interactions can be ignored.

The most general superpotential can be written in the form:

$$W = \mu^2 M\left(\beta + \left(\frac{\phi}{M}\right) + \frac{\Delta}{2}\left(\frac{\phi}{M}\right)^2 + \frac{\Delta\lambda}{2}\left(\frac{\phi}{M}\right)^3 + \Delta\sum_{n=4}^{\infty}\frac{a_n}{n}\left(\frac{\phi}{M}\right)^n\right) . \tag{1.5}$$

All the parameters in W are chosen to be real and we will consider only phase transitions in which ϕ slow rolls from ϕ=0 along the positive real ϕ-axis. (Very little generality is lost from these assumptions, as we will comment later.) ϕ can always be shifted so ϕ=0 corresponds to the beginning of the phase transition.) The potential, V, generated from Eq. (1.1) can be expanded around ϕ=0 (which corresponds to the high temperature phase and the inflationary false vacuum).

$$V = \mu \sum_{n=0}^{\infty} c_n \left(\frac{\phi}{M}\right)^n . \tag{1.6}$$

Denote by σ the minimum of V corresponding to the true vacuum state where the transition ends ($\sigma > 0$). We will now consider the constraints on the coefficients in the expansion of V and what they imply for the coefficients of W.

The constraints for a successful inflationary universe scenario that we have considered fall into three categories:

1. Slow Rollover Constraints: these are the constraints on V that insure: (a) that the potential be flat enough near $\phi=0$ to obtain sufficient inflation to solve the horizon, flatness, monopole and domain wall problems;[1-3] and (b) that the potential be flat over the correct range of ϕ so that the energy density perturbation amplitude generated by the inflation is $\delta\rho/\rho \lesssim 10^{-4}$ on all observable scales (which is what cosmologists require to explain galaxy formation and yet to be consistent with the observed isotropy of the microwave background).[4] The constraints have been codified into a precise prescription for a completely general $V(\phi)$.[18] Almost all these constraints can be met simply in Eq. (1.6) by setting μ to be of $O(10^{-(3-4)}M)$ (and all the c_n's can be O(1)). Notice that the coefficients of ϕ are suppressed by powers of (μ/M); thus, couplings can be made very small by this one choice of μ without any continual fine-tuning of dimensionless parameters. This is the property that makes Supersymmetric Inflation attractive. The only additional condition necessary to satisfy the slow rollover constraints are

$$c_1 \lesssim O(10^{-6}) \ , \quad c_2 \lesssim O(10^{-3}) \ , \tag{1.7}$$

where all the other c_n's have been taken to be of O(1). Tuning c_1 to be zero eliminates the linear term in ϕ from the potential, which may or may not be considered a tuning per se. The tuning of c_2 cannot be passed off so easily, but the constraint is not all that horrible either. We are satisfying more constraints than was possible in the Coleman-Weinberg new inflationary models[2,3] with less fine-tuning.

2. Cosmological Constant Constraint: This constraint requires $V(\sigma)$ be zero. (the constraint is necessary because the cosmological constant is measured to be nearly zero in the present phase of our universe which is supposed to correspond to the true vacuum $\phi=\sigma$.) In GUTs this is a trivial constraint that can be satisfied by simply adding an overall constant to V and adjusting it so that V is equal to zero at the state corresponding to our present universe. In supersymmetric models the constraint is much more subtle, because one cannot change just the constant term in V without explicitly breaking supersymmetry (which destroys the whole purpose of considering supersymmetric models in the first place). A means of setting the cosmological constant to zero is to add an overall constant to W and use it to adjust the cosmological constant. However, from Eq. (1.1) we see that adding a constant term to W not only changes the constant term in V, c_0, but also changes an infinite number of other terms in the expansion of V (e.g. because of the exponential prefactor). Setting the cosmological constant to zero can interfere with attempts to make a flat potential. Thus, supersymmetry connects two apparently unrelated sets of constraints: a constraint on the potential near $\phi=0$ (the high temperature phase) and a constraint on the potential at large $\phi \simeq \sigma$ (our present low temperature phase). In terms of V, the cosmological constant constraint requires fine-tuning, no doubt, but the same fine-tuning as required in nearly any model to explain the observed value of the cosmological constant.

3. Thermal Constraint: From the constraints that have already been

applied, it has been insured that the effective potential has the right shape between $\phi=0$ and $\phi=\sigma$ for a successful inflationary universe scenario. However, one must explain how the universe entered the false vacuum $\phi=0$ phase in the first place. The conventional explanation has been that at high temperatures the $\phi=0$ phase might be favored and then the universe supercools in that phase to T=0 as inflation commences. If this is the case, one must check that the favored state at high T is $\phi=0$ so that a transition will occur in which ϕ rolls across the flat portion of the potential that has been designed according to the slow rollover constraints. If ϕ near σ is favored at high temperatures (which is possible, especially in cases of scalar fields coupled only to themselves), then as the temperature decreases ϕ will never roll over the flat region of the potential. There will be no inflation even though the slow rollover constraints have been satisfied. Using Eq. (1.4), we find that for a completely general W:

$$\Delta V = \frac{k^2 T^2}{12} \mu^4 [\Delta(1+\lambda)-\beta](\phi+\phi^+)+.... \tag{1.8}$$

From the leading term in ΔV we see that $\phi<0$ is favored if:

$$\Delta(1+\lambda)-\beta\geq 0 \quad . \tag{1.9}$$

Although the constraint is not terribly stringent by itself, we shall see that, added to the other constraints, the thermal constraint is important in eliminating many possible models. (Olive and Srednicki[17] have pointed out that Eq. (1.8) is not quite correct for the supersymmetric models we have studied, but their more precise expression leads to the same constraint, Eq. (1.9).)

We hope that the important role that the thermal constraint plays in our analysis leads to a re-evaluation of the issue of how the inflationary phase transition begins. There is a potentially dangerous problem which may not only affect the analysis of Supersymmetric Inflation, but may also threaten the entire inflationary universe program. The problem is that the usual assumption that the universe is driven to the false vacuum phase by high temperature effects is not tenable. Given the very weak couplings required for the inflation fields in order to obtain small amplitude density fluctuations, it is straightforward to show that the inflation fields cannot have been in thermal equilibrium before the inflationary transition began. It was previously shown that interaction rates for GUT fields (with couplings of order the gauge coupling constant) are so slow that they could not have reached thermal equilibrium until a temperature of $0(10^{15}$ GeV)[19].

For inflation the couplings have to be nearly ten orders of magnitude smaller. The inflation fields could not have reached thermal equilibrium before $T=0(10^6$ GeV). Yet, the critical temperature for the Supersymmetric Inflation transition should be $0(\mu) = 0(10^{14}$ GeV). Although the detailed numerics may change somewhat, the same conflict should occur in any inflationary model, supersymmetric or not, because of the requisite weak couplings.

What is required is a new approach to beginning the inflationary phase transition. A first guess might be to suppose that when our observable universe is contained within a region of less than Planck radius, gravitational interactions are so strong that they thermalize the particle distribution. As the universe expands beyond the Planck radius, the gravitational interactions weaken and the particles fall out of equilibrium (until their non-gravitational interactions would bring them into equilibrium again, but not until after the inflationary transition). The distribution of particles would nevertheless be nearly thermal, and perhaps the expression for the high temperature potential would still be a good approximation.

Unfortunately, it does not make any sense to even consider the notion of particles being in thermal equilibrium when the universe is within a Plack radius.[20] The reason is that there is less than one particle per causal horizon during that epoch, so the notion of thermal equilibrium is hopeless. (The particle number increases as T^3 at high temperatures, but the causal horizon volume decreases as T^6; the number of particles within the horizon becomes less than unity at a temperature slightly below the Planck scale, depending on the number of particle species.) To assume that the strong gravitational fields can act over distances greater than a causal horizon distance to thermalize the particles would defeat one of the major purposes of inflation.

Four solutions to the dilemma appear possible, but none has been made rigorous: (a) Fluctuating field approach[21]: although almost no particles lie within a causal horizon, the scalar field is continuously defined everywhere. It might be reasonable to suppose that, due to gravitational interactions, say, the field is uncorrelated over distances greater than a causal horizon distance as the universe expands beyond the Planck radius. Depending upon the signs of the couplings of the fields to themselves and others, it may be energetically favorable for the fields to relax to a false vacuum state within the causal horizon because this may be the state of least gradient field energy even though it is a state of higher potential energy. A preliminary study of this possibility shows that the theories for which this approach might work would be identical to those that obey the thermal constraint, Eq. (1.9), above. (b) Fluctuating curvature approach[21]: Suppose that the gravitational interactions produce randomly varying curvature which, in many regions, is very high. Quantum field theory in highly curved backgrounds are very similar to field theories in high temperature backgrounds. If the thermal constraint, Eq. (1.9), is satisfied, the high curvature regions should be driven to the false vacuum phase. (c) Chaotic inflation[22]: Random fluctuations for an arbitrary theory may drive rare regions into the false vacuum phase. Only such rare regions inflate and that happens to be where we live; it is not clear what fraction of the total universe such regions occupy. No thermal constraint is required. (d) Universe Springs from Nothing[23]: Our universe tunnels from a state with no space or time into a false vacuum state. It then inflates, reheats, etc. No thermal constraint is required.

For the moment, approaches (a) and (b) seem the most attractive to us. since both they and the conventional approach require the same thermal constraint to be satisfied, we will remain with it for the rest of our analysis. However, when the reader sees the important role the thermal constraint plays, he may wish to re-evaluate the feasibility of the constraint.

Let us now consider how the constraints on V affect the choice of parameters in the superpotential, W. We first consider W with a constant and linear term only, encounter failure, and gradually add terms to see if and how matters improve.

Case 1: The superpotential is

$$W = \mu^2 M(\beta + (\phi/M) , \qquad (1.10)$$

which corresponds to a Polonyi model[13], the type of superpotential that has been employed in so-called "hidden sector" theories. For our purposes, the potential is completely analogous to the more general O'Raifeartaigh models discussed above[10]; both models spontaneously break supersymmetry and in both cases the inflaton field appears at most linearly in the superpotential. For $(\phi/M) = R$ = real,

$$V = \mu^4 e^{R^2} \{[1-3\beta^2]-[4\beta]R + [\beta^2-1]R^2 + [2\beta]R^3 + R^4\} . \qquad (1.11)$$

There are two possibilities. First, it is possible to have a minimum at $\sigma = (\sqrt{3}-1)M$ with $V(\sigma) = 0$ if $\beta = 2-\sqrt{3}$. In this case, however $|c_1| = 4\beta = 1.07$ which violates constraint Eq. (1.7) by six orders of magnitude. Second, take $c_1=0$ to satisfy the constraint. Then the minimum has non-zero cosmological constant. Thus, the problem with O'Raifeartaigh potentials is not that they don't reheat, as was originally believed. The problem is that, once one takes careful account of all constraints, one discovers that the models could never really inflate (assuming the global supersymmetry O'Raifeartaigh model can only be part of a local super-symmetry theory).

Case 2: The superpotential is

$$W = \mu^2 M(\beta + (\frac{\phi}{M}) + \frac{\Delta}{2}(\frac{\phi}{M})^2 + \frac{\Delta\lambda}{3}(\frac{\phi}{M})^3). \qquad (1.12)$$

For $(\phi/M) = R$ = real,

$$V = \mu^4 e^{R^2}\{[1-3\beta^2] + [2(-2\beta+\Delta)]R$$

$$+ [\beta^2-1 + \Delta(\Delta-\beta) + 2\lambda]R^2 + \ldots\} . \qquad (1.13)$$

The constraint on $c_1 = 2(-2\beta+\Delta)$ is satisfied if

$$\Delta = 2\beta . \qquad (1.14)$$

The constraint on $c_2 = 2\lambda$ is satisfied if

$$\lambda = 0 . \qquad (1.15)$$

Setting $\lambda=0$ guarantees that the curvature of V near $\phi=0$ is positive in the imaginary ϕ direction. this justifies the assumption in this paper that the slow rollover transition occurs in the real direction. The thermal constraint is satisfied if

$$\beta > 0 . \qquad (1.16)$$

In this case however, V has no minimum for R>0. (If one is willing to suspend the thermal constraint, we showed that a superpotential of the form[7]

$$W = \mu^2 M\left(-\frac{1}{2} + \left(\frac{\phi}{M}\right) - \frac{1}{2}\left(\frac{\phi}{M}\right)^2\right) \tag{1.17}$$

is the simplest which satisfies all other constraints. This superpotential was recently used by the Ross, et al.[25] in their Supersymmetric Inflation models.)

Case 3: the superpotential is

$$W = \mu^2 M\left(\beta + \left(\frac{\phi}{M}\right) + \beta\left(\frac{\phi}{M}\right)^2 + \frac{\beta}{2} a_4 \phi^4\right). \tag{1.18}$$

For $(\phi/M) = R = \text{real}$

$$V = \mu^4 e^{R^2}\{[1-3\beta^2] + [3\beta^2 - 1]R^2 + [2\beta(1+2a_4)]R^3 + \ldots\} \quad . \tag{1.19}$$

There are two kinds of minima, σ. The first kind preserves supersymmetry. Taking $V(\sigma) = 0$ implies

$$\beta = \frac{-3\sigma}{2(2+\sigma^2)} \quad . \tag{1.20}$$

For $(\sigma) > 0$, Eq. (1.20) requires $\beta < 0$. This violates the thermal constraint. The second kind of minimum spontaneously breaks supersymmetry. Since $c_3 = 2\beta(1+a_4) < 0$ and $\beta>0$ it follows that $a_4 < -1/2$. For these values of a_4, one can show that $V(\sigma)$ must be non-zero which violates the cosmological constant constraint.

Case 4: The superpotential is

$$W = \mu^2 M\left(\beta + \left(\frac{\phi}{M}\right) + \beta\left(\frac{\phi}{M}\right)^2 + \frac{\beta}{2} a_4\left(\frac{\phi}{M}\right)^4 + \frac{2\beta}{5} a_5\left(\frac{\phi}{M}\right)^5 + \ldots\right). \tag{1.21}$$

There are two types of minima with $V(\sigma) = 0$. The first preserves supersymmetry. Assuming, without loss of generality, that $\sigma = M$, we find:

$$a_5 = -\frac{15}{16} a_4 \quad , \quad \beta = \frac{-1}{2(1+a_4/16)} \quad . \tag{1.22}$$

All constraints are seemingly satisfied if

$$-\infty < a_4 < -16\left(\frac{\sqrt{3}}{2} + 1\right) \quad . \tag{1.23}$$

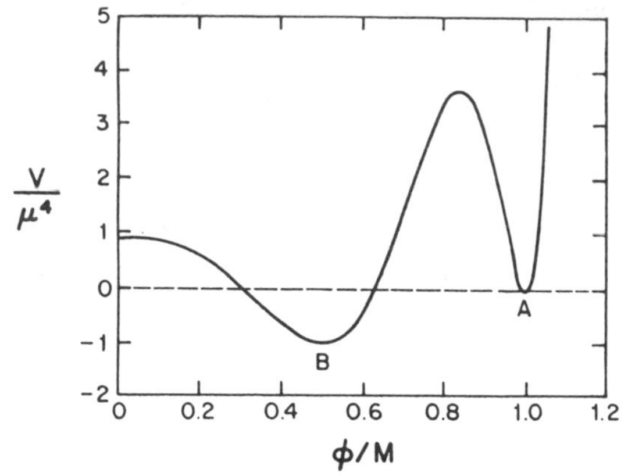

Figure 1: Graph of the potential energy density for a model satisfying slow rollover constants which also has a supersymmetry preserving minimum (A) with V=0. Minimum B preserves supersymmetry but has V<0. Unfortunately, rollover takes the universe to B, not A, which is unacceptable.

However, consider the graph of the potential one obtains for typical parameters in Figure 1. Minimum A is the supersymmetry preserving minimum with zero cosmological constant. Minimum B is also supersymmetry preserving but has $V < 0$. Minimum A is separated from R=0 by a large potential barrier. The result is a cosmological disaster because the universe slow rolls from ϕ to a state with a large negative cosmological constant, not the desired $V=0$ state. Amazingly enough, one can show that this situation occurs for almost any reasonable polynomial superpotential, no matter how many additional terms and no matter the choice of parameters.[7] the idea of rolling from $\phi=0$ to a supersymmetry preserving state was an important part of the "inflaton" approach, and now we see that the notion is in deep trouble. (There are a few tiny loopholes, besides giving up the thermal constraint: (a) Complex valued couplings in W: Although possible, this approach makes the analysis of the thermal constraint and the slow rollover so complicated that new technologies will be required to study them. (b) Regenerative Universe: Arrange a potential where the universe is trapped in a phase that is metastable even at T=0, i.e. a barrier remains separating $\phi=0$ from $\phi=\sigma$. Rarely, a bubble will be nucleated inside of which ϕ has a value on the $\phi=\sigma$ side of the barrier, but suppose the potential is very flat for that value of ϕ. Then the interior of the bubble will undergo the usual inflationary scenario. See Ref. 26 for further implications. Unfortunately, this approach requires a very artificial potential at best, and a supersymmetric example has not been demonstrated as of this writing.[27] (c) Add more fields to the superpotential. This appears to us to be the most likely approach to work, although no example has been demonstrated as of this writing. We hope, however, to convince the reader that an alternative approach is much more workable and much, much more attractive.)

The second kind of minimum spontaneously breaks supersymmetry. The potential for $\beta=.2$, $a_4 = -2.191$ and $a_5 = 1$ is shown in Figure 2. We find that the Kahler derivative $D_\phi W|_\sigma = 2.2\mu^2 \neq 0$. Hence supersymmetry is spontaneously broken even though the superpotential, Eq. (1.21)is not of the O'Raifeartaigh type.[11] This "new type" of supersymmetry breaking is possible in locally supersymmetric theories only; although it is a simple consequence of the form for V in Eq. (1.1), its significance had not been fully appreciated prior to our analysis.[7,28] (An additional advantage of the new supersymmetry breaking mechanism we hope to exploit in the future is that, unlike O'Raifeartaigh superpotentials, a gauge singlet field which appears only linearly in W is not required. This could play an important role in our eventual hope of combining inflation in a supersymmetric GUT theory in which the inflation fields would be Higgs fields already present in the theory.)

For the example given, there are more than 10^6 e-foldings of inflation, the energy density amplitude is $\delta\rho/\rho \sim 10^{-4}$, and the reheating temperature is of $O(10^{10}GeV)$, marginally sufficient to account for the observed baryon asymmetry. We find the same qualitative results for a wide range of parameters and additional terms in W.

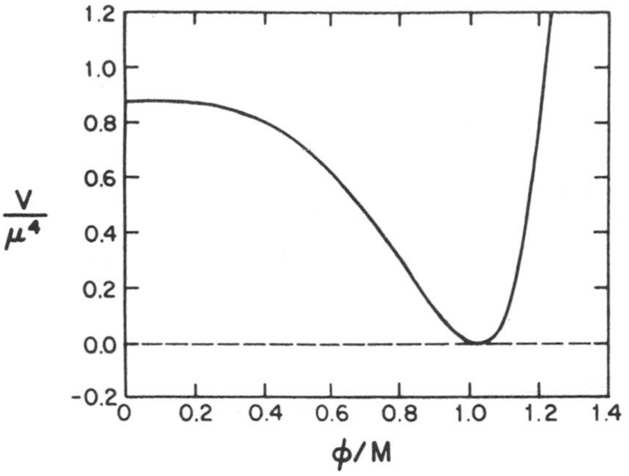

Figure 2: Graph of the potential energy in which the minimum satisfies V=0, but spontaneously breaks supersymmetry, Eq. (1.20).

Thus, we have demonstrated a class of supersymmetric models that can satisfy all the constraints for inflation, and we have been forced to stumble across a new way of spontaneously breaking supersymmetry in the process. Perhaps the simplest way to incorporate the models in particle physics is to use these W's to replace the "hidden sector" superpotentials employed in the standard approaches to supersymmetric phenomenology.[24] However, those already familiar with supersymmetry phenomenology have already observed the impending disaster. Our examples have a value of μ of $O(10^{-(3-4)}M) \simeq O(10^{14}GeV)$ which implies a supersymmetry breaking scale, m_{SB}, of $O(\mu)$. In natural supersymmetric phenomenological models,[29] the electroweak unification scale is $O(m_{SB}^2/M)$ which, to obtain a value of $O(10^2\ GeV)$, requires $m_{SB} \approx 10^{10}GeV << O(\mu)$. We now observe that there is a startling conflict between the constraints of inflationary cosmology (the constraint of small density fluctuation amplitudes in particular) that demand $\mu = O(10^{14}\ GeV)$, and the constraints of the mass hierarchy, which demand $m_{SB} = O(10^{10}GeV)$, in order to obtain the correct electroweak unification scale.

II. RESOLVING THE CONFLICT BETWEEN INFLATIONARY COSMOLOGY AND THE MASS HIERARCHY[8]

In Section I a special class of superpotentials was demonstrated which meets all the constraints of inflationary cosmology and which requires a supersymmetry scale, μ of $O(10^{-(3-4)}M)$. A simple way of incorporating these models into realistic particle theories is to replace the Polonyi superpotentials[1] in standard hidden sector approaches to supersymmetric phenomenological models, since the special class of potentials also spontaneously breaks supersymmetry. However, in order to use the same superpotentials in a model that also describes particle physics, a theory must be found in which the supersymmetry scale, m_{SB} is not $O(\mu) \simeq (10^{-(3-4)}M)$, as in most cases, but is instead $O(10^{-8}M)$. In natural models, the electroweak unification is generated dynamically at a scale $m_{ew} \approx O(m_{SB}^2/M)$ and a value of $m_{SB} \approx O(10^{-8}M)$ is required to obtain $m_{ew} \approx O(10^{-16}M) \approx O(10^2GeV)$.[29] In this section we will demonstrate a subclass of models which employ the "new" supersymmetry breaking scheme which can satisfy all constraints for inflationary cosmology and also yield the correct mass hierarchy ($m_{ew} \simeq 0(10^2 GeV)$).

The key to resolving the conflict between inflationary cosmology and the mass hierarchy is (1) to allow more than one chiral superfield in the inflationary (hidden) sector; and (2) to employ a mechanism (henceforth referred to as "hierarchy mechanism") that sets the supersymmetry breaking scale, m_{SB}, to be much smaller than the fundamental supersymmetry scale, μ, in the superpotential. In Section I we excluded the possibility of resolving the conflict for the case of just one superfield in the hidden sector; but by adding one or more fields and using the hierarchy mechanism, μ can maintain the value desired for inflationary cosmology, $0(10^{-(3-4)}M)$, while the supersymmetry breaking scale, and, consequently, the electroweak scale, can be made small enough to be consistent with observations.

The important mechanism can be illustrated by the following simple superpotential:[30]

$$W = -\mu^2 \phi + \frac{\phi^2 \psi}{2} \quad . \tag{2.1}$$

The supersymmetry breaking scale is given by $m_{SB} \simeq \max \{\sqrt{D_{\phi_i} W}\}$ and the induced electroweak scale is $0(m_{SB}{}^2/M)$. In the limit that gravity is turned off ($M \to \infty$), the minimum of V corresponds to a supersymmetric state with $D_\phi W = D_\psi W = 0$ and

$$\phi = \frac{\mu^2}{\psi} \quad , \quad \psi \to \infty \quad . \tag{2.2}$$

When gravity is turned on (M is finite), the minimum of V cannot occur beyond ψ of $0(M)$. The solution for ϕ remains unchanged to leading order and, hence,

$$\phi \simeq 0 \left(\frac{\mu^2}{M}\right) \quad , \quad \psi \simeq 0\ (M) \quad . \tag{2.3}$$

Both Kahler derivatives are now non-zero at the minimum of V and, therefore, supersymmetry is broken. Since $D_\psi W = \phi^2/2 + \ldots$

$$D_\psi W \simeq 0\ (\mu^4/M^2) \tag{2.4}$$

which is μ^2/M^2 times smaller than is expected in a model with supersymmetry scale μ. $D_\phi W$ is smaller than $D_\psi W$ and can be ignored. It follows that m_{SB} is $0(\mu^2/M)$ and that the electroweak scale is $0(\mu^4/M^3)$. If we set $\mu \simeq 0(10^{-(3-4)}M)$ to satisfy the constraints of inflationary cosmology, the electroweak scale is of $0(10^2 GeV)$, which is compatible with observations.

An example of a superpotential which incorporates all constraints for a successful inflationary cosmology and employs the hierarchy mechanism is:[8]

$$W = -\mu^2\phi + \frac{a_4}{4M}\phi^4 + \frac{a_5}{5M^2}\phi^5 + \frac{\lambda'}{2M}\phi^2\psi^2 + \frac{\lambda}{2M^2}\phi^2\psi^3 + \frac{\delta}{3}\psi^3 \quad (2.5)$$

where

$$\mu = 10^{-4}M \qquad a_4 = .74\,\mu^2/M \qquad a_5 = -18.5\mu^2/M^2$$
$$\lambda' = 16.7\mu^2/M^2 \qquad \lambda = 1 \qquad \delta = .04\,\mu^4/\lambda \quad . \quad (2.6)$$

All parameters in W are chosen to be real. Simpler superpotentials of this type can be found, but Eq. (2.5) is presented because it is very easy to analyze. The model was specifically designed so that the inflation and the reheating occur in very different parts of the potential. The inflation occurs while (ϕ,ψ) roll along the real ϕ-axis and reheating occurs after (ϕ,ψ) roll off the ϕ-axis to the true minimum. The theory has the following properties:

(1) At high temperature, T, the energetically favored state is $(\phi,\psi) = (0,0)$. The universe supercools in this state to T near zero and (0,0) serves as the false vacuum state of the inflationary scenario. As T approaches zero, this state becomes unstable and the slow rollover transition begins. Hence, the thermal constraints are satisfied.

(2) Even at zero temperature the curvature in ψ is positive near (0,0). The zero temperature potential is unstable along the real ϕ-axis; the potential projected along the real ϕ-axis is given by:

$$V = \mu^4\left[1 - \frac{2a_4}{\mu^2 M}\phi^3 + \frac{1}{2M^4}\left(1 - \frac{4a_5M^2}{\mu^2}\right)\phi^4 + \ldots\right] \quad . \quad (2.7)$$

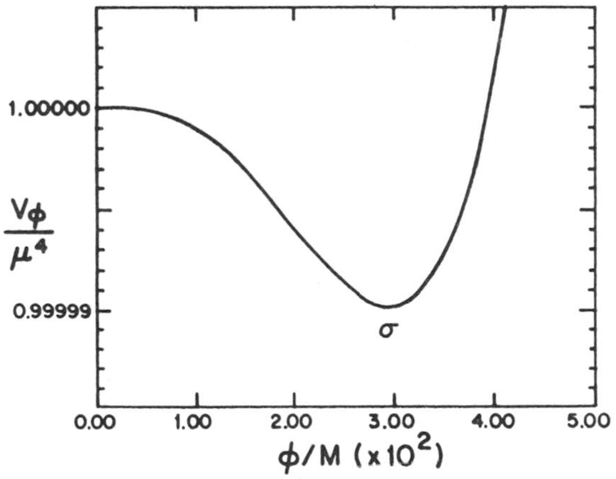

Figure 3: Graph of the slow rollover path down positive ϕ-axis for example of Eq. (2.5).

Therefore, the slow rollover proceeds directly along the positive real ϕ-axis. Because the slow rollover depends on essentially only one of the fields, the analysis developed in Section I could be applied to determine the choice of parameters in Eq. (2.6) necessary to meet the slow rollover constraints. As ϕ rolls from (0,0) to an extremum, σ, at $(\phi,\psi) = (3 \times 10^{-2}M,0)$ there is more than sufficient inflation ($>10^6$ e-foldings) and the density perturbation amplitude is $\delta\rho/\rho \sim 10^{-4}$ on observable scales. The one difference with the analysis of Section I is that we do not demand that the slow rollover constraint be satisfied at σ. The point, σ, is an unstable (in the ψ direction) extremum of the potential (saddle point) and once (ϕ,ψ) reaches σ, fluctuations drive the system off the ϕ-axis in the positive ψ direction. The roll then proceeds to a true minimum well off the real ϕ-axis and it is this point, not σ, which must have $V = 0$. A graph of V/μ^4 is shown in Figure 3.

(3) Once the system rolls off the real ϕ-axis near σ, it rapidly rolls to the true minimum of the potential. No further inflation occurs, and hence, the number of e-foldings and the value of $\delta\rho/\rho$ are determined solely by the slow roll down the ϕ-axis. The minimum occurs at $\tau = (.5\ \mu^2/2M,\ 1.3M)$. Coefficient δ is fine-tuned so that this minimum has zero cosmological constant. We have checked the curvatures in the complex ϕ,ψ directions to be sure τ is a minimum and not a saddle point. The same mechanism exemplified in the superpotential of Eq. (2.1) has been used (terms 1 and 5 in Eq. (2.5)) so that the electroweak scale is of $O(\mu^4/M^3) = O(100\ \mathrm{GeV})$, the observed value. A graph of V/μ^4 along the path of steepest descent from σ to the true minimum is shown in Figure 4. At the true minimum, τ, we find that:

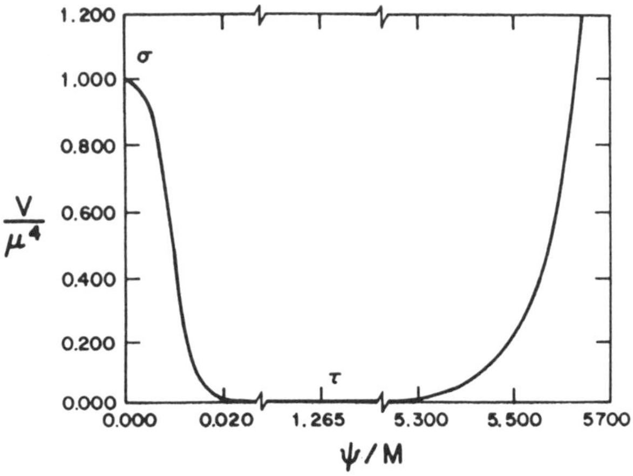

Figure 4: Graph of path from extremum σ, to true minimum, τ, projected along ψ-axis for example in Eq. (2.5).

$$D_\phi W = .37\ \frac{\mu^4}{\lambda M^2}\ ,\quad D_\psi W = 1.1\ \frac{\mu^6}{\lambda^2 M^4} \tag{2.8}$$

and the masses of the scalar field are:

$$M_\psi = 1.4\ \frac{\mu^4}{\lambda M^3}\ ,\quad M_\phi = 6.4\ \lambda\ M\ . \tag{2.9}$$

The surprising disparity in the masses is a characteristic feature of models which employ the hierarchy mechanism, a characteristic that will play a very important role in Section III.

III. BARYON ASYMMETRY AND THE GRAVITINO PROBLEM(S)[9]

Finding a mechanism that resolves the conflict between inflationary cosmology and the mass hierarchy, as illustrated in Section II, is significant. However, we have subsequently discovered that the same hierarchy mechanism may help resolve two additional cosmological problems: the generation of baryon asymmetry after inflation and the "gravitino entropy crisis". The resolution of both problems depends crucially on how reheating takes place after the inflationary epoch. As a result of the hierarchy mechanism, the reheating process is strikingly different in the new models compared to any previous models of new inflation. After analyzing the reheating, we find that a baryon asymmetry can be generated that is consistent with astrophysical observations and the gravitino problem(s) can be avoided for $\mu \simeq O(10^{14-15}\text{GeV})$. Remarkably, the resolution of the baryon asymmetry and gravitino problems comes as a free byproduct of our attempts to resolve inflation with the mass hierarchy.

Reheating following the inflationary epoch is determined by the curvature of the potential near the global minimum, τ, and the couplings of the inflation fields. For the case under study, the superpotential defines a hidden sector which couples only through gravitationally induced interactions to ordinary matter. The decay width for hidden sector fields (particles or coherent oscillations) is given by:

$$\Gamma_\phi \sim m_\phi^3/M^2 , \tag{3.1}$$

where m_ϕ is the mass of the hidden sector field, which also depends on the curvature at the global minimum.

The key feature of models which employ the hierarchy mechanism is that the curvature of the potential at the global minimum varies by many orders of magnitude, as we see from Eq. (2.9). One field, ψ, has a small mass of $O(\mu^4/M^3)$, while the other, ϕ, has a large mass of $O(M)$. This mass hierarchy is directly related to the fact that the supersymmetry breaking scale is $O(\mu^2/M)$. The "small mass" ψ-field has a small decay width according to Eq. (3.1). The "large mass" ϕ-field has a very large decay width, $O(M)$, implying a lifetime shorter than the Hubble expansion time during or after inflation. The ϕ-fields will be referred to as "reheating" fields because their decay is primarily responsible for the conversion of oscillation energy to radiation energy after inflation.

Because of the anomalously small ψ mass, $V(\phi,\psi)$ is very anharmonic in the ψ-direction. For deviations in the ψ-direction greater than $\Delta\psi > O(\mu^2/M)$, the quartic terms in V are greater than the quadratic ones. The effect of the quartic terms is that the small mass oscillation direction deviates slightly from the ψ-direction, corresponding to the curve $\phi = (\mu^2/\lambda)\,\psi^{-3}$ near τ.

The equations of motion for the evolution of the (ϕ,ψ) fields as they oscillate about τ are given by:

$$\ddot{\phi} + \partial V/\partial\phi = -3H\dot{\phi} - \Gamma_\phi\dot{\phi}$$

$$\ddot{\psi} + \partial V/\partial\psi = -3H\dot{\psi} - \Gamma_\psi\dot{\psi} \tag{3.2}$$

$$\rho_r = -4H\rho_r + \Gamma_\phi\dot{\phi}^2 + \Gamma_\psi\dot{\psi}^2$$

where $H^2 = 1/(3M^2)(1/2\dot{\phi}^2 + 1/2\dot{\psi}^2 + V + \rho_r)$ is the Hubble constant, ρ_r is the radiation energy density; and $\Gamma_\phi = M$ and $\Gamma_\psi \simeq \mu^{12}/M^{11}$ are the rates at which oscillations in the (ϕ,ψ) directions decay to radiation (see Eq. (3.1)).

The solution of these equations can be understood qualitatively as follows: A plot of the rollover to the true minimum, τ, shows that the (ϕ,ψ) first approach τ along the ψ-direction. If the oscillations proceed exactly along this small mass direction, the decay to radiation would occur at the rate Γ_ψ. For $\mu \simeq O(10^{-(3-4)}M)$ width is so small that most of the oscillation energy would be redshifted away (via the H-dependent terms in Eq. (3.2)), rather than be converted to radiation. The reheat temperature would be[18]

$$T^{o}_{rh} \simeq (\Gamma_\psi M)^{1/2} \sim O(\mu^6/M^5) \sim O(10^{-(0-5)}\ \mathrm{GeV}). \tag{3.3}$$

However, because the anharmonicity of the potential causes the minimum V" (small mass) oscillation path to bend away from the ψ direction away from τ, each oscillation results in a small fraction, $O(\mu^4/M^4)$ of the small mass oscillation energy being converted to reheating field (ϕ) oscillation energy. This latter energy decays in less than one ψ-oscillation time to radiation because Γ_ϕ is so large.

The resulting effective decay rate of the oscillation energy created by the anharmonicity is $\Gamma_e \simeq \mu^4/M^3$. The reheating temperature is then given by:

$$T_{rh} \simeq (\Gamma_e M)^{1/2} \simeq O(\mu^2/M) \approx O(10^{10-12}\mathrm{GeV}) \quad . \tag{3.4}$$

This analysis is confirmed by exact numerical solution of Eq. (3.2). <u>Baryon Asymmetry:</u> Although the reheating temperature in Eq. (3.4) represents a significant improvement over the naive value, Eq. (3.3), it appears at first to still be too low for baryon asymmetry to be generated after inflation. In the conventional scheme[31], baryon asymmetry is generated when particles with baryon-violating decays (e.g. color triplet Higgs, X meson) fall out of thermal equilibrium at some time after the temperature of the universe cools below their mass. From limits on proton decay,[32] it is at best marginal and probably ruled out for these masses to be as small as T_{rh}, so these particles are probably not returned to thermal equilibrium after inflation.

Nevertheless, a baryon asymmetry can be generated after inflation using a more unconventional scheme. The key point is that at the end of inflation the universe is as far from thermal equilibrium as possible - all the energy density of the universe is in the form of coherent field oscillations. A baryon asymmetry can be generated while the universe is very far from equilibrium (as opposed to the conventional slightly-out-of-equilibrium scheme) as the oscillating fields first begin to decay.[18,33] If the mass of the fields is greater than that of the color triplet Higgs or X meson, the oscillating fields can decay to the color triplet Higgs or X mesons. However, the reheating fields (ϕ) have a mass O(M), which certainly exceeds the mass of the color triplet Higgs or X meson. Thus the mass hierarchy scheme which produces such different mass scales turns out to be crucial in allowing baryon asymmetry after inflation.

The baryon-to-entropy ratio, n_B, produced by the unconventional very-far-from-equilibrium scheme is given by:

$$n_B = .75 \left(\frac{T_{rh}}{M}\right) \varepsilon \quad , \tag{3.5}$$

where ε is the baryon excess per ϕ decay. (ε can be enhanced by increasing the coupling of the color triplet Higgs to ϕ). For our choice of parameters, this is compatible with the observed baryon asymmetry $O(10^{-10})$, if ε is $O(10^{2-4})$ or greater for $\mu \simeq O(10^{-(3-4)})M$. This limit on ε may not be so severe once factors of O(1) are taken into account for particular models.

Gravitino Problems: The "gravitino problems" common to almost all supersymmetric unified theories can lead to the dilution of any baryon asymmetry produced just after inflation. The gravitino problems involve the production of large amounts of entropy later in the history of the universe when the temperature is too low for further baryon asymmetry generation: thus, the baryon-to-entropy ratio is reduced to a value incompatible with observations.

The gravitino problems are extremely difficult to avoid because there are three different ways the undesired entropy can be produced. As the name suggests, the undesired entropy can result from the decay of gravitinos [24] (mass = $m_g \simeq O(\mu^4/M^3) \simeq 10^{2-3}$GeV) whose lifetime is given by:

$$t_g = \Gamma_g^{-1} \simeq (m_g^{\,3}/M^2)^{-1} \simeq O(M^{11}/\mu^{12}) \simeq O(10^{18-30}\text{GeV}^{-1}) \quad . \tag{3.6}$$

The lifetime is roughly the time for primordial nucleosynthesis ($t_N \simeq O(10^{24}\text{GeV}^{-1})$ or $T \simeq 1$ MeV). If there are a sufficient number of non-relativistic gravitinos at $t = t_N$ though, the universe will expand at a matter dominated $R{\sim}t^{2/3}$ rate, instead of the expected radiation dominated rate $R{\sim}t^{1/2}$, and the successful predictions of nucleosynthesis will not be obtained. Alternatively, from the decay of so many gravitinos the universe may reheat to a temperature above that of nucleosynthesis. The universe will cool and undergo nucleosynthesis again after the gravitinos decay, but the entropy produced in the decay will dilute the baryon-to-entropy ratio. The best hope is to dilute the number of gravitinos to such a degree that they produce negligible entropy when they decay and they never dominate the energy density.

Gravitinos produced before inflation are diluted to insignificant numbers by the exponential expansion[35] but there are two ways gravitinos can be produced after inflation: Too many gravitinos can be produced by thermal equilibrium processes if T_{rh} exceeds $O(10^{12}$ GeV).[34] Too many gravitinos can also be produced via the initial decay of the reheating fields.[36]

Because gravitinos are so light ($m_g \approx 10^{2-3}$ GeV), they are highly relativistic just after reheating at time $t=t_{rh}=O(M/T_{rh}^{\,2}) \simeq O(M^3/\mu^4)$ for our models. Thus, the energy density in gravitinos, ρ_g, and the radiation energy density, ρ_r, both scale as R^{-4}, where R is the Robertson-Walker scale factor. (Recall, relativistic matter energy density scales as R^{-4}, non-relativistic as R^{-3}. $R{\sim}t^{1/2}$ if the universe is dominated by highly relativistic particles but $R{\sim}t^{2/3}$ is the universe is matter dominated, say, by non-relativistic gravitinos.) When the universe cools enough for the gravitinos to become non-relativistic, at $t = t_{nr}$, ρ_g scales a R^{-3} and at a later time

$$\rho_g/\rho_r|_t = \rho_g/\rho_r|_{t_{nr}} \quad (R(t)/R(t_{nr})) \ . \qquad (3.7)$$

R(t) continues to grow as $t^{1/2}$ until at some later time, $t=t_d$, the energy density becomes dominated by the mass of non-relativistic gravitinos. The ratio, ρ_g/ρ_r continues to grow until the gravitinos decay at $t = t_g$ (see Eq. (3.6)); ρ_g is then converted to radiation (entropy) and the baryon-to-entropy ratio is diluted by a factor:

$$\Delta = [\rho_g/\rho_r|_{t_g}]^{3/4} \qquad (3.8)$$

Given the barely acceptable baryon asymmetry generated just after inflation, Eq. (3.5), a value of Δ that is O(1) is required for successful nucleosynthesis. This is accomplished by making t_g/t_d as small as possible so that the gravitinos dominate the energy density for at most a short time.

These problems are difficult enough to avoid, but there is yet a third way too much entropy can be produced after inflation: Energy density in the form of coherent oscillations of the fields associated with supersymmetry breaking scales with R in the same way as non-relativistic matter ($\sim R^{-3}$), and can eventually dominate the energy density of the universe.[37] This last problem is classified under gravitino problems because the supersymmetry breaking fields must have, by construction, the same mass and decay rate as the gravitino. Therefore, the same difficulties can arise for the coherent oscillations of the scalar fields as for the gravitinos themselves.

Previous attempts to find inflationary models which avoid all three gravitino problems have involved theories in which the scalar fields responsible for inflation are completely different from the fields responsible for supersymmetry breaking. All such attempts required, in the final analysis, some degree of fine-tuning.[36] In the inflationary models we are considering, the same fields that produce inflation also break supersymmetry. thus, a special new analysis is required. We find that inflationary models which employ the mass hierarchy scheme automatically avoid all the gravitino problems.[9]

The gravitinos produced by thermal processes can be avoided because the models do not reheat above $T \simeq 10^{12}$ GeV. The mass hierarchy scheme is crucial here because normally such a low reheating temperature would rule out baryon asymmetry production after inflation altogether, as we pointed out above. However, with the very-far-from-equilibrium baryon asymmetry generation mechanism, a reheating temperature low enough to avoid the gravitino problem can be tolerated.

The gravitinos produced directly through the decay of the inflation fields can be avoided because the reheating fields have mass much greater than T_{rh}. Gravitinos will be produced directly at time $t = t_{rh}$ through the decay of the reheating fields. The reheating field has mass O(M) so the gravitinos produced have momentum $p_g(t_{rh}) \simeq M$. The ratio, $\rho_g/\rho_t|t_{rh}$, should be O(1) since the reheating (hidden sector) fields couple roughly equally to all particle species through gravitationally induced couplings. However, because gravitinos couple too weakly to come into thermal equilibrium, the gravitino momentum is not reduced by subsequent scatterings; it simply redshifts with the expansion of the universe:

$$p_g(t) = M\,(R(t_{rh})/R(t)) \,. \tag{3.9}$$

Ordinary radiation does equilibrate and it has an initial mean momentum $p_r(t_{rh}) \simeq T_{rh}$, which also redshifts with the expansion:

$$p_r(t) = T_{rh}(R(t_{rh})/R(t) \,. \tag{3.10}$$

As a result, when the temperature cools below the mass of the gravitino after reheating, the typical radiation particle will have momentum less than the gravitino mass, $p_r \simeq T \lesssim m_g$, but the gravitino momentum will still be much greater than m_g. Thus, the gravitinos remain highly relativistic even below $T = m_g$ and the ratio ρ_g/ρ_r does not increase. Not until time $t = t_{nr}$ where

$$p_g = m_g \text{ or } R(t_{rh})/R(t_{nr}) \sim m_g/M \tag{3.11}$$

should the gravitinos become non-relativistic. Since ρ_g/ρ_r has not changed from O(1) since reheating, just after t_{nr} gravitinos dominate the energy density of the universe; i.e. in our notation $t_d = t_{nr}$. Then t_d can be computed using $t_{rh} \simeq O(M^3/\mu^4)$:

$$t_d = t_{rh}(M/M_g)^2 = O(M^{11}/\mu^{12}) \quad . \tag{3.12}$$

However, this also corresponds to the lifetime of the gravitino, t_g (from Eq. (3.8)), or the time at which the gravitinos are converted into entropy. Since $t_g/t_d \simeq O(1)$, we conclude that the gravitinos decay just before they come to dominate the energy density of the universe and no significant entropy is produced by their decay ($\Delta \sim 1$).

Avoiding the third gravitino problem caused by the energy density trapped in coherently oscillating supersymmetry breaking fields depends crucially on the form of the potential, V, for our models and the fact that the inflation fields are the same as the supersymmetry breaking fields. As we outlined before, the coherent oscillations after exponential expansion are along the small mass direction.

Recall that the oscillations in the small mass direction decay by first being converted to oscillations in the large mass direction by the anharmonic terms in the potential, and then the large mass oscillations decay quickly. While the small oscillations are dominated by the quartic terms in V, the oscillation energy decreases as R^{-4}, due to the anharmonicity of the potential.[38] However, some harmonic oscillation energy density remains after reheating. Once the amplitude in the small mass direction becomes less than $\Delta \lesssim O(\mu^2/M)$, the anharmonic terms become irrelevant compared to the quartic ones, and the remaining small mass oscillation energy is not converted to large mass oscillation energy. Thus, there remains a harmonic oscillation energy density

$$\rho_0(t_{rh}) \sim O(\mu^{12}/M^8) \ll \rho_r(t_{rh}) \simeq O(T_{rh}^4) \tag{3.13}$$

which scales with R in the same way as non-relativistic matter, $\propto R^{-3}$. Given that $T_{rh} \simeq O(\mu^2/M)$, one finds that

$$\rho_0/\rho_r|_t = \rho_0/\rho_r|_{t_{rh}}(R(t)/R(t_{rh})) \simeq O(\mu^4/M^4)[\frac{R(t)}{R(t_{rh})}] \quad . \qquad (3.14)$$

Although this ratio is small at $t=t_{rh}$, eventually, at time $t=t_d$, the oscillation energy can dominate the radiation energy density (up to that time $R \sim t^{1/2}$). Then t_d can be computed from Eqs. (3.3, 3.14) to be:

$$t_d = t_{rh}(M^8/\mu^8) \simeq O(M^{11}/\mu^{12}) \quad . \qquad (3.15)$$

Once again, we find that this time, t_d, equals the lifetime of the gravitino. Thus, t_g/t_d is O(1) and no significant entropy is produced ($\Delta \sim 1$).

Although it appears as if these results depend sensitively on the reheating temperature, it is worth mentioning that this is not the case. If the reheating temperature were higher than our estimate, more baryon asymmetry would be generated just after inflation, Eq. (3.5). However, $t_g/t_d > O(1)$ (unlike above) and gravitinos do dominate the energy density for a short time. One finds that $\Delta > 1$, but just enough entropy is produced by the decay of the gravitinos to bring the baryon asymmetry back down to the value we have already found, Eq. (3.5).

CONCLUSIONS

Although most supersymmetric models are found to be incompatible with the constraints of inflationary cosmology, a special new class of models has been demonstrated to be an excellent candidate for Supersymmetric Inflation. The class of models meet all the constraints for inflationary cosmology, including producing small density fluctuations, spontaneously breaking supersymmetry, setting an electroweak unification scale consistent with observations, producing a baryon asymmetry after inflation and avoiding all the gravitino problems. The examples constructed so far are rather artificial and require tuning of the parameters in the superpotential, but this has been done to simplify the analysis of the inflationary constraints. A better technology is being developed which will allow us to analyze simpler, more natural models of the same class.

Should one be optimistic about Supersymmetric Inflation? It is hard to say. The route has proven to be much more difficult than was originally supposed. On the other hand, each time an obstacle has been overcome, some extra benefit has come for free. We set out to find supersymmetry models that satisfied the constraints for inflationary cosmology, and we were forced to stumble across a new supersymmetry breaking scheme. We had to reconcile the new models with the mass hierarchy, and we discovered a class of models that also generates a baryon asymmetry and avoids the gravitino problem. We can only hope that the extra benefits are signposts indicating that we are heading in the right direction.

We would like to thank S. Raby and M. Turner for their many valuable comments. One of us (PJS) would like to thank Charles Louis and Joseph Solomon for their super contributions and criticisms, and for making it possible to stay awake late enough to complete the manuscript. This work was supported in part by DOE Grant DE-AC02-81ER 40033 B.000 (BAO) at Rockefeller University, by DOE Grant EY-76-C0203071 BAO and PJS) at the University of Pennsylvania, and the Alfred P. Sloan Foundation (PJS).

REFERENCE

[1]A. Guth, Phys. Rev. **D23**, 347 (1981).
[2]A. D. Linde, Phys. Lett. **108B**, 389 (1982).
[3]A. Albrect and P. J. Steinhardt, Phys. Rev. Lett.**48**, 1220 (1982).
[4]A. A. Guth and S.-Y. Pi, Phys. Rev. Lett. **49**, 1110 (1982); S. W. Hawking, Phys. Lett. **115B**, 395 (1982); A. A. Starobinskii, Phys. Lett. **117B**, 175 (1982); J. M. Bardeen, P. J. Steinhardt and M. S. Turner, Phys. Rev. **28**, 579 (1983).
[5]M. Sher, Phys. Lett. **135B**, 52 (1984).
[6]Q. Shafi and A. Vilenkin, Phys. Rev. Lett. **52**, 691 (1984).
[7]B. A. Ovrut and P. J. Steinhardt, Phys. Lett. **133B**, 161 (1983).
[8]B. A. Ovrut and P. J. Steinhardt, to appear in Phys. Rev. Lett; B. A. Ovrut and P. J. Steinhardt, Rockefeller University Preprint RU84/B/** (1984), submitted to Phys. Rev. D.; Rockefeller University Preprint RU84/B/91 (1984).
[9]B. A. Ovrut and P. J. Steinhardt, Penn Preprint UPR-0164T (1984).
[10]A. Albrecht, S. Dimopoulos, W. Fischler, E. Kolb, S. Raby and P. Steinhardt, Nucl. Phys. **B229**, 528 (1983); for an impersonal discussion of reheating, see A. Albrecht and P. Steinhardt, Phys. Lett. **131B**, 45 (1983).
[11]L. O'Raifeartaigh, Nucl. Phys.**B96**, 331 (1975).
[12]S. Dimopoulos and S. Raby, Nucl. Phys. **B219**, 479 (1983).
[13]J. Polonyi, Budapest:Preprint KFKI-1977-93, unpublished. See also E. Cremer, B. Julia, J. Scherk, P. van Nieuwenhuizen, S. Ferrara, L. Girardello, Phys. Lett. **79B**,231; Nucl. Phys. **B147**, 105 (1979);(1979).
[14]D. V. Nanopoulos, K. A. Olive, M. Srednicki and K. Tamvakis, Phys. Lett. **123B**, 41 (1983); D. V. Nanopoulos, K. A. Olive, and M. Srednicki, Phys. Lett. **127B** 30 (1983); D. V. Nanopoulos and M. Srednicki, Phys. Lett. **133B**, 287 (1983). See also M. Srednicki's contribution to these proceedings.
[15]E. Cremmer, B. Julia, J. Scherk, P. van Niewenhuizen, S. Ferrara and L. Girardello, Phys. Lett. **79B**, 23 (1978); Nucl. Phys. **B147**, 105 (1979); J. Bagger and E. Witten, Phys. Lett. **118B**, 103 (1982).
[16]D. Kirshnitz and A. Linde, Phys. Lett. **42B**, 471 (1972); L. Dolan and R. Jackiw, Phys. Rev. **D9**, 3320 (1974); S. Weinberg, Phys. Rev. **D9**, 3357 (1974).
[17]K. Olive and M. Srednicki, UCSB preprint 84-0519 (1984). See also G. B. Gelmini, D. V. Nanopoulos, K. A. Olive, Phys. Lett. **131B**, 53 (1983).
[18]P. J. Steinhardt and M .S. Turner, Phys. Rev. **D29**, 2162 (1984).
[19]J. Ellis and G. Steignman, Phys. Lett. **89B**, 186 (1980).
[20]We are grateful to A. Guth for his discussions on this point.
[21]S. Nussinov and P. J. Steinhardt, unpublished (1984).

[22]A. Linde, Phys. Lett. **129B**, 177 (1983).
[23]A. Vilenkin, Phys. Lett. **117B**, 25 (1982).
[24]H. Nilles, Nucl. Phys. **217B**, 366 (1983); R. Arnowitt, A. Chamsedine, and P. Nath, Phys. Rev. **49**, 970 (1982); L. Ibanez, Phys. Lett. **118B**, 73 (1982); R. Barbieri, S. Ferrara, and C. Savoy, Phys. Lett. **119B**, 343 (1982).
[25]R. Holman, P. Ramond, G. G. Ross, Phys. Lett. **137B**, 343 (1984).
[26]See P. J. Steinhardt, "Natural Inflation" in The Very Early Universe, ed. by G. Gibbons, S. Hawking, and S. Siklos, (Cambridge University Press, 1983).
[27]See M. Srednicki's contribution to these proceedings.
[28]B. A. Ovrut and S. Raby, Phys. Lett. **31B**, 461 (1984). See also B. Gato, J. Leon, M. Guiros, M. Ramon-Medrano, CERN-TH 3538 (1983).
[29]L. Ibanez, Madrid Preprint FIUAM/82-8 (1982); J. Ellis, D. Nanopoulos, and K. Tamvakis, CERN preprint TH. 3418 (1982); L. Alvarez-Guame, J. Polchinski, and M. Wise, Nucl. Phys. **B221**, 495 (1983);H. P. Nilles, Phys. Lett. **115B**, 193 (1982); B. Ovrut and S. Raby, Phys. Lett. **130B**, 277 (1983).
[30]For related examples, see C. Oakley and G. Ross, Phys. Lett. **125B**, 159 (1983); L. Ibanez and G. Ross, Preprint RI-83-060/T-326 (1983); B. Ovrut and S. Raby, Phys. Lett. **134B**, 51 (1984).
[31]For a review see E. W. Kolb and M. S. Turner, Ann. Rev. Nucl. Part. Sci. **33**, 645 (1983).
[32]See J. S. Kim, A. Masiero, D. V. Nanopoulos, CERN-TH-3866 (1984), for a recent discussion.
[33]L. Abbott, E. Farhi, and M. Wise, Phys. Lett. **117B** 29 (1982); A. Dolgov and A. Linde, Phys. Lett. **118B**, 329 (1982).
[34]See, for example, L. Krauss, Nucl. Phys. **B227**, 556 (1983). Recently, in J. Ellis, J. Kim, D. V. Nanopoulos, CERN-TH-3839 (1984), a limit of $T_{rh} \lesssim 10^{10}$ GeV was derived. This is barely within the allowed region for our parameters.
[35]A. Linde, J. Ellis, D. V. Nanopoulos, Phys. Lett. **118B**, 59 (1982).
[36]See, for example, G. D. Coughlan, R. Holman, P. Ramond, G. G. Ross, U. of Florida preprint UFTP-83-24 (1984).
[37]G. Coughlan, W .Fischler, E.W. Kolb, S. Raby, and G. G. Ross, Phys. Lett. **131B**, 59 (1983).
[38]M.S. Turner, Phys. Rev.**D28**, 1243 (1983).

Primordial Inflation in N=1 Supergravity: A Brief Review

Mark Srednicki

The new inflationary scenario[1] is a compelling model of very early cosmology due to its ability to resolve the horizon, flatness, isotropy, homogeneity, and rotation problems. In addition, it is possible to generate the correct spectrum of primordial density fluctuations which can act as seeds of galaxy formation.[2] The problem is to construct an equally compelling (and, one hopes, correct) model of microphysics which produces an acceptable history of the universe. The general requirements of such a model do not seem difficult to satisfy. The key ingredient is the inflaton, a scalar field whose potential leads to a phase transition at some time during the very early universe. Bubbles or regions of the new phase must form with the inflaton field still far from its value at the absolute minimum of its potential. The energy stored in the inflaton field causes each bubble to grow exponentially, producing an island universe with $\Omega = 1$. Meanwhile, quantum fluctuations in the inflaton field result in the needed primordial density perturbations. Finally, the inflaton converts its stored energy into a hot gas of ordinary particles.

It is possible to elucidate the general features of potentials which yield acceptable scenarios along these lines.[3,4] The results are somewhat disturbing: the needed potential bears no obvious resemblance to any of those usually encountered in particle physics. The scalar self-couplings must all be very small. The upper limit on some of them grows with the value of the inflaton field at its absolute minimum; thus, we would like this to be as large as possible. The general shape cannot be achieved with a fourth degree or lower polynomial if the original phase is metastable at zero temperature. Higher order polynomials lead to nonrenormalizable theories; such theories generally make sense only if all possible couplings are included, and even then only below some characteristic energy scale (the 'cutoff').

What are we to make of this? We need to introduce very small couplings and, probably, nonrenormalizability in some natural way. Fortunately there is an existing framework in which this can be done without drastic revision of our usual ideas of particle physics. This framework is supergravity.

Supersymmetry relates bosons to fermions, and has been popular for several years now due to the possibility of resolving[5] the notorious hierarchy problem: why is the scale of weak interactions (~100 GeV) so much less than that of grand unification (~10^{15} GeV) or gravity (~10^{19} GeV)? Here, too, the main requirement is essentially a very small effective coupling of the electroweak Higgs scalars to superheavy particles. Ordinarily, radiative corrections produce large couplings even if they were small initially. In supersymmetric theories, there are remarkable cancellations which prevent this. In this sense, very small couplings are natural in supersymmetric theories. If the solution to the hierarchy problem indeed lies in supersymmetry, then clearly models for the inflaton should also take supersymmetry into account. As a bonus, we get a natural framework for the necessarily very small inflaton self-couplings.[3]

Anyone who believes in supersymmetry and gravity automatically believes in supergravity.[6] The spin-2 graviton is related by supersymmetry to a spin-3/2 particle, the gravitino. In the simplest form of supergravity (called N=1) there is one gravitino (more complicated forms of supergravity have up to eight different gravitini; I will not consider these theories here). The amount of supersymmetry breaking can be characterized by the mass of the gravitino (exact supersymmetry would yield a massless gravitino). Gravitational interactions alone will give the spin-0 partners of quarks and leptons masses of order the gravitino mass. These particles have not yet been observed, and so must weigh at least some tens of GeV. If the gravitino has a mass of this order, so do the squarks and sleptons, automatically! This is an elegant and simple way to explain the absence of superpartners in high energy experiments to date. If it is true, it means gravity plays an essential role in particle physics.[7]

Furthermore, N=1 supergravity is a nonrenormalizable theory. The necessary cutoff can be no larger than the Planck mass ($M_{Planck} = G^{-1/2}_{Newton} = 1.2 \times 10^{19}$ GeV), which is the natural scale of gravitational phenomena.

We see that the introduction of supersymmetry, plus gravity, has led us to a theoretical framework incorporating small couplings and nonrenormalizability (with a cutoff at the Planck scale) in a natural way. This is what we need for a successful inflationary scenario.[8]

In N=1 supergravity, all couplings among spin-0 fields and their spin-1/2 partners are specified by giving a single function of these fields, $G(\phi_i, \phi^*_i)$. (All scalar fields must be complex in supergravity.) The scalar potential is then given by

$$V = e^G[G^i(G^i_j)^{-1}G_j - 3] \quad (1)$$

where $G_i = \partial G/\partial\phi_i$, $G^i = \partial G/\partial\phi^*_i$, etc., and I have set $M = M_{Planck}/\sqrt{8\pi} = 1$. The kinetic energy terms are

$$G^i_j \partial^\mu \phi^*_i \partial_\mu \phi^j \quad (2)$$

The kinetic energy assumes its usual form only if $G^i_j = \delta^i_j$; in this case, the couplings are said to be "minimal", and we can write

$$G = \phi^*_i \phi^i + \log f(\phi) + \log f^*(\phi^*) \quad (3)$$

where the superpotential $f(\phi)$ is not a function of the ϕ^*_i. Now the potential becomes

$$V = e^{M^{-2}\phi^*_i\phi^i}[|\frac{\partial f}{\partial \phi_j} + M^{-2}\phi^*_j f|^2 - 3M^{-2}|f|^2] \quad (4)$$

where I have restored factors of the Planck mass (f has dimension mass cubed). Let us focus on the superpotential for a single field ϕ, to be identified with the inflaton.[8] We take the inflaton to be a singlet under the grand unifying gauge group. As already mentioned, we will want the inflaton to have as large an expectation value as possible, possibly inconsistent with the unification scale. We can write the superpotential as a power series in the field:

$$f(\phi) = \mu^3[\lambda + \sum_{n=0}^{\infty}(n+1)^{-1}\lambda_n(\phi/M)^{n+1}] \quad (5)$$

Here I have scaled the field with factors of M, and the overall superpotential with a different (and as yet unspecified) mass parameter μ. The λ_n will be chosen to be in the range $0(10^{-3})$ to $0(1)$. The most natural value for μ is, of course, of order M; this, however, will not lead to the needed small inflaton self-couplings, and so we must appeal to the more technical aspect of naturalness provided by supersymmetry. Since μ enters only in the overall scale of f, the only scale entering in the determination of the value of ϕ at the minimum of the potential is M. Thus, we expect ϕ to be of order M at the minimum. Since only the scale of gravity (and not of grand unification) is relevant, this scenario is called primordial inflation.

We will construct f so that there is a metastable minimum at $\phi = 0$ (it need not be exactly at $\phi = 0$, but should differ by order M from the true minimum). Thus we get the large value of ϕ which is needed to minimize fine-tuning among the λ_n. By appropriate choices of the λ_n and μ (μ must be about $10^{-2}M$), an acceptable shape can be achieved for the inflaton potential. One difficulty[9] is that finite temperature corrections[10] will shift the metastable minimum towards the true minimum, possibly causing the inflaton field to end up in the true minimum before any inflation occurs. This could be avoided if the shifted metastable minimum never moves byond the barrier between the two minima.

Our region of the universe will form according to the standard new inflationary scenario. We must require that the inflaton field convert itself into a hot gas of ordinary particles, and so it must be coupled somehow to ordinary fields. We could, of course, simply include some sort of coupling in the complete superpotential. A more elegant scheme[11] is to leave the inflaton separate, so that the complete superpotential is the sum of $f(\phi)$ as given in Eq. (5) and terms involving only the usual fields of a grand unified theory (and not the inflaton). In this case, all couplings of the inflaton will be suppressed by inverse powers of M. An inflaton particle will decay to ordinary particles at a rate $\Gamma_\phi \simeq m_\phi^3/M^2$, where m_ϕ is the inflaton mass in the stable minimum and should be of order μ^3/M^2, or about 10^{12} GeV. This rate is quite small, and so the universe becomes matter dominated by ϕ particles which decay when the Hubble parameter H is of order Γ_ϕ. Since the energy density of the ϕ particles is $3H^2M^2$, the decay products will have a temperature roughly given by

$$T \simeq 1.7N^{-1/4}(m_\phi/M)^{3/2}M \simeq 10^9 \text{ GeV} \tag{6}$$

where N is the number of degrees of freedom in the decay products. If $m_\phi > m_{H_3}$, the decay products will include Higgs color triplets. If $T < m_{H_3}$, they will be out of equilibrium,[4,12] and their subsequent decays will generate the needed baryon asymmetry. If $T > m_{H_3}$, some other mechanism is needed to get out-of-equilibrium decay.[11]

Although supergravity provides a useful framework for addressing the naturalness problems of both inflationary cosmology and particle physics, it introduces new problems. Gravitinos, with a mass $m_{3/2}=0(100)$ GeV and a decay rate $\Gamma_{3/2} \simeq m_{3/2}^3/M^2$, can prove to be a cosmological embarrassment.[13] In order to avoid a universe dominated by gravitinos either today or during the time of helium synthesis, we must make sure that not too many of them were produced after inflation.[14] We can kinematically forbid inflaton decay to gravitino plus inflatino (which would otherwise produce about as many gravitinos as any other particle species), but ordinary scattering events[15,11] will still produce gravitinos at a rate $\Gamma_p \simeq \alpha N T^3/M^2$, yielding an abundance relative to photons of order $\Gamma_p/H \simeq 10^{-9}$ using the previous parameters. This is a safe value.

Another possible problem is associated with the scalar field responsible for spontaneous breakdown of local supersymmetry. Such a field has a mass of order the gravitino mass, and can be far from its minimum after inflation. Energy stored in this field must be reduced to an acceptable value.[16] Known solutions of this problem are very model dependent.[17] An elegant possibility is that supersymmetry breaking is dynamical in nature (i.e., due to instantons or fermion condensates),[5,18] and so there simply is no problematic scalar field.

I have not yet discussed how the universe manages to end up in an SU(3) x SU(2) x U(1) symmetric phase after inflation. This problem also has a variety of model dependent solutions.[19,20] Typically, it is necessary to make the potential for the field which breaks the grand unification symmetry (i.e., SU(5)) very flat. This can be achieved with a judicious use of nonrenormalizable coupling, so that it is not necessary to introduce very small coupling constants (i.e., of order 10^{-12}) by hand.[19]

Are there any traces left today of supersymmetric inflation? One of the new particles predicted by supersymmetry is stable. Naturally, this particle is a good candidate for the dark matter needed to have $\Omega = 1$, provided it has zero electric charge. The photino is a likely possibility.[21] Surprisingly, a galactic halo of ~3 GeV photinos would produce observable fluxes of cosmic ray antiparticles through photino-photino annihilation.[22] Indeed, the predicted number of low energy antiprotons is comparable to the experimental result,[23] a result which is difficult to explain via secondary production in the standard cosmic ray propagation models. Confirmation of this exciting possibility would imply that supersymmetry has played a role in astrophysics from the inflationary era to the present day.

I would like to thank John Ellis, John Hagelin, Demetri Nanopoulos, Keith Olive, Joe Silk, and Kyriakos Tamvakis for very enjoyable collaborations.

REFERENCES

[1]A. Guth, Phys. Rev. D**23**, 347 (1981); A. D. Linde, Phys. Lett. **108B**, 389 (1982); A. Albrecht and P. J. Steinhardt, Phys. Rev. Lett. **48**, 1220 (1982).

[2]A. Guth and S.-Y. Pi, Phys. Rev. Lett. **49**, 1110 (1982); S. Hawking, Phys. Lett. **115B**, 395 (1982); A. A. Starobinski, Phys. Lett. **117bb**, 175 (1982); J. M. Bardeen, P. J. Steinhardt, and M. S. Turner, Phys. Rev. D**28**, 579 (1983).

[3]J. Ellis, D. V. Nanopoulos, K. A. Olive, and K. Tamvakis, Phys. Lett. **118B**, 335 (1982); Nucl. Phys. **B221**, 524 (1983); Phys. Lett. **120B**, 331 (1983).

[4]P. J. Steinhardt and M. S. Turner, Phys. Rev. D**29**, 2162 (1984).
[5]E. Witten, Nucl. Phys. **B185**, 513 (1981); M. Dine, W. Fischler, and M. Srednicki, Nucl. Phys. **B189**, 575 (1981); S. Dimopoulos and S. Raby, Nucl. Phys. **B192**, 353 (1981).
[6]E. Cremmer, S. Ferrara, L. Girardello, and A. Van Proeyen, Nucl. Phys. **B212**, 413 (1983), and references therein; J. Basser, Nucl. Phys. **B211**, 302 (1983), and references therein.
[7]For a review and references, see J. Polchinski, these proceedings.
[8]D. V. Nanopoulos, K. A. Olive, M. Srednicki, and K. Tamvakis, Phys. Lett. **123B**, 41 (1983).
[9]B. A. Ovrut and P. J. Steinhardt, Phys. Lett. **133B**, 161 (1983).
[10]G. B. Gelmini, D. V. Nanopoulos, and K. A. Olive, Phys. Lett. **131B**, 53 (1983); K. A. Olive and M. Srednicki, UCSB preprint (1984); P. Binetruy and M. K. Gaillard, UC Berkeley preprint UCB-PTH-84/19 (1984).
[11]D. V. Nanopoulos, K. A. Olive, and M. Srednicki, Phys. Lett. **127B**, 30 (1983).
[12]R. Holman, P. Ramond, and G. G. Ross, Phys. Lett. **137B**, 343 (1984).
[13]S. Weinberg, Phys. Rev. Lett. **48**, 1303 (1982).
[14]J. Ellis, A. D. Linde, and D. V. Nanopoulos, Phys. Lett. **118B**, 59 (1982).
[15]S. Weinberg, unpublished, quoted in Ref. 11.
[16]G. D. Coughlan, W. Fischler, E. Kolb, S. Raby, and G. G. Ross, Phys. Lett. **131B**, 59 (1983).
[17]D. V. Nanopoulos and M. Srednicki, Phys. Lett. **133B**, 287 (1983); M. Dine, W. Fischler, and D. Nemeschansky, Phys. Lett. **136B**, 169 (1984); G. D. Coughlan, R. Holman, P. Ramond, and G. G. Ross, Phys. Lett. **140B**, 44 (1984); B. A. Ovrut and P. J. Steinhardt, Rockefeller Univ. preprints (1984), and P. J. Steinhardt, these proceedings.
[18]I. Affleck and M. Dine, Phys. Rev. Lett. **52**, 1677 (1984), and references therein.
[19]D. V. Nanopoulos, K. A. Olive, M. Srednicki, and K. Tamvakis, Phys. Lett. **123B**, 41 (1983).
[20]A. D. Linde, Phys. Lett. **131B** 330 (1983); Phys. Lett. **132B**, 312 (1983); K. Enqvist and D. V. Nanopoulos, CERN preprint TH.3847 (1984).
[21]S. Weinberg, Phys. Rev. Lett. **50**, 387 (1983); H. Goldberg, Phys. Rev. Lett. **50**, 1419 (1983); J. Ellis, J. S. Hagelin, D. V. Nanopoulos, K. Olive, and M. Srednicki, Nucl. Phys. **B238**, 453 (1984).
[22]J. Silk and M. Srednicki, Phys. Rev. Lett. **53**, 624 (1984).
[23]A. Buffington, S. Schindler, and C. Pennypacker, Astrophys. J. **248**, 1179 (1981).

Supersymmetric Inflationary Cosmology

R. Holman

ABSTRACT

We use N = 1 supergravity coupled to hidden sectors to construct a completely viable inflationary scenario. Constraints on the gravitino abundance allow us to predict that $(\delta\rho/\rho)_{gal} \sim 10^{-4}$ as required for galaxy formation.

SUPERSYMMETRIC INFLATIONARY COSMOLOGY

Inflation is an attractive solution to the flatness, horizon and monopole problems of the standard cosmology (and GUTs for the latter).[1] It would be even more attractive if it could be implemented successfully! This has not been possible for many reasons; in Guth's original version the universe could not exit from the inflationary era gracefully.[1] The "New" inflationary scenario[2] could do so, but fell into the wrong vacuum[3] and, more seriously, generated dangerously large density fluctuations.[4] Models based on adding extra scalars to the new inflationary scenario,[5] on global supersmmetry (SUSY)[6] and on N = 1 supergravity (SUGRA)[7] have been constructed but none of these are completely satisfactory.

We believe we can remedy this state of affairs with what we (P. Ramond, G. Ross and myself) call the Supersymmetric Inflationary Cosmology (S.I.C.).[8] This model satisfies all known cosmological constraints and constrains the particle physics sector to a large extent. What is more surprising is that, in essence, it is a one-parameter model. We will see that this one parameter is bounded rather tightly by cosmological constraints and predicts the amplitude of density fluctuations to be $\sim 10^{-4}$ on galactic scales, which is just the value required to create galaxies! This is very different from previous models where this was used as an input parameter.

We follow Ref. (7) and use the N = 1 supergravity coupled to hidden sectors as the basis of our model. Our superpotential consists of three sectors: the inflaton sector (I) responsible for driving inflation, the O'Raifertaigh sector (S) which breaks SUSY and the GUT sector (G) which is self explanatory. These sectors will be hidden from one another except for gravity strength interactions. We write our total superpotential, P, as the sum:

$$P = I + S + G, \tag{1}$$

where I and S are given by:

$$I = \frac{\Delta^2}{M}(\Phi - M)^2, \quad \left(M = \frac{M_P}{\sqrt{8\pi}} \simeq 2.4 \times 10^{18}\ \text{GeV}\right) \tag{2a}$$

$$S = A(\lambda B^2 + \mu^2) + mBC + q\ . \tag{2b}$$

We will discuss the GUT sector later. The superfields Φ,A,B,C are gauge singlet, chiral superfields, Δ,μ,M,m are mass parameters, λ is a Yukawa coupling and q is a constant, needed to cancel the cosmological constant of the ground state of the theory. When we couple P to the N = 1 SUGRA multiplet, an effective scalar potential V is generated:[9]

$$V = \exp \sum_i \frac{|\phi_i|^2}{M^2} \sum_i \left|\frac{\partial P}{\partial \phi_i} + \frac{\phi_i^+ P}{M^2}\right|^2 - \frac{3}{M^2}|P|^2 , \tag{3}$$

where the sum is over all scalars in the theory.

We now turn to the inflaton sector I; the effective potential for the inflaton (scalar piece of Φ) is

$$V(\phi) = \Delta^4 e^{\tilde{\phi}^2}\left[1 - \tilde{\phi}^2 - 4\tilde{\phi}^3 + 7\tilde{\phi}^4 - 4\tilde{\phi}^5 + \tilde{\phi}^6\right] , \tag{4}$$

where $\tilde{\phi}$ is the real part of ϕ scaled by M. The scale for inflation is set by Δ^4. $V(\phi)$ has a minimum at $\phi = M$ and a long rollover region.[10] From Eq. (4) we find that ϕ has mass $\sim O(\Delta^2/M)$, self interaction strength $(\Delta/M)^4$ and interaction strength $(\Delta/M)^2$ with other sectors. This brings us to a major difference between our model and previous inflationary scenarios. For the allowed values of Δ/M, ϕ cannot be in thermal equilibrium with the surrounding heat bath until $T \lesssim 10^3$ GeV. Hence the finite T formalism is inapplicable until this temperature. In particular, we cannot use it to stabilize the origin. This could allow ϕ to be too near M to allow sufficient inflation to occur. We can deal with this in two ways. We can add an extra field ψ to the inflaton sector with a strong coupling to ϕ, e.g.:

$$I' = (\bar{m} + h\phi)\psi^2 . \tag{5}$$

For a certain region in $(\bar{m},h)$ space, this will generate a thermal barrier near $\phi = 0$ while leaving the rest of the scenario intact. We have shown that the ϕ_R ridge is the preferred one locally, but there may be additional minima waiting to swallow up ϕ_R. The second way out of this problem, is to use the recent work of Linde[11] and assume that the values of $\phi_{initial}$, $\dot{\phi}_{initial}$ satisfy some probability distribution in various regions of space; those regions with $\phi_{initial} \lesssim H_0$, $\dot{\phi}_{initial} \lesssim H_0^2$ will inflate generating bubble universes (here $H_0^2 = V(0)/3M^2 \approx \Delta^4/3M^2$). We will in fact see that we have a large amount of freedom in the choice of value of $\phi_{initial}$ that can generate sufficient inflation.

We will also need to know that ϕ couples most strongly to the heaviest scalars of the theory.

We next examine the O'Raifertaigh sector S. The F-term $\partial S/\partial A$ breaks SUSY so that the spinor part of A is the Goldstino which gives the gravitino its mass via the super-Higgs mechanism. The scalar part of A can thus decay into two gravitinos ($\tilde{G}$).

The gravitino mass, $m_{\tilde{G}}$, is μ^2/M so that, for SUSY to be useful for particle physics μ must be $O(10^{10-11}$ GeV) [see e.g. Polchinski's contribution in these proceedings]. Now the gravitino will decay into photinos and photons after nucleosynthesis; in order that the predictions of the He abundance of the standard cosmology remain undisturbed we must demand that the ratio of energy densities $\rho_{\tilde{G}}/\rho_{\gamma}$ at decoupling be $\lesssim$ O(1-10) which implies [see the 2nd reference in[8] for the calculation]

$$\frac{\Delta}{M} > 10^{-4} - 10^{-5} , \tag{6}$$

When the inflaton oscillates about its minimum and reheats the universe to a temperature T_R, gravitini can be created. In order to not produce too many we must demand that[8]

$$\frac{\Delta}{M} < 10^{-3.5} , \ \alpha_{\lambda} \lesssim 10^{-3} - 10^{-4} , \tag{7}$$

where

$$\alpha_{\lambda} = \lambda^2/4\pi . \tag{8}$$

Hence the SUSY breaking sector puts stringent bounds on the scale of inflation $\Delta = 10^{-4} \lesssim \Delta/M \lesssim 10^{-3.5}$ (we choose 10^{-4} in Eq. (8) so that ϕ will be massive enough to decay into the Higgs color triplet which can then create baryon asymmetry via its decays).

While we have not yet constructed a specific GUT sector, we are now in a position to list some constraints on it:

a) It must generate acceptable particle physics(!),

b) GUT breaking must occur at scales greater than Δ to solve the monopole problem,

c) There must be no cosmological constant at the minimum of the full theory,

d) There must be enough Higgs color triplets to generate a $\Delta B \sim O(10^{-5})$,

e) In order that ϕ decay predominantly into Higgs triplets, H_3, we must have a mechanism which splits up the SU(2) doublet from H_3. Such mechanisms exist, i.e. sliding singlet or incomplete multiplet (we prefer the latter).[12]

Condition a) is, of course, highly non-trivial; in particular, the GUT sector must have light triplets [$m(H_3) \sim 10^{10} - 10^{11}$ GeV] so that ϕ may decay into H_3's. However, one must make sure that proton decay is not too fast!

The scenario is shown in Table 1. We now turn to the results of the model:

a) The number of e-folds, N, is given by

$$N \simeq \int_{\phi_{initial}}^{\phi_e} H_0 \, dt \ . \tag{9}$$

If we take $\phi_{initial} \sim H_0$, then, using ϕ_e(e = end of inflation)

~ 1/8 M we find $N = 10^8$! In order to secure just the right amount of inflation (N = 60) we could have taken $\phi_{initial} \sim 1/10\ M = 10^7\ H_0$. Hence there is a <u>very</u> large range of initial conditions that allows inflation in our model.

b) By using conservation of energy, we find that the reheat temperature T_R is given by Δ^3/M^2 or 10^6 GeV for $\Delta/M \sim 10^{-4}$;

c) Using the Abbott, Farhi, Wise[13] results for baryosynthesis via ϕ decays into H_3 we find:

$$\frac{n_b}{n_\gamma} \sim \left(\frac{T_R}{m_3}\right) \Delta B \sim 10^{-4}\ \Delta B\ . \tag{10}$$

(c.f. condition c) on the GUT sector). Finally, we compute the spectrum of density fluctuations in our model using reference (4) to find:

$$\frac{\delta\rho(\lambda)}{\rho} \sim \left(\frac{\Delta}{M}\right)^2 \left[1 + \frac{3}{2}\ \ell n \frac{\lambda}{H_0^{-1}}\right]^2 , \tag{11}$$

where λ is the scale of the fluctuation. For galactic sized fluctuations we find

$$\left(\frac{\delta\rho}{\rho}\right)_{gal} \cong 10^4 \left(\frac{\Delta}{M}\right)^2 \sim 10^{-4}\ (!!), \tag{12}$$

for $\Delta/M \sim 10^{-4}$. Thus we have <u>predicted</u> the (correct!) value of $(\delta\rho/\rho)_{gal}$.

In conclusion, we have constructed an inflationary model which is not only consistent with all known cosmological requirements but is in fact highly <u>predictive</u> and <u>testable.</u>

This model should be viewed as a phenomenological one since we have no idea where the inflaton field comes from; it could perhaps be some bound state of the N = 8 theory.

Finally, we must now construct a viable GUT sector for the theory to prove its internal consistency. This work is now in progress.

REFERENCES

[1] A.H. Guth, Phys. Rev. **D23**, 347 (1981).

[2] A.D. Linde, Phys. Lett. **108B**, 389 (1982); A. Albrecht and P.J. Steindardt, Phys. Rev. Lett. **48**, 1220 (1982).

[3] J.D. Breit, S. Gupta and A. Zaks, Phys. Rev. Lett. **51**, 1007 (1983); J. Kodaira and J. Okada, Phys. Lett. **131B** 287 (1983); C.W. Kim and J.S. Kim, Johns Hopkins preprint JHU-HET 8310.

[4] S.W. Hawking, Phys. Lett. **115B**, 295 (1982); A. Guth and S.Y. Pi, Phys. Rev. Lett. **49** 1110 (1982); J. Bardeen, P.J. Steinhardt and M.S. Turner, Phys. Rev. **D28** 679 (1983); A. Starobinski, Phys. Lett. **117B**, 175 (1982).

[5] Q. Shafi and A. Vilenkin, Phys. Rcv. Lett. **52**, 691 (1984); S.Y. Pi, Phys. Rev. Lett. **52** 1725 (1984).

[6] P. Steinhardt in <u>The Very Early Universe</u>, ed. by G. Gibbons, S. Hawking and S. Siklos (Cambridge Univ. Press, 1983); A. Albrecht, S. Dimopoulos, W. Fischler, E. Kolb, S. Raby and P.J. Steinhardt, Los Alamos preprint LA-UR 82-2947 (1982).

[7]J. Ellis, D.V. Nanopoulos, K.A. Olive, K. Tamvakis, Phys. Lett. **120B**, 331 (1983); B.A. Ovrut and P.J. Steinhardt, Phys. Lett. **133B**, 161 (1983).
[8]R. Holman, P. Ramond and G.G. Ross, Phys. Lett. **137B**, 343 (1984); G.D. Coughlan, R. Holman, P. Ramond and G.G. Ross, University of Florida preprint, UFTP-83-24.
[9]E. Cremmer, B. Julia, J. Scherk, S. Ferrara, L. Giradello and P. Van Nieuwenhuizen, Phys. Lett. **79B**, 231 (1978).
[10]P.J. Steinhardt and M.S. Turner, Phys. Rev. **D29** 2126 (1984).
[11]A.D. Linde, Phys. Lett. **129B** 177 (1983).
[12]B. Grinstein, Nucl. Phys. **B206**, 387 (1982); S. Dimopoulos and F. Wilczek, Santa Barbara Preprint UM-HE81-71 (1982); A. Masiero, D.V. Nanopoulos, K. Tamvakis and T. Yanagida, Phys. Lett. **117B** 380 (1982).
[13]L.F. Abbott, E. Fahri and M.B. Wise, Phys. Lett. **117B** 29 (1982).

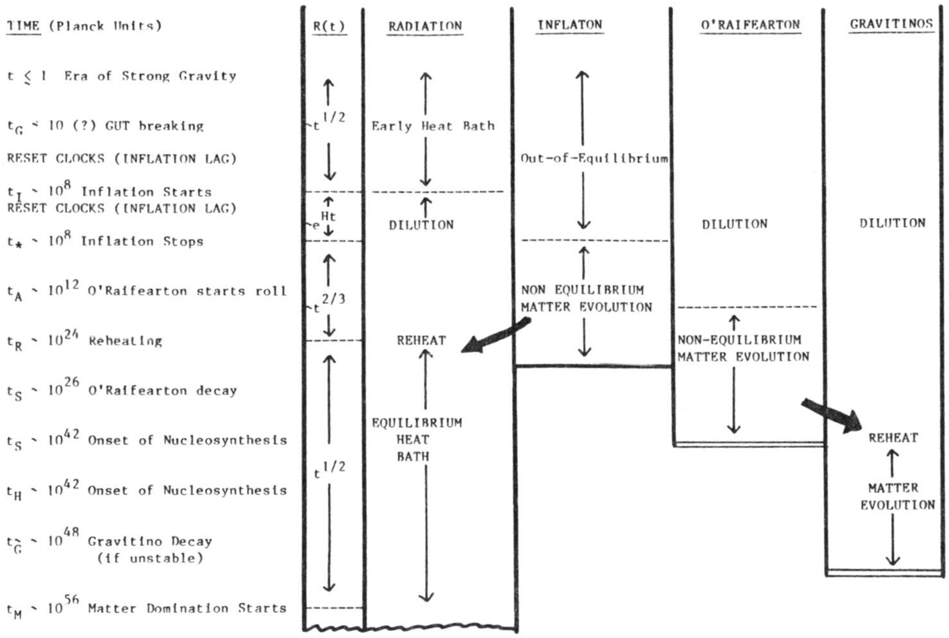

Note: In the above $\Delta = 10^{-4}P$, $\mu = 10^{-8}P$, $\alpha_\lambda = 10^{-4}$

Table 1. Important dates in supersymmetric inflationary cosmology.

Generation of Fluctuations in Cosmological Models

Robert H. Brandenberger

ABSTRACT

It is shown how spatial correlations in the vacuum state wave functional of a scalar quantum field in curved space-time generate classical matter perturbations. The primordial energy density fluctuation spectrum at initial Hubble radius crossing is a scale invariant Zeldovich spectrum.

In inflationary universe models[1,2] there is a causal mechanism which generates the primordial energy density fluctuations required as initial conditions in theories of galaxy formation. The aim of this presentation is to discuss how classical matter fluctuations are initially produced when considering a scalar quantum field theory in curved space-time.

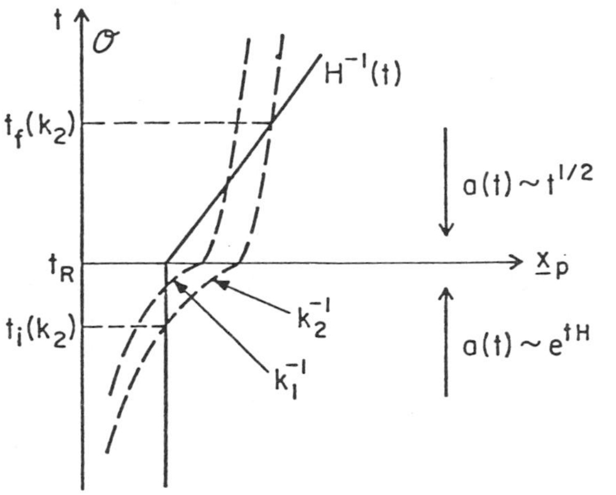

Figure 1: Sketch of the evolution of fluctuations on two scales in physical coordinates x_p. The evolution inside the Hubble radius in the de Sitter phase is time translation invariant.

As sketched in Fig. 1, there are two periods in the evolution of cosmological perturbations in inflationary universe models. Initially in the de Sitter period scales of cosmological interest are inside the Hubble radius $H^{-1}(t)$ (we are considering spatially flat Robertson-Walker metrics with scale factor $a(t)$ and $H(t) = \dot{a}(t)a(t)^{-1}$). Microphysics can act coherently and will generate classical energy density perturbations. Outside the Hubble radius perturbations will then evolve classically, according to the linearized Einstein equations.[3,4]

We will show how quantum fluctuations generate classical matter inhomogeneities, and that the resulting spectrum is scale invariant,[5] i.e. that the amplitude is independent of scale k when measured at the time $t_i(k)$ the scale leaves the Hubble radius. Time translation invariance of the de Sitter phase is the heuristic reason. The evolution of perturbations of different scales are related by time translation.

We will study the initial period $t < t_i(k)$ and will make two crucial assumptions. First, we will neglect metric perturbations.

Since it can be shown[4] that metric perturbations remain very small up to $t_i(k)$, this approximation is well justified. We also neglect quantum field nonlinearities, i.e. we treat the generation of classical fluctuations as a problem of free field theory in the de Sitter phase of a Robertson-Walker universe. This simplification is reasonable since in new inflationary universe models[2] the scalar field potential near the origin is virtually flat. The scalar field $\phi_o(t)$ remains in this flat section of the potential until long after all scales of cosmological interest have crossed the Hubble radius.

The basic question is how inhomogeneities can develop given a homogeneous and isotropic background geometry and a quantum state which does not break space translational invariance. The mechanism must be nontrivial as the following incorrect analysis shows: The Einstein equations are

$$G_{\mu\nu} = 8\pi G T_{\mu\nu}^{cl} \; . \tag{1}$$

$T_{\mu\nu}^{cl}$ is a classical energy-momentum tensor for matter. Given a quantum field theory as matter source, the question arises how to define a classical energy-momentum tensor. The most obvious approach is to take the expectation value of the quantum operator $T_{\mu\nu}$ in the prescribed quantum state $|s\rangle$. But since the state is space-translation invariant, the expectation value will be x independent. The approach predicts the absence of fluctuations.

It is not hard to see what is missing in the above argument: although the state $|s\rangle$ is translationally invariant, it contains spatial correlations. The main idea of this work is to demonstrate how these spatial correlations give rise to classical matter fluctuations.

The usual argument for fluctuations is based on Hawking radiation.[7] An observer with a particle detector traveling along a world line in de Sitter space will detect a thermal flux of particles corresponding to a temperature $T_H = H/2\pi$. It is, however, incorrect to conclude from this that there will be thermal matter fluctuations. After all, the energy-momentum tensor of Hawking radiation is not that of a thermal bath. It is de Sitter invariant[8] ($p = -\rho$) or at least (in the case of an approximate de Sitter phase of a FRW universe) approximately de Sitter invariant.[9]

A second method to predict the magnitude of fluctuations is based on the statistics of correlation functions. The mean deviation of a stochastic function f(x) on wave number k is related to its autocorrelation function by

$$(\delta\tilde{f}(k))^2 = \frac{k^3}{(2\pi)^3} \int d^3x \; e^{i\mathbf{k}\cdot\mathbf{x}} \langle f(\mathbf{x})f(0)\rangle \; . \tag{2}$$

Guth and Pi[5] and others obtain initial fluctuations in the classical matter field $\phi(\mathbf{x},t)$ by applying Eq. (2) and using the two-point Green's function in place of the classical autocorrelation function. In this work we attempt to justify the correlation function method from "first principles".

To motivate the physical arguments for a scalar quantum field theory, we first consider a quantum mechanical harmonic oscillator with one degree of freedom q in an expanding universe. The ground state wave function is a Gaussian with width inversely proportional to the frequency of the oscillator. In an expanding universe the frequency scales as $a(t)^{-1}$. Hence the width increases as a(t). The width of the same packet is given by the two-point function. Hence the classical position of the oscillator is

$$q^{cl}(t) = (\langle\psi|Q^2|\psi\rangle)^{1/2} . \tag{3}$$

A scalar quantum field theory is a quantum system with an infinite number of degrees of freedom. A state wave functional contains not only information about one degree of freedom, but also about correlations between different degrees of freedom, i.e. spatial correlations. These correlations correspond to classical field perturbations. Thus the basic idea of our approach is to define a classical field $\phi_{cl}(\mathbf{x},t)$ which contains both information about the expected value of a single degree of freedom and about spatial correlations.[10,11] We define

$$\phi_{cl}(\mathbf{x},t) \equiv \phi_o(t) + \delta\phi(\mathbf{x},t) . \tag{4}$$

$\phi_o(t)$ gives the space-averaged spread of the vacuum state wave functional, and $\delta\phi(\mathbf{x},t)$ describes spatial correlations. Hence in analogy to the case of a single degree of freedom, Eq. (3).

$$\phi_o(t) \equiv (\langle\psi_o|\Phi^2(\mathbf{x})|\psi_o\rangle)^{1/2} . \tag{5}$$

$|\psi_o\rangle$ is the vacuum state. We work in the Schrödinger representation in which the state wave functionals carry the time dependence. We define $\delta\phi^2$ in momentum space as the Fourier transform of the spatial correlation function

$$\delta\tilde{\phi}(\mathbf{k},t) = (V\int d^3r\ e^{i\mathbf{kr}}\langle\psi_o|\Phi(\mathbf{r})\Phi(0)|\psi_o\rangle)^{1/2}. \tag{6}$$

The square of the right-hand side is the expectation value of $\tilde{\Phi}(\mathbf{k})^2$ in the vacuum state, given a volume cutoff V. $\delta\tilde{\phi}(\mathbf{k},t)$ is simply related to the two-point Green's function in momentum space

$$\langle\psi_o|\tilde{\Phi}(\mathbf{k})^*\tilde{\Phi}(\mathbf{e})|\psi_o\rangle = V^{-1}\delta^3(\mathbf{k}-\mathbf{e})\delta\tilde{\phi}^2(\mathbf{k},t) . \tag{7}$$

$\phi_{cl}(\mathbf{x},t)$ can be used to compute a classical energy-momentum tensor $T_{\mu\nu}^{cl}(\mathbf{x},t)$ by applying the usual formulas for classical field theory. In particular, the classical energy density fluctuation becomes

$$\frac{\tilde{\delta\rho}}{\rho}(\mathbf{k},t) = \dot{\phi}_o(t)\delta\tilde{\dot{\phi}}(\mathbf{k},t)\rho^{-1} . \tag{8}$$

Our ideas lead to the following procedure for computing initial energy density fluctuations at Hubble radius crossing.[10] We consider a free massless scalar field in the de Sitter phase of the cosmological model. The first step is to define what we mean by the vacuum state (we will return to this point). We then compute the two-point Green's functions in position and momentum space and use Eqs. (4)-(7) to define a classical field which is not homogeneous in space. The energy density fluctuations are then determined by evaluating Eq. (8) at Hubble radius crossing.

We now turn to a few technical points, the most important of which is the correct definition of the vacuum state. The vacuum $|\psi_0\rangle$ is defined by

$$\alpha_{\mathbf{k}}|\psi_0\rangle = 0 \ \forall\ \mathbf{k} \tag{9}$$

where the $\alpha_{\mathbf{k}}$ are the annihilation operator coefficients of the expansion of the quantum field $\Phi(\mathbf{x},t)$ in terms of positive and negative frequency solutions $g_{\pm}^{\mathbf{k}}(t)\exp(i\mathbf{k}\cdot\mathbf{x})$ of the classical field equations:

$$\Phi(\mathbf{x},t) = (2\pi)^{-3}\int d^3k\{\alpha_{\mathbf{k}} g_{+}^{\mathbf{k}}(t)e^{i\mathbf{k}\cdot\mathbf{x}} + \text{h.c.}\} . \tag{10}$$

This prescription relies on having a predetermined space-time slicing, a time direction in order to define positive and negative frequencies and a flat spatial section in order to be able to Fourier transform. In special relativity there is a unique class of inertial frames; in general relativity there is not. Hence there is no unique vacuum state. Given a frame, we can define a state by the above second quantization procedure. The state we obtain is empty of scalar particles when viewed from the chosen frame.

Our choice of frame is motivated by physical considerations. In the usual inflationary universe models a hot FRW period precedes the de Sitter phase. Matter is a thermal bath of relativistic particles in the FRW phase. The bath is at rest with respect to the FRW coordinate frame. After a few Hubble times into the de Sitter phase the particles initially present will have redshifted away exponentially and the resulting state can be described as empty of particles in the coordinate frame. Hence our vacuum state is defined by Eq. (9) in the FRW coordinate frame.

We calculate the vacuum state wave functional $\psi_0(\phi(\mathbf{x}),t)$, the quantum field theory analog of the ground state wave function in quantum mechanics, defined by the formal expansion.

$$|\psi_0\rangle = \int[d\phi]\psi_0(\phi(\mathbf{x}),t)|\phi(\mathbf{x})\rangle . \tag{11}$$

$|\phi(\mathbf{x})\rangle$ is an eigenstate of the field operator $\Phi(\mathbf{x})$ with eigenvalue $\phi(\mathbf{x})$; the functional integral runs over all classical field configurations.

Our result is not surprising. The ground state wave function of a quantum mechanical harmonic oscillator is a Gaussian with width inversely proportional to its frequency. A free scalar field theory is an infinite assembly of uncoupled harmonic oscillators, one for each wave number $\mathbf{k}$. Hence we expect the ground state wave functional to be the product of the harmonic oscillator wave functions for each $\mathbf{k}$. The effect of curved space-time is to render the frequencies time dependent. The result is[10]

$$\psi_0(\tilde{\phi}(\mathbf{k}),t) = N \exp\left\{-\frac{1}{2}\, a^3(t)(2\pi)^{-3}\int d^3k\,\omega(\mathbf{k},t)|\tilde{\phi}(\mathbf{k})^2|\right\} . \tag{12}$$

N is the normalization constant and $\omega(\mathbf{k},t)$ is the frequency of the oscillator labeled by comoving coordinate $\mathbf{k}$. $\omega(\mathbf{k},t)$ will depend on the solutions $g_+^{\mathbf{k}}(t)$ which in turn depend on the coupling of gravity to matter. For minimal coupling we obtain[10]

$$\mathrm{Re}\,\omega(k) = ke^{-tH}\frac{(-k\tau)^2}{1+(-k\tau)^2}$$

where $\tau(t) = -H^{-1}\exp(-Ht)$ is conformal time.

Since the vacuum state wave functional is a Gaussian in momentum space, the two-point Green's function in momentum space is simply

$$\langle\psi|\tilde{\Phi}(\mathbf{k})^*\tilde{\Phi}(\mathbf{e})|\psi_o\rangle = \delta^3(\mathbf{k}-\mathbf{e})(2\pi)^3a^{-3}(t)\,\frac{1}{2}\,[\mathrm{Re}\,\omega(k)]^{-1}. \tag{13}$$

The two-point function in position space computed by Fourier transforming Eq. (13) is both infrared and ultraviolet divergent. The infrared divergence is handled by imposing a cutoff $k_{min} = H$. This means that perturbations on wavelengths larger than the Hubble radius at the beginning of the de Sitter phase are set to zero (or taken as part of a local background). The ultraviolet divergence is treated by renormalization.[10,11] For minimal coupling we recover the result of Vilenkin and Ford[12] and Linde[13]:

$$\langle\psi_o|\Phi^2(\mathbf{x})|\psi_o\rangle = (2\pi)^{-2}H^3t. \tag{14}$$

For non-minimal coupling there is an effective gravitationally induced potential barrier which keeps the expectation value of $\Phi^2(\mathbf{x})$ constant in time, but for minimal coupling the wave functional will spread in time.

The physical interpretation of the two-point function in momentum space is interesting. The Green's function scales as in Minkowski space for wavelengths inside the Hubble radius ($-k\tau>1$) but like in de Sitter space for wavelengths with ($-k\tau<1$).

The Green's functions (13) and (14) determine the classical field. By Eqs. (5) and (6)

$$\phi_o(t) = (2\pi)^{-1}H^{3/2}t^{1/2} \tag{15}$$

and

$$\delta\tilde{\phi}(\mathbf{k},t) = V^{1/2}(2\pi)^{3/2}a^{-3/2}(t)\,\frac{1}{\sqrt{2}}\,[\mathrm{Re}\,\omega(k)]^{-1/2}. \tag{16}$$

Hence by Eq. (8), at initial Hubble radius crossing $t_i(k)$

$$\frac{\delta\tilde{\rho}}{\rho}\,(\mathbf{k},t_i(k)) = \mathcal{O}(1)V^{1/2}k^{-3/2}\left(\frac{H}{\sigma}\right)^4 . \tag{17}$$

The crucial factor $k^{-3/2}$ stems from evaluating $a(t)$ at Hubble radius crossing $H^{-1} = k^{-1}a(t_i(k))$. ρ is of order σ^4 (σ is the scale of grand unification) and since H is the only dimensional parameter in the quantum problem, the factor H^4 is required by dimensional arguments (it is easy to explicitly trace its origin).

Equation (17) is our final result. It represents a scale invariant Zeldovich spectrum with amplitude $(H/\sigma)^4$ at initial Hubble radius crossing. We recall that the quantity which characterizes the physical size of perturbations on a length scale k^{-1} is the average mass excess $\delta M/M$ inside a ball of radius k^{-1}. Fluctuations in the energy density on scales smaller than k^{-1} average to zero in the ball and hence do not contribute to $\delta M/M$ on scale k^{-1}. Thus[14]

$$\left(\frac{\delta M}{M}\right)^2(k) = \mathcal{O}(1)V^{-1}k^3\left(\frac{\delta\tilde{\rho}}{\rho}\right)^2(k) , \tag{18}$$

which at initial Hubble radius crossing $t_i(k)$ is of order $(H/\sigma)^4$.

ACKNOWLEDGMENTS

For useful discussions, I thank Jim Bardeen, Doug Eardley, Lenny Susskind, Nick Tsamis, and Tony Zee. This material is based upon research supported in part by the National Science Foundation under Grant No. PHY77-27084, supplemented by funds from the National Aeronautics and Space Administration.

REFERENCES

[1]A. Guth, Phys. Rev. D **23**, 347 (1981).
[2]A. Linde, Phys. Lett. **108B**, 389 (1982); A. Albrecht and P. Steinhardt, Phys. Rev. Lett. **48**, 1220 (1982).
[3]J. Bardeen, P. Steinhardt, and M. Turner, Phys. Rev. D **28**, 679 (1983); A. Guth and S. Y. Pi, Phys. Rev. Lett. **49**, 1110 (1982); S. Hawking, Phys. Lett. **115B**, 295 (1982); A. Starobinsky, Phys. Lett. **117B**, 175 (1982).
[4]R. Brandenberger and R. Kahn, Phys. Rev. D **29**, 2172 (1984).
[5]E. Harrison, Phys. Rev. D **1**, 2726 (1970); Ya. Zeldovich, Mon. Nat. R. Astr. Soc. **160**, 1p (1972).
[6]W. Press, Physica Scripta **21**, 702 (1980).
[7]S. Hawking, Comm. Math. Phys. **43**, 199 (1975); G. Gibbons and S. Hawking, Phys. Rev. D **15**, 2738 (1977).
[8]T. Bunch and P. Davies, Proc. R. Soc. Lond. **A360**, 117 (1978).
[9]R. Brandenberger, Phys. Lett. **129B**, 397 (1983).
[10]R. Brandenberger, Nucl. Phys. **B245**, 328 (1984).
[11]Similar ideas have been discussed by J. Bardeen, unpublished (1983).
[12]A. Vilenkin and L. Ford, Phys. Rev. D **26**, 1231 (1982).
[13]A. Linde, Phys. Lett. **116B**, 335 (1982).
[14]J. Peebles, <u>The Large Scale Structure of the Universe</u> (Princeton University Press, Princeton 1980).

Inflation and Phase Transitions

Robert M. Wald

ABSTRACT

We comment briefly on the cosmological problems and issues that inflation proposes to solve. Then we briefly summarize the arguments, which Mazenko, Unruh, and I have given in detail elsewhere, that a phase transition in the early universe should not, as a general rule, be expected to produce an era of inflation.

Recently, Mazenko, Unruh, and I[1] have analyzed what happens during a phase transition of the type considered in the "new inflationary scenario". We concluded that, in general, such phase transitions will not lead to an era of inflation in the early universe. In this note, I shall briefly summarize the main point of our arguments, referring the reader to Ref. (1) for further details.

Before doing so, however, I would like to comment upon the cosmological problems that inflation proposes to solve: (1) The horizon "problem": Why is the microwave background radiation so isotropic in view of the fact that in the standard Robertson-Walker model (without inflation) photons arriving from angular separations of more than several degrees last interacted with matter in regions that could not have causally communicated with each other. (ii) The flatness "problem": Why was the spatial curvature in the very early universe so small compared with the energy density of matter; equivalently, why is the time at which spatial curvature dominates over energy density (namely, $\tau \gtrsim 10^{10}$ years) so large compared with the natural timescales associated with quantum gravity ($t_P \sim 10^{-43}$sec) or particle physics. (iii) The monopole problem: Why does our universe have many fewer magnetic monopoles than are predicted to be produced in the standard Robertson-Walker model (without inflation) using grand unified models of particle physics. (iv) The galaxy perturbation "problem": From where did the departures from homogeneity required to produce galaxies arise?

I have put the word "problem" in quotes in (i), (ii), and (iv) above because I do not feel that the use of that word is justified; the word "issue" would be far more appropriate. Here I define a problem to be a conflict between theory and observation or an internal inconsistency of a theory. I define an issue to be a fact which one believes should be explained in an ultimate theory but which is neither explained nor contradicted by the present theory. Thus, for example, I consider the question "Why is $\alpha \approx 1/137$?" to be an issue, not a problem, in quantum electrodynamics. Questions (i), (ii), and (iv) involve no contradiction with the standard model; they relate directly to cosmological initial conditions. Since present theory says nothing about these initial conditions (i.e., these initial conditions are part of the input rather than the output of present theory), I will henceforth apply the term "issue" to them.[2] On the other hand, question (iii) is a genuine problem, although at present it is far from clear as to whether it is a problem associated with the standard cosmological models (at $\tau \sim 10^{-35}$sec) or with the grand unified theories.

Issues (i), (ii), and (iv) are of considerable importance in giving us guidance toward searching for an ultimate theory which will explain our initial conditions. The observational facts associated with (i), (ii), and (iv) suggest that the initial conditions of the universe were very "special". If interpreted in this manner, they would lead us to search for an ultimate theory which has such special initial conditions built in; some proposals in that direction have been advanced by Penrose.[3] However, if an era of inflation occurred in the very early universe, then the observational facts mentioned in (i), (ii), and (iv) may become consistent with "generic" initial conditions. We would then have little guidance toward an ultimate theory which accounts for initial conditions, but on the other hand, we may be able to make many more predictions without access to such an ultimate theory, since the dynamical evolution of the observable universe would be far less sensitive to cosmological initial conditions.

Thus, the existence of an era of inflation would dramatically effect the implications of issues (i), (ii), and (iv) for the initial state of the universe. I do not consider the changes in our viewpoint of the initial state of the universe introduced by inflation as intrinsically "good" or "bad", but they clearly are important. Thus, it is of great interest to determine whether it is plausible that an era of inflation occurred in the early universe.

In order for inflation to occur, it is necessary that, for a sufficiently long era, the stress-energy tensor of the matter field have the form

$$T_{ab} = -\lambda g_{ab} \tag{1}$$

where λ is a positive constant (using metric signature $-+++$). It has been proposed that this could occur in a theory with spontaneous symmetry breaking when the fields are cooled below the phase transition temperature. In the simple case of a single Higgs field ϕ, the idea is to "supercool" ϕ by keeping it in the high temperature mimimum of the free energy at $\phi = 0$ for a sufficiently long time as the temperature drops below the phase transition temperature. If during such supercooling the field has negligible kinetic and spatial derivative energy compared with potential energy, its stress-energy will be of the required form, eq. (1). Guth's original proposal[4] involved cooling to $T = 0$ with the field remaining in a metastable state, but it suffered from difficulties in obtaining a "graceful exit" from inflation. The "new inflationary scenario" of Linde[5] and Albrecht and Steinhardt[6] addressed that problem by considering the case where the "$\phi = 0$ minimum" becomes unstable at low temperatures. It is assumed that the field then classically "rolls down the hill" to its true minimum and that inflation occurs during the "roll-down".

Most current work on inflation is aimed at obtaining (zero temperature) potentials from particle physics models which are sufficiently "flat" and have a sufficiently large distance from the "high temperature minimum" at $\phi = 0$ to the true minimum at $\phi = \phi_c$ so that the "roll-down" time is long enough to produce the desired degree of inflation. However, a key point which we emphasized in ref. (1) is that it takes much more than a (zero temperature) potential with a long "roll-down" time to produce the conditions needed for inflation. In order to get the field to "roll down the hill", it certainly is necessary to get the field to the top of the hill initially! Now, it might seem that this is automatically satisfied for the models considered in new inflation, since at high temperatures the free energy has a minimum at $\phi = 0$ (and, indeed, this minimum gets more narrow with increasing temperature). Thus, one has "$\phi = 0$" initially if one starts in thermal equilibrium at high temperatures. However, this argument is not correct. The "ϕ" of which the free energy is a function is the thermal expectation value of the quantum field ϕ averaged over all space. The narrow minimum at $\phi = 0$ of the free energy at very high temperatures does not correspond to a state where the field is ordered at $\phi = 0$ and has only small fluctuations about the value 0. Rather, it corresponds to a highly disordered state, with ϕ rarely being close to the value zero (although, of course, its spatial average is zero and the fluctuations in its spatial average become small since the correlation length becomes small at high temperatures). In other words, the field fluctuates away from zero more, not less, at high temperatures. The field does not, in any meaningful sense, start at the "top of the hill" at high temperatures.

The above remarks have immediate implications for the new inflationary scenario. In order for the dynamics of the field to resemble the "ball rolling down the hill" picture[7] it is necessary that initially the field ϕ be ordered near $\phi = 0$. If ϕ is highly disordered, with only its spatial average being zero, it will locally have direct dynamical access to the true minima of the potential and will immediately break up into domains. The dynamics of the phase transition then will be governed by the growth and coalescence of these domains. Most importantly, there will be no era of "supercooling" where the stress-energy tensor will be given by eq. (1). Hence, there will be no era of inflation.

Thus, in order to produce a viable model yielding inflation, it is essential to have a mechanism which orders the field at $\phi = 0$. I know of no fundamental reason which would prevent one, in principle, from constructing a model which does this and sufficient ordering may well occur in some models, such as the Coleman-Weinberg model with $g^2 \ll 1$. However, for many of the models which have been proposed in the new inflationary scenario, there is little or no evidence for the existence of an effective means of locally ordering the field near zero. In particular, there appears to be no such mechanism at all in the case where ϕ is not directly coupled to other fields. The coupling of ϕ to another field as in the gauge-field coupling of the Coleman-Weinberg model can provide an ordering mechanism, but for a coupling constant of order unity as given by grand unified models, it is far from obvious that it is sufficient to prevent immediate breakup into domains of the true minima as the field is cooled. In this regard, it should be emphasized that the existence, below the phase transition temperature, of a metastable state with ϕ ordered near zero does not suffice to produce inflation; it is essential, in addition, that the field dynamically cool into that metastable state.

In summary, the local ordering of a quantum field (at a value other than at the minimum of its potential) as it is cooled from high temperature is an essential ingredient of any model proposing to produce an era of inflation in cosmology by means of phase transitions. The dynamical behavior of the field during cooling - as opposed to merely the shape of the zero-temperature potential - must be carefully analyzed before one can conclude that an inflationary model is viable. There appears to be little evidence presented thus far that the models which have been considered would yield an era of inflation as described in the standard scenarios.

REFERENCES

[1]G. F. Mazenko, W. G. Unruh, and R. M. Wald, Phys. Rev. D. (in press).

[2]Issues (i), (ii), and (iv) could be considered as problems if one appends to standard theory the additional assumption that the initial conditions of the universe were "chosen at random" since then one could argue that our universe is an "unlikely" one. However, until a viable theory of initial condition selection is given, I consider it premature to declare what initial conditions are "unlikely" or "unnatural".

[3]R. Penrose, in General Relativity: An Einstein Centenary Survey, eds. S. W. Hawking and W. Israel (Cambridge University Press, 1979).

[4]A. Guth, Phys. Rev. D**23**, 347 (1981).

[5]A. D. Linde, Phys. Lett. **108B**, 389 (1982).

[6]A. Albrecht and P. J. Steinhardt, Phys. Rev. Lett. **48**, 1220 (1982).

[7]It should be emphasized that even if the field ϕ is locally ordered near zero initially, this does not mean that mean field dynamics will adequately describe its subsequent evolution. The dynamical evolution of ϕ is presently being investigated by A. Guth and S.-Y. Pi and by W. Bardeen and C. Hill.

The Behavior of the Higgs Field In the New Inflationary Universe

Alan H. Guth and So-Young Pi

In this talk I would like to summarize some research[1] which I have been doing in collaboration with So-Young Pi, concerning the quantum-mechanical treatment of the Higgs field in the context of the new inflationary universe[2,3,4]. Our research in this area is continuing, but so far we have constructed and solved a toy model which I would like to describe in this lecture.

I would like to begin by summarizing what can be regarded as the standard picture for the behavior of the Higgs field. The generic behavior of the finite temperature effective potential is shown in Figure 1. The curves have been shifted vertically so that they all coincide at $\phi = 0$. At zero temperature the potential is extremely flat near $\phi = 0$, and has a minimum at $\phi = \phi_c$. At extremely high temperatures ($T >> T_c$) the effective potential has a unique minimum at $\phi = 0$. As the universe cools the Higgs field ϕ gets caught at $\phi \approx 0$. The Higgs field hovers for a while at $\phi \approx 0$, and then eventually begins to roll classically down the hill of the potential energy diagram. The inflationary era occurs during this slow rollover. The homogeneous solution to the classical equations of motion will be called $\phi = \phi_0(t)$.

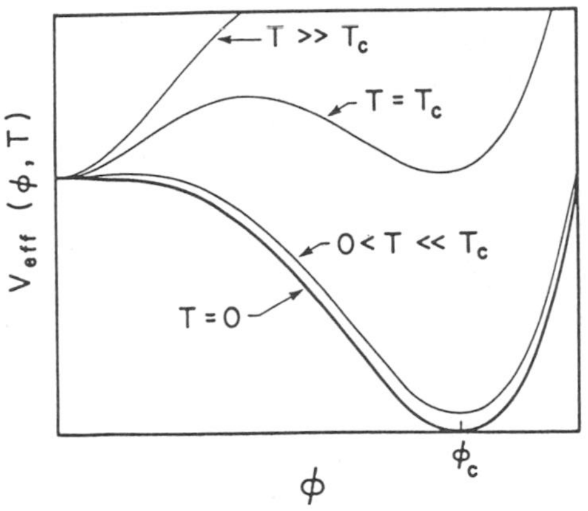

Figure 1: The generic behavior of the finite-temperature effective potential for the Higgs field in the new inflationary universe model.

Of course the Higgs field is not entirely classical, so quantum fluctuations about the classical solution must be taken into account.[5] These quantum fluctuations produce inhomogeneities in the Higgs field

$$\phi(\vec{x},t) = \phi_0(t) + \delta\phi(\vec{x},t), \tag{1}$$

which in turn give rise eventually to inhomogeneities in the mass density of the universe.

This standard picture relies on a number of assumptions, and various questions have been raised about its validity:

1. Is the picture of a classical slow rollover valid? It has been pointed out by Mazenko, Unruh, and Wald[6] that at high temperatures, when one says that $\phi \approx 0$, one really means that the spatial or time average of ϕ is about equal to zero -- the field itself is undergoing large fluctuations. As the system cools, they

argue, it is possible that these fluctuations, which extend initially out to $\phi = \pm\phi_c$ or more, will cause the Higgs field to settle quickly into small regions with $\phi = \pm\phi_c$ in each of these regions. When one looks at the spatial average one might see what appears to be a rolling motion, but the actual local dynamics could be quite different.

2. What is the physical significance of the classical function $\phi_0(t)$? Hawking and Moss[7] point out that the system begins in a thermal ensemble which possesses an exact symmetry, $\phi \to -\phi$. The dynamics is also consistent with this symmetry, and it therefore follows that $\langle \phi(\vec{x},t) \rangle$ remains zero for all time -- the field presumably does roll down the hill, but since it is equally likely to roll in any direction, the expectation value remains zero. Hawking and Moss developed an alternative method of calculation in which only the operator $\phi^2(\vec{x},t)$ appears. Using this method, they calculated[7] a value for the fluctuation amplitude which was in significant disagreement with calculations done using the standard picture.[5] These authors have since retracted their calculation,[8] but the question which they raise still requires an answer.

3. The standard picture described above gives no prediction for how long the Higgs field will hover about $\phi = 0$ before it begins rolling down the hill of the effective potential. This question is of course crucial, since it determines whether or not sufficient inflation is obtained. The question has been studied by Linde[9] and by Vilenkin and Ford[10], who calculated the behavior of $\langle \phi^2(t) \rangle$. This operator is infinite and requires a subtraction, and it was never completely clear to us how this expectation value is related to the behavior of ϕ itself. Our methods will address this question in what I feel is a more precise way, but I should say at the beginning that our results are in qualitative agreement with the results of these previous authors.

4. It has been pointed out by Börner and Seiler[11] that the effective potential shown in Figure 1 is actually ill-defined. In particular, the behavior shown for the zero temperature potential at $\phi < \phi_c$ violates rigorous convexity theorems. The shape of the true effective potential depends on choices made in its definition. One choice leads to a Maxwell construction, $V(\phi) = 0$ for $|\phi| < \phi_c$. Another choice, based on analytic continuation in the parameters of the theory, leads to an effective potential which is complex when $|\phi| < \phi_c$. In the light of these statements, it is not clear that the standard picture is reliable.

The goal of the present work is to provide answers to these questions. In order to do this we have constructed an exactly soluble toy model for the behavior of the Higgs field in the new inflationary universe. Before describing this toy model I will first describe a baby toy problem which contains what we consider to be the most important features of the real problem. This baby problem is the study of a single particle moving in one dimension, under the influence of a potential with the form of an upside-down harmonic oscillator:

$$V(x) = -\tfrac{1}{2}kx^2. \qquad (2)$$

We will track the behavior of an initial wave packet having the Gaussian form:

$$\psi(x,t = 0) \sim \exp\{-\tfrac{1}{2}x^2/h_o^2\} \tag{3}$$

The solution is expressed most simply by introducing the new variables:

$$a^2 = \frac{\hbar}{\sqrt{mk}} \tag{4a}$$

$$\omega^2 = k/m. \tag{4b}$$

Note that a corresponds to the characteristic quantum mechanical length scale of the problem, analogous to the Bohr radius of the hydrogen atom. More precisely, a corresponds to the width of the ground state wave function of the upside-right harmonic oscillator potential with the same absolute value of k. Similarly, ω describes the natural frequency of the corresponding upside-right harmonic oscillator. The solution maintains the Gaussian form $\psi(x,t) = A(t)\exp\{-B(t)\}$ for all time, where A(t) and B(t) are complex functions. For large times the solution has the form:

$$\psi(x,t) \sim e^{-x^2/2h^2(t)}e^{ix^2/2a^2} \tag{5a}$$

where

$$h(t) = \frac{\sqrt{a^4+h_o^4}}{4h_o}e^{\omega t} \quad . \tag{5b}$$

The behavior of equation (5b) is easily understood -- the width of the wave packet at large times is minimized for a specific value of h_o, which turns out to be a. If h_o were chosen very small, then the width would become large due to the h_o in the denominator, corresponding to the fact that the uncertainty principle would require the initial wave packet to have a large spread in momentum. On the other hand, a large value of h_o means that the initial wave packet is already part way down the hill, and therefore spreads quickly.

Our main point in introducing this model is to show that the quantum mechanical wave function at large times is accurately described by classical physics. Most textbooks speak of the classical limit of quantum mechanics in terms of sharply peaked wave packets, which is certainly not the case for equation (5). However, classical physics also applies whenever the length over which the phase of the quantum mechanical wave function changes by 2π (i.e., the DeBroglie wavelength) is much smaller than any other length in the physical problem. However, in such cases the system cannot be described by a single classical trajectory. It must instead be described by a classical probability distribution, which in this case is given by:

$$f(x,p,t) \sim e^{-x^2/h^2(t)}\delta(p - x\sqrt{mk}). \tag{6}$$

This claim is justified because the following two facts can be demonstrated:

a) $f(x,p,t)$ describes classical physics - that is, it obeys the classical equations of motion.

b) For any dynamical variable (i.e., any function $Q(x,p)$), the expectation value of Q can be computed by using either the quantum mechanical wave function ψ or the classical probability distribution f. It can be shown that:

$$\langle Q\rangle_f = \langle Q\rangle_\psi \{1 + O(e^{-2\omega t})\}.$$

The probability distribution f describes an ensemble of particles, each of which rolls from rest at $x = 0$. The various classical trajectories $x_i(t)$ contained in the ensemble differ from each other only by an offset of the time variable:

$$x_i(t) = x(t - \delta t_i),$$

where the function $x(t)$ is a solution to the classical equations of motion which approaches 0 as $t \to -\infty$. Thus, at large times the particle rolls classically down the hill, but the time at which it started is described by a probability distribution. In this situation an ensemble average (whether it be an average of x or x^2) can obscure the physics, since it averages over systems which are in very different stages of their evolution. In our opinion, it was an overly naive interpretation of ensemble averages which led to the error in the calculation of Ref. (7).

Having discussed the baby model I will now move on to describe the toy model which we feel provides a reasonable approximation to the behavior of the Higgs field in the new inflationary universe. The virtue of this toy model is that it is a free field theory, and therefore exactly soluble. It is of course unrealistic, but we believe that the model qualitatively describes the correct physics. Furthermore, we hope that it will serve as a valid zero order of approximation to a more complicated calculation in which interactions are taken into account perturbatively. I will not have time to present the calculations here, but I would like to describe the assumptions and the conclusions.

The dynamics of our toy model is described by two assumptions. First, we assume a perfect background deSitter space:

$$ds^2 = -dt^2 + e^{2\chi t}d\vec{x}^2. \tag{7}$$

This is a reasonable approximation for most stages of the rollover. Corrections due to the previous Friedmann-Robertson-Walker phase have not yet been considered, but they are presumably significant only for the behavior of the very long wavelength modes. Second, we treat the Higgs field as a free scalar quantum field in deSitter space, governed by the time-dependent potential

$$V(\phi) = -\frac{1}{2}\mu^2\phi^2 + cT^2\phi^2 \quad , \tag{8a}$$

where $\mu^2 > 0$ is a fixed parameter, $c > 0$ is a constant which contains terms of order λ (the quartic scalar self-coupling) and g^2 (the gauge coupling), and T represents a background temperature obeying the relation

$$T = T_0 e^{-\chi t}. \tag{8b}$$

The value of the parameter T_0 depends of the arbitrary origin of t, and hence has no physical significance. The second term on the right-hand-side of equation (8a) describes the leading high-temperature behavior of the Higgs field interactions, including the quartic self-interaction as well as the interactions with the other fields in the theory. As I mentioned before, the potential of our toy model describes a free field theory which is only an approximation to the truth. In particular, it is stable for early times and becomes unstable for late times, so ϕ will roll indefinitely to larger and larger values. The toy model will clearly not describe how ϕ settles into its minimum, but we believe that it will give a good description of the slow rollover.

In order to complete the model we must also assume an initial state. Our choice is motivated by the fact that, at very short length scales, deSitter space is essentially indistinguishable from Minkowski space. For example, it is conceivable that our universe has a positive cosmological constant, and we might today be living at the beginning of a deSitter phase with an exponential time constant of the order of 10 billion years. However, that fact would in no way affect experiments being done at Fermilab today. The exponential expansion simply has no effect at short distance scales. To take advantage of this fact, imagine describing the spatial variation of the Higgs field as a Fourier integral:

$$\phi(\vec{x},t) = \int d^3\vec{k}[\sigma_1(\vec{k},t)\sin(\vec{k}\cdot\vec{x}) + \sigma_2(\vec{k},t)\cos(\vec{k}\cdot\vec{x})]. \tag{9}$$

(For clarity we have used real spatial eigenfunctions rather than the usual $\exp(i\vec{k}\cdot\vec{x})$.) As $t \to -\infty$, each $\sigma(\vec{k},t)$ obeys an equation of motion which approaches that of a simple harmonic oscillator. The frequency of the harmonic oscillator shifts as the universe exponentially expands, with a rate of order χ. However, as $t \to -\infty$, the harmonic oscillator frequency $\omega \to \infty$, and thus for early times $\omega >> \chi$. Although the frequency of the harmonic oscillator is changing, the change is adiabatic; each mode then behaves precisely as a mode of a scalar field in Minkowski space, and we treat it accordingly. We assume that each of these harmonic oscillator variables is described at asymptotically early times by a thermal equilibrium density matrix at temperature T. Since T is changing with time (equation (8b)), one might think that thermal equilibrium can hold at one time only. However, ω is also changing, and the ratio $\omega/T \to$ constant as $t \to -\infty$. This means that thermal equilibrium can be maintained as $t \to -\infty$.

Since the model is a free field theory, everything about it can be calculated exactly once the dynamics of the system and the initial state have been specified. The calculations will be shown elsewhere[1], but I will now summarize our results:

1. Monte Carlo calculations. It is impossible to directly generate Monte Carlo configurations for the quantum field $\phi(\vec{x},t)$, because a quantum field at a single point is invariably described by a probability distribution with an infinite width. To obtain a measurable operator one must "smear" the quantum field over some finite volume. One therefore defines a smeared field

$$\hat{\phi}_\ell(t) = \frac{1}{(2\pi)^{3/2}\ell^3} \int d^3\vec{y}\; e^{-(\vec{x}-\vec{y})^2/2\ell^2}\, \phi(\vec{y},t). \tag{10}$$

Since the dynamical variables $\sigma(\vec{k},t)$ introduced in equation (9) are described by a classical probability distribution at large times, it is possible to use Monte Carlo procedures to generate a typical set of values for these variables. One can then draw pictures which illustrate typical configurations of the Higgs field at any given time, to get some feeling for what the field is actually doing. One can also calculate $\langle \hat{\phi}_\ell^2(t) \rangle$, but one must bear in mind the caveat that such an expectation value averages over systems in different stages of evolution. For potentials which are extremely flat (i.e., $c \ll 1$), the calculation of $\langle \hat{\phi}_\ell^2(t) \rangle$ for $\ell \approx 1/T$ shows that $\hat{\phi}_\ell(\vec{x},t)$ has a high probability of hovering at $\hat{\phi}_\ell \approx 0$ for a long time before beginning to roll down the hill of $V(\phi)$. (There is no real contradiction with Ref. (6), which makes definitive statements only for $c = O(1)$.)

2. The meaning of $\phi_0(t)$. In the new inflationary universe scenario, the observed universe evolves from part of a much larger region which has evolved into the false vacuum. Let b denote the coordinate diameter of the region which evolves into the observed universe, and let $k_U = b^{-1}$. It is then reasonable to separate the Fourier integral of equation (9) into two parts:

$$\phi(\vec{x},t) = \int_{|\vec{k}|<k_U} d^3\vec{k}[\ldots] + \int_{|\vec{k}|>k_U} d^3\vec{k}[\ldots]. \tag{11}$$

The first term on the right-hand-side describes fluctuations with wavelengths longer than the diameter of the observed universe, and thus can be considered homogeneous for astronomical purposes. It is this term which we identify with $\phi_0(t)$. In our formalism it obeys a classical equation of motion, but the time at which the rolling begins is described by a probability distribution. The second term represents inhomogeneities on scales less than that of the observed universe, and these correspond to the term $\delta\phi(\vec{x},t)$ in equation (1).

3. How long will $\phi_0(t)$ hover around $\phi \approx 0$? In our model this question can be answered unambiguously by calculating the probability distribution of $\phi_0(t)$. Our results agree approximately with the qualitative results derived earlier by Linde[9] and by Vilenkin and Ford[10].

4. Calculation of density perturbations. In the toy model, the probability distribution for density perturbations $\delta\rho/\rho$ can be calculated exactly. The results are completely in accord with those obtained earlier[5] using the standard picture.

In conclusion, let me say that our calculations are certainly not definitive - they are calculations in a toy model based on free field theory. However, we believe that this toy model does contain the important features of the real problem; and we think that we have found strong evidence that the basic features of the standard picture are correct.

After this work was completed, we learned that similar work is being carried out by R. Brandenberger[12], J. M. Bardeen[13], W. Bardeen and C. Hill[14].

A. H. G. was supported in part through funds provided by the U. S. Department of Energy (DOE) under contract DE-AC02-76ER03069, in part by the National Aeronautics and Space Administration (NASA) under grant NAGW-553, and in part by an Alfred P. Sloan Fellowship. S.-Y. P. was supported in part through funds provided by the Department of Energy under the Outstanding Junior Investigator Program.

REFERENCES

[1] A. H. Guth and S.-Y. Pi, manuscript in preparation.

[2] A. H. Guth, Phys. Rev. D**23**, 347 (1981).

[3] A. D. Linde, Phys. Lett. **108B**, 389 (1982).

[4] A. Albrecht and P. J. Steinhardt, Phys. Rev. Lett. **48**, 1220 (1982).

[5] A. A. Starobinsky, Phys. Lett. **117B**, 175 (1982); A. H. Guth and S.-Y. Pi, Phys. Rev. Lett. **49**, 1110 (1982); S. W. Hawking, Phys. Lett. **115B**, 295 (1982); J. M. Bardeen, P. J. Steinhardt, and M. S. Turner, Phys. Rev. D**28**, 679 (1983); R. Brandenberger, R. Kahn, W. H. Press, Phys. Rev. D**28**, 1809 (1983); R. Brandenberger, R. Kahn, Phys. Rev. D**29**, 2172 (1984).

[6] G. F. Mazenko, W. G. Unruh, and R. M. Wald, Does a Phase Transition in the Early Universe Produce the Conditions Needed for Inflation?, University of Chicago preprint (1984); see also the contribution by Wald in these proceedings.

[7] S. W. Hawking and I. G. Moss, Nucl. Phys. B**224**, 180 (1983).

[8] I. G. Moss, Phys. Lett. **128B**, 385 (1983); Nucl. Phys. B**238**, 436 (1984); S. W. Hawking, private communication.

[9] A. D. Linde, Phys. Lett. **116B**, 335 (1982).

[10] A. Vilenkin and L. H. Ford, Phys. Rev. D**26**, 1231 (1982); A. Vilenkin, Nucl. Phys. B**226**, 504 (1983).

[11] G. Börner and E. Seiler, Some Remarks Concerning the Inflationary Universe, Max-Planck-Institut preprint, (1983).

[12] R. H. Brandenberger, Quantum Fluctuations as the Source of Classical Gravitational Perturbations in Inflationary Universe, Santa Barbara ITP Preprint (1984); see also the contribution by Brandenberger in these proceedings.

[13] J. M. Bardeen, private communication.

[14] W. Bardeen and C. Hill, private communication.

Dynamical Evolution of a Classical Higgs Field and the Fate of the Inflationary Universe

Katsuhiko Sato and Hideo Kodama

I. INTRODUCTION

The most interesting consequence of the grand unified theories (GUTs) on cosmology is the possible appearance of an exponentially expanding phase, namely the so-called inflationary stage, in the early universe.[1,2] After very active investigations within a surprisingly short period of a couple of years, people now realize that the inflationary universe model has as many difficulties as fascinating features.

In the original model the inflation was considered to be given rise to by the supercooling, namely the sticking of the Higgs field to the symmetric state due to a large barrier of the effective potential. The inevitable strongly first-order nature of the phase transition in this model, however, entailed fatal difficulties that the phase transition does not terminate in general,[3] and even if it terminates, an extremely inhomogeneous universe would be left after the phase transition.[4]

The central source of difficulties in this model resided in that the presently observed part of the universe should be formed from a collection of coherent domains (or bubbles), each of which has strongly spatial inhomogeneity itself. Noticing this point, Linde[5] and Albrecht and Steinhardt[6] proposed independently a new version of the inflationary universe model. Its essential idea was that if a model is constructed such that each coherent domain formed by the phase transition continues inflation, the observed part of the present universe can be contained in one coherent domain and the difficulties of the original scenario disappear. In fact, they explicitly showed that if the Coleman-Weinberg type potential is used to break symmetry, a model with the expected properties can be constructed. In this new model the Higgs field spends a lot of time near the metastable symmetric point, where the potential energy is still very large, because of the flatness of the potential. As a result each coherent region in which the Higgs field has a nearly uniform nonvanishing expectation value expands exponentially for a sufficiently long time while the Higgs field rolls down to the absolute minimum point of the potential.

Unfortunately, however, there is one unexpected pitfall in this new inflation model. In most of the investigation until now it has been assumed that the vacuum expectation value of the Higgs field evolves directly to the SU(3)×SU(2)×U(1) direction from the start of rolling down. Recently, however, some people have pointed out[7-9] that the Higgs field goes toward SU(4)×U(1) state, which may give rise to serious difficulties in the new inflationary scenario. The purpose of the present work is to investigate this point in more detail by examining the evolution of the vacuum expectation value Φ of a Higgs field belonging to the adjoint representation of SU(5) in the full 24-dimensional space by numerical simulation and to elucidate whether the new inflationary scenario is consistent with cosmological observations or not.

II. BASIC EQUATIONS AND ASSUMPTIONS

For the convenience of the numerical computation, we first assume that the time derivative of the Higgs field vanishes, i.e., $\dot{\Phi} = 0$, just when the Higgs field acquires non-vanishing expectation values in the first stage of the phase transition. This assumption is reasonable if a region with a non-vanishing expectation value of the Higgs field is created by quantum tunneling even in the new inflationary universe model, as discussed by Hawking and Moss,[10] and recently investigated by Albrecht et al.[11] and Jensen and Steinhardt[12] in detail. On the other hand, Vilenkin and Ford[13] and Hawking and Moss[14] discussed another mechanism for the Higgs field to acquire non-vanishing expectation values, i.e., the destabilization of the state $\Phi = 0$ by quantum fluctuations. In this point of view, however, it is not well clarified when the Higgs field acquires a non-vanishing expectation value, though when and how the phase transition terminates should be determined by the expectation value eventually. Though the mutual relation between these two points of view has not been investigated well yet at present, it is certain that Φ must be sufficiently small in both scenarios at the beginning of the stage during which the classical description of the Higgs field is valid, in order for a sufficient inflation to occur. Hence our assumption on Φ is not so restrictive, since the essential part of the result in the present work holds even if Φ is non-vanishing initially so far as it is sufficiently small.

Under this assumption that $\dot{\Phi} = 0$ at the initial point, the Higgs field Φ represented by an arbitrary 5×5 hermitian traceless matrix can be diagonalized in each coherent region by a global gauge transformation, keeping $\dot{\Phi} = 0$. The equation of motion of the Higgs field guarantees that Φ remains diagonal in the course of its evolution if Φ is diagonal and $\dot{\Phi} = 0$ at the start. Thus in the calculation of evolution we can restrict the form of the Higgs field without loss of generality as

$$\Phi = \mathrm{diag}\,[\phi_1, \phi_2, \phi_3, \phi_4, \phi_5] \tag{1}$$

with the constraint

$$\mathrm{Tr}\,\Phi = \sum_{i=1}^{5} \phi_i = 0.$$

In this representation the Coleman-Weinberg potential with the one-loop correction by the gauge boson (the Higgs boson contribution to the one-loop correction is not included) is given by[10]

$$V(\phi) = \frac{3g^4}{256\pi^2}\Big[C\,\Big\{\sum_{i=1}^{5} \phi_i{}^4 - \frac{7}{30}\Big(\sum_{i=1}^{5} \phi_i{}^2\Big)^2\Big\} + \sum_{i,j=1}^{5} (\phi_i - \phi_j)^4 \Big\{\ln\Big(\frac{\phi_i - \phi_j}{\mu}\Big) - \frac{1}{2}\Big\}\Big], \tag{2}$$

where C is an arbitrary parameter of this potential and μ is a renormalization parameter related to the vacuum expectation value of the

Higgs field at the $SU(3)\times SU(2)\times U(1)$ minimum point $\phi = \sigma_3\times$ diag $[1,1,1,-3/2,-3/2]$ ($\sigma_3 \sim 4.5\times 10^{14}$GeV) as $\mu = 5\sigma_3/2$. In the present work we neglect the effect of temperature on the potential because the inflation begins after the cosmic temperature becomes less than the GUT temperature ($\sim 10^{14}$GeV) and the essential fate of the Higgs field is determined before the universe is heated up again after the Higgs field has fully developed its vacuum expectation value.

In general the potential $V(\Phi)$ has two kinds of local minimum points. The first type is twenty $SU(3)\times SU(2)\times U(1)$ minimum points with $V(\Phi) = V_0 = 5625g^4/2048\pi^2$, and the second type is ten $SU(4)\times U(1)$ ones where Φ is represented by $\Phi = \sigma_4 \times$ diag $[1,1,1,1,-4]$ ($\sigma_4 = (\sigma_3/2)\exp(-C/60)$) and the expressions obtained by permutating the components with $V(\Phi) = V_0\{1-(2/3)\exp(-C/15)\}$. Depending on the value of C, the relative height of $V(\Phi)$ at these two types of local minimum points changes, and in some range of C, either type of points cease to be local minimum points and become mere saddle points. More concretely speaking, the following four situations occur:

i) $C>15$

In this range only the $SU(3)\times SU(2)\times U(1)$ points are the local minimum points, hence at the same time they are the global minimum points.

ii) $-15\ln(1.5) < C < 15$

In this range both types of points are the local minimum points, but the $SU(3)\times SU(2)\times U(1)$ points are the global minimum points.

iii) $-15 < C < -15\ln(1.5)$

In this range both types of points are the local minimum points again, but the $SU(4)\times U(1)$ points become the global minimum points now.

iv) $C<-15$

In this range the $SU(3)\times SU(2)\times U(1)$ points become saddle points and the $SU(4)\times U(1)$ points are the local and global minimum points.

Thus in order that the symmetry is broken to $SU(3)\times SU(2)\times U(1)$ eventually, the parameter C must be larger than $-15\ln(1.5)$ at least. Further for $C>15$ the contribution of the Higgs field to the effective potential becomes important as pointed out by Breit et al.[8] Hence we limit the consideration to the range ii) in this paper.

Because of the traceless condition

$$\sum_{i=1}^{5} \phi_i = 0,$$

five components of the Higgs field ϕ_i, $i = 1,2,\cdots 5$, are not independent. This makes the numerical computation complicated if we calculate the time evolution of these components directly. For the convenience of numerical computation, we introduce the following four components fields which are completely independent of each other,

$$\psi_i = \phi_i + \sum_{j=1}^{4} \phi_j/(1 + \sqrt{5}) \ . \ (i = 1,2,3,4) \tag{3}$$

In Fig. 1, contours of the potential on a plane ($x = \sqrt{3}\psi_1 = \sqrt{3}\psi_2 = \sqrt{3}\psi_3$, $y = \psi_4$) are displayed for the case of the potential parameter $C = 1$. On this plane there are two SU(3)×SU(2)×U(1) minima and four SU(4)×U(1) minima.

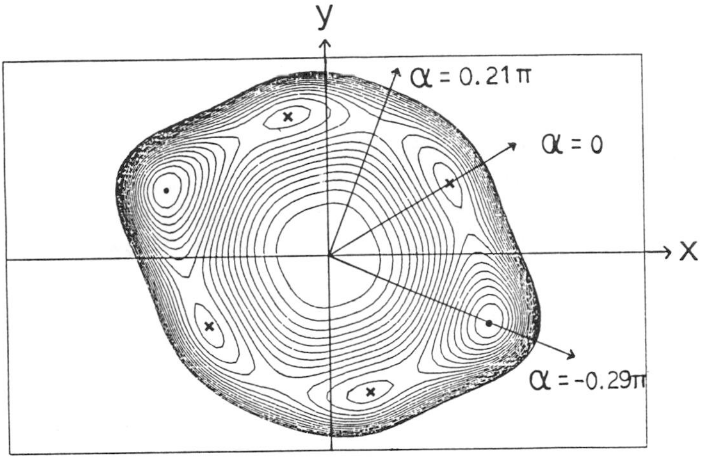

Figure 1: Contour map of the Coleman-Weinberg potential for the case $C = 1$ is displayed on the plane ($x = \sqrt{3}\psi_1 = \sqrt{3}\psi_2 = \sqrt{3}\psi_3$, $y = \psi_4$). On this plane, there exist four SU(4)×U(1) minima (x) and two SU(3)×SU(2)×U(1) minima (•). In this plane, $\alpha = 0$ means SU(4)×U(1) direction, $\alpha = -0.29\pi$ SU(3)×SU(2)×U(1) direction and $\alpha = 0.21\pi$ the direction vertical to the SU(3)×SU(2)×U(1) direction. Numerical computation is carried out in the range $-0.29\pi < \alpha < 0.21\pi$.

The equations of motion for the fields ψ_i are given by

$$\ddot{\psi}_i + 3(\dot{R}/R)\dot{\psi}_i + \partial V/\partial\psi_i + C_{vis}|\psi_i|\dot{\psi}_i = 0 \tag{4}$$

where a viscosity term $C_{vis}|\psi_i|\dot{\psi}_i$ is introduced in order to convert the energy of the Higgs field into thermal energy. The scale factor of the universe R is calculated by the expansion equation of the universe. (The absolute units $c = \hbar = G = 1$ are used throughout this paper.)

$$(\dot{R}/R)^2 = 8\pi\ (\rho_r + \rho_\phi)/3 \tag{5}$$

where we have assumed that the universe is spatially flat, which is adequate in the early universe even if it is not flat exactly. The change of radiation energy density ρ_r is described by

$$d(\rho_r R^4)/dt = \sum_{i=1}^{4} C_{vis}|\psi_i|\dot{\psi}_i^2 R^4, \tag{6}$$

and the energy density of the Higgs field ρ_ϕ is directly expressed in terms of ψ_i as

$$\rho_\phi = \frac{1}{2}\sum_{i=1}^{4} \dot{\psi}_i^2 + V. \tag{7}$$

The initial value of the Higgs field has four degrees of freedom: one is the norm of Higgs field

$$\|\psi\| = (\sum_{i=1}^{5} \phi_i^2)^{1/2} = (\sum_{j=1}^{4} \psi_j^2)^{1/2}$$

in the four dimensional ψ space. In the present investigation, we take $\varepsilon = \|\psi\|_0 = 8\times10^{-6}$ as the initial value of $\|\psi\|$, which is about 0.2H, where $H = (8\pi GV(0)/3)^{1/2}$. In order to parametrize the initial direction of ψ we utilize three angles α, θ and ϕ. α represents the deviation angle from the SU(4)×U(1) direction $\psi_1=\psi_2=\psi_3=\psi_4$ (>0) on the plane $\psi_1=\psi_2=\psi_3$ as shown in Fig. 1, and θ and ϕ represent the deviation angles off this plane. Though we do not limit the range of α essentially, we restrict θ and ϕ in the very narrow range $|\theta|<10^{-4}$ and $|\phi|<10^{-4}$ for the convenience of the analysis of the numerical computation as a first step.

III. RESULTS OF NUMERICAL SIMULATION

In Figs. 2a-2e, some results of numerical computation for the case of the potential parameter C = 1 are shown. As demonstrated in these figures, the Higgs field goes to the SU(4)×U(1) direction $\psi_1=\psi_2=\psi_3=\psi_4>0$ at first independent of the values of viscosity parameter C_{vis} and the initial angle α, provided that $-0.29\pi<\alpha<0.21\pi$. This is obviously a natural consequence of the fact that the potential has the steepest gradient along this direction as far as the norm $\|\psi\|$ is small[8] (see Fig. 1).

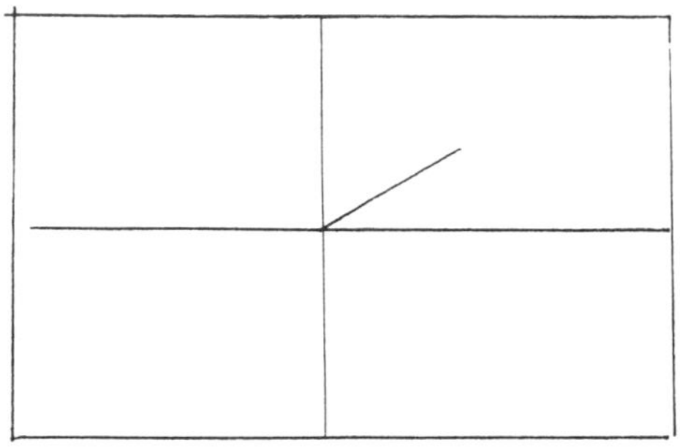

Figure 2a: Time evolution of the Higgs field for the case C = 1, C_{vis} = 0.1 and $\alpha = 0.1\pi$ is projected on the same plane as shown in Figure 1.

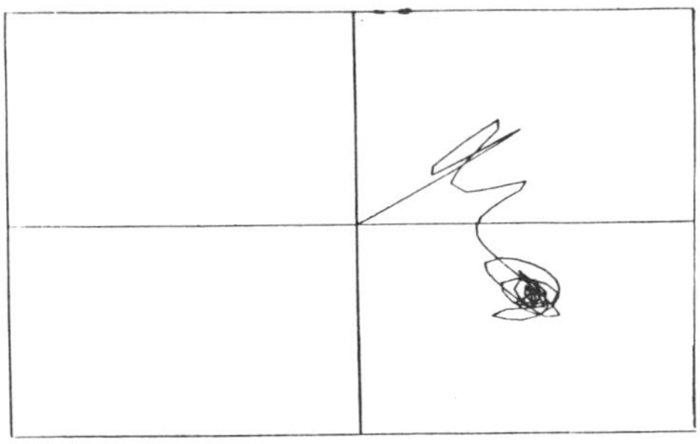

Figure 2b: Same as Figure 2a, but for the case C = 1, C_{vis} = 0.01 and $\alpha = -0.2\pi$.

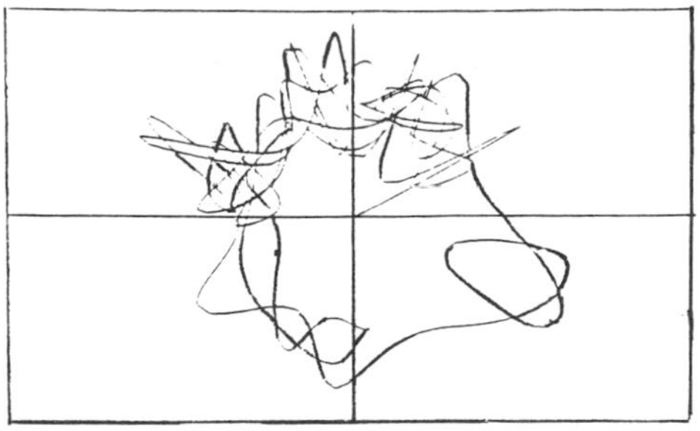

Figure 2c: Same as Figure 2a, but for the case $C = 1$, $C_{vis} = 0.001$ and $\alpha = 0.2\pi$.

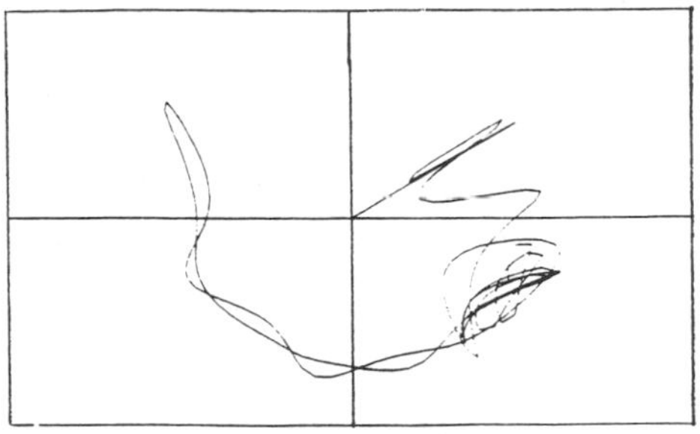

Figure 2d: Same as Figure 2a, but for the case $C = 1$, $C_{vis} = 0.003$ and $\alpha = -0.2\pi$.

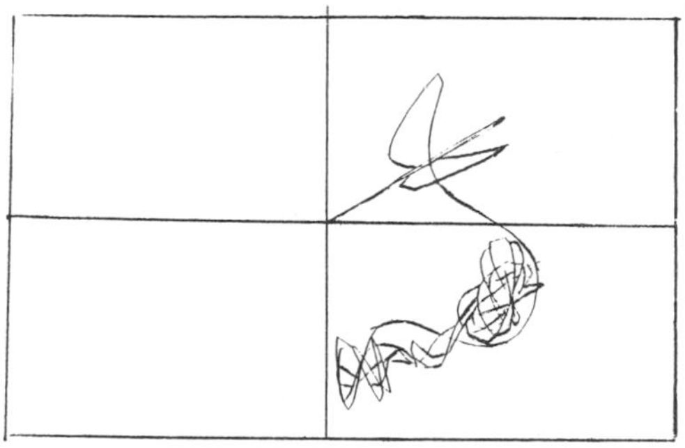

Figure 2e: Same as Figure 2a, but for the case $C = 1$, $C_{vis} = 0.003$ and $\alpha = -0.1\pi$.

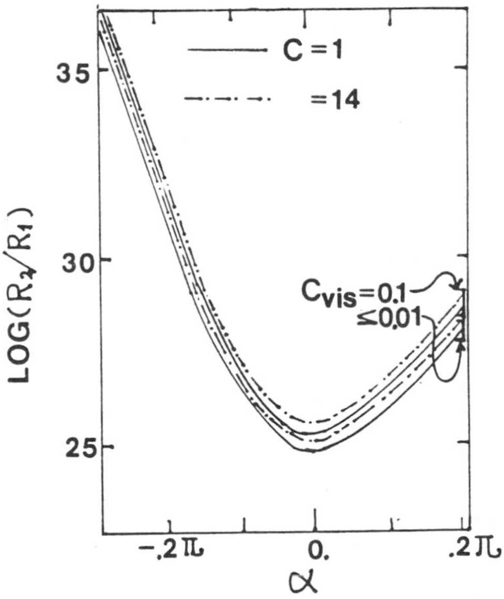

Figure 3: The dependence of the degree of inflation on the initial angle α. Initial absolute value of the Higgs field is assumed to be $(\psi_1{}^2+\psi_2{}^2+\psi_3{}^2+\psi_4{}^2)^{1/2} = 8\times10^{-6}\sigma_3$.

In Fig. 3, the dependence of the degree of inflation on the initial angle α is shown. Here we define the degree D as the ratio of the cosmic scale factors, $D = R_2/R_1$, where R_1 is the value when the inflation begins, i.e., when the vacuum energy density becomes greater than that of the radiation, and R_2 is the value when the Higgs field arrives near the SU(4)×U(1) minimum. Note that inflation begins again when the Higgs field settles down at an SU(4)×U(1) state because of the remaining vacuum energy density. Of course this inflation is not taken into account in this definition. As shown in Fig. 3, the degree of inflation is very sensitive to the initial angle α, but almost independent of the potential parameter C and the viscosity parameter C_{vis} provided that $C_{vis}<0.1$. The evolution after the arrival to this minimum, of course, depends on the value of C_{vis}. Generally speaking, the Higgs field settles down in a very direct way to the SU(4)×U(1) minimum after short time oscillation independent of the angle α provided that $C_{vis}>1$ as is illustrated in Fig. 2a. For the smaller values of the viscosity parameter C_{vis}, however, it can depart from this local minimum and further evolve to an SU(3)×SU(2)×U(1) minimum state. As shown in Fig. 2b ($C_{vis} = 10^{-2}$ and $\alpha = 0.2\pi$), the Higgs field evolves to the nearest SU(3)×SU(2)×U(1) state via the SU(4)×U(1) state and settles down to this state after oscillation around it. Note that, however, details of the evolution of the Higgs field are different for the different initial angles even if the value of the viscosity parameter C_{vis} is the same. For example, if we take $\alpha = -0.1\pi$, the Higgs field settles down to the SU(4)×U(1) state after large amplitude oscillations. When we take the smaller values for C_{vis}, the Higgs field begins to circulate in this (x,y)-plane and wanders around a lot of SU(3)×SU(2)×U(1) states and SU(4)×U(1) states. After a few time circulations, the Higgs field goes out from this plane and begins to wander around more numbers of SU(3)×SU(2)×U(1) and SU(4)×U(1) states in the four dimensional space of the Higgs field. This result suggests that the eventual state of the Higgs field is changed greatly by the very small deviation of the initial direction of the Higgs field.

In Fig. 4, final states of the Higgs field are summarized in the plane of the initial angle α and the viscosity parameter C_{vis}. As has been discussed, final states of the Higgs field depend on the value of the viscosity parameter C_{vis} strongly. The final state is an SU(4)×U(1) minimum if C_{vis} is greater than a critical value. Although the critical value depends on the initial angle α as shown in Fig. 4, we may conclude that the final state is SU(4)×U(1) if $C_{vis}>10^{-2}$. This state is, however, unstable if we take into account the quantum tunneling effect. Although the Higgs field can reach a nearby SU(3)×SU(2)×U(1) state by this tunneling, the same difficulties as appeared in the original inflation scenario[1-4] arise because this phase transition is of first order, i.e., large-scale inhomogeneity is created by bubbles formed by this phase transition as pointed out by Breit, Gupta and Zacks.[8]

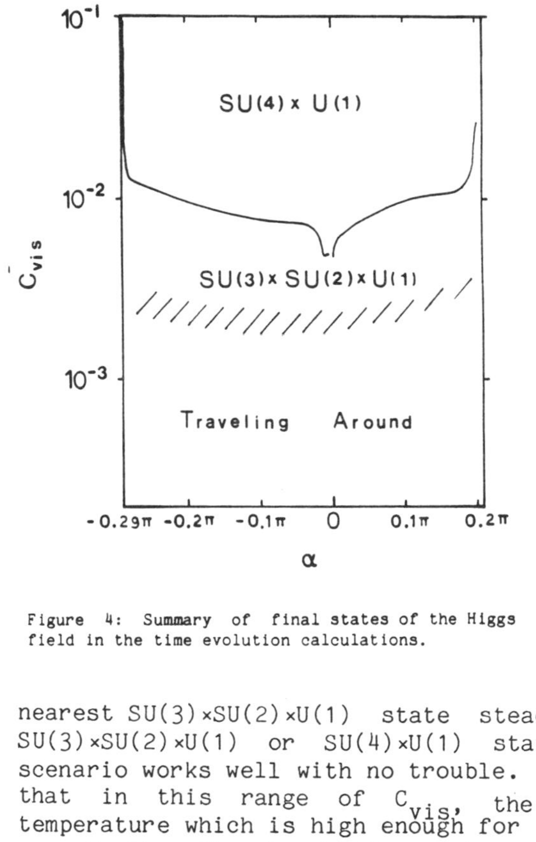

Figure 4: Summary of final states of the Higgs field in the time evolution calculations.

When we take the value of the viscosity parameter in the range $2\times10^{-3}<C_{vis}<10^{-2}$, the Higgs field can settle down to the nearest SU(3)×SU(2)×U(1) state steadily without traveling to other SU(3)×SU(2)×U(1) or SU(4)×U(1) states. In this case, the inflation scenario works well with no trouble. Our computational results show that in this range of C_{vis}, the universe is heated up again to a temperature which is high enough for the baryosynthesis to occur.

On the other hand, if we take values $C_{vis}<2\times10^{-3}$, the Higgs field travels around a lot of minimum states as illustrated in Fig. 2c. Eventual states to which the Higgs field settles down are changed by fluctuations of initial values of the Higgs field. This result strongly suggests that a coherent region, which is formed by nucleation of bubbles or spinodal decomposition in the early stage $\|\phi\|<H$, will be fragmented into many SU(3)×SU(2)×U(1) and SU(4)×U(1) states by the fluctuations of the Higgs field associated with the initial state. Even if the classical fluctuations of the Higgs field are extremely small, the quantum ones might fragment the original coherent region into small pieces of different state.[8] Thus this result suggests that large-scale inhomogeneities also appear for the too small value of C_{vis}, which conflict with present observation.

We have carried out numerical computations for the different values of the potential parameter C (Eq. (2)) in the range $0<C<14$. The result displayed in Fig. 1 (the case C = 1) does not change qualitatively for the other values of C in this range except that the critical value of the viscosity parameter C_{vis}, which decides whether the final state is SU(4)×U(1) or not, depends more sensitively on the initial angle α. We have also carried out the simulation by using the viscosity of the form $C_{vis}\ \|\psi\|\dot{\psi}_i$ instead of $C_{vis}|\psi_i|\dot{\psi}_i$. The results were essentially the same both qualitatively and quantitatively except for the small change of the critical value of C_{vis}.

IV. ESTIMATE OF THE VISCOSITY PARAMETER

The result of the numerical computation given in Section 3 clearly shows that the viscosity term representing the conversion of the vacuum energy or the energy of the classical Higgs field into the energy of thermal particles and radiation essentially determines the fate of the universe after the end of the GUT phase transition in the new inflation model.

As has been shown by Abbott et al.,[15] the gross estimate of the energy conversion rate in the oscillatory phase of the Higgs field can be easily estimated and roughly given by the decay rate of Higgs particles $\sim h^2 m_H$, where h is a coupling constant between the Higgs particles and fermions and m_H is a typical mass of the Higgs particle. For the form of the viscosity term employed in the present paper, the energy conversion rate in the oscillatory phase is roughly expressed as

$$C_{vis}\phi\dot{\phi}^2/m^2\phi^2 \sim C_{vis} m/g^2 \ . \tag{8}$$

Hence Abbott et al.'s estimate corresponds to the value of C_{vis} of $O(g^2h^2)$, which can be in the respected range obtained in Section 3 if $h = O(1)$.

This argument is too crude in two respects. First, Abbott et al.'s estimate neglects the effect of the temperature on the dissipation rate. Hosoya, Sakagami and Takao, and Hosoya and Sakagami[16] studied this effect by evaluating the friction of the motion of classical Higgs field with Higgs particles and fermions which exist as thermal gases. They considered a small linear deviation in the distribution of these particles from the thermal equilibrium, which is induced by the change of the Higgs mass, $m_H = (\partial^2 V(\phi)/\partial\phi^2)^{1/2}$ and fermion mass $m_f = h\phi$, where h is a Yukawa coupling constant. They evaluated the viscosity due to the friction with Higgs particles for the high temperature limit, $T \gg m$, as

$$C_{vis}\phi\dot{\phi} \sim \frac{3\times 2^5 (V^{(3)}(\phi))^2}{\pi\lambda^2 T} \log (T/m)\ \dot{\phi}\ , \tag{9}$$

where $V^{(3)}(\phi)$ is the third derivative of the potential with respective to ϕ and λ is the four body self-coupling constant of ϕ field. This result suggests that the viscosity parameter becomes very large if the universe is reheated up to $T \gg m$.

The second point is that Abbott et al.'s argument is limited to the oscillatory phase. As the argument in Section 3 indicates, how much fraction of the Higgs energy is dissipated into radiation before the oscillatory phase has a comparable importance as the dissipation in the oscillatory phase to determine the critical value of the viscosity coefficient. Furthermore, since the dissipation rate during the monotonically changing phase of the Higgs field determines the temperature of the universe when the Higgs field begins to oscillate, Hosoya et al.'s result also strengthens the importance of the estimate of the dissipation rate during the monotonic phase in the zero temperature limit.

The first investigation of this problem has been done recently by Morikawa and Sasaki.[17] They determined the structure of the viscosity term and evaluated the dissipation coefficient by calculating the decaying part of $\langle\phi(t)^2\rangle$ based on the idea that the dissipation occurs through the creation of Higgs particles due to the change of the vacuum expectation value ϕ and their subsequent decay into fermions. They found that the viscosity term is given as

$$C_{vis}\phi\dot{\phi} = \frac{2^{1/2}\lambda^{3/2} f^2}{3072\pi^2}\ \phi\dot{\phi}\ , \tag{10}$$

where f is the Yukawa coupling constant between fermions and Higgs bosons. Their result suggests that the viscosity parameter C_{vis} is less than 10^{-4} at least if we assume $f, \lambda < 1$. Though this value looks a little smaller than the respected value, we think that the situation is not so serious. First the right hand side of Eq. (10) might be multiplied by the number of helicity states to which the Higgs particles can decay, which can amount to $O(10^2)$. Second it should be noticed that the above estimate can be only applied to the case when rolling over speed is much less than the mass of quasi Higgs boson, i.e., $\dot{\phi}/\phi \ll m(\phi)$. This is the case long before Higgs field rolls over to the $SU(4)\times U(1)$ minimum, but it is no longer valid when the Higgs field comes near the minimum point. It should also be noticed that since Morikawa and Sasaki's method relies on the time dependent Fock representation of the Higgs field in a crucial way, their result seems to depend on how the quasi-particles are defined at each time.

V. CONCLUSION

In the present work, we have found that in order for an inflationary universe model consistent with observation to be constructed, the value of the viscosity must be in an adequate range, otherwise large-scale inhomogeneities which conflict with observations arise. At present, it is hard to determine whether the actual value of viscosity lies in the range adequate for the new inflationary scenario or not, because the result of recent works[16,17] are rather qualitative. In order to make clear whether a consistent scenario can be constructed or not, more precise evaluation of viscosity is necessary.

ACKNOWLEDGEMENTS

The authors thank A. Hosoya, M. Sasaki and D. W. Sciama for valuable discussion. This work is supported in part by the Grant in Aid for Science Research Fund of the Ministry of Education, Science and Culture No. 59740126 and also by Asahi Scholastic Promotion Fund. Numerical computations were carried out on the FACOM M-190 of LICEPP.

REFERENCES

[1]K. Sato, Phys. Lett. **99B**, 66 (1981); Month. Notices Roy. Astron. Soc. **195**, 467 (1981).
[2]A. H. Guth, Phys. Rev. **D23**, 347 (1981); A. H. Guth and E. J. Weinberg, Phys. Rev. **D23**, 876 (1981).
[3]A. H. Guth and E. J. Weinberg, Nucl. Phys. **B212**, 321 (1983).
[4]M. Sasaki, H. Kodama and K. Sato, Prog. Theor. Phys. **68**, 1561 (1982); H. Kodama, M. Sasaki and K. Sato, Prog. Theor. Phys. **68**, 1979 (1982).
[5]A. D. Linde, Phys. Lett. **108B**, 389 (1982).
[6]A. Albrecht and P. J. Steinhardt, Phys. Rev. Lett. **48**, 1220 (1982).
[7]I. G. Moss, Phys. Lett. **128B**, 385 (1983).
[8]J. D. Breit, S. Gupta and A. Zacks, Phys. Rev. Lett. **51**, 1007 (1983).
[9]J. Kodaira and J. Okada, Phys. Lett. **133B**, 291 (1983).
[10]S. W. Hawking and I. G. Moss, Phys. Lett. **110B**, 35 (1982).
[11]A. Albrecht, L. G. Jensen and P. J. Steinhardt, Preprint UPR-0229T (1983).
[12]L. G. Jensen and P. J. Steinhardt, Nucl. Phys. **B237**, 176 (1984).
[13]A. Vilenkin and L. H. Ford, Phys. Rev. **D26**, 1231 (1982).
[14]S. W. Hawking and I. G. Moss, Nucl. Phys. **B224**, 180 (1983).
[15]L. F. Abbott, E. Farhi and M. B. Wise, Phys. Lett. **117B**, 29 (1983).
[16]A. Hosoya and M. Sakagami, Phys. Rev. **D29**, 2228 (1984); A. Hosoya, M. Sakagami and M. Takao, Ann. Phys. **154**, 229 (1984).
[17]M. Morikawa and M. Sasaki, Prog. Theor. Phys. **72**, 782 (1984).

Conditions for the Formation of Bubble Universes

J. Richard Gott, III

ABSTRACT

The causal structure associated with the formation of bubble universes in the "new inflationary" scenario is analyzed and a Penrose diagram for this is presented. Requirements on the bubble formation rate and the amount of inflation within bubbles are discussed.

This paper will discuss the geometrical implications of bubble formation in the new inflationary scenario for the early universe.

The advantages for cosmology if the early universe contained a de Sitter phase of exponential expansion have been outlined clearly by Englert[1] and Guth.[2] In particular all the material we are able to see can have been causally connected in the past and the large radius of curvature of the universe ($\gtrsim$6000 Mpc) is explained. Guth[2] pointed out how an inflationary de Sitter phase would arise naturally in SU(5). In this original inflationary scenario the Higgs potential $V(\phi)$ for $T < T_{CRIT}$ would have a metastable local minimum at $\phi = 0$, $V(\phi)$ then reaches a maximum at $\phi = \phi_{MAX}$ before dropping precipitously to its true minimum $V(\phi_0) = 0$ at ϕ_0. The symmetric vacuum $\phi = 0$ has a density $\rho_{VAC} = V(0)$ and $P_{VAC} = -\rho_{VAC}$. The geometry in this phase approximates a piece of de Sitter space with radius of curvature $r_0 = (3/8\pi\ \rho_{VAC})^{1/2}$. At late times this space has an inflationary expansion $a(t) \sim r_0 \exp(t/r_0)$. The normal (asymmetric) vacuum state which we encounter today is one in which $\phi = \phi_0$ and $\rho = P = 0$. Coleman[3] pointed out that the metastable false vacuum state ($\phi = 0$) would decay by the formation of low density ($\rho = 0$) bubbles in the same kind of first order phase transition which occurs when we boil water. Coleman[3] found that bubble formation would occur by tunnelling from $\phi = 0$ through the barrier at $\phi = \phi_{MAX}$ directly to $\phi = \phi_0$. At an event E a bubble of radius $\sigma_1 < r_0$ forms spontaneously by tunnelling. Inside this sphere $\phi = \phi_0$ and $\rho = 0$. The material which would have filled the bubble (with $\rho = \rho_{VAC}$) is now all concentrated in a thin bubble wall. Since the pressure inside the bubble $P = 0$ is greater than the pressure outside $P = -\rho_{VAC}$ the bubble wall is accellerated outward. The bubble wall traces out a hyberboloid of one sheet which is the set of points at a spacelike separation of σ_1 from the event E. The geometry in the interior of the bubble is that of Minkowski space; the entire future light cone of E is contained within the bubble and one could think of this as an empty open $\Omega = 0$ cosmology as noted by Coleman and De Luccia.[4]

Let us examine the geometrical structure of this model by constructing a Penrose diagram of this type of bubble formation. A complete de Sitter space may be embedded as the hyperboloid

$$-V^2 + W^2 + X^2 + Y^2 + Z^2 = r_0^2 \tag{1}$$

in a flat five-dimensional space with a metric

$$ds^2 = -dV^2 + dW^2 + dX^2 + dY^2 + dZ^2 \tag{2}$$

This can be mapped conformally onto a static three sphere with metric

$$ds^2 = r_o^2 \, [-d\eta'^2 + d^2\chi + \sin^2\chi \, (d\theta^2 + \sin^2 \theta d\phi^2)] \tag{3}$$

(cf. Hawking and Ellis[5]) where $0 \leq \chi \leq \pi$ and $0 < \eta' < \pi$. Thus the Penrose diagram of a complete de Sitter space is a square in the (η',χ) plane. If the event E for example is located at $W = r_o$, $X = Y = Z = V = 0$, the bubble wall is the set of points on the de Sitter hyperboloid with $W = (r_o^2 - \sigma_1^2)^{1/2}$. Now for convenience pick E to have coordinates $X = Y = Z = 0$, $V = r_o$, $W = (2)^{1/2} r_o$, so $\chi = 0$, $\eta' = 3\pi/4$. Assume that the de Sitter phase begins at $V = 0$ ($\eta' = \pi/2$). This may be the result of tunnelling from some pre-geometric state as suggested by Vilenkin[6], may be preceeded by a hot Friedmann stage as suggested by Guth[2] or may be the result of chaotic initial conditions. In any case the de Sitter phase quickly forgets the initial conditions that produced it (cf. Hawking and Moss[7]). The formation of Coleman bubbles can be seen by consulting Figure 1. The bubble forms at E with finite radius EG = σ_1. The outwardly accellerating bubble wall is shown by the line Gi°. The region DEGi°E'BC has a de Sitter geometry and $\phi \approx 0$. Gott[8] noted that the Coleman bubble is surrounded by an event horizon H. Guth[2] showed that the bubbles do not in general percolate. This creates a problem in that our universe does not resemble an unpercolated froth of bubbles.

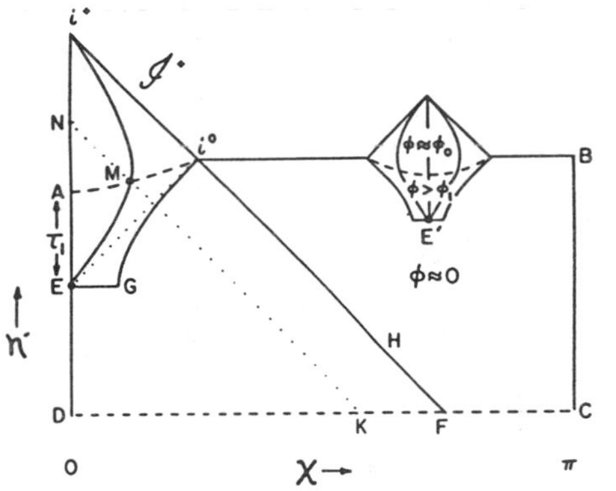

Figure 1: Penrose diagram for bubble formation in inflationary models.

Gott[8] approached this problem from the general relativity viewpoint by trying to find solutions which looked like standard Friedmann solutions at late times and de Sitter space at early times. At large V de Sitter space resembles a cone and just as a cone can be cut to produce a circle, parabola or hyperbola, phase transitions can be arranged to produce either $k = +1$, $k = 0$, $k = -1$ cosmologies. A phase transition out of the de Sitter phase for example at $V = \text{const} > 0$ gives a $k = +1$ cosmology, at $V = \text{const} - W$ gives a $k = 0$ cosmology while at $W = \text{const} > r_o$, $V > 0$ gives a $k = -1$ cosmology. Gott found that the only one to be causally connected (i.e., where the entire phase transition can be signaled from a single event) and to have event horizons was the $k = -1$ case. In this case the phase transition occurs along a negatively curved hyperboloid which is the set of events in de Sitter space at a timelike separation τ_1 from an event E. The geometry is shown in figure 1. The line Ai° is the hyperboloid. The length of AE is τ_1. The region Ai°BCD has a de Sitter geometry while the region Ai°i^+ has the geometry of a standard $k = -1$ Friedmann model with $0 < \Omega < 1$. Gott[8] noted that the Penrose diagram for this is exactly the same as for a Coleman bubble, and proposed that our universe is simply one of the bubbles. The main difference with the Coleman picture is that the bubbles must form slowly over a timescale τ_1 so that there is some inflation within the bubble. Setting $t = 0$ at E, during the period $0 < t < \tau_1$, $a(t) = r_o \sinh (t/r_o)$. Gott found that the value of τ_1 determined Ω_o with $\tau_1 \sim 69 \, r_o$ giving $\Omega_o \sim 0.1$, if $\tau_1 \gg 69 \, r_o$ then

$\Omega_0 \simeq 1 - 0.9 \exp[2(69 - \tau_1 r_0^{-1})] \simeq 1$. Our bubble universe is surrounded by an event horizon H and other bubble universes could form at events E' beyond the event horizon which we would never see. What was needed was a slow forming single bubble model. Interestingly, shortly after this Linde[9], and Albrecht and Steinhardt[10] independently produced detailed particle physics models ("new inflationary" models) based on the Coleman-Weinberg potential in SU(5) which led in fact to slow forming single bubble models (cf. Figure 2). In this potential there is a small barrier at ϕ_{MAX} (greatly exaggerated in Figure 2 for clarity) a long flat shelf beyond the barrier where $V(\phi) \sim V(0)$ and a final drop to $V(\phi_0) = 0$. The small barrier is naturally produced by terms in the potential of the form $m^2\phi^2$, $\xi R\phi^2$, $c_1 T^2\phi^2$ at small ϕ. (cf. Hawking and Moss[7]) where m, ξ, and c_1 are constants, R is the Riemann curvature scalar and T is the temperature. In the de Sitter phase $R = 12/r_0^2$, and $T = 1/2\pi r_0$ is the Hawking temperature. After tunnelling through the barrier (forming a bubble) one rolls down the flat shelf for a period of time τ_1. During this period the value of ρ_{VAC} remains nearly equal to the symmetric vacuum value ($V(\phi) \simeq V(0)$) so the universe inflates by a factor of approximately $\exp(\tau_1/r_0)$ allowing the expanding bubble to encompass the observable universe. Finally one rolls very quickly down the cliff to ϕ_0, dumping the vacuum energy density into ordinary particles with $P \sim 1/3\ \rho$. After this stage is reached the model behaves exactly like a standard big bang model.

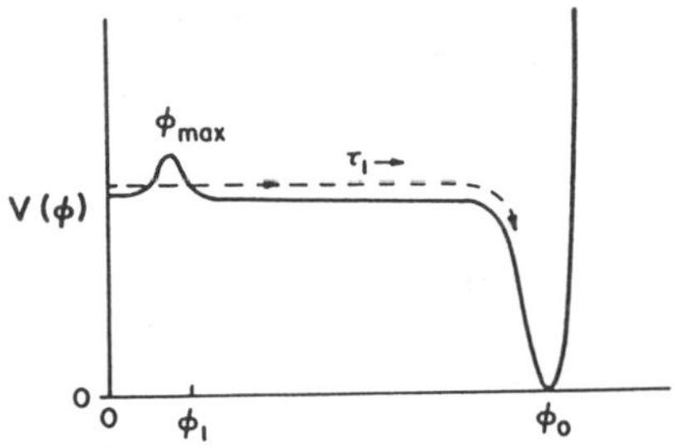

Figure 2: Higgs potential $V(\phi)$ in the new inflationary scenario where ϕ is some measure of the magnitude of the Higgs field. The height of the barrier at ϕ_{max} has been exaggerated for clarity.

The geometrical picture of what happens in the new inflationary scenario has been discussed by Gott and Statler.[11] The region EGi°E'BCD is in the metastable $\phi \simeq 0$ state with a de Sitter geometry. A bubble of radius σ_1 = EG forms at E when we tunnel across the barrier to $\phi = \phi_1$. Since $V(\phi_1) \simeq V(0)$, the value of the vacuum density inside the bubble is almost exactly the same as outside. Thus the amount of material in the bubble wall Gi° is vanishingly small. It is accellerated outward as before by the small pressure differential between the inside and the outside. Since $V(\phi) \simeq V(0)$ for $\phi_1 < \phi \lesssim \phi_0$ we know furthermore that the geometry inside the bubble for a time τ_1 is also well approximated by the same de Sitter geometry that applies outside. Thus the region EGi°A has a de Sitter geometry and $\phi_1 < \phi < \phi_0$. At Ai° we fall off the cliff to reach $V(\phi) = 0$ at $\phi = \phi_0$, we dump the vacuum energy into normal particles and enter the Friedmann phase. These changes are illustrated explicitly in the E' bubble.

Whether the bubbles percolate depends on the dimensionless bubble formation rate ε which is the probability of forming a bubble in a four-volume r_0^4. Guth and Weinberg[12] found that there is a critical bubble formation rate ε_{cr} such that if $\varepsilon < \varepsilon_{cr}$ the bubbles will not percolate. They show that $5.8 \times 10^{-9} < \varepsilon_{cr} < 0.24$. Our present position in the diagram is illustrated by the point N. The distance AM is $a(\tau_1)\eta_o$ where η_o is quite accurately given by the co-moving particle horizon radius in the standard big bang model:

$$\eta_o = \ln \frac{1 + (\Omega_{r,o})^{1/2} + (1 - \Omega_o)^{1/2}}{1 + (\Omega_{r,o})^{1/2} - (1 - \Omega_o)^{1/2}} \tag{4}$$

where $\Omega_{r,o}$ is the current ratio of the radiation density to the critical density (cf. Gott and Statler[11]). If any other bubbles E' formed within our past light cone (region NDK) then we would see them as hot spots in the sky and this would spoil the isotropy of the microwave background. The probability that no other bubbles form in this region is

$$P = \exp\,[-4\pi\varepsilon\{2 \ln [\exp(\eta_o) + 1] + \exp(2\eta_o)\}/3] \tag{5}$$

(cf. Gott and Statler[11] for details). If we demand that $P > 0.05$ we can set upper limits on ε, i.e., $\varepsilon < 7.6 \times 10^{-4}$ if $\Omega_o \sim 0.1$ and $\varepsilon < 0.30$ if $\Omega_o \sim 1$. We expect that $\varepsilon \propto \exp[-B]$ where B is the action involved in crossing the potential barrier at ϕ_{max}. So it is easy in principle to produce grand unified theories with exponentially small values of ε which will be in agreement with our observations of the isotropy of the microwave background and in which the bubble universes do not percolate. In this case the bubbles form an infinite number of isolated bubble clusters each containing an infinite number of bubbles. If we wait till $t = \infty$ (the point i+) we will be able to see all the other bubbles in our cluster but we will never be able to see the other clusters. Our bubble cluster is surrounded by an event horizon (cf. Gott[13]).

Of course we are a long way from the correct grand unified theory. The Coleman-Weinberg potential has already been superseded because it produces fluctuations that are too large in amplitude (cf. Bardeen, Steinhardt, Turner[14]) but promising supersymmetric models are now available which solve this problem. Any theory which has the general features of Figure 2 will produce bubble universes as shown in Figure 1. The requirements are (1) a de Sitter phase with a finite beginning, (2) a metastable false vacuum state, (3) a barrier high enough to produce a small bubble formation rate, (4) a flat shelf allowing enough inflation ($\tau_1 \gtrsim 69\ r_o$) within the bubble. Bubble universes have negative curvature. If $\tau_1 \gg 69\ r_o$, $a(t_o) \gg 6000$ Mpc and $\Omega_o \sim 1$ but with minimal inflation $\tau_1 \sim 69\ r_o$, $a(t_o) \sim 6000$ Mpc and $\Omega_o \sim 0.1$. If the true value of Ω_o is significantly less than 1 the length of the inflationary period τ_1 can be established with rather high accuracy setting a strong constraint on allowed grand unified theories.

REFERENCES

[1] F. Englert, <u>Physical Cosmology, Les Houches 1979 Session XXXII</u> (North-Holland Publishing Company, R. Balian <u>etal.</u> eds., 1979).

[2] A. H. Guth, Phys. Rev. D**23**, 347 (1981).

[3] S. Coleman, Phys. Rev. D**15**, 2929 (1977).

[4] S. Coleman, F. De Luccia, Phys. Rev. D**21**, 3305 (1980).

[5] S. W. Hawking and G. F. R. Ellis, <u>The Large Scale Structure of Space-Time</u> (Cambridge University Press, 1973).

[6] A. Vilenkin, Phys. Rev. D**27**, 2824 (1983).

[7]S. W. Hawking and I. G. Moss, Physics Lett. **110B**, 35 (1982).
[8]J. R. Gott, Nature **295**, 304 (1982).
[9]A. D. Linde, Phys. Lett. **108B**, 389 (1982).
[10]A. Albrecht and P. J. Steinhardt, Phys. Rev. Lett. **48**, 1220 (1982).
[11]J. R. Gott and T. S. Statler, Phys. Lett. **136B**, 157 (1984).
[12]A. H. Guth and E. J. Weinberg, Nucl. Phys. **B212**, 321 (1983).
[13]J. R. Gott, In *Early Evolution of the Universe and its Present Structure*, I.A.U. Symposium 104, (Reidel, Dordrecht, G. Abell ed., 1982).
[14]J. M. Bardeen, P. J. Steinhardt, M. S. Turner, Phys. Rev. D**28**, 679 (1983).

Inflation in a Wall Dominated Universe

David Seckel

ABSTRACT

We examine the possibility that the Universe was, at some time in its past, dominated by topological domain walls. In particular, we discuss the possibility that such an epoch could solve the flatness and horizon problems. We conclude that such a model would not work. The problems of baryogenesis and density perturbations are discussed.[1]

I. MOTIVATION

In 1981, Guth[2] invented inflation. Since then many problems have been discovered to plague the inflationary model builder.[3] Although all these problems may be solved one at a time, no one has demonstrated an existence proof of a working inflationary model. Therefore, it makes sense to look for alternatives to inflation. What we have in mind is a Universe with an equation of state such that a rapid expansion is effected. However, the expansion does not have to be exponential, as in normal inflation. We refer to such scenarios as generic inflation. We will examine a particular case of generic inflation, a Universe dominated by topological domain walls. We will also look at the problems of baryogenesis and density perturbations in the wall dominated scenario.

II. GENERIC INFLATION

Let us call the cosmic scale factor, S, and the cosmic time, t. Then a necessary condition to improve the flatness and horizon problems is that the ratio S/t be an increasing function of time.[4] To solve these problems such behavior must occur for a sufficiently long period of expansion. At the end of that time the Universe must reheat, i.e. whatever stress energy was involved in the inflationary epoch must be converted into radiation in such a way as to not lose the successes of the standard big-bang model. For simplicity, we consider power law expansion, $S \sim t^a$. Then if $a > 1$ we will have generic inflation in that the horizon and flatness problems are improved.

First consider the horizon problem. Suppose the inflationary epoch begins at some time t_0 and continues until t_1, and for the purposes of illustration assume that the Universe is radiation dominated after t_1. Then the particle horizon at some time $t > t_1$ is given by

$$\ell = S \int \frac{1}{S}\, dt \simeq \frac{t_0}{a-1} \frac{S}{S_0} + 2t \ . \tag{1}$$

The first term in eq.(1) represents the stretching out of the particle horizon that existed at the begining of the inflationary epoch, whereas the second term is due to particle propagation in the post-inflationary radiation dominated regime. It is possible for the first term to exceed the second, in which case there is an apparent horizon problem, i.e.

the Universe is homogeneous on scales larger than the apparent particle horizon. However, no problem exists since at early times a comoving region which will subsequently grow to be much larger than the apparent horizon was in causal contact with itself.

Now consider the flatness problem. We start by writing down the Einstein equation for a Friedman model.

$$\left(\frac{\dot{S}}{S}\right)^2 = \frac{8\pi G\rho}{3} + \frac{k}{S^2}, \tag{2}$$

where ρ is the density and k is the curvature. We can parameterize the importance of curvature by defining the quantity

$$\alpha \equiv \frac{3k}{8\pi G\rho S^2}. \tag{3}$$

During a radiation dominated era, $\rho \sim S^{-4}$, and therefore α increases with the scale factor. An observer living in a radiation dominated era with $\alpha \sim 1$ will wonder at the miraculous initial conditions necessary to produce his observed Universe. However, the problem can be solved if the Universe previously went through a phase of generic inflation. To show this it is sufficient to consider the Einstein equations without the curvature term. For a power law expansion the density develops as $\rho \sim t^{-2}$. Then α behaves as t^2/S^2. If (S/t) increases then the importance of curvature decreases. This argument is valid even if the curvature term is large at the beginning of inflation, as long as the Universe is open. In the absence of curvature, an equation of state that produces a scaling law $S \sim t^a$, must have an energy density that scales as $\rho \sim S^{-2/a}$. With this scaling law for ρ we find that α behaves as $\alpha \sim S^{(2/a)-2}$. As a result, an equation of state that reproduces generic inflation will eventually achieve negligible curvature.

It should be clear that in order to solve the horizon problem it is necessary to have the observable Universe today be contained within the particle horizon at the beginning of the inflationary epoch. The same condition applies for solving the flatness problem. To show this we could write down scaling laws for the various epochs (inflationary, radiation, matter) and apply matching conditions. A more direct method is to think of the whole Universe as being flat with some local perturbations in the curvature. The parameter α is then related to the growth of those perturbations while outside the horizon. The recent work on density perturbations tells us that the amplitude of a curvature perturbation will be the same (up to factors of order unity) at both horizon crossings.[5] Then, to solve the flatness problem we want to have α of order unity at the beginning of inflation and have todays horizon be roughly the scaled up horizon from the beginning of inflation. (Caution, this argument is not valid in the case of exponential inflation where the quantity (ρ + p) is very small)

Before proceeding, we note that the monopole problem is solved in any generic inflation model. If enough inflation occurs to solve the horizon and flatness problems then the monopole to entropy ratio becomes negligible. This assumes that no monopoles are produced during reheating.

III. WALL DOMINANCE

In a wall dominated era the density scales as $\rho \sim 1/S$. This follows since the density of walls goes as $1/S^3$ whereas the area in any given wall goes as S^2. Using $\rho \sim S^{-2/a}$ we find that $a = 2$; so, $S \sim t^2$. Therefore, a wall dominated Universe undergoes generic inflation. The important question is then, can this inflation solve the horizon and flatness problems.

To answer this question it is necessary to understand how the Universe reheats after wall dominance. The walls must dissipate their energy into radiation. One way in which this might happen is if the domain walls are not absolutely stable. An example of such walls are the walls produced in many axion models.[6] Now, such walls may decay via hole formation, a process similar to the decay of the false vacuum as described by Coleman and Callan.[7] In the case of axion walls the rate of hole formation is hopelessly low, however it is not difficult to make a model where hole formation proceeds at an interesting rate. (We note two attractive features of this decay scenario. Since tunneling phenomena depend exponentially on parameters of the Lagrangian it is only necessary to tune a model to order one percent. Also, since the decay is exponential and the expansion of the Universe is a power law the inflationary epoch is guaranteed to end.)

Now, just after the walls decay the Universe will be composed of hot thin sheets that are quite far apart. One must wait until these sheets thicken and overlap before the Universe will be homogeneous again. But the sheets thicken at the speed of light so their width is just the second term in eq.(1). On the other hand the distance between walls is scaling with the expansion of the Universe. It is usually thought that walls on smaller scales than the horizon will find ways to annihilate due to the strong stresses involved. So, the natural scale for the distance between walls is just the horizon length at the beginning of inflation expanded by the growth in scale factor. This is just the first term in eq.(1). Therefore, walls cannot merge until the two contributions to the particle horizon are comparable. This is just the opposite of the condition that ensures a solution to the horizon and flatness problems, so it is improbable that a wall dominated era could solve the horizon and flatness problems.

The situation is actually more restrictive than presented above. In order to achieve nucleosynthesis in the standard manner it is required that the Universe be homogeneous at the time of nucleosynthesis. Therefore, wall dominated inflation cannot be expected to solve the flatness and horizon problems past the age of nucleosynthesis. To have a marginally flat world today, would require a fine tuning of initial conditions to one part in 10^{16}. There is a possible way out of this dilemma, which is to have many walls per horizon at the beginning of inflation. However, one must suppose that wall annihilation is peculiarly inefficient.

IV. BARYOGENESIS AND DENSITY PERTURBATIONS

At first thought baryogenesis[8] might seem to be difficult in wall dominated scenarios because the reheating temperature is quite low. However, one does not have to have a homogeneous Universe for baryogenesis, so one can form baryons directly after the walls decay. We can estimate the baryon to photon ratio by treating the domain walls as a coherent state of the scalar field with a large occupation number. The decay of the walls can be treated as the out of equilibrium decay of the scalars. If this decay violates CP and baryon number then baryons may be produced. Let each scalar produce ε baryons. If the thickening of the walls is an adiabatic process then the number of photons produced per scalar may be calculated from the average energy density at the time the walls decay. If one requires a minimal solution to the horizon and flatness problems then the baryon to photon ratio is of order $(10^{-7}-10^{-8})\varepsilon$. Assuming maximal CP violation it is possible to generate enough baryons.

The question of density perturbations[9] can also be dealt with in a fairly simple fashion. There will be two types of density fluctuations depending on whether the scale in question is larger or smaller than the horizon at the time the walls form. For larger scales, fluctuations in the background temperature at the time the walls form will induce adiabatic perturbations in the wall density. On small scales, fluctuations come from the fact that the walls will be randomly located. This results in a fluctuation spectrum which scales as $\delta \sim (\lambda/\ell)^{1/2}$ at horizon crossing. To account for galaxies and measurements of the 3°K background radiation the amplitude must be normalized so that $\delta \sim 10^{-4}$ at recombination. This implies that $\delta \sim 1$ at a temperature of about 1 Gev. This must correspond to the time when the walls are just starting to merge. One should worry that a large number of solar mass black holes might form at this time. In order to not overclose the Universe with such objects the energy fraction trapped in blackholes should be no greater than 10^{-8}.

Part of this work was done at the University of Washington.

REFERENCES

[1]The contents of this contribution are discussed in more detail in a paper under preparation.
[2]A. Guth, Phys. Rev. **D23**, 347 (1981).
[3]P.J. Steinhardt, these proceedings.
[4]Similar considerations have been made by L. Abbott and M. Wise, see L. Abbott, these proceedings.
[5]J. Bardeen, P.J. Steinhardt, and M.S. Turner, Phys. Rev. **D28**, 679 (1983); J.A. Frieman and M.S. Turner, Univ. of Chicago preprint (1983).
[6]J. Preskill, these proceedings.
[7]C. Callan and S. Coleman, Phys. Rev. **D16**, 1762 (1977).
[8]For a review of baryogenesis see E. Kolb and M.S. Turner, Ann. Rev. Nuc. Part. Sci. **33**, 645 (1983).
[9]See talks by S. White and J. Silk, these proceedings.

PART V

MASSIVE MAGNETIC MONOPOLES

Interest in magnetic monopoles was revived in 1974 with the discovery by 't Hooft and Polyakov that certain classes of non-Abelian gauge theories admit monopole solutions. Grand Unified Theories of the strong and electroweak interactions are examples of spontaneously broken gauge theories with magnetic monopoles. Before gauge theories, the mass of the magnetic monopole was a free parameter, but in gauge theories the mass is related to the scale of symmetry breaking. In Grand Unified Theories the mass of the magnetic monopoles is expected to be of order 10^{16} GeV. This enormous mass makes the production of GUT monopoles by terrestrial accelerators impossible, and the only site with enough energy for monopole production is the very early Universe. The first predictions for the abundance of GUT monopoles produced in the big bang were embarrassingly large, and efforts to reduce the predicted number started Guth on the path to the inflationary Universe. Inflationary scenarios, on the other hand, predict a dearth of monopoles -- typically less than one in the observable Universe. The present-day detection of a massive monopole would be exciting because it would offer a glimpse into physics at high energies, and it would be direct evidence that the Universe was once at temperatures of the order of 10^{16} GeV.

An overview of magnetic monopoles in particle physics and cosmology is given by Preskill. Further aspects of monopoles in cosmology are discussed in papers by Weinberg, who carefully investigates annihilation as a mechanism to rid the Universe of monopoles, and Bracci, et al., who discuss the formation of monopole-proton bound states. At present there are many groups searching for cosmic ray magnetic monopoles. Stone reviews the status of the overall effort for monopole detection, and short reports from several experimental groups are presented. One of the most exciting recent developments in monopole physics is the observation by Rubakov and Callan that monopoles may catalyze nucleon decay at a rate typical of the strong interactions. Harvey gives a summary of the effect and discusses some of its implications for particle physics and cosmology.

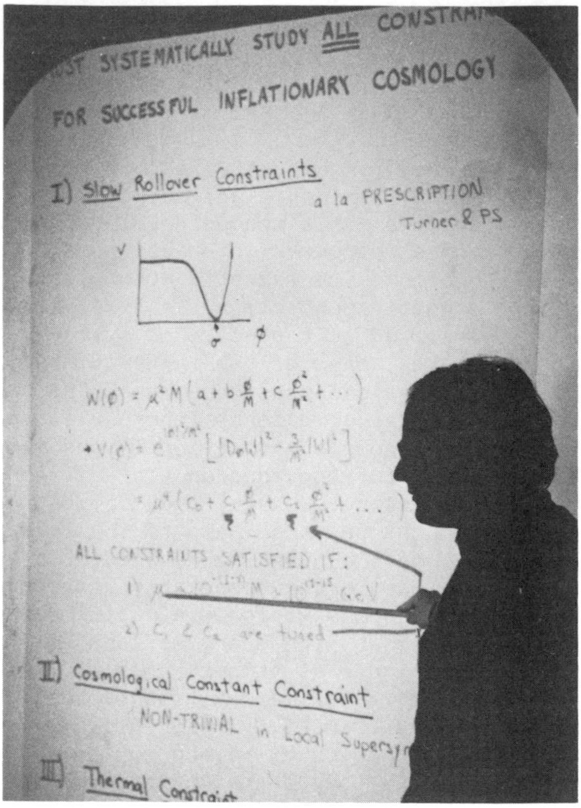
SYSTEMATICALLY STUDY ALL
FOR SUCCESSFUL INFLATIONARY COSMOLOGY
I) Slow Rollover Constraints
a la PRESCRIPTION
Turner & PS
V
σ
ϕ
ALL CONSTRAINTS SATISFIED IF:
II) Cosmological Constant Constraint
NON-TRIVIAL in Local
III) Thermal

Magnetic Monopoles in Particle Physics and Cosmology

John Preskill

I. INTRODUCTION

Hardly any topic better illustrates the connection between particle physics and cosmology than the topic of magnetic monopoles. While there is no persuasive evidence that a monopole has ever been detected, the existence of monopoles is implied by deeply cherished beliefs about the structure of matter at extremely short distances. And the fact that monopoles are so rare as to have escaped detection has profound implications concerning the very early history of the universe. In this talk I give a brief overview of the theory of magnetic monopoles and its relevance to cosmology. In Section II, I explain the connection between monopoles and the unification of the fundamental interactions. In Section III, I describe how monopoles might have been produced in the very early universe. Theoretical limits on the abundance of monopoles derived from astrophysical considerations are the subject of Section IV. Section V contains conclusions.

A much more detailed discussion of these and related matters may be found in Ref. 1.

Those of us who do theoretical particle physics have tended more and more, in recent years, to speculate about the structure of matter at exceedingly short distances, distances as small as 10^{-29} cm or smaller. Such speculations can be discouraging, because to explore this extremely short-distance physics experimentally would require that we do accelerator experiments at correspondingly high energies, energies as high as 10^{15} GeV or more. Such experiments won't be done in the foreseeable future. It's partly for this reason that some particle physicists have become interested in cosmology. We believe that the universe was once exceedingly hot, so hot that particle collisions with energies of 10^{15} GeV or above were occurring copiously, and we can therefore hope to use the very early universe as our accelerator and look today for a relic of our distant, very energetic past.

Figure 1 is a summary of the history of the universe; the age of the universe in seconds and its temperature in GeV are indicated on a logarithmic scale. Traditional cosmology begins when the universe was about 1 second old and temperature was about 1 MeV. At that time the universe was a nuclear physics laboratory. Nuclear reactions were occurring with copious rates, and it's around that time that the story of helium synthesis began. We now believe that we understand the structure of matter down to distances of around 10^{-16} cm or up to energies of about 100 GeV, and therefore we can extrapolate back the big bang cosmology to much earlier times, times of the order of the first billionth of a second. But to extrapolate back further than that, we must speculate about the structure of matter at still shorter distances. There's no shortage of speculation. What's indicated in the figure is one possible speculation, according to which essentially no new particles or interactions arise between present-day energies and energies of the order of 10^{15} GeV. The history of the universe in such a scenario would be extremely dull for many decades of time before the first billionth of a second, but even in this extreme, radically conservative scenario, eventually, at sufficiently high energies, or sufficiently early times, qualitatively new particles and interactions do come into play. On the basis of this picture, we should look today for some kind of relic of the physics of the first 10^{-35} sec. Most of what I'm going to say in this talk, however, does not rely on any specific picture of physics at distances much less than 10^{-16} cm. Most of it follows, in fact, from a very general hypothesis about the unification of the fundamental interactions which I'll spell out more specifically as I proceed.

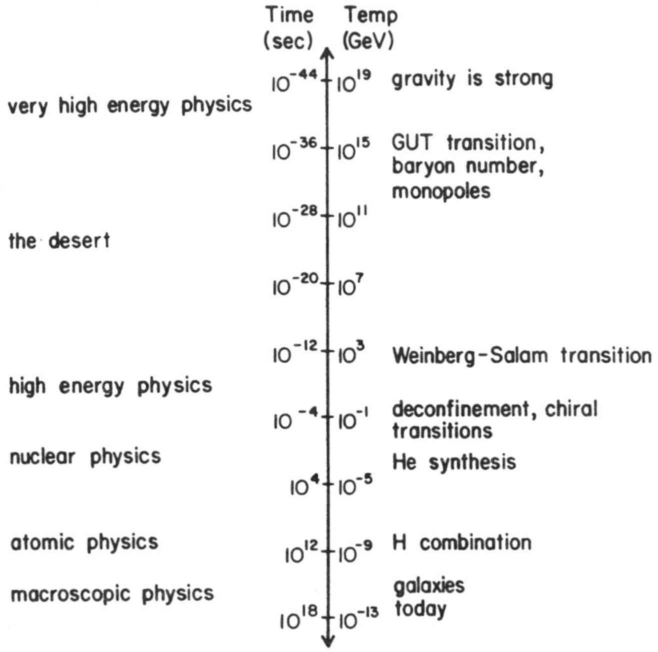

Figure 1: History of the universe.

II. MONOPOLES AND UNIFICATION

The relic of the very early universe which I want to consider in this talk is the magnetic monopole, the existence of which is predicted in a very wide class of theories. Among the theories which predict that monopoles should exist are those known as grand unified theories (GUTs). These are theories in which the standard low-energy gauge group $SU(3)_{color} \times [SU(2)\times U(1)]_{electroweak}$ is embedded in some simple group G which becomes spontaneously broken at a large mass scale M_X. The simplest and most familiar example of the unifying gauge group is SU(5), as proposed by Georgi and Glashow.[2] But that's just one of many possibilities. In the case of the SU(5) model, the unification scale M_X can be computed to be about 10^{14} GeV[3], but in a different model it could be different.

Grand unified theories have three properties which are especially relevant when we consider their cosmological implications. First of all, we expect that at sufficiently high temperatures, the full grand unified symmetry group G became restored; therefore, there was a phase transition as the universe cooled, in which the symmetry first became spontaneously broken. It is possible that the primordial density fluctuations from which galaxies evolved were a relic of such a phase transition. Second, these theories are expected to have new interactions with range of the order of M_X^{-1} which can change baryon number. This leads, on the one hand, to the prediction that the proton should be unstable and, on the other hand, to a basis for understanding, in microscopic terms, the baryon number of the universe. And third, in grand unified theories the quantization of electric charge is explained in a very natural way. In these theories, the electric charge operator obeys nontrivial commutation relations with other operators in the theory, and just as the angular momentum algebra requires J_z to have eigenvalues which are integer multiples of $(1/2)\hbar$, so in these theories the non-trivial commutation relations obeyed by the electric charge operator require electric charge to be a multiple of a fundamental unit.

Very closely connected to this natural explanation of quantization of electric charge is the prediction that grand unified theories contain stable particles which carry magnetic charges, that is magnetic monopoles. The observation that grand unified theories necessarily contain magnetic monopoles was made some ten years ago by 't Hooft and Polyakov,[4] but the idea that there is a deep connection between quantization of charge and magnetic monopoles is much older. It was first proposed by Dirac[5] over 50 years ago.

Dirac envisioned a monopole as an extremely long and extremely thin solenoid. One end of this solenoid viewed in isolation appears to be a magnetic charge (Fig. 2). But it makes sense to identify the end of the solenoid as a magnetic monopole only if no conceivable experiment can detect the solenoid (which is also called the Dirac string), in the limit in which it becomes infinitesimally thin. We can imagine trying to detect the solenoid by doing an electron interference experiment. But that experiment gives a null result if the phase picked up by the electron wave function when it is transported around the string is a trivial phase. This phase is

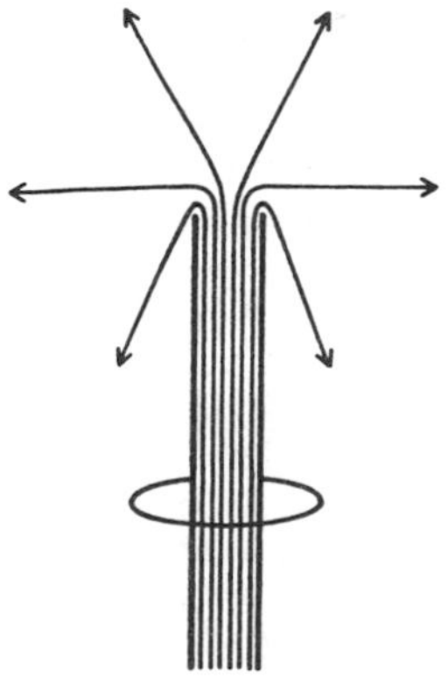

Figure 2: The end of a long solenoid.

$$\exp\left[-ie \oint \vec{A} \cdot d\vec{r}\right] = \exp\left[-i\, 4\pi\, eg\right] = 1 \ , \tag{1}$$

where $-e$ is the electron charge and g is the magnetic charge of the monopole. (The solenoid carries magnetic flux $4\pi g$.) Thus, the magnetic charge is required to satisfy Dirac's quantization condition

$$eg = n/2 \ , \tag{2}$$

where n is an integer. The minimum allowed magnetic charge $g_D = 1/2e$ is called the Dirac magnetic charge.

Furthermore, one can turn this argument around, as Dirac observed. If there exists any particle with electric charge q and there exists a magnetic monopole somewhere in the world, then the requirement that the string of the monopole is undetectable is equivalent to the condition that the electric charge must be an integer multiple of 1/2g where g is the magnetic charge of the monopole. So the existence of a magnetic monopole explains quantization of electric charge.

We can restate this result in a mathematically equivalent way, which we'll find useful in a moment. One can say that the Dirac string, the solenoid, is a "gauge artifact;" the vector potential in the vicinity of the string can be removed by doing a gauge transformation. It is a "pure gauge,"

$$A_\phi = -\frac{1}{ie}(\partial_\phi \omega)\,\omega^{-1} , \tag{3}$$

where $\omega(\phi)$ is the gauge function close to the string, and ϕ is the azimuthal angle around the string. Requiring $\omega(\phi)$ to be single valued, we obtain by integrating Eq. (3),

$$\omega(0) = \omega(2\pi) = \exp\left[-ie \int_0^{2\pi} A_\phi d\phi\right] \omega(0) , \tag{4}$$

which is equivalent to Eq. (1). In this way, we see that the integer n in Eq. (2) is the number of times the phase ω winds around the unit circle as the azimuthal angle ϕ changes from 0 to 2π. So we see that the Dirac quantization condition for the magnetic charge has a topological interpretation; it can be interpreted as the statement that the winding number of a circle around the circle must be an integer.

A natural question to ask, is why is it the electron charge which appears in the Dirac quantization condition? We believe that quarks exist, and the electric charge of a down quark, for example, is $-e/3$. Will not the same argument as before, applied to the down quark instead of the electron, lead to the conclusion that the minimal allowed magnetic charge is three times the Dirac magnetic charge?

Well, in fact, if we want to apply to quarks the argument we applied earlier to electrons, we have to allow for the fact that quarks are permanently confined inside hadrons, and therefore it makes sense to speak of doing a quark interference experiment only over distances less than the size of a hadron, of the order of 10^{-13} cm. It's true that if a down quark, say, which has charge $-e/3$, is transported around the string of a monopole with the Dirac magnetic charge, its wave function picks up a non-trivial phase. But we have to recall that a quark, in addition to its electric charge, carries another degree of freedom, namely color. The string will be undetectable if the monopole, in addition to its ordinary magnetic charge, carries also a color magnetic charge such that the phase acquired by the down quark due to the interaction of its color charge with the color magnetic charge of the monopole compensates for the phase associated with its electric charge. The conclusion, then, isn't that a monopole with the Dirac magnetic charge is forbidden because of the existence of quarks, but rather that the monopole with Dirac magnetic charge must also carry a color magnetic charge of appropriate strength.[6] The color magnetic field becomes screened by nonperturbative color effects at distances greater than the size of a hadron; it can't be detected in a macroscopic experiment.

Again we can restate this conclusion in the winding number language used earlier. Although quarks have fractional electric charge, objects which can be isolated have trivial color triality; they can be made out of a number of quarks minus anti-quarks which is a multiple of three, and such objects, with trivial triality, have integer electric charge. A mathematical way of restating this is that $\exp[i2\pi\, Q_{quark}]$, where Q_{quark} is the charge operator acting on a quark, is an element of the center of the color group SU(3); it's just the unit matrix acting on quarks of all three colors times a non-trivial phase, $e^{-i2\pi/3}$.

Well, that means that a phase which winds one third of the way around the electromagnetic gauge group goes from the element 1 to an element of the center of SU(3). And therefore the gauge transformation which can remove the vector potential of the monopole in the vicinity of the Dirac string, needs to wind only one third of the way around the electromagnetic gauge group and it can then return to its starting point, the identity, by going through the $SU(3)_{color}$ group (Fig. 3). But that means that this gauge transformation, in addition to having winding number one third in the electromagnetic gauge group, has also an $SU(3)_{color}$ winding number, which is to say that the monopole carries both kinds of magnetic charge, ordinary magnetic charge and color magnetic charge.

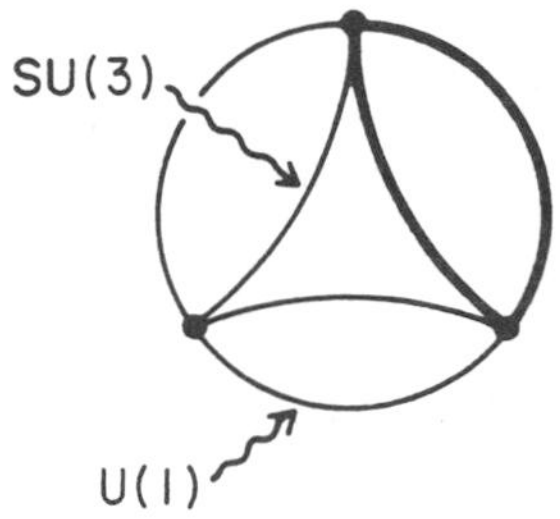

Figure 3: A minimal nontrivial loop in $[SU(3)\times U(1)]/Z_3$.

Dirac's reasoning demonstrated the consistency of the existence of the magnetic monopoles in quantum mechanics. But 't Hooft and Polyakov reached a more powerful conclusion. They found that the existence of monopoles is actually necessary in spontaneously broken unified gauge theories. To illustrate their result, let's consider a simple example in which the gauge group SO(3) is spontaneously broken to the subgroup U(1). This example is easy to visualize because we can think of the order parameter of the spontaneous symmetry breakdown, which is some scalar field, as being an arrow which points somewhere in a three-dimensional internal symmetry space. The SO(3) gauge symmetry rotates the arrow and the unbroken U(1) symmetry consists of rotations around the axis picked out by the arrow. This order parameter, the scalar field Φ, has an expectation value in the vacuum state; more precisely, its modulus is required by energetic considerations to have an expectation value which we'll call **v**. But the direction in which it points in the internal symmetry space is undetermined, and therefore there are many degenerate vacua, as many vacuua as there are points on a two-dimensional sphere.

In this model in which spontaneous symmetry breaking occurs, let us ask the following question: Are there stable, time-independent solutions of finite energy to the classical equations of motion in this theory, other than the trivial vacuum solution? If there are such solutions, they behave like particles in the classical theory. They're called solitons, and we can expect them to survive in the spectrum of the quantum theory. In fact, the answer to the question is 'yes', and to see that we can seek such a solution by the following strategy.

First of all, let's determine what conditions must be satisfied in order for a field configuration to have finite energy. It's evidently necessary for the modulus of the scalar field Φ to approach its energetically preferred value **v** at spatial infinity.

$$|\Phi| \xrightarrow[r\to\infty]{} v \ . \tag{5}$$

There is also a gradient term in the energy which must be finite and that means that the gradient of the scalar field Φ must get small at large **r** sufficiently rapidly (faster than $\mathbf{r}^{-3/2}$). Actually, because this is a gauge theory, it's a covariant derivative of Φ which occurs, the sum of an ordinary gradient and a term in which the gauge field acts on the scalar field,

$$|D_\mu \vec{\Phi}| = |\partial_\mu \vec{\Phi} - e\,\vec{A}_\mu \times \vec{\Phi}| \xrightarrow[r\to\infty]{} 0 \ . \tag{6}$$

Consider, now, a scalar field configuration of finite energy such that as $\vec{\mathbf{r}}$ goes to spatial infinity along any fixed direction, the scalar field Φ goes to a point in the internal symmetry space that has modulus **v**, but points in a direction in the internal symmetry space which corresponds to the direction in actual space along which $\vec{\mathbf{r}}$ goes to infinity,

$$\Phi^a \xrightarrow[r\to\infty]{} v\hat{r}^a \ . \tag{7}$$

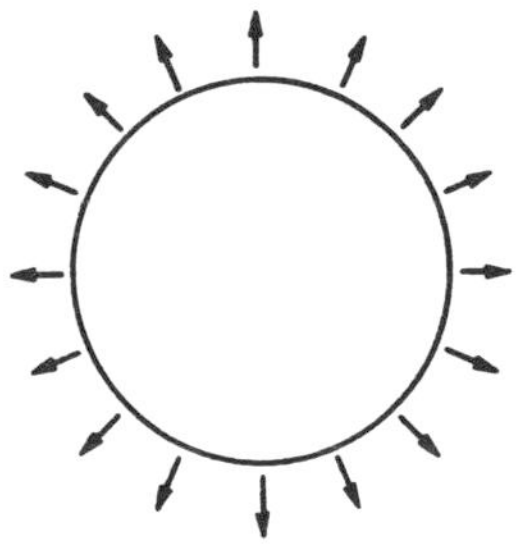

Figure 4: A hedgehog.

Such a configuration looks like a hedgehog (Fig. 4). In order for it to have finite energy, since there is a spatial gradient term in the circumferential direction which is falling off only like $1/\mathbf{r}$, it's necessary for the gauge field to also fall off like $1/\mathbf{r}$, so that the covariant gradient of the scalar field falls off faster than $1/\mathbf{r}$.

Once we've constructed this configuration with finite energy we can deform it continuously until we reach a stationary point of the energy functional. When we find such a stationary point we have a solution to the classical equations of motion which is stable and time independent. So we have succeeded in constructing a soliton unless the solution we have obtained is the vacuum solution. But in this case we know it cannot be the vacuum solution because it is impossible to continuously deform the hedgehog configuration with which we began, while the energy remains finite, to the vacuum solution; the vacuum solution, at least in a particular gauge, is a configuration in which the scalar field points in the same direction at each point at spatial infinity rather than in a direction which is correlated with the direction in space. You can't comb the hair on a billiard ball without making a part, you can't peel an orange without breaking the skin, you can't continuously deform the hedgehog configuration to a vacuum configuration. Thus we have succeeded in constructing a soliton in this theory, and we expect it to survive in the spectrum of the quantum theory.

In fact, the configuration which we have constructed is a magnetic monopole. To see this, we perform an SO(3) gauge transformation which rotates the scalar field Φ to a standard direction, so that it points in the same direction everywhere on the sphere at spatial infinity (Fig. 5). As I've just said, there is no such transformation without a singularity, so the gauge transformation is singular at some point on the sphere, let's say the south pole. Therefore, the gauge field, which you recall falls off like $1/r$, acquires a string singularity in this gauge. Obviously, this singularity is a gauge artifact; we produced it by performing a gauge transformation, so we can remove it by performing a gauge transformation.

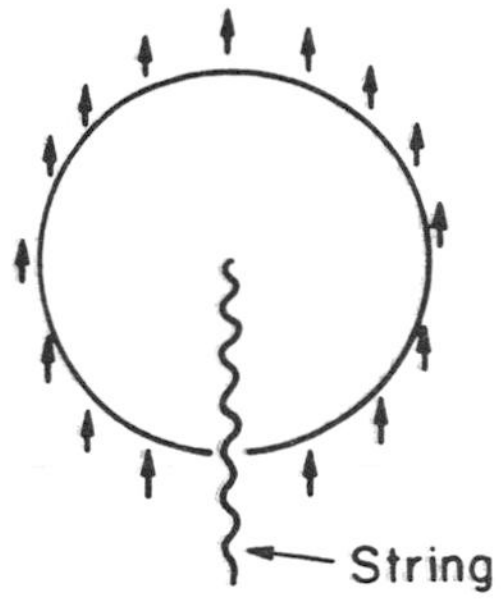

Figure 5: Hedgehog in the "string gauge."

Consider, now, this gauge transformation in more detail. On the sphere at spatial infinity, it takes the value

$$\omega(\theta,\phi) \in SO(3) \ . \tag{8}$$

Here ω is singular at the south pole, $\theta = \pi$, but since we know that both Φ and $\omega\Phi$ are smooth, it must be that

$$\omega(\theta=\pi,0)\omega^{-1}(\theta=\pi,\phi) \in U(1) \ , \tag{9}$$

so that the discontinuity in ω is an element of the unbroken U(1) subgroup, which acts trivially on Φ. Thus the gauge transformation ω, in the vicinity of the south pole, (the string) defines a loop in U(1). That loop must have a winding number; if it did not, there would be no need for the string. Then we would be able to continuously deform a hedgehog to a vacuum configuration, which we know is impossible. We've already seen that this winding number can be identified with the magnetic charge carried by the long-range gauge field. Thus, the "topological charge" of our soliton, which prevents it from being deformed to a vacuum configuration, is precisely the same as its magnetic charge.

To generalize this observation, we first note that the loop defined by Eq. (9) can be smoothly shrunk to a point through the group SO(3); we need only let θ range from π to 0. (The loop can simply slip off the top of the sphere, without crossing the string. See Fig. 6.) Furthermore, given a nontrivial loop in U(1) which can be shrunk to a point in SO(3), we can associate with it a (singular) gauge transformation $\omega(\theta,\phi)$ on the sphere, and a corresponding topologically nontrivial Φ configuration on the sphere. There is, in general, a one-to-one correspondence between the topological charge and magnetic charge of a soliton.

Figure 6: A loop around the south pole slips off the sphere.

Generalizing further, given any model in which the gauge group **G** is spontaneously broken to a subgroup **H**, such that there are loops with a nontrivial winding number in **H** which can be shrunk to a point in **G**, we can construct solitons with **H**-magnetic charges. In particular, such loops always exist if **G** is simple and **H** contains a U(1) factor. All grand unified theories contain monopoles.

I emphasize again the generality of this conclusion. The prediction that magnetic monopoles exist does not depend on the mechanism of the symmetry breakdown: in particular, it does not matter whether the Goldstone bosons associated with the symmetry breakdown are elementary or composite. Nor does it matter whether gravitation becomes unified with the other particle interactions at the unification scale.

At spatial infinity in the monopole solution, as I've said, the scalar order parameter approaches its vacuum value. But deep inside the monopole there is a core in which the scalar order parameter departs by a significant amount from its vacuum value, and in which non-abelian gauge fields are excited. The size R_c of the core is determined by competition between two energetic effects. It is chosen to minimize the sum of the energy stored in the magnetic field outside the core and the energy due to the scalar field gradient inside the core. In order of magnitude these are

$$E_{mag} \sim 4\pi g^2 R_c^{-1} \sim (4\pi/e^2) R_c^{-1}$$

$$E_{scal} \sim 4\pi v^2 R_c \sim (4\pi/e^2) M_X^2 R_c \qquad (10)$$

where M_X is the typical mass of a heavy vector boson. The core size and monopole mass are determined to be

$$R_c \sim M_X^{-1}$$

$$m \sim (4\pi/e^2) M_X \; . \qquad (11)$$

We see that, if the gauge coupling e is small, then the Compton wavelength of the monopole is much smaller than the size of its core. It is a nearly classical object, and quantum mechanics plays a minor role in determining the structure of the core.

In general, we can't predict the mass M_X of the heavy vector bosons, but we can predict it if we make the assumption (the desert hypothesis) that there are essentially no new particles or interactions between present day energies of order 100 GeV and the unification scale. If we make that assumption we predict the M_X is about 10^{14} GeV; thus the size of the monopole core would be about 10^{-28} cm, and its mass around 10^{16} GeV. Of course the desert hypothesis could be wrong and the mass of the monopole could turn out to be something different. But it's interesting to know that the mass of the monopole can plausibly be expected to be enormous for the mass of a stable elementary particle;

10^{16} GeV is about 10^{-8} grams, roughly the mass of a bacterium, or 10^{6} joules, roughly the kinetic energy of a charging rhinoceros.

While most of the mass of the monopole is concentrated in its tiny core, it has interesting structure also on the other distance scales (Fig. 7). In particular, it has some structure at the weak interaction

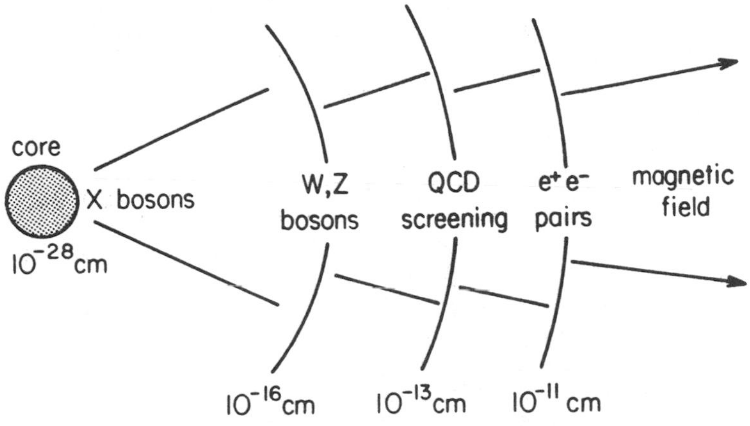

Figure 7: Structure of a grand unified monopole.

scale associated with its coupling to **W** and **Z** bosons, it has a color magnetic field which is screened at a distance of about one Fermi, and it's strongly coupled to a surrounding cloud of electron-positron pairs. At distances greater than the Compton wavelength of the electron only its long-range magnetic field can be detected.

To put monopoles in perspective, it helps to consider the particle spectrum of a typical grand unified theory. Nothing is stable in these theories unless there is a good reason. The photon and graviton are stable because, in fact, they are expected to be exactly massless. The electron is stable because it's the lightest particle which carries electric charge, and electric charge is exactly conserved. The lightest neutrino in the theory, is stable, because it's the lightest particle with half-odd-integer angular momentum, and therefore its decay is forbidden by conservation of angular momentum. And the magnetic monopole is stable, because it's the lightest particle which carries magnetic charge, and magnetic charge is also exactly conserved.

The monopoles have other interesting properties. I'll emphasize only one because it will be relevant in the ensuing discussion. The scattering of low-energy fermions off of a monopole has a very peculiar property. Namely, the outcome of such a scattering process, even when the energy of the incoming fermion is very low, is strongly dependent on the structure of the core of the monopole. One might think that a very long wavelength incoming electron would be unable to detect the structure of the very tiny monopole core, but in fact it can.

To begin to understand this phenomenon, one observes that even at the classical level, the angular momentum of a particle with electric charge e in the field of a monopole with magnetic charge g has a peculiar term,

$$\vec{J}_{em} = - eg\,\hat{r} \tag{12}$$

where $\hat{r}$ is the unit vector pointing from the monopole to the charged particle. Thus, an electron is forbidden (by conservation of angular momentum) to pass through the monopole unless it's able to flip its helicity or its charge as it does so. This statement has a quantum mechanical counterpart. If I solve the Dirac equation in the background of a monopole in the zero angular momentum partial wave, I find that there is a solution of negative helicity which is an incoming wave only, and a solution of positive helicity which is an outgoing wave only. And furthermore, that the Dirac equation itself provides no information about how to link up the incoming and outgoing wave.

A boundary condition is therefore needed in order for the problem of an electron scattering off a monopole to be well-defined.[7] This is the key to understanding what's going on. A deeply cherished principle in quantum field theory, which we tend to regard as an inviolable principle, sometimes called the decoupling principle, says that very short distance physics has effects which must be suppressed in processes at low energy by some power of the short-distance scale. Therefore, in the case of electron scattering off the monopole, one would expect, at energies which are very small compared to the inverse size of the core of the monopole, to be able to regard the monopole as effectively point-like and analyze scattering off a point-like monopole. But that's wrong because the problem of a fermion scattering off a point-like monopole is not well-defined. The outcome of the scattering experiment isn't determined until we specify the boundary condition which must be satisfied by the electron at the pole. It is short-distance physics, the structure of the core, that determines that boundary condition, and through the boundary condition exerts a strong influence on low-energy physics. In other words, low-energy fermions are able to penetrate to the core of the monopole and thus provide us with a unique window on the structure of the monopole core, and hence on the structure of all matter at extremely short distances. And, in particular, if the heavy gauge bosons associated with the monopole core have baryon number violating interactions, as is typically the case in a grand unified theory, we should expect that in the $J = 0$ partial wave the cross section for baryon number violation in the scattering of a fermion off a monopole will be unsupressed; it is independent of the size of the core in the low-energy limit.[8]

III. MONOPOLES AND COSMOLOGY

So far I have argued that monopoles should exist as a consequence of a very general notion of unification of the fundamental interactions at short distances. But that doesn't mean that there need be any monopoles around to be observed today. If monopoles are extremely heavy as we expect, they can't be produced in accelerator experiments, or by any process going on in the current universe. The only way we could detect monopoles is by observing ones which were produced in the very early universe, when much higher energies were available. So the abundance of monopoles is a cosmological issue. We must ask, were monopoles produced in the very early universe? How copiously should we expect them to have been produced?

Well, first of all, to discuss cosmological monopole production, we have to recall that we expect that, at sufficiently high temperature in the early universe, the full grand unified gauge symmetry was restored. We've seen that the existence of monopoles is a consequence of spontaneous symmetry breaking. Therefore, at sufficiently high temperature, monopoles could not exist. They could only form when the universe cooled below the critical temperature at which the grand unified symmetry became spontaneously broken.

Once the phase transition occurred, monopoles could be produced. Once produced, because they are stable particles, their abundance per comoving volume in the expanding universe could be reduced only by annihilation of monopole anti-monopole pairs. But we can plausibly expect that, because of the rapid expansion of the universe, monopoles and anti-monopoles were unable to find each other in order to annihilate completely, so that a potentially significant number of monopoles could be left over today. So let's first consider how the monopoles might have been produced, and then how effectively they could have annihilated once produced.

A reasonably general argument leads to the expectation that production of a significant number of monopoles is virtually unavoidable in the early universe. The point is that when the universe was so hot that the grand unified symmetry was restored, though there were no monopoles, there were thermal fluctuations in the scalar order parameter. As the universe rapidly cooled, and passed through the critical temperature, it became quenched, and some of these fluctuations of the order parameter became frozen in. Different regions, in which the fluctuating order parameter points in different directions, can't, in general, consistently be brought together without trapping a topological defect, a hedgehog, where the regions meet. Such a hedgehog we've seen, is a magnetic monopole. And therefore one expects that in the phase transition, an abundance of monopoles would be produced which is of the order of ξ^{-3}, where ξ is some typical correlation length.[9] That is, the abundance of monopoles should be comparable to the abundance of fluctuation regions which were frozen in during the phase transition.

Well, in the vicinity of the critical temperature, it's possible that correlations grew to be very large, but in the early universe there's a limit imposed by causality on the distances over which things could have become correlated. We don't expect correlations to be established over distances greater than the particle horizon, the distance that light could travel since the initial singularity.[10,11] Hence we obtain a lower bound on the abundance of monopoles,

$$n/s \gtrsim O(10)\ (T_c/m_{pl})^3. \tag{13}$$

Here T_c is the critical temperature, $m_{pl} \sim 10^{19}$ GeV is the Planck mass, which enters because it determines the expansion rate of the universe, and n/s is the ratio of the monopole density to the entropy density, a dimensionless measure of the monopole abundance. (The ratio of monopoles to entropy is a convenient quantity to consider because as the universe evolved, it remained constant, as long as the expansion was adiabatic.) When we put in plausible numbers, a critical temperature of the order of 10^{15} GeV say, we obtain the result that the ratio of monopole density to entropy, would have to be greater than about 10^{-11}.

What about the annihilation of monopoles, once formed? The annihilation of monopoles and anti-monopoles is a rather complicated process. The monopole and anti-monopole must first capture each other by emitting photons or gluons. Once capture occurs the pair cascades down by emitting many photons and gluons until finally the monopole and anti-monopole explode in a final burst of scalar particles and heavy vector bosons. It's clear, though, that the rate of annihilation can't be any greater than the rate at which capture occurs. And capture is a relatively inefficient process. It's plausible, and in fact can be shown on the basis of a more careful analysis, that once the monopole abundance is sufficiently small, the annihilation rate is small compared to the expansion rate of the universe, and therefore annihilation is quite ineffective in reducing the abundance per co-moving volume further.[10,12] And in fact one finds that if the monopole abundance is less than the order of 10^{-11} annihilation is a negligible effect and the monopole abundance isn't further reduced at all. Hence we arrive at the statement of what is sometimes called the 'monopole problem.' We conclude, using for the mass of the monopole that expected in a grand unified theory respecting the desert hypothesis, that the abundance of monopoles should be comparable to the abundance of baryons. That's obviously absurd, because the mass of the monopole is about 10^{16} times the mass of a baryon and we would conclude that the universe is dominated by monopoles by many many orders of magnitude.

The fact that monopoles are actually observed in our universe to be very rare is thus a quite significant piece of information. It tells us something about either particle physics at very short distances, or the very early evolution of the universe. One way to resolve the problem is simply by saying that there is no unification; the particle interactions which are in effect at arbitrarily high energies are just the ones that we're already familiar with. Then there need be no monopoles, and there is no cosmological monopole problem. If we wish to cling to the idea of unification, we must find some way of modifying either the cosmological scenario or the particle physics, so as to explain the suppression of the monopole abundance.

The preferred and most elegant way of suppressing monopoles is based on the inflationary universe scenario.[13] In this scenario there was a large effective cosmological constant in the very early universe, which caused it to expand exponentially. If this inflation occurred subsequent to the production of monopoles, it could have rapidly reduced the monopole abundance to a negligible value. Eventually the cosmological constant thermalized; it was turned into radiation and the universe reheated. After reheating, the baryon number of the universe could have been produced by baryon-number-violating interactions, which were brought into thermal equilibrium, or nearly into equilibrium, in the reheated universe. But it's quite plausible that during and after the reheating, monopole production was a small or negligible effect; that the baryon number of the universe was created after reheating but the monopole abundance produced after reheating was very small.[14] On the other hand, it is possible in some models, that the monopole abundance produced after reheating, while small enough to be acceptable, might be large enough to be interesting, might, at least in principle, be detectable.

I'll briefly describe one other scenario which is based more on a modification of the particle physics than of the underlying cosmology. In this scenario, as the universe cooled, it entered a high temperature superconducting phase.[15] In a superconductor, magnetic flux is expelled and therefore tends to collapse to a flux tube. Monopoles and anti-monopoles would be connected by such tubes of flux, so there would be in effect a linear potential between monopoles and anti-monopoles, which would greatly enhance the annihilation rate. Eventually, as the universe continued to cool, it re-entered a normal phase, and this annihilation mechanism then shut off. Although the flux tubes are very effective at making monopoles and anti-monopoles annihilate, some monopoles would be unable to find an anti-monopole to pair up with so that when the universe re-entered the normal phase a small but, again, perhaps detectable abundance of monopoles might have been left over.[16]

So the conclusion seems to be that there is no definite cosmological prediction for the monopole abundance. Thus, theoretical cosmology should not discourage the prospective monopole hunter. It seems quite conceivable that the monopole abundance could be both small enough to be acceptable, and yet large enough to be detectable.

IV. LIMITS ON THE MONOPOLE ABUNDANCE

I have remarked that our failure to observe a monopole is itself a significant piece of information, but it would be even more interesting to see one. In judging the prospects for detecting a monopole, certain astrophysical considerations are very relevant. The first important point is that monopoles in cosmic rays are likely to be relatively slowly moving. A monopole in our galaxy would be accelerated by the galactic magnetic field. That field has a magnitude of the order of 3 microgauss and a coherence length of roughly 10^{21} cm. Thus a monopole of 10^{17} GeV would be accelerated by the field to a velocity of about 10^{-3} c. For monopoles more massive than 10^{17} GeV, just gravitational effects would be expected to accelerate them to a velocity of the order of 10^{-3} c, that being the typical galactic in-fall velocity.

A clever argument due to Parker[17] puts a very severe limit on the flux of magnetic monopoles in the galaxy. The galaxy has a magnetic field which tends to accelerate monopoles, and if the field accelerates magnetic monopoles, that drains energy from the field. By demanding that the field survive for at least of the order of 10^8 years, roughly the time that's required for the galactic dynamo to regenerate the field, we obtain a limit on the flux, F,

$$F \lesssim 10^{-15}\ \mathrm{cm}^{-2}\,\mathrm{s}^{-1}\,\mathrm{sr}^{-1}\ . \tag{14}$$

Actually this argument has to be reconsidered slightly if the monopole is heavier than 10^{17} GeV.[18] Such heavy monopoles will enter a coherent domain of the galactic magnetic field with an energy larger than the energy they would pick up if accelerated from rest in that domain. Thus, in first approximation they are as likely to put energy into the magnetic field as take it out. But as they cross the domain they tend to be deflected so that there is a second order effect in which they do drain energy from the magnetic field. There is still a limit on the flux, but the limit rises linearly with the monopole mass if the mass is larger than 10^{17} GeV,

$$F \lesssim 10^{-15}(m/10^{17}\ \text{GeV})\ \text{cm}^{-2}\text{s}^{-1}\text{sr}^{-1}\ . \tag{15}$$

When a monopole becomes sufficiently heavy, the best limit on the flux comes from considering not its magnetic charge, but just its enormous mass. We can limit the abundance of monopoles by demanding that the monopoles in the galaxy not exceed the galactic mass of about 10^{12} solar masses. And then if the velocity is taken to be about 10^{-3} c we obtain a limit on the flux,

$$F \lesssim 10^{-10}(m/10^{17}\ \text{GeV})\text{cm}^{-2}\text{s}^{-1}\text{sr}^{-1}\ . \tag{16}$$

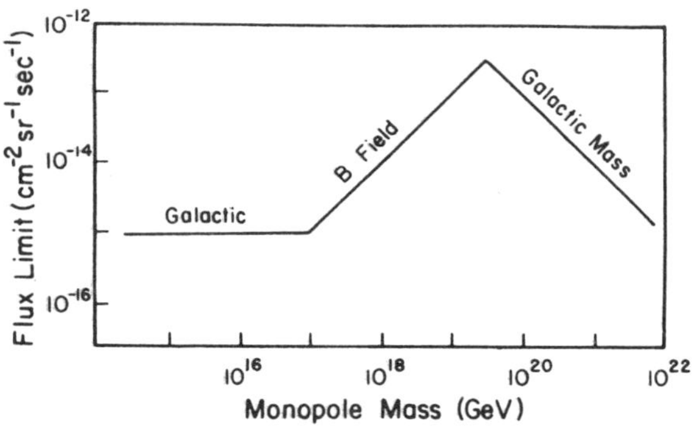

Figure 8: Astrophysical limits on the monopole flux as a function of monopole mass.

The two limits obtained from the energetics of the magnetic field and the mass of the galactic halo cross for a monopole mass of the order of 10^{19} GeV (Fig. 8). And there is a mass-independent bound of about 10^{-13}. It's interesting that the limit is weakest for a mass of the order of 10^{19} or 10^{20} GeV, since that's actually what we would expect the mass of the monopole to be in a certain class of unified theories, the Kaluza-Klein theories, in which gravity is unified with the other particle interactions.

Actually Parker's limit is subject to one potentially serious criticism, which is that we have ignored the back reaction on the magnetic field of the monopoles.[19] If the monopole abundance is sufficiently large, then the plasma frequency of the monopole anti-monopole plasma might be greater than the inverse time required for a monopole to cross a domain of the galactic field. And hence the field might excite magnetic plasma oscillations rather than irreversible currents. In fact, we obtain a lower bound on the flux by demanding that the monopole abundance be such that the plasma frequency is sufficiently large for such considerations to apply. I'm not sure how seriously to take this objection. It seems to me rather implausible that these magnetic plasma oscillations could be maintained coherently through many oscillations, as would be required for Parker's limit to be exceeded by orders of magnitude, in the presence of the inevitable magnetic and gravitational inhomogeneities which would tend to destroy coherence of the oscillations.

An even stronger limit on the monopole flux can be obtained if we consider the astrophysical consequences of the catalysis by monopoles of nucleon decay.[20] These astrophysical consequences are most spectacular in neutron stars. Monopoles in cosmic rays tend to be captured by neutron stars. Once the star has gathered a significant number of monopoles, they, catalyzing nucleon decay at a furious rate, would heat up the star so that it would emit x-rays. There are good observational limits on the x-ray luminosity of neutron stars and from these we obtain limits on the product of the monopole flux and the cross section for catalysis of nucleon decay.

If we assume the cross section is of the order of 10^{-28} cm^2, some sort of strong interaction cross section, this flux limit is six orders of magnitude better than Parker's limit. And in fact it can be improved by yet another six orders of magnitude for monopoles which are sufficiently light, if we take into account the capture of monopoles by the star while it is on the main sequence before it becomes a supernova. Well, this result is very discouraging, Parker's limit says that we would need about a football field of detector to have a chance of seeing a few monopoles in a year. This improved limit says that in all of Cook County at most a few monopoles per year could be detected. It's possible, of course, that there exist monopoles which don't catalyze nucleon decay, to which this much stronger limit would not apply, but it's especially discouraging to those who hope to detect monopoles through their catalysis effect.

This neutron star limit is subject to a few criticisms; you might argue that we don't really have enough understanding of the detailed structure of neutron stars to confidently say how monopoles in the interior of a star would affect it. I won't go into the details of what could go wrong. Instead we'll just say, that even if you insist that we disregard these bounds based on neutron stars, we can still obtain a bound, using similar reasoning, by considering the luminosity of white dwarfs;[21] the interior of white dwarfs we think we understand very well. And this limit is only weaker by about three orders of magnitude than the one obtained from neutron stars, and still, therefore, three orders of magnitude better than Parker's limit. Is there any point then, to doing a terrestrial experiment, looking for the catalysis effect. Well, one can imagine one possible scenario in which there might be a point, although the plausibility of this scenario is difficult to judge. We do expect the cross section for capture of a nucleon by a monopole to be large. But suppose the rate for catalysis is much slower; suppose it takes a microsecond or so for the catalysis to occur after capture. Then it still might be possible to observe the catalysis phenomenon in a terrestrial experiment, but the much lower rate, the catalysis rate rather than the capture rate, would be the relevant one in an astrophysical context, and the bounds obtained in this way on the monopole flux would be insignificant.[22]

V. CONCLUSIONS

We've seen that our failure to observe a monopole in our universe is a significant piece of information about very-short-distance physics and the very early universe. Naturally it would be much more exciting to see a monopole than not see one. The discovery of a monopole would, for one thing, confirm that the universe was once extremely hot, so hot that production of monopoles was possible. And the abundance of monopoles, if measured, would have to be explained by any realistic cosmological scenario; it would be a severe constraint on cosmological model building.

From the point of view of particle physics, discovery of the monopole would confirm the basic idea of unification of the fundamental interactions. If, as should eventually prove possible, the mass of the monopole can be measured, we would determine for the first time the unification scale at which electrodynamics becomes truly united with the other particle interactions. Perhaps, by scattering fermions off the monopole, we could obtain some more information about particle physics at extremely short distances, information to which we could gain access in no other way.

I don't know whether a monopole will ever be seen, but already the theoretical study of magnetic monopoles has led to marvelous insights into the dynamics of gauge theories, the subtleties of quantum field theory involving the interactions of fermions, and the connection between particle physics and cosmology. I feel that even if a magnetic monopole never is seen, those of us who have studied its properties will have been amply rewarded.

REFERENCES

[1] J. Preskill, Ann. Rev. Nucl. Part. Sci. **34**, 461 (1984).
[2] H. Georgi and S.L. Glashow, Phys. Rev. Lett. **32**, 438 (1974).
[3] H. Georgi and H. Quinn, and S. Weinberg, Phys. Rev. Lett. **33**, 451 (1974).
[4] G. 't Hooft, Nucl. Phys. **B79**, 276 (1974); A.M. Polyakov, JETP Lett. **20**, 194 (1974).
[5] P.A.M. Dirac, Proc. R. Soc. London **A133**, 60 (1931).
[6] G. 't Hooft, Nucl. Phys. **B105**, 538 (1976); E. Corrigan, D. Olive, D. Fairlie, and J. Nuyts, Nucl. Phys. **B106**, 475 (1976).
[7] A.S. Goldhaber, Phys. Rev. **D16**, 1815 (1977).
[8] V. Rubakov, JETP Lett. **33**, 644 (1981); Nucl. Phys. **B203**, 311 (1982); C.G. Callan, Phys. Rev. **D25**, 2141 (1982); **D26**, 2058 (1982); Nucl. Phys. **B212**, 391 (1983).
[9] T.W.B. Kibble, J. Phys. **A9**, 1387 (1976).
[10] J. Preskill, Phys. Rev. Lett. **43**, 1365 (1979).
[11] A.H. Guth and S.-H. Tye, Phys. Rev. Lett. **44**, 631 (1980); M.B. Einhorn, D.L. Stein, and D. Toussaint, Phys. Rev. **D21**, 3295 (1980).
[12] Ya.B. Zeldovich and M.Y. Khlopov, Phys. Lett. **79B**, 239 (1978).
[13] A.H. Guth, Phys. Rev. **D23**, 347 (1981); A.D. Linde, Phys. Lett. **108B**, 389 (1982); **114B**, 431 (1982); A. Albrecht and P.J. Steinhardt, Phys. Rev. Lett. **48**, 1220 (1982).
[14] M.S. Turner, Phys. Lett. **115B**, 95 (1982); G. Lazarides, Q. Shafi, W.P. Trower, Phys. Rev. Lett. **49**, 1756 (1982); J. Preskill, in "The Very Early Universe," ed. S.W. Hawking, et al., Cambridge, Cambridge University Press (1983); W. Collins and M.S. Turner, Phys. Rev. **D29**, 2158 (1984).
[15] P. Langacker and S.-Y. Pi, Phys. Rev. Lett. **45**, 1 (1980).
[16] E. Weinberg, Phys. Lett. **126B**, 441 (1983).
[17] E.N. Parker, Astrophys. J. **160**, 383 (1970).
[18] M.S. Turner, E.N. Parker, and T. Bogdan, Phys. Rev. **D26**, 1296 (1982).
[19] E.E. Salpeter, S.L. Shapiro, and I. Wasserman, Phys. Rev. Lett. **49**, 1114 (1982); J. Arons and R. Blandford, Phys. Rev. Lett. **50**, 544 (1983).
[20] E.W. Kolb, S.A.Colgate, and J.A. Harvey, Phys. Rev. Lett. **49**, 1373 (1982); S. Dimopoulos, J. Preskill, and F. Wilczek, Phys. Lett. **119B**,

320 (1982); K. Freese, M.S. Turner, and D.N. Schramm, Phys. Rev. Lett. **51**, 1625 (1983).
[21]K. Freese, University of Chicago Print-84-0573 (1984).
[22]A.S. Goldhaber, in "Magnetic Monopoles," ed. R.A. Carrigan and W.P. Trower, New York, Pelnum (1983).

Can The Primordial Monopole Problem Be Solved By Annihilation?

Erick J. Weinberg

It has been known for some time that grand unified theories, when combined with standard hot big bang cosmology, appear to predict a density of primordial superheavy magnetic monopoles which is far in excess of the known bounds on the present monopole abundance.[1] A possible solution to this "monopole problem" is offered by the inflationary universe scenario. However, it remains to be established that the universe passed through an inflationary period; certainly most GUTs do not imply inflation. Even in an inflationary cosmology, a problem can arise from monopoles produced after the inflationary era; this is particularly true in examples (e.g., SO(10)models) with relatively low mass monopoles. It is therefore still worth examining non-inflationary approaches to the problem. I will argue in this talk that the most natural of these are those which invoke some particularly efficacious annihilation scheme, and will then derive a rather general constraint on such schemes.[2]

Monopoles are first produced when the GUT is broken to a subgroup which allows non-trivial topology; typically this occurs at a critical temperature $T \sim 10^{15}$ GeV. Arguments[3] outlined below indicate that the initial value of the monopole to entropy ratio, r, must then be at least 10^{-12}. Once produced, the monopoles and antimonopoles may interact and annihilate. However, examination of a wide class of electromagnetic and other annihilation mechanisms shows that these are ineffective once r is less than $10^{-10}(M_{mon}/10^{16}\text{GeV})$. In the absence of annihilation, r remains constant as long as the expansion of the universe is adiabatic.

For comparison, the bounds on the total mass density of the universe imply that at present $r \lesssim 10^{-24}\ (10^{16}\ \text{GeV}/M_{mon})$. Other limits can be inferred from astrophysical bounds on the galactic monopole flux. The most stringent of these[4], based on catalysis of nucleon decay in neutron stars, give bounds in the range 10^{-32} to 10^{-37} if monopoles are uniformly distributed throughout the universe and a factor of 10^5 lower if they cluster with galaxies.

There are essentially three ways of modifying the above scenario so as to reconcile the cosmological predictions with the bounds on the present value of r. One can either lower the initial monopole production, increase the entropy density, or enhance the monopole-antimonopole annihilation. In a non-inflationary context the first requires a reduced critical temperature; the more stringent astrophysical limits can be met only with an implausibly low T_c. Additional entropy could be generated if at some time after the intial monopole production there was a first order transition with supercooling followed by reheating. The chief difficulties are that parameters must be very finely tuned to give sufficient supercooling and yet have the transition eventually completed and that there must be a mechanism for regenerating the baryon number excess.

The third possibility, enhanced annhiliation, seems more promising. It could be implemented by superimposing on the GUT a particularly efficacious annihilation mechanism which was not tied to the unification mass scale and which did not require fine-tuning of parameters. An example is the Langacker-Pi[5] model, in which Higgs structure is such that before the transition to the final SU(3)xU(1) phase the universe passes through a phase with only SU(3) symmetry. During this phase the universe is essentially a superconductor and any previously produced monoples and antimonopoles are linked by magnetic flux tubes. These flux tubes rapidly contract, leading to rather efficient monopole-antimonopole annihilation. However, this annihilation cannot be complete; instead, it is subject to a constraint imposed by causality and magnetic charge conservation.

In deriving this constraint I will not make use of any particular annihilation mechanism, but will instead consider an idealized scenario in which monopoles are formed during a phase transition at a temperature T_c and subsequently, during a temperature range $T_1>T>T_2$, are subject to a "perfect" annihilation mechanism (i.e., one with the maximum possible efficiency). The aim will be to place a lower bound on the value of r at the end of the annihilation era. To begin, note that for any system a knowledge of charge fluctuations can be used to obtain a lower bound on the density. Suppose that the system is divided into many regions, each of volume L^3. If $\Delta N(L)$ is the average magnitude of the net charge in these regions, then the density certainly satisfies

$$n \geq \frac{\Delta N(L)}{L^3} \tag{1}$$

for any value of L. Next, note that while annihilation processes will tend to wipe out any initial charge fluctuations, causality prevents them from doing so on scales greater than the horizon length. Taking into account the expansion of the universe, we see that the magnetic charge fluctuations at T_2 are related to the initial fluctuations at T_c by

$$\begin{aligned} \Delta N(L,T_2) &\approx \Delta N\left(\frac{R(T_c)}{R(T_2)} L, T_c\right) , \\ &\approx \Delta N\left(\frac{T_2}{T_c} L, T_c\right) \end{aligned} \tag{2}$$

To estimate the magnitude of the initial charge fluctuations, recall how monopoles are produced during the phase transition. As the universe cools below T_c, the Higgs field develops a non-zero value throughout space. The direction of this field is not uniform but instead is correlated only over some characteristic length ξ, leading to a sort of domain structure. (In a first-order transition bubbles play the role of domains.) At the intersection of several domains there is a probability $p \sim 1/10$ of a topologically non-trivial Higgs field, and thus a monopole. While the evaluation of ξ is a difficult dynamical question, causality implies that it should be no greater than the horizon length; this leads to the lower bound $r_{init} \gtrsim (T_c/M_p)^3$.

Now recall that the total topological charge in a region can be expressed as a surface integral of a current J_i which depends on the orientation of the Higgs field. Since the latter is random over distances much greater than ξ, the root mean square value of the integral will be proportional to the square root of the surface area for regions large relative to ξ. Consequently,

$$\Delta N(L,T_c) \sim n_{init}^{1/3} L \qquad (3)$$

(Note that this implies a certain degree of correlation between monopoles and antimonopoles, since independent random distributions of the two species would give $\Delta N \sim L^{3/2}$.) Eqs.(1)-(3), with $L=d_H(T_2)$, lead to

$$r(T_2) \gtrsim 2\pi\left(\frac{2\pi^2\mathcal{N}}{45}\right)^{1/2} r_{init}^{1/3} \left(\frac{T_2}{M_P}\right)^2 \qquad (4)$$

where $\mathcal{N}$ counts the number of effectively massless degrees of freedom. Taking $T_2 \sim 1$ TeV, as in the Langacker-Pi scheme, gives $r(T_2) \gtrsim 10^{-34}$; given the uncertainties in the flux limits based on nucleon decay catalysis in neutron stars, this may be barely acceptable. Clearly T_2 cannot be many orders of magnitude greater than this, or else the mass density bound would be violated. Values of T_2 much less than 1 TeV would give an acceptably small monopole density, but would be hard to reconcile with known low energy physics.

The validity of this bound has been questioned. For example, Vilenkin[6] has argued that flux tube annihilation might proceed much more rapidly than Eq.(4) would allow. The essential point is that with fluctuations described by Eq.(3) it is possible to link the monopoles and antimonopoles using only short flux tubes; contracting these would lead to rapid annihilation. The flaw in this argument can be seen by considering a line (of length L) of alternating monopoles and antimonopoles, with the leftmost a monopole and the rightmost an antimonopole. If each monopole is linked with the antimonopole to its right, then only short flux tubes are needed. However, if it were not for the end points it would be equally advantageous for flux tubes to run from a monopole to the antimonopole on its left; to have the correct pattern established all along the line clearly takes a time $t \gtrsim L/c$. Consistency with Eq. (4) is restored once this additional time is taken into account.

Finally, these ideas can be tested by computer simulations of systems of particles and antiparticles.[7] The results are consistent with the causality bound and provide further evidence that the existence of spatial fluctuations in the initial magnetic charge density leads to a severe constraint on attempts to solve the primordial monopole problem without inflation.

REFERENCES

[1]J. Preskill, Phys. Rev. Lett. **43**, 1365 (1979); Ya. B. Zeldovich and M. Y. Khlopov, Phys. Lett. **79B**, 239 (1978).
[2]E. Weinberg, Phys. Lett. **126B**, 441 (1983).
[3]M. Einhorn, D. Stein, and D. Toussaint, Phys. Rev. **D21**, 3295 (1980)
[4]S. Dimopoulos, J. Preskill, and F. Wilczek, Phys. Lett. **119B**, 320 (1982); E. Kolb, S. Colgate, and J. Harvey, Phys. Rev. Lett. **49**, 1373 (1982); K. Freese, M. Turner, and D. Schramm, Phys. Rev. Lett. **51**, 1625 (1983).

[5]P. Langacker and S.-Y. Pi, Phys. Rev. Lett. **45**, 1 (1980).
[6]A. Vilenkin, Phys. Lett. **136B**, 47 (1984).
[7]K. Lee and E. Weinberg, Nucl. Phys. **B246**, 354, (1984).

50

Monopole-Proton Bound States in the Hot Universe

L. Bracci, G. Fiorentini, G. Mezzorani, and P. Quarati

The interaction of slow magnetic monopoles with matter is a problem which has to be understood for devising efficient detection techniques. In this respect it is interesting that the hamiltonian of the monopole-proton system has bound states: there is an infinite number of bound states with $J = 0$, the binding energies being in an exponential sequence, $E_n = E_o e^{-7.8n}$ [1]. The energy of the ground state, E_o, is estimated to be in the range 100÷1000 KeV. This poses the following question: are monopoles free or are they bound to protons when reaching a detector? In order to answer the question one has to calculate the cross section for the formation of monopole-proton systems and for the inverse (detachment) process and to follow the life history of the monopole interacting with its environment. The hot universe era, say from kT ≃ 1 MeV down to the recombination temperature, is particularly important for the formation of bound states since the proton density was very high at those times. On the other hand, bound states could be destroyed by the energetic photons present in the hot plasma. Thus one has to follow a delicate balance between formation and dissociation in an expanding and cooling Universe. The solution of the problem, which is quite similar to the discussion of the primordial nucleosynthesis, depends on the value of the energy scale E_o and on the protons to photons ratio, $\eta = N_p/N_\gamma$: the smaller the binding energy is the less stable the bound states are and, similarly, the more abundant the photons are the more likely dissociation occurs.

We have calculated the cross section for formation and dissociation of monopole-proton bound states and we have studied the kinetics of the reactions in the hot era[2,3]. The results are shown in figs. 1 and 2. We find that at the end of the hot era monopoles are bound to protons most likely, i.e. for values of E_o and η in a widely accepted range.

We have also studied the life history of the monopole after the recombination time. We find that the monopole-proton system (a positive dyon) captures an electron, thus forming atomic systems similar to atomic Hydrogen. The rate of the dyon-electron system approaching a detector onto Earth depends on its velocity. If the velocity is larger than $10^{-4}c$ the electron will be lost in collisions with atoms in the atmosphere or in the portion of Earth crossed by the monopole. On the other hand, the proton stays bound to the monopole for velocity

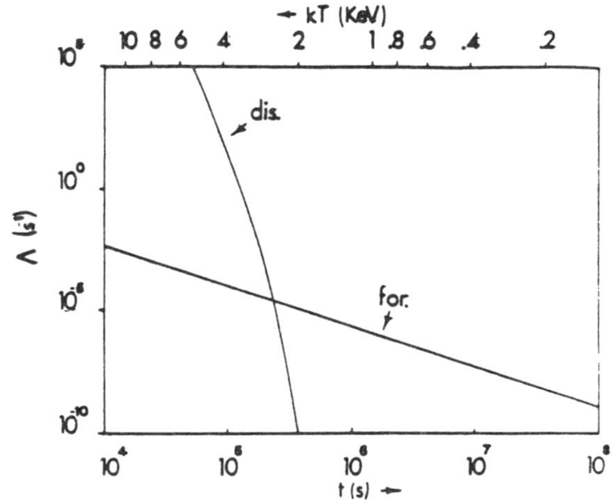

Figure 1 - The formation and dissociation rates as a function of time t (lower scale) and temperature T_9 (upper scale) for E_o = 100 keV and $\eta = 10^{-9}$.

smaller than 10^{-2}. Thus in the range $10^{-4} < v/c < 10^{-2}$, which is the most likely range for heavy monopoles, one has to do with a monopole-proton bound system. This system behaves quite differently than a bare monopole. For example, the energy loss resulting from adiabatic transitions between Zeeman levels is an order of magnitude smaller for the dyon than it is for the bare monopole[4].

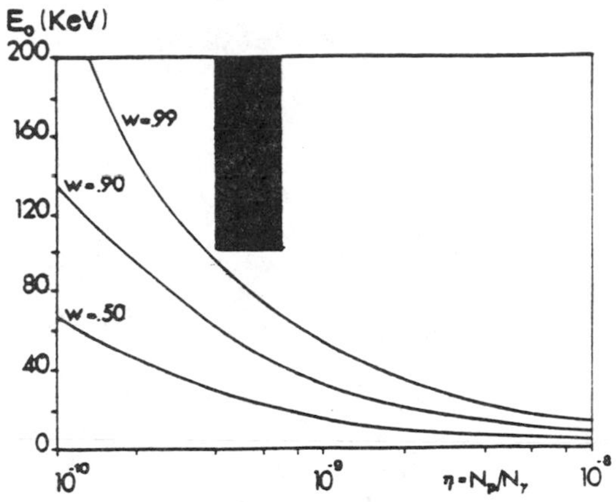

Figure 2 - Curves of equal binding probability w. η is the proton to photon ratio and E_o is the binding energy of the most tightly bound (Mp) state. The dashed area corresponds to the most likely region in the (η, E_o) plane.

REFERENCES

1 See G. Fiorentini, in Proceedings of Monopole '83, J. Stone Editor, to be published by Plenum Press, New York.

2 L. Bracci, G. Fiorentini, G. Mezzorani and P. Quarati, Pisa Preprint IFUP TH-84-3, to appear in Physics Letters B.

3 L. Bracci and G. Fiorentini, Pisa Preprint IFUP-TH-84-05.

4 N. M. Kroll, S. Parke, V. Ganapathi and S. D. Drell, in Proceedings of Monopole '83, J. Stone Editor, to be published by Plenum Press, New York.

51

Magnetic Monopole Search Experiments

James L. Stone

INTRODUCTION

Experimental searches for the elusive magnetic monopole have extended over 50 years -- ever since Dirac[1] showed that the existence of isolated magnetic charges could account for the observation of quantized electric charge. Early searches for Dirac's monopole were hampered since there was no indication of the expected monopole mass, velocity, or origin in the theory. Consequently, numerous experiments searched more or less blindly for relativistic low mass monopoles in the cosmic rays and in particle collisions at high energy accelerators. No viable candidate monopole events were observed in these early efforts and enthusiasm for performing monopole search experiments waned. However, in 1974 G. 't Hooft[2] and A. Polyakov[3] working independently provided new guidance for monopole hunters and huntresses. In a completely general analysis, their work showed that superheavy magnetic monopoles were a natural consequence of any Grand Unified Theory (GUT) in which the unifying group contains a U(1) factor as a subgroup. These proposed GUTs provided exciting new prospects for particle physics and cosmology. For example, a GUT analysis in an expanding, cooling universe yields a phase transition at an energy ~ 10^{15} GeV and at a cosmic time ~10^{-35} seconds after the big bang. General consequences of GUTs and this phase transition are the prediction of proton decay, the production of superheavy magnetic monopoles, a consistent explanation for the apparent excess of matter over anti-matter in the universe, and the possibility that monopoles in proximity to nuclear matter may "catalyze" nucleon decay reactions. Attempts to provide the necessary experimental verification for GUTs has grown into a major industry over the past 10 years. Massive proton decay detectors have been built in deep underground laboratories and currently very large area detectors are being developed to search for GUT monopoles.

A discussion of magnetic monopoles in the context of theoretical particle physics and cosmology was presented by J. Preskill at this conference. In this paper, I will review briefly the monopole properties which are relevant for their detection and summarize current experimental efforts using induction, ionization/excitation, and catalysis techniques. A complete listing of monopole references can be found in the bibliography by Craven and Trower.[4]

PHYSICAL PROPERTIES OF GUT MONOPOLES

Charge: The magnetic field due to an isolated magnetic charge, g, in the formulation of electromagnetism is given by

$$B = g \frac{\hat{r}}{r^2} .$$

The magnetic charge, g, and electric charge, q, obey the Dirac relation

$$g = ng_{Dirac} = \frac{\hbar c}{2q} n = 68.5\ nq ,$$

where $n = \pm 1, \pm 2, \pm 3, \ldots$. In general, the values $n = 1$ and $q = e$ are assumed, although some theories favor higher values of n. If the fundamental electric charge $q \neq e$, the Dirac relation is still valid.

Mass: The monopole mass, M_m, in GUTs is related to the unification mass, M_x, and the coupling strength of the interaction, α_{GUT}, at the mass scale of the theory, i.e.

$$M_m \propto \frac{1}{\alpha_{GUT}} M_x .$$

In minimal SU(5), for example, $M_m \sim 10^{16}$ GeV/c^2 although it is possible to construct theories with monopole masses as low as 10^4 GeV/c^2. However, in the recently revived Kaluza-Klein theories which attempt to bring gravity into the unification scheme, monopole masses can exceed 10^{19} GeV/c^2. The experimentally excluded mass region from particle accelerators is < 150 GeV/c^2.

Velocity: Because of their large mass, the velocity of GUT monopoles in our galaxy should be of the same order as the orbital velocity of stars, i.e. ~300 km/s corresponding to $\beta_m \sim 10^{-3}$. Through gravitational interactions or accelerations by the galactic magnetic field, however, monopole velocities in the range $10^{-4} \leq \beta_m \leq 10^{-2}$ are expected. For monopoles bound to our solar system $\beta_m \leq 10^{-4}$, whereas monopoles bound to the earth would have $\beta_m \leq 10^{-5}$.

Structure: Unlike point Dirac monopoles, GUT monopoles are expected to have a complex internal structure consisting of layers composed of various particle anti-particle pairs depending on the distance from the monopole core. Since this structure is somewhat model dependent, I will not attempt to give a detailed description here. However, an outer layer in which fermion antifermion pairs as well as baryon and lepton number violating condensates reside, gives rise to the possibility that monopoles could catalyze nucleon decay reactions. At large distances, a GUT monopole appears identical to the Dirac monopole.

Bound States: The electromagnetic interaction of a monopole's magnetic charge with the magnetic dipole moment of a nucleus may yield bound systems. Monopolic atoms have been studied in detail by Fiorentini,[5] Bracci,[6] and Goebel.[7] For a monopole passing through the earth, the cross section for capturing an Al^{27} nucleus could be as large as 0.3 mb due to the relative abundance of Al^{27} and it's strong magnetic moment. In designing detectors, it is important to consider the possibility that <u>all</u> monopoles may come with a proton or heavier nucleus attached.

Astrophysical Limits: The most widely accepted astrophysical limit on the monopole flux is due to an argument first suggested by E. Parker.[8] Massive monopoles moving slowly through the galactic magnetic field (~3μG) will undergo deflections and absorb energy from the galactic field. The maximum allowable isotropic monopole flux consistent with the regeneration of the galactic field by dynamo action has become known as the Parker limit. For monopoles with mass $\sim 10^{16}$ GeV/c^2 and velocity $\beta_m \leq 5 \times 10^{-2}$, the derived upper limit for an isotropic monopole flux is $\leq 6 \times 10^{-16}$ cm^{-2}sr^{-1}s^{-1}. For higher mass or higher velocity monopoles, the Parker limit is less restrictive.

Another astrophysical flux limit can be derived by assuming that monopoles provide the missing mass of the universe. In this case, an isotropic monopole flux of $\leq 10^{-15}$ cm^{-2}sr^{-1}s^{-1} is obtained for monopoles with mass $\sim 10^{16}$ GeV/c^2 and velocity $\beta_m \sim 10^{-3}$. These limits are plotted in the summary in Fig. 12.

The message provided to experimenters by these astrophysical limits is that very large area detector arrays will be required to observe monopoles or to place significant new limits on their abundance.

INDUCTION EXPERIMENTS

In the late 1960's L. Alvarez et al.[9] pioneered the induction technique in searching for monopoles trapped in moon rocks, meteorites, and other materials by passing them through a superconducting coil. In an ideal sense, detecting the current induced in a coil by a passing magnetic charge (Faraday induction) is the most reliable method of monopole detection since it measures magnetic charge directly. In a superconducting coil, the signal arises solely from the long range electromagnetic interaction between the moving magnetic charge and the macroscopic quantum state of the superconducting loop. For a given coil configuration, not only is the induced current persistent, it also has a predictable magnitude for a given value of the charge. In addition, the interaction is independent of the monopole's velocity, mass, and electric charge.

For a detector coil with inductance, L, the current change induced by a passing monopole is given by $\Delta I = 2n\Phi_0/L$, where n = 1 for a single magnetic charge and $\Phi_0 = 2.07 \times 10^{-15}$ webers is the quantum of magnetic flux. In the example where the loop area is ~1 m^2 and the inductance L ~ 10 μH, a singly charged monopole induces a current $\Delta I \sim 8 \times 10^{-10}$ amperes.

Alvarez's experiment employed signal averaging of repeated passages of the samples through the detection coil in order to amplify the induced current to a measurable level. However, this very small current must be detected and measured directly if the induction technique is used to search for in-flight monopoles which make a single pass though the superconducting loop. The development of a Josephson junction device in recent years called a SQUID (Superconducting QUantum Interference Device) provides this capability. The intrinsic noise in the SQUID circuitry is limited only by Johnson noise in the Josephson device, however, the dominant noise source will be due to variations of the ambient magnetic field in the vicinity of the detection coil and SQUID.

To achieve the required signal-to-noise ratio in a coil-SQUID circuit, superconducting shields and gradiometer detection coil configurations are used. In general, the technique involves lowering the ambient magnetic field as much as possible initially by using soft-iron and/or mu-metal shielding. Inside this lowered field region, superconducting shields made of lead or niobium foils are used to impose the Meissner effect which pins or freezes the remaining ambient field inside the foil shields. In addition, detection coils in the shape of twisted loops (gradiometers) serve to cancel long range magnetic field variations. These techniques have achieved shielding factors of $\sim 10^8$. Figure 1 shows schematically how the induction technique utilizes SQUIDs and superconducting shields. Typically, the induced current is recorded as an offset on a strip chart recorder.

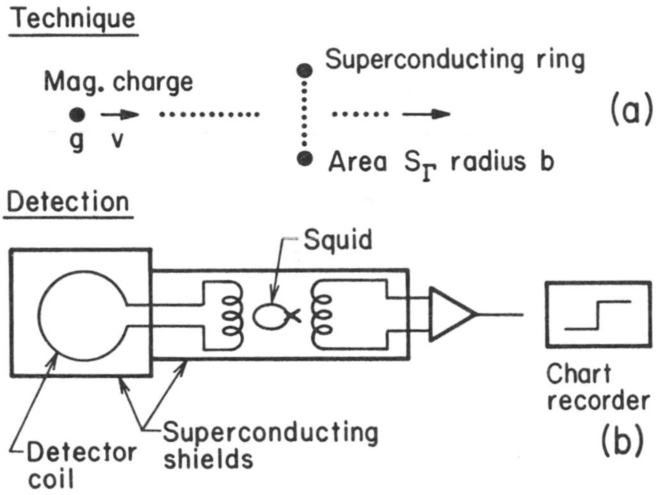

Figure 1: A schematic diagram showing the experimental arrangement for the superconducting coil-SQUID induction technique for monopole detection. The s.c. detection coil, SQUID, and shields are enclosed in a cryostat to maintain the operating temperature of ~4.2°K.

Mechanical, acoustic, and/or thermal disturbances of induction detectors can produce spurious monopole-like signals, although generally they do not have the correct amplitude and are usually not clean offsets. Nevertheless, current experiments address the spurious signal problem by using mechanically precise and rigid detector assemblies which are carefully monitored by magnetometers, accelerometers, sound alarms, etc. In addition, two or more independently instrumented detection coils are used by most experiments to require coincident monopole signals in order to nullify spurious events.

In 1981-82, B. Cabrera et al.[10] at Stanford were operating a single 4-turn, 5 cm diameter superconducting coil-SQUID detector. The detector was inside an ultra-low field region ($\sim 5 \times 10^{-8}$ gauss), however, no devices were being used to monitor possible disturbances. After 151 days of running, an event was observed on February 14, 1982, with precisely the signature expected for a monopole of charge $g = g_D$. The laboratory was unoccupied at the time and the event was not correlated to liquid helium or nitrogen transfers. The event corresponded to a monopole flux $F_m \simeq 1.4 \times 10^{-9}$ $cm^{-2}sr^{-1}s^{-1}$. Subsequent running of the original Stanford detector as well as an improved 3 coil detector has produced no additional monopole events. At this time, the most probable explanation is that Cabrera's Valentine Day event was instrumental in nature, however, numerous attempts to induce similar signals by mechanically disturbing the apparatus yielded no clean monopole-like offsets.

Following Cabrera's observation, experimental groups at IBM, Chicago-Fermilab-Michigan, Kobe University, Imperial College, and the National Bureau of Standards built induction monopole detectors. These groups employed various new concepts in shielding, coil configuration, SQUID-coil inductance matching, cryostat design, etc. which has resulted in vastly improved detector systems. Coil areas on the order of a square meter have been achieved by some groups. Table I summarizes the properties of the superconducting induction detectors and reports the 90% C.L. flux limits obtained as of May, 1984. The global flux limit derived by combining the results from all of the operating induction detectors is $F_m < 3.7 \times 10^{-12}$ $cm^{-2}sr^{-1}s^{-1}$ (90% C.L.).

Table I. Summary of superconducting induction detector properties and the 90% C.L. flux limits obtained as of May 1984.

Group	Arrangement	Ave. Area over 4π (cm^2)	Live Time (days)	Flux Limit (90% C.L.) ($cm^{-2}sr^{-1}s^{-1}$)
Stanford I[10]	Simple Loop	10	151	1.4×10^{-9}
Stanford II[12]	3 ⊥ loops	476	365	2.1×10^{-11}
IBM I[13]	2 Planar Gradiometers	25	165	5.1×10^{-10}
IBM II[14]	6 Planar Gradiometers	~1000	150	1.4×10^{-11}
Chicago-FNAL-Michigan[15]	Twisted loops	2100	110	9.2×10^{-12}
Kobe U.[16]	Simple loop	25	185	4.6×10^{-10}
Imperial College[17]	Twisted loop	~300	-	-

Global flux limit: $F_m < 3.7 \times 10^{-12}$ $cm^{-2}sr^{-1}s^{-1}$ (90% C.L.)

Presently, the Stanford, IBM, and Chicago-Fermilab-Michigan groups are studying ways of building modular superconducting induction detectors which could be expanded into large area arrays, e.g. > 100 m^2. A large area detector under design at IBM is shown in Fig. 2. The effective coincident area of the IBM detector module is 3.2 m^2. Also, Stanford and C-F-M are in the process of building modules with comparable coincident areas. The only inherent limitation on the size of induction arrays is governed by the cost associated with the number of cryostats and SQUIDs required. As lower cost techniques are developed, large area induction arrays may very well be feasible. A good review of the induction experiments and future possibilities can be found in the paper by Frisch in Ref. 11.

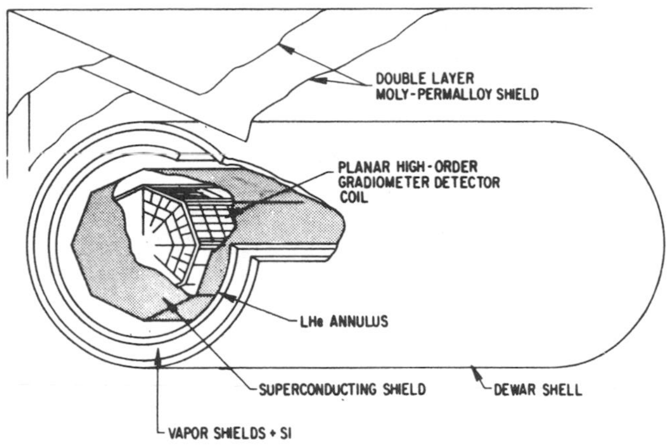

Figure 2: The large area detector under design at IBM consists of 8 rectangular and 2 octagonal planar gradiometer coils monitored by 10 independent thin film dc SQUIDs. The dewar is roughly 7 ft. in diameter by 14 ft. long. Two closed ferromagnetic shields surround the entire dewar.

SCINTILLATOR EXPERIMENTS

An important experimental technique for GUT monopole searches relies on detection of ionization or excitation energy loss (dE/dX) as monopoles pass through matter. Scintillation counters and gaseous proportional counters have been used extensively at accelerators for detection of charged particles. The response of these detectors is well understood for relativistic particles. However, for slow projectiles (e.g. massive charged particles or GUT monopoles) the energy loss mechanism and detector signals are much less understood.

Ahlen[18] has pointed out that energy loss cannot be equated directly to detector signal. For example, water and scintillator exhibit similar nuclear stopping powers. However, a particle with velocity below the Cherenkov threshold (~0.75 c) would not produce a detectable signal in water, but would in scintillator if its velocity is above a threshold of $\sim 6 \times 10^{-4}$ c. In general, a monopole signal in an ionization/excitation detector depends critically on the nature of the monopole. For example, does the monopole have bare magnetic charge? Does it come with an attached proton? Is it a positive or negative dyon? Does it capture and drag along a heavy nucleus? Ahlen and Tarlé[19] have calculated the scintillation yield for monopoles, dyons, and monopoles with a bound Al nucleus for various types of scintillator. Their result for PS-10 acrylic scintillator is shown in Fig. 3.

The actual technique for monopole detection using a typical scintillation counter telescope is sketched in Fig. 4. For standard scintillators, a minimum ionizing particle yields only a few photoelectrons (p.e.). For a prompt particle such as a relativistic muon, the few p.e.'s clump together in a narrow pulse (width ≈ few ns), however, for a slow monopole, the few p.e.'s are spilled out over several microseconds depending on the thickness of the counter. Hence, several-fold redundancy can be achieved for the monopole signature by requiring broad pulses from each scintillator coupled with a time delay between layers consistent with the passage of a low velocity particle.

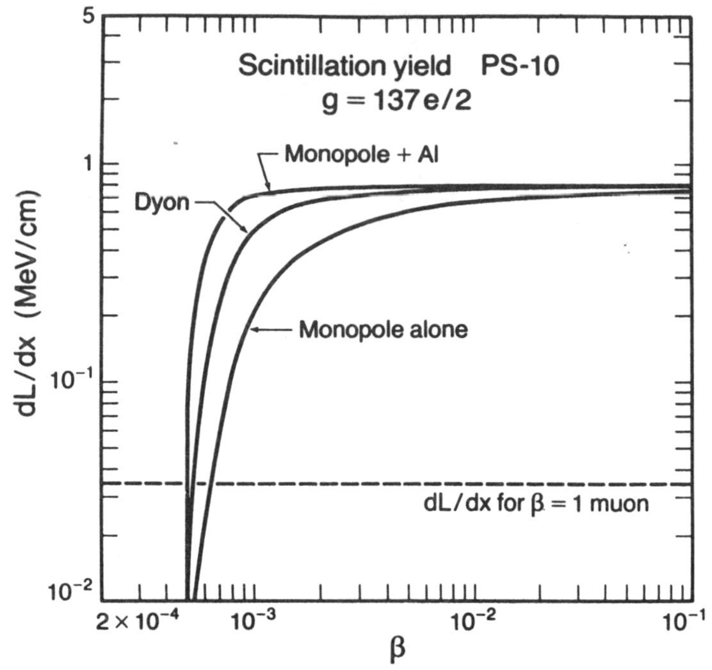

Figure 3: Light yield for PS-10 acrylic scintillator. The dashed line is the yield for a minimum ionizing singly charged particle.

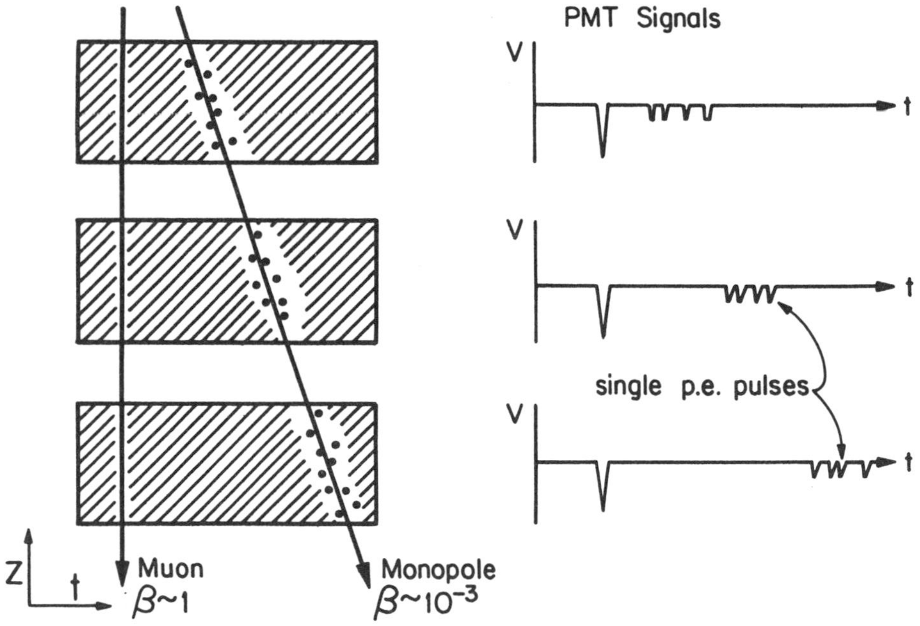

Figure 4: A Comparison of monopole and muon signals for a typical scintillation counter telescope.

A large number of experiments have already been performed to search for superheavy, slowly moving projectiles, (e.g. monopoles) using the scintillator technique. No monopole candidate events have been observed. Table II summarizes the 90% C.L. monopole flux upper limits derived from these efforts. A few of the ongoing experiments are described in more detail below.

Table II. Summary of Scintillation Counter Monopole Search Experiments.

Experiment	A·Ω (m^2sr)	dE/dX (x min. ion.)	Velocity Range	Flux Limit (90% CL) ($cm^{-2}sr^{1}s^{-1}$)
Baksan[20]	1800	0.25	10^{-3} - 5×10^{-2}	$< 8.5\times10^{-15}$
Berkeley-Indiana[21]	17.5	0.6	6×10^{-4} - 3×10^{-3}	$< 4.1\times10^{-13}$
Bolonga[22]	33.0	10-25	2×10^{-3} - 6×10^{-1}	$< 2.0\times10^{-13}$
BNL ν-expt.[23]	14.5	0.33	10^{-3} - 2×10^{-1}	$< 5.2\times10^{-12}$
Kobe University[24]	20.0	0.1	3×10^{-3} - 1.0	$< 9.0\times10^{-13}$
Tokyo-ICRR[25]	11.0	0.05	6×10^{-4} - 10^{-1}	$< 1.8\times10^{-12}$
Tokyo-Kamioka[26]	22.0	0.06	3×10^{-4} - 1.0	$< 1.5\times10^{-12}$
Tokyo University[27]	2.0	1.25	10^{-2} - 10^{-1}	$< 1.5\times10^{-11}$
	6.0	0.13	3×10^{-4} - 5×10^{-3}	$< 3.0\times10^{-12}$
Utah-Stanford[28]	2.7	0.13	10^{-3} - 3×10^{-2}	$< 8.2\times10^{-12}$
Homestake[29]	1200	0.1	6×10^{-4} - 1.0	Taking data
Texas A&M[30]	400	0.1	6×10^{-4} - 1.0	Taking data
Gran Sasso	12000	0.1	6×10^{-4} - 1.0	Proposed

Baksan Underground Scintillator Telescope

The most impressive scintillator experiment so far to look for monopoles is the Baksan detector located 850 m.w.e. under the Anderchi mountain in North Caucasus, Baksan Valley, USSR. The detector consists of ~3200 liquid scintillator modules arranged in a nearly cubic geometry 16 m × 16 m × 10.8 m. There are 4 horizontal planes (spaced 3.6 m apart) and 4 vertical planes of modules. The planes are separated by low background concrete absorbers (~180 gm/cm^2). Altogether there are 6 external and 2 internal planes of scintillator modules which gives a total acceptance of 1800 m^2sr. A cross sectional view of the detector is illustrated in Fig. 5.

One external scintillator plane and one internal plane firing within a 100μs time interval constitutes a monopole trigger. About 50% of the background triggers come from $\mu \to e$ decays inside the detector and about 50% from random cosmic ray muons. For an event to be considered a monopole candidate, ≥ 3 scintillator planes are required with the time delays between layers corresponding to a slowly moving

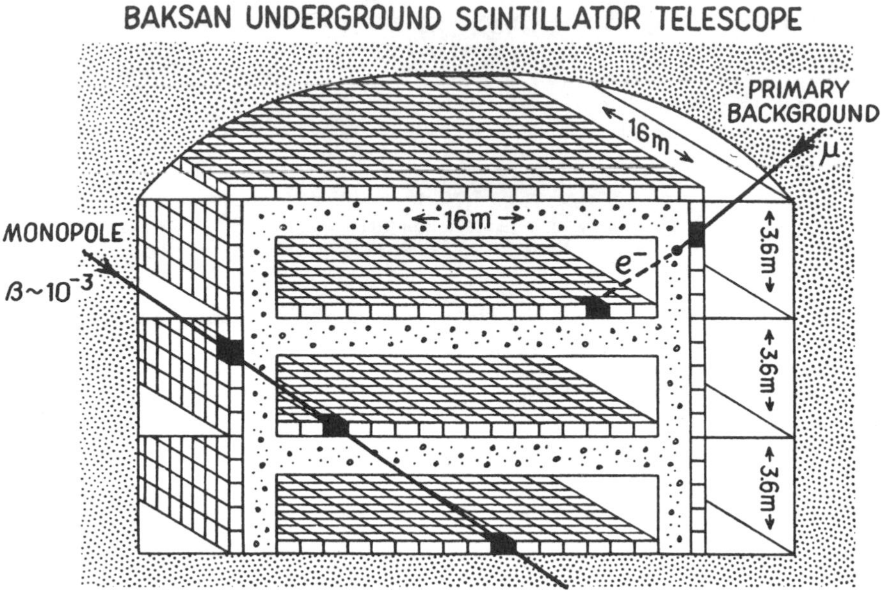

Figure 5: Cross sectional view of the Baksan underground scintillator telescope.

particle. After 425 days of detector livetime, no monopole candidates have been found. The 90% C.L. monopole flux upper limit derived from the Baksan detector is plotted below in Fig. 12.

A deficiency of the Baksan detector arises from the scheme of triggering on anode signals with a typical pulse width of 50 ns. Since a slowly moving particle can have a time-of-flight through a scintillator plane greater than the 50 ns integration time, the effective monopole velocity threshold is increased to $\sim 10^{-3}$. Previously reported limits at lower velocity were based solely on the time-of-flight sensitivity of the detector. However, if the monopole velocity is below the trigger threshold, time-of-flight measurements would not be possible.

Homestake Large Area Scintillator Detector

A large area detector array is now under construction deep underground (4200 m.w.e.) in the Homestake Gold Mine near Lead, South Dakota. The detector consists of 200 modules containing 140 tons of liquid scintillator. Each module has dimensions 30 cm × 30 cm × 8 m. The modules are arranged in the shape of a hollow box 8 m × 8 m × 16 m (with one end open) yielding an aperture of 1200 m^2sr. For slow ionizing particles, the detector has a sensitivity ≈0.1 × minimum ionizing (or 0.2 MeV cm^2/g) and time-of-flight electronics sensitive to particle velocities down to $\sim 1.5 \times 10^{-4}$c. A slow monopole ionizing at a level of ~0.1 × minimum would have an energy loss of ~5 MeV in traversing a module. The expected pulse width for a slow monopole is ~1 ns/β_m and the expected time delay between top and bottom layers (or side to side) is ~25 ns/β_m.

The monopole trigger requires that ≥ 2 scintillator modules fire. Flash ADCs digitize each pulse into 100 ns bins providing 6 bits of pulse height and pulse shape information. Hence, the following additional constraints can be imposed: (1) the two pulse heights must be consistent with each other; (2) the pulse widths in entering and exiting counters should be ~1 ns/β_m; and (3) the time delay between entering and exiting counters should be consistent with the passage of a slow particle, i.e. 25-250 μs. Hence, a monopole signal in the Homestake detector has multiple redundancy. The only background triggers expected are due to random cosmic ray muons. These triggers will be easily rejected at the analysis stage by the pulse height and shape criteria.

Texas A & M GUT Monopole Detector

The Texas A & M group has deployed a monopole detector underground in a salt mine at a depth of 500 m.w.e. The detector is built up from 64 PS-10 acrylic scintillator counter elements, each with dimensions 6' × 6' × 3/8". For triggering, the detector is segmented into 4 groups of 16 counters. Three groups comprise horizontal layers 24' × 24' in size separated by 2'. The fourth group covers the sides of the detector box. A monopole trigger requires a majority coincidence of 3 of the 4 groups. Pulse heights and relative timing of the counters firing in a trigger are recorded. The electronics for the detector is designed to be sensitive to prompt light resulting from electronic ionization energy loss and to possible delayed fluorescence from level crossing interactions. Although the Texas group digitizes the pulse heights into 50 ns bins over periods of 22.5 μs using a flash ADC system, their thin scintillators are not sufficient to constrain the monopole signal by pulse shape information. Hence, monopole candidate events will be selected on the basis of the relative firing times of the 3 counters forming the trigger.

Gran Sasso Monopole, Astrophysics, and Cosmic Ray Observatory

A very large area GUT monopole detector has been proposed by an American-Italian collaboration.[31] The detector would be located in an experimental hall excavated deep underground adjacent to a road tunnel through the Gran Sasso mountain east of Rome. The hall has dimensions 17 m × 17 m × 125 m at a depth of 4000 m.w.e. As now proposed, the detector will consist of liquid scintillator tanks 50 cm × 25 cm × 12 m stacked together in the shape of a hollow box 12 m × 5 m × 110 m. The interior of the box will be filled with 10 layers of He+n-pentane filled limited streamer tubes 3 cm × 3 cm × 12 m separated by low background concrete absorbers, 50 cm thick. In addition, multiple sheets of CR-39 and Lexan plastic track detectors will be placed across the center plane of the array.

The general philosophy guiding this collaboration is to design and build a detector which will generate redundant and independent signatures for a passing monopole. The scintillator system alone has > 3-fold redundancy for monopole detection by requiring that the pulse widths, pulse shapes, and relative time delay between 2 layers be consistent with the passage of a slowly moving, weakly ionizing particle. Independently, the He+n-pentane limited streamer tubes can be triggered by monopoles via the level crossing mechanism and Penning effect (see section on Gaseous Detectors below). The redundancy of the gaseous detector system is several-fold since a monopole trajectory could be followed through several layers in a slow time sequence. And finally, if monopole events are observed in either the scintillator or streamer tube systems, the corresponding panels of plastic track detectors would be removed for etching and scanning.

The acceptance of the Gran Sasso detector is ≈12000 m^2sr yielding a sensitivity to an isotropic flux of monopoles an order of magnitude below Parker limit in a few years of livetime. Likewise, the experiment will be able to place significant new limits on whether superheavy monopoles make up the missing mass of the universe. Other physics objectives planned for the detector include searches for evidence of neutrino oscillations, supernova explosions, neutrino point sources, exotic particle, muon bundles, etc. The layout of the proposed detector is shown in Fig. 6.

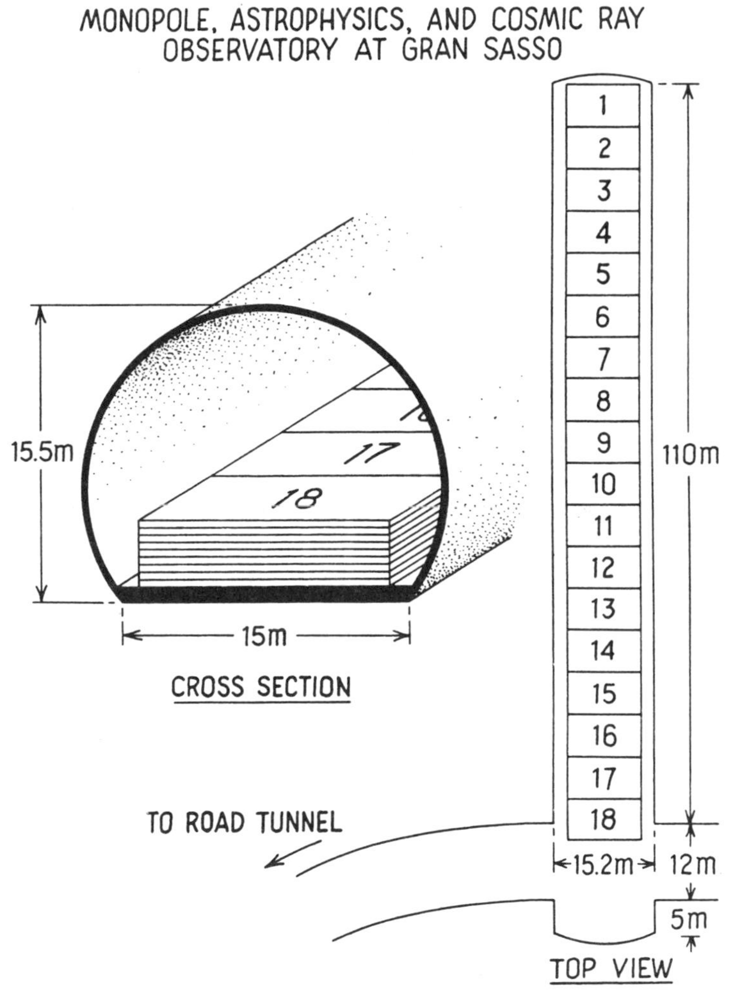

Figure 6: Layout of the proposed Monopole, Astrophysics, and Cosmic Ray Observatory (MACRO) at Gran Sasso.

GASEOUS DETECTORS

Drell et al.[32] have done a significant calculation demonstrating a mechanism for atomic and molecular excitation at extremely low velocities. By considering the interaction of the atomic electrons of H and He with the intense magnetic field of the monopole, they found that Zeeman effect level mixings can result in a high cross section for leaving a simple atom in a metastable excited state. Using this technique it may be possible to detect the passage of a monopole by 1) observing the photons emitted in the de-excitation of the excited atom, or 2) by observing ionization caused by energy transfer from excited atoms to complex molecules with low ionization potential (Penning effect).

For a detector using a mixture of He-CH_4 gas, the passage of a bare monopole through the detector would result in excited helium atoms with metastable energy levels of 20 eV. However, the ionization potential of CH_4 is 13 eV, hence, collisional de-excitation of the He atom can result in ionization of the CH_4 molecule. For typical He-CH_4 gas mixtures, the mean collision time is much less than the lifetime of the metastable He state.

Also, Kroll et al.[33] have evaluated the Zeeman effect interaction of negative and positive dyons as well as Dirac monopoles of higher charge. Although the monopole velocity thresholds and detection means differ for the various cases, their general conclusion is that gaseous He detectors will detect the passage of superheavy magnetic monopoles of arbitrary charge in the velocity range $10^{-4} < \beta_m < 10^{-3}$. Whether or not heavier noble gases such as argon can detect slowly moving massive monopoles by this mechanism is still uncertain. However, using two-body kinematics, Ahlen and Tarle[19] have calculated the threshold monopole velocity for excitation of argon gas to be $v_0 \approx 2 \times 10^{-3}c$.

Experiments which have used gaseous detectors to search for slowly moving monopoles are listed in Table III. Values listed in the table

Table III. Summary of Monopole Search Experiments Using Gaseous Detectors.

Experiment	Detection Mode	Gas Mixture	A·Ω (m^2sr)	Velocity Range, β_m	Flux Limit ($cm^{-2}sr^{-1}s^{-1}$) (90% C.L.)
BNL[34]	Proportional	90% Ar 10% CH_4	1.9	4×10^{-4} - 10^{-3}	< 3.5×10^{-11}
KGF[35]	Proportional	90% Ar 10% CH_4	250	$>2\times10^{-3}$	< 1.9×10^{-14}
Mont Blanc[36]	Streamer	29% Ar 48% CO_2 23% n-pentane	9.5	$>2\times10^{-3}$	< 6.4×10^{-13}
Soudan I[37]	Proportional	91% Ar 9% CO_2	73	10^{-3} - 10^{-2}	< 1.3×10^{-13}
Tokyo-ICRR[38]	Proportional	90% He 10% CH_4	24.7	3×10^{-4} - 1.0	< 7.2×10^{-13}
Frejus Tunnel[39]	Geiger	98% Ar 2% Ethylene	1000	10^{-3} - 10^{-1}	Taking Data
Gran Sasso	Streamer	25% He 75% n-pentane	12000	10^{-4} - 1.0	Proposed Experiment

are tabulated from the published references. Caution should be exercised when interpreting velocity thresholds less than 2×10^{-3}c for gases other than helium. At this time, the Tokyo group has performed the most significant experiment using a He-CH_4 gaseous detector array. Their result is plotted in the summary found in Fig. 12.

THE BERKELEY MICA EXPERIMENT

The Price group at Berkeley which pioneered the development of track sensitive plastic detectors such as CR-39, Lexan, etc., recently reported results from a monopole search using ancient muscovite mica.

Track-etch detectors record the passage of charged particles by submicroscopic damage trails in the crystalline lattice of the detector material. These trails can be amplified by chemical etching. Layers of the track sensitive solids can be scanned and measured to trace particle trajectories. For given plastic and etchant conditions, there is a well defined response function for the rate at which etchpits develop versus Z/β. The use of mica as a track sensitive detector has been calibrated using low velocity ions with $10 \leq Z \leq 90$. A slowly moving bare monopole would not produce a detectable track in mica, however, a monopole bound to a heavy nucleus would produce a track provided that the nucleus is at least as heavy as Aluminum. The damage rate in mica for the monopole-nucleus composite is related to that of the nucleus alone if one assumes an infinitely massive nucleus and neglects the small magnetic interaction of the monopole with the mica crystal.

The mica experiment scenario of Ahlen, Fleischer, Guo, and Price[40] is illustrated in Fig. 7. The basic idea is that a "bare" monopole will

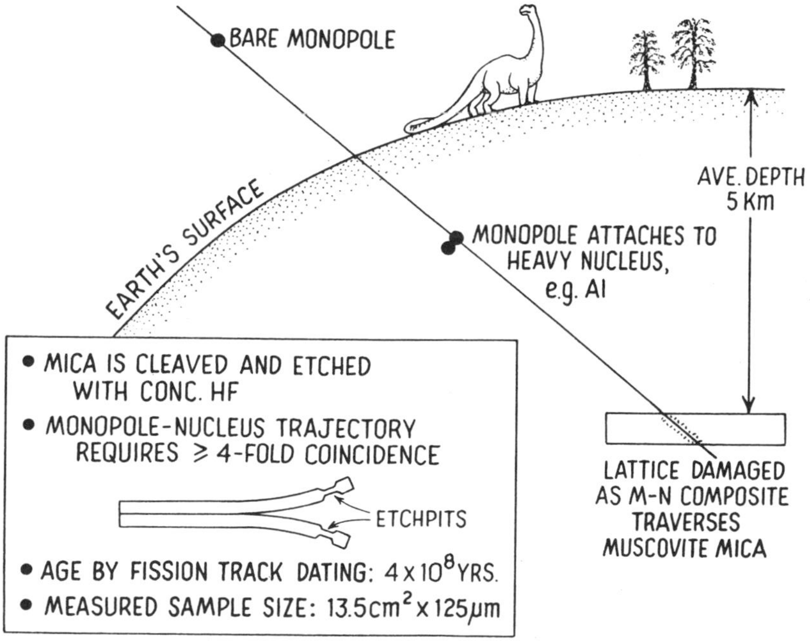

Figure 7: Scenario of the Berkeley Mica Experiment.

very likely capture a heavy nucleus as it enters the earth. Aluminum is the most likely nucleus to be captured because it is very abundant and has a large magnetic moment. If the monopole drags the nucleus through a piece of mica buried in the earth, elastic nuclear collisions result in a trail of lattice defects in the mica. As long as the mica is not reheated, the damage trail will survive. The authors have estimated that the mean free path for a monopole to capture an aluminum nucleus is ~10 Km for a monopole velocity in the range 10^{-3} to 10^{-4} c.

The piece of mica used in this experiment has a total size of 1000 cm^2 × 1 cm, although only 13.5 cm^2 × 125 μm were analyzed. Two techniques were used to determine the age of the mica sample. These included fission track dating which gives the time since the mica cooled and started recording tracks (~4.6×10^8 yrs.) and Ru-Sr dating. Hence, although the measured sample is relatively small, it has been recording tracks for more than 400 million years. The absence of 4-fold etchpit coincidences, which would be indicative of a monopole-nucleus trajectory, places an upper limit on the cosmic monopole flux well below the Parker limit. This result is shown in the summary in Fig. 12.

Although the mica experiment offers a clever way of challenging the Parker limit, it should not be considered a definitive result since there are plausible arguments for evading the limit: e.g., (1) if monopoles catalyze nucleon decay with a large cross section, then monopole-nucleus bound states could be short lived; (2) Arafune and Fukugita[41] have argued that a long range force due to extra angular momentum carried by a monopole-electric charge system could reduce the probability for a monopole to capture a nucleus; and (3) if the monopole is a positive dyon or comes with a proton attached as suggested by Bracci and Fiorentini,[6] then coulomb repulsion would inhibit the monopole from capturing a nucleus.

MONOPOLE CATALYSIS OF NUCLEON DECAY

One of the more exciting ideas to emerge from Grand Unified Theories (GUTs) is the suggestion by Callan,[42] Rubakov,[43] and Wilczek[44] that superheavy magnetic monopoles may catalyze nucleon decay via the interaction,

$$M + N \rightarrow M' + e^+ + \text{meson(s)}, \text{ where } N = n \text{ or } p.$$

The cross section for this process is expected to be of the order typical for strong interactions, i.e., $\sigma \sim 0.1 - 10$ mb for monopoles in the velocity range $10^{-4} \lesssim \beta_m \lesssim 10^{-2}$. Hence, the possibility exists that the large nucleon decay detectors now on the air could detect a time sequence of multiple ($\gtrsim 2$) catalyzed nucleon decays with essentially no background. Such an observation would be a tremendous boost for GUTs in that it would represent the simultaneous discovery of monopoles and nucleon decay. Data have now been reported from 5 nucleon decay experiments (Kolar Gold Field, Soudan I, Mont Blanc, Kamioka, and IMB). In addition, a small deep underwater detector at Lake Baikal, USSR, which will eventually grow into a large volume DUMAND-type array, has reported preliminary catalysis results.

The experimental parameters of the various detectors will limit their range of sensitivity for observing multiple catalyzed reactions.

For a given detector, the rate of expected monopole catalyzed events is given by

$$N_0 = F_m \times A \times \Omega \times \Delta t \times \varepsilon(\text{geom.}, \beta_m, \lambda_c) \times \varepsilon_d ,$$

where F_m is the monopole flux, $A \times \Omega$ is the effective area-solid angle for an isotropic monopole flux, Δt is the detector livetime, ε and ε_d are efficiency factors. In addition to detector geometry, the efficienty factor $\varepsilon(\text{geom.}, \beta_m, \lambda_c)$ depends on the monopole velocity, β_m, and the catalysis interaction length, λ_c. From the outset, nucleon decay detectors and their associated electronics were designed to observe single spontaneous nucleon decays. Hence, the electronic livetime and deadtime for most of the detectors restricts the range of sensitivity in β_m and λ_c for observing multiple decays during the passage of a slowly moving monopole. This effect is demonstrated graphically in Fig. 8. The parameter ε_d in the above rate equation is the triggering and software filtering efficiency for single nucleon decay events. Once again for the nucleon decay detectors this efficiency is high, typically > 70%. However, for detectors with experimental objectives other than nucleon decay such as the Lake Baikal detector, this factor can be small, e.g. $\sim 10^{-3}$.

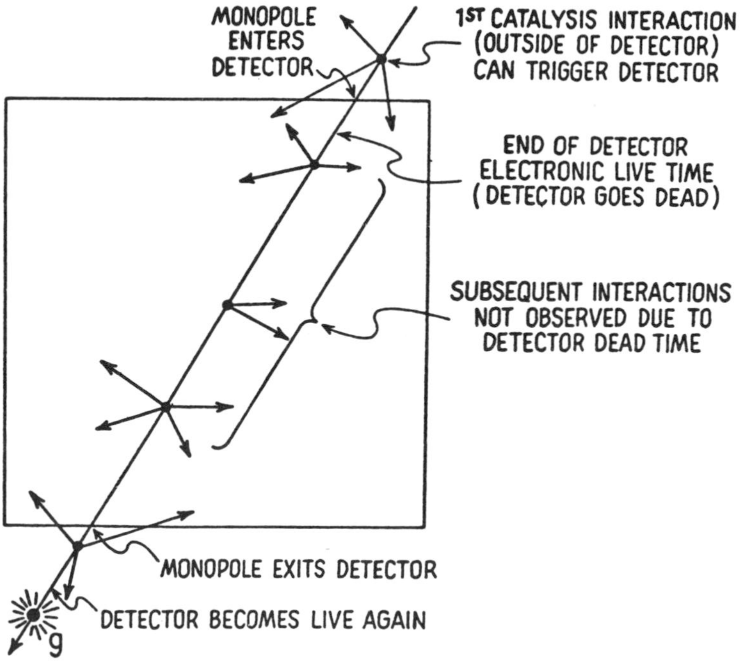

Figure 8: Technique for searching for ≥ 2 monopole catalysis of nucleon decay events. The deficiencies of the technique due to detector electronic livetime and deadtime are indicated.

Table IV summarizes the results from experiments which have reported 90% C.L. limits on the monopole flux based on the non-observation of multiple catalyzed decays. A few of the experiments are described in more detail below.

Table IV. Summary of Monopole Flux Limits from Nucleon Decay Experiments

Detector	Live Time (days)	Interaction Length, λc (meters)	Velocity Range (β)	Monopole Flux Limits ($cm^{-2}sr^{-1}s^{-1}$) ≡ 1 Interaction	≥ 2 Interactions
KGF[45]	561	λ_c=4 m		2×10^{-13}	
	561	σ_c>10 mb	$\beta_m > 10^{-3}$		2×10^{-14}
Soudan I[46]	285	$10^{-2}<\sigma_0<10^{2}$	$\beta_m > 10^{-3}$		1.5×10^{-13}
Mont Blanc[47]	317	$\lambda_c \approx 4$ m	$\beta_m > 10^{-4}$	6.1×10^{-13}	
	317	λ_c<1 m	$\beta_m > 10^{-4}$		2.3×10^{-14}
Kamioka[48]	135	λ_c<1.7 m	$10^{-4} < \beta_m < 10^{-3}$		7.6×10^{-14}
	135	λ_c<17 m	$10^{-5} < \beta_m < 10^{-3}$		6.4×10^{-15}
IMB[49]	200	$\lambda_c \approx 16.7$ m		1.7×10^{-12}	
	300	λ_c<1.0 m	$5\times10^{-3} < \beta_m \times 10^{-2}$		2.4×10^{-15}
	300	λ_c<0.1 m	$10^{-3} < \beta_m < 10^{-1}$		2.1×10^{-15}

Irvine-Michigan-Brookhaven

The IMB detector is the largest of the nucleon decay experiments. The 8000 metric tonne ring-imaging water Cherenkov detector has 3300 tonne fiducial mass viewed by 2048 5" diameter hemispherical photomultiplier tubes (PMTs). There are 2 time scales associated with the detector electronics: a T1 scale extending to 0.5 μs and a T2 scale which keeps the detector live for an additional 7.5 μs. Following a trigger on the T1 scale, the T2 scale is activated to enable the detection of $\mu \rightarrow e\nu\nu$ decays for muons which stop in the detector. For monopoles, the T1-T2 timing enables the detection of a sequence of multiple catalyzed nucleon decays in an 8 μs interval. The IMB monopole trigger requires at least one event with ≥ 30 PMTs on the T1 time scale and at least one additional event with ≥ 50 PMTs within any 300 ns window on the T2 time scale. The measured event rate is 4.6 ± 0.3 coincidences/day which is consistent with the expected rate of 4.7 events/day due to random two-fold coincidences from the 2.7 muons per second passing through the detector. All coincident events were scanned by physicists and rejected as entering random cosmic ray muons. With no observed catalysis events, the IMB collaboration has reported 90% C.L. upper limits on the monopole flux based on 300 days of analyzed data for various values of the interaction mean free path λ_c. The IMB results are plotted as a function of monopole velocity, β_m in Fig. 9.

Kamioka

The Kamioka nucleon decay collaboration has reported directly measured limits on the monopole flux by searching for multiple catalyzed

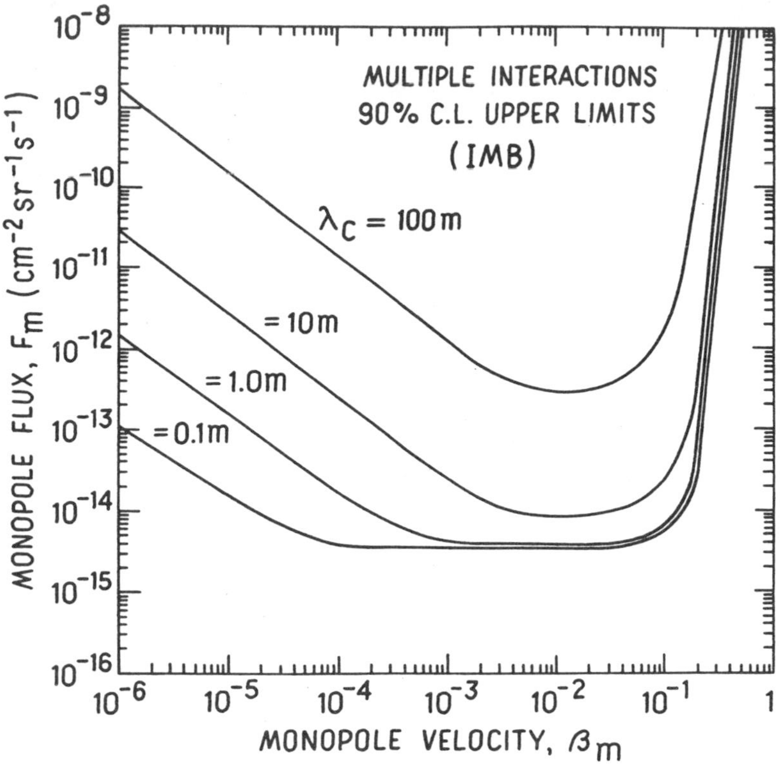

Figure 9: Monopole flux limits based on the non-observation of multiple (≥ 2) catalyzed nucleon decay events in the IMB detector.

decays in a manner similar to IMB. For 201 days of data, these results are shown by the solid lines in Fig. 10.

In addition to a direct search for catalyzed nucleon decays occurring in their detector, the Kamioka group has followed an analysis suggested by Arafune et al.[50] to search for indirect evidence of monopoles trapped in the sun. In summary, the analysis proceeds as follows: Monopoles which become trapped in the sun and catalyze nucleon decay could produce 0.45 ν_e/proton decay, ($p \rightarrow \pi^+ X$, 57%; $\pi^+ \rightarrow \mu^+ \rightarrow \nu_e$, 80%). The average ν_e energy is $\langle E\nu_e \rangle \sim 35$ MeV. For a neutrino intensity of 1.2×10^4 cm^{-2}s^{-1}, they estimate that ~40 $\nu_e e^- \rightarrow \nu_e e^-$ interactions will occur in their detector for each kilotonne-year of exposure. For a 201 day exposure with a 450 tonne fiducial mass, they find 3 events passing the event selection criteria, viz. a single showering track with $29 \leq E_e \leq 52$ MeV. The observed event rate places a 90% C.L. limit on $I\nu_e < 2.5 \times 10^4$ cm^{-2}s^{-1}. The Kamioka indirect monopole flux limits derived from this analysis are shown as dashed curves in Fig. 10.

Lake Baikal

A deep underwater experiment located at Lake Baikal[51] in the USSR began to operate its first string of PMTs in April, 1984. The present detector consists of 6 photomultiplier (PMT) modules and one electronics

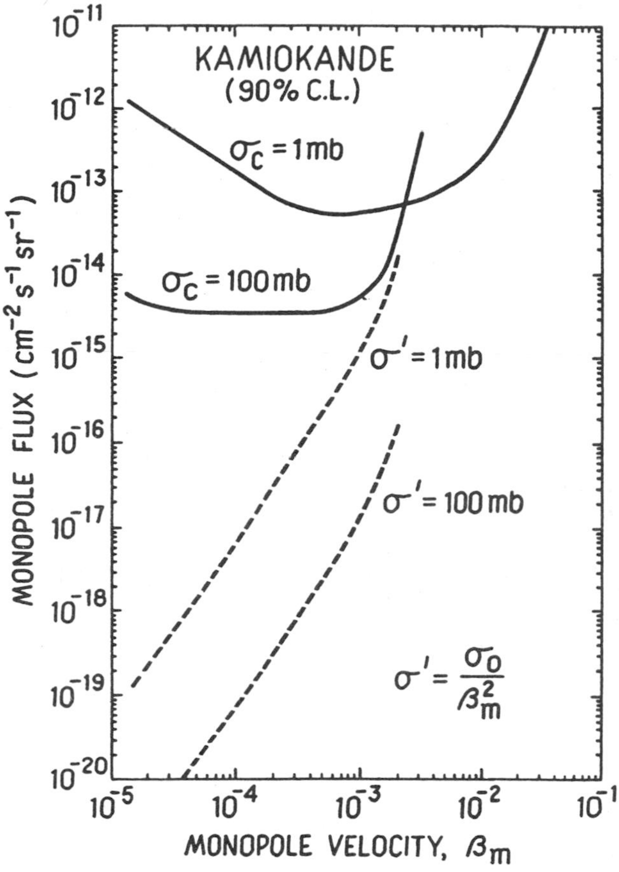

Figure 10: Monopole flux limits based on the non-observation of ≥ 2 nucleon decay events in the Kamioka detector (solid curves). The dashed curves are indirect limits derived from low energy neutrino interactions in the detector resulting from monopole catalysis in the sun.

module deployed at a depth of ~1 Km. Each PMT module consists of 2 phototubes with a 15 cm diameter flat photocathode. The electronics module contains the coincidence logic, power supplies, buffer memory, and communication electronics. The PMT string is connected to a shore station nearly 5 Km away by a geophysical well-logging cable. Figure 11 shows schematically the Lake Baikal detector and its deployment. In the future, the Baikal collaboration will deploy additional strings of PMTs, thus building up to a large volume array for studying very high energy muon and neutrino astrophysics.

The present single stationary string of PMTs has a sensitivity to monopole catalysis of nucleon decay for monopoles in the velocity range of $10^{-5} \leq \beta_m \leq 10^{-4}$ and having catalysis cross section $\geq 10^{3}$ barns. The Baikal monopole catalysis trigger requires a two-fold coincidence of greater than eleven 0.25 × minimum ionizing pulses within a 500 μs time interval. The triggering efficiency for a single nucleon decay at a distance of 20 m is $\sim 10^{-3}$. This low efficiency restricts the sensitivity

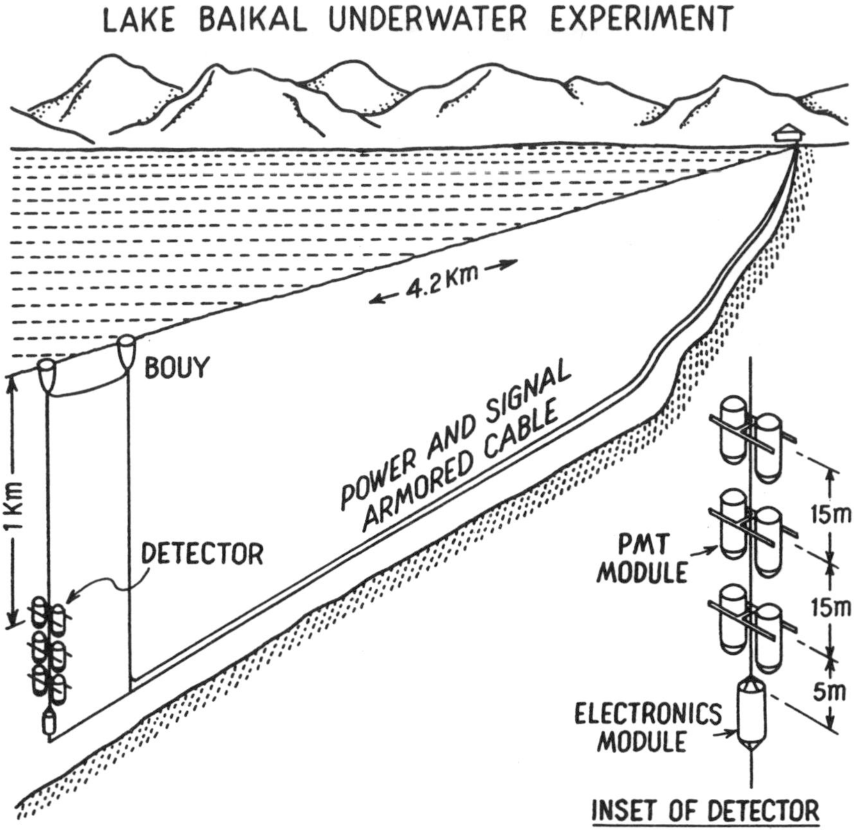

Figure 11: Deployment of the deep underwater PMT string at Lake Baikal.

of the Baikal detector to very low monopole velocities and extremely large catalysis cross sections. After 50 days of operation, the Baikal group reports the following 90% C.L. limits on the monopole flux for velocity, $\beta_m \leq 10^{-4}$:

1) For $\sigma_c \geq 10^3$ barns , $F_m \leq 7 \times 10^{-15}\ cm^{-2}s^{-1}sr^{-1}$

2) For $\sigma_c \geq 10^5$ barns , $F_m \leq 6 \times 10^{-16}\ cm^{-2}s^{-1}sr^{-1}$.

Astrophysical Catalysis Limits

In addition to the terrestrial experimental measurements described above, very stringent monopole flux limits have been derived using monopole catalysis of nucleon decay inside neutron stars and old radio pulsars. These limits are based on observations of x-ray luminosities and spindown ages and fall generally in the range 10^{-20} to $10^{-28}\ cm^{-2}sr^{-1}s^{-1}$. The derivations have inherent uncertainities due to assumptions regarding models for neutron star interiors, the number density of old neutron stars, the effect of interstellar absorption, possible m-$\bar{m}$ annihilations, etc. A comprehensive description of these astrophysical monopole flux limits can be found in the papers by Harvey, Turner, Kolb, Freese, and Drukier in Ref. 52 along with the references found therein.

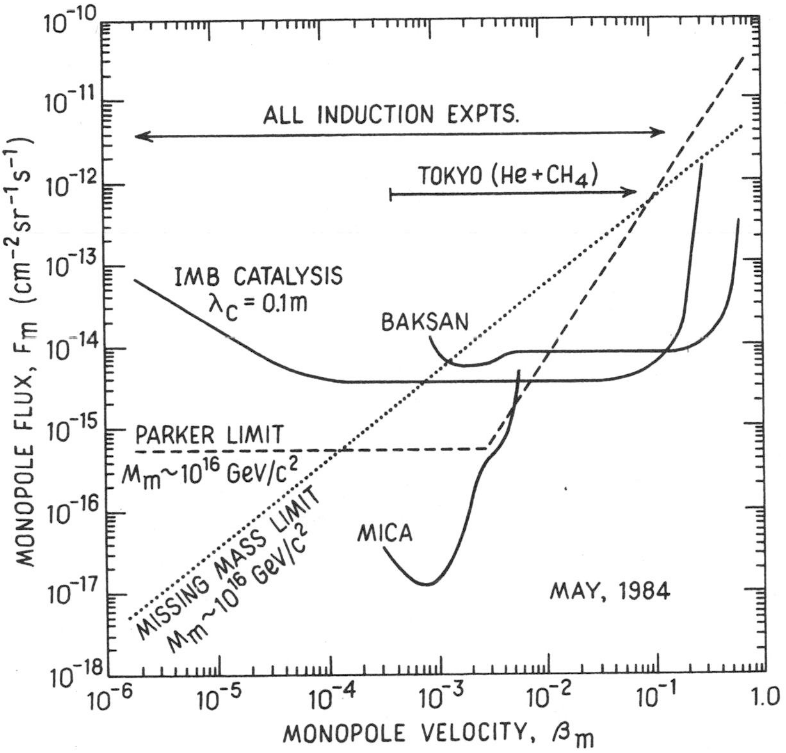

Figure 12: Summary of monopole flux limits versus monopole velocity. The solid curves are the most restrictive experimental limits from each of the various techniques discussed in this paper. The broken curves are limits derived from astrophysics.

CONCLUSION

A large number of theoretical and experimental developments in monopole physics have followed the work of 't Hooft and Polyakov in 1974. The marriage of particle physics and cosmology in the context of grand unified theories has yielded a wealth of explanations and new speculations about our universe. Although these analyses give no definite prediction of the monopole abundance, they suggest monopole properties which provide new guidance to experiments. In addition, arguments based on the persistence of the galactic magnetic field and the missing mass of the universe give a clear message, viz., detectors of unprecedented size (>10000 $m^2 sr$) will be required to detect monopoles or to place significant new limits on their abundance. Progress has been made recently in understanding the response of detectors for slow moving magnetically charged particles. Ahlen's calculation has shown that scintillators can detect monopoles above a threshold velocity of $\sim 6 \times 10^{-4}c$. Likewise, calculations by Drell, Parke, Kroll, et al. have shown that helium gaseous detectors can detect monopoles through Zeeman effect level crossing interactions. Large detectors using these techniques are now running and even larger detectors have been proposed. Also, several groups have made advances in techniques for building large superconducting induction detectors. Ultimately, any observed monopole flux would require confirmation by direct measurement of the monopole charge, hence, induction detector R and D efforts are complimentary to ionization/excitation techniques and should be continued.

Magnetic monopoles have shown up repeatedly in the development of physics: classically, they provide symmetry to Maxwell's equations; quantum mechanically, they explain charge quantization; and most recently they occur as a natural consequence of grand unified theories. The existence of monopoles is so fundamental and important to physics, this author believes that every possible experimental means for observing them should be pursued vigorously.

REFERENCES

[1] P.A.M. Dirac, Proc. R. Soc., London, Ser. **A133**, 60 (1931).

[2] G. 't Hooft, Nucl. Phys. **B79**, 276 (1974).

[3] A.M. Polyakov, Pis'ma Zh. Eksp. Teor. Fiz. **20**, 430 (1974). [JETP Lett. **20**, 194 (1974)].

[4] R. Craven and W.P. Trower, Fermilab-Pub-82/96-THY (1982).

[5] G. Fiorentini, "The Coupling Between Magnetic Charges and Magnetic Moments," MONOPOLE '83, ed. by J.L. Stone, (Plenum, New York, 1984) pg. 317.

[6] L. Bracci and G. Fiorentini, Phys. Lett. **129B**, 29 (1983).

[7] C.J. Goebel, "Binding of Nuclei to Monopoles," MONOPOLE '83, ed. by J.L. Stone, (Plenum, New York, 1984) pg. 333.

[8] E.N. Parker, Astrophys. J. **163**, 225 (1971); E.N. Parker, "Galactic Magnetic Fields and Magnetic Monopoles," MONOPOLES '83, ed. by J.L. Stone, (Plenum, New York, 1984) pg. 125.

[9] L.W. Alvarez, P.H. Eberhard, R.R. Ross, and R.D. Watt, Science **167**, 701 (1970); Phys. Rev., **D4**, 3260 (1971); Phys. Rev. **D8**, 698 (1973).

[10] B. Cabrera, Phys. Rev. Lett. **48**, 1378 (1982); B. Cabrera, in Magnetic Monopoles, ed. by R.A. Carrigan and W.P. Trower, (Plenum, New York, 1983) pg. 175.

[11] H. Frisch, "A Summary of Induction Detectors and Techniques," MONOPOLE '83, ed. J.L. Stone, (Plenum, New York, 1984) pg. 515.

[12] B. Cabrera et al., "Report on Stanford Superconductive Monopole Detectors," MONOPOLE '83, ed. by J.L. Stone, (Plenum, New York, 1984) pg. 439.

[13] J.F. Ziegler et al, Phys. Rev. **D28**, 1793 (1983); see also C.C. Tsuei, in Magnetic Monopoles, ed. by R.A. Carrigan and W.P. Trower, (Plenum, New York, 1983) pg. 201.

[14] C.C. Chi et al., "Monopole Searches at IBM: Present Status and Future Plans," MONOPOLE '83, ed. J.L. Stone, (Plenum, New York, 1984) pg. 451.

[15] J.R. Incandela et al., "First Results from the Chicago-Fermilab-Michigan Cosmic Ray Magnetic Monopole Detector," MONOPOLE '83, ed. J.L. Stone, (Plenum, New York, 1984) pg. 461.

[16] T. Ebisu and T. Watanabe, J. Phys. Soc. Japan **52**, 2616 (1983).

[17] J.C. Schouten et al., "The Imperical College Monopole Detector," MONOPOLE '83, ed. J.L. Stone, (Plenum, New York, 1984) pg. 471.

[18]S. Ahlen, invited talk presented at Monopole '83 Conference, Ann Arbot, MI, October 6-9, 1983.
[19]S.P. Ahlen and G. Tarle, Phys. Rev. **D27**, 688 (1983).
[20]E.N. Alexeyev, M.M. Boliev, A.E. Chudakov, B.A. Makoev, S.P. Mikheyev, and Yu. V. Sten'kin, Lett. Nuovo Cimento **35**, 413 (1982), and private communication from A.E. Chudakov, May 1984.
[21]G. Tarle, S.P. Ahlen, and T.M. Liss, "First Results from a Sea Level Search for Supermassive Magnetic Monopoles," MONOPOLE '83, ed. J.L. Stone, (Plenum, New York, 1984) pg. 551.
[22]R. Bonarelli et al., Phys. Lett. **126B**, 137 (1983); Phys. Lett. **112B**, 100 (1982).
[23]P.L. Connolly, "A Monopole Search Using an Accelerator Detector," MONOPOLE '83, ed. J.L. Stone, (Plenum, New York, 1984) pg. 617.
[24]S. Higashi et al., Proc. of 18th ICRC, Bangalore, ed. by P.V. Ramana Murthy (Tata Institute 1983) Vol. 5, pg. 52.
[25]F. Kajino et al., Phys. Rev. Lett. **52**, 1373 (1984).
[26]T. Mashimo et al., Phys. Lett. **128B**, 327 (1983).
[27]T. Mashimo, K. Kawagoe, and M. Koshiba, J. Phys. Soc. Japan **51**, 3067 (1982).
[28]D.E. Groom, E.C. Loh, H.N. Nelson, and D.M. Ritson, Phys. Rev. Lett. **50**, 573 (1983).
[29]R.I. Steinberg et al., "Monopole Search and Neutrino Astrophysics with Liquid Scintillation Detectors," MONOPOLE '83, ed. J.L. Stone, (Plenum, New York, 1984) pg. 567.
[30]R.C. Webb, "Status of the Texas A & M Monopole Search," MONOPOLE '83, ed. J.L. Stone, (Plenum, New York, 1984) pg. 581.
[31]Gran Sasso Monopole Collaboration: Bari, Bologna, Frascati, Milano, Roma, Torino, CERN, Indiana University, Caltech., The University of Michigan, Drexel University, Virginia Tech., Texas A & M.
[32]S.D. Drell, N.M. Kroll, M.T. Mueller, S.J. Parke, and M.A. Ruderman, Phys. Lett. **50**, 644 (1983).
[33]N.M. Kroll, V. Ganapathi, S.J. Parke, and S.D. Drell, "Excitation of Simple Atoms by Slow Magnetic Monopoles," MONOPOLE '83, ed. J.L. Stone, (Plenum, New York, 1984) pg. 295.
[34]J.D. Ullman, Phys. Rev. Lett. **47**, 289 (1981).
[35]M.R. Krishnaswami et al., Proc. of the 18th ICRC, Bangalore, ed. by P.V. Ramana Murthy (Tata Institute, 1983).
[36]M. Price, "A Search for Magnetic Monopoles in the Mont Blanc Nucleon Decay Apparatus," MONOPOLE '83, ed. J.L. Stone, (Plenum, New York, 1984) pg. 605.
[37]J. Bartelt et al., Phys. Rev. Lett. **50**, 655 (1983).
[38]F. Kajino et al., Phys. Rev. Lett. **52**, 1373 (1984).
[39]P. Eschstruth, "Monopole Search with the Frejus Tunnel Nucleon Lifetime Experiment," MONOPOLE '83, ed. J.L. Stone, (Plenum, New York, 1984) pg. 611.
[40]S.P. Ahlen, P.B. Price, S. Gue, and R.L. Fleischer, "A Search for GUT Monopoles with Mica Track Etch Detectors," MONOPOLE '83, ed. J.L. Stone, (Plenum, New York, 1984) pg. 383; also submitted to Phys. Rev. Lett.
[41]J. Arafune and M. Fukugita, Phys. Rev. Lett. **50**, 1901 (1983).
[42]C.G. Callan, Jr., Phys. Rev. **D26**, 2058 (1982).
[43]V.A. Rubakov, Nucl. Phys. **B203**, 322 (1982), and Pis'ma Zh. Eksp. Teor. Fiz. **33**, 658 (1981), [JETP Lett. **33**, 644 (1981)].

[44]F. Wilczek, Phys. Rev. Lett. **48**, 1146 (1982).
[45]M.E. Krishnaswami et al., Proc. of 18th ICRC, Bangalore, ed. by P.V. Ramana Murthy (Tata Institute, 1983).
[46]J. Bartelt, et al., Phys. Rev. Lett. **50**, 655 (1983).
[47]G. Battistoni, et al., Phys. Lett. **133B**, 454 (1983).
[48]M. Koshiba, private communications, April, July 1984.
[49]S. Errede, et al., Phys. Rev. Lett. **51**, 245 (1983).
[50]J. Arafune, et al., preprint RIFP-531, Kyoto, Japan, (1983).
[51]L.B. Bezrukov, et al., "Preliminary Results of the Search for Slow Monopoles in a Deep Water Experiment at the Baikal Lake," contributed paper at the XXII Inter. Conf. on HEP, Leipzig, July 1984).
[52] MONOPOLE '83, ed. J.L. Stone, (Plenum, New York, 1984).

A Monopole Flux Limit From A Large Area Induction Detector

J. Incandela

ABSTRACT

The design and performance during seven months of data taking of a superconducting induction detector with two 60-cm-diameter superconducting loops is presented. No candidate events were observed and an upper limit on the flux of cosmic ray magnetic monopoles $F_m \lesssim 6.2 \times 10^{-12} cm^{-2} sr^{-1} sec^{-1}$ (90% C.L.) is set. The detector demonstrates the possibility of operating large induction detectors in ambient magnetic fields greater than 1 mgauss.

Our detector has two identical gradiometer loops, each of diameter 60 cm and cell width 5 cm, (see figure1) separated by 21 cm. They are located inside cylindrical shields of the same diameter. The endcaps of the shields are domed for mechanical stability. The distance between the gradiometer and the endcaps thus varies between 4.25 cm and 9.50 cm. Each loop is coupled, via an impedance matching transformer,[1] to a SHE Corporation Model 30 SQUID.[2]

The CFM (Chicago-Fermilab-Michigan) detector is shown in figure 2. The detectors operate in a magnetic field of ~1-10 mgauss which is trapped by the superconducting shields at the time of the transition to the superconducting state and is obtained by reducing the earth's field with the aid of a 183 cm diameter, 218 cm tall, steel pipe.

Mechanical isolation is obtained by operating the detectors in vacuum and cooling by conduction. This eliminates vibration caused by bubbling and pressure changes which are encountered when cooling in a liquid helium bath. This also ensures that thermal time constants are large enough that no abrupt or localized temperature changes can occur, and that all thermal gradients (which cause currents to flow) are stable.

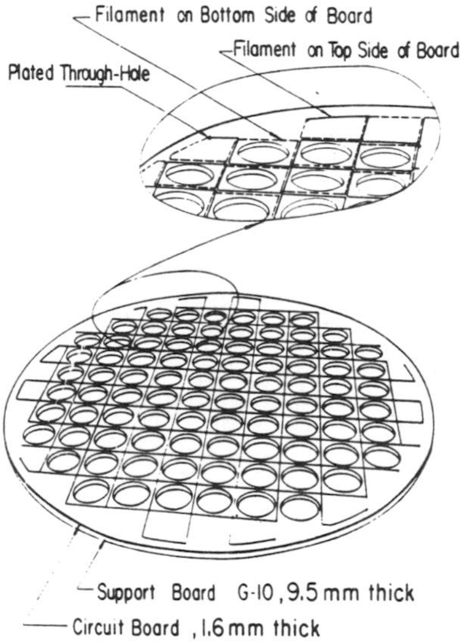

Figure 1. A perspective view of the gradiometer circuit board mounted to a 1 cm thick G-10 support plane. The gradiometer pattern uses both sides of the circuit board. Holes are cut to reduce mass for economy in cooling to 4°K.

We have found that disturbances in the environment of the apparatus can cause d.c. offsets[3] in the signal from the detector. The most common sources of these signals are magnetic field fluctuations and mechanical vibration of the SQUID systems. Radio-frequency (RF) radiation can also cause offsets by interfering with the RF biasing of the SQUID. To distinguish these background signals from true events, we have installed a variety of monitoring devices. These measure ambient magnetic field, vibration, RF radiation, sound, pressure of the nitrogen system, and the temperature and pressure of the cryostat. The data from the detector

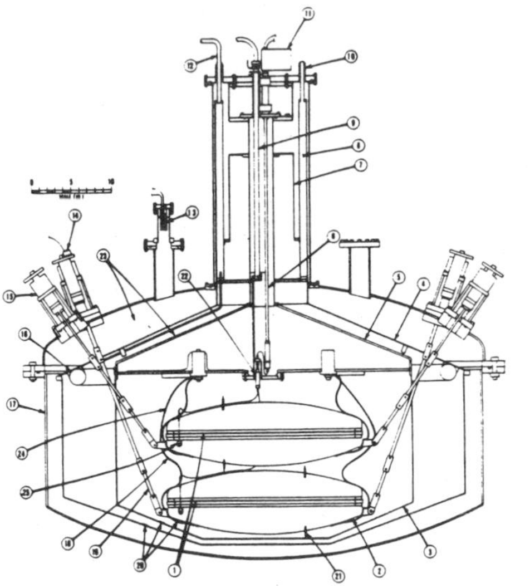

Figure 2. The apparatus: (1) Detector plan; (2) Lead lined copper shield; (3) Helium temperature shield; (4) Nitrogen temperature shield; (5) Helium reservoir; (6) SQUID tube; (7) Heat pick-up; (8) Nitrogen jacket; (9) Helium fill tube; (10) Nitrogen vent; (11) SQUID electronics; (12) Nitrogen fill line; (13) Vacuum gauge; (14) Accelerometer; (15) Vibration isolator assembly; (16) Nitrogen cooling ring; (17) Vacuum vessel; (18) Cooling straps; (19) Suspension rod; (20) Thermometers; (21) Pseudopole; (22) Vacuum feedthrough; (23) Super insulation; (24) Cooling straps; (25) SQUID matching transformer.

and some of the monitors is recorded by an 8-channel strip chart recorder. A PDP-11 with a 32 channel analog-to-digital system samples all of the monitors and the SQUIDS at 10.1 Hz and records the data onto magnetic tape.

The detector sensitivity is calibrated indirectly to better than 10% by measuring the self-inductance of the gradiometer. A direct calibration, good to better than 5%, is obtained by exciting any of six slender solenoids ("pseudopoles"), which thread each shield and dectector loop, to the level of a monopole penetrating the detector along the axis of the cylindrical shield. To evaluate the effect of the mutual inductance of the shield and the detection loop, we have calculated the response function of our detector by solving in closed form the boundary value problem of a monopole passing through a cylindrical shield with top and bottom endcaps. The response function was obtained by generating events with random trajectories through the shield, and calculating the detector response for each. The resulting probability curve is shown in figure 3.

There were no coincident offsets observed. A histogram of unexplained[5] offsets greater than 1 mV appearing in either of the detector loops is shown in figure 3. Although the long tail of the detector response curve extends down to the region of the small offsets, the total integrated probability below 3.2 mV is only 1.2%. The complete absence of offsets in the high probability region of the curve indicates that the unexplained single offsets are part of a spurious signal background and are not monopole candidates at the 90% C.L. One large offset of 12.7 mV, not shown in figure 3, was observed. This offset is not consistent with a singly charged Dirac monopole, or even an integral number of Dirac charges.[6]

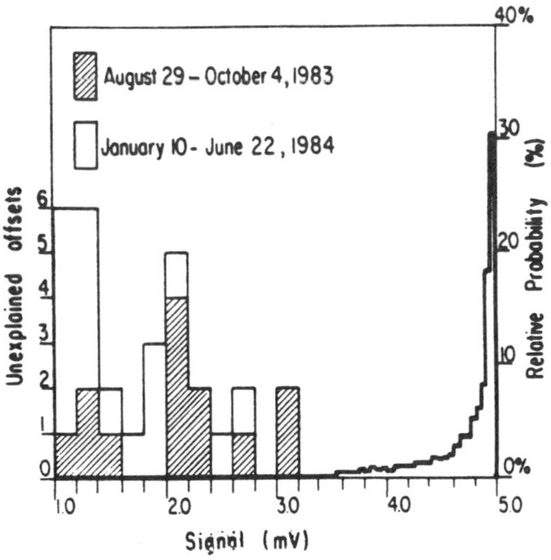

Figure 3. The measured spectrum of events (left-hand scale), and the calculated response function of the dector to a single-Dirac charge monopole (right-hand scale). The cross-hatched events are from the first 'shake-down' run.

In conclusion, we have demonstrated that inexpensive, large area superconducting detectors can be stably operated in easily obtained ambient magnetic fields. No monopole candidate events have been recorded for either detector loop in 163 days of operation. The total sensitive area for a monopole passing through either or both of the loops is 2100 cm² averaged over all angles. This exposure thus sets a 90% C.L. limit, $F_m \lesssim 6.2 \times 10^{-12}\ cm^{-2} sr^{-1} sec^{-1}$ on the flux of cosmic ray magnetic monopoles at the earth's surface, independent of monopole velocity. This experiment corresponds to 230 times the sample of B. Cabrera's first experiment. The probability that these two results are compatible is less than 0.5%.

I thank my collaborators: M. Campbell, H. Frisch and S. Somalwar from the University of Chicago; M. Kuchnir from Fermilab and H. R. Gustafson from the University of Michigan.

We thank R. Armstrong, L. Fiscelli, R. Gabriel, P. McGoff, and R. Northrop of the University of Chicago for their excellent mechanical and design work; R. Smith of Fermilab for the loan of a dewar; and A. Guthke and C. Kendzione of Fermilab for the loan and patient repair of a turbo pump. This research was supported by the National Science Foundation, the Department of Energy, and the University of Chicago Physical Sciences Divsion.

REFERENCES

[1]Impedance matching between the 21μH loop and the 2μH input of the SQUID is necessary to obtain the maximum signal. In this detector we used a Niobium wire superconducting transformer - we have since realized that one can link the gradiometer element in a series - parallel arrangement to make the loop impedance equal to the SQUID impedance. (EFI preprint 84-10 to be published in Nucl. Inst. and Meth.).

[2]S.H.E. Corporation, 4174 Sorrento Valley Blvd., San Diego, CA, 92121.

[3]The dominant source of offsets we see is in the SQUIDS themselves. A detailed study of this has been done by A. Clark, M. Cromar, and F. Fickett of NBS, Proc. Monopole '83, Ann Arbor, Michigan, J. Stone ed.

[4]The signal-to-noise ratio is $\geq$ 20 at 0.1 Hz bandwidth.

[5]An 'unexplained' offset is one for which no disturbance in the environment of the detector was measured by our monitors. The most common 'explained' disturbances are large changes in the magnetic field due to occasional trucks in the alley next to our building, people moving large iron objects in the building, vibrations and occasional RF interference.

[6]This 12.7 mV step occurred one hour after recovering from a catastrophe in which the power cord to the computer caught fire. This was prior to the installation of the RF interference monitor. We suspect that this offset was due to RF interference.

Future IBM-BNL Large-Area Superconducting Inductive Monopole Detectors

S. Bermon, C.C. Chi, C.C. Tsuei, P. Chaudhari,
M. Ketchen, C.D. Tesche, and A. Prodell

INTRODUCTION

The observation of massive moving magnetic monopoles would have extremely important implications for grand unification theories and cosmological models for the creation of the universe. Among detection methods, the superconducting induction technique is unique in that it directly and unambiguously measures the sole property of the monopole of which we are certain--its magnetic charge--the detector response being independent of all other characteristics such as the monopole mass, its velocity, the presence of a companion electric charge, or the detailed nature of its interaction with matter, in particular the degree of ionization/excitation for heavy monopoles moving at $v_m < 10^{-3}c$, or the size of the cross section for monopole induced baryon decay. Since Cabrera's observation[1] of a monopole candidate event in February 1982, however, none of the subsequently set-up induction detectors[2] have registered any additional candidates. At the time of this conference, the area-time exposure for all induction experiments was approximately 300 times the original Stanford exposure, indicating that the actual monopole flux is much lower than implied by the single candidate event. Indeed, bounds based on astrophysical constraints suggest rather small upper limits for such a flux. Parker and colleagues,[3] e.g., have calculated the maximum monopole flux consistent with the existence of the ~3μG galactic magnetic field to be roughly $F_m \cong 10^{-15}$ $cm^{-2}sr^{-1}s^{-1}$ corresponding to one monopole per year through a planar area of about 500 m^2, some five orders of magnitude lower than the Stanford value. Described herein are our plans for constructing an induction detector sufficiently large to reach the Parker bound in several years of operation.

THE PROTOTYPE DETECTOR

The design of the large-area detector is based in large part on the superconducting induction detector presently in operation at the IBM T.J. Watson Research Center. That detector, which consists of six independent planar gradiometer induction coils each connected to an individual RF SQUID (Superconducting QUantum Interference Device), mounted on the six surfaces of a rectangular parallelopiped, is the largest area, fully coincident induction detector currently running, with a total <u>coincident</u> area for isotropic flux (integrated over 4π sr) of 1000 cm^2. The detector has been in operation since early October, 1983. After 210 days of running (as of June 30, 1984) no candidate events were observed, setting an upper limit for the monopole flux (90% confidence level) of $1.2\times10^{-11} cm^{-2}sr^{-1}s^{-1}$. A brief description of the detector is contained in the Proceedings of the Monopole '83 Conference,[4] with complete details contained in a forthcoming publication.

One of the major problems presented by large-scale detectors already addressed in the design of the 1000 cm^2 detector is the sensitivity of large area induction coils to fluctuations in the ambient magnetic field. For example, the inductance of a single circular loop of one meter diameter made of 0.25 mm dia. wire is approximately 10μH. A monopole passing through such a loop induces a current of $2\Phi_0/L \cong 4\times10^{-10}$A (where $\Phi_0 = 2.07\times10^{-7}$G-cm^2 is the superconducting flux quantum). For a S/N ratio of 10, the required field stability would have to be better than 4×10^{-12}G or about 1 part in 10^{11} th of the earth's field. An effective technique to reduce the coil sensitivity to ambient field fluctuations[5,6] is to twist the coil into a set of coplanar loops of alternating polarity cells to cancel those field fluctuations, while maintaining full sensitivity to the passage of a monopole whose path threads only one of the cells. The design of the planar "gradiometer" developed at IBM[4,5,6] is unique in that the cell areas are varied in such a way as to systematically cancel out successively higher-order terms in the Taylor-series expansion of the field. Such is not the case with other schemes which simply use equal-area cells.[7] A configuration insensitive to the (n-1)st spatial derivative of the field is termed an nth order gradiometer. For a given loop area, the higher the order of the gradiometer, the smaller the cell size, and the greater is the insensitivity to external fields. The corresponding increase in the loop inductance, however, which is roughly proportional to the length of wire in the gradiometer, causes the induced current to be reduced and eventually seriously degrades the electrical S/N ratio. The optimum design is a compromise between the magnetic and electrical sensitivities.

A very important additional advantage of the gradiometer scheme is that it allows optimum usage of the expensive cryogenic volume within which the detector is enclosed. The primary method for obtaining adequate field stability is to enclose the detector in a superconducting shield. Such a shield does not exclude magnetic flux, but simply traps it. Possible flux jumps in the shield can simulate a monopole signal. More importantly, the flux emanating from the doubly-quantized vortices (at both entrance and exit) left in the shield walls by the passage of a monopole, generates an opposing current in the detector loop which degrades the monopole signal, with the degree of degradation dependent upon the particular monopole trajectory, which cannot be independently known. The gradiometer greatly reduces the interaction between the coil and shield allowing the coil to be placed very close to the shield wall--a distance approximately equal to an average cell size, rather than the entire loop linear dimension. The detector can thus be expanded to fill almost the entire inner dimensions of the dewar.

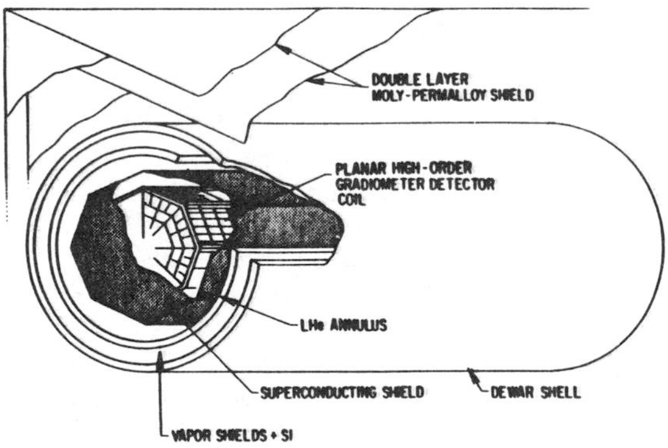

Figure 1: Schematic drawing of 4 m^2 prototype monopole induction detector showing octagonal prism gradiometer, surrounding superconducting shield, LHe dewar (7 ft. dia. by 16 ft. long) and enclosing ferromagnetic shield.

A schematic drawing of the large-area prototype detector-dewar-ferromagnetic shield system is shown in Fig. 1. The cylindrical dewar is 7 ft (2.1 m) in external diameter and over 16 ft. (4.9 m) in length. Liquid helium is contained in a 5 cm wide annular region surrounding the 4.2 K pumpable He exchange gas space, 64 inches (1.6 m) in diameter and 12 ft (3.6 m) long, which houses the detector and surrounding superconducting shield. Access is gained to the sample space through a 5 ft dia. crushed-In seal cold flange. The closed-box detector, in the shape of an octagonal prism, is made up of ten independent gradiometer plates, each connected to a separate SQUID--eight lateral rectangular plates of area 1.68 m^2 each, and two octagonal end plates of 1.28 m^2, for a total surface area of 16 m^2. The effective collection area for isotropic flux integrated over 4π sr is then 1/4 of this area, or 4.0 m^2--40 times the area of the current detector. An important feature of the above design is that any monopole passing through the closed-box detector must excite exactly two and only two of the 10 induction coils. The 4 m^2 area thus represents a 100% coincidence area, but in addition, the geometry discriminates against those types of spurious disturbances that might cause more than two (in particular--all) of the detector circuits to fire.

Associated with the detector are a full complement of "counter insurgency" devices to identify spurious events that might mimic a monopole signal. These include a 3-axis fluxgate magnetometer to monitor external field changes, a 3-axis accelerometer mounted on the dewar and strain gauges mounted on the detector itself to monitor mechanical disturbances, an RF interference meter, an electrical power line monitor, and a proportional counter to measure cosmic-ray activity. Data taking will be automated using an IBM Series One computer. The large area gradiometer coils, with their consequently high inductance, require sensitive SQUID's to achieve an adequate electrical S/N ratio. Ultra-low noise tunnel junction dc SQUID's with tightly-coupled planar input coils[8] have been fabricated at IBM[9] with the specific intention of employing them in this and future detectors.

The problem of flux jumps in the superconducting shields and the coil wires themselves can be minimized by cooling the assembly in low field. The surrounding moly-permalloy ferromagnetic shield enclosure (8.5 ft × 8.5 ft × 17.5 ft) is designed to achieve an ambient field level of several milligauss over the entire superconducting shield volume using degaussing techniques developed as part of the Josephson computer program at IBM. Although a two-layer structure is pictured, a single-layer enclosure is initially being constructed in an attempt to achieve the required field levels.

The LHe dewar is doubly vapo-shielded with provision for circulating LN_2 through the outer thermal shield in place of the LHe evaporant should this significantly improve the boil-off performance, which is estimated to be between 1.5 and 2.5 l/hr. The dewar is specially designed to minimize temperature gradients in the thermal shields, which can produce appreciable magnetic fields through thermoelectrically generated currents. The system will be installed at the Brookhaven National Laboratory and is expected to be operating by late 1985.

As an aide to design, and in order to do preliminary testing of large-area coil and shield performance, a "pre-prototype" detector of effective area 1.0 m^2 consisting of six independent gradiometer plates mounted on the surfaces of a rectangular parallelopiped and housed in a square cross-section dewar is presently under construction with completion expected in mid 1985.

THE ARRAY

To pursue the Parker bound, an array of detectors is planned. Although studies are still proceeding on determining the optimum combination of size and number of dewars to reach a given total detection area, the present projection calls for 16 dewars, each 7 ft in dia. (similar to the prototype), but 25 ft (7.6 m) long, each housing an octagonal prism detector with an averaged isotropic area of 6.7 m^2. The dewars would be arranged in a 2 × 8 array spread over an area of 60 ft (18 m) by 120 ft (36 m). The solid angle overlap between nearby detectors (a monopole can pass through more than one detector) reduces the effective total area from the sum of the individual areas--107 m^2, to about 100 m^2. Such a 100 m^2 detector would reach the Parker bound in about 3 years of operation.

Liquid helium consumption is estimated to be about 50 l/hr. with LHe being supplied by a closed-cycle helium liquifier of capacity 200 l/hr. presently existing at BNL. The present ferromagnetic shield enclosure, which is of modular construction, would be enlarged to accommodate the longer dewars. Only one enclosure is needed. Each dewar can be cooled in turn through the superconducting T_c, thereby trapping within the SC shield the low field existing in the enclosure, and then removed for operation in the earth's field. The cost of the ferromagnetic shielding then becomes a small fraction of the total cost of the project.

ACKNOWLEDGMENTS

We should like to thank S. Kirkpatrick and M. Gutzwiller for valuable insights concerning the high-order planar gradiometers.

REFERENCES

[1] B. Cabrera, Phys. Rev. Lett. **48**, 1378 (1982).

[2] See papers by Cabrera, et al., Incandela, et al., and Stone - this conference.

[3] E. N. Parker, Astrophys. J. **160**, 383 (1970); M. S. Turner, E. N. Parker, and T. J. Bogdan, Phys. Rev. D **26**, 1296 (1982).

[4] C. C. Chi, C. D. Tesche, C. C. Tsuei, P. Chaudhari and S. Bermon, in Monopole '83, J. Stone, ed., Plenum, NY (1984, to be published).

[5] C. C. Tsuei, Magnetic Monopoles, R. A. Carrigan, Jr. and W. P. Trower, eds., Plenum, NY, p. 201 (1983).

[6] C. D. Tesche, C. C. Chi, C. C. Tsuei and P. Chaudhari, Appl. Phys. Lett. **43**, 384 (1983); C. D. Tesche, Fourth Workshop on Grand Unification, H. A. Weldon, P. Langacker, and P. J. Steinhardt, eds., Birkhauser, p. 121 (1983).

[7] J. Incandela, et al., in Monopole '83, op. cit.

[8] M. B. Ketchen and J. M. Jaycox, Appl. Phys. Lett. **40**, 736 (1982).

[9] C. D. Tesche, K. H. Brown, A. C. Callegari, M. M. Chen, J. H. Greiner, H. C. Jones, M. B. Ketchen, K. K. Kim, A. W. Kleinsasser, H. A. Notarys, G. Proto, R. H. Wang and T. Yogi, Proc. of the 17th Inter. Conf. on Low Temp. Phys., North-Holland, p. 263 (1984).

Cosmic-Ray Monopole Searches at Stanford Using Superconducting Detectors

M. Taber, B. Cabrera, R. Gardner and M. Huber

I. INTRODUCTION

A Dirac magnetic charge g passing through a superconducting ring changes the magnetic flux threading the ring by exactly 2 ϕ_0, where $\phi_0 = 2.07 \times 10^{-7}$ G cm^{-2} is the magnetic flux emanating from the monopole. Superconductive detectors based on this property directly measure magnetic charge of a passing particle independently of its velocity, mass, electric charge, and higher magnetic multipole moments. The detector response follows from simple and fundamental theoretical considerations.[1] Such detectors are thus natural choices in the search for any flux of cosmic-ray magnetic charges, particularly the nonrelativistic and supermassive monopoles predicted by grand unification theories.[2]

A 5-cm diameter single-loop prototype detector based on this principle was operated for a total time of 151 days and produced a single candidate event.[1] The existence of one inconclusive candidate motivated the construction of a new detector with significantly improved sensing area, redundancy, mechanical stability, bandwidth, and signal-to-noise ratio. A three-loop detector satisfying these requirements has been in low-noise operation for over a year now, and we present here an update of initial results reported in more detail earlier.[3]

II. CURRENT RESULTS FROM THE THREE-LOOP DETECTOR

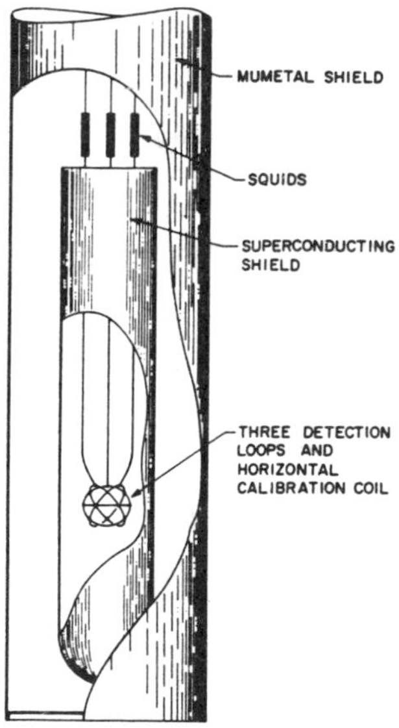

Figure 1: Schematic of the three-loop detector.

The detection apparatus consists of three superconducting coils, each made of two turns NbTi wire wound on a 10.2 cm diameter Pyrex sphere (Fig. 1). They form mutually orthogonal concentric circles oriented with threefold rotational symmetry about the vertical axis. A two-turn calibration coil, also concentric with the loops, is mounted on the same sphere with its axis vertical. This loop assembly is mounted in an evacuated Pyrex chamber and connected to shielded SQUID sensors located above the vacuum compartment. During cryogenic operation, the detector loops are located in a 20 cm diameter by 100 cm long cylindrical superconducting shield which is closed at the bottom and open at the top. Ambient magnetic field in the detection region is ~20 nG. Sensitivity to external field variations is further reduced by a high-permeability magnetic shield outside the Dewar yielding a combined attenuation of ~160 dB.

To guard against spurious signals, additional instruments monitor parameters known to affect the detector. A fluxgate magnetometer monitors external field variations, a pressure transducer monitors pressure above the liquid helium bath, an accelerometer detects motion of the apparatus, and a powerline monitor detects line noise or fault conditions.

The output signals from the fluxgate, the accelerometer, and the three independent SQUID systems are sampled continuously at 200 Hz (100 Hz bandwidth) with a computer data-acquisition system. These data are temporarily stored in a circular buffer and digitally filtered to 0.1 Hz bandwidth in real time for permanent storage along with data from the other instruments. If an offset in the filtered SQUID data occurs which corresponds to a current change in excess of 0.1 ϕ_0/L in any one of the three sensor loops, where L is the total inductance of the sensor loop plus the SQUID input inductance, the higher bandwidth data both preceding and following the offset can likewise be permanently stored. This technique is useful for detecting mechanically induced disturbances.

The response of our three-loop detector has been characterized by a detailed Monte Carlo calculation which simulates an isotropic distribution of monopole trajectories through the apparatus. A trajectory which passes through a loop changes the flux threading it by exactly 4 ϕ_0(2 ϕ_0 per turn). However, in computation of the induced supercurrent, the cylindrical superconducting shield surrounding the loops must be considered since a Dirac charge would leave doubly quantized supercurrent vortices of opposite vorticity at the intersection of its trajectory with the shield. For each of the simulated trajectories the changes induced in each loop were found by combining the direct coupling of the particle to the loop (either 4 ϕ_0 or zero) together with the flux coupled to that loop from the field generated by the vortices induced in the shield.

This calculation accomplishes two things: it allows discrimination between the class of detector offsets which could correspond to real Dirac charges from those which could not, and secondly it provides an estimate of the average detector sensing area. The detector sensing area has two components, one which corresponds to trajectories which intercept at least one loop and the other which corresponds to trajectories which miss the loops but intercept the shield near enough to the sensing region to be detected. The magnitude of both these components, but most particularly the latter, is a function of the signal-to-noise and the consequent threshold and coincidence criteria that are set to qualify potential candidate events. In our case we have set the criterion that a potential candidate must produce an offset greater than 0.1 ϕ_0/L in at least two loops. With this criterion, the direct loop sensing area (averaged over 4π solid angle) is 70.5 cm^2 and the near-miss sensing area is 405 cm^2 for a total of 476 cm^2.

Since cooldown on 25 January 1983 the three-loop apparatus has been operating continuously with rms noise levels of approximately 0.02 ϕ_0/L. For comparison, trajectories which intercept a detector loop will generally produce a signal in the 3-4 ϕ_0/L range and a near miss will be less than 1 ϕ_0/L. As of 20 June 1984, we have accumulated 441.5 days of live data with no candidates meeting our coincidence criterion. This result represents approximately 140 times the exposure of the original prototype detector and sets an upper limit of 1.0×10^{-11} cm^{-2} sec^{-1} sr^{-1} (90% confidence level) for an isotropic flux of cosmic-ray monopoles passing through the Earth's surface at any velocity. If the initial five days of generally noisy data are excluded, we have observed 18 unexplained single events above threshold (Fig. 2), all of them having a magnitude less than 1 ϕ_0/L. If we assume uncorrelated rates obtained from data after the first five days, accidental unexplained double coincidences would occur at a mean rate of about once in every 47,000 years.

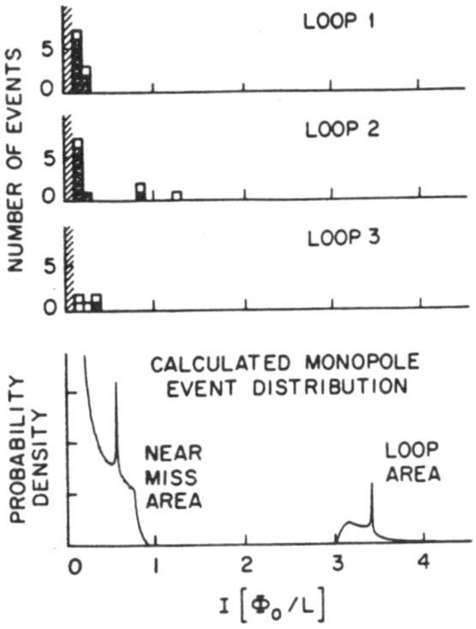

Figure 2: Histograms of events above threshold prior to elimination by coincidence veto. Unshaded events occurred within the first 5 days after cooldown. The single-loop distribution function for monopole events is included for reference.

III. LARGE SCALE SUPERCONDUCTIVE MONOPOLE DETECTOR

In view of the negative results from the three-loop detector, further substantial progress can be made only by devising a detector of significantly larger cross section. Because SQUID systems are expensive, this generally requires the sacrifice of signal level. Spurious signal discrimination would normally suffer as a consequence, but this can be avoided or minimized if lower noise SQUID systems are used, coincidence requirements are maintained, and if statistical consistency checks are used.

Work done at IBM[4] and by a University of Chicago-University of Michigan-Fermilab collaboration[5] has demonstrated the utility of planar gradiometer configurations for achieving large sensing areas. This type configuration consists of a two dimensional array of superconducting loops interconnected with alternating polarities so that sensitivity to uniform field changes is eliminated. This has two advantages: ambient field and mechanical stability constraints are considerably relaxed, and the interaction with the surrounding superconducting shield is significantly reduced. The latter advantage allows a smaller separation between the detection loops and the shield without an unacceptable loss of sensitivity or broadening of the response distribution due to the monopole-induced vortices in the shield.

The detector which we are currently building will consist of eight gradiometer panels arranged in an octagonal cylinder with each panel being coupled to a separate SQUID (Fig.3). It will have a double-coincidence sensing area of ~1.5 m^2 averaged over solid angle

(~30 times the three-loop detector area) and will be enclosed in a cylindrical superconducting shield measuring 50 cm diameter by 6 m long. Each panel will be approximately 16 cm wide by 6 m long and will consist of two distributed parallel gradiometer arrays connected to the SQUID in parallel in order to achieve optimum coupling efficiency.[6] Although the distributed parallel arrangement offers no significant improvement in signal strength over other optimal interconnection schemes, it does have the advantage of being realizable on one side of a circuit board. We anticipate a signal-to-noise ratio of 20 using rf SQUID technology with additional improvement being available with dc SQUIDs. The detector system will be housed in a currently available dewar which will be augmented by external high-permeability magnetic shield, for the purpose of providing an ambient field in the milligauss range.

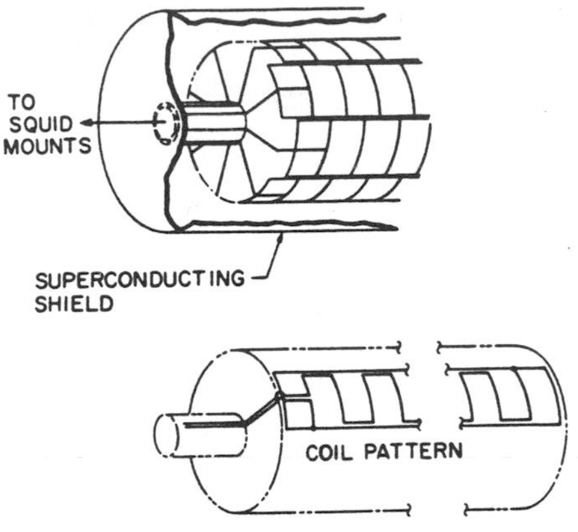

Figure 3: Schematic of the 1.5 m^2 monopole detector.

It is anticipated that this system will serve as a prototype for a yet larger detector that can be made using present dewar and SQUID technology. For example, a 100 m^2 array could be built using six dewars, each 8 feet in diameter by 40 feet long. An array of this size could reach the Parker bound[2] in one year of operation and could be built at a cost comparable to that of typical high energy experiments.

REFERENCES

[1]B. Cabrera, Phys. Rev. Lett. **48**, 1378 (1982).
[2]See Magnetic Monopoles, eds. R.A. Carrigan and W.P. Trower (Plenum, New York, 1983).
[3]B. Cabrera, M. Taber, R. Gardner and J. Bourg, Phys. Rev. Lett. **51**, 1933 (1983).
[4]C.D. Tesche, et al., Appl. Phys. Lett. **43**, 384 (1983).
[5]H. Frisch, Monopole 83, ed. J. Stone (Plenum, New York, in press).
[6]S. Somalwar, et al., Enrico Fermi Institute preprint 84-10 (see also the paper from the CMF group, this conference).

Monopoles and the Homestake Underground Scintillation Detector

M. L. Cherry, S. Corbato, D. Kieda, K. Lande,
C. K. Lee, and R. I. Steinberg

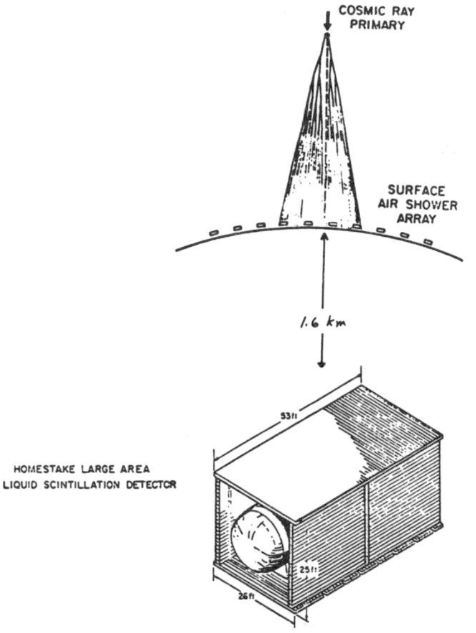

Fig. 1: Surface and underground Homestake detectors.

The Homestake Large Area Scintillation Detector (Fig. 1) consists of 140 tons of liquid scintillator in a hollow 8m × 8m × 16m box surrounding the Brookhaven ^{37}Cl solar neutrino detector. The experiment is located at a depth of 4850 ft. (4200 m.w.e.) in the Homestake Gold Mine. Half of the detector is currently running; the full detector will be taking data by the fall of 1984. An extensive air shower array is currently under construction on the earth's surface above the underground detector, consisting of 100 scintillators, each 3m^2, covering approximately 0.8 km^2; the first portion of the surface array will also be providing data this fall. Together, the two new Homestake detectors will be used to search for slow, massive magnetic monopoles; to study the zenith angle distribution of neutrino-induced muons; to search for neutrino bursts from the gravitational collapse of massive stars; to measure the multiplicity and transverse momentum distributions of cosmic ray muons; and to study the composition of the primary cosmic rays. Both instruments have been described in detail elsewhere[1]; we discuss here the capabilities of the underground device as a detector of slow magnetic monopoles.

Arguments based on the Parker limit[2] suggest that the mean flux of monopoles in the galaxy can be no more than about 10^{-15} cm^{-2} sec^{-1} sr^{-1}, and other arguments[3] suggest that even this value may be high. Experimentally[4], the limits set by induction experiments are now down below 4×10^{-12} cm^{-2} sec^{-1} sr^{-1}. The expected velocities of monopoles caught in the solar or galactic gravitational fields are in the range 10^{-4} -10^{-3} c. The Homestake Large Area Scintillation Detector has therefore been built with as large an area and as low a velocity threshold as possible: the detector has an aperture for isotropic monopoles of 1200 m^2 sr, large enough to detect one event in 3 years at a flux level of 9×10^{-16} cm^{-2} sec^{-1} sr^{-1}, and can detect signals as low as 0.1 times minimum ionizing (corresponding to velocities 1.5-6 × 10^{-4}c).

The calculation of expected scintillation yield from a slowly moving monopole is somewhat uncertain. Ahlen and Tarle[5] suggest that the yield from a monopole at $10^{-3}c$ is approximately 6 times minimum ionizing, and that excitation cuts off for velocities below $6 \times 10^{-4}c$. Earlier calculations[6] have been more optimistic. Given the uncertainty in the calculation, we have designed the Homestake scintillator to be sensitive to as low a signal as possible. With a scintillator thickness of 25 g cm^{-2}, a signal of 0.1 times minimum ionizing corresponds to a pulse height of 5 MeV, well above the level of typical energy deposits due to background radioactivity.

The Large Area Scintillation Detector consists of a hollow 8m × 8m × 16m box of 30 cm × 30 cm × 8m liquid scintillation detectors surrounding the existing ^{37}Cl solar neutrino tank. Each of the 200 scintillator elements is a PVC box, lined with teflon (for total internal reflection), containing a low-cost mineral oil-based liquid scintillator developed to have excellent light collection and transmission characteristics, a light attenuation length greater than 8m, long-term stability, a high flash point, and low toxicity. Each detector element is viewed by two 5-inch photomultiplier tubes in coincidence, one at each end. Fast muons passing through the midpoint of one of the modules produce an average of 350 photoelectrons at each photomultiplier. A particle ionizing even at 0.01 times minimum would thus produce 3-4 photoelectrons at each photomultiplier and still be visible. The low energy threshold is therefore set not by the scintillator light output, but rather by the background produced by the ambient radioactivity (primarily MeV gamma rays) from the rock walls. The individual detector elements have ±1.3ns time resolution, spatial resolution of ±15 cm, and a very low muon background flux (1100 m^{-2} yr^{-1}, a factor of 10^7 below the surface flux).

Cosmic ray muons and neutrino-induced muons will typically produce two pairs of coincident photomultiplier tube pulses, one pair as the muon enters the detector and one delayed pair as the muon leaves the detector. Each photomultiplier tube pulse will have a rise time of approximately 10 ns. The delay between the entering and exiting pulses will be about 25 ns. From the time difference between the pulses at each end of a given module, and the corresponding ratio of pulse heights, we can locate the position of a muon to ±15 cm. We can also recognize multiple muons passing through a given module by the increased pulse height and the mismatch between time differences and pulse height ratios. Since each module is an independent muon detector, we can recognize and count very high muon multiplicities associated with a given cosmic ray interaction. In addition, our depth (corresponding to a muon threshold at the surface of 2.7 TeV) and our transverse dimensions (8×16 m^2) allow us to contain multiple muon events with transverse momentum greater than 2 GeV/c. From the location of the entering and exiting points (± 15cm), the muon direction can be determined to ±3°.

For a monopole, we expect a pair of slow pulses with width 1 ns/β as the monopole enters the detector and, after a delay of 25 ns/β, a second pair of slow pulses as the monopole leaves (Fig. 2). For the velocities of interest, the delay time between the two entering and exiting pulses will be 25-250μs. Such long delays can only be correlated in a very low background environment such as that available in a deep mine. The most severe background will be due to two independent, traversing muons, each of which is detected in only one of the two counters through which it passes. The accidental coincidence rate for independent muons mimicking a slow particle of velocity β is $4 \times 10^{-5}/\beta$

yr^{-1}. At $\beta = 10^{-4}$, this corresponds to 0.4 accidental coincidences per year. This rate will be further reduced by requiring that there be no fast (25 ns) coincidence and that the four individual photomultiplier pulses be wide (1 ns/β). The monopole position in each box will be determined from the ratio of pulse heights at each end of the box; the individual pulse heights are then corrected for the position, and the monopole pulse height is determined.

The detector array provides an aperture of 1200 m^2 sr, corresponding to one event in 3 years at a flux level of 9×10^{-16} cm^{-2} sec^{-1} sr^{-1}. This sensitivity limit is compared to the limits obtained with other detectors in Fig. 3. The Homestake sensitivity is shown as a solid line for $\beta \geq 6 \times 10^{-4}$ and a dashed line down to $\beta = 1.5 \times 10^{-4}$, reflecting the uncertainty in the estimates of energy loss at low velocities.

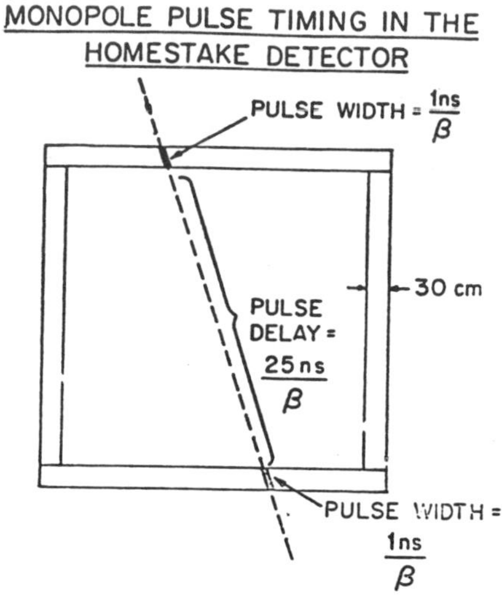

Fig. 2: Monopole pulse timing in the Homestake detector.

The electronics are designed to permit us to look at both fast muons and slow monopoles. An amplifier/discriminator is mounted outside the module at each photomultiplier tube. A fast discriminator on this unit gives the fast muon timing pulse. The photomultiplier pulse is then integrated (with an integration time currently set at 2μs) and fed to a flash ADC and a slow discriminator. A coincidence of either fast or slow discriminator pulses from two ends of a single box provides the system trigger. For each photomultiplier whose discriminator (either fast or slow) fires, the integrated pulse height is then sampled by the ADC. The system is currently set to sample the pulse height once every 300 ns for a total of 2μs; these time intervals can be increased if desired. A muon pulse then appears as a rapidly-rising (10 ns) pulse, whereas a monopole rises more slowly (1/β ns).

The range of interesting times is from 1 ns to 250μs. The system is therefore equipped with a fast clock (2.5 ns resolution) which covers the first 500 ns and a slow UTC clock (0.1 μs resolution) to cover the time span thereafter. A 16 word deep memory buffer is associated with each photomultiplier so that multiple pulses for each event can be recorded.

An individual high voltage supply is also located at each phototube. The supply is fed by 15V DC power, and provides an adjustable 0-2.5 kV at 2 mA. An external control line permits measurement and adjustment of the high voltage from the computer in the main electronics area.

The mechanical work on the underground detector is essentially finished. The southern half of the detector has been filled with liquid and turned on. We are currently testing and calibrating this portion of the detector while installing the north-side electronics and filling the remaining detector modules with oil. The detector is expected to be fully operational by the fall of this year.

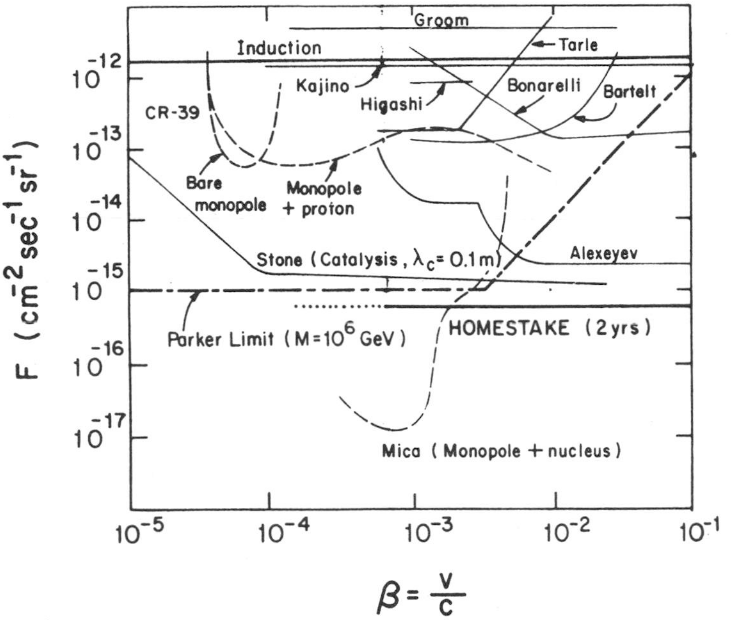

Fig. 3: Slow monopole flux limits. The Parker limit is that given for monopoles of $M=10^{16}$ GeV by Turner et al.[2] The induction limit is the global limit suggested by Stone.[4] The CR-39 limit is the Price result.[7] The mica limit comes from Price et al.[7] The Homestake line is the result of running for three years with the detector described here. Other limits are referenced in Ref. 7.

This work has been supported in part by U. S. Department of Energy contract DE-AC02-81-ER-40012. The assistance and generous cooperation of Mr. A. Gilles and the Homestake Mining Company are deeply appreciated, as are the advice and assistance of R. Davis, B. Cleveland, R. Reid, I. Davidson, and K. Brown.

REFERENCES

[1] K. Brown et al., Proc. MONOPOLE '83, Ann Arbor (1983); M. L. Cherry et al., Proc. Workshop on Science Underground, Los Alamos (1982).
[2] See M. S. Turner et al., Phys. Rev. D**26**, 1296 (1982) for a recent discussion.
[3] See J. Preskill, this meeting, and R. A. Carrigan and W. P. Trower, Nature **305**, 673 (1983) for reviews.
[4] J. Stone, this meeting.
[5] S. Ahlen and G. Tarlè, Phys. Rev. D**27**, 688 (1983).
[6] D. M. Ritson, SLAC Report No. SLAC-Pub-2950 (1982); cf. also R. A.

Carrigan and W. P. Trower, eds., Magnetic Monopoles, Plenum, N.Y. (1983), and ref. 3.

[7] P. B. Price, CERN preprint EP/84-28 (1984); D. E. Groom et al., Phys. Rev. Lett. **50**, 573 (1983); F. Kajino et al., 18th Intl. Cosmic Ray Conf., Bangalore, **5**, 56 (1983); S. Higashi et al., 18th ICRC, Bangalore, **5**, 69 (1983); J. Bartelt et al., Phys. Rev. Lett. **50**, 655 (1983); R. Bonarelli et al., Phys. Lett. **126B**, 137 (1983); P. B. Price et al., Phys. Rev. Lett. **52**, 1265 (1984); G. Tarlè et al., Phys. Rev. Lett. **52**, 90 (1984); E. N. Alexeyev, 18th ICRC, Bangalore, **5**, 52 (1983).

Progress Report on the Texas A & M Monopole Experiment

Michael J. Shepko

ABSTRACT

The Texas A & M University high energy physics research group has just completed construction of a large scintillation telescope at the United Salt Corporation mine in Hockley, Texas to search for superheavy magnetic monopoles.[1] Measurements include both the time of flight and specific ionization of particles passing through the detector. A GUT monopole event would be characterized by a rather long transit time between layers of the telescope coupled to the observation of fluorescence (either prompt or delayed) in the scintillation counters.

The Texas A & M monopole detector was designed to achieve a sensitivity of 10^{-15} $cm^{-2}sec^{-1}sr^{-1}$ for a monopole velocity in the range of $8 \times 10^{-4} < \beta < 10^{-1}$.

THEORY

Arguments based on survival of the interstellar magnetic field place an upper bound of 10^{-13} to 10^{-15} $cm^{-2}sec^{-1}sr^{-1}$ on the primordial monopole abundance.[2] Clearly, it is advantageous to build a detector of large areal coverage to reach this astrophysical flux limit within a reasonable period of time.

Detection of monopoles via the scintillation mechanism is primarily limited by a lower velocity threshold below which a monopole would not be observed. Within the past few years progress has been made on theoretical predictions of scintillation response due to the passage of a slowly moving magnetic monopole. Modeling the valence electrons of the scintillator as a fermi gas with a band gap of 5 ev, Ahlen and Tarlé suggest a lower detectable velocity limit of $\beta = 6 \times 10^{-4}$ using conventional plastic scintillation counters.[3] Above this velocity cutoff, magnetic monopoles would be detected through the emission of scintillation light with relative light yield greater than that of ordinary cosmic ray muons.

DETECTOR DESIGN

The scintillation telescope consists of a series of 64 (6' × 6' × 3/8") acrylic based napthalene scintillation counters arranged in three horizontal layers with 16 (4×4) counters to a given layer. Interplane separation is approximately three feet. A final set of 16 vertical counters completely enclose the sides yielding an effective surface area of 576 ft^2 with an angular acceptance of 9.4 sr. For triggering purposes, each horizontal layer and vertical side comprise a trigger face. In total, seven faces constitute the trigger logic.

Scintillation light from a panel is collected along each counter edge using BBQ doped (90mg/l) wave shifter bars. The resultant light is channeled by total internal reflection to either one of a pair of fast photomultiplier tubes placed an opposite corners of the plastic panel.

Light output from each pair of photomultiplier tubes is analog summed and the time structure of the light pulse is clocked into a 455 element CCD array at a 20 megahertz rate for storage and eventual waveform digitizing. This provides an observational window of 22 microseconds with time resolution of 50 nanoseconds.

To produce a trigger, signals from each phototube are also routed into an integration circuit. At this point, pulses are discriminated upon at a threshold of one photoelectron. A face trigger occurs when the discriminator outputs from a pair of phototubes on a panel are coincident within a resolving time of one microsecond (expected lifetime of the triplet excitation). The resulting trigger is a coincidence between any three faces within a time window of 25 microseconds. Events satisfying these requirements will initiate CCD readout of all detectors containing pulse height information.

CALIBRATION OF THE DETECTOR

Several methods of calibrating the experiment are carried out during on-line data acquisition.

One method employs the observation of penetrating cosmic ray muons. Monitoring muons through a separate muon trigger determines detector efficiency due to the passage of minimum ionizing particles. A second source of calibration consists of a centrally located LED on each of the 64 scintillator panels. Counter sensitivity to the deposition of a low light level is determined by firing the LED's near the one photoelectron level.

Lastly, an externally produced test pulse, similar in amplitude to that of a photomultiplier tube signal, is input directly into the data acquisition system, testing electronic hardware independent of the light collection scheme.

BACKGROUND RATE

Above the one photoelectron level, photomultiplier tube noise rates average 3 kHz. Corresponding to a face-face trigger once every ten seconds.

Observationally, the average panel coincidence rates are 91Hz, 115Hz and 149Hz for the three horizontal planes top, middle and bottom respectively. This rate is over an order of magnitude larger than that expected from purely cosmic ray muon backround events.

Using a sodium iodide crystal, the backround energy spectrum revealed the presence of two strong peaks at energies of 1.4 MeV and 2.6 MeV due to gamma radiation from nuclear decays of ^{40}K and ^{232}Th. The primary source of radiation was found to originate from the supporting concrete floor.[4] Covering the floor with ten inch thick salt blocks attenuated the backround radiation by approximately a factor of four independent of counter position above the floor.

PRESENT EXPERIMENTAL STATUS

Presently, the detector has been on-line for a total of 575 hours.

Taking into account counter efficiency, the physical three fold muon rate averages 1.6Hz in close agreement with the expected flux.[5]

Raw three fold monopole trigger rates occur at a frequency of 0.1Hz. Approximately one three fold event per hour produces a combined pulse height spectrum at the one photoelectron level (summing both phototubes). During off-line analysis we require the analog sum of both phototubes on a given panel to be greater than twice minimum ionizing

(4 pe). Primarily, the dominant form of backround rate stems from Compton scattered photons observed below a two photoelectron amplitude. Consequently, during the off-line analysis, we do not expect a contribution to the three fold monopole rate from this source of radiation. A more detailed discussion of backround events simulating monopole passage can be found in a previously referenced paper.

To date we have observed no monopole events above the minimum ionization threshold.

REFERENCES

[1]P.M. McIntyre and R.C. Webb, "Searching for the GUT Monopole," Texas A & M preprint DOE-ER40039-4 (1982).

[2]E.N. Parker, "Cosmical Magnetic Fields," (Claredon, Oxford, 1979).

[3]S.P. Ahlen and G. Tarlé, Phys. Rev. **D27**, 688 (1983), (Rapid Communications).

[4]S.L. White, "A Determination of the Backround Gamma Radiation in the Hockley Salt Mine," Texas A & M preprint DOE-ER40039-8 (1984).

[5]R.K. Adair and H. Kasha, Cosmic Ray Muons, in "Muon Physics," ed. C.S. Wu and V.W. Hughes, Academic Press (1977).

Properties of Exotic Monopoles or Fermions Never Lie

J. A. Harvey

ABSTRACT

I discuss some of the properties of magnetic monopoles other than the minimal SU(5) monopole. The topics covered include stability of higher charge monopoles, the weak structure of monopoles, and the connection between fermion boundary conditions and dyon quantum numbers.

I. INTRODUCTION

In this talk I will discuss some of the properties of "non-minimal" monopoles which occur in grand unified theories other than SU(5) and also in SU(5) for a certain range of parameters in the Higgs potential. These monopoles are of experimental interest because in some scenarios they could be more abundant than minimal charge monopoles and also because in some models they do not catalyze nucleon decay at a strong interaction rate. They are of theoretical interest because they shed some light on the generality of the Callan/Rubakov effect and because their semi-classical quantization is a bit more involved than for minimal charge monopoles. I will start by discussing various theories in which stable multiple-charge monopoles rise. I will then discuss the weak structure of both minimal and non-minimal monopoles. I will conclude by discussing the relationship between fermion boundary conditions and dyon quantum numbers with particular emphasis on the spin and color charges.

II. MULTIPLE-CHARGE MONOPOLES

There are two ways that I know of obtaining stable monopoles with magnetic charge greater than the minimal Dirac charge $g_D = 1/(2e)$. The first involves models with two stages of symmetry breaking. A well known example[1] is a SO(10) theory with

$$SO(10) \underset{M_1}{\rightarrow} SO(6) \times SO(4) \underset{M_2}{\rightarrow} SU(3) \times SU(2) \times U(1) \qquad (1)$$

where the first stage of symmetry breaking occurs at a mass scale M_1 and the second stage at a mass scale M_2. The first stage of symmetry breaking produces Z_2 monopoles (monopoles whose magnetic charge is only conserved mod 2) with mass $\approx M_1/\alpha$. In the second stage of symmetry breaking these monopoles are converted into ordinary unit Dirac charge monopoles and in addition monopoles with twice the Dirac charge and mass $\approx M_2/\alpha$ are produced. Since $M_2 < M_1$ these monopoles are clearly stable. If something dramatic happens between M_1 and M_2 (i.e. inflation or a superconducting phase) it is quite possible that the abundance of double charge monopoles could be much larger than the abundance of single charge monopoles. Z_n monopoles and their cosmological implications have been studied in detail by Weinberg, London, and Rosner[2].

The second mechanism for obtaining multiple charge monopoles applies to a much larger class of theories but depends on the values of parameters in the Higgs potential[3]. The basic idea is very simple. If one imagines bringing two single charge monopoles together they will orient themselves in group space so as to minimize their interaction energy. In this orientation they are repelled due to Abelian gauge boson exchange and attracted due to non-Abelian gauge boson exchange while for Higgs boson exchange they are attracted in the Abelian case and repelled in the non-Abelian case. By adjusting the masses of the Higgs bosons one can make configurations which are attractive at separations less than the inverse mass of any of the Higgs bosons but greater than the size of the monopole core. One can show that the energies of the multiple-charge bound states that are constructed this way coincide with the first corrections to the mass away from the Prasad-Sommerfield limit for spherically symmetric multiple charge monopoles[3]. In this way it is possible to construct stable monopoles in SU(5) with magnetic charges of two, three, four, and six times the Dirac charge. I would now like to explore some of the ways that higher-charge monopoles differ from Dirac charge monopoles.

III. WEAK STRUCTURE

One can see that the weak structure of higher-charge monopoles must be non-trivial by the following argument. For concreteness I will consider the double charge monopole. Imagine probing the monopole at distances $r \ll M_W^{-1}$ or alternatively consider high temperatures in the early universe when SU(3) × SU(2) × U(1) was still unbroken. Then the Dirac quantization condition plus the stability conditions of Brandt and Neri[4] and of Coleman[5] determine the long range magnetic field to be

$$\vec{B} = \frac{\hat{r}}{gr^2} Q_2 \qquad (2)$$

where Q_2 is a certain matrix in group space. In SU(5) with the weak hypercharge given by
$Y = \mathrm{diag}\,(1/3,1/3,1/3,-1/2,-1/2)$, the color hypercharge by
$Y^c = \mathrm{diag}\,(1/3,1/3,-2/3,0,0)$, the weak isospin by
$T_3 = \mathrm{diag}\,(0,0,0,1/2,-1/2)$ and the generator of electromagnetism by
$Q_{em} = Y - T_3$ we have

$$Q_2 = Y + 1/2\, Y^c = Q_{em} + 1/2\, Y^c + T_3 = 1/2\, \mathrm{diag}\,(1,1,0,-1,-1) \qquad (3)$$

Clearly this cannot describe the magnetic field of the monopole when SU(2) × U(1) is broken since the gauge bosons corresponding to T_3 are then massive. One might be tempted to conclude in analogy with the confinement of monopoles in a superconductor that these double monopoles would be confined after the SU(2) × U(1) transition. However if we pursue this analogy a little further we can see that this is not correct[6]. In analogy with a superconductor the massive components of $\vec{B}$ will be screened by surface currents. Since the analog of the Cooper pair is the Higgs field ϕ^0 these currents will screen the Z^0 field but leave the purely electromagnetic part alone. Therefore for $r \gg M_W$ the magnetic field should look like

$$\vec{B} = \frac{\hat{r}}{gr^2} (Q_{em} + 1/2\, Y^c) \qquad (4)$$

On a scale of $\approx M_W$ there must be a rearrangement of the weak magnetic field. Although one could try to determine the radial behavior of different components of the magnetic field by minimizing the energy it is only necessary to know the long range and short range behavior of $\vec{B}$ to draw some general conclusions. First, it is clear that these monopoles are not confined by flux tubes since the formation of flux tubes (with energy/length $\approx M_W^2$) costs more energy than the rearrangement of the $\vec{B}$ field which only affects the energy by an amount of order M_W. Another consequence of this rearrangement of the weak fields is that it destroys the rotational invariance of the monopole. As a result there will be collective coordinates to describe the orientation of the monopole and these monopoles should have a rotational spectrum with the energy spacing of order M_W. I should mention that for single charge monopoles there is no such problem since their stable long range magnetic fields lie completely in the color and electromagnetic directions and are thus unaffected by the weak symmetry breaking.

Superconducting systems in which the U(1) factor is "realigned" rather than completely broken as occurs in the Weinberg-Salam transition might occur in other contexts. For example if quark stars exist it is reasonable to assume that the dense quark matter would be superconducting[7]. Quarks which are color triplets and have say charge -1/3 should form Cooper pairs in the most attractive channel i.e. color antitriplets with charge -2/3. A condensate of these Cooper pairs breaks U(1) of electromagnetism and SU(3) of color down to SU(2) but leaves a linear combination of electromagnetism and a diagonal SU(3) generator unbroken. As a result there can exist long range magnetic fields which are linear combinations of ordinary magnetic fields and chromomagnetic fields.

While I am on the subject of the weak structure of monopoles I would like to make a few comments that also pertain to single charge monopoles. It is sometimes assumed that fermion masses and weak symmetry breaking can be ignored when discussing monopole catalysis of nucleon decay since this involves the short distance effective core boundary conditions where the long range effects of weak breaking are presumably negligible. However this need not be the case, even at the classical level. Consider the expectation value of the weak Higgs doublet ϕ around the minimal SU(5) monopole. In vacuum the expectation value of ϕ is fine tuned to be small by adjusting parameters in the Higgs potential that involve the coupling of ϕ to the adjoint Higgs field which breaks SU(5). However around a monopole the adjoint Higgs field has radial dependence with some of its components going to zero at the monopole core. It is not hard to see that this will disturb the cancellation mechanism leading to either a very large positive or very large negative effective mass term for the ϕ field around the monopole. In the latter case the ϕ field expectation value near the monopole will have the form $\langle\phi\rangle = c/r$ for some constant c. This is just due to the fact that the core acts like a source for the ϕ field while outside the core the field is effectively massless (for distances $\ll 1/M_W$). This behavior can also be induced by the coupling of ϕ to fermions.[8] It is therefore likely that the effective fermion mass behaves like ε/r around the monopole where ε is the relevant Yukawa coupling. This repulsive potential is still weak compared to the attractive monopole potential and has little effect on say the behavior of the s-wave fermion wave function near the origin. It does make the neglect of fermion masses problematical, particularly for the heavier generations.

Dyon Quantum Numbers

The semiclassical quantization of single charge monopoles involves subtleties which have been only recently appreciated.[9] In particular the dyonic excitations of monopoles that carry color magnetic fields do not form definite representations of SU(3) as would be naively expected but can only be labeled by their quantum numbers under the subgroup of SU(3) which commutes with the long range magnetic field. The dyons associated with multiple charge monopoles have some additional interesting properties which were first suggested by an analysis of fermion scattering.[10]

As an example consider the double charge monopole which is produced in the second stage of symmetry breaking of the SO(10) model mentioned earlier:

$$SO(6)\times SO(4)\approx SU(4)\times SU(2)_L\times SU(2)_R \xrightarrow{M_2} SU(3)\times SU(2)\times U(1) \qquad (5)$$

under $SU(4)\times SU(2)_L\times SU(2)_R$ the fermions transform as (4,2,1) + (4,1,2). Specifically the fermions are

$$\begin{pmatrix} \underset{\sim}{u}_L & \nu_L \\ \underset{\sim}{d}_L & e_L \end{pmatrix} ; \begin{pmatrix} \underset{\sim}{u}_R & \nu_R \\ \underset{\sim}{d}_R & e_R \end{pmatrix} \qquad (6)$$

At energies large enough to excite the dyons the fermion boundary conditions allow the following transitions:

$$\begin{pmatrix} u_{3L} \\ \nu_L \end{pmatrix} \begin{pmatrix} d_{3L} \\ e_L \end{pmatrix} \begin{pmatrix} u_{1R} \\ d_{1R} \end{pmatrix} \begin{pmatrix} u_{2R} \\ d_{2R} \end{pmatrix} \begin{pmatrix} u_{3R} \\ d_{3R}+\nu_R \\ e_R \end{pmatrix} \qquad (7)$$

ΔY	2/3	2/3	1	1	5/3
ΔY^c	-2/3	-2/3	0	0	-2/3

where the change in the value of the weak hypercharge and color hypercharge is as shown. Since these quantities are conserved the dyons must be able to carry (Y,Y^c) charges of (2/3,-2/3), (1,0), (5/3,-2/3), etc. Note that the massive SU(4) X-bosons have charge (2/3,-2/3) and the massive $SU(2)_R$ W^+ bosons have charge (1,0) so that as for the minimal SU(5) monopole the dyon states can be thought of as bound states of monopoles and massive vector bosons except that here there are two independent sets of massive vector bosons. From the collective coordinate point of view this is only possible if there are two U(1) symmetries which generate independent rotations of the classical monopole solution.

To see that this is the case we construct a spherically symmetric Ansatz for the double charge monopole. The Ansatz is symmetric under $\vec{L} + \vec{T}$ where $\vec{T} = \vec{T}_2 + \vec{T}_4$ and

$$\vec{T}_2 = \vec{\tau}/2 \;\; ; \;\; \vec{T}_4 = \begin{pmatrix} 0 & 0 \\ 0 & \vec{\sigma}/2 \end{pmatrix} \qquad (8)$$

are generators of $SU(2)_R$ and a SU(2) subgroup of SU(4) respectively. The long range part of the magnetic field is determined by the matrix

$$Q_2 = \vec{z} \cdot (\vec{T}_2 + \vec{T}_4) = Y + Y^c/2 \tag{9}$$

The Higgs field Φ which transforms as (4,1,2) under $SU(4) \times SU(2)_L \times SU(2)_R$ is given by

$$\Phi = \begin{pmatrix} 0 & 0 \\ 0 & 0 \\ F(r) & + H(r)\vec{r}\cdot\vec{t} \end{pmatrix} \tag{10}$$

where $\vec{t}$ are certain two by two matrices. The $SU(2)_R$ and SU(4) gauge fields are

$$W_i = (\vec{T}_2 \times \vec{r})_i \frac{K_2(r)-1}{gr}$$
$$A_i = (\vec{T}_4 \times \vec{r})_i \frac{K_4|r|-1}{gr} \tag{11}$$

respectively.

There are potentially 12 collective coordinates corresponding to the 12 generators of $SU(3) \times SU(2) \times U(1)$. However $SU(2)_L \times SU(2)_I$ where $SU(2)_I$ is the isospin subgroup of $SU(3)_c$ acting on the upper two components of Φ is an exact symmetry of the Ansatz. Furthermore[9] we know that the $SU(3)_c$ generators λ^4, λ^5, λ^6, λ^7 that do not commute with the long range part of the gauge field are not legitimate collective coordinates. We are thus left with 12 - 10 = 2 collective coordinates corresponding to Y and Y^c transformations. Since no linear combination of Y and Y^c leaves the Ansatz invariant both of these are independent collective coordinates. The dyon quantum numbers which follow from semiclassical quantization agree with those deduced from fermion scattering - hence fermions never lie.

In some cases fermions tell us something more interesting - they tell us that certain dyons must carry angular momentum even though the monopoles from which they are built are spherically symmetric! As an example consider the triple charge monopole in $SU(5) \to SU(3) \times SU(2) \times U(1)$ with magnetic field

$$\vec{B} = \frac{\hat{r}}{gr^2} Q_3 \quad ; \; Q_3 = \frac{3}{2} Y - \frac{1}{2} T_3 \tag{12}$$

Wilkinson and Goldhaber[11] showed that there is an Ansatz spherically symmetric under $\vec{L} + \vec{T}$ whenever the monopole charge matrix Q has the form

$$Q = I_3 - T_3 \tag{13}$$

where $\vec{I}$ generates another SU(2) subgroup and commutes with Q and the long range Higgs field. For the triple monopole we have

$$I_3 = 1/2 \text{ diag } (1, -1, 0, 0, 0)$$
$$T_3 = 1/2 \text{ diag } (2, 0, 1, -2, -1) \tag{14}$$

Under the SU(2) subgroup generated by $\vec{T}$ the 5 of fermions transform as a doublet $(\bar{d}_3, \nu)$ and a triplet $\bar{d}_1, \bar{d}_2, e^-)$. The fermion boundary conditions allow the charge exchange process $\bar{d}_1 \to e^-$ which corresponds to $\Delta Q_3 = 1/2$. Since Q_3 measures the field angular momentum this process must leave a dyon which carries not only color and electric charge but also angular momentum 1/2! Since the full SU(3) color group commutes with the long range magnetic field these dyons will transform as definite representations of SU(3). The semiclassical quantizations of dyons when the monopole Ansatz has $\vec{T} \neq 0$ is quite interesting and has been carried out in detail by Dixon[12]. He has shown that spherically symmetric monopoles which obey an Ansatz with the SU(2) subgroup generated by $\vec{T}$ embedded nontrivially in the gauge group have dyonic excitations which carry angular momentum determined by the representation of $\vec{T}$. To my knowledge the coupling of fermions to these non-Abelian rotators has not been carried out. Monopole catalysis in these theories might have some additional interesting features.

CONCLUSIONS

I have tried to convince you that higher charge monopoles can exist in a wide class of theories and that the properties of these monopoles are quite interesting. They provide an interesting theoretical laboratory to test current ideas regarding semiclassical quantization of monopoles and monopole catalysis of nucleon decay. The connection between dyon quantum numbers as determined by fermion boundary conditions and as determined by semiclassical quantization is clearly required for physical consistency but exactly how this connection works is not clear to me and deserves further study.

REFERENCES

[1] G. Lazarides and Q. Shafi, Phys. Lett. **94B**, 149 (1980); G. Lazarides, M. Magg, and Q. Shafi, Phys. Lett. **97B**, 87 (1980).
[2] E. J. Weinberg, D. London, and J. Rosner, Nucl. Phys. **B236**, 90 (1984).
[3] C. L. Gardner and J. Harvey, Phys. Rev. Lett. **52**, 879 (1984).
[4] R. A. Brandt and F. Neri, Nucl. Phys. **B161**, 253 (1979).
[5] S. Coleman, unpublished.
[6] F. Bais, Phys. Lett. **98B**, 437 (1981).
[7] B. Barrois, Ph.D Thesis, unpublished (1979); D. Bailin and A. Love, J. Phys. **A12**, **L283** (1979); Nucl. Phys. **B190**, 175 (1981); Nucl. Phys. **B190**, 751 (1981).
[8] E. Witten, unpublished.
[9] A. Abouelsaood, Nucl. Phys. **B226**, 309 (1983).
[10] C. G. Callan Jr., Princeton preprint (1984).
[11] D. Wilkinson and A. S. Goldhaber, Phys. Rev. D **16**, 1221 (1977).
[12] L. Dixon, Nucl. Phys. **B248**, 90 (1984).

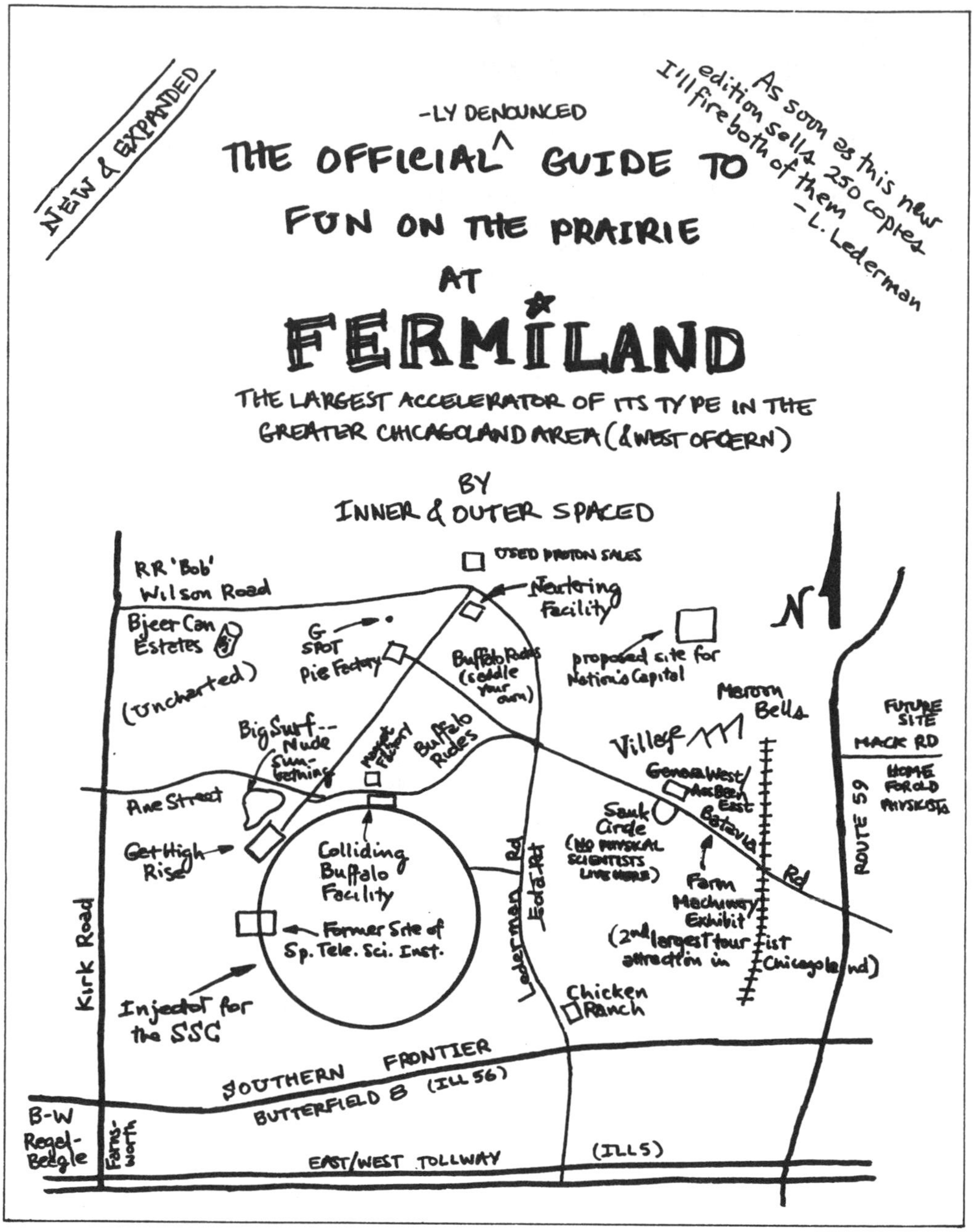
NEW & EXPANDED
-LY DENOUNCED
THE OFFICIAL GUIDE TO
FUN ON THE PRAIRIE
AT
FERMILAND
THE LARGEST ACCELERATOR OF ITS TYPE IN THE
GREATER CHICAGOLAND AREA (& WEST OF CERN)
BY
INNER & OUTER SPACED
As soon as this new
edition sells 250 copies
I'll fire both of them
-L. Lederman
USED PROTON SALES
RR 'Bob'
Wilson Road
Neutering
Facility
Bjeer Can
Estates
G
SPOT
Pie Factory
Buffalo Rides
(saddle
your
own)
proposed site for
Nation's Capitol
N
(uncharted)
Meson
Bells
Big Surf--
Nude
Sun-
bathing
Buffalo
Rides
Village
FUTURE
SITE
MACK RD
Pine Street
Geneva West/
East
Batavia
Rd
HOME
FOR OLD
PHYSICISTS
Sauk
Circle
(NO PHYSICAL
SCIENTISTS
LIVE HERE)
Get High
Rise
Colliding
Buffalo
Facility
Farm
Machinery
Exhibit
(2nd largest tourist
attraction in
Chicagoland)
ROUTE 59
Lederman Rd
Eola Rd
Former Site of
Sp. Tele. Sci. Inst.
Kirk Road
Injector for
the SSC
Chicken
Ranch
SOUTHERN FRONTIER
BUTTERFIELD 8 (ILL 56)
B-W
Regel-
Beagle
Farns-
worth
EAST/WEST TOLLWAY
(ILL 5)

PART VI

SUPERSYMMETRY, SUPERGRAVITY, AND QUANTUM GRAVITY

The standard model of particle physics described in Chapter I by Langacker provides a remarkably successful picture of physics up to energies of 100 GeV. However, there are reasons to suspect that the theory is incomplete at energies well above 100 GeV. In particular, these theories suffer from a problem of naturalness; they contain at least two vastly different energy scales -- the weak scale and the Planck scale. The introduction of a supersymmetry relating bosons and fermions, in addition to its intrinsic beauty, offers a possible explanation for this naturalness problem. As discussed in Chapter IV on inflation, supersymmetric theories tend to have flat potentials which are what is needed to implement inflation. Supersymmetric theories have 'R-parity', which, if unbroken, requires the lightest R-odd particle to be stable. This stable superparticle could provide the dark matter in the Universe.

A review of supersymmetry and supergravity is presented by Polchinski, with particular attention to possible masses and interaction strengths of stable superpartners. The possibility that the lightest superpartner survives annihilation in the early Universe in sufficient numbers to make an important contribution to the overall mass density of the present Universe is discussed by Hagelin. The role of superpartners in dark matter and galaxy formation was discussed in Chapter III. Finite temperature effects in supersymmetry may have played an important part in the evolution of the early Universe and the breaking of supersymmetry. Some of the possible effects are reviewed by Goldberg and by Kullab.

It has long been realized that quantum gravity should play an important role in the evolution of the Universe at the Planck time and before. Hartle presents the results of recent work on the construction of a quantum mechanical wave function of the Universe. This approach is a possible way to understand the initial conditions of the big bang. The chapter concludes with papers on other aspects of Quantum Cosmology.

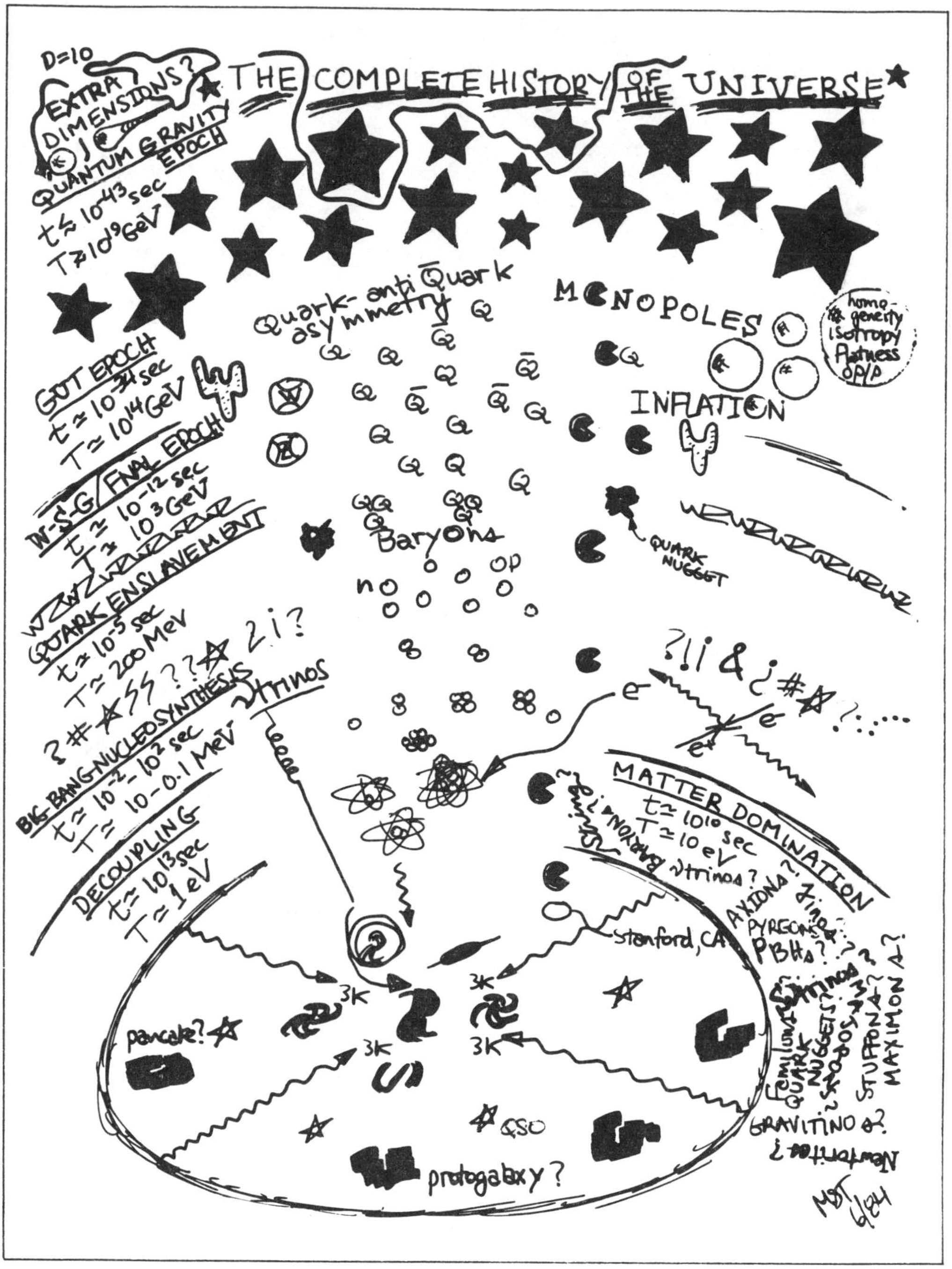
THE COMPLETE HISTORY OF THE UNIVERSE
D=10
EXTRA DIMENSIONS?
QUANTUM GRAVITY EPOCH
t ≲ 10^-43 sec
T ≳ 10^19 GeV
Quark-anti Quark asymmetry
MONOPOLES
homogeneity
isotropy
flatness
GUT EPOCH
t ≈ 10^-34 sec
T ≈ 10^14 GeV
INFLATION
W-S-G/FINAL EPOCH
t ≈ 10^-12 sec
T ≈ 10^3 GeV
Baryons
QUARK NUGGET
QUARK ENSLAVEMENT
t ≈ 10^-5 sec
T ≈ 200 MeV
BIG BANG NUCLEOSYNTHESIS
t ≈ 10^-2-10^2 sec
T ≈ 10-0.1 MeV
νtrinos
MATTER DOMINATION
t ≈ 10^10 sec
T ≈ 10 eV
DECOUPLING
t ≈ 10^13 sec
T ≈ 1 eV
Stanford, CA
AXIONS?
PYRGONS?
PBHs?
3K
pancake?
QSO
protogalaxy?
GRAVITINOS?
MAXIMON?
MST 6/84

Supersymmetry and Supergravity

Joseph Polchinski

ABSTRACT

The role of supersymmetry and supergravity in particle physics is reviewed. Particular attention is given to the properties of stable superpartners in the various models.

I. THE HIERARCHY PROBLEM: MOTIVATION FOR SUPERSYMMETRY

In 1967, the Weinberg-Salam model of the weak interactions was discovered.[1] There were earlier weak interaction theories which correctly described the physics at the energies then accessible to experiment. However, these theories had a problem: at somewhat higher energies, a few hundred GeV (the "weak interaction scale" or Λ_W) they no longer made sense. A new theory was needed to describe the physics at and above the weak interaction scale. The Weinberg-Salam model does this, remaining consistent up to much higher energies.

Many of the predictions of the Weinberg-Salam model have since been beautifully confirmed. At the same time, though, the suspicion has grown that, while the Weinberg-Salam model is consistent well above the weak interaction scale, it becomes incomplete at or just above that energy. It suffers from a more subtle problem, a naturalness problem. The problem arises if we suppose that at some energy far above Λ_W, call it Λ_F, we have a more fundamental theory, and that the Weinberg-Salam theory is valid from Λ_W up to Λ_F. In this more fundamental theory, could we understand why $\Lambda_W << \Lambda_F$? The answer is no. The weak interaction scale is fixed by the mass of the Higgs scalar, $\Lambda_W \sim M_H$. Theories with a very small ratio M_H/Λ_F occupy a tiny and undistinguished region in the parameter space of the more fundamental theory. The word "undistinguished" is important here: if the theory with $M_H = 0$ were distinguished by having a larger symmetry, for example, we could understand the small value of M_H/Λ_F as the consequence of an approximate symmetry. In theories in which the standard model (that is, Weinberg-Salam plus the strong interaction, color) is all there is at the weak interaction scale, this is not the case, and the region in which M_H/Λ_F is small really is tiny and undistinguished. Either we have to believe that physics lies in this tiny region by incredible luck, or we have to look for a more natural theory.

The specific technical problem is this: the mass of the Higgs boson is a sum of many contributions, including, for example, the Feynman graph of Fig. 1a, in which the Higgs emits and reabsorbs a gauge boson. This graph is quadratically divergent at high energy and gives a contribution to M_H^2 proportional to the scale, Λ_F, squared. In order that M_H^2, and thus Λ_W^2, be much smaller than this we must somehow remove or cancel this contribution.

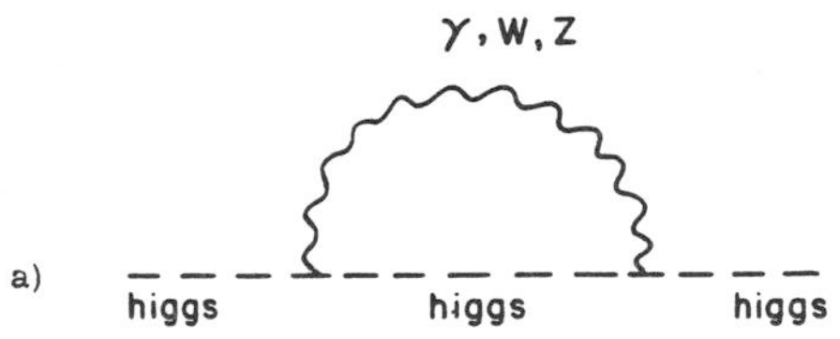

Figure 1a: A graph giving a large contribution to the Higgs mass squared.

Theories in which this is done by fine tuning of the parameters of the theory are unnatural in the sense described above; we would like to have some way in which the contribution of Fig. 1a is cancelled automatically. A great deal of effort has gone into this problem, because it seems to be one of the few weak points of the standard model. One possible solution is supersymmetry.[2] In a supersymmetric theory, every boson has a fermonic partner and vice versa. In addition to the graph of Fig. 1a, we now also have Figs. 1b and 1c. Supersymmetry relates the various couplings in these graphs in such a way that the dangerous parts cancel:

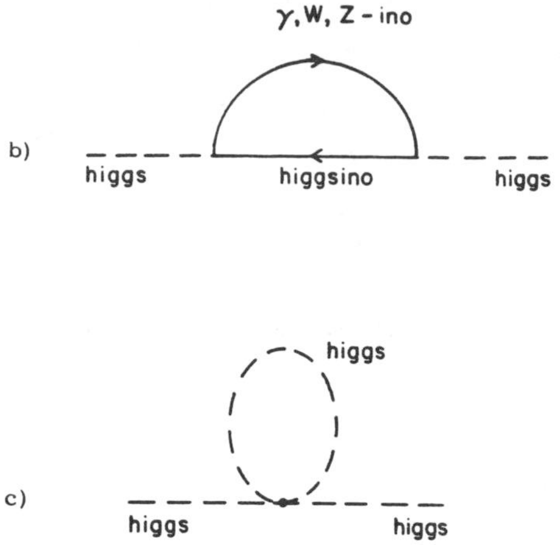

Figure 1b,c: Additional graphs in a supersymmetric theory.

$$\text{Fig. 1a + Fig. 1b + Fig. 1c} \approx \Lambda_W^2 \quad , \qquad (1)$$

even though the individual terms in (1) are of the much larger order Λ_F^2. The fact that such cancellations occur in supersymmetric theories makes very attractive the idea that the correct theory at the weak interaction scale is a supersymmetric version of the standard model, and this might then be complete up almost to the Planck scale, 10^{19} GeV.

II. THE SUPERSYMMETRIC STANDARD MODEL

A supersymmetric version of the standard model must have at least the following particles:

TABLE I. The Supersymmetric Standard Model

A. The Standard Model		B. The Superpartners	
Quarks	spin 1/2	Quarks	spin 0
Leptons		Sleptons	
Photon	spin 1	Photino	spin 1/2
W,Z		W,Z-ino	
Gluon		Gluino	
(two)Higgs	spin 0	Higgsinos	

The first column lists the particles of the standard model (plus an extra Higgs), while the second lists the respective superpartners. It is natural to ask whether we have double-counted here. That is, it would be nice if some of the known particles in column A were superpartners of each other. This was tried at first and does not work; the superpartners are all new, unseen, particles. Supersymmetry at least doubles the particle content of the standard model.

Supersymmetry requires superpartners to have the same gauge couplings (color, charge, and weak isospin) and mass. Because there is no charged boson degenerate with the electron, supersymmetry must be a broken symmetry. The particles listed in column B must be given masses of the order of 1 to 10^3 GeV. The charged superpartners, for example, must weigh at least 20 GeV, or they would have been seen at PETRA. The upper bound of roughly 10^3 GeV on the masses comes about because the cancellation in Fig. 1 is no longer exact once supersymmetry is broken. In order that Eq. (1) not leave a remainder larger than Λ_W^2, the superpartners themselves must not be much heavier than Λ_W. This is the importance of the naturalness argument: it tells us that supersymmetry is not merely a nice mathematical structure which might appear at the Planck scale, but that evidence of supersymmetry (or some other new physics) must appear at energies not far above those now being explored.

Here, then, is a recipe for a realistic supersymmetric model: take the supersymmetric standard model and add extra masses for the unwanted column B particles. Supersymmetry is now explicitly broken--it is only an approximate symmetry. That is,

$$[Q^\alpha, H] \neq 0 \; ; \qquad (2)$$

the supersymmetry generator does not commute with the Hamiltonian. As long as the breaking of supersymmetry is soft, which means, roughly speaking, that all supersymmetry breaking effects are proportional to the scale Λ_W, the desired cancellation (1) still occurs and the smallness of M_H/Λ_F is still natural.[3] However, breaking the symmetry by hand like this is ugly and arbitrary. Further, since the supersymmetry is no longer exact, the theory can no longer be coupled to gravity in a consistent and supersymmetric way.

It is better if the supersymmetry is broken spontaneously:

$$[Q^\alpha, H] = 0 \quad \text{but} \quad Q^\alpha|\text{vacuum}\rangle \neq 0. \qquad (3)$$

That is, supersymmetry is still an exact symmetry of the interactions, but the vacuum is not supersymmetric. Several types of system with spontaneous supersymmetry breaking are known. Their common feature is a massless fermion, the goldstino:

$$Q^\alpha|\text{vacuum}\rangle = F_G|1 \text{ goldstino}\rangle \; . \qquad (4)$$

F_G, with units of $(\text{mass})^2$, is known as the goldstino decay constant and measures how badly supersymmetry is broken. When supersymmetry is gauged, that is, coupled to supergravity, the massless spin-3/2 gravitino and the massless spin-1/2 goldstino combine to form the massive spin-3/2 gravitino. The properties of the gravitino are largely determined by F_G. Its mass, denoted $m_{3/2}$, is

$$m_{3/2} = \left(\frac{4\pi}{3} \right)^{1/2} \frac{F_G}{M_P} , \tag{5}$$

(M_P = Planck scale, 1.2×10^{19} GeV) and its couplings are proportional to $1/F_G$.

Before discussing supersymmetric models in more detail, we need to discuss a certain combination of symmetries which has important cosmological consequences in supersymmetric theories. This is[4]:

$$R = (-1)^{3B} (-1)^{L} (-1)^{F} . \tag{6}$$

Here, B is baryon number, which is known to be very well conserved. L is lepton number. It is likely that L is violated by small neutrino masses, but these change L by 2 units and leave $(-1)^L$ unchanged. Effects which change L by an odd number of units are as small as those which violate B in most unified models. $(-1)^F$ is fermion number mod 2, and it is exactly conserved as a consequence of rotation invariance. Thus, R is a product of 3 symmetries, each of which is exact or nearly so, and so it is almost exactly conserved. R is a multiplicative quantum number: the product of the R's of the particles in the initial state equals the product in the final state. A particle with R = -1 cannot decay entirely into R = 1 particles, and so the lightest R = -1 particle is quite stable, probably as long-lived as the proton.

What is R? For quarks it is $(-1)^1 (-1)^0 (-1)^1 = 1$; for leptons, $(-1)^0 (-1)^1 (-1)^1 = 1$; for Higgs and gauge bosons and the graviton it is $(-1)^0 (-1)^0 (-1)^0 = 1$: it is 1 for all the particles of the standard model. All the superpartners, including the gravitino, have the same B and L as their partners but opposite F and so they have R = -1. Thus, in supersymmetric models there will be a new stable particle, the lightest superpartner (LSP).

III. SCHEME NO. 1: LOW-ENERGY SUPERSYMMETRY (LESS)

A realistic model with spontaneously broken supersymmetry can be obtained in the following way.[5,6] Take the supersymmetric standard model and add to it a system which has spontaneously broken supersymmetry. Let some of the fields of the supersymmetry breaking sector carry SU(3) × SU(2) × U(1) gauge quantum numbers. The gauge couplings now carry the supersymmetry breaking to all the fields of the model. For example, Fig. 2a gives the gauginos (photino, gluino, Wino,

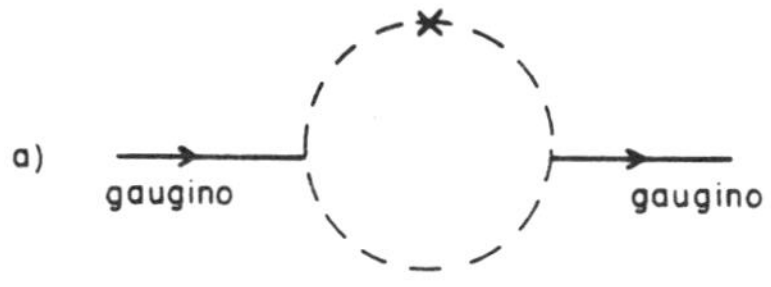

Figure 2a: Graph giving mass to the gauginos. The dashed line is a field coupling both to the supersymmetry breaking field (X) and to the gauge fields.

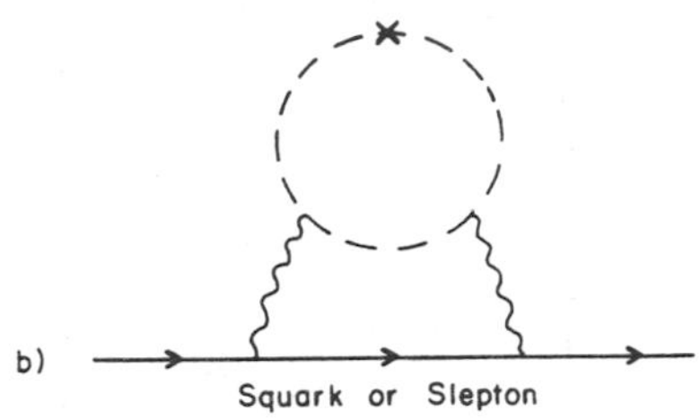

Figure 2b: Graph giving mass to the scalars, via gauge field exchange.

Zino) masses of order $(10^{-2}-10^{-4}) \times \sqrt{F_G}$. Figure 2b gives scalar quarks, leptons, and Higgs masses of the same order. (Figure 2b gives a two-loop mass-squared, which is comparable to a one-loop mass.) Thus, for $\sqrt{F_G}$ of order $10^4 - 10^6$ GeV, the unseen gauginos, squarks and sleptons will have masses of order 10^2 GeV, just as we want. Actually, this should make you suspicious. Why are the quarks, leptons, and gauge bosons not also this heavy? There is a simple reason, in fact. The squarks, sleptons, gauginos (and Higgsions and Higgs) can all be given $SU(3) \times SU(2) \times U(1)$ gauge invariant mass terms, and as soon as supersymmetry breaks these masses are typically generated. The quarks, leptons, and gauge bosons, on the other hand, some of which are much lighter than 10^2 GeV, cannot be given gauge-invariant masses, even when supersymmetry is broken. Thus, there is a natural reason why we have seen just the particles we have seen and not the other half of the spectrum; the supersymmetric doubling of the spectrum is not as troubling as it at first seems. A second question is, why does the Higgs get a negative mass-squared, causing $SU(2) \times U(1)$ to break, while the scalar quarks and leptons have positive mass-squared? At least one very appealing explanation for this has been found.[5] If there is a heavy top or 4th generation quark, graphs with a loop of this quark will have precisely the desired effect. So, LESS gives a good picture of why the world is the way it is.

For $\sqrt{F_G} \sim 10^4 - 10^6$ GeV, we have $m_{3/2} \sim 10^{-2}$ eV - 10 keV. The gravitino is similar to a neutrino in its properties but more weakly coupled, so that its helicity 1/2 components decouple at $T \sim (F_G^2/M_P)^{1/3} \sim 10^{-1} - 10^2$ GeV. Thus, its present number density (including only 2 of its 4 spin states, as the other decouple at $T \sim M_P$) is ~0.1 times the density of a neutrino species (unless F_G is very near its minimum value). One sees that a gravitino with a mass of order 100 eV represents an interesting amount of dark matter, while a gravitino too much heavier than this is forbidden.[7] The ordinary superpartners, in column B of Table I, are expected to be much heavier than this. They can decay to the gravitino plus their R = 1 partner at a rate $\sim M^5/F_G^2$. This is $\gtrsim 1\,\mathrm{sec}^{-1}$ for typical numbers $m \geq 1$ GeV, $F_G \leq 10^6$ GeV. A few models may have trouble with a very light higgsino decaying after nucleosynthesis, but most have no problem.

LESS models were the first supersymmetric models to produce a realistic spectrum, and the first to show that it could be done in a natural way. They are not in fashion now--the idea I will discuss next has received much more attention. However, LESS may still be the way nature works, and these models deserve more study.

III. SCHEME NO. 2: SUPERGRAVITY AS THE MESSENGER OF SUPERSYMMETRY BREAKING

In scheme No. 1, the $SU(3) \times SU(2) \times U(1)$ gauge interactions connected the supersymmetry breaking sector to ordinary matter. We might also imagine turning off all gauge and matter interactions between the two sectors, so they couple only through (super) gravity.[8] This sounds like a radical idea, but these turn out to be the simplest realistic supersymmetric models. The main effects come about from the exchange of auxiliary fields, related by supersymmetry to the graviton and gravitino, as shown in Fig. 3. Although we now have propagating gravitational lines, most of the problems of quantum gravity can be ignored because we are at energies far below the Planck scale. In particular, all effects of incalculable gravitational radiative corrections can be absorbed in a few free parameters. Actually, the number of free parameters, and therefore the amount of predictive power, depends on assumptions about physics at the Planck scale. See[9] for a more thorough discussion. The auxiliary field exchange of Fig. 3 produces effective supersymmetry breakings in the low-energy theory, namely

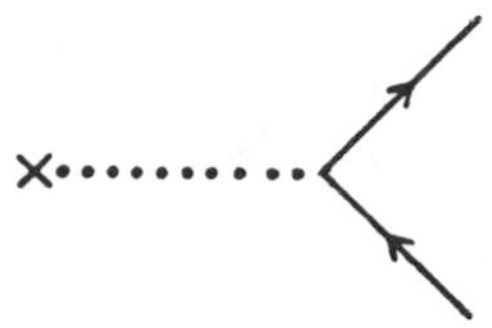

Figure 3: Graph in which a gravitational auxiliary field (dotted line) couples ordinary matter (solid line) to the supersymmetry breaking (X).

(a) A mass term, of order the gravitino mass $m_{3/2}$, for all scalars (squarks, sleptons, higgs),
(b) Additional scalar interactions proportional to $m_{3/2}$, and
(c) Possibly a mass of order $m_{3/2}$ for all gauginos.

If (c) is absent, loop graphs generate a gaugino mass of order $\alpha m_{3/2}$, $\alpha \sim 10^{-1} - 10^{-2}$.

As in LESS, normal particles tend to be light and their partners heavy, and for the same reason. The scale of all the supersymmetry breakings is set by $m_{3/2}$. This determines $m_{3/2} \sim 10^2$ GeV: much lighter and the superpartners would have been seen; much heavier and the cancellation in Fig. 1 is spoiled. As in LESS, a heavy quark is a natural source of SU(2) × U(1) breakdown.[10] How heavy? A mass of at least 60-80 GeV works best. The reported value of 30-50 GeV[11] can work, but it requires other parameters to be adjusted rather finely.

The LSP in thse models might be an ordinary superpartner, from column B of Table I, or the gravitino, or something else. We will consider these possibilities in order. Even the simplest supergravity models have several free parameters, so that it is difficult to make firm statements about which is the lightest ordinary superpartner (LOSP). Table II lists the possibilities in approximate order of theoretical likelihood, with the first two substantially more likely than the others. Note that all the particles in Table II are colorless, a feature on which almost all models agree. The models allow a wide range for the mass of the LOSP, from ~1 GeV up to ~100 GeV.

TABLE II. LOSP Candidates

photino	or a mixture of these
neutral higgsino	
charged slepton	
charged higgsino	
neutral slepton (sneutrino)	

Having summarized the theoretical expectations for the LOSP, let us consider the cosmological consequences if the LOSP is also the LSP and therefore stable. Reference 12 gives a thorough review of this question; see also the talk by Hagelin. Stable charged particles, if they existed, would be mixed with ordinary matter at levels far above the observed upper limits.[12,13] This is a severe problem for models in which the LSP is charged. If instead one of the neutral particles in Table II is the LSP, it would be seen today through its contribution to the mass density of the universe. In order that this not be too great, it must reduce its number density by a factor of roughly 10^8 through annihilation. The neutral particles in Table II annihilate through the exchange of heavier particles, including Z, W, and various superpartners. The situation is thus very similar to the Lee-Weinberg calculation of the limits on a heavy stable neutrino,[14] and gives a lower bound on the mass of the LSP. For the photino or higgsino, this bound is ~2 GeV if the squarks and sleptons weigh 40 GeV, and ~15 GeV if they weigh 100 GeV.[15,12] (If a sneutrino is the LSP, the limit is much weaker.[16]) This is an interesting result, for it is significantly stronger than any particle physics limits on these masses. It is also a lower limit indirectly on the gluino mass, as $m_{gluino} = (3\alpha_{strong}/8\alpha_{em})m_{photino} \sim 6m_{photino}$ is a prediction of most models. A photino or higgsino with a mass near the lower limit would lead to a significant amount of dark matter today (see the talk by Srednicki for a discussion of the possible detection of photinos in the galactic halo).

We now consider the gravitino, whose mass, as discussed earlier, is of order 10^2 GeV. Working backwards from Eq. (5), this gives a supersymmetry breaking scale which is the geometric mean of the weak scale and the Planck scale,

$$F_G \sim \Lambda_W M_P \sim (10^{10\cdot5}\ \mathrm{GeV}) \quad . \tag{7}$$

Also, all couplings of the gravitino involve $1/M_{Planck}$ and are of gravitational strength. In most cases the gravitino will be heavier than the LOSP and can therefore decay, for example into photino + photon. However, its lifetime is rather long,

$$\tau_{gravitino} \sim M_P^2/m_{3/2} \gtrsim 10^7\ \mathrm{sec}\ , \tag{8}$$

so that it decays well after nucleosynthesis, increasing n_γ/n_B by an unacceptable amount.[17] In order to decay before nucleosynthesis, the gravitino must weigh at least 10^4 GeV,[17,18] which is impossible if supersymmetry is to solve the hierarchy problem. The gravitino couples far too weakly for annihilation to effectively reduce its number density. On the other hand, because it couples so weakly, its density can be reduced by inflation.[19] After inflation, processes such as

quark + gluon → squark + gravitino

will restore the gravitino density at a rate[20]

$$\frac{d}{dt}\left(\frac{n_G}{T^3}\right) \sim \alpha\, N\, T^3/M_P^2 \quad , \tag{9}$$

where $\alpha \sim 10^{-1}$ and N, the number of species present, is $\sim 10^2$. Comparing this with $\dot{R}/R \sim N^{1/2}T^2/M_P$, one sees that the final gravitino density is reduced relative to the thermal density by

$$Y = \frac{\alpha N^{1/2} T_R}{M_P} \quad , \tag{10}$$

where T_R is the reheating temperature after inflation. Limits on entropy production after nucleosynthesis require $Y < 10^{-6}$, while limits on the mass density of the decay products (assuming the gravitino decay products involve a superpartner with a mass of at least a few GeV) require $Y < 10^{-7}$, which translates into $T_R < 10^{12}$ GeV. Stronger bounds come from considering the effect of the gravitino decay products on the existing D and ^{3}He after nucleosynthesis (see[21] and the talk by Kim), leading to

$$T_R < 10^9 - 10^{10}\ \text{GeV} . \tag{11}$$

This is quite a low temperature, below the energies at which baryon number violating processes are normally expected, and so is a potential problem for baryonsynthesis. There are still several ways to generate enough asymmetry, but the low reheating temperature is a severe constraint on the models.

Finally, there might be some new R = -1 particle, call it $\tilde{X}$, lighter than the LOSP, so that decays such as photino → photon + $\tilde{X}$ would be allowed. If this decay occurred fast enough, it would remove all constraints on the mass of the LOSP.[22] In order that the mass density of $\tilde{X}$ not be a problem, it must weigh less than a few hundred eV. It was suggested that the axino, superpartner of the axion, might have these properties. However, it turns out that the axino is too heavy to be the $\tilde{X}$[23]:

$$10\ \text{MeV} \lesssim m_{axino} \lesssim 100\ \text{GeV}\ . \tag{12}$$

Axino masses in the lower end of this range are a potential problem: the cosmic axino mass density will be far too large. Axinos with masses near the upper end can decay into ordinary superpartners and have little cosmological effect. While the axino cannot be $\tilde{X}$, some other light weakly coupled particle may play this role. A related possibility is that the R symmetry may not be as good as we have assumed, so that decays such as photino → photon + neutrino could occur rapidly, again eliminating the constraint on the LOSP.[24] Note that neither R violation nor $\tilde{X}$ affects the problem with the gravitino, as its decay rate is fixed by its gravitational strength coupling; the upper bound (11) is hard to evade.

V. OUTLOOK

We have discussed two schemes for producing realistic supersymmetric models. In the first, LESS, ordinary matter couples to the supersymmetry breaking as strongly as possible, through the low-energy gauge interactions. In the second, the coupling was as weak as possible, through gravity alone. There are also intermediate possibilities, such as coupling through superheavy matter fields.[25] All these models give similar results at the weak scale. In particular, as discussed in Section III, it is rather natural that the observed world would be as it is, with superpartners not yet seen and with SU(2) × U(1) broken due to a heavy quark. Supersymmetry also passes other tests (flavor changing neutral currents) which killed technicolor, an earlier solution to the hierarcy problem. Thus, the outlook is good that physics up to 1 TeV, and perhaps well beyond, is described more or less by Table I.

The models we have discussed suggest several sources of interesting amounts of dark matter. In LESS, this would be a gravitino with a mass in the hundred eV range and an effective temperature of about 1K. In supergravity, it would be a photino or other neutral superpartner with a mass of order 10 GeV. Conversely, cosmology puts constraints on the models. Most notable is the low post-inflation reheating temperature, $T_R < 10^9 - 10^{10}$ GeV, required in supergravity models (or in any model with $\sqrt{F_G} \sim 10^{10} - 10^{11}$ GeV).

Thus far, we have only discussed the physics of supersymmetric models near the weak interaction scale. The picture at higher energies is much less clear. There are no supersymmetric grand unified models as simple and appealing as the low-energy models we have been discussing. One of the problems is to explain why the Higgs scalars which are needed to break SU(2) × U(1) are light, while their colored unified partners, which can cause baryon decay, are heavy. A second problem is to explain why the low-energy theory has the gauge symmetry SU(3) × SU(2) × U(1), since supersymmetric grand unified theories typically have several vacua, some with SU(4) × U(1) or SU(5) symmetry. Various solutions to these problems have been suggested, but all have some difficulties, and it may be that important ideas are still missing. A related subject that we have not discussed is supersymmetric inflationary cosmology. Several other speakers will discuss possible scenarios in detail (see talks by Srednicki, Steinhardt, and Holman).

Supersymmetry makes it natural that the ratio M_H/M_P should be very small rather than of order 1. The important question remains, why this small number, or the related ratio $m_{3/2}/M_P$, is not exactly zero. Two ambitious ideas have recently been put forward. It has been shown that in some models $m_{3/2}$ vanishes (supersymmetry is unbroken) to all orders in perturbation theory in the coupling constants but is in fact non-vanishing due to non-perturbative effects.[26] Because these non-perturbative effects are small, the resulting ratio $m_{3/2}/M_P$ is very small. It remains to be seen whether this mechanism fits into a realistic model. In the other proposal, the "no scale" idea, there are many vacua, with different values of $m_{3/2}/M_P$; it has been argued that quantum effects can choose a vacuum in which this ratio is very small.[27] This idea is very ambitious but it is questionable whether it actually works in its present form: the quantum effects in fact seem to lead to $m_{3/2}/M_P$ exactly zero or of order 1.[28]

Finally, mention should be made of recent experimental developments. The CERN collider has seen events which just possibly could be due to supersymmetry.[29] Even if this is not the case, this accelerator and others now under construction will be probing deeper into the energy range where superpartners are expected. If the theoretical arguments of Section I are justified and supersymmetry is found, there will be much more to say. With some of the parameters of the theory experimentally determined, it will be possible to make much more precise statements about the properties of the LSP. Further, knowing more about the parameters of the theory will be a great help in extrapolating the physics up to much higher energies, perhaps to the Planck scale, and thus extrapolating the history of the universe back in time.

REFERENCES

[1]S. Weinberg, Phys. Rev. Lett. **19**, 1264 (1967); A. Salam, in Elementary Particle Theory, ed. N. Svartholm (Almquist and Forlag, Stockholm, 1968).

[2]E. Witten, Nucl. Phys. **B185**, 513 (1981); M. Dine, W. Fischler, and M. Srednicki, Nucl. Phys. **B189** 375 (1981); S. Dimopoulos and S. Raby, Nucl. Phys. **B192**, 353 (1981); N. Sakai, Z. Phys. **C11**, 153 (1981); S. Dimopoulos and H. Georgi, Nucl. Phys. **B193**, 150 (1981). For a thorough recent review, see H.P. Nilles, to appear in Phys. Reports.

[3]L. Girardello and M. Grisaru, Nucl. Phys. **B194**, 65 (1981).

[4]G. Farrar and P. Fayet, Phys. Lett. **76B**, 575 (1978).

[5]L. Alvarez-Gaume, M. Claudson, and M. Wise, Nucl. Phys. **B207**, 16 (1982); J. Ellis, L. Ibanez, and G. Ross, Phys. Lett. **113B**, 283 (1982); K. Inoue, A. Kakuto, H. Komatsu, and S. Takeshita, Prog. Theor. Phys. **68**, 927 (1982).

[6]M. Dine and W. Fischler, Phys. Lett. **110B**, 227 (1982); C. Nappi and B. Ovrut, Phys. Lett. **113B**, 175 (1982).

[7]H. Pagels and J.R. Primack, Phys. Rev. Lett. **48**, 223 (1982).

[8]A. Chamseddine, R. Arnowitt, and P. Nath, Phys. Rev. Lett. **49**, 970 (1982); R. Barbieri, S. Ferrara, and C. Savoy, Phys. Lett. **119B**, 343 (1982).

[9]J. Polchinski, in Proceedings of the Fourth Workshop on Grand Unification, ed. H.A. Weldon, P. Langacker, and P. Steinhardt (Birkhäuser, Boston, 1983).

[10]H.P. Nilles, Phys. Lett. **115B**, 193 (1982); L. Ibanez, Phys. Lett. **118B**, 73 (1982); J. Ellis, D.V. Nanopoulos, and K. Tamvakis, Phys. Lett. **125B**, 275 (1983); L. Ibanez, Nucl. Phys. **B218** 514 (1983); L. Alvarez-Gaume, J. Polchinski and M. Wise, Nucl. Phys. **B221**, 495 (1984); J. Ellis, J. Hagelin, D.V. Nanopoulos, and K. Tamvakis, Phys. Lett. **124B**, 484 (1983).
[11]UA1 Collaboration, G. Arnison et al., CERN press release.
[12]J. Ellis, J. Hagelin, D.V. Nanopoulos, K. Olive, and K. Tamvakis, Nucl. Phys. **B238**, 453 (1984).
[13]S. Wolfram, Phys. Lett. **82B**, 65 (1979); P.F. Smith and J.R.J. Bennett, Nucl. Phys. **B149**, 525 (1979).
[14]B.W. Lee and S. Weinberg, Phys. Rev. Lett. **39**, 165 (1977).
[15]H. Goldberg, Phys. Rev. Lett. **50**, 1419 (1983).
[16]L. Ibanez, Phys. Lett. **137B**, 160 (1984).
[17]S. Weinberg, Phys. Rev. Lett. **48**, 223 (1982).
[18]L. Krauss, Nucl. Phys. **B227**, 556 (1983).
[19]J. Ellis, A.D. Linde, and D.V. Nanopoulos, Phys. Lett. **118B**, 59 (1982); S. Dimopoulos and S. Raby, Nucl. Phys. **B219**, 479 (1983).
[20]S. Weinberg, unpublished.
[21]M. Yu. Khlopov and A.D. Linde, Phys. Lett. **138B**, 265 (1984); J. Ellis, J.E. Kim, and D.V. Nanopoulos, CERN preprint TH-3839 (1984).
[22]J.E. Kim, A. Masiero, and D.V. Nanopoulos, Phys. Lett. **139B** 346 (1984).
[23]J. Polchinski, in preparation.
[24]L. Hall and M. Suzuki, Nucl. Phys. **B231**, 419 (1984).
[25]T. Banks and V. Kaplunovsky, Nucl. Phys. **B206**, 45 (1982); S. Dimopoulos and S. Raby, Nucl. Phys. **B219**, 479 (1983); M. Dine and W. Fischler, Nucl. Phys. **B204**, 346 (1982); R. Barbieri, S. Ferrara, and D.V. Nanopoulos, Z. Phys. **C13**, 267 (1982); J. Polchinski and L. Susskind, Phys. Rev. **D26**, 3661 (1982).
[26]I. Affleck, M. Dine, and N. Seiberg, Phys. Lett. **137B**, 187 (1984); Y. Meurice and G. Veneziano, CERN preprint TH-3803 (1984).
[27]J. Ellis, A.B. Lahanas, D.V. Nanopoulos, and K. Tamvakis, Phys. Lett. **134B**, 429 (1984).
[28]T. Banks and J. Polchinski, unpublished.
[29]G. Arnison et al., CERN preprint EP/84-42; J. Ellis and H. Kowalski, DESY preprint 84-045.

Supersymmetric Relics from the Big Bang

John S. Hagelin

ABSTRACT

We discuss cosmological constraints on supersymmetric theories with a new stable particle. Circumstantial evidence points to a neutral gauge/Higgs fermion as the best candidate for this particle, and we give bounds on the parameters in the Lagrangian which govern its mass and couplings. One favored possibility is that the lightest supersymmetric particle (LSP) is predominantly a photino $\tilde{\gamma}$ with mass above 1/2 GeV, while another is that the lightest neutral supersymmetric particle is a Higgs fermion $\tilde{H}$ with mass above 5 GeV or less than O(100) eV. It is also conceivable that the LSP is a scalar neutrino $\tilde{\nu}$, which could have interesting phenomenological implications.

In the past few years, supersymmetric extensions of the standard model have received a great deal of attention from theorists. All these models predict an exciting variety of new particles. Alas, there are few experimental constraints of their masses and couplings, particularly on those of neutral particles. To guide future work, both experimental and theoretical, we must learn all we can about the parameters of these theories. Here, as in many other cases, useful information can be obtained from a study of early cosmology.

The essential feature of supersymmetric theories which makes them amenable to cosmological study is that in many models one of the new particles is absolutely stable. In the very early universe, all these particles will be present in thermal equilibrium. As the temperature falls, the heavier new particles decay into the lighter ones. Eventually only the lightest supersymmetric particle (LSP) will be left. It can disappear only by pair annihilation and we must require that this pair annihilation is efficient enough to reduce the present day number density of the LSP to an acceptable level.

Here we present the results of a detailed analysis of this question for the class of models we favor: a minimal supersymmetric extension of the standard model, with supersymmetry broken by gravitational couplings to spontaneously broken supergravity. For details, see refs. 1 and 2.

First we consider the likely possibility that the LSP is a colorless, neutral fermion. The mass matrices for the neutral SUSY fermions -- gauginos and Higgsinos -- are determined by the Lagrangian terms.

$$L \sim \varepsilon\varepsilon_{ij}\tilde{H}_1^i\tilde{H}_2^j - M_2\tilde{W}_a\tilde{W}_a - M_1\tilde{B}\tilde{B}$$

where W_a and B denote SU(2) and U(1) gauge superfields respectively, the tildes denote fermionic components and i,j (a) are doublet (triplet) SU(2) indices. We generally expect the parameters M_1, M_2 to be $O(M_W)$, while it is conceivable that ε might be very much smaller. The diagonalization of the resulting mass matrix leads to four neutral Majorana fermions whose masses and couplings depend on ε and the gaugino masses.

The constraints on the LSP can be summarized as follows. If the LSP is predominantly a photino then $M_{\tilde{\gamma}}$ is required to be $\gtrsim$(1/2,2,5) GeV for fermion masses $\gtrsim$(20,50,100) GeV respectively. If the LSP is predominantly a Higgsino then $M_{\tilde{H}} \geq 5$ GeV, or else $M_{\tilde{H}} \lesssim 100$ eV in which case the $\tilde{H}$ behaves cosmologically like a neutrino.

It is also possible that the LSP is scalar neutrino.[2] Since sneutrinos annihilate very efficiently through gaugino exchange there is no lower bound on $M_{\tilde{\nu}}$ provided ε is not too small.[2] (If ε and $M_i \to 0$, the $\tilde{\nu}$'s cannot annihilate via gaugino exchange which requires a Majorana mass term.) Because the $\tilde{\nu}$ has suitable characteristics as a dark matter candidate, it is interesting that ε can be tuned so that the present cosmological mass density of the $\tilde{\nu}$ can have practically any value. If the $\tilde{\nu}$ were in fact the LSP, one could hope to observe them in experiments such as $e^+e^- \to \tilde{\nu}\tilde{\nu}\gamma$ and $Z \to \tilde{\nu}\tilde{\nu}$.[2]

REFERENCES

[1]J. Ellis, J. S. Hagelin, D. V. Nanopoulos, K. Olive and M. Srednicki, Nucl. Phys. B238, 453 (1984).
[2]J. S. Hagelin, G. L. Kane and S. Raby, Los Alamos preprint LA-UR-83-3711 (1983).

Strongly-Coupled Nature of the Primordial SUSY SU(5) Plasma

H. Goldberg

ABSTRACT

It is shown that for a large class of SU(5) SUSY GUTs, (i) the Debye free energy of the plasma is about 2 times larger than the free energy of the SUSY SU(5) ideal gas, (ii) the plasmon $(\text{mass})^2$ is about 2 times the mean $\langle p^2 \rangle$ of gluons in the plasma, (iii) the mean free path of an SU(5) gluon in the plasma is about 1/3 of the average interquantum spacing of the ideal gas.

The successful implementation of inflation[1,2] seems to require[2,3] a considerable gap between the GUT mass scale ($\sim 10^{16}$ - 10^{18}GeV) and the completion of the SU(5) phase transition ($T \sim 10^9$ - 10^{12}GeV). During this time the energy density and pressure are presumed to be dominated by the stress tensor of some quasistatic field (either gauge singlet or not) providing the false vacuum, while the essentially massless SU(5) symmetric plasma behaves as a negligible dilute background.

In this talk I will outline the results of some calculations[4] which indicate that in a large class of supersymmetric models (global or local), the SU(5)-symmetric plasma cannot be regarded as even approximately dilute without much more elaborate study. The class of models referred to are those in which $\alpha_{SU(5)}(m_{GU}) \approx \alpha_{QCD}(m_W) \sim 0.14$; as explained later, such a relation results from the suppression of quadratic and cubic self-couplings for the adjoint Higgs field in the superpotential. A large suppression is desirable for cosmological reasons.

I will come to the conclusion concerning the non-dilute nature of the plasma in two entirely separate ways. The first is through a calculation of the Debye screening energy, the second involves a calculation of the transport properties of the supersymmetric SU(5) plasma.

I. SCREENING

The perturbative finite temperature QCD calculation of Kapusta[5] can be extended to the case of a SUSY SU(N) gauge theory with arbitrary chiral matter fields. In the present work, I will concentrate on the $O(g^3)$ (Debye) corrections, and note here two principal differences between Kapusta's calculation and the present case:

(1) Many more fields contribute to the sum of ring diagram contributions to the free energy.

(2) For each scalar field, an additional zero mode is present in the finite temperature propagator, so that ring diagram sums must be performed for all the scalar fields.

Using the methods of Refs. 5 and 6, I have calculated the electric and scalar plasmon masses for a SUSY SU(N) theory, and find

$$m_{el}^2 = \frac{1}{3} g^2T^2 \left(\frac{3}{2} N + \sum_R n_R\, T(R)\right) \tag{1}$$

$$m_{sc}^2(R) = \frac{3}{4} g^2T^2\, C_2(R) + \text{non-gauge contributions} \tag{2}$$

In Eqs. 1 and 2, R is the SU(N) label of the complex chiral matter field which is light before SU(5) breaking, T(R) is the Dynkin number ($\mathrm{Tr}T^aT^b = \delta^{ab}T(R)$, T(fund) = 1/2), and $C_2(R)$ is the quadratic Casimir invariant ($C_2(R) = (N^2 - 1)T(R)/d(R)$, d(R) = dim (R)).

Thus we have for the $\alpha^{3/2}$ contribution to the free energy density[5,6]

$$\mathcal{F}^{(3)} = -(T/12\pi)\left[(N^2-1)m_{el}^3 + 2\sum_R n_R d(R)\, m_{sc}^3(R)\right] \tag{3}$$

(The factor of 2 comes from the complexity of the scalar field in SUSY.)

The ideal gas free energy density for a SUSY SU(N) theory is

$$\mathcal{F}^{(0)} = -(\pi^2T^4/24)\left(N^2 - 1 + \sum_R n_R d(R)\right) \tag{4}$$

In order to proceed, a model must be specified. Acceptable values of $\delta\rho/\rho$ and a smooth completion of the SU(5) phase transition are most naturally obtained in models which leave light SU(3) color octet and SU(2) weak triplet supermultiplets after SU(5) breaking.[3] Renormalization group equations then imply $\alpha_{GU}^{-1} \sim \alpha_3^{-1}(m_W) \sim 7$.

Examples of models giving reasonable values of $\sin^2\theta_W$ are:

$$\{R\}_1 = \underset{\sim}{24} + 3(\underset{\sim}{10} + \underset{\sim}{\bar{5}}) + 2(\underset{\sim}{5} + \underset{\sim}{\bar{5}})$$

$$M_{GU} = 3.6 \times 10^{15}\ \text{GeV}$$

$$\alpha_{GU}^{-1} = 5.8 \qquad \text{(First paper, Ref. 3)} \tag{5a}$$

$$\{R\}_2 = \underset{\sim}{24} + 3(\underset{\sim}{10} + \underset{\sim}{\bar{5}}) + (\underset{\sim}{5} + \underset{\sim}{\bar{5}}) + 2(\underset{\sim}{10} + \underset{\sim}{\overline{10}})$$

$$M_{GU} = 2 \times 10^{18}\ \text{GeV}$$

$$\alpha_{GU}^{-1} = 7.1 \qquad \text{(Ref. 7)} \tag{5b}$$

Fine tuning or missing partner mechanisms are used to prevent the $5 + \bar{5}$ or $10 + \overline{10}$ pairs from attaining $O(m_{p\ell})$ masses.

On processing (5a) and (5b) through Eqs. 1-4, I find[8]

$$\text{for } \{R\}_1 \quad \mathcal{F}^{(3)}/\mathcal{F}^{(0)} \approx 31\, \alpha^{3/2} \approx 2.2 \tag{6a}$$

$$\text{for } \{R\}_2 \quad \mathcal{F}^{(3)}/\mathcal{F}^{(0)} \approx 34\, \alpha^{3/2} \approx 1.8 \tag{6b}$$

We may also conveniently assess the importance of possible quasiparticle effects by comparing m_{el}^2 with the mean $(\text{momentum})^2$ of a gluon ($(2.7T)^2$). I find

$$\text{for } \{R\}_1 \qquad M_{el}{}^2/\langle p^2\rangle \approx 12\alpha \approx 2.0 \tag{7a}$$

$$\text{for } \{R\}_2 \qquad M_{el}{}^2/\langle p^2\rangle \approx 15\alpha \approx 2.0 \tag{7b}$$

Other plasmons similarly acquire large masses (see Eq. 2).

In order to gauge just how non-perturbative such ratios are, we may compare with the case of pure QCD (SU(3)), no quarks, where[5,6]

$$\mathcal{F}^{(3)}/\mathcal{F}^{(0)} \approx 5.4\ \alpha_{QCD}^{3/2} \tag{8}$$

$$M_{el}{}^2/\langle p^2\rangle = 1.7\ \alpha_{QCD} \tag{9}$$

With no quarks, $\alpha_{QCD}{}^{-1} \approx (11/2\pi)\ln(4T/\Lambda_{MOM})$ so that ratios comparable to Eqs. 6-7 are obtained in QCD only for $T \lesssim \Lambda_{MOM}$, definitely in a non-ideal gas domain.

I conclude that a large class of cosmologically acceptable SUSY GUT models lead to strongly coupled SU(5) plasmas even at $T \sim m_{GU}$. Before commenting about the implications of this result, I will show how one comes to a similar conclusion from a different direction.

II. MEAN FREE PATH

Let λ_a be the mean free path of a typical quantum a in the SU(5) plasma, and let d be the typical spacing between quanta. Then the plasma can be considered dilute only if $\lambda_a/d >> 1$. If the number density of quantum species i is n_i, and the spin-averaged scattering cross section of a from i is σ_{ai}, then $\lambda_a{}^{-1} = \sum_i n_i\sigma_{ai}$, $d = (\Sigma n_i)^{-1/3}$, and

$$\lambda_a/d = (\sum n_i)^{1/3} / (\sum n_i\sigma_{ai}) \tag{10}$$

To illustrate, I will choose a = vector gluon = g. In calculating total scattering cross sections for various channels (e.g., g + g → g + g), there are small-angle divergences which are in practice cut off by the Debye screening length. To obtain a plausible lower bound for σ_{gi} (and simplify the algebra at the same time), I have calculated $(d\sigma/d\cos)_{90°}$ for the various processes, and have set $\sigma_{gi} \approx 2(d\sigma_{gi}/d\cos\theta)_{90°}$.

The various approximate values of σ_{gi} are shown in Table I of Ref. 4, where can be found the remaining details of the calculation. I give here the result:

$$(\lambda_g/d)_{SUSY\ SU(N)} = \frac{4.15}{\alpha^2} \cdot C^{-1} \cdot (N^2-1 + \sum_R n_R d(R))^{1/3}$$

$$= 0.24 \text{ for } \{R\}_1$$

$$= 0.31 \text{ for } \{R\}_2 \tag{11}$$

Here

$$C = 38\tfrac{1}{4}N^2 + 15 \sum n_R C_2(R)T(R) + 16\tfrac{3}{8} N \sum n_R T(R) \tag{12}$$

For quarkless, ordinary QCD[4],

$$(\lambda_g/d)_{QCD} = 0.028/\alpha^2_{QCD} \tag{13}$$

The value of (λ_g/d) found for $\{R_1\}$ corresponds to $\alpha_{QCD} = 0.34$, while for $\{R_2\}$ it corresponds to $\alpha_{QCD} = 0.30$. I find that these values of α_{QCD} correspond to $T = 1.2\Lambda_{QCD}$ $1.4\Lambda_{QCD}$, respectively. Both are in regions of sizeable deviations from perturbative behavior of the SU(3) Yang-Mills gas.

III. SUMMARY AND SPECULATIONS

1. In a class of cosmologically acceptable SU(5) SUSY GUTs, I have found that, prior to the SU(5) phase transition, (1) the Debye screening energy is larger than the Stefan-Boltzmann free energy, (2) the plasmon masses are substantially larger than the mean momenta of the particles, and (3) the mean free path for gluon collisions in the plasma is substantially smaller than the average interquantum spacing of a dilute plasma at the same temperature.

Comparison with QCD_3 indicates that under any of these circumstances, the SUSY-SU(5) plasma is in a strongly-coupled phase. In all three cases, the effect is due to a combination of the proliferation of quantum species (especially the presence of the adjoint representation), and the large size (~1/7) of the gauge coupling at $T \sim M_{GU}$. The last is a feature of models which lead to an acceptable completion of the SU(5) →3.2.1 phase transition after inflation.

2. Lack of space prevents presentation of a heuristic speculation illustrating how a strongly coupled plasma in a metastable phase could by itself drive a "new" inflation. The reader is referred to Ref. 4 for this discussion.

REFERENCES

[1] A. H. Guth, Phys. Rev. D**23**, 347 (1981).

[2] A. D. Linde, Phys. Lett. **108B**, 389 (1982); A. Albrecht and P. J. Steinhardt, Phys. Rev. Lett. **48**, 1220 (1982).

[3] J. Ellis, D. V. Nanopoulos, K. A. Olive, and K. Tamvakis, Nucl. Phys. B Phys. Lett. **120B**, 331 (1983); D. V. Nanopoulos, K. A. Olive, M. Srednicki, and K. Tamvakis, Phys. Lett. **123B**, 41 (1983); Phys. Lett. **124B**, 171 (1983); A. Albrecht, L. G. Jensen, and P. J. Steinhardt, Univ. of Penn. preprint VPR-0229 (July 1983); B. Orvut and S. Raby, Phys. Lett. **125B**, 270 (1983).

[4] H. Goldberg, Phys. Lett. **139B**, 45 (1984).

[5] J. Kapusta, Nucl. Phys. **B148**, 461 (1979).

[6] D. J. Gross, R. D. Pisarski, and L. G. Yaffe, Rev. Mod. Phys. **53**, 43 (1981).

[7] L. Ibañez, Phys. Lett. **126B**, 196 (1983).

[8] Eqs. (6a) and (6b) are numerically corrected versions of Eqs. (8) and (9) of Ref. 4. My thanks to J. Kapusta and D. Reiss for pointing out

these corrections (which do not change any of the conclusions of the work).

Is a Spontaneously Broken Supersymmetry Restored at Positive Temperature

M.K. Kullab

We consider an O'Raifeartaigh type model which has no other symmetry than supersymmetry (SUSY).[1] The SUSY is spontaneously broken at T = 0. We find that there is a phase transition at some critical temperature. Such a model has a massless Goldstone fermion associated with the T = 0 supersymmetry breaking. We want to know if this Goldstino acquires a mass at T > 0 when the one-loop self-energy corrections are taken into account.

The model considered by Girardello et al.[2] consists of three chiral superfields ϕ_a, a = 0, 1 and 2, and has a discrete chiral symmetry. This chiral symmetry is spontaneously broken at T = 0, and is restored by a phase transition at a critical temperature

$$T_c^2 = \frac{4\xi}{g} - \frac{4m^2}{g^2} .$$

We add a term $(-\eta\ \phi_1)$ to the superpotential which explicitly breaks the discrete chiral symmetry and leaves the model with SUSY as the only symmetry. We calculate and minimize the effective potential as in the model of Girardello et al. We find that the mean value, A_2, of the scalar part of the a = 2 superfield is now determined by:

$$A_2^3 + \left[T^2 - \left(\frac{4\xi}{g} - \frac{4m^2}{g^2} \right)\right] A_2 - \frac{8m\eta}{\sqrt{2}\ g^2} = 0 .$$

We find that there is a second-order phase transition at

$$T_c^2 = \frac{4\xi}{g} - \frac{4m^2}{g^2}\ 3 \left(\frac{4m\eta}{\sqrt{2}\ g^2} \right)^{2/3} .$$

At the one-loop level we find that the fermion mass matrix is

$$M_f = \begin{pmatrix} \Sigma_{00} & \Sigma_{01} & \Sigma_{02} \\ \Sigma_{10} & (\Sigma_{11}+\mu) & \Sigma_{12} \\ \Sigma_{20} & \Sigma_{21} & (\Sigma_{22}-\mu) \end{pmatrix} .$$

Here,

$$\mu = \sqrt{\frac{2m^2 + g^2 A_2^2}{2}}$$

and Σ's are the self-energy corrections. We are interested in the zero-frequency mode at zero momentum, which is the Goldstino at $T = 0$. Therefore, it is sufficient to evaluate the Σ's in the infrared limit, $P = 0$, $P_0 \to 0$. In this limit we find that the only contributions come from the bosonic part of ϕ_2. These calculations lead to the result

$$\det M_f = 0 \ .$$

This means that there is a zero-eigenvalue. The Goldstino is still present at all temperatures.

We conclude that SUSY is not restored at any temperature, not even by the phase transition which this model has. It could be, though, that in more complicated models the SUSY is restored (in the sense that there is no Goldstino) and this should be looked at. Furthermore, cosmological implications still need to be explored. Otherwise, the absence of the SUSY at $T > 0$ is a fact of life we must learn to live with.

ACKNOWLEDGMENTS

It is a pleasure to thank J. Kapusta for advising me on this problem. This work was supported in part by the U.S. Department of Energy under Contract No. DOE/DE-AC02-79ER-10364, and by Yarmouk University, Jordan.

REFERENCES

[1]M. Kullab, University of Minnesota preprint (to be published).
[2]L. Girardello, M.T. Grisaru and P. Salmonson, Nucl. Phys. **B178**, 331 (1981).

Initial Conditions

James B. Hartle

ABSTRACT

The role initial conditions play in cosmological theories is discussed. The construction, in quantum gravity, of the wave function corresponding to the state of minimum excitation is described. There is evidence that this state could be the state of our universe.

The subject of cosmology presents us with a problem which is of a fundamentally different kind from that encountered elsewhere in physics. The familiar type of problem in physics is to use local laws to calculate the evolution of a system. For example, we use Maxwell's equations and Newton's laws of mechanics to predict the evolution of a plasma. We use Schrödinger's equation to predict the evolution of an atomic state. The local laws of physics require boundary conditions: sometimes initial conditions, sometimes spatial boundary conditions, sometimes radiative boundary conditions, and often a combination of these. These boundary conditions are set by the physical conditions of those parts of the universe which are not part of the physical system under consideration. There are no particular <u>laws</u> determining these conditions, they are specified by <u>observations</u> of the rest of the universe.

The situation is different in cosmology. Boundary conditions, and in particular initial conditions, are still required to solve the local laws governing the evolution of the universe. They are needed, for example, to solve Einstein's equation. There is, however, no "rest of the universe" to pass their specification off to. If there is a general specification of these initial conditions it must be part of the laws themselves.[1]

I would like to review and discuss some attitudes towards this problem and describe some ideas in quantum gravity which provide a framework in which at least one of these attitudes may be discussed.

To make the discussion concrete let me focus on just one aspect of the observable universe-the observed large scale homogeneity and isotropy. In the 3° cosmic background radiation we see the light from approximately 100,000 years after the big bang. This radiation, and the universe at the time it was emitted, is isotropic to a part in a few thousand.[2] If we are not at a preferred position in the universe, and there is no evidence from galaxy surveys today that we are, then the universe must have also been approximately homogeneous at that time. Extrapolating this data backward in time we can find initial conditions, themselves approximately homogeneous and isotropic, which give rise to this state of affairs. What attitude are we to take to these initial conditions?

A number of attitudes <u>have</u> been taken. Many of them can be summarized in the following <u>four</u> rough categories:

Attitude 1: That's the way it is.

The universe might have been in any one initial state as well as any other. It happens that the one it is in is homogeneous and isotropic on the scales we observe. That's as far as physics can go. Its not the proper subject of physics to explain these initial conditions only to discover what they were.

This is a reasonable but not very adventurous attitude. It certainly has no predictive power about what we will see as with increasing time we are able to observe larger regions of the universe. I believe we will only be able to say it is correct when all attempts to explain the initial conditions have failed.

Attitude 2: The boundary conditions which determine the universe are not initial conditions but present conditions and in particular the fact that we exist.

This attitude is related to the set of ideas called the anthropic principle.[3] The universe must be such as to allow galaxy condensation, star formation, carbon chemistry and life as we know it. This is indeed a restriction on the structure of the universe. Perhaps, if one were given a choice of three or four very different cosmologies one could identify our own using the anthropic principle. As stressed by Penrose,[4] however, the anthropic principle does not seem strong enough to single out the observed universe from among all possibilities. Suppose, for example, the sun had been located in a cloud near the galactic center and we had not been able to make observations of the large scale structure. Would we have been able to predict the large scale homogeneity and isotropy using the anthropic principle? I think not.

Attitude 3: Initial conditions are not needed; dynamics does it all.

Interesting features like the large scale homogeneity and isotropy will arise from any reasonable initial conditions through the action of physical processes over the course of the universe's history. Even if it started in an inhomogeneous and anisotropic state the universe would evolve towards a homogeneous and isotropic one over the scales we can observe it. This is an attractive idea not least because we can achieve determinism with the existing dynamical laws of physics. A variety of physical mechanisms have been proposed to implement this idea beginning with the work of Misner and his coworkers.[5] The most successful mechanism is inflation.[6]

Any dynamical explanation of the large scale structural features of the universe has to confront the problem of cosmological horizons. I might illustrate this in the following way: As we look at the microwave radiation on the sky we are seeing a picture of the universe as it was about 100,000 years after the big bang. It is remarkably isotropic. If we are to have a dynamical explanation of this isotropy then the different regions from which the radiation was emitted must have been able to communicate with each other sometime during this history. However, there is only a finite time since the big bang for them to do so. Whether they were able to communicate depends on the history of the universe prior to the time the radiation was emitted. The size of the region which at any one time could have been in causal contact is the horizon size at that time. A prerequisite for any dynamical explanation of the observed isotropy of the background radiation is that the horizon size at decoupling be larger than the size of the universe we can observe then.

The most naive extrapolation of the history of the universe to times earlier than decoupling is to assume that the spacetime geometry remains approximately homogeneous and isotropic (the Friedmann-Robertson-Walker model), that the matter energy density is dominated by the density of approximately free radiation and that the evolution is governed by Einstein's equation. With this early history only regions at the time of decoupling now subtending a few degrees on the sky could have been in causal contact since the big bang and no dynamical explanation of the observed isotropy would be possible. This extrapolation, however, is too naive. The horizon can be much larger than the observable universe in models where the geometry is significantly anisotropic before decoupling,[7] as a consequence of quantum effects at the Planck epoch[8] or as a consequence of an inflationary de Sitter-like expansion arising from a matter phase transition at the GUT time. In the inflationary history with only conservative assumptions on the matter physics, the horizon at decoupling would be at least 10^4 times larger than the observable universe and could be very much larger. There is thus opportunity for dynamical processes to act to drive the universe towards isotropy and the inflationary expansion itself provides a mechanism to do this.[9] Further, the region that becomes the observed universe at decoupling is so much smaller than the horizon size at the end of inflation that one might suppose that any reasonable initial conditions which are inhomogeneous on the scale of the horizon will appear smooth on the scale of the observed universe. With an inflationary history no special assumptions on initial conditions are required to explain the observed isotropy. We see the universe as homogeneous and isotropic simply because we do not see a very big part of it.

If the universe is inhomogeneous on a large scale eventually we will find this out. The size of the universe we can observe grows with every second. No dynamical explanation of homogeneity and isotropy can thus ever be an explanation for all time. Eventually we will see outside the horizon and have to face up to the problem of initial conditions. However, if the inflationary history is correct, under even the conservative assumptions mentioned above we may be able to postpone this discussion for 10^{22} years.

Even today, even in the limited region we can observe, no dynamical explanation can ever be a complete explanation for the features of the universe we see. We can always imagine a present state of the observable universe which is highly inhomogeneous. Whatever the laws of geometry and matter, whether they be classical or quantum, whether there are phase transitions or not, these laws can be used to extrapolate this state backward in time and reach some initial condition.[10] No dynamical explanation can, therefore, ever completely exclude a present state of inhomogeneity without some restrictions on the initial conditions. As impressive as they may be in broadening our choice for initial conditions compatible with our present observations, dynamical explanations of these observations can never be complete.

Attitude 4: There is a law of physics specifying the initial conditions.

Specification of the boundary conditions is just as much a law of physics as are the dynamical equations governing their evolution. In this view, the question for physics is whether there exists a compelling, simple, and predictive principle which will single out the initial state of our universe. Any search for such a principle is likely to involve the quantum theory of gravity in an essential way. This is because the classical theory of gravity by itself strongly suggests that quantum gravitational phenomenon will be important for the early universe. One of the important theoretical advances in general relativity in the last several decades is the singularity theorems. Applied to cosmology[11] these theorems suggest that if we extrapolate the presently observed universe back into the past according to the classical laws we will eventually reach a singularity - a time when the density of matter goes to infinity, the temperature goes to infinity and the spacetime curvature goes to infinity. This is the big bang. Only very special configurations of matter which violate the assumptions of these theorems can prevent this singularity from being in our past.

Quantum gravitational effects will be important when the curvature varies significantly over a Planck length, $(hG/c^3)^{1/2} \approx 10^{-33}$ cm. The singularity theorems suggest that such curvatures will inevitably be reached near the big bang. A theory of initial conditions must therefore be essentially quantum gravitational.

There is not today a complete and manageable quantum theory of gravity. The difficulties lie not only with such technical issues as finiteness which exist for any field theory but more fundamentally with the question of what Lagrangian to start from and even more fundamentally with issues like finding the correct Hilbert space of states and giving the theory a meaningful interpretation.[12] How are we to proceed in the absence of a complete theory? One reasonable answer is to give up. An equally reasonable answer is to extrapolate the theoretical ideas we have, understanding that we will not be able to answer every question.

In this talk I want to describe some of the quantum gravitational framework for discussing the issue of initial conditions and a recent proposal about what this initial condition might be.

In quantum mechanics if one knows the wave function of a system one is said to know its state and everything that one can about it. In the case of a single particle we write

$$\Psi = \Psi(\vec{x},t) \quad . \tag{1}$$

Ψ is the probability amplitude that the particle is found at $\vec{x}$ on the surface of constant time t and nowhere else on that surface. In the field theory of a scalar field, the wave function is a functional of scalar field configurations and the time so that we would write

$$\Psi = \Psi[\phi(\vec{x}),t] \quad , \tag{2}$$

Ψ is the probability that the field at each $\vec{x}$ on the surface of constant time t has the value $\phi(\vec{x})$ and no other value.

In the quantum mechanics of closed cosmologies the wave function is a functional of the three geometry on a spacelike surface. Denoting the three metric by $h_{ij}(\vec{x})$ we would write

$$\Psi = \Psi[h_{ij}(\vec{x})] \quad . \tag{3}$$

No additional time label is needed because a generic three geometry will fit as a spacelike surface in at most a few places in a generic spacetime. The three geometry itself carries the information about its location in spacetime; that is about time. For example, in a closed Friedman universe, if one knows the radius of the spatial geometry then one knows one must be at one of two ages. The counting of degrees of freedom is correct: Three of the six components of h_{ij} are gauge degrees of freedom corresponding to transformations of the three coördinates in the spacelike surface. Of the remaining three components, two represent the physical degrees of freedom of the gravitational field and one represents the time.

An important state for any physical system is the ground state or state of minimum excitation. In the quantum mechanics of a single particle moving in a potential $V(\vec{x})$ there are two ways of calculating the ground state wave function, $\Psi_0(\vec{x})$. One can find the lowest energy eigenfunction of the Hamiltonian

$$H\Psi_0(\vec{x}_0) = E_0\Psi_0(\vec{x}_0) \quad . \tag{4}$$

Completely equivalently, the wave function of the ground state may be calculated as a Euclidean functional integral

$$\Psi_0(\vec{x}_0) = \int \delta\vec{x}(\tau)\exp(-I[\vec{x}(\tau)]/h) \quad . \tag{5}$$

Here, $I[\vec{x}(\tau)]$ is the Euclidean action functional

$$I[\vec{x}(\tau)] = \int d\tau\left[\frac{1}{2}\, m\dot{\vec{x}}^2 + V\,(\vec{x})\right] \quad . \tag{6}$$

The sum is over all paths which start at $\vec{x}_0$ at time $\tau = 0$ and proceed in the infinite past to a configuration of minimum action.

It is not difficult to sketch the demonstration of the equivalence of (4) and (5). One begins with the path integral for the propagator

$$\langle \vec{x}'', t'' | \vec{x}', t' \rangle = \int \delta x(t)\exp(iS[\vec{x}(t)]/\hbar) \quad . \tag{7}$$

Here, S is the usual action [i.e. (6) with the opposite sign for $V(\vec{x})$] and the sum is over paths which start at $\vec{x}'$ at time t' and wind up at $\vec{x}''$ at time t". Consider the particular propagator $\langle \vec{x}_0, 0 | 0, t \rangle$ and expand it in a complete set of energy eigenstates as follows:

$$\langle\vec{x}_0,0|0,t\rangle = \sum_n\langle\vec{x}_0,0|n\rangle\langle n|0,t\rangle$$

$$= \sum_n e^{+iE_nt}\,\Psi_n(\vec{x}_0)\Psi_n^*(0) \quad , \tag{8}$$

where $\Psi_n(\vec{x})$ are the wave functions of the energy eigenstates. Equate the last line of (8) to the right hand side of (7) and rotate the time to imaginary values, $t\rightarrow -i\tau$, on both sides of the equation. One has

$$\sum_n e^{E_n\tau}\,\Psi_n\,(\vec{x}_0)\Psi_n^*(0) = \int\delta x(\tau)\exp(-I[\vec{x}(\tau)]/h) \quad . \tag{9}$$

Then take the limit $\tau\rightarrow-\infty$. If one normalizes the energy so that the lowest eigenvalue is zero, only the ground state term survives in the sum on the left hand side of (9). The sum over paths on the right hand side becomes the sum described above and one recovers after a normalization eq.(5).

In the quantum gravity of closed cosmologies one does not expect to find a notion of ground state as the lowest eigenfunction of a Hamiltonian. There is no natural spacelike slicing for general closed spacetimes, correspondingly no natural notion of time, correspondingly no natural Hamiltonian, and therefore no natural ground state as the lowest eigenfunction of a Hamiltonian. Even if one chooses a time slicing in an ad hoc fraction the corresponding Hamiltonian will in general be time dependent and not yield a unique ground state. Thus one of the methods for constructing a ground state wave function in the quantum mechanics of a particle does not work in the quantum mechanics of closed cosmologies.

The construction of the wave function of the state of minimum excitation using a Euclidean functional integral can be generalized to the quantum mechanics of closed cosmologies.[13,14] Schematically, one would write

$$\Psi_0[h_{ij}] = \int\delta g\exp(-I[g]/\hbar) \quad . \tag{10}$$

The sum is over a class of Euclidean four geometries which have a boundary on which the induced three metric is h_{ij}. [This is the analog of the paths starting at $\vec{x}_0$ in (5)]. The action is the Euclidean action for general relativity on a manifold M with boundary ∂M (we use units were $c = G = 1$)

$$I[g] = -2\int_{\partial M} Kh^{1/2}d^3x - \int_M (R-2\Lambda)g^{1/2}d^4x \quad , \tag{11}$$

where, since the subject is cosmology, we have included a cosmological constant Λ. To completely specify the wave function Ψ_0 there remains the complete specification of the class of geometries summed over [the analog of the paths going to minimum action at $\tau\rightarrow-\infty$ in (5)]. The proposal is that one should sum over compact Euclidean four geometries.[13] The remaining boundary condition for geometries contributing to the state of minimum excitation is that there is no other boundary.

The wave function so naturally identified by the Euclidean functional integral prescription (10) displays many properties one would associate with a state of minimum excitation when analyzed in simple models.[13,15] As our own universe is not in a state of very high excitation, and as this state emerges so simply in the theory, it is a natural conjecture[15,16] that this wave function is the wave function of our universe and that the law (10) is the law specifying the initial conditions. It may seem paradoxical that the complexity which we do see could be represented by a state of minimum excitation but in reflecting on this it is important to remember two things: (1) We are part of this system, and (2), complex results can emerge from simple laws.

We have gone only a short distance towards a complete analysis of this proposal for a wave function of the universe. The proposal has been tested in simple models which contain only some of the degrees of freedom in our universe. These models are constructed by restricting the three geometries on which the wave function is evaluated to be geometries of high symmetry and the sum in the functional integral (10) to be over compatible four geometries of high symmetry. These restricted sums are more tractable than the general case. The restriction on the three geometries means that the wave function is evaluated on only a subspace of the superspace of all three geometries. For this reason these models are called minisuperspace models.

The simplest restriction with which to illustrate the construction of minisuperspace models is to allow only three geometries which are homogeneous and isotropic and four geometries which have homogeneous and isotropic spatial sections. A cosmological spacetime with these symmetries has the closed Robertson-Walker line element

$$ds^2 = -dt^2 + a^2(t)d\Omega_3^2 \quad . \tag{12}$$

Here, $d\Omega_3^2$ is the metric on the unit three sphere and $a(t)$ is the scale factor. These cosmologies may be viewed as an evolving three sphere whose radius is $a(t)$. A closed Friedmann universe is a geometry of this class which solves the Einstein equation. In the Friedmann universe the three sphere evolves from zero radius at the big bang, through a maximum radius and back to zero radius at the big crunch. Closed, homogeneous, isotropic three geometries are characterized by a single number - the three sphere radius a_0. The wave function in the minisuperspace model with this symmetry is therefore a function of this single number

$$\psi_0 = \psi_0(a_0) \quad . \tag{13}$$

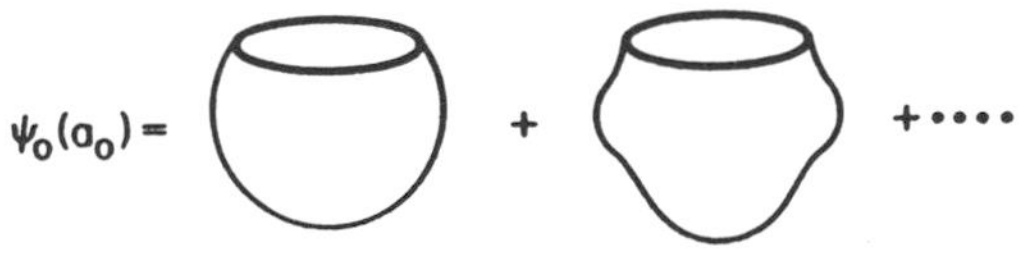

FIG. 1 A schematic representation of the Euclidean functional integral prescription for the wave function of the state of minimum excitation in a minisuperspace model. In a minisuperspace model where the three geometries are restricted to be homogeneous isotropic three spheres like the spatial sections of a closed Robertson-Walker model, the wave function is a function only of the radius of the three sphere, a_0. The wave function is given by a sum of exp(-action) over the Euclidean four geometries represented here with two of their dimensions suppressed. The four geometries have three sphere symmetry and can be thought of as composed of three sphere sections of varying radii. The wave function in the minisuperspace model is defined as the sum of exp(-action) over the four geometries with this symmetry which are compact and have the three sphere of radius a_0 as their only boundary.

The minisuperspace is thus the positive real line. The functional integral (10) which defines $\psi_0(a_0)$ is over all compact Euclidean four geometries which have homogeneous and isotropic spatial sections one of which is the boundary with radius a_0 (Figure 1). These four geometries have the line element

$$ds^2 = d\tau^2 + a^2(\tau)d\Omega_3^2 \quad , \tag{14}$$

where τ is like a polar angle. The functional integral thus reduces to one over a single function of one variable - $a(\tau)$,

$$\psi_0(a_0) = \int \delta a \exp\,(-I[a]) \quad , \tag{15}$$

where $I[a]$ is the gravitational action restricted to geometries of the form (14). (From now on we put $h = c = G = 1$.)

Even with this significant reduction in degrees of freedom the functional integral is not evaluable explicitly. One can get a handle on the qualitative behavior of the resulting ψ_0 by evaluating the integral semiclassically i.e. in the steepest descents approximation. In the steepest descents approximation, the functional integral (15) is dominated by the compact four geometry or geometries of the form (14) which have a three sphere boundary radius a_0 and make the action stationary i.e. solve the Euclidean Einstein equation

$$R_{\alpha\beta} = \Lambda g_{\alpha\beta} \quad . \tag{16}$$

If there is one dominant stationary geometry we would write for the semiclassical approximation

$$\psi_0(a_0) \approx \exp[-I_{c\ell}(a_0)] \quad , \tag{17}$$

where $I_{c\ell}(a_0)$ is the action evaluated at the stationary geometry. (There is, of course, a prefactor but we shall ignore it for a qualitative discussion.)

The unique compact solution of the Euclidean Einstein equation (16) with the symmetries of (14) is the Euclidean four sphere with radius $(3/\Lambda)^{1/2}$. For not too big a_0, the appropriate compact geometry with boundary is a part of the 4-sphere bounded by a three sphere of that radius (Figure 2). The action of the 4-sphere is negative. As a_0 is increased from zero and more and more of the four sphere increasingly is included in the stationary geometry, the exponent in (17)

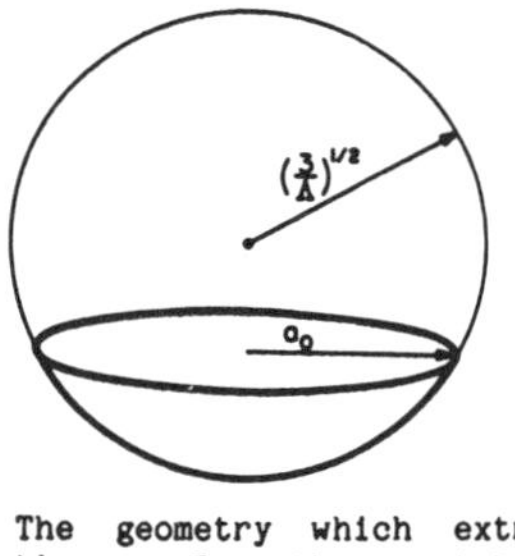

FIG. 2 The geometry which extremizes the action. In the semiclassical approximation the dominant contribution to the minisuperspace wave function $\psi_0(a_0)$ when $a_0<(3/\Lambda)^{1/2}$ is given by the Euclidean geometry represented here. The four sphere of radius $(3/\Lambda)^{1/2}$ is the Euclidean geometry with the symmetries of the model which extremizes the action. The extremizing compact geometry with a three sphere boundary of radius a_0 is a part of this 4-sphere. As a_0 increases from zero the part of the four sphere increases. For $a_0>(3/\Lambda)^{1/2}$, the three sphere will not fit in the four sphere and the extremizing geometry becomes complex.

becomes increasingly positive and the wave function ψ_0 (a_0) increases.

When a_0 passes beyond the radius $(3/\Lambda)^{1/2}$ the three sphere will no longer fit in the four sphere and there is no longer a real stationary four geometry. There are, however, complex stationary geometries. Since the original integral (10) is real these contribute to the steepest descent approximation in complex conjugation pairs. Thus, for $a_0 > (3/\Lambda)^{1/2}$ the wave function oscillates

$$\psi_0(a_0) \approx \cos\left(|I_{cl}(a_0)| - \pi/4\right), \tag{18}$$

while for $a_0 < (3/\Lambda)^{1/2}$ we have the damped behavior of (17).

These qualitative features are reproduced in numerical calculations such as that shown in Figure 3. The wave function is damped for three sphere radii below $(3/\Lambda)^{1/2}$ and oscillating and slowly decreasing above this radius. I shall now argue that this is the appropriate behavior for a state of minimum excitation for this model.

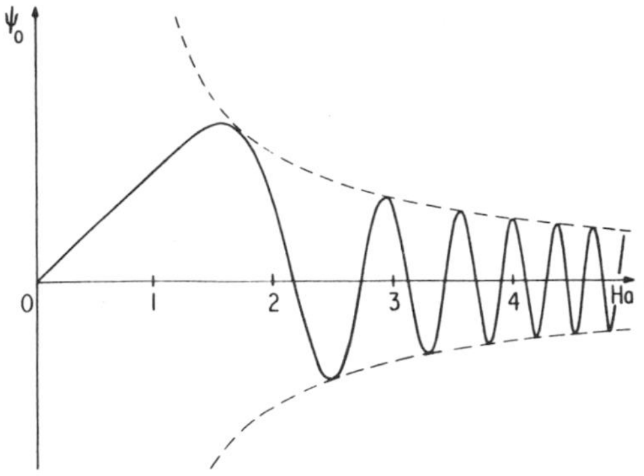

FIG. 3 The wave function $\psi_0(a_0)$. The wave function is damped for $a_0 < H^{-1}$ where $H^2 = \Lambda/3$. Semi-classically this corresponds to the absence of three spheres with these radii in Lorentzian de Sitter space. The wave function is oscillatory for $a_0 > 1/H$. Its envelope corresponds to the distribution of three spheres in Lorentzian de Sitter space.

This model contains only gravitational degrees of freedom. The classical solution of Einstein's equation with cosmological constant having the symmetries of (12) is de Sitter space. The geometry of de Sitter space is the geometry of a Lorentz hyperboloid in a flat five dimensional Lorentz signatured spacetime (Figure 4). de Sitter spacetime may be viewed as the time evolution of a homogeneous isotropic three sphere. The sphere starts from a large radius, collapses until a minimum radius $(3/\Lambda)^{1/2}$ is reached and then reexpands. Thus classical de Sitter space contains no three spheres of radius less than $(3/\Lambda)^{1/2}$ and contains three spheres of all radii greater than $(3/\Lambda)^{1/2}$. The wave function constructed from the Euclidean functional integral reflects these properties. It is damped [eq. (17)] for $a_0 < (3/\Lambda)^{1/2}$ and only slowly varying for $a_0 > (3/\Lambda)^{1/2}$. In fact, the slow variation of the envelope of the wave function reflects exactly the distribution of three spheres in de Sitter space.[13] Thus, in this simplest of models the Euclidean functional integral prescription yields a wave function which reflects the properties of the classical solution of highest symmetry or minimum excitation.

In some rough sense de Sitter space is the spacetime closest to our own universe were we limited to the degrees of freedom in this model. However, to explore seriously the proposal that the wave function constructed from the Euclidean functional integral is the wave function of our universe one clearly needs models like that described above but whose degrees of freedom more nearly reflect those of the real universe. At least three questions must be addressed of the wave functionss resulting from these models.

1. Can the universe get as large as we see it today? If gravity were the only force in the universe one would expect the scales of size and age to be set by the Planck length. Our universe is about 10^{60} times larger and 10^{60} times older. The wave function must have a significant amplitude for a large universe.

2. Is there matter in the universe? We certainly do not live in a state of minimum excitation as far as the matter content of the universe is concerned. The wave function must have a significant amplitude for the appropriate matter content.

3. Is the universe nearly homogeneous and isotropic? The wave function should be peaked about homogeneous and isotropic universes yet have the fluctuations which eventually give rise to galaxies.

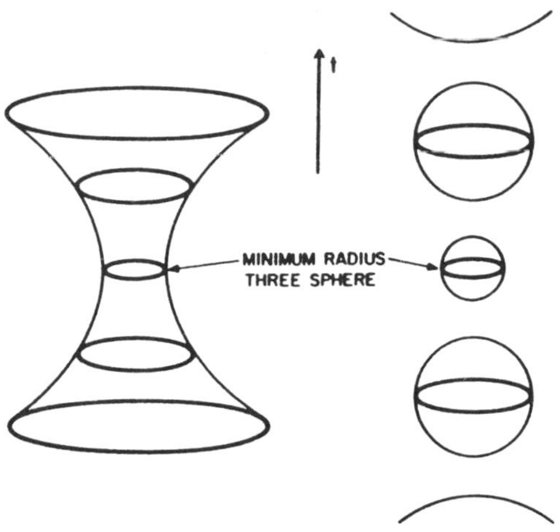

FIG. 4 Lorentzian de Sitter space. The most symmetric solution of Einstein's equation with cosmological constant is de Sitter space. It is a Lorentz hyperboloid in a flat five dimensional Lorentz signatured spacetime. It may be thought of as the evolution of a three sphere which collapses to a minimum radius $(3/\Lambda)^{1/2}$ and then reexpands.

Hawking and Hawking and Luttrell[15-18] have constructed a series of models, only slightly more complex than the one described above which give some hope that these questions will be answered affirmatively and that the features of the real universe will emerge from the state of minimum excitation. In the simplest model gravity is coupled to a scalar field of mass m representing the matter content of the universe. The geometry is again restricted to have three sphere symmetry. The minisuperspace has two dimensions-the radius of the universe, a, and the amplitude of the scalar field ϕ. Hawking finds that in a certain region of minisuperspace the semiclassical wave function constructed from the Euclidean functional integral prescription may be viewed as a superposition of states which correspond in the classical limit to cosmologies which do get big and do contain matter. These classical cosmologies have an initial regime in which the scalar field varies only slowly. The $m^2\phi^2$ term in the Lagrangian of the scalar field acts as an effective cosmological constant in this regime and the universe inflates. Universes with the appropriate m and initial ϕ can reach the present size. Eventually the field $\phi(t)$ varies significantly and begins oscillating in time. The matter field thus acquires energy and the universe evolves in a matter dominated way. Thus Hawking's model answers affirmatively two of the questions described above. [For a model which bears on the isotropy see Ref. 17.]

The Euclidean functional integral prescription defines the wave function which is naturally called the state of minimum excitation in the quantum gravitational mechanics of closed cosmologies. Moreover, it yields this wave function in a way which is accessible to approximation and generalization. In its study we shall certainly learn more about approaches to a quantum theory of gravity. We may even be studying the wave function of our own universe.

ACKNOWLEDGMENT

Preparation of this talk was supported in part by the National Science Foundation under grant PHY 81-07384.

REFERENCE

[1]Alternatively, the laws of physics may be consequences of the initial conditions. See, J. A. Wheeler, e.g. in Foundational Problems in the Special Sciences, R. Butts and J. Hintikka (eds.) (D. Reidel, Dordrecht, 1977) or in The Physicists Conception of Nature, J. Mehra (ed.) (D. Reidel, Dordrecht, 1973).

[2]See, for example, the review of D. W. Wilkinson in these proceedings.

[3]See, for example, R. H. Dicke, Nature **192**, 440 (1961), B. Carter in Confrontation of Cosmological Theories with Observational Data (IAU Symp. 63), M. S. Longair (ed.), (D. Reidel, Dordrecht, 1974) and B. J. Carr and M. J. Rees, Nature **278**, 605 (1979). It should probably be pointed out that not all authors advocating "an anthropic principle" would advocate one as strong as the extreme version presented here nor endorse its application to the determination of initial conditions.

[4]R. Penrose in General Relativity: An Einstein Centenary Survey, S. W. Hawking and W. Israel (eds.), (Cambridge University Press, Cambridge 1979).

[5]C. W. Misner, Ap. J. **151**, 431 (1968), R. A. Matzner and C. W. Misner Ap. J. **171**, 415 (1972), R. A. Matzner, Ap. J. **171** 433 (1972),

[6]See, for example, the review of A. Guth in these proceedings.

[7]C. W. Misner, Phys. Rev. Lett. **22** 1071 (1969), D. M. Chitre, Investigations of the Vanishing Horizon for a Bianchi IX (Mixmaster) Universe, Ph.D. dissertation, Univesity of Maryland (1972).

[8]See, e.g. A.A. Starobinsky, Phys. Lett. **91B**, 99 (1980) and P. Anderson, Phys. Rev. **D28**,271 (1983), Phys. Rev. **D29**, 615 (1984).

[9]See, e.g. W. Boucher and G. W. Gibbons, in The Very Early Universe, G. W. Gibbons, S. W. Hawking and S. T. C. Siklos (eds.), (Cambridge University Press, Cambridge, 1983).

[10]J. M. Stewart, Mon. Not. R. Astron. Soc. **145**, 347 (1969), C. B. Collins and J. M. Stewart, Mon. Not. R. Astron Soc. **153**, 419 (1971).

[11]See, e.g.,S. W. Hawking and G. F. R. Ellis, Ap. J. **152**, 25 (1968) and R. Geroch and G. T. Horowitz, in General Relativity: An Einstein Centenary Survey, S. W. Hawking and W. Israel (eds.), Cambridge University Press, Cambridge (1979).

[12]For some interesting reviews of the current status of the search for a quantum theory of gravity see the articles by B. S. DeWitt and S. W. Hawking in General Relativity: An Einstein Centenary Survey, S. W. Hawking and W. Israel (eds.), (Cambridge University Press, Cambridge 1979), and by K. Kuchar, in Quantum Gravity 2, C. J. Isham, R. Penrose and D. W. Sciama (eds.) (Clarendon Press, Oxford, 1981).

[13]J. B. Hartle and S. W. Hawking, Phys. Rev. **D28**, 2960 (1983).

[14]Related ideas have been put forward by A. Vilenkin, Phys. Lett. **117B**, 25 (1983), Phys. Rev. **D27**, 2848 (1983), preprint (1984); by D. Atkatz and H. Pagels, Phys. Rev. **D25**, 2065 (1982); by T. Padmanabhan Int. Jour. theor. Phys. **22**; 1023 (1983) J. V. Narlikar and T. Padmanabhan, Physics Reports **100**, 151 (1983) and very possibly by many others.

[15]S. W. Hawking, in Relativity, Groups and Topology II, ed. by B.S. DeWit and R. Stora (North Holland, Amsterdam, 1984).
[16]S. W. Hawking, Nucl. Phys. **B239**, 257 (1984).
[17]S. W. Hawking and J. C. Luttrell, The Isotropy of the Universe (preprint).
[18]S. W. Hawking and J. C. Luttrell, Higher Derivatives in Quantum Cosmology (preprint).

Notes on Cosmological Phase Transitions

B. L. Hu

Phase transitions in the early universe from the Planck time (10^{-44}sec) to the Grand Unification (GUT) time (10^{-34}sec) are believed to have played an important role in the development of recent cosmological theories,[1] most notably the inflationary models[2] - old or new, primordial or GUT. Despite the large amount of work on cosmological phase transitions done and the many interesting implications drawn therefrom, we find the current level of understanding still somewhat unsatisfactory. This is because much of the existing work on GUT phase transitions from which the standard inflationary scenario evolved are based on results derived from flat-space finite temperature quantum field theory, e.g., the Coleman-Weinberg effective potential in the new inflationary universe. We find the theoretical foundation for these models not too well established, especially in the treatment of the gravitational and statistical aspects of these problems. In this note, we comment on a number of issues which call for more careful consideration and try to identify areas which need improvement or new development. To gain a more thorough understanding of these processes we find it necessary to carry out a more systematic and complete investigation into the effects of spacetime curvature, topology, dynamics and finite temperature under general cosmological conditions. An earlier discussion of these general issues and a partial summary of our results can be found in Ref. 3.

I. EFFECT OF SPACETIME CURVATURE

It is often remarked that gravitational effects at times later than the Planck time are unimportant. This is an overstatement. It is undeniably true that quantum gravitational effects are important at and before the Planck time. However, gravitational effects can influence a particle-field system at all times as it is coupled universally. Classically the scalar curvature acts as an effective mass of the field which can influence the symmetry behavior of the system. In the same vein, how the field couples to curvature is also relevant. Semi-classically, gravitational 1-loop corrections to the quantum fields can also be important on scales associated with the field parameters, not just at the Planck scale. This is borne out by our study of the symmetry behavior of a $\lambda\phi^4$ field coupled arbitrarily to the Einstein universe[4]. We find that the near-minimally coupled fields behave very differently from the conformally coupled fields in that they tend to inhibit rather than enhance symmetry restoration. Curvature anisotropy can also alter the symmetry behavior of a system. We calculated the effective potential for a static Taub universe[5] and found that deformation into an oblate configuration in general enhances symmetry breaking while a prolate deformation tends to restore symmetry. This result can be used to study possible effects of curvature anisotropy on the suppression of primordial inflation and the quantum instability of Kaluza-Klein cosmologies. In realistic settings, factors associated with dynamic spacetimes (extrinsic), such as expansion and shear may be more dominant than factors associated with static spacetimes (intrinsic), such as curvature and anisotropy, in influencing the phase transition. (See Section II).

A special class of spacetime in which the effect of curvature has been studied in detail is the de Sitter universe[6], which forms the basis of all descriptions of inflationary cosmology. As there exist Euclidean sections in these maximally symmetric spacetimes, quantum processes in de Sitter universes can be conveniently analyzed by Euclidean quantum field theory techniques. An important consequence is the existence of a minimum temperature, the Hawking temperature, associated with the event horizon (related to the vacuum energy density or cosmological constant). Its role in GUT phase transitions and primordial galaxy formation theories is well-known. However, the results obtained for de Sitter universes are rather unique to itself (e.g. Hawking effect) and the methodology used therein (e.g. Euclidean theory) is not readily generalizable to arbitrary dynamic spacetimes. Effective potentials obtained for the de Sitter universe are useful for analyzing processes during the inflation but not before or after. Thus it cannot address the question of how the universe evolved from the initial Robertson-Walker (or chaotic) universe to the de Sitter universe, or how the system settled and remained in the false vacuum for a long time to make inflation possible. To address these questions one needs a more general framework in which both the gravitational (e.g. curvature and dynamics of spacetime) and statistical (e.g. formation of fluctuation domains, approach to critical point) effects can be dealt with in greater generality. One should try to develop statistical field theories applicable under general cosmological conditions (see Section IV and V), or, in approximate form, construct finite temperature effective Lagrangians for model field theories.[7] This is the level we aim at in our current program of investigation. (It is perhaps worth pointing out that a "finite temperature theory" in a de Sitter universe used to depict the Hawking temperature associated with an event horizon is very different from the finite temperature theory usually associated with a classical ambient radiation field - the former is intrinsically in the nature of vacuum fluctuations and is observer-kinematics dependent. See references in Section III.

II. EFFECT OF SPACETIME DYNAMICS

In adapting an effective potential for the description of symmetry breaking (as in most discussions of the new inflationary universe), one tacitly makes the assumption that the background Higgs field $\hat{\phi}$ (the order parameter) is a constant field. But this is in general inconsistent with a dynamical background spacetime as assumed in any evolutionary cosmology. The correct way is to work with an effective Lagrangian $L[\hat{\phi}, g_{\mu\nu}]$, solve the wave equations governing the background field $\hat{\phi}$ and the fluctuation field ϕ together with the Einstein equation governing the background metric $g_{\mu\nu}$ self-consistently. D. J. O'Connor and I have recently devised a method to treat cases where the background field varies slowly in time or space. By means of a Riemann normal coordinate expansion for the metric and a momentum space representation for the Green function, we are able to derive in analytic form the one-loop effective Lagrangian for a $\lambda\phi^4$ theory in curved spacetime which is exact to all orders in λ and includes variation of the background field up to the second order. Ultraviolet divergences are removed by small proper time expansion and dimensional regularization. We obtain a generalized expression for the a_2 Minakshisundaram-DeWitt coefficient of the scalar wave operator with a spacetime dependent background field. A set of renormalization group equations for the coupling constants of the theory is also obtained, which can be used for analyzing their curvature and energy dependence. We also presented an alternative derivation of the effective Lagrangian via the heat kernel technique for anisotropic harmonic oscillators.

Our result is useful for the study of quantum processes in the early universe or black holes under conditions where spacetime curvature and dynamical field effects are important. When generalized to gauge theories, the effective Lagrangian obtained here in the form of a quasi-potential should replace the flat-space Coleman-Weinberg potential assumed in most discussions of the new inflationary universe. By analyzing the infrared spectrum of the scalar wave operator in curved space in the form of anisotropic harmonic oscillators, one can study the symmetry behavior of such a system as a function of varying background curvature and fields. We find that the second derivative (in time and space) of the scalar curvature R and the order parameter Φ contribute to the effective mass of the system in such a way that symmetry tends to be broken for negative values (deceleration) and restored for positive values (acceleration). This suggests a form of dynamical symmetry breaking due to the changing background curvature and fields alone, without invoking particle interactions. When realistic cosmological and field theoretical parameters for the standard models are used, this should yield interesting new results.[9]

One can extend the present method to include higher order variations of the background curvature and fields. This will enable one to analyze situations where the spacetime changes more rapidly, as in phase transitions closer to the Planck time. Although exact forms of the effective Lagrangian including higher order non-local terms do not always exist, one can nevertheless carry out perturbation expansions (similar to the adiabatic expansion) using the 2nd order "exact" Lagrangian as a starting point. The corresponding model is that of an anisotropic anharmonic oscillator. When the non-locality of the background field is viewed as possessing spatial inhomogeneity rather than being dynamic, the present problem is analogous to assuming next-nearest-neighbor spin interactions in a Ginsburg-Landau theory. It is also similar to the multipole expansion in QCD for the description of soft-gluon processes, where large distance (or low momentum) behavior becomes more important.

III. THERMAL AND STATISTICAL EFFECTS

In most considerations of GUT phase transitions, finite temperature (T) corrections to the effective potential are usually assumed to be the same as in flat space, i.e., proportional to $T^2/24$. Our investigation for general curved spacetime[10] shows that this is true only under rather restrictive conditions which can preserve the adiabaticity of the system, e.g., for conformal fields in conformally flat spacetimes. Any deviation from these conditions introduces correction terms of order, T^{-2}, T^{-4}, etc., with coefficients proportional to curvature, coupling and the non-adiabatic expansion rate of the system. Valid at high temperatures, these results are derived from an adiabatic expansion of the finite temperature energy momentum tensor. At low temperature or for the consideration of long wavelength modes of interest in phase transition studies, alternative approximations should be devised.[11] For these problems a real-time Green function approach may prove to be more appropriate.

Under conditions where a finite temperature field theory is approximately well-defined, one can define thermal Green functions by imposing quasi-periodic conditions on the imaginary time and adopt a Riemann normal coordinate expansion for the background spacetime to include the effect of spacetime curvature on the thermal properties of the system. Combining the results from the present calculation one can in principle derive an "exact" finite temperature effective (quasi-) potential including variations of the background field up to the second order. It will be useful for studying finite temperature quantum effects in the early universe. The effective Lagrangian for non-abelian gauge theories at finite temperature derived in this way should provide a significant improvement over the existing flat-space Coleman-Weinberg potentials for the discussion of quantum processes in GUT phase transitions.

IV. CRITICAL DYNAMICS IN THE EARLY UNIVERSE

The very condition of thermal equilibrium which allows one to use finite temperature techniques is not always present in the early universe. A more complete treatment of phase transitions in dynamic spacetimes which can include large thermal fluctuations and departures from thermal equilibrium is by way of the Langevin equation (or time-dependent Ginsburg-Landau equation in the absence of noise), as in critical dynamics.[12] The change in the order parameters is determined by two factors: one of dynamic nature driven by the wave equation, the other of stochastic nature, accounting for the interaction of the system with its environment. A two-time approximation will be useful to separate the slowly varying background field and the rapid interaction and relaxation. One would expect to see characteristic behavior like critical slowing down or the like occurring in these phase transitions. The theoretical framework developed here will be useful for analyzing how the system approaches the critical point, a problem of critical importance to the working of inflationary cosmologies.

Concerning the viability of the standard GUT inflationary scenario, we take the attitude that all relevant factors should be examined carefully before one can draw any lasting conclusions, as subtleties are abound in these problems. The existing framework from which the many impressive results are deduced is, in our view, too crude to warrant any definitive claim. It is partly towards the resolution of these doubts[13] and partly for understanding the physics of the many interesting aspects of these phenomena that our present program of research is undertaken.

V. PHASE TRANSITIONS NEAR THE PLANCK TIME

The quasi-local approximation adopted in Ref. 8 which allows one to derive an exact form for the effective Lagrangian can only account for gradual variations of the background curvature and fields. Non-local approximations can be carried out to higher orders to account for more rapid changes, but are of limited validity. When the spacetime changes so rapidly that particle production becomes important, or when global properties of the spacetime (e.g. boundary or non-trivial topology) are involved, non-perturbative methods will have to be invoked. These conditions are prevalent for (primordial) phase transitions near the Planck time, which may play an important role in quantum, super-, or induced gravity theories. In such cases, not only is the effective potential ill-defined, the effective Lagrangian one should deal with can acquire imaginary terms. The process of phase transition from the false vacuum to the true vacuum will take a form similar to the problem of quantum tunneling with dissipation in certain problems in hydrodynamics and condensed matter physics.[14] As in the case without dissipation, this problem can be analyzed first in quantum mechanics language and then in quantum field theory. Ultimately one needs to combine these results with previous studies on particle production in order to deal with Planck time critical phenomena, which in many ways form the last frontier of our knowledge of phase transitions in the early universe.

Supported in part by the National Science Foundation under grant PHY81-07387.

REFERENCES

[1]For a recent review, see G. W. Gibbons, S. W. Hawking and S. T. Siklos (ed.) The Very Early Universe (Cambridge University Press, 1983).
[2]A. H. Guth, Phys. Rev. D**23**, 347 (1981); A. Linde, Phys. Lett. **108B**, 383 (1982); A. Albrecht and P. Steinhardt, Phys. Rev. Lett. **48**, 1220 (1982).
[3]B. L. Hu, in Proceedings of the Tenth International Conference on General Relativity and Gravitation, Vol. II, pp. 1086-1089, edited by B. Bertotti, F. de Felice and A. Pascolini (Padova, Italy 1983).
[4]D. J. O'Connor, B. L. Hu and T. C. Shen, Phys. Lett. **130B**, 31 (1983).
[5]T. C. Shen, B. L. Hu, and D. J. O'Connor, Phys. Lett. B (1984).
[6]G. Shore, Ann. Phys. (N.Y.) **117**, 121 (1980); S. W. Hawking and I. G. Moss, Phys. Lett. **110B**, 35 (1982); B. Allen, Nucl. Phys. B**226**, 228 (1983); A. Vilenkin, Nucl. Phys. B**226**, 504 (1983); and articles by A. Linde and A. A. Starobinski in Ref. 1.
[7]B. L. Hu, in Proceedings of the Third Marcel Grossman Meeting on Recent Developments in General Relativity, ed. H. Ning (Science Press, Beijing, 1983).
[8]B. L. Hu and D. J. O'Connor, Phys. Rev. D**30**, 743 (1984).
[9]B. L. Hu and D. J. O'Connor, "Effects of Spacetime Dynamics on Phase Transitions in the Early Universe" (Maryland preprint 1984).
[10]B. L. Hu, Phys. Lett. **103B**, 331 (1982); **108B**, 19 (1983).
[11]B. L. Hu, in Ref. 1.
[12]P. C. Hohenberg and B. J. Halperin, Rev. Mod. Phys. **49**, 435 (1977).
[13]G. F. Mazenko, W. G. Unruh and R. M. Wald, Enrico Fermi Institute preprint (1984).
[14]A. O. Caldeira and A. J. Leggett, Ann. Phys. (N.Y.) **149**, 374 (1983).

Path Integral Quantum Cosmology

Beverly K. Berger

The effect of the Hamiltonian constraint on the quantization of gravity is explored by application of Feynman path integral formalism to the quantum mechanics of spatially homogeneous cosmologies.[1] The reduction method of Arnowitt, Deser, and Misner (ADM) has been implemented within the path integral by a canonical transformation in the subphasespace of the non-dynamical degree of freedom to variables which are the Hamiltonian constraint and a time coordinate condition. The Jacobian of this transformation introduces an extra factor in the measure. Minisuperspace transition amplitudes have been constructed for several cosmological models. Closed form expressions have been obtained for the Kasner and fluid-filled Kasner cosmologies. The infinitesimal transition amplitude has been found for Robertson-Walker with scalar field and mixmaster near the minimum of the potential.

All the models may be described by combinations of the minisuperspace variables Ω related to the logarithm of the spatial volume and $\vec{\beta} \equiv (\beta_+, \beta_-)$ describing the spatial anisotropy, a parameter N which allows a change of time coordinate, and other variables to describe the spatially homogeneous scalar field, dust, radiation, spatial curvature, and cosmological constant. For the Kasner model the classical equations of motion may be obtained from

$$S(f,i) = \int_{t_i}^{t_f} dt[\vec{p}\cdot(d\vec{\beta}/dt) + p_\Omega\, d\Omega/dt - N/2\ (p^2 - p_\Omega^2)]. \quad (1)$$

The p's are the conjugate momenta and the term in parentheses is the Hamiltonian constraint H. Since $p_\Omega^2 = e^{6\Omega} h^2$ where h is the Hubble parameter, the other models are described by additional energy terms in H.

Classically, set $\Omega = t$ and solve $H = 0$ for $p_\Omega \equiv -H_{ADM}$. Quantum mechanically, the transition amplitude in the Kasner minisuperspace is the path integral

$$\langle \vec{\beta}_f, \Omega_f | \vec{\beta}_i, \Omega_i \rangle = \int D(\vec{p}, p_\Omega, \vec{\beta}, \Omega, N) e^{iS(f,i)}. \quad (2)$$

The measure is chosen to perform the canonical transformation mentioned previously to express Eq. (2) as

$$\langle \vec{\beta}_f, \Omega_f | \vec{\beta}_i, \Omega_i \rangle = \int D(\vec{p}, \vec{\beta}) \exp [i\int_{\Omega_i}^{\Omega_f} d\Omega(\vec{p}\cdot d\vec{\beta}/\ H_{ADM})]. \quad (3)$$

The factors in the measure required to go from Eq. (2) to Eq. (3) are retained for all generalizations of the Kasner model. Integrating out all momenta in Eq. (3) yields the infinitesimal (Lagrangian) reduced transition amplitude. This can also be constructed by integrating out all momenta in Eq. (2) before performing $\int$ dN to avoid difficulties associated with the non-polynomial character of H_{ADM}.

Supported in part by NSF Grants RII-83-10331 and PHY-82-13411. Part of this work formed the Master's thesis of C. N. Vogeli at Oakland University.

REFERENCES

[1] B. S. DeWitt, Phys. Rev. **160,** 1113 (1967); C. W. Misner, Phys. Rev. **186,** 1319 (1969).

Effective Coupling Constants and GUT's in the Early Universe

Leonard Parker

I would like to discuss here some new consequences of grand unified theories (GUTs) that depend in an essential way on spacetime curvature. This is work done in collaboration with David Toms.[1,2] We find that at high curvature, such as existed in the early universe when particle energies were at the GUT scale, the effective values of certain gravitationally significant coupling constants are largely determined by the particle content of the GUT under consideration. These coupling constants appear in the part of the Lagrangian containing the curvature tensor $R^{\alpha}{}_{\beta\gamma\delta}$:

$$L_{curv} = \Lambda + \kappa R + \alpha_1 R^{\mu\nu\rho\sigma} R_{\mu\nu\rho\sigma} + \alpha_2 R^{\mu\nu} R_{\mu\nu} + \alpha_3 R^2$$

$$- \xi_\phi R \, tr(\phi^2) - \xi_H R H^{\dagger} H \; . \qquad (1)$$

Here Λ, κ are related to the cosmological constant Λ_c and Newtonian gravitatinal constant G by

$$\kappa = (16\pi G)^{-1} \; , \qquad \Lambda = -(8\pi G)^{-1} \Lambda_c \qquad (2)$$

and ϕ, H are Higgs fields. The GUTs we consider include Georgi-Glashow SU(5)[3] and a related theory[4] in which the scalar self-interactions are asymptotically free.

We use the effective action, background field method[5,6] and dimensional regularization.[7] We calculate the counterterms required in the bare coupling constants to cancel the poles coming from the one-loop part of the effective action. Let $\varepsilon = (4\pi)^2(d-4)$, where d is the spacetime dimension. For $\delta\xi_\phi$, the counterterm appearing in ξ_ϕ, we find

$$\delta\xi_\phi = \varepsilon^{-1}(\xi_\phi - \frac{1}{6}) \, f_1 \text{ (g, Higgs self-couplings)}$$

$$+ \varepsilon^{-1}(\xi_H - \frac{1}{6}) \, f_2 \text{ (g, Higgs self-couplings)} \qquad (3)$$

where f_1 and f_2, functions of the gauge coupling constant g and the Higgs self couplings, are given explicitly in Ref. 2. There is a similar expression for $\delta\xi_H$. (These results hold for arbitrary gauge.) Note that only $(\xi_\phi-(1/6))$ and $(\xi_H-(1/6))$ appear. The other counterterms are

$$\delta\Lambda = \varepsilon^{-1} \, [\frac{1}{2} N_0{}^R \mu_\phi{}^4 + \frac{1}{4} N_0{}^C \mu_H{}^4] \qquad (4)$$

$$\delta\kappa = -\varepsilon^{-1}[N_0{}^R(\xi_\phi - \frac{1}{6}) \mu_\phi{}^2 + N_0{}^C(\xi_H - \frac{1}{6})\mu_H{}^2] \qquad (5)$$

$$\delta\alpha_1 = (720\varepsilon)^{-1} \, [4N_0{}^R + 8N_0{}^C - 52N_1 + 7N_{1/2}{}^W + 14N_{1/2}{}^D] \qquad (6)$$

and similar expressions for $\delta\alpha_2$ and $\delta\alpha_3$. Here μ_ϕ and μ_H are the Higgs boson masses. Also $N_0{}^R$ = no. of real scalars, $N_0{}^C$ = no. of complex scalars, N_1 = no. of vectors, $N_{1/2}{}^W$ = no. of Weyl (2-comp.) spinors, $N_{1/2}{}^D$ = no. of Dirac spinors. The values of the N's depend on the particular GUT considered.

Using these expressions for the counterterms, we write down and integrate the renormalization group equations, which give the dependence of the effective coupling constants on the renormalization point μ. It can be shown that scaling of μ by a dimensionless parameter s, $\mu \to \mu s$, corresponds to a scaling of the curvature invariants of the spacetime. The large s limit corresponds to the high curvature limit. Thus, the renormalization group equations give information about the values of the effective coupling constants at high curvature.

Solving the renormalization group equations for the effective cosmological constant $\Lambda_c(s)$ and setting its present value equal to zero, we find that in the GUT era $\Lambda_c(s)$ is large and positive:

$$\Lambda_c(s) \sim (m_{Planck})^{-2}(m_{Higgs})^4. \tag{7}$$

This may provide a complementary or alternative mechanism for producing an inflationary expansion.[8]

In contrast to Λ_c, the solution to the renormalization group equation for $G^{-1}(s)$ is dominated by the constant of integration, so that the gravitational constant G(s) remains essentially constant at its present value through the GUT era. Only as the Planck scale is approached does G(s) change significantly.

For the $\alpha_i(s)$ appearing in the terms in L_{curve} which are quadratic in the curvature, we find

$$\alpha_i(s) = b_i \ln s + \text{constant}\ , \tag{8}$$

where the b_i are constants depending on the numbers of particles in the theory. In the GUT era the $b_i \ln s$ terms are dominant, so that one can calculate the form of the curvature squared terms in the Lagrangian. We find that if ξ_ϕ and ξ_H (the Higgs-curvature coupling constants) approach the conformal value of 1/6 at high curvature, then the part of the gravitational Lagrangian quadratic in the curvature becomes

$$L_{quad} = A\, C^{\alpha\beta\gamma\delta} C_{\alpha\beta\gamma\delta} + B\, E \tag{9}$$

where $C_{\alpha\beta\gamma\delta}$ is the Weyl tensor, and

$$E = R^{\alpha\beta\gamma\delta} R_{\alpha\beta\gamma\delta} - 4\, R^{\alpha\beta} R_{\alpha\beta} + R^2 \tag{10}$$

is the integrand of the Euler characteristic, a topological invariant. Also

$$A = -\frac{1}{120(4\pi)^2}(N_0{}^R + 2N_0{}^C + 12N_1 + 3N_{1/2}{}^W + 6N_{1/2}{}^D)\ln s \tag{11}$$

and

$$B = \frac{1}{720(4\pi)^2}(2N_0{}^R + 4N_0{}^C + 124N_1 + 11N_{1/2}{}^W + 22N_{1/2}{}^D)\ \ln s \qquad (12)$$

have the same signs for any GUT. (This result is quite general.) Note that no independent term proportional to R^2 appears in L_{quad}.

Finally, integration of the renormalization group equations for the effective couplings ξ_ϕ and ξ_H shows that, under natural assumptions, in the fully asymptotically free theories ξ_ϕ and ξ_H do indeed approach the value of 1/6. This means that at high curvature the effective gravitational and matter equations of motion are dominated by conformally invariant terms. In more general GUTs, the fact that ξ_ϕ and ξ_H appear in one-loop counterterms only in the combinations $(\xi_\phi - (1/6))$ and $(\xi_H - (1/6))$ implies that it is possible that ξ_ϕ and ξ_H will approach 1/6, but their behavior ultimately depends on the values of the Higgs self-interactions in the GUT era.

In summary, we have found that the effective values of certain gravitationally important coupling constants in the GUT era are largely determined by the particle content of the theory.

REFERENCES

[1]L. Parker and D. J. Toms, Phys. Rev. Lett. **52**, 1269 (1984).
[2]L. Parker and D. J. Toms, Phys. Rev. **D29**, 1584 (1984).
[3]H. M. Georgi and S. L. Glashow, Phys. Rev. Lett. **32**, 438 (1974).
[4]N. Chang, A. Das, and J. Perez-Mercader, Phys. Rev. **D22**, 1829 (1980).
[5]B. S. DeWitt, Dynamical Theory of Groups and Fields (Gordon and Breach, New York, 1965).
[6]R. Jackiw, Phys. Rev. **D9**, 1686 (1974).
[7]G. 't Hooft and M. Veltman, Nucl. Phys. **B44**, 189 (1972).
[8]A. Guth, Phys. Rev. **D23**, 347 (1981).

Unification of Gravitation With Particle Physics Via Metric-Connection Theories

John Dell and Lee Smolin

In the small space available here we would like to briefly discuss some progress which has been made towards the construction of perturbatively sensible quantum field theories based on a new proposal for the unification of general relativity with Yang-Mills theories. These new theories are called metric-connection theories because they result from a generalization of Einstein-Yang-Mills theory in which, for both spacetime and the internal geometries, the relevant metric and connection variables are both dynamical and unconstrained. As we shall discuss, this program has led to the discovery of a new class of gauge theories which are renormalizable and perturbatively unitary and causal, and there is some reason to believe it will lead to the construction of a class of gravitational theories with these properties.

The invention of the metric connection theories was originally motivated by the desire to construct a unified theory in which the dynamics of gravitation and the other forces could be understood as arising from a single structure.[1] One question which such a unified theory must answer is why the dynamics of general relativity and Yang-Mills theory, which seem to be the relevant dynamics at low energies, are so different from each other. These differences arise from the fact that, while kinematically general relativity and Yang-Mills theory each involve a metric and a connection field, in general relativity the metric is the dynamical field while the connection is constrained, by the condition of metric compatibility, to be a function of the metric. On the other hand, in Yang-Mills theory the connection is the dynamical field while the metric, which is defined on the associated bundle (or internal vector space) in which the matter fields live, is fixed a-priori, and non-dynamical. This internal metric, while non-dynamical, plays two roles, first it provides an inner product on the space of matter fields which allows an invariant matter lagrangian to be defined, and second, it specifies the preferred SU(N) subgroup of SL(N,C) which is gauged in Yang-Mills theory. Interestingly enough, the fact that the same metric plays both roles may be expressed by saying that, while differing as to whether it is the metric or connection which is dynamical, in both Yang-Mills theory and general relativity the connection enjoys a property of metric compatibility.

The basic idea on which the construction of the metric connection theories was based was then to find a class of theories in which both metrics and both connections are initially dynamical and unconstrained which would naturally reduce to Einstein-Yang-Mills theory in a suitable low energy limit. We thus seek a class of theories in which the dynamics governing the internal geometry and the spacetime geometry arise from a simple unified structure. The striking differences between the dynamics of general relativity and Yang-Mills theory are then to be explained dynamically rather than being put in by hand. Indeed the fact that in both Yang-Mills theory and general relativity the connection is metric-compatible suggests that this can be accomplished by dropping, in each case, the condition of metric compatibility, and then finding a dynamical mechanism for freezing out the metric-noncompatible parts of the connections at low energies.

A class of unified metric-connection theories which achieves these ends was constructed in 1978, using a formulation involving general connections over frame bundles.[1] These theories contain a gravitational theory, which generalizes Einstein's theory in that both the spacetime metric and connection are dynamical, and an internal sector, which generalizes Yang-Mills theory. While it is not possible in the space available to describe in detail this unified theory, we can mention two important features which govern the recovery of the Einstein-Yang-Mills theory in the low energy limit. The important point is that the low energy limit is dominated by the invariant dimension two terms which may be formed from the variables of the theory. The freezing out of the non-metric compatible degrees of freedom at low energies is in both cases a consequence of the presence of dimension two terms of the form $M^2(\nabla g)^2$, where g may be either the spacetime or internal metric and ∇ is the relevant covariant derivative. Moreover, all of the differences between the dynamics of the spacetime variables and the internal variables at low energies are consequences of the fact that in the spacetime case it is possible to form a dimension two term linear in the curvature, $M^2 g^{\mu\nu} R(\Gamma)_{\mu\nu}$, whereas no such term is possible for the internal variables.

In the past few years we have been engaged in trying to see if these theories can be made into sensible quantum field theories. In the remaining space we would like to describe progress in the construction of the generalized Yang-Mills theory, which has led to the discovery of a new class of renormalizable quantum field theories which are unitary and ghost-free in perturbation theory, and have asymptotically free coupling constants.

The internal sector, which is what we call a metric-connection gauge theory is constructed from an SL(N,C) connection field, ω_μ, and an internal metric field q, which is assumed to be hermitian and positive definite. From these we may construct the SL(N,C) curvature, or field strength, $F_{\mu\nu} = \partial_\mu \omega_\nu - \partial_\nu \omega_\mu + [\omega_\mu, \omega_\nu]$, and the covariant derivative of q, given by $\nabla_\mu q = \partial_\mu q - \omega_\mu^+ q - q\omega_\mu$. The theory is then given by the following Lagrangian,[1,2]

$$L = -\frac{1}{4g^2} \mathrm{Tr}(q^{-1}F^+_{\mu\nu} q F^{\mu\nu}) + \frac{1}{8g'^2} \mathrm{Tr}(F_{\mu\nu}F^{\mu\nu} + F^+_{\mu\nu}F^{\mu\nu +}) + \frac{M^2}{8} \mathrm{Tr}(q^{-1}\nabla_\mu q)^2$$
$$+ \frac{1}{8h} \mathrm{Tr}(\nabla^2 q^{-1})(\nabla^2 q) - \frac{\lambda_1}{32} [\mathrm{Tr}(q^{-1}\nabla_\mu q)(q^{-1}\nabla^\mu q)]^2 \qquad (1)$$
$$- \frac{\lambda_2}{32} [\mathrm{Tr}(q^{-1}\nabla_\mu q)(q^{-1}\nabla_\nu q)]^2$$

This is the most general action containing terms of dimension four or less, invariant under SL(N,C) gauge transformations, and a certain discrete symmetry, which is defined by, $P_B : \omega_\mu \rightarrow -\omega_\mu^+$; $P_B : q \rightarrow q^{-1}$.[3]

In order to better understand the physics of this theory it is convenient to decompose the connection into two pieces, $\omega_\mu = iA_\mu + B_\mu$. The A_μ is defined by $iA_\mu = (1/2)(\omega_\mu - q^{-1}\omega_\mu^+ q)$ and gauges the SU(N) subgroup which commutes with q at a given point. The B_μ then generates the remaining SL(N,C)/SU(N) components of the group, and for this reason is called the metric-noncompatible part of the connection. It is further often convenient to choose the gauge $q = 1$, which reduces the explicit gauge invariance to an SU(N) subgroup. The field strength $F_{\mu\nu}$ may also be decomposed in this manner, and in this gauge its SU(N) component is given by $G_{\mu\nu} = f(A)_{\mu\nu} + [B_\mu, B_\nu]$, and its SL(N,C)/SU(N) component by $W_{\mu\nu} = D_\mu B_\nu - D_\nu B_\mu$, where $f(A)_{\mu\nu}$ is the usual Yang-Mills field strength, and D_μ is the usual Yang-Mills covariant derivative. Further in this gauge we have the identity $q^{-1}\nabla_\mu q = -2B_\mu$. The lagrangian (1) then takes the form

$$L|_{q=1} = -\frac{1}{4e^2}G_{\mu\nu}\cdot G^{\mu\nu} - \frac{1}{4f^2}W_{\mu\nu}\cdot W^{\mu\nu} + \frac{M^2}{2}B_\mu\cdot B^\mu - \frac{1}{2h}\mathrm{Tr}(D_\mu B^\mu)^2 - \frac{\lambda'_1}{4}(B_\mu\cdot B^\mu)^2 - \frac{\lambda_2}{4}(B_\mu\cdot B_\nu)(B^\mu\cdot B^\nu) \qquad (2)$$

where the coupling constants are given by, $1/e^2 = 1/g^2 + 1/g'^2$, $1/f^2 = 1/g^2 - 1/g'^2$ and $\lambda'_1 = \lambda_1 - 8h^{-1}$. In this gauge the theory looks like an SU(N) gauge theory coupled to a multiplet of massive vector fields in the adjoint representation. Note, however, three features of the dynamics of the B' s not present in any other gauge theory: the invariant mass, the dynamical longitudinal mode and the independently renormalized four point couplings, all arising from gauge invariant terms in the fundamental lagrangian.

This theory enjoys all of the desirable features of perturbative quantum field theory. It is renormalizable, asymptotically free in at least two of the coupling constants (e^2 and f^2), and is perturbatively free of ghosts and tachyons.[2] The perturbative stability of the theory is due to the presence of the four point couplings among the B' s. To see this, we look at the Hamiltonian for constant B fields, which has the form[2]

$$U = \frac{-M^2}{2}(B_\mu\cdot B^\mu) + \frac{\lambda'_1}{4}(B_\mu\cdot B_\mu)^2 + \frac{\lambda_2}{4}(B_\mu\cdot B_\nu)(B^\mu\cdot B^\nu). \qquad (3)$$

It is clear that for $M^2 \neq 0$ the usual perturbative vacuum is not even locally a minimum of the energy. In order to find the ground state one must expand around a mean field description of the vacuum, and look for states around which all perturbations result in an increase of energy. The result of this calculation is that the ground state contains a condensate of gauge and Lorentz invariant pairs of B' s such that $\langle\psi|B^a_\mu|\psi\rangle = 0$ and $\langle\psi|B^a_\mu B^b_\nu|\psi\rangle = \delta^{ab}\eta_{\mu\nu}F$, where F is a computable constant of order M^2/λ_1. One can show that the perturbation theory around this state is free of ghosts and tachyons, and that for the small fluctuations the B_μ are massless, while the A_μ acquire a mass of order $(e/f)F^{1/2}$.[2,4,5]

The associated gravitational theories are even more complex, and their quantization is presently under investigation. Given the successful quantization of (1) there is reason to believe that there does exist a perturbatively sensible class of metric-connection gravitational theories, particularly because the decoupling of the metric and connection at high energy may allow the theory to scale at high energies without the need for higher derivative instabilities. Two other topics under investigation are the role of second class constraints (which arise in many of these theories) in the functional integral formulation,[6] and the role of fermions in the condensate discussed above. If the condensate does contain a component of bound states involving a pair of fermions and a B, then flavor symmetries may be dynamically broken.[4]

ACKNOWLEDGEMENTS

This research was supported in part by N.S.E.R.C. grant 9391/9047-8279 to the University of Windsor and by N.S.F. grant PHY 80-26043 and D.O.E. grant AC0282ER-40073 to the University of Chicago.

REFERENCES

[1] J. C. Dell, University of Maryland Preprint (1979) and dissertation (1980).

[2] L. Smolin, Phys. Rev. D **30**, 2159 (1984).

[3] Theories of this type, but only containing terms written in the first line of eqn. (1), and without reference to any connection with gravity, have been proposed independently by Cahill and by Zee. Please see K. Cahill, Phys. Rev. D**18** 2930 (1978); Phys. Rev. D**20**, 2636 (1979); J. Math. Phys. **21**, 2676 (1980); Phys. Rev. D**26**, 1916 (1982); and J. E. Kim and A. Zee, Phys. Rev. D**21**, 1939 (1980); A. Zee, Institute for Advanced Study Preprint (1984).

[4] J. Dell and L. Smolin, Enrico Fermi Institute Preprint, April 1984.

[5] The parameters of the theory can be chosen such that $m_A \ll \Lambda_{SU(N)} \ll F^{1/2}$ so that the low energy spectrum is, to leading order in $\Lambda_{SU(N)}/F^{1/2}$, the normal confined SU(N) spectrum, plus some additional singlet states involving B'_s. However at energies above $\Lambda^2_{SU(N)}/m_A$ the physics will differ drastically from that predicted by pure Yang-Mills theory.

[6] J. Dell and L. Smolin, Institute for Advanced Study Preprint, August 1983; L. Smolin, to appear in Nucl. Phys. B (1984).

PART VII

COSMOLOGICAL CONSTRAINTS ON PARTICLE PHYSICS

The relationship between particle physics and cosmology is a symbiotic one. The cosmologist uses particle physics as a tool to understand the forces and particles present in the early Universe. Without an understanding of the microphysics it is impossible to understand the evolution of the Universe. The particle physicist uses the unique astrophysical environments of the contemporary Universe, and the high temperatures of the early Universe, as non-traditional laboratories to study the interaction of particles under conditions that cannot be duplicated in terrestrial laboratories. Cosmology has placed constraints on the properties of neutrinos, 'invisible' axions, photinos, cosmic strings, and many other exotic particles. This chapter reviews the astrophysical and cosmological constraints that have been placed on particle properties.

The existence of a small neutrino mass of order 10-100 eV would have profound cosmological consequences. These same cosmological consequences result in a bound on the mass of light, stable neutrinos that is much better than laboratory limits for the muon and tau neutrinos. In this chapter Sciulli reviews the status of laboratory experiments on the properties of neutrinos. Some cosmological implications of a neutrino mass are discussed in papers by Madsen and Epstein. Measurements of the background radiation provide limits on the radiative lifetimes of neutrinos that are much stronger than laboratory bounds, as is discussed by Bowyer. Numerical simulations of structure formation in a Universe with massive neutrinos were discussed in Chapter III. Sikivie presents a review of axions in astrophysics and cosmology, and Iwamoto discusses the role of axions in the cooling of neutron stars. Chapter III also included numerical simulations of galaxy formation with axions and a review by Vilenkin of cosmic strings and their astrophysical and cosmological implications. In this chapter a further astrophysical consequence of cosmic strings, gravitational lensing, is discussed by Turner and by Frieman. Kim and Salati discuss some cosmological implications of the supersymmetric particles discussed in Chapter VI. The chapter concludes with papers on cosmic rays, Higgs particles, mirror fermions, and fractionally-charged particles.

Review of Neutrinos and Neutrino Mass from Particle Physics Experiments

Frank Sciulli

I. INTRODUCTION

Our mission today is to describe our present (June 1984) understanding of the nature of neutrinos as gleaned from experimental particle physics. Sometimes this information comes indirectly, as was the case when the neutrino was first proposed as a particle; quite often, it comes directly from the observation of neutrinos interacting with matter. Table I shows some typical sources of neutrinos available for such direct observation, together with typical detector sizes and event rates.

TABLE I. Source of Neutrinos for Experimentation

Source	Energy Range	Source Process	Distance from Source	Detector Mass	Event Rates	Detection Process	Beams
Sun	0.2-8 MeV	β decay $p+p\to D+e^{+}+\nu_e$	1.5×10^{11}m	600 tons	1/2 cnt/day	$\nu_e+{}^{37}Cl \to e^{+}+{}^{37}Ar$	isotropic
Reactor Bombs	0.2-10 MeV	β decay	10-40m >500m	10 "	50/day 10/explosion	$\bar{\nu}_e+p\to e^{+}+n$	"
LAMPF (800 MeV)	150 MeV	$\mu^{+}\to e^{+}+\nu_e+\bar{\nu}_\mu$ $\pi^{+}\to\mu^{+}+\nu_\mu$	200m	20 "	70/day	$\nu_\mu+n\to\mu^{-}+p$ $\bar{\nu}_e+p\to e^{+}+n$	wide band focus
BNL PS (30 GeV)	1 GeV	$\{^{\pi}_{K}\}\to\mu+\nu_\mu$	100m	30 "	~10/min	$\nu_\mu+N\to\mu+N'+\pi$	"
SPS FNAL (400→1000 GeV)	100→ 300 GeV	$\{^{\pi}_{K}\}\to\mu+\nu$	1000m	600 "	~10/min to ~50/min	$\nu_\mu+N\to\mu+X$	narrow band focus
FNAL Beam Dump	~50 GeV	$F\to\tau+\nu_\tau$	100m	20 "	~15/day	$\nu_\tau+N\to\tau^{-}+X$	beam dump

II. HOW MANY NEUTRINOS (LEPTON NUMBER)

The first question we should address is "How many different kinds of neutrinos (and antineutrinos) are there?" We are absolutely sure of two distinct types: electron-type (ν_e) and muon-type (ν_μ). They have been observed to directly interact and preserve their lepton quality through birth and death.[1] There is a high level of confidence in one more, the tau-type (ν_τ), whose existence is inferred from the demonstrated existence and decay properties of the τ-lepton.[2] Hence, there are at least three. We are told by cosmologists[3] that there should be no more than four. Clearly it is important to make direct measurements to find how many there really are. Such information should be forthcoming when the resonant width of the Z^0 can be measured with sufficient precision.

The neutrinos interact weakly through the V-A coupling. Admixtures of other couplings are small, if they exist at all. This means that, to a good approximation, all neutrinos have left-handed spin, and all anti-neutrinos are right-handed. All observed interactions are consistent with the "standard" Lepton Number assignment, shown in Table II. Antiparticles have equal and opposite assignments.

TABLE II. Conventional Lepton Number Assignments

Charged Lepton	L_e	L_μ	L_τ	Neutrino
e^-	1	0	0	ν_e
μ^-	0	1	0	ν_μ
τ^-	0	0	1	ν_τ

Alternatives to the strict conservation of these lepton numbers have been:

(1) Multiplicative Rule: $\Sigma(L_e+L_\mu+L_\tau)$ = constant and $\pi(-1)^{L_i} = C_i$ would permit, for example, $\mu^+ \rightarrow e^+ + \bar{\nu}_e + \nu_\mu$ as well as $\mu^+ \rightarrow e^+ + \nu_e + \bar{\nu}_\mu$. This has been ruled out at the 10% level or better.[4]

(2) Majorana Neutrinos: In this scheme, the lepton assignment is retained, but for example, the right-handed neutrino is the <u>same</u> particle as the left-handed antineutrino. We will have more to say on this subject later.

(3) Lepton Number Violation among the families occurs at a low level. Typical limits on such violations are shown in Table III.

TABLE III. Limits on Forbidden Transitions Relative to Allowed

Reaction	Upper Limit	Reference
$\nu_\mu + N \rightarrow e^- + X$	3×10^{-3}	5
$\nu_\mu + N \rightarrow \tau^- + X$	6×10^{-3}	6
$\nu_e + N \rightarrow \tau^- + X$	0.35	7

III. NEUTRINO MASS DIRECTLY MEASURED

Of equal importance to numbers of lepton species to the cosmologists are the masses of the neutrinos. The measurement technique is to use momentum and energy conservation in the decay process. Clearly, reactions in which the energy release is smallest permit the most sensitivity. This is illustrated in Table IV, which shows some typical limits as of a few years ago and the reactions from which these limits are inferred.

TABLE IV. Typical Limits on Neutrino Mass

Neutrino	Mass Limit	Reaction	p_ν	Reference
ν_e	55 eV	$H^3 \rightarrow He^3 + e^+ + \nu_e$	<18 keV/c	8
ν_μ	500 keV	$\pi \rightarrow \mu + \nu_\mu$	37 MeV/c	9
ν_τ	250 MeV	$\tau^- \rightarrow \nu_\tau + e^- + \bar{\nu}_e$	<750 MeV/c	10

A. Electron Neutrinos

Figure 1 shows the limits on direct measurement of electron neutrino mass as a function of time. As a rule of thumb, the measurements have improved by about a factor of five per decade. The present emphasis on this measurement should provide even better prospects over the next few years.

Much of this emphasis comes from the 1980 positive result by the ITEP Group for finite electron neutrino mass.[11] This measurement made on electrons from the decay $^3H \rightarrow {}^3He + e^+ + \nu_e$ (with the tritium contained in the valine compound), resulted in a quoted mass range

$$14 < M_{\nu_e} < 46 \text{ eV} .$$

This range was not in contradiction to existing limits.

There were several criticisms of this result:

(1) The positive result depends on the resolution function used. The resolution function, in turn, depends on the intrinsic line width of the calibration source.[12]

(2) There is some dependence on the assumed branching fraction into the 2s and 1s atomic states of the valine molecule.

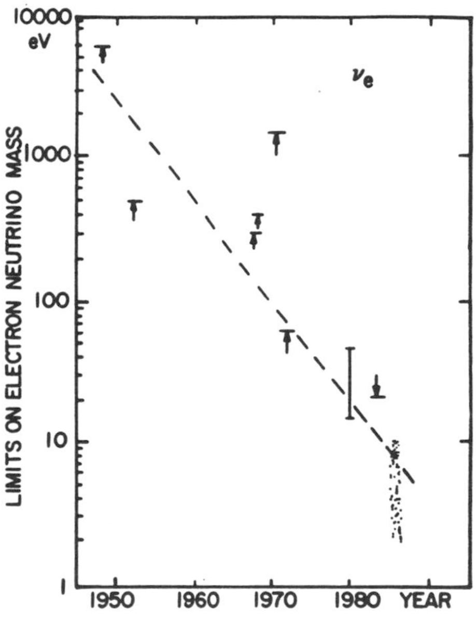

Figure 1: Limits (measurements) on ν_e mass as a function of time.

Recently, new results[13] have come from this group with an apparatus that is substantially improved. The improvements include:

(a) an electrostatic grid which removes background and permits the spectrometer to be set constant;

(b) focusing to provide higher signal rate;

(c) improved resolution, with the calibration line width measured in a separate experiment and effect of ionization losses measured separately.

Figure 2 shows the rate as a function of electron energy near the endpoint from the new ITEP experiment. The old background level is shown; the new background level is smaller by a factor of about 20. The fits to this data still indicate a finite neutrino mass under most assumptions for the valine final state. The experimenters conclude that the data indicate that $M_{\nu_e} > 20$ eV, with the endpoint parameter

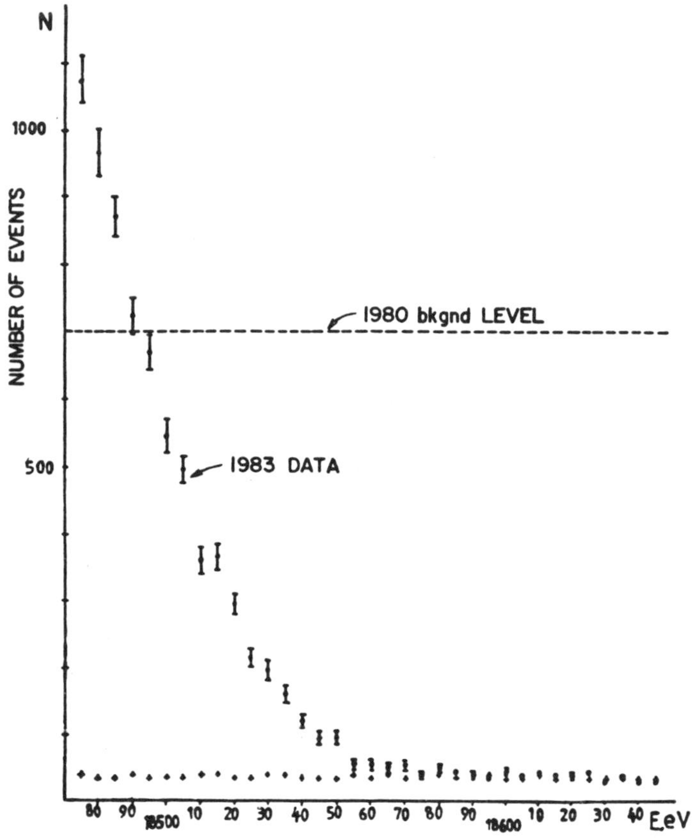

Figure 2: Spectrum of electrons from ^3H decay, showing the reduction in background in the 1983 ITEP data.

$E_0 > 15580$ eV. Figure 3 shows a Curie plot of the data with two mass hypotheses. Although the lower energy data together with the high endpoint seems to require finite mass, the higher energy data do not fit this hypothesis very well. It has been pointed out[14,22] that the ITEP result for the endpoint is not in good agreement with an independent measurement of the ^3H - ^3He mass difference,[15] which gives $E_0 = 18549 \pm 7$ eV.

Hence, the substantially improved ITEP experiment still indicates a non-zero neutrino mass, though this interpretation remains controversial. There are many new experiments expected to come on line this year to resolve the controversy. Table V shows a partial list of

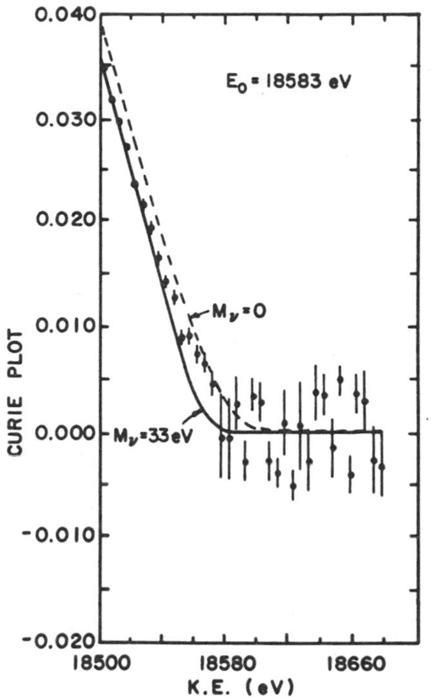

Figure 3: Curie plot of tritium decay electron (ITEP) data. Curves assume endpoint value (E_o) shown. Lower endpoint values would shift curves to the left.

new experiments, most with resolutions at least that of the ITEP experiment and many with free ^3H. Corroboration or contradiction from these experiments can be expected soon.

New techniques involving electron capture by heavy nuclei are being considered for the future.[16] While not yet competitive with the traditional β-decay endpoint method, it shows promise for studying reactions with very small energy release.

TABLE V.

Experiment	Source	Resolution (rms)	Sensitivity
Fackler et al. Rock-FNAL-LLL	Solid Molecular ^3H	1-2 eV	$m_\nu > 4$ eV
Boyd Ohio State	"	10 eV	$m_\nu > 10$ eV
Bowles et al. LAMPF	Atomic ^3H	40 eV	$m_\nu > 10$ eV
Clark IBM	Solid ^3H	5 eV	-
Heller et al. UC Berkeley	^3H in Semi-conductor	100 eV	$m_\nu > 30$ eV
Graham et al. Chalk River	-	10 eV	$m_\nu > \sim 20$ eV
Bergkvist	^3H in valine	~ 25 eV	$m_\nu > 19$ eV
Kundig Zurich	-	5 eV	$m_\nu > 10$ eV
INS Japan	-	13 eV	$m_\nu > 25$ eV

B. Muon Neutrinos

The precision of measurements of mass of the muon neutrino (ν_μ) is limited by the energy release of reactions in which it partakes, notably that of π decay. The history of these measurements is illustrated in Fig. 4, where we see approximately a factor of 3 improvement per decade. Most recent measurements give limits at or below the mass of the electron, as shown in Table VI.

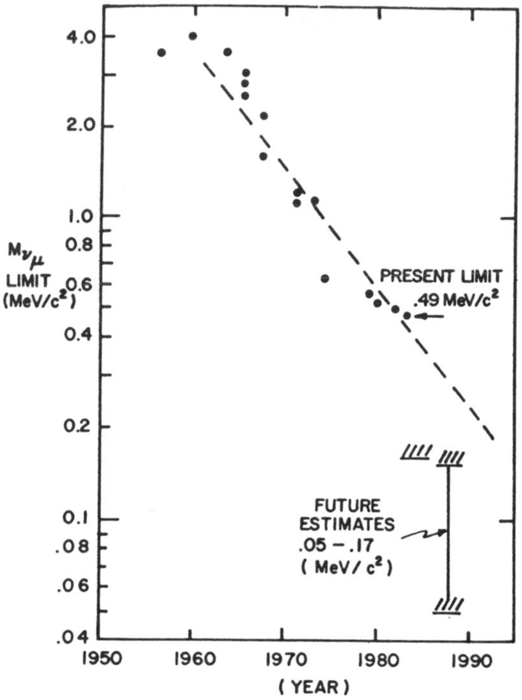

Figure 4: Limits on ν_μ mass as a function of time.

TABLE VI. Most Recent Limits from π Decay

Reference	m_{ν_μ}
17 (1980)	< 0.52 MeV/c^2
9 (1982)	< 0.50
18 (1983)	< 0.49

C. Tau Neutrinos

The tau-neutrino (ν_τ) mass is even more difficult to determine accurately. Early measurements of the mass gave[10] $m_{\nu_\tau} < 250$ MeV/c^2 from $\tau^- \rightarrow \nu_\tau + e^- + \bar{\nu}_e$. A very recent report,[19] using the low Q-value decay $\tau^+ \rightarrow \pi^+\pi^+\pi^-\pi^0\nu_\tau$ reports:

$$m_{\nu_\tau} < 164 \text{ MeV/c}^2 .$$

Figure 5 shows the invariant mass of the $3\pi^{\pm}\pi^{0}$ system with the curves for $m_{\nu_\tau} = 0$ (solid) and $m_{\nu_\tau} = 250$ MeV/c^2 (dashed). High statistics, reactions involving even smaller energy release (e.g. involving F's), and innovative approaches could reduce this limit even further.

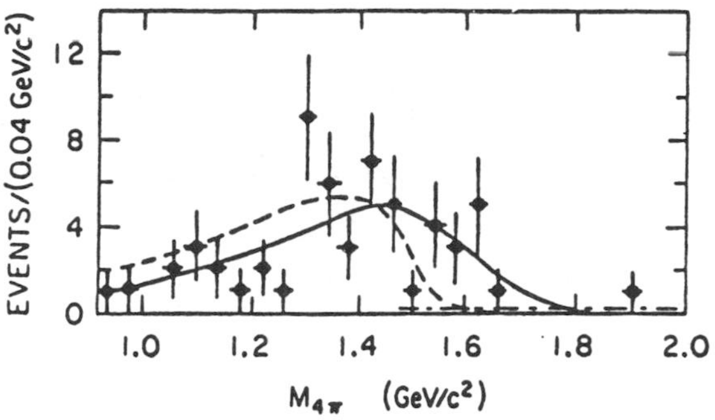

Figure 5: Invariant mass[10] of the $3\pi^{\pm} - \pi^{0}$ system together with the curves expected for $m_{\nu_\tau} = 0$ (solid) and $m_{\nu_\tau} = 250$ MeV/c^2.

D. Conclusions

In summary, the masses of the three known types of neutrinos are limited to values of

$m_{\nu_e} \lesssim 50$ eV (possibly positive result!)

$m_{\nu_\mu} < 490$ keV

$m_{\nu_\tau} < 164$ MeV

where the scale of the limit is set by the energy release obtainable in the reaction. With cleverness, dedication, lots of hard work, and some luck, we might see improvements in these limits by a factor of ten over the next decade.

IV. ALTERNATIVE EXPERIMENTAL HANDLES ON NEUTRINO MASS

As we have seen, a straightforward measurement of neutrino mass is difficult and limited by nature's available decay modes. Alternatively, one can search for effects of finite mass much more sensitively if the properties of neutrinos are appropriate. In particular, if conventional lepton number is violated, even at a low level, there are approaches to mass determination which are very different.

A. Majorana Neutrino

In the Standard Model, the V-A weak interaction causes neutrinos born in the reaction $\pi^+ \rightarrow \mu^+ + \nu_\mu$ to be left-handed (ν_μ^L). The subsequent charged current process is $\nu_\mu + N \rightarrow \mu^- + X$. But if the mass of the neutrino is finite, there is a small but finite amplitude for a right-handed neutrino to be made (ν_μ^R). (This would also happen for a finite V+A amplitude.) This becomes important if neutrinos are Majorana particles, for which right-handed neutrinos are equivalent to antineutrinos ($\nu_\mu^R = \bar{\nu}_\mu^R$). Under these circumstances, there would be a small, but finite, probability for the subsequent charged-current interaction of a ν_μ^R created from π^+ decay to give $(\bar{\nu}_\mu^R = \nu_\mu^R) + N \rightarrow \mu^+ + X$.

Unfortunately, the precision of the test just described is not very sensitive; the limit is only about 10^{-3} due to backgrounds. However, limits from double β-decay process, which tests precisely the same hypothesis for electron neutrinos, is at the level of 10^{-9} to 10^{-10}.

A normal second-order weak process expected for a nucleus of atomic weight, A, and atomic number, Z, is

$$(2\nu)\ [A,Z] \rightarrow [A,Z+2] + e^- + e^- + \bar{\nu}_e + \bar{\nu}_e \ .$$

If the neutrinos are of the Majorana type, they can be internal as shown in Fig. 6. In that case, the double β-decay can occur without external neutrinos:

$$(0\nu)\ [A,Z] \rightarrow [A,Z+2] + e^- + e^- \ .$$

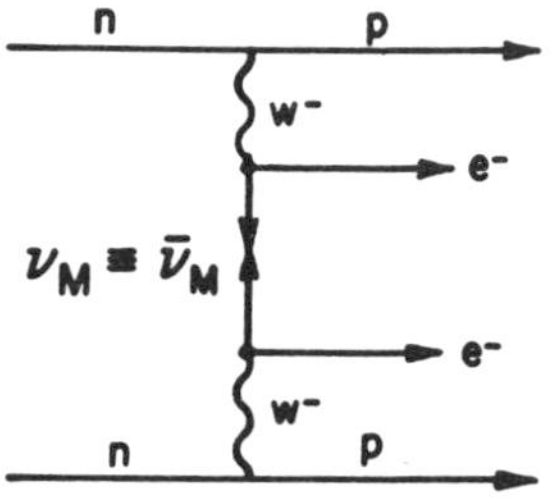

Figure 6: Second order weak process with internal Majorana neutrinos (0_ν) resulting in no final state neutrinos.

One historical method has been to search for ^{130}Te and ^{128}Te decays as evidenced by the appropriate xenon isotope in ore (geochemical); these would be daughters of the double β-decay process. If the (2ν) process alone takes place, it is expected that the ratio should be $R_{T_e} = T^{130}_{1/2}(2_\nu)/T^{128}_{1/2}(2_\nu) = 2 \times 10^{-4}$. Any contribution from the (0ν) process would make this ratio larger. One measured value[20] gave $R_{T_e} = (6.3 \pm 0.2) \times 10^{-4}$, which could imply Majorana neutrinos with mass in the range of 10 eV. A more recent measurement[21] gives $R_{T_e} = (1.0 \pm 1.1) \times 10^{-4}$, which is in contradiction to the earlier number, and is consistent with no (0ν) process. There is no firm evidence as yet for finite Majorana neutrino mass from such experiments. There have been other measurements, both in laboratory and geochemical, of processes involving (0ν) double β-decay.[22] None has shown evidence for the (0ν) process. Present limits[23] are approaching 10 eV. New laboratory measurements by groups at Milano, Caltech, Batelle and Irvine are expected to reduce these limits even further.

B. Non-Distinct Neutrino Species

Even if neutrinos are not Majorana particles, there is another way in which greater sensitivity to mass might be available to experimenters. This would be true if there were couplings among neutrino-types with new neutrinos, or with other species of neutrinos. With such mixing, there would be mass eigenstates of neutrinos, designated $|\nu_1\rangle, |\nu_2\rangle, \ldots, |\nu_i\rangle$. The neutrinos created in the weak interactions would be superpositions of these.

$$|\nu_e\rangle = \sum_i U_{ei} |\nu_i\rangle$$

$$|\nu_\mu\rangle = \sum_i U_{\mu i} |\nu_i\rangle$$

$$\vdots \qquad \vdots$$

where U is a unitary matrix (barring decays).

One case[24] would be if mass differences among the states were very large. A two-body decay (e.g. $\pi \to \mu + \nu$) would show this as a double peak in the momentum distribution of the charged secondary. For three-body decays, it would be manifest in a kink in the spectrum. No evidence is seen for such a phenomenon as yet. The limits[25] on the matrix elements are shown in Fig. 7 (ν_μ processes) and Fig. 8 (ν_e processes).

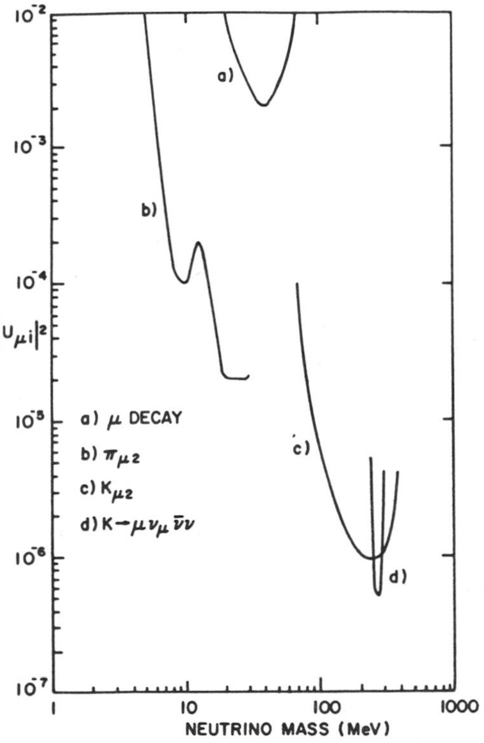

Figure 7: Limits on couplings to more massive neutrinos versus neutrino mass for muon type neutrinos.

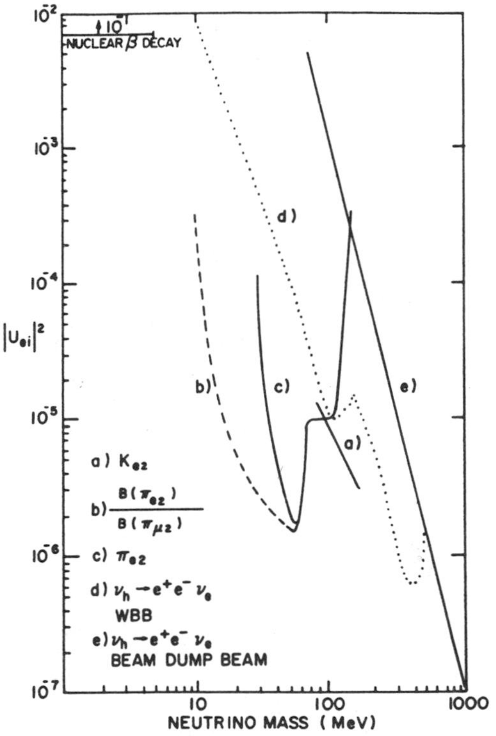

Figure 8: Limits on couplings to more massive neutrinos for electron-type neutrinos.

A more interesting case for such mixing of massive neutrino states occurs if the mass differences are small: the phenomenon of neutrino oscillations. Consider the simplest 2-component case, in which the muon neutrino mixes with one other species. A beam of muon neutrinos born from π-decay is described at the instant of birth by

$$|\psi(0)\rangle = |\nu_\mu\rangle = \cos\theta|\nu_1\rangle + \sin\theta|\nu_2\rangle ,$$

where $|\nu_1\rangle$ and $|\nu_2\rangle$ are the mass eigenstates, and θ is the single parameter describing the 2×2 unitary mixing matrix. At a later time, the state has evolved to

$$|\psi(t)\rangle = \cos\theta\, e^{-i\phi_1}|\nu_1\rangle + \sin\theta\, e^{-i\phi_2}|\nu_2\rangle ,$$

where the time-dependent phases are of the form, e.g.

$$\phi_1(t) = E_1t - p_1L \approx L(E_1 - p_1) \approx L\frac{m_1^2}{2E} ,$$

where L is the distance traversed in the lab and we assume that the mass is small compared to the lab energy. The phase difference,

$$\Delta\phi = \frac{L}{2E}\left(m_2^2 - m_1^2\right) ,$$

is the relevant parameter in the description of the state. The probability for the state $|\nu_x\rangle$, orthogonal to the $|\nu_\mu\rangle$, to appear is then

$$P_{\nu_x}(\ell) = \sin^2 2\theta \sin^2\left[\frac{L\left(m_2^2 - m_1^2\right)}{4E}\right] .$$

Clearly the effect requires some mixing between states ($\theta \neq 0$), as well as a finite mass difference between eigenstates. The sensitivity to the mass difference scales with L/E. Figure 9 shows the scales in this parameter together with the neutrinos sources useful in various ranges. Clearly, it is possible to investigate mass difference values in the range

$$0.01 < \Delta m^2 < 1000 \text{ eV}^2 ,$$

through the use of various facilities.

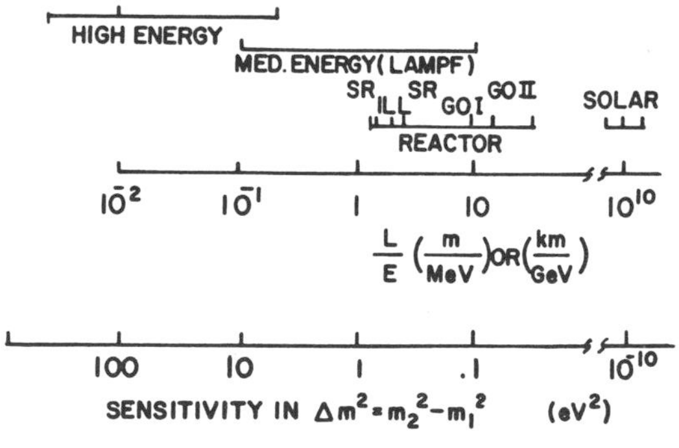

Figure 9: Ranges in L/E covered by various neutrino facilities and the corresponding sensitivity range in mass square.

There are two general types of oscillation experiments: exclusive (or appearance) experiments and inclusive (or disappearance) experiments. The former class consists of looking for the appearance of a different type than created; for example, the reaction $\nu_e + N \to e^- + X$ appearing from a beam of pure ν_μ would signal an oscillation. It has the advantage that it can be done with a single detector, and has sensitivity out to large Δm^2. It has the disadvantage that it is sensitive only to a single coupling among known neutrinos. The inclusive experiments, which measure the distance dependence of the ν_μ flux, are sensitive to couplings of ν_μ to anything. They require either more than one detector or independent knowledge of the flux, neutrino cross sections, detector efficiencies, etc. They are limited at high Δm^2 by experimental resolution on distance (L) and energy (E).

Figure 10 shows the present exclusive limits for (a) $\nu_\mu \to \nu_\tau$ and (b) $\nu_\mu \to \nu_e$ as extracted for a recent review.[25] No evidence for a positive effect has been seen. Future experiments are being planned which might reduce these limits by about an order of magnitude in each variable.

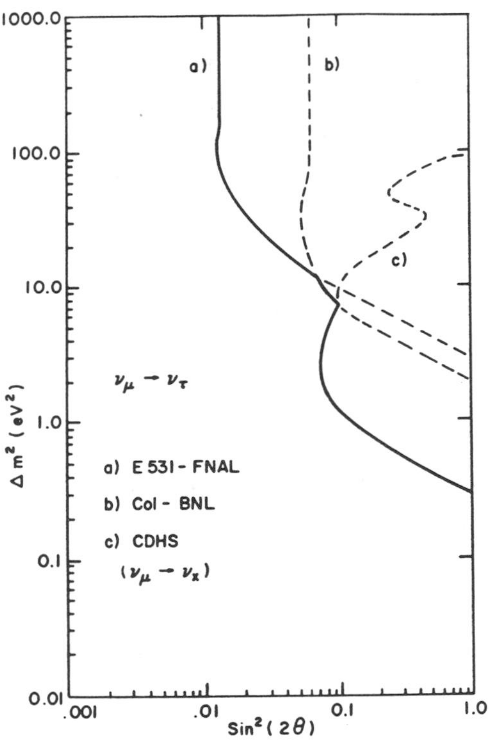

Figure 10a: Limits on $\nu_\mu \to \nu_\tau$ mass square differences versus coupling.

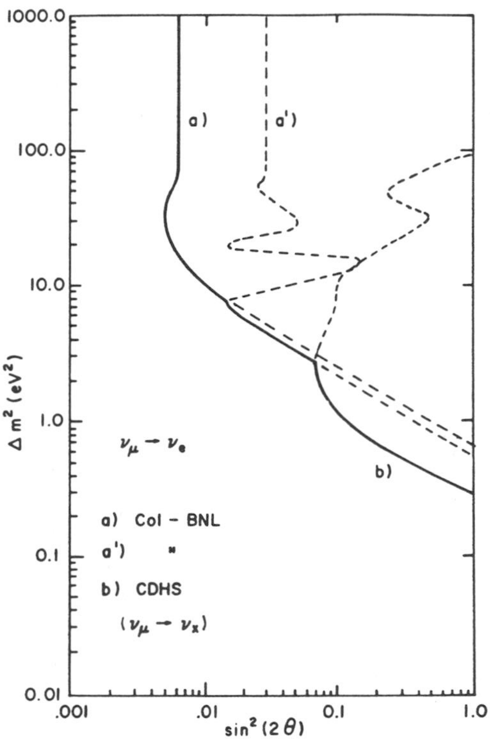

Figure 10b: Limits on $\nu_\mu \to \nu_e$ mass square differences versus coupling.

The best inclusive ν_μ limits come from two experiments which cover complementary mass ranges. The CDHS group[26] used the CERN PS with two detectors at distances of 150 m and 880 m from the effective source point in a beam of neutrinos with typical energy of 3 GeV. The CCFR group,[27] working here at Fermilab, had two detectors at 715 m and 1116 m from the source in a beam of neutrinos with typical energy 100 GeV. Both experiments look for events of the type $\nu_\mu + N \to \mu^- + X$.

Figure 11 shows the ratio of events from the CDHS two detectors as a function of the projected range of the final state muon. The curve corresponds to full mixing with $\Delta m^2 = 2$ eV^2 (i.e. for one mass dominant, $m_2 \approx 1.4$ eV).

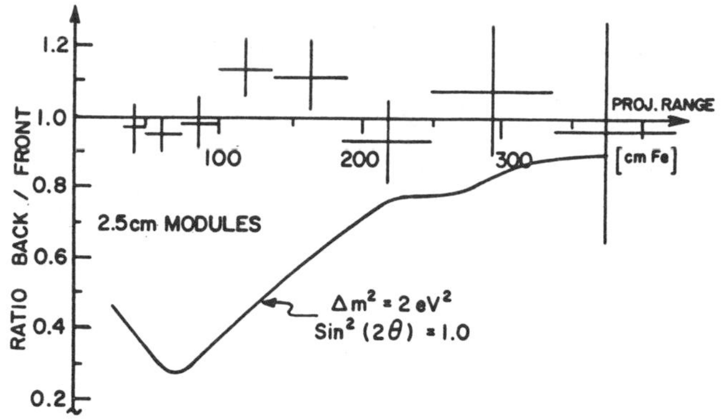

Figure 11: Ratio of events from two detectors in the CDHS experiment as a function of projected range. A curve for a typical Δm^2 value and maximum mixing is shown.

Figure 12 shows the ratio of events from the CFRR two detectors as a function of neutrino energy. The dashed smooth curve corresponds to full mixing with $\Delta m^2 = 162$ eV^2 (i.e. for single mass dominant,

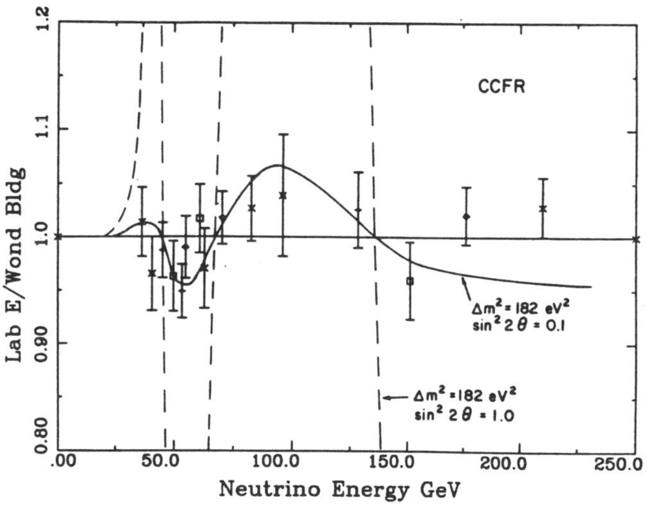

Figure 12: Ratio of events in two detectors from the CCFR experiment. A curve (solid) for a typical δm^2 value and 10% mixing is shown. The curve for maximal mixing (dashed) at the same Δm^2 is largely off-scale.

$m_2 = 13$ eV). In both cases, the data are consistent with no oscillations. The 90% confidence levels are shown in Fig. 13. Taken

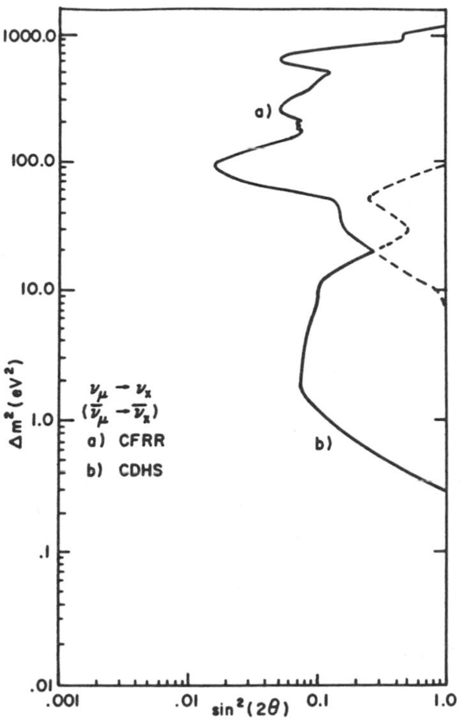

Figure 13: The 90% confidence limits on exclusive oscillations as a function of $\sin^2 2\theta$ and Δm^2 are shown for (a) CCFR experiment, and (b) CDHS experiment. The area to the right of the curves is excluded.

together, the two experiments cover the mass square range $1.0 < \Delta m^2 < 1000\ eV^2$, with mixing angles ruled out for $\sin^2 2\theta \gtrsim 0.09$ at lower masses and $\sin^2 2\theta \gtrsim 0.05$ at higher masses. Future experiments are planned which could extend to lower masses and smaller mixing angles in the low mass regions. Such improvement might improve the low mass limits by as much as a factor of five along each axis.

Inclusive ν_e experiments have been very important in the oscillation question. An early result[28] by the UCI group at the Savannah River reactor indicated a positive result with maximum mixing and $\Delta m^2 \sim 0.7\ eV^2$. By 1982, data taken[29] at the Goesgen reactor had limits which excluded this effect. Recently, this group has taken data at another position; no hint of an effect is evident. Figure 14 shows the ratio $\bar{\nu}_\mu + p \to e^+ + n$ events at two distances as a function of the positron energy. The 90% confidence limits for (a) calculated flux and (b) two

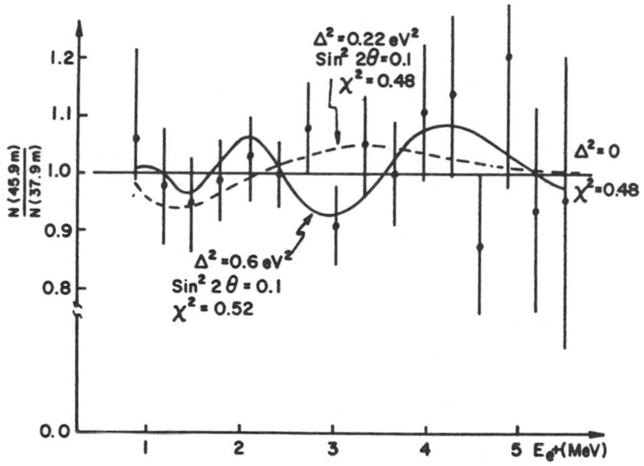

Figure 14: Ratio of events (corrected for solid angle, etc.) at two locations from Goesgen reactor data.

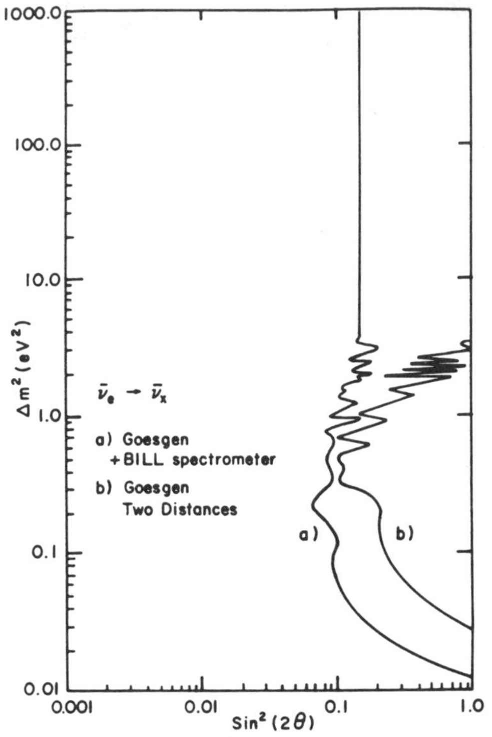

Figure 15: Limits on mass square difference versus coupling for $\bar{\nu}_e$ inclusive (Goesgen data). Curve (a) uses calculated flux, while curve (b) uses the two detector ratio.

detector ratio are shown in Fig. 15. Other groups have been experimenting along same lines. New results from these groups are expected soon.

V. CONCLUSIONS

Experimental Particle Physics has strong evidence for three neutrino types, though one has yet to be observed to interact. Collider experiments at the Z^0 should ultimately give a firm determination of the number of low mass neutrinos. There are several experimental approaches to determine whether some or all neutrinos have finite mass; although hints exist, there is no unassailable evidence from particle physics experiments that neutrinos do have finite mass. Cosmological inputs have been extremely stimulating. In many cases, they have oriented searches (e.g. in stimulating searches in certain mass ranges). In many instances, the cosmological framework has been more specific than that from theoretical physics. It is hoped that the benefits of mutual stimulation among these fields will continue to the point where much more about neutrinos is understood, and this understanding fits into all of our ideas about the nature of particles and the structure of the universe.

REFERENCES

[1]G. Danby et al., Phys. Rev. Lett. **9**, 36 (1962).
[2]M.L. Perl et al., Phys. Rev. Lett. **35**, 1489 (1975).
[3]See review by M.S. Turner, Proceedings of the 1981 Conference on Neutrino Physics, ed. R.J. Cence, E. Ma, A. Roberts, U. of Hawaii, Vol. I, p. 95.
[4]S.E. Willis et al., Phys. Rev. Lett. **44**, 522 (1980); erratum PRL **45**, 1370 (1980).
[5]N.J. Baker et al., Phys. Rev. Lett. **47**, 1576 (1981).
[6]N. Ushida et al., Phys. Rev. Lett. **47**, 1694 (1981).
[7]P. Fritze et al., Phys. Lett. **96B**, 427 (1980).
[8]K. Bergkvist, Nucl. Phys. **B39**, 317 (1972).
[9]H.B. Anderhub et al., Phys. Lett. **114B**, 76 (1982).
[10]W. Bacino et al., Phys. Rev. Lett. **42**, 749 (1979); C.A. Blocker et al., Phys. Lett. **109B**, 119 (1982).
[11]V.A. Lubimov et al., Phys. Lett. **94B**, 266 (1980).

[12]J.J. Simpson, paper presented at ICOMAC 83, Frascati, Italy (1983).
[13]S. Boris et al., Proceedings of HEP 83, ed. J. Guy, C. Costain, Rutherford Appleton Lab.
[14]J.J. Simpson, P. Vogel, Low Energy Tests of Conservation Laws, AIP Conference Proceedings, New York (1984).
[15]L.G. Smith, E. Koets, A.H. Wapstra, Phys. Lett. **102B**, 114 (1981).
[16]See, for example, review by M. Shaevitz, Proceedings of 1983 Symposium on Lepton and Photon Interactions, ed. D.G. Cassel, D.L. Kreineck, Cornell University.
[17]D.C. Lu et al., Phys. Rev. Lett. **45**, 1066 (1980).
[18]R. Abela et al., SIN Newsletter **15**, 26 (1983).
[19]C. Matteuzzi et al., Phys. Rev. Lett. **52**, 1869 (1984).
[20]E. Hannecke et al., Phys. Rev. **C11**, 1378 (1975).
[21]T. Kirsten et al., Phys. Rev. Lett. **50**, 474 (1983).
[22]For a recent review of double β-decay experiments, see F. Boehm, P. Vogel, Annual Review of Nuclear and Particle Science, Vol. 34 (1984).
[23]E. Bellotti et al, Proceedings of HEP 83, ed. J.B. Giuy, C. Costain, Rutherford Appleton Lab.
[24]R.E. Shrock, Phys. Lett. **96B**, 159 (1980); Phys. Rev. **D24**, 1232, 1275 (1981).
[25]M. Shaevitz, Proceedings of the 1983 Symposium on Lepton Photon Interactions, ed. D. Cassel, D.L. Kreineck, Cornell University.
[26]F. Dydak et al., Phys. Lett. **134B**, 281 (1984).
[27]I.E. Stockdale et al., "Search for ν_μ and $\bar{\nu}_\mu$ Oscillations at Large Values of Δm^2," to be submitted to Zeit. für Physik.; I.E. Stockdale et al., Phys. Rev. Lett. **52**, 1384 (1984).

Lower Limits on m_ν from the Distribution of Dark Matter in Galaxies

Jes Madsen and Richard I. Epstein

If one flavour of neutrino and antineutrino of mass m_ν is responsible for the non-luminous matter in the outer parts of galaxies, one can obtain firm lower limits on the value of m_ν from the observed distribution of dark matter.

Consider a galaxy where dark matter dominates outside a radius r_0, and the neutrino density is limited by

$$\rho_\nu(r) \leq \rho(r_0 \left(\frac{r}{r_0}\right)^{\beta-3} \quad \text{for } r > r_0$$

where β is a number between 0 and 2. If the mass interior to r_0 satisfies $M(r_0) \geq (4\pi/\beta)\ \rho(r_0)r_0^3$ (this constraint is for instance satisfied for a spiral with flat rotation curve), it follows that

$$M(r) \leq M(r_0) \left(\frac{r}{r_0}\right)^{\beta} \quad \text{for } r > r_0 \,.$$

Assume that the neutrinos for $r \geq r_0$ are in spherically symmetric pressure equilibrium. The neutrino pressure at r_0 is then bounded from above by

$$P_\nu(r_0) = \int_{r_0}^{\infty} \frac{GM(r)}{r^2}\,\rho_\nu(r)dr \leq \int_{r_0}^{\infty} \frac{G}{r^2}\,\rho(r_0)M(r_0) \left(\frac{r}{r_0}\right)^{2\beta-3} dr$$

or

$$P_\nu(r_0) \leq \frac{GM(r_0)\rho(r_0)}{[4-2\beta]r_0} \,.$$

On the other hand, the neutrino pressure is bounded from below by the ground state degeneracy pressure

$$P_\nu(r_0) \geq \frac{8\pi}{15}\left(\frac{3}{8\pi}\right)^{5/3} h^2 \,\frac{1}{m_\nu^{8/3}}\,\frac{\rho_\nu(r_0)^{5/3}}{f^{2/3}}$$

where the occupation number, f, must be less than 0.5 for relic neutrinos from the Big Bang.

Taken together, these pressure equations imply a lower bound on m_ν

$$m_\nu \geq 32.2 \text{ eV}\,\frac{[4-2\beta]^{3/8}}{f^{1/4}}\left(\frac{\rho(r_0)}{10^{-24}\text{g cm}^{-3}}\right)^{1/4}\left(\frac{r_0}{10 \text{ kpc}}\right)^{3/8}\left(\frac{M(r_0)}{10^{10}M_\odot}\right)^{-3/8}$$

For the giant elliptical M87 application of the formula above gives $m_\nu > 8$ eV.[1] Utilizing data from surveys of spiral galaxies,[2] we find that the highest bound on m_ν is obtained for NGC 4605, which like most other spirals in these surveys has $\rho \propto r^{-1.7}$ at the optical limit of the galaxy. Using therefore $\beta = 1.3$ and data for the density and mass at the optical radius, we get $m_\nu > 39\ \mathrm{eV} \cdot h_0^{1/2}$, where h_0 is the Hubble parameter in units of 100 km $s^{-1}Mpc^{-1}$. Several other spirals give m_ν-limits above 30 $\mathrm{eV} \cdot h_0^{1/2}$.

Note that the method described here does not depend on the actual coarse-grained phase-space distribution of neutrinos in galaxies (except for assuming isotropy). Another approach to the problem with the same advantage is presented in (1).

REFERENCES

[1] J. Madsen and R. I. Epstein, Ap. J. **282** (in press), (1984).

[2] D. Burstein, V. C. Rubin, N. Thonnard, and W. K. Ford, Ap. J. **253**, 70 (1982); V. C. Rubin, W. K. Ford, N. Thonnard, and D. Burstein, Ap. J. **261**, 439 (1982).

J. M. was supported by the Danish Natural Science Research Council.

Astrophysical Data on 5 eV to 1 keV Radiation from the Radiative Decay of Fundamental Particles: Current Limits and Prospects for Improvement

Stuart Bowyer and Roger F. Malina

ABSTRACT

Line emission from the decay of fundamental particles, integrated over cosmological distances, can give rise to detectable spectral features in the diffuse astronomical background between 5 eV and 1 keV. Spectroscopic observations may allow these features to be separated from line emission from the numerous local sources of radiation. We review the current observational status and existing evidence for such features. No definitive detections of non-galactic line features have been made. Several local sources of background mask the features at many wavelengths and confuse the interpretation of the data. No systematic spectral observations have been carried out to date. We review upcoming experiments which can be expected to provide significantly better constraints on the presence of spectral features in the diffuse background from 5 eV to 1 keV.

I. INTRODUCTION

The diffuse astronomical background between 5 eV and 1 keV (~ 2400 - 12Å) is of significant interest to particle physics and cosmology. In regard to particle physics, the decay radiation from particles such as photinos, massive neutrinos, axions, and possibly others[1] may fall within this bandpass. In regard to cosmology, a number of processes may produce observable features; one example is the signature of the reheating of primordial gas in galaxy formation.

Astrophysical observations of the decay radiation may provide the best tests of many theories of the radiative decay of fundamental particles because of the large path-lengths available. Depending on the history and spatial distribution of the emission, this radiation will be seen as line emission or as an emission edge starting at the wavelength of the minimum redshift. Thus, for example, the decay of particles trapped in the halo of a galaxy will appear as a narrow spectral line.

The spatial distribution of this radiation is expected to be similar to, and thus masked to some extent by, other cosmological sources of radiation. These include the integrated light from galaxies and quasars and emission from the intergalactic medium. Local sources of line radiation are also known which will mimic any decay radiation. These include emission from hot gas in our own galaxy and emitted light from galactic sources such as fluorescing molecular hydrogen.[2] Also prominent in the diffuse background spectrum are line emission from within the solar system and from the upper atmosphere (in the case of earth orbiting instruments). Thus spectral rather than spatial observations will be required for such studies.

Another factor in the ability to detect cosmological line radiation is the opacity of local matter: at 250 eV (44Å) the opacity of the interstellar medium becomes significant over galactic distances; at 40 eV (~ 300Å) even the thickness of the galactic disk is sufficient to attenuate any extra-galactic flux. Thus in this energy band only decay radiation from particles sufficiently massive to be trapped in the potential well of our own galaxy can be probed with high sensitivity.

Observations of the diffuse background in this band now have been carried out for some fifteen years beginning with the first detection of the soft X-ray background[3] and the first observations of the diffuse far ultraviolet background. Very little spectral data in any of these bands exists, as almost all measurements have been broad band.

In spite of the limited observational status, existing observations have been used to provide significant limitations on the predictions of current particle theories. For decay photons between 5 and 50 eV, for instance, a number of arguments have been used[4] to derive lower limits on the neutrino lifetimes which are much greater than the age of the universe. Since theoretical estimates range from 10^{12} to as high as 10^{37} seconds, large portions of parameter space are excluded by these astronomical observations. More limited constraints for neutrino masses between 4 and 10 eV[5,6] have also been provided by observations of the Coma and Virgo clusters of galaxies. It is clear that additional observations can be applied to constraining current particle theories.[7]

Several upcoming experiments will provide significant improvements. In the next few years several new space experiments will be studying the background in a number of different ways using new types of instrumentation and carrying out both imaging and spectral measurements. Since spectral observations promise to be the most useful to particle theorists, we confine our discussion here to a summary of the current observational situation and the improvements which can be expected from forthcoming experiments.

II. SUMMARY OF EXISTING BROAD BAND OBSERVATIONS

It is useful to discuss separately observations made with broad bandpasses ($\lambda/\Delta\lambda < \sim 5$-10) and those carried out with moderate spectral resolution ($\lambda/\Delta\lambda > \sim 20$); the former achieve sufficient sensitivity to map out the overall distribution of the background, while the latter are crucial to establishing the nature of the observed flux through spectral analysis. In Figure 1 we summarize available data on the intensity of the diffuse far ultraviolet, extreme ultraviolet, and soft X-ray backround.

The intensity and spatial distribution of the soft X-ray band (.1 - 1 keV) has been reviewed by Tanaka and Bleeker;[8] the observations, carried out with wide field of view broad band instruments, reveal a ubiquitous background with large spatial anisotropies. This background can be explained by the emission from hot plasma in the local interstellar medium, but there may well be a component due to emission in the galactic halo[9], and it is possible that an extra-galactic component of this radiation is present. The nature of the dominant source of the background is confirmed by the detection of a blend of OVII, OVIII and CV lines[10,11]. Several all-sky surveys now exist: the most recently published are the surveys carried out from sounding rockets by the Wisconsin group[12] and from the SAS-3 satellite by the M.I.T. group[9].

In the EUV, an astronomical backround has been detected only to 140Å[13,14]. The Apollo-Soyuz measurements[14] covered 10% of the sky and established the existence of a background from 100-150Å which was shown to be diffuse rather than produced by point sources. These observations indicated the presence of a component of the interstellar medium with temperatures cooler than the million degree plasma detected in the soft X-ray background. This plasma should exhibit a rich spectrum of collisionally excited lines in the 80-900Å bandpass[15].

From ~ 150 to 912Å, emission from the very local interstellar medium is expected to dominate any observable diffuse flux, as the interstellar opacity attenuates emission from further than a few hundred parsecs. In addition, line emission is produced by solar resonance scattering with the interplanetary medium (HeI 584Å) and the plasmasphere (HeII 304Å). In this band only upper limits have been obtained on flux from outside the solar system.

At the Lyman edge of hydrogen at 912Å the interstellar opacity opens up again, permitting lines of sight out of the galaxy. However, all instrumentation in earth orbit must contend with contamination from resonant scattering of solar hydrogen Lyman α radiation at 1216Å by geocoronal hydrogen. Even in the interplanetary medium this line is present, due to resonant scattering of solar radiation by interstellar material entering the solar system. Hence instrumentation is normally designed to observe either shortward or longward of 1216Å. There have been limited broad band observations between 912-1216Å[16,17].

The flux from 1200-2000Å has been the subject of much controversy for several years, both as to its intensity and to its spatial distribution[18,19]. It is now established that the diffuse flux is anisotropic, varying from ~ 300 to ~ 3000 ph/cm^2/s/Å/ster. The fluxes are correlated with neutral hydrogen column density. This flux has been attributed at least in part to galactic starlight scattered into the field of view by high latitude dust clouds. Recent detailed studies of this correlation[20] also indicate that in some directions an additional component may be present, perhaps due to molecular hydrogen fluorescence.

III. SUMMARY OF SPECTROSCOPIC OBSERVATIONS OF THE DIFFUSE BACKGROUND

No spectral ($\lambda/\Delta\lambda \gtrsim 20$) observations have been carried out of the diffuse background between 1 keV and ~ 25 eV. This is primarily due to the difficulty of achieving sufficient sensitivity; instrument designs require grazing incidence optics to obtain sufficient collecting area, and obtaining sufficient field of view requires dedicated large instrumentation. From 25 eV to 10 eV (~ 500-1200Å), available spectroscopic observations with significant sky coverage are not numerous but those that are available are shown in Figure 1. Paresce and Bowyer[21] scanned near the galactic plane with a resolution of 40Å. Sandel, Shemansky and Broadfoot[22] obtained data with spectrometers on the Voyager mission; the fields of view were .1 × .9° with a resolution of 25Å. These observations specifically avoided bright stars. Kimble[23] has shown that this background cannot be isotropic since the resulting ionization of the local ISM would exceed measured values.

Kimble and Bowyer[24] reported coverage of 12% of the sky, obtained with a 7 × 9° field of view; a spectral resolution of 8 Å was achieved from 300-1400Å. The upper limits derived are significantly more stringent than other high galactic latitude measurements. The measurements are still above theoretical predictions for those directions. At low latitudes the signal detected longward of 912Å is in excellent agreement with stellar model atmosphere predictions, suggesting that the FUV radiation field is dominated by a small number of bright early type stars in this band.

IV. PROSPECTS FOR IMPROVED MEASUREMENTS

As summarized above, the observational situation is quite rudimentary. There is a clear need for additional observations with sufficient resolution to identify spectral lines. A number of experiments with this capability have been approved by NASA and are currently being readied for flight.

The University of California at Berkeley[25] is currently readying a diffuse extreme ultraviolet spectrometer for launch by the Shuttle in 1985 on the UVX payload. The instrument is housed in a single Get Away Special Can and is the first utilizing the GAS PLUS concept. It will remain attached to the Shuttle; approximately six hours of data will be accumulated. The spectrograph is a Rowland mount and is fed by a normal incidence off-axis paraboloid. Two imaging microchannel plate detectors record the spectrum of the Lyman hydrogen α line. The entrance slit of the spectrometer is 0.1 × 4°, permitting a resolution of 2-5Å; for the first flight the instrument will be configured to provide a resolution of 10Å from 600-1950Å.

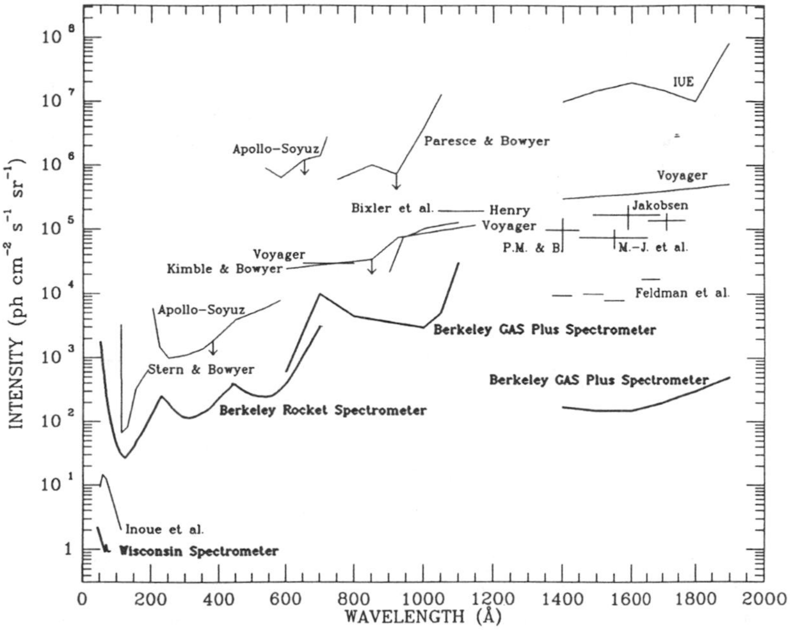

Figure 1: Current data on the intensity of the diffuse far ultraviolet, extreme ultraviolet and soft X-ray background. Existing data are indicated by light curves; they are primarily broad band measurements. These results are plotted as upper limits on any line flux that could exist in the band indicated. The heavy lines are measurements that are expected to be carried out in the next few years. Most of these will have moderate (~ 10Å) resolution or better. See text for details.

The instrument sensitivity as a function of wavelength is shown in Figure 1. As can be seen, the instrument, in one nighttime pass of the Shuttle, will have a limiting flux one to two orders of magnitude fainter than the measured broad band flux. The instrument is designed to avoid sources of contamination which have affected previous instrumentation. The imaging detectors, for instance, will allow faint B and A stars which enter the slit to be removed from the data. The grating has very low intrinsic scattering and is kept in a sealed vacuum box until observations are initiated in flight. The instrument is equipped with a number of filters to allow accurate determination of the internal background, as well as verifying the expected low level of contamination by geocoronal Lyman hydrogen α. Recovery of the instrument will permit post-flight calibration.

The University of Wisconsin has built a Bragg spectrometer for studies of the diffuse soft X-ray backround; the instrument will be part of the OSS Shuttle payload to be flown in 1986. The instrument is based on the use of a curved Bragg crystal which images the same wavelength from within the field of view onto a single focus. Different wavelengths are imaged by sections of the crystal array with different curvatures. A plane position sensitive proportional counter is used to image the spectrum. The instrument uses 1800 cm^2 of lead stearate crystals to cover the range of 42 to 84Å with throughput of about 0.02 cm^2 sr. Spectral resolving power ranges from 35 to 50. A second similar unit equipped with thallium acid pthallate crystals will cover the range of 11-24Å with about half the above throughput and twice the resolving power. In Figure 1[26] we show the predicted response to a two temperature thermal spectrum.

A sounding rocket is under construction by Labov, Bowyer, and Martin at the University of California, Berkeley for launch in 1985 which will provide the first spectral measurements of the extreme ultraviolet background from 80-600Å. This instrument will have a resolution of 5-10Å. The instrument uses a new optical design which combines a grazing incidence collector with a conical diffraction grating set. The spectra are recorded by two-dimensional imaging microchannel plate arrays. The instrument sensitivity is shown in Figure 1.

ACKNOWLEDGEMENTS

This work was performed under NASA grants JPL-956650 and NGR 05-003-450. We would like to thank Christopher Martin, Simon Labov, Michael Hettrick and Robert McCammon for material contributed to this paper.

REFERENCES

[1]J. Audouze and J. Tran Van Thanh, eds. Formation and Evolution of Galaxies and Large Structures in the Universe (Reidel, London): 1983.
[2]P. Jakobsen, Astron. Astrophys. **106**, 375, 1982.
[3]S. Bowyer, G. B. Field, and J. E. Mack, Nature **217**, 32, 1968.
[4]R. Kimble, S. Bowyer, and P. Jakobsen, Phys. Rev. Lett. **46**, 80, 1981.
[5]H. L. Shipman, and R. Cowsik, Astrophys. J. Lett. **247**, L111, 1981.
[6]R. C. Henry and P. D. Feldman, Phys. Rev. Lett. **47**, 618, 1981.
[7]F. Paresce, Proc. of the First Workshop on Galactic and Extragalactic Dark Matter, Il Nuovo Cimento, 1983.
[8]Y. Tanaka, and J. A. M. Bleeker, Space Sci. Rev. **20**, 815, 1977.
[9]F. J. Marshall and G. W. Clark, submitted to Astrophys. J., 1984.
[10]H. Inoue, et al., Astrophys. J. Lett. **227**, L85, 1980.
[11]H. W. Schnopper, et al., Astrophys. J. **253**, 131, 1981.
[12]D. McCammon, Astrophys. J. **269**, 107, 1983.
[13]W. Cash, R. F. Malina, and R. Stern, Astrophys. J. Lett. **204**, L7, 1976.
[14]R. Stern and S. Bowyer, Astrophys. J. **230**, 755, 1979.

[15]R. Stern, E. Wang, and S. Bowyer, Astrophys. J. Suppl. **37**, 195, 1978.
[16]R. C. Henry, et al., Astrophys. J. **223**, 437, 1978.
[17]J. Bixler, S. Bowyer, and M. Grewing, Astron. Astrophys., in press, 1984.
[18]A. Davidsen, S. Bowyer, and M. Lampton, Nature **247**, 513, 1974.
[19]F. Paresce and P. Jakobsen, Nature **288**, 119, 1980.
[20]P. Jakobsen, S. Bowyer, R. Kimble, P. Jelinsky, M. Grewing, G. Krämer, and C. Wulf-Mathies, Astronomy and Astrophysics in press, 1984.
[21]F. Paresce and S. Bowyer, Astrophys. J. **207**, 432, 1976.
[22]B. R. Sandel, D. E. Shemansky, and A. L. Broadfoot, Astrophys. J. **227**, 808, 1979.
[23]R. Kimble, Ph. D. Thesis, Univ. Cal. Berkeley, 1983.
[24]R. Kimble and S. Bowyer, submitted to Astrophys. J., 1984.
[25]C. Martin and S. Bowyer, Proceedings of the IAU Colloquium No. 81, "The Local Interstellar Medium," 1984 (in press).
[26]D. McCammon, private communication.

Axions in Astrophysics and Cosmology

P. Sikivie

I. INTRODUCTION

The axion[1,2] was postulated approximately seven years ago to explain why the strong interactions conserve P and CP.[3] The parameter that sets the amount of P and CP violation in QCD is

$$\bar{\theta} = \theta - \arg\det m \ , \qquad (1)$$

where m is the quark mass matrix and θ is the coefficient of $(g^2/32\pi^2) G^a_{\mu\nu} \tilde{G}^{a\mu\nu}$ in the action density for QCD. Using the Adler-Bell-Jackiw anomaly,[4] one readily shows that QCD depend on θ and arg det m only through the combination (1). Because $G\tilde{G}$ is a four-divergence, the $\bar{\theta}$ dependence of QCD is due purely to quantum effects. Quantum effects are most important when the coupling constant is large, i.e., in the case of QCD, at energies below a few GeV. They can be more or less reliably calculated using instanton and current algebra techniques.

The present upper limit on the neutron electric dipole moment requires[5] $\bar{\theta} \lesssim 10^{-8}$. If the CP violation necessary to explain $K_L \to 2\pi$ introduced into the standard $SU_L(2) \times U_Y(1) \times SU^c(3)$ model of particle interactions in the manner of Kobayaski and Maskawa,[6] then arg det m is an arbitrary (random) angle and there is absolutely no reason why $\bar{\theta} \lesssim 10^{-8}$. Other methods of introducing CP violation into the standard model also suffer from this difficulty[7] which is believed to be quite general and which has been given the name of "strong CP problem".

Peccei and Quinn[1] proposed the following simple and elegant solution to the problem. Let us postulate a $U_{PQ}(1)$ symmetry for the classical action density under which the quark fields and scalar fields transform generically as follows:

$$q \to e^{i\beta\gamma_5} q, \quad \phi \to e^{-2i\beta} \phi \ . \qquad (2)$$

The Yukawa interactions and scalar self-interactions have the following general form

$$- K\, q_L^{+} q_R \phi + \text{h.c.} - V(\phi^{+}\phi) \ . \qquad (3)$$

$V(\phi^\dagger \phi)$ has the shape of a "Mexican hat" and hence

$$\langle\phi\rangle = v e^{i\alpha}$$

$$m = K v e^{i\alpha}$$

$$\bar{\theta} = \theta - N(\delta + \alpha) \ , \qquad (4)$$

where δ is the overall phase of the matrix K of Yukawa couplings, and N is a model dependent integer.

Because of the $U_{PQ}(1)$ symmetry, the bottom of the Mexican hat potential is degenerate and the value of α is indifferent at the classical level. But the quantum effects (instantons...) which make the physics of QCD $\bar{\theta}$-dependent will lift this degeneracy and align α in a particular direction. The most straightforward way to determine the direction of alignment is by minimizing the Yukawa interaction energy and using the fact that QCD produces quark-antiquark condensates $\langle q_L^+ q_R \rangle$ which are CP conserving.[8] One readily finds that α aligns in such a way that $\bar{\theta} = 0$. The strong CP problem is thus solved.

Weinberg and Wilczek[2] independently pointed out that the Peccei-Quinn solution to the strong CP problem implies the existence of a light pseudoscalar particle, which they called the axion. The axion is the pseudo-Nambu-Goldstone boson associated with the spontaneous breaking of the $U_{PQ}(1)$ quasi-symmetry; i.e., it is the degree of freedom corresponding to rolling at the bottom of the Mexican hat potential. The axion would be massless if $U_{PQ}(1)$ were not broken by QCD instanton effects. One can compute the axion mass using the same considerations as those which determine the alignment of α discussed above. The result is

$$m_a \simeq \frac{f_\pi m_\pi}{v}\left(\frac{N}{6}\right) \simeq 50 \text{ keV } \frac{250 \text{ GeV}}{v}\left(\frac{N}{6}\right) , \tag{5}$$

where v and N are the quantities that appear in Eq. (4). In Eq. (5), v and N are normalized to the values they had in the earliest axion models. The coupling of the axion to quarks is

$$\sim i \frac{a}{v} m_q \bar{q} \gamma_5 q . \tag{6}$$

The coupling of the axion to the electromagnetic field is

$$- \frac{4\alpha}{3\pi} N \frac{a}{v} \vec{E} \cdot \vec{B} . \tag{7}$$

The value of the coupling strength[9] given in Eq. (7) holds for grand unified theories in which the unrenormalized value of the electroweak angle is $\sin^2\theta_w^0 = 3/8$. Note that both the axion mass, Eq. (5), and its coupling to the electromagnetic field, Eq. (7), are proportional to N/v. We will call the combination $f_a = v(6/N)$ the axion decay constant. The presence of the factor 6 is due to historical considerations.

It was first thought that the breaking of $U_{PQ}(1)$ occurred at the electroweak scale; i.e., $v \simeq 250$ GeV. The corresponding axion was searched for in K, J/ψ and Υ decays and in reactor and beam dump experiments, but it was not found. Soon, however, it was discovered[10] how to construct axion models with arbitrarily large values of v. These were called "invisible" axion models because for v >> 250 GeV, the axion is so weakly coupled that the event rates in the axion search experiments mentioned above are hopelessly small. For a while, it was thought that the strong CP problem was solved without any presently observable consequences whatsoever.

Fortunately, astrophysics and cosmology came to the rescue. As we will see in Section III, they provide us with arguments that imply the axion decay constant should lie in the range 10^8 GeV $\lesssim f_a \lesssim 10^{12}$ GeV. A second cosmological constraint arises because axion models have, as a rule, multiple degenerate vacua and hence domain walls. In Section II we will describe the properties of these domain walls, the cosmological catastrophe they produce and the ways in which this catastrophe may be avoided. In Section IV we give the reasons why axions are an excellent candidate to constitute the dark matter of galactic halos. In Section V we describe detectors to look for axions floating about in the halo of our galaxy and for axions emitted by the sun.

II. AXIONIC DOMAIN WALLS

Axion models often have a spontaneously broken exact discrete symmetry.[8] In that case, they have discretely degenerate vacua and hence domain walls. The domain walls are the soliton-like boundaries between regions which happen to be in different vacua.

The exact discrete symmetry in question is the overlap of the group of anomaly free global flavor symmetries of the colored fermions (quarks,...) with $U_{PQ}(1)$. For example, in the Dine-Fischler-Srednicki model[10] with n quarks

$$[SU_L(n) \times SU_R(n) \times U_V(1)] \cap U_{PQ}(1) = Z(2n) . \tag{8}$$

Z(2n) is an exact discrete symmetry of that model. Indeed, as a subgroup of $U_{PQ}(1)$ it is a symmetry of the classical action, and as a subgroup of the group of **anomaly-free** global symmetries of QCD, it is respected by the quantum effects as well. Z(2n) is spontaneously broken down to Z(2) by the vacuum expectation value $\langle\phi\rangle = ve^{i\alpha}$ that breaks $U_{PQ}(1)$ see Eq. (4)]. Hence the Dine-Fischler-Srednicki model has n degenerate vacua, as many as there are quarks (e.g., n=6). In other axion models, however, the number N of degenerate vacua is different from the number of quarks. In general N is given by the formula[11,12]

$$2N = \frac{2\pi}{T_\theta} \sum_f Q_f t_f . \tag{9}$$

Here the sum is over the colored left-handed fermions in the model, Q_f is their Peccei-Quinn charge, t_f is their "color-anomaly" defined by $Tr(T_f^a T_f^b) = 1/2\ t_f\ \delta^{ab}$ where T_f^a are the generators of $SU^c(3)$ for the color representation to which the fermions f belong, and T_θ is the period of θ. $T_\theta = 2\pi$ for QCD, the standard $SU^c(3) \times SU_L(2) \times U_Y(1)$ model and the SU(5) grand unified theory (GUT), but $T_\theta = 4\pi$ for the O(10) GUT and $T_\theta = 6\pi$ for the E_6 GUT.[9] For example and for reasons that will soon become clear (see the cosmological domain wall problem below), Georgi and Wise[11] build a three generation SU(5) grand unified axion model with the fermion representation content $3(10)_1+3(\bar{5})_1+5(5)_{-1}+5(\bar{5})_{-1}$ where the subscripts indicate the Peccei-Quinn charge of the corresponding multiplet. Using Eq. (9), one readily verifies that this model has a unique vacuum (N=1).

To derive the kinematic properties of the domain walls in axion models, one uses the effective action for the axion a

$$S_a = \int d^4x \left[\frac{1}{2}\,\partial_\mu a \partial^\mu a - \frac{m_a^2 v^2}{N^2}\, f\left(N\,\frac{a}{v}\right)\right]$$

$$= v^2 \int d^4x \left[\frac{1}{2}\,\partial_\mu \alpha \partial^\mu \alpha - \frac{m_a^2}{N^2}\, f(N\alpha)\right] , \qquad (10)$$

where $\alpha = a/v$ denotes collectively all the phases that rotate under a $U_{PQ}(1)$ transformation and f is a periodic function of period 2π, whose Taylor expansion begins with $f(x) = 1/2\, x^2 + \ldots$; for example, $f(x) = 1 - \cos x$. The axion self-interaction potential is then Z(N) symmetric and m_a is the axion mass. A domain wall, in the x-y plane for example, is the static classical solution $\alpha(z)$ obtained by minimizing the energy associated with Eq. (10) with the boundary conditions $\alpha(z) \to 0$ as $z \to -\infty$ and $\alpha(z) \to (2\pi)/N$ as $z \to +\infty$. One readily finds that the axionic domain walls have thickness of order m_a^{-1} and energy per unit surface $\sigma \simeq 8m_a v^2 \simeq 8f_\pi m_\pi v$. The tension in the domain wall equals its surface energy density σ. This follows from energy conservation and the fact that σ is a constant. The energy momentum tensor of a thin domain wall in the x-y plane is thus

$$(T_{\mu\nu}) = \sigma\delta(z)\,\mathrm{diag}(1, -1, -1, 0) \, . \qquad (11)$$

Domain walls are a very unusual source of gravity. They are in fact gravitationally repulsive.[13-15] To clarify this statement, let us first remark that the Newtonian limit of Einstein gravity is valid only when T_{00} is much larger than the other components of $T_{\mu\nu}$. Hence, intuition derived from Newtonian gravity is inapplicable to the gravity of domain walls. Einstein's equations for planar domain walls have been solved exactly.[14,15] There is a unique reflection symmetric solution which is free of curvature singularities. It corresponds to a uniform gravitational field in which observers on either side are repelled by the domain wall with constant acceleration $2\pi G_N\sigma$, where G_N is Newton's gravitational constant. More generally it has been shown that,[14] for a wall of arbitrary shape and motion and with arbitrary tension τ and surface energy density σ, the sum of the accelerations towards the wall on both sides, as measured by observers hovering just off the wall, is $4\pi G_N(\sigma-2\tau)$. For a dust wall (τ=0) one recovers the Newtonian result. For a domain wall ($\tau=\sigma$), the acceleration has equal magnitude as for a dust wall but opposite direction!

Axionic domain walls also have unusual electromagnetic properties.[9] To investigate these, one writes down the effective action density for photons and axions

$$= -\frac{1}{4} F_{\mu\nu}F^{\mu\nu} + \frac{1}{2} \partial_\mu a \, \partial^\mu a - \frac{m_a^2 v^2}{N^2} f\left(\frac{Na}{v}\right) + \frac{a}{3\pi} \frac{Na}{v} F_{\mu\nu}\tilde{F}^{\mu\nu} . \tag{12}$$

The strength of the $a\gamma\gamma$ coupling given in Eq. (12) is obtained by assuming that there is grand unification with the unrenormalized valuc of the electroweak angle $\sin^2\theta_W^0 = 3/8$. The equations of motion derived from Eq. (12) are

$$\vec{\nabla} \cdot \left(\vec{E} - \frac{e^2}{3\pi^2} \frac{Na}{v} \vec{B}\right) = 0$$

$$\vec{\nabla} \times \left(\vec{B} + \frac{e^2}{3\pi^2} \frac{Na}{v} \vec{E}\right) - \frac{\partial}{\partial t} \left(\vec{E} - \frac{e^2}{3\pi^2} \frac{Na}{v} \vec{B}\right) = 0$$

$$\vec{\nabla} \cdot \vec{B} = 0$$

$$\vec{\nabla} \times \vec{E} + \frac{\partial \vec{B}}{\partial t} = 0$$

$$\Box\, a = -\frac{e^2 N}{3\pi^2 v} \vec{E} \cdot \vec{B} - \frac{m_a^2 v}{N} \sin \frac{Na}{v} . \tag{13}$$

Consider an axionic domain wall of arbitrary shape and motion. Across the domain wall, $(Na)/v$ changes by 2π. The boundary conditions across the domain wall surface implied by Eq. (13) in the thin wall approximation are

$$\Delta B_\perp = \Delta \vec{E}_\parallel = 0, \quad \Delta E_\perp = \frac{2e^2}{3\pi} B_\perp , \quad \Delta \vec{B}_\parallel = -\frac{2e^2}{3\pi} \vec{E}_\parallel . \tag{14}$$

We see that an axionic domain wall becomes electrically charged when traversed by magnetic flux. The electric surface charge density is

$$\sigma = \frac{2e^2}{3\pi} \vec{n} \cdot \vec{B} , \tag{15}$$

where $\vec{n}$ is the unit normal in the direction of increasing $(Na)/v$. Similarly, an electric field parallel to an axionic domain wall induces a surface current density

$$\vec{K} = -\frac{2e^2}{3\pi}\vec{n} \times \vec{E} \ . \tag{16}$$

These unusual effects are necessary to make sense of the Witten dyon charge[16] in the presence of axionic domain walls. Witten has shown that in a θ-vacuum, magnetic monopoles acquire electric charge $q_\theta = C\theta g$ where g is the magnetic charge on the monopole and C is a model dependent constant. When a magnetic monopole traverses an axionic domain wall, the local value of θ changes by 2π and hence the electric charge on the monopole changes by one unit. One may well wonder whence that extra unit of electric charge came or what happens to electric charge conservation. The answer[9] is that the magnetic field of the monopole induces an electric charge density onto the domain wall, Eq. (15). When the monopole approaches the wall, the induced electric charge becomes concentrated near the impact point. It jumps onto the monopole when the monopole traverses the wall. The Witten dyon charge on the magnetic monopole plus the electric charge, Eq. (15), induced onto the axionic domain wall is conserved.

Domain walls exist in any theory in which a discrete symmetry is spontaneously broken. In 1974, Zel'dovich, Kobzarev and Okun[17] pointed out that because of these domain walls the spontaneous breaking of an exact discrete symmetry is incompatible with standard cosmology. Their argument is very simple. The universe starts off at some very high temperature at which the discrete symmetry is unbroken. At some critical temperature, the spontaneous breakdown does occur and the order parameter chooses among several equally probable values (or directions), corresponding to the various vacua of the theory. Different regions of the universe will in general settle into different vacua and hence be separated by domain walls. In particular, regions which are outside each other's horizon are causally disconnected and thus totally uncorrelated. Hence, there will be at least on the order of one domain wall per horizon at any given time. The energy density in domain walls today would be

$$\rho_{d.w.}(t_0) \simeq \frac{\sigma}{t_0} = \rho_{crit}\left(\frac{\sigma}{10^{-5}\ \mathrm{GeV}^3}\right), \tag{17}$$

where $t_0 \simeq 10^{10}$ years is the age of the universe today and $\rho_{crit} \simeq 10^{-29}$ gr/cm^3 is its present critical energy density for closure. Since $\sigma \simeq f_\pi m_\pi v >> 10^{-5}$ GeV^3, it is clear that if axions exist and N > 1, our present universe would be domain wall dominated many times over. But this can not be. A domain wall dominated universe would be expanding like $R \sim t^2$ (R is the cosmological scale parameter) and at a much higher rate than we observe today.

The cosmological domain wall problem just described can be avoided in a number of ways. Below are the three types of evasion which I am aware of:

1. The inflationary universe scenario [18] provides a solution if the inflationary epoch comes after the Peccei-Quinn phase transition at

$T_{PQ} \simeq v$ where $U_{PQ}(1)$ is spontaneously broken. Indeed, inflation will align the phase α of $\langle\phi\rangle = ve^{i\alpha}$ over enormous distances. Later, when the QCD instanton effects turn on at ~ 1 GeV temperatures, each enormous region will fall entirely into the same vacuum and hence be free of domain walls. For this to work, it is of course necessary that the post-inflation reheating temperature T_{reheat} be less than the temperature $T_{PQ} \simeq v$ at which the $U_{PQ}(1)$ symmetry is restored. We will see in the next section that v should be less than about 10^{12} GeV. On the other hand T_{reheat} must be sufficiently large for the baryon number asymmetry to be produced after inflation, since inflation wipes out any previous baryon number asymmetry. Hence, the set of constraints

$$T_{\text{baryo-genesis}} < T_{reheat} < T_{PQ} \simeq v \lesssim 10^{12} \text{ GeV} , \tag{18}$$

which may be difficult to satisfy in practice.

2. It is possible to construct axion models which have a unique vacuum.[8,19,11,20,21] One way is to build the model in such a way that N = 1 where N is the integer given by Eq. (9) [Note: if N = 0, the Peccei-Quinn mechanism is inoperative; see Eq. (4)]. When N = 1, the model only has a discrete Z(2) symmetry which is not spontaneously broken. Hence the vacuum is unique. Many N = 1 models have been constructed, e.g., Kim's original "invisible" axion model[10] and the grand unified axion model of Georgi and Wise[11] mentioned above. Another way to construct axion models with a unique vacuum is to embed the discrete Z(N) symmetry into a gauged[19,20] or an exact global continuous symmetry.[21] In that case, the N vacua are either gauge equivalent and hence not distinct or they are part of a larger continuous degeneracy and hence can be rotated into each other by adding coherent states of massless Nambu-Goldstone bosons.

The argument leading to the cosmological domain wall problem discussed above clearly does not apply to axion models with a unique vacuum. It is not immediately obvious, however, that such models are entirely free of cosmological difficulties because they, in fact, have domain walls, too.[23,12] When one traverses these domain walls one moves away from the unique vacuum and back to it along some topologically nontrivial path. This path is most readily visualized as one turn along the bottom of the Mexican hat potential $V(\phi^+\phi)$ of Eq. (3) from the unique vacuum at $\alpha = 0$ through $\alpha = \pi$ and back to the vacuum at $\alpha = 2\pi$. These domain walls are quantum-mechanically unstable[24,12] because holes can be poked in them through some tunneling process. The rate for this process is very much smaller than the $(\text{age})^{-1}$ of the universe, however, so that the domain walls are in fact stable for cosmological purposes. What saves axion models with a unique vacuum from the cosmological disaster of one domain wall per horizon at temperatures $\lesssim$ 1 GeV is the earlier appearance of strings.[23,12] When, at temperature $T_{PQ} \simeq v$, the phase transition occurs where $U_{PQ}(1)$ becomes spontaneously broken by $\langle\phi\rangle = ve^{i\alpha(x)}$, strings appear because $\pi_1[U(1)] = Z$. When one moves around the string once, the local value of $\alpha(x)$ varies from 0 to 2π. From the usual causality arguments one expects at least on the order of one string per horizon from T_{PQ} onward until QCD temperatures when the domain walls appear. Each string then becomes the edge of a domain wall. The typical size of a domain wall bounded by a string or of a closed domain wall is the horizon size ($\simeq 10^{-4}$ sec) at QCD temperatures. The probability of finding a domain wall much larger than that is exponentially small.[12] The finite size domain walls oscillate and dissipate away long before they dominate the cosmological energy density.

3. The last evasion of the domain wall problem is based on the observation[8,22] that a **tiny** explicit breaking of the Z(N) symmetry is sufficient to make the domain walls disappear before they dominate the energy density. A soft explicit breaking of $U_{PQ}(1)$ will introduce shifts $\langle\Delta\mathcal{V}\rangle$ in energy density amongst the various vacua, and it will also introduce a finite value of $\bar{\theta}$. When the domain bubbles have average size $\tau_B = \sigma/\langle\Delta\mathcal{V}\rangle$, the differences in volume energy among bubbles is of order their surface energy, the Z(N) breaking effects become important and the true vacuum takes over. One can show that the $\bar{\theta} \lesssim 10^{-8}$ constraint can be made compatible with the requirement that the domain walls disappear before they dominate the energy density provided $v \lesssim 10^{15}$ GeV. Finally, we note that an explicit breaking of the Z(N) symmetry is of course very artificial if done by hand. On the other hand, this evasion of the domain wall problem is a natural property of the ultimate theory of the world if the latter has in its low energy effective theory an automatic $U_{PQ}(1)$ which is then explicitly broken by higher order corrections.

III. ASTROPHYSICAL AND COSMOLOGICAL CONSTRAINTS ON THE AXION DECAY CONSTANT

The astrophysical constraint[25] arises because stars emit the weakly coupled axions from their whole volume whereas they emit photons only from their surface. Axions are produced in Compton, Primakoff and bremstrahlung type processes when photons collide with nuclei and electrons in stellar interiors. Because the axions are so weakly coupled, they can leave the star without further collisions. It has been shown[25] that if 250 GeV $\lesssim f_a \lesssim 10^8$ GeV, axion emission by stars is too copious to be consistent with our understanding of stellar evolution. If $f_a \lesssim 250$ GeV, the axion is too heavy to be produced in stars; if $f_a \gtrsim 10^8$ GeV, it is too weakly coupled to be produced overabundantly. Since $f_a \lesssim 250$ GeV appears to be ruled out by the unseccessful laboratory searches, it follows that f_a should be larger than about 10^8 GeV.

The cosmological bound[26] ($f_a \lesssim 10^{12}$ GeV) arises because axions are abundantly produced in the early universe when the temperature $T \simeq 1$ GeV. The argument is as follows. When T falls below $T_{PQ} \simeq v$, $U_{PQ}(1)$ becomes spontaneously broken by $\langle\phi\rangle = ve^{i\alpha(x)}$. The values of $\alpha(\vec{x})$ are at that time randomly chosen since the QCD instanton effects which lift the degeneracy at the bottom of the Mexican hat potential are negligible when T is larger than a few GeV. $\alpha(\vec{x})$ is spatially inhomogeneous. However, all wiggles in $\alpha(\vec{x})$ which fall within the horizon at any given time will start to oscillate thenceforward and thus red-shift away.[27] The result of this is that, at any given time t, $\alpha(x)$

is approximately homogeneous over the horizon scale t. [Of course, if there is inflation with $T_{reheat} < T_{PQ}$, $\alpha(\vec{x})$ is perfectly homogeneous over distances much larger than t. So much the better.] When QCD instanton effects become important at temperatures of order 1 GeV, the axion mass switches on and $\alpha(x)$ begins to oscillate with frequency m_a about the CP conserving minimum at $\alpha = 0$. Thus, at about 1 GeV temperature, a coherent state of nonrelativistic axions suddenly appears. The axions are nonrelativistic because their momenta are of order the inverse of the horizon scale at QCD temperatures $t_{QCD}^{-1} \simeq (10^{-4}\ \text{sec})^{-1} \simeq 10^{-11}$ eV, which is smaller than the axion mass (for $v \lesssim M_{Planck}$). The axion energy density just after the axion mass has switched on is

$$\rho_a(t_{QCD}) \simeq f_\pi^2\, m_\pi^2\, \alpha^2(t_{QCD}) \simeq \rho_{rad}(t_{QCD})\, \alpha^2(t_{QCD}) , \tag{19}$$

where ρ_{rad} is the energy density in radiation. But the nonrelativistic energy density ρ_a decreases with time as R^{-3} (R is the cosmological scale factor), whereas $\rho_{rad} \sim R^{-4}$. Moreover, for $v \gtrsim 10^8$ GeV, the axion fluid is effectively decoupled. It can be shown that the axions do not reheat nor convert into radiation.[26] Thus unless $\alpha^2(t_{QCD})$ is very small, the universe will become axion matter dominated too soon. If we require the axion energy density today too be less than ten times ρ_{crit}, we need $\alpha^2(t_{QCD}) \lesssim 10^{-6}$.

How can $\alpha(t_{QCD})$ be so small? If the switch-on of the axion mass were **sudden,** we would have $\alpha(t_{QCD}) \sim O(1)$ and all axion models would be ruled out. "Sudden" means that the switch-on rate $(1/m_a)(dm_a/dt)$ is large compared to the frequency m_a at which $\alpha(t)$ oscillates. The opposite of "sudden" is "adiabatic": $(1/m_a)(dm_a/dt)$ small compared to m_a. The latter regime is characterized by the adiabatic invariant

$$\oint p\, dq = \pi A^2(t) m_a(t) \simeq \text{time independent} , \tag{20}$$

where A(t) is the amplitude of the oscillation $\alpha(t) = A(t)\cos(m_a t+\delta)$. Equation (20) tells us that, in the adiabatic regime, the oscillation amplitude decreases while the axion mass is being switched on. Hence, provided the switch-on is sufficiently adiabatic, an excessive axion energy density may be avoided. Let us define a time t_1, such that $m_a^{-1}(t)\,(dm_a/dt) > m_a(t)$ for $t < t_1$, and $m_a^{-1}(t)\,(dm_a/dt) < m_a(t)$ for $t > t_1$. Before t_1, the switch-on is sudden. Hence $A(t_1) = O(1)$. After t_1, the switch-on is adiabatic. Hence

$$A^2(t_{QCD}) = \frac{A^2(t_1) m_a(t_1)}{m_a(t_{QCD})} = O(1)\ \frac{m_a(t_1)}{m_a} . \tag{21}$$

The time dependence of the axion mass follows from its temperature dependence which has been calculated[28]: $m_a(t) = m_a[T(t)]$. Using this, the following result was obtained[26] for the axion energy density today

$$\rho_a(t_0) \simeq 5\rho_{crit}\left(\frac{f_a}{10^{12}\ \text{GeV}}\right)^{7/6} . \tag{22}$$

Hence, the constraint $f_a \lesssim 10^{12}$ GeV which applies to all axion models independently of their vacuum structure and of the history of the universe before the temperature reached O(10) GeV.

Steinhardt and Turner[29] have considered entropy production when the temperature of the universe is between 1 GeV and 1 MeV, by out-of-equilibrium decays of a relic particle species or by a first-order QCD phase transition, as a means to dilute the axion energy density and thus to weaken the $f_a \lesssim 10^{12}$ GeV bound.

IV. AXIONS AND GALAXY FORMATION

There is good evidence[30] that individual galaxies possess dark halos with masses exceeding that of the luminous galactic matter by a factor ~10. These galactic halos could be made of axions.[31] First, if $f_a \gtrsim 2 \times 10^{10}$ GeV, axionic matter is abundant enough to make up the halos [cf., Eq. (22)]. Second, since axions are effectively decoupled for such large values of f_a, axionic halos are automatically dark. Neutrinos are similar to axions in these two respects and they have indeed been a very popular candidate for the halo matter. However, neutrino halo models have run into rather serious difficulties because the neutrino phase space density tends to be too small to allow them to cluster into galactic halos and because neutrio free streaming greatly inhibits the growth of all matter density perturbations on all mass scales less than about 10^{15} $M_\odot$. Axions, on the other hand, because they are nonrelativistic from the moment of their first appearance at ~1 GeV temperatures, have enormous phase space density and vanishingly small free streaming distance.[31] The large phase space density allows them to cluster easily into galactic halos, whereas the absence of free streaming allows the growth of primordial density perturbations to proceed on all scales. Recent computer simulations[32] of the growth of density perturbations in the early universe have shown that indeed cold dark matter (e.g., axions, photinos, gravitinos...) appears preferable to hot dark matter (e.g., neutrinos).

V. "INVISIBLE" AXION DETECTORS

One of the exciting aspects of the hypothesis that galactic halos are made of axions is the fact that it can be tested experimentally. How can this be done? One exploits the coupling, Eq. (7), of the axion to the electromagnetic field and the fact that we have available in the laboratory large oscillating electric and magnetic fields with frequencies or wave-vectors of order the axion mass ($\hbar$=c=1)

$$m_a \simeq 1.24\ 10^{-5}\ \text{eV}\left(\frac{10^{12}\ \text{GeV}}{f_a}\right)$$

$$= \frac{2\pi}{10\ \text{cm}}\left(\frac{10^{12}\ \text{GeV}}{f_a}\right)$$

$$= (2\pi)\ 3\ \text{GHz}\left(\frac{10^{12}\ \text{GeV}}{f_a}\right) \quad . \tag{23}$$

The general idea[33] is that an externally applied magnetic or electric field will stimulate the conversion of an axion to a photon through the coupling, Eq. (7). The outgoing photon is relatively easy to detect. We will see below that this process can be used both for the detection of axions floating about in the halo of our galaxy and for the detection of axions emitted by our sun. In addition, one may attempt to observe the static forces with range of order m_a^{-1} due to virtual axion exchange. We refer the reader to the work of Moody and Wilczek[34] who have discussed these effects in detail.

1. Axion Haloscope

If the Milky Way halo is composed of axions, their number density in our vicinity is approximately

$$\rho_a \simeq \frac{10^{-24}\text{gr}}{\text{cm}^3} \simeq \frac{(0.5)10^{14}\ \text{axions}}{\text{cm}^3}\left(\frac{f_a}{10^{12}\ \text{GeV}}\right) \quad . \tag{24}$$

These axions have energies

$$\varepsilon_a = m_a\left(1 + \frac{1}{2}\beta^2\right) = m_a\ [1+O(10^{-6})] \quad . \tag{25}$$

where $\beta \simeq 10^{-3}$ is the galactic virial velocity. Consider an electromagnetic cavity permeated by a strong static magnetic field $\vec{B}_0$. When the frequency ω of one of certain appropriate cavity modes equals the axion mass, there will be resonant conversion of Milky Way halo axions into quanta of excitation (photons) of that mode. For a rectangular cavity, the appropriate modes are $TM_{n\ell o}$ with n and ℓ odd (the longitudinal direction is that of $\vec{B}_0$). The power on resonance into such a mode is[33]

$$P_{n\ell} \simeq (0.8)\, 10^{-19}\ \text{Watt} \left(\frac{V}{5\times10^{4}\ \text{cm}^{3}}\right)\left(\frac{B_0}{8\ \text{Tesla}}\right)^{2}$$

$$\times \frac{1}{n^{2}\ell^{2}} \left(\frac{m_a}{1.24\times10^{-5}\ \text{eV}}\right) \text{Min}\,(1, 10^{-6}\, Q_{n\ell}) \quad , \tag{26}$$

where V is the volume of the cavity and $Q_{n\ell}$ is the quality factor for that mode. Because the axion energies have spread of order 10^{-6} m_a, the power [Eq. (26)] is not increased by having quality factors $Q \gtrsim 10^{6}$. [However, if the axion were to be found and its mass were known, one could use superconducting cavities with $Q >> 10^{6}$ to **resolve** the spectrum of galactic halo axion energies.] To scan the allowed range of axion masses in a reasonable amount of time, a very sensitive detector of microwave radiation is required. A typical state of the art detector[35] today has a noise temperature T_N of order 10-20°K in the 1-40 GHz frequency range. The noise equivalent power of such detectors over a bandwidth set by the quality of the cavity [$B \simeq (2f/Q)$] is

$$\text{NEP} = 1.2 \times 10^{-20}\ \frac{\text{Watt}}{\sqrt{\text{Hz}}} \left(\frac{f}{\text{GHz}} \cdot \frac{10^{6}}{Q}\right)^{1/2} \left(\frac{T_N}{20°\text{K}}\right) \quad . \tag{27}$$

Comparison with Eq. (26) suggests that the experiment may be feasible. To keep the thermal noise below the signal, the cavity must be cooled to less than 0.1°K. Also, to distinguish the signal from fluctuations in thermal and detector noise, it will be necessary to modulate the resonant frequency of the cavity at some audio frequency and carry out phase-sensitive detection of the cavity output.

Rather detailed feasibility studies have been carried out for this experiment. It appears that a set of large cavities (say 150 cm long and 50 cm wide) placed in the bore of an 8 Tesla solenoidal superconducting magnet and equipped with state of the art microwave detectors should be able to cover, in one or two year continuous running time, the range of axion masses between 1 GHz and 30 or 40 GHz [10^{11} GeV $\lesssim f_a \lesssim 3\times10^{12}$ GeV] with a signal to noise ratio of three. It is very conceivable that design improvements will extend this range in both directions. Note that this experiment is capable of exploring that special range of values of the axion mass for which axions may provide the critical energy density for closing the universe.

2. Axion Helioscope

The solar axion flux on earth is approximately $0.8 \times 10^{13}\ \text{sec}^{-1}\text{cm}^{-2}(10^{8}\ \text{GeV}/f_a)^{2}$. The solar axions have a broad spectrum of energies centered about 1 keV, the temperature in the sun's interior. Using the coupling given in Eq. (7), one finds[33] the following general cross-section for axion → photon conversion in a volume V in which there is a static inhomogeneous magnetic field $\vec{B}_0(\vec{x})$

$$\sigma = \frac{1}{16\pi^2|\vec{\beta}_a|}\left(\frac{e^2N}{3\pi^2 v}\right)^2 \sum_\lambda \int d^3k_\gamma\, \delta(k_\gamma - E_a)$$

$$\times \left| \int_V d^3x\, e^{i\vec{q}\cdot\vec{x}}\, \vec{B}_0(\vec{x}) \cdot \vec{\varepsilon}(\vec{k}_\gamma,\lambda) \right|^2 , \qquad (28)$$

where $\vec{q} = \vec{k}_\gamma - \vec{k}_a$ and the sum is over photon polarizations. Consider then a detector of length L in the directions $\vec{n}$ of the sun, inside of which there is a transverse magnetic field $\vec{B}_0 = B_0\hat{\varepsilon}\cos[(2\pi/d)\vec{n}\cdot\vec{x}]$. The cross section for $a \to \gamma$ conversion for axions coming from the direction of the sun is

$$\sigma = \frac{1}{16|\vec{\beta}_a|}\left(\frac{e^2NB_0}{3\pi^2 v}\right)^2 VL\left[\frac{\sin\left(\frac{2\pi}{d} - q_z\right)\frac{L}{2}}{\left(\frac{2\pi}{d} - q_z\right)\frac{L}{2}} + \frac{\sin\left(\frac{2\pi}{d} + q_z\right)\frac{L}{2}}{\left(\frac{2\pi}{d} + q_z\right)\frac{L}{2}}\right]^2 \qquad (29)$$

where V is the detector volume and

$$q_z = (\vec{k}_\gamma - \vec{k}_a)\cdot\vec{n} = E_a - \sqrt{E_a^2 - m_a^2} = \frac{1}{2}\frac{m_a^2}{E_a}$$

$$\simeq \frac{2\pi}{16\ \mathrm{cm}}\left(\frac{10^8\ \mathrm{GeV}}{f_a}\right)^2\left(\frac{\mathrm{keV}}{E_a}\right) . \qquad (30)$$

On resonance ($q_z \simeq 2\pi/d$) the event rate in the detector is[33]

$$\frac{\#\text{x-ray}}{\text{time}} \simeq \frac{10^{-2}}{\mathrm{sec}}\frac{Vd}{(\mathrm{meter})^4}\left(\frac{B_0}{8\ \mathrm{Tesla}}\right)^2\left(\frac{10^8\ \mathrm{GeV}}{f_a}\right)^4 . \qquad (31)$$

Assuming that a signal of one x-ray/ten days can be distinguished from background, it appears that a cubic meter detector can detect solar axions if 10^8 GeV $\lesssim f_a \lesssim 10^9$ GeV. If f_a were to be in this range, axions would provide us with a powerful tool to study the solar interior.

ACKNOWLEDGMENTS

I am grateful to SLAC for its hospitality while this talk was written up.

REFERENCES

[1]R.D. Peccei and H. Quinn, Phys. Rev. Lett. **38**, 1440 (1977) and Phys. Rev. **D16**, 1791 (1977).
[2]S. Weinberg, Phys. Rev. Lett. **40**, 223 (1978); F. Wilczek, Phys. Rev. Lett. **40**, 279 (1978).
[3]A.A. Belavin, A.M. Polyakov, A.S. Schwartz and Y.S. Tyuplin, Phys. Lett. **59B**, 85 (1975); G. 't Hooft, Phys. Rev. Lett. **37**, 8 (1976) and Phys. Rev. **D14**, 3432 (1976); R. Jackiw and C. Rebbi, Phys. Rev. Lett. **37**, 172 (1976); C.G. Callan, R.F. Dashen and D.J. Gross, Phys. Lett. **63B**, 334 (1976).
[4]S.L. Adler, Phys. Rev. **117**, 2426 (1969); J.S. Bell and R. Jackiw, Nuovo Cimento **A60**, 47 (1969).
[5]V. Baluni, Phys. Rev. D**19**, 2227 (1979); R.J. Crewther, P. Di Vecchia, G. Veneziano and E. Witten , Phys. Lett. **88B**, 123 (1979).
[6]M. Kobayashi and K. Maskawa, Progr. Theor. Phys. **49**, 652 (1973).
[7]See, however: A. Nelson, Phys. Lett. **136B**, 387 (1984); S. Barr, Univ. of Washington preprint 40048-8P4 (1984).
[8]P. Sikivie, Phys. Rev. Lett. **48**, 1156 (1982).
[9]P. Sikivie, Phys. Lett. **137B**, 353 (1984).
[10]J. Kim, Phys. Rev. Lett. **43**, 103 (1979); M. Shifman, A. Vainshtein and V. Zakharov, Nucl. Phys. **B166**, 493 (1980); M. Dine, W. Fischler and M. Srednicki, Phys. Lett. **104B**, 199 (1981); H.P. Nilles and S. Raby, SLAC-PUB-2743 (1981).
[11]H. Georgi and M. Wise, Phys. Lett. **116B**, 123 (1982).
[12]P. Sikivie, "Axions in Cosmology," U. Florida preprint UFTP-83-6, published in the Proc. of the Gif-sur-Yvette Summer School in Particle Physics, Sept. 1982.
[13]A. Vilenkin, Phys. Rev. **D23**, 852 (1981).
[14]J. Ipser and P. Sikivie, "The Gravitationally Repulsive Domain Wall," U. Florida preprint (Oct 1983), to be published in Phys. Rev. D.
[15]A. Vilenkin, Phys. Lett. **133B**, 177 (1983).
[16]E. Witten, Phys. Lett. **86B**, 283 (1979).
[17]Y.B. Zel'dorich, I.Y. Kobzarev and L.B. Okun, Zh. Eksp. Teor. Fiz. **67**, 3 (1974) [Sov. Phys. JETP **40**, 1 (1975)].
[18]A. Guth, Phys. Rev. **D23**, 347 (1981); A. Linde, Phys. Lett. **108B**, 389 (1982); A. Albrecht and P. Steinhardt, Phys. Rev. Lett. **48**, 1220 (1982).
[19]G. Lazarides and Q. Shafi, Phys. Lett. **115B**, 21 (1982).
[20]S. Dimopoulos, P. Frampton, H. Georgi and M. Wise, Phys. Lett. **117B**, 185 (1982); S. Barr, D. Reiss and A. Zee, Phys. Lett. **116B**, 227 (1982); S. Barr, X.C. Gao and D. Reiss, Phys. Rev. **D26**, 2176 (1982); T. Kephart, Phys. Lett. **119B**, 92 (1982); D. Dominici, R. Holman and C.W. Kim, Phys. Lett. **130B**, 39 (1983); K. Kang, C.K. Kim and J.K. Kim, Phys. Lett. **133B**, 79 (1983). K. Kang, S. Ouvry, Phys. Rev. **D28**, 2662 (1983); A. Yildiz, Phys. Rev. **D29**, 297 (1984).
[21]F. Wilczek, Erice Lectures, ITP (Santa Barbara) Report No. NSF-ITP-84-14 (1984).
[22]B. Holdom, Phys. Rev. **D28**, 1419 (1983).
[23]A. Vilenkin and A.E. Everett, Phys. Rev. Lett. **48**, 1867 (1982); T.W.B. Kibble, G. Lazarides and Q. Shafi, Phys. Rev. **D26**, 435 (1982).
[24]S. Coleman, unpublished.
[25]D. Dicus, E. Kolb, V. Teplitz and R. Wagoner, Phys. Rev. **D18**, 1829 (1982) and Phys. Rev. **D22**, 839 (1980); M. Fukugita, S. Watamura and M. Yoshimura, Phys. Rev. Lett. **48**, 1522 (1982); see also N. Iwamoto, in the proceedings of this conference.

[26]L. Abbott and P. Sikivie, Phys. Lett. **120B**, 133 (1983); J. Preskill, M. Wise and F. Wilczek, Phys. Lett. **120B**, 127 (1983); M. Dine and W. Fischler, Phys. Lett. **120B**, 137 (1983).

[27]A. Vilenkin, Phys. Rev. Lett. **48**, 59 (1982).

[28]D. J. Gross, R.D. Pisarski and L.G. Yaffe, Rev. Mod. Phys. **53**, 43 (1981).

[29]P.J. Steinhardt and M. Turner, Phys. Lett. **B129**, 51 (1983).

[30]J.P. Ostriker, P.J.E. Peebles and A. Yahil, Astrophys. J. Lett. **193**, L1 (1974); J. Einasto, A. Kaasik and E. Saar, Nature **250**, 309 (1974).

[31]J. Ipser and P. Sikivie, Phys. Rev. Lett. **50**, 925 (1983); F.W. Stecker and Q. Shafi, Phys. Rev. Lett. **50**, 928 (1983); M.S. Turner, F. Wilczek and A. Zee, Phys. Lett. **125B**, 35 (1983); M. Axenides, R. Brandenberger and M. Turner, Phys. Lett. **126B**, 178 (1983); M. Fukugita and M. Yoshimura, Phys. Lett. **127B**, 181 (1983).

[32]A.L. Melott, J. Einasto, E. Saar, I. Suisalu, A. Klypin and S.F. Shandarin, Phys. Rev. Lett. **51**, 935 (1983); C.S. Frenk, S.D.M. White and M. Davis, Astrophys. J. **271**, 417 (1983); G.R. Blumenthal, S.M. Faber, J.R. Primack and M.J. Rees, preprint SLAC-PUB-3307 (1984).

[33]P. Sikivie, Phys. Rev. Lett. **51**, 1415 (1983); ibid., **52**, 695 (1983). Note: All axion $\rightarrow$ photon conversion rates given there contain an error of a factor 4π. To obtain the correct expected rates, divide by 4π.

[34]J.E. Moody and F. Wilczek, ITP (Santa Barbara) Report No. NSF-ITP-83-177 (1984).

[35]P.L. Richards and L.T. Greenberg in "Infrared and Millimeter Waves," Vol. 6, ed. K.J. Button, Acad. Press, New York, 1982.

Role of Axions in Neutron Star Cooling

Naoki Iwamoto

We report on our study[1] of how axions affect neutron star cooling within the current astrophysical[2] and cosmological[3] bounds on the mass scale F at which the global chiral symmetry $U_{PQ}(1)$ spontaneously breaks down,[4] 10^8 GeV < F < 10^{12} GeV. We have calculated the energy loss rates due to axion bremsstrahlung from degenerate neutron-neutron collisions in the interior, $n + n \rightarrow n + n + A$; and the axion bremsstrahlung by the relativistic degenerate electrons colliding with the nuclei with atomic number Z and mass number A in the crust, $e^- + (Z,A) \rightarrow e^- + (Z,A) + A$. They are (in erg cm^{-3} s^{-1}) $\varepsilon_{ANN} \simeq 5 \times 10^{32}\, g_{ANN}^2 (m_n^*/m_n)^2 (n_b/n_o)^{1/3} F(x) T_8^6$ and $\varepsilon_{Ae} \simeq 4 \times 10^{22}\, g_{Aee}^2 (Z^2/A) \rho\, (n_b/n_o)^{-2/3} Y_e^{-2/3} [2\, \ell n(2\gamma) - 1] T_8^4$, respectively, where g_{ANN} is the axion-nucleon coupling, m_n^* the neutron effective mass, n_b the baryon number density, $n_o = 1.7 \times 10^{38}$ cm^{-3} the nuclear matter density, F(x) a function of density of order unity, $T_8 = T/10^8$K, g_{Aee} the axion-electron coupling, ρ the mass density, Y_e the electron number per baryon, and γ the Lorentz factor of the electron. These axion energy loss rates are compared with the neutrino rates given by Soyeur and Brown.[5] In order to allow for the uncertainties in the neutron star matter equation of state and the superfluid transition which the nucleons are expected to undergo in the interior, we have chosen two types of equation of state (medium-soft and stiff) and considered the cases with or without superfluidity separately. We find that the conventional neutrino cooling scenario of neutron stars must be modified, namely, neutron stars would cool predominantly due to axion emission rather than neutrino emission unless the following conditions are satisfied: (1) In the absence of superfluidity, $g_{ANN} < 4 \times 10^{-10}$ or $F > 3 \times 10^9$ GeV, where $g_{ANN} = C_A\, m_N/F$ with $C_A \simeq 1.25$ the axial vector coupling constant and m_N the nucleon mass. (2) In the presence of superfluidity, $g_{ANN} < (4\text{-}6) \times 10^{-10}$ or $F > (3\text{-}2) \times 10^9$ GeV and $g_{Aee} < (9\text{-}6) \times 10^{-13}$ or $F > (6\text{-}9) \times 10^8$ GeV, where $g_{Aee} = m_e/F$ with m_e the electron mass.

The conclusion from our analysis is two fold: (1) In spite of the uncertainties in the neutron star matter equation of state, axion emission could be the dominant cooling mechanism of neutron stars if 10^8 GeV < F $\lesssim 3 \times 10^9$ GeV. This parameter range of F has not been ruled out from any previous considerations.[2,3] (2) Since the cooling calculations based on the standard neutrino scenario[6] are found to be consistent with the recently observed X-ray flux from three point sources in the centers of supernova remnants (Crab, Vela and RCW 103)[7] within the theoretical uncertainties, one may tentatively conclude that there should not be any cooling mechanism in addition to neutrino emission, in which case one can obtain a better bound on the axion parameters. However, this conclusion is contingent on the confirmation through further spectral observations that these X-rays are thermal and that there are no heating mechanisms inside neutron stars.

This work was supported in part by the NSF under Grant Numbers PHY-77-27084 (supplemented by funds from the NASA) at ITP, UC Santa Barbara and DMR-83-04213 together with the NASA Grant NAGW-122 at Washington University.

REFERENCES

[1]N. Iwamoto, Phys. Rev. Lett. **53**, 1198 (1984).
[2]M. Fukugita, S. Watamura, and M. Yoshimura, Phys. Rev. D**26**, 1840 (1982); J. Ellis and K. A. Olive, Nucl. Phys. B**223**, 252 (1983) and the references therein.
[3]J. Preskill, M. B. Wise, and F. Wilczek, Phys. Lett. **120**B, 127 (1983) and other papers in the same issue.
[4]P. Sikivie, in these Proceedings; F. Wilczek, Erice Lecture Notes (1983); ITP Preprint NSF-ITP-84-14 (to be published).
[5]M. Soyeur and G. E. Brown, Nucl. Phys. A**324**, 464 (1979).
[6]K. Nomoto and S. Tsuruta, Astrophys. J. Lett. **250**, L19 (1981).
[7]D. J. Helfand and R. H. Becker, Nature (London) **307**, 215 (1984).

Gravitational Lenses and Particle Properties

Edwin L. Turner

ABSTRACT

The potential of observations of gravitational lens systems for the determination of cosmological constants and for tests of the nature and distribution of dark matter is illustrated. The advantages and disadvantages of gravitational lenses as cosmological probes are evaluated.

I. INTRODUCTION

Gravitational lenses are the latest addition to the list of astronomical objects predicted by theorists during the birth of modern physics in the first half of this century and discovered by observers during the renaissance of observational techniques in the second half. The history of the subject is now fifty years of intermittent theoretical attention and five years of fairly intensive observation and modeling ignited by the discovery of 0957+561 in 1979 and fed by nearly annual discoveries of additional systems since then. No comprehensive review of either theoretical or observational work will be attempted here.

The relevance of gravitational lenses to the themes of this meeting is that they provide a potentially very powerful new tool for the determination of cosmological parameters and the properties of dark matter. I emphasize "potentially" and "new" because while on the one hand no major application of lenses has yet been realized or even appears to be an immediate prospect, on the other hand, lenses allow the possibility of novel approaches to long standing and formidable problems including some which are otherwise completely unassailable observationally. The reason is that lensing depends on such simple and physically relevant properties of the Universe as the spatial mass distribution and the geometry of random alignments but not on such complex and difficult to interpret properties as emissivity distributions or radiation mechanisms. In addition, gravitational lens systems, like pulsars, have the pleasant property that much of fundamental significance can be learned from detailed observations of a single particularly simple and well understood or otherwise special case.

In order to illustrate the potential of gravitational lens observations, sections II through V below describe possible future lens experiments, including the type of lens system required, the observational requirements, and what would be learned. Section VI returns to a more general discussion and attempts to evaluate some of the advantages and disadvantages of lenses as cosmological probes from a broader perspective.

II. TIME DELAYS

Consider a quintuple image of a high redshift quasar produced by an intervening galaxy of moderate redshift (e.g., $z\sim0.5$). Suppose that the galaxy is a quite ordinary spiral of type Sb and with a luminosity near

L* and that it is resident in the field or in a small group but not in a rich cluster. This last requirement ensures that the lensing effect is totally dominated by the galaxy's own mass distribution so that no effects due to an associated cluster need be considered. Further suppose that the quasar, like many other quasars, varies irregularly in its optical continuum brightness on a timescale of months.

Deep ST Wide Field/Planetary Camera images of this system will conveniently resolve the galaxy and separate the five quasar images near or within the galaxy's image; the angular scale of the galaxy will be of order 3 to 4 arc sec, and that of the quasar image separations, of order 2 arc sec. This data will allow an accurate determination of the galaxy's Hubble type and inclination to the line of sight as well as a measurement of the relative positions and brightnesses of the five quasar images. Spectroscopy (probably ground based) will give the redshifts of the galaxy and quasar. The galaxy's rotational velocity could be inferred from its luminosity via the Tully-Fischer relation or, just possibly, measured by very difficult VLA 21 cm. line observations. Given this data and our extensive knowledge of the rather regular and standardized mass distributions in the inner regions of low redshift spirals of the same Hubble type,[1] it should be possible to construct a detailed and reliable model of the lensing galaxy's mass distribution and of the geometrical alignment of the background quasar and the foreground galaxy.

For such a well determined lensing situation the differential light travel time back to the quasar along each of the five paths; these will be of order of weeks to years and will be directly proportional to the inverse of Hubble's constant.[2] In addition to the delay produced by the lens itself, there will be differential delays introduced by mass concentrations outside of the beams which will bend the paths along which each image reaches us. These additional "racetrack" delays depend on the rms amplitude and scale of fluctuations in the total density about the path out to the quasar and will be a source of confusion for H_o determinations. Fortunately, these two effects can be separated because the "racetrack" delays will vary linearly with position (in some direction) across the sky. Thus, a determination of all four independent time delays in such a well understood lens system would give a direct measure of H_o, independent of all local distance measures, on a cosmological scale and would give some constraint on ρ_{rms} in the Universe. This second datum would perhaps be even more valuable than the H_o measurement since it would test the hypothesis that the large scale mass distribution is like the large scale galaxy distribution, a critical but unchecked assumption in many cosmological arguments.

III. BRIGHTNESS VARIATIONS DUE TO MINI-LENSES

One of the most exciting ideas for a potential gravitational lens application is to test the hypothesis that the dark matter in galaxies and clusters is in the form of condensed objects of very roughly stellar mass.[3] If so, such objects will occasionally strongly lens one of the images of a quasar in a galaxy or cluster lens system resulting in a variation of order unity in its apparent brightness. These events can be distinguished from intrinsic variations in the background quasar because they will be time symmetric and absent from the other quasar images (at the appropriate time delay). The timescale of such brightness variations would be of order decades for dark objects of order one solar mass and scales as the square root of the objects' masses. The "optical depth" for such mini-lensing events depends only on (and is strictly proportional to) the surface mass density of the objects and can thus be estimated from standard mass distribution models. A lower limit on the mass of objects detectable by this means is set by the unknown true angular size of quasar emission regions; it could be as small as $10^{-3}M_{\odot}$ for plausible quasar models.

In a typical galaxy lens situation, there will frequently be an image on the outskirts of the galaxy with $\tau \ll 1$, an image near the critical radius with $\tau \approx 1$, and an image near the galaxy core with $\tau \gg 1$. This situation would afford us the maximum opportunity to estimate the masses and velocities of the objects making up the bulk of the surface density and to disentangle the effect from any other sources of variability.

Observation of mini-lensing events in sufficient numbers could effectively rule out the possibility that galaxy and cluster masses are dominated by any species of individual elementary particles. Conversely their absence in a suitably well studied sample of lenses could lend the elementary particle model important support by ruling out a strong competing possibility.

IV. DARK LENSES

If, as is widely hypothesized,[4] the baryonic matter in galaxies and clusters was accreted by pre-existing objects made of some form of dark matter, then it is obviously possible that some of these dark condensed objects never accreted any significant amount of ordinary baryonic matter and thus remained dark. One can also entertain speculations of dark, dead galaxies in which all of the stars either never shone (due to low masses) or have already burned out (due to high masses). Such a potentially important class of objects could be detected by their lensing effects even if their contribution to Ω_o were as small as 10^{-3} given that their individual masses and surface densities were high enough.

In order to claim detection of such a dark lens, one would need to locate a convincing example of a multiply imaged quasar in which lensing by any ordinary galaxy or cluster could be ruled out by their absence in very deep images of the system. This is a challenging but by no means overwhelmingly difficult observational task in a favorable case. The lens mass can be calculated for any assumed lens redshift as can an upper limit on its size based on the observed angular separation of the images. The a priori probability distribution for lens redshifts may also be calculated and is strongly peaked at moderate redshifts (say 0.2 to 0.8), especially for large angular separations and/or low quasar redshifts. Careful and deep imaging (particularly with ST) should essentially always detect the lensing object if it is a normal galaxy or cluster in large splitting cases.[5] The failure to detect a lensing object should allow us to put lower limits on the lens mass-to-light ratio of order 10^3 to 10^4 in the best cases.

Two of the five known lenses, 1635+267 and 2345+007, have no reported detection of a lensing object despite serious efforts to find one. For these two cases, the indicated minimum lens masses are 0.7 and 2.0 times $10^{12}M_{\odot}$, respectively with upper limits on their sizes of 10 and 18 kpc assuming a most probable lens redshift near 0.5. Available observations probably already rule out such a low lens redshift and require either an even more massive galaxy at a substantially greater redshift or a dark lens.

V. STATISTICAL STUDIES

The preceding sections have concentrated on possible applications of individual lens systems. It is also obviously possible and desirable to consider the statistical properties of samples of lenses.[5] Such studies can address a large number of cosmological problems including the determination of cosmological parameters and the large scale distribution of matter. The primary observational requirement is a reasonably large sample of similarly observed lenses selected in some well understood way. Obviously, some lens parameters (e.g., image separations) will be far more easily determined for large samples than others (e.g., differential time delays). Fortunately, the easily determined parameters are of some interest; even a good determination of the frequency of lensing events would be quite valuable.

One possible example of such a statistical application would be to plot image separation versus lens redshift. The difference in the logarithmic slope of this relation between lens redshifts of 0.2 and 0.8 (the likely range) is 20% for $\Omega_0 = 1$ strongly clustered versus uniformly distributed models.

VI. DISCUSSION

It should first be emphasized that the preceding sections have only scratched the surface of potential lens applications. Lenses also may have important implications for the cosmological constant Λ, the global homogeneity and isotropy of the Universe, neutrino objects, Zel'dovich pancakes and filaments, primordial black holes, Population III objects, the intergalactic medium, the quasar luminosity function evolution, quasar emission region models, black pits, discordant redshifts, cosmic strings, etc.[6] It seems unlikely that they have anything to tell us about the solar neutrino problem, however, an exception I mention only to prove the rule.

Despite these inviting possibilities, the skeptic will note that 1) lenses are quite rare with only five cases known among the roughly 2500 known quasars, 2) the Universe is a dirty laboratory and real world complications may ruin the proposed applications in much the way they have stymied the in-principle elegant standard candle tests (for instance), and 3) the observations will be exceedingly difficult since they will require high angular resolution and in some cases synoptic study of faint and rare sources. These objections must be admitted to have some validity but also some rebuttal. First, the modest discovery rate of about one per year can be expected to increase as extensive systematic searches[7] get under way. Second, despite the various real world complications, lenses offer us a fundamentally new tool with which to attack problems against which the old tools (e.g., the standard candle tests) have already been beaten dull. I think that it is particularly important that one can imagine learning a great deal from individual lens systems which turn out to be particularly free from complicating factors; the problems are not all inherently statistical.

Third and finally, the observational difficulties can be met with the new generation of astronomical instruments such as the VLA, ST, the VLB array, and the NNTT. In any case, they may be regarded as a source of job security.

Discussions with J. R. Gott, J. E. Gunn, and J. P. Ostriker have greatly informed my understanding of the issues discussed in this paper. The support of NASA grant NAGW-626 is gratefully acknowledged. This work was performed under the auspices of an Alfred P. Sloan Research Fellowship.

REFERENCES

[1]V. C. Rubin, W. K. Ford, and N. Thonnard, Astrophys. J. Lett. **225**, L107 (1978).
[2]S. Refsdal, M.N.R.A.S. **128**, 307 (1964).
[3]K. Chang and S. Refsdal, Nature **282**, 561 (1979); J. R. Gott, Astrophys. J. **243**, 140 (1981).
[4]See the discussions of galaxy formation and large scale structure elsewhere in this volume, for example.
[5]E. L. Turner, J. P. Ostriker, and J. R. Gott, Astrophys. J. **284**, 1 (1984).
[6]Many, but not all, of these potential applications have been discussed in the literature in references too numerous to list here. Unfortunately, no comprehensive review has appeared. The lensing effects of cosmic strings are described elsewhere in this volume by Vilenkin.
[7]C. R. Lawrence, D. P. Schneider, M. Schmidt, C. L. Bennett, J. N. Hewitt, B. F. Burke, E. L. Turner, and J. E. Gunn, Science **223**, 46 (1984).

Gravitational Lensing by Strings

Joshua A. Frieman

Several years ago, Vilenkin[1] pointed out the unusual gravitational properties of vacuum strings, among them the possibility of gravitational lensing, which has recently been explored in detail.[1-3] Gravitational lenses are distinguished by two sorts of properties: their image characteristics (e.g., the number of images formed, their angular separation, distortion, and magnification), and their statistics (e.g., the fraction of QSO's which would be lensed). We would like to know if this combination of characteristics distinguishes string lenses sufficiently from other types of lens (either compact objects or galaxies) that one could conceive of using QSO surveys to search for, or rule out, evidence of strings.

For brevity, we consider static strands of string produced in the spontaneous breakdown of a gauge symmetry. (Lensing by loops, not discussed here, is considered in refs. 1 and 3.) Strings form with linear mass density and tension $\mu \sim \langle\Phi\rangle^2 \sim T_c^2$, where $\langle\Phi\rangle$ is the expectation value of a Higgs field component and T_c is the critical temperature of the phase transition. Gravitational effects are determined by the dimensionless parameter $G\mu \sim (T_c/m_{pl})^2$ ($\approx 10^{-6}$ for $T_c \approx 10^{16}$ GeV), where m_{pl} is the Planck mass. Note that since μ is quadratic in $\langle\Phi\rangle$, the effects are very sensitive to the SSB scale. Since the spacetime around a static string is flat, it exerts no gravitational force on nearby matter. The 2-surface orthogonal to the string is, however, conical (a flat surface missing a wedge of angle $8\pi G\mu$), so points on opposite sides of the string can be connected by 2 geodesics, i.e., objects can appear double. Light rays are deflected through an angle $\Delta\phi = 4\pi G\mu \approx 2.6$ arc sec (4.3 arc min) for $T_c \approx 10^{16}$ GeV ($T_c \approx 10^{17}$ GeV).

Since $\Delta\phi$ is independent of impact parameter, the image of an object behind the string will be neither distorted nor amplified; the characteristic signature of string lensing is the appearance of two equal brightness images of identical shape, with angular separation $\lesssim 2\Delta\phi$. For $G\mu \approx 10^{-6}$, corresponding to a few arc sec separation, the differential time delay due to the difference in path length is $\lesssim 3$ years, so the images can have different brightnesses if the QSO source is variable. Observation over an extended period, however, should be able to distinguish this delay-induced difference from a true asymmetry in the imaging. By contrast, black holes and galaxies tend to form distorted images of unequal brightness even in the absence of time delay. For example, for a point mass lens, the ratio of the brightness of the two images $\sim \theta^4$, where θ is the angular separation of the source and lens; for strings, this ratio is unity as long as $\theta \lesssim \Delta\phi$. Further distinctions between strings and conventional lenses arise: galaxies form an odd number of images, not two (although in practice they might not all be resolved); an angular separation of a few arcsec to a few arcmin, reasonable for strings, requires a point mass of $10^{12} - 10^{16}\ M_\odot$, more massive than typical galaxies (for a point mass, $\Delta\phi \approx 3(M/10^{12}M_\odot)^{0.5}$ arcsec); for galaxies, the lens count should increase with QSO apparent magnitude m_Q as[4] $N_L \sim 10^{1.3m_Q}$, while for a string, the lensing probability just scales with the QSO distribution, $N_L \sim N_Q \sim 10^{0.9m_Q}$.

To date, five lenses have been observed,[5] with an average

separation $\langle\Delta\phi\rangle \approx 5$ arcsec. Since strings naturally produce separations in this range, while for a model population of isothermal galaxies[6] $\langle\Delta\phi\rangle \lesssim 1$ arcsec, it is tempting to speculate that strings may have been seen already. Two of the observed lenses, however, show an odd number of images, while in a third, the brightness ratio is 4.3/1, almost certainly an imaging asymmetry. In Q0957 + 561, two images of fairly equal brightness are seen (the ratio has varied above and below 1/1), but a candidate galaxy lens in a rich cluster has been identified. Although the galaxy/cluster model for the lens must invoke dark matter, the spatial asymmetry of the images, particularly in the radio structure, rules out a simple string model here as well. For the fifth lens, Q2345 + 007, a double with the largest observed separation (7.3 arcsec), no candidate galaxy lens has been found, although there is evidence for a rich cluster. The brightness ratio is large, 2.5/1, but the time delay (for a string) could be up to 24 years, so this remains a possible string lens if the source is highly variable. It is clear that strings cannot explain the unexpected large angular separations of the observed lenses. Fortunately, these separations can be obtained by embedding galaxy lenses in rich clusters.[6]

This less exotic explanation of the observations is also borne out by considering statistics. In most string scenarios, a typical observer sees one strand stretched across the horizon, all objects within a band of angular width $\approx 2\Delta\phi$ around the string appearing twice. The fraction of QSO's (or anything else) lensed by a strand is $P_L = 2\pi\Delta\phi/4\pi = 2\pi G\mu \approx 6 \times 10^{-6}(T_c/10^{16}\ \mathrm{GeV})^2$. (For comparison, for galaxy lenses $P_L \approx 10^{-3} - 10^{-2}$.) In units of the closure density, the density of string is $\Omega_s \approx 2\pi G\mu$, so we recover the familiar result of Press and Gunn[7] that $P_L \approx \Omega_s$. Since roughly 2000 QSO's have been searched for lensing, the chances are poor ($\sim 10^{-2}$) that a strand lens with $G\mu \sim 10^{-6}$ (which would give a separation in the range so far observed) would have been seen yet. More complete surveys would certainly improve this, but they would have to go very deep in magnitude: the surface density inferred from past surveys[4] indicates of order 10^5 QSO's brighter than 19th magnitude spread over the sky, so $\sim (T_c/10^{16}\ \mathrm{GeV})^2$ of them would be lensed by a strand. (This assumes of course that a large portion of the strand is at smaller redshift than most QSOs, but a strand placed at random in our Hubble volume has a less than 20% chance of having its minimum redshift $z < 2$.)

Turning this argument around, the fact that at most 1 out of the 1900 QSOs surveyed is lensed by a string gives an upper limit on the actual lensing fraction due to strings of $P_L < 2.9 \times 10^{-3}$ at 95% confidence level (assuming a fair sample and that no sub-arcmin lenses were missed). Since $P_L \approx 2\pi NG\mu$, where N is the number of strands with small redshift, this implies $NG\mu < 4.6 \times 10^{-4}$, which provides a few interesting constraints on string scenarios: (i) strands obviously do not close the universe, $\Omega_s < 1.7 \times 10^{-2}$; (ii) either $N = 0$ or $T_c < 2 \times 10^{17}$ GeV; (iii) for GUT-scale strings, N cannot be much larger than unity. As a consequence, intercommutation, which reduces the strand density and keeps the universe from becoming string-dominated, must remain efficient up to the present epoch. [Suppose intercommutation is efficient down to a temperature T_a; for $T < T_a$, the strand density grows as $\rho_s \sim 1/R^2 \sim T^2$, so that at the present, $\Omega_s = 2\pi G\mu(T_a/2.7\mathrm{K})^2$. The lensing constraint, with $G\mu > 10^{-6}$, implies $T_a < 140$K.] For relativistic strings, additional constraints may come from the microwave anisotropy[2] and from gravitational radiation by loops.[8]

This report is a direct product of discussions with Arlin Crotts. I thank D. N. Schramm for encouragement and acknowledge support by DOE at Chicago.

REFERENCES

[1]A. Vilenkin, Phys. Rev. D**23**, 852 (1981); this volume; Tufts preprint (1984).
[2]J. R. Gott III, Princeton preprint (1984).
[3]C. Hogan and R. Narayan, unpublished.
[4]J. A. Tyson, Astrophys. J. **272**, L41 (1983).
[5]For a review, see B. F. Burke, Comments on Astrophys. **10**, 75 (1984).
[6]E. L. Turner, J. P. Ostriker, J. R. Gott III, Princeton preprint (1984).
[7]W. H. Press and J. E. Gunn, Astrophys. J. **185**, 397 (1973).
[8]C. Hogan and M. Rees, Caltech preprint (1984).

The Photino, the Axino and the Gravitino in Cosmology

Jihn E. Kim

ABSTRACT

I review cosmological constraints of the photino, the axino and the gravitino which have been obtained in collaboration with Ellis, Masiero and Nanopoulos. The photino mass constraint is revised for the case of the Peccei-Quinn symmetry, and the gravitino problem remains even in the inflationary universe scenario.

I. INTRODUCTION

One of the naturalness problems is the strong CP problem. QCD introduces[1]

$$\mathcal{L}_{eff} = \frac{\bar{\theta}g^2}{32\pi^2} F^a_{\mu\nu} \tilde{F}^{a\mu\nu} , \tag{1}$$

which violates both P and CP. Therefore, from the experimentally measured electric dipole moment, we estimate[2] the bound on $\bar{\theta}$: $|\bar{\theta}| = |\theta + \text{Arg.det.}M_q| < 10^{-8}$. The strong CP problem is the naturalness problem of why $\bar{\theta}$ is so small. The only known mechanism in the sense of t'Hooft[3] is due to the Peccei-Quinn symmetry[4] which is the axial symmetry in the quark sector. But it is known that a massless quark does not exist, implying that the global Peccei-Quinn symmetry is broken and there exists a Goldstone boson. This is the axion.[5] For phenomenological reasons, the axion must be very light. This can be achieved by introducing a complex SU(2) × U(1) singlet Higgs field developing a large vev. The new particle residing in the phase of the complex Higgs singlet is the invisible axion,[6] and its mass is

$$m_a \cong \frac{F_\pi m_\pi}{v_{PQ}} \frac{\sqrt{Z}}{1+Z} \frac{1}{\Gamma(\sigma)} \cdot$$

$$\cdot \left| \sum_{i=\text{light } q_L, \bar{q}_R} \Gamma(i) + \sqrt{2} \sum_{j=\text{heavy } Q_L, \bar{Q}_R} \frac{\alpha_c^2}{\pi^2} \Gamma(j) \ln \frac{m_j^2}{m_u m_d} \right| , \tag{2}$$

where $Z = m_u/m_d$, v_{PQ} is the Peccei-Quinn symmetry breaking scale, and Γ is the Peccei-Quinn quantum number. Since the coupling of the invisible axion to light quarks is also inversely proportional to v_{PQ}, the invisibility is reached by making $v_{PQ} \gg M_W$.

Another naturalness problem is the gauge hierarchy problem, which, it has been suggested, is solved by supersymmetry. Supersymmetry, and recently supergravity, have been applied to particle physics[7] with interesting predictions, such as the t quark mass. Supersymmetrization of the invisible axion scheme introduces[8] the axino, the common superpartner of the invisible axion and the saxino. In this talk, we take the viewpoint that the axino is the lightest sypersymmetric particle.

Let us now discuss the photino, axino and the gravitino in cosmology, which I have obtained in collaboration with Ellis, Masiero and Nanopoulos.[9,10] These particles are invented out of the naturalness problem.

II. PHOTINOS AND AXINOS

If the R parity is unbroken, there must be an absolutely stable R odd particle. [Models with a broken R parity, e.g., with a broken lepton number, can be constructed, but they are, in general, very contrived.[7]] Then there must be a cosmologically forbidden region for the mass of this stable particle.[11,12] Since the gluino is known to be heavy, one also expects the photino to be heavy.

If the photino is the lightest R odd particle, and if the scalar electron mass is of several tens of GeV, the Lee-Weinberg bound[12] is roughly applicable. Indeed, Goldberg[13] and Ellis et al.[14] have confirmed this by explicit calculations. A conservative bound for the stable photino mass is

$$m_{\tilde{\gamma}} \geq 0.5 \text{ Gev} , \tag{3}$$

for the scalar electron mass of 20 GeV. Increasing the scalar electron mass to 100 GeV also increases the photino mass bound to 15 GeV since the photinos are decoupled much earlier.

As discussed in the Introduction, it is very tempting to solve the strong CP problem by the Peccei-Quinn mechanism with an invisible axion. In supersymmetry/supergravity theories,[7] this implies the existence of the axino.[8] Because the axion can decay to two photons by the interaction

$$\mathcal{L}_{a\gamma\gamma} = \frac{e^2 N}{12\pi^2 v_{PQ}} \frac{1}{2} a F_{\mu\nu} F_{\rho\sigma} \varepsilon^{\mu\nu\rho\sigma} , \tag{4}$$

the photino can also decay into an axino and a photon by

$$\mathcal{L}_{\tilde{\gamma}\tilde{a}\gamma} = \frac{e^2 N}{12\pi^2 v_{PQ}} \tilde{\gamma}(\sigma^m \bar{\sigma}^n - \sigma^n \bar{\sigma}^m) \tilde{a} F_{mn} . \tag{5}$$

In the remainder of this section, we describe the cosmology based on Eq. (5).

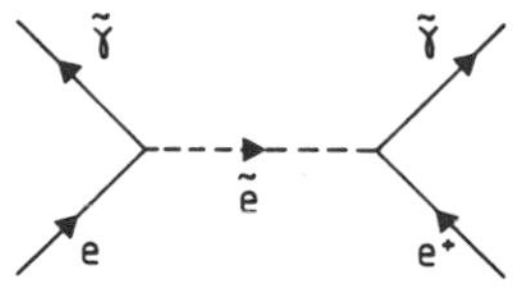

Fig. 1

The process $\tilde{\gamma}\tilde{\gamma} \rightleftarrows f\bar{f}$ (f = ordinary fermion) proceeds through Z^0 and $\tilde{f}$ exchange, such as shown in Fig. 1. When this reaction rate starts to be smaller than the expansion rate of the Universe, the photino decouples. The decoupling temperature is

$$T_D = m_{\tilde{\gamma}} \left[\ell n \left(\tilde{a}\, m_{\tilde{\gamma}}\, M_P \sqrt{\frac{45}{\pi^5 N_F}}\, x_D^{1/2} + \tilde{b}\, m_{\tilde{\gamma}}\, M_P \sqrt{\frac{45}{\pi^5 N_F}}\, x_D^{3/2} \right) \right] , \qquad (6)$$

where $x_D = T_D/m_{\tilde{\gamma}}$ and $\tilde{a}$ and $\tilde{b}$ are related to the scalar electron mass.[14] For a given photino mass, the decoupling temperature slightly increases as the scalar electron mass increases. By augmenting the photino mass, one also augments the decoupling temperature. For $m_{\tilde{e}} \gtrsim 30$ GeV and $m_{\tilde{\gamma}} > 10$ MeV, the decoupling temperature is always greater than 2 MeV, which is above the nucleosynthesis temperature (see Table of Ref. 9).

Therefore, it is interesting to see whether the photino can dominate the mass density of the Universe at the time of nucleosynthesis. For three light neutrino species, we have the following T*, at which the Universe becomes photino-dominated[15]:

$$T^* = \frac{m_{\tilde{\gamma}}}{(1.68)(2.75)} \exp(-m_{\tilde{\gamma}}/T_D) \quad . \qquad (7)$$

For most interesting values of the photino mass and the scalar electron mass, the photino dominates the mass density of the Universe after nucleosynthesis (see Table of Ref. 9). For unstable neutrinos, the requirement of the correct amount of He^4 has been obtained by Kolb and Scherrer,[16] which means in our case that the photino dominates the mass density of the Universe at $T^* < 3.3 \times 10^{-2}$ MeV. This is satisfied for most values of the Table of Ref. 9. Therefore, the stringent condition is that it decays before deuterium is formed, which is shown as the dash-dotted line in Fig. 2. In fact, this is the most restrictive bound.

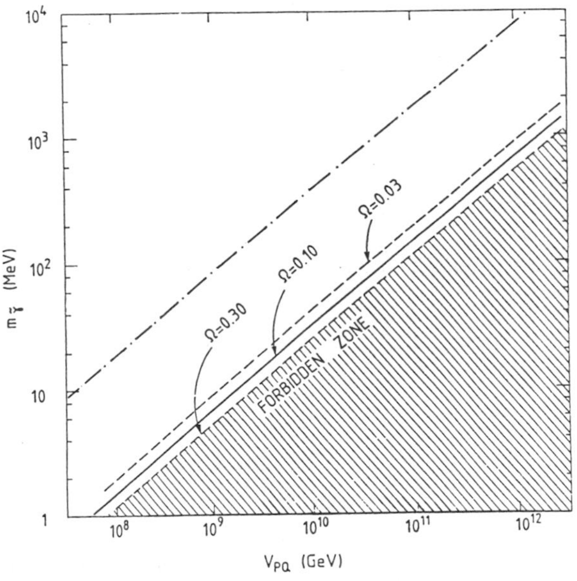

Fig. 2

If deuterium production has another primordial source, we can look for other cosmological constraints for the photino mass. For this purpose, we study the thermalization requirement[17] of the decay photon, which should decay before $\tau_{\tilde{\gamma}} < 3 \times 10^5$ sec.[15] Since the decay rate is given by

$$\Gamma_{\tilde{\gamma}} = \frac{\alpha_{em}^2 N^2}{9\pi^3} \left(\frac{m_{\tilde{\gamma}}}{V_{PQ}}\right)^2 m_{\tilde{\gamma}} \quad , \tag{8}$$

we obtain the bound given as solid and dotted lines (depending on Ω_B) in Fig. 2. If the invisible axion exists, the photino mass bound is a function of the Peccei-Quinn symmetry breaking scale, which is summarized in the same figure.

III. THE GRAVITINO PROBLEM

The gravitino problem is that the mass of the gravitino in the region 20-400 GeV advocated in supergravity-particle physics falls in the forbidden zone (1 keV ~ 10 TeV) of Pagels and Primack[18] and of Weinberg.[19] Ellis, Linde and Nanopoulos[20] suggested "inflation" for dilution of the gravitino number density, but it became apparent that gravitinos are regenerated after inflation.[21]

Recently, Ellis et al.[10] considered this problem seriously, and calculated the number density of gravitinos using N = 1 supergravity[22] and supersymmetric gauge Lagrangians. The result is expressed as a function of T_{max}, the reheating temperature after inflation. Certainly, this is a lower bound of the gravitino number density, since there may be other sources of gravitino production.[23] The result of this calculation is shown in Table 1 of Ref. 10. Summing the entries of the right-hand column of this table, we obtain

$$\sigma_t^{MSSM} = \frac{1}{M^2} (15.59\alpha_3 + 5.25\alpha_2 + 1.65\alpha') \quad , \tag{9}$$

where $M = M_P/\sqrt{8\pi} = 2.43 \times 10^{18}$ GeV, and

$$\alpha_3(T) \cong 0.0635(1-0.024 \ \ln T_9) \ , \tag{10}$$

$$\alpha_2(T) \cong 0.040(1+0.00675 \ \ln T_9) \ , \tag{11}$$

$$\alpha'(T) \cong 0.0115(1+0.0237 \ \ln T_9) \ , \tag{12}$$

with $T_9 = T/10^9$ GeV. Equation (9) has been obtained by multiplying the following fermion number densities on the cross-section and factoring out the common factor n_b:

$$n_b = \frac{N_b \zeta(3)}{\pi^2} T^3, \ n_f = \frac{3}{4} \frac{N_f \zeta(3)}{\pi^2} T^3 \ . \tag{13}$$

Here N_b and N_f count the boson and fermion degrees respectively. Therefore, the gravitino number density is

$$\Delta n_{\psi_\mu} (\text{at } T_f) = \sum_{ab} \sigma_t(ab \to c\ \psi_\mu) \cdot M \frac{\zeta(3)^2}{\pi^4} N_a N_b \sqrt{\frac{90}{\pi^3 N}}\ T_{max}\ T_f^3$$

$$= 3.35 \times 10^{-12}\ T_f^3\ T_9^{max}\ (1 - 0.018\ \ell n\ T_9^{max}) \ . \qquad (14)$$

Applying the estimate (14) at the time of nucleosynthesis does not restrict T_{max} very much. For example, by requiring that the gravitino dominate the mass density of the Universe only after $T < 3.3 \times 10^{-2}$ MeV, we obtain

$$T_{max} \lesssim 0.9 \times 10^{16} \text{ GeV} \ . \qquad (15)$$

The most stringent bound comes from the deuterium destruction of the decay photon. The temperature at the decay time of the gravitino is, if both $\psi_\mu \to$ gluino + gluon and $\psi_\mu \to$ photino + photon are allowed,

$$T_\Gamma = \left(\frac{90}{32\pi^3 N}\right)^{1/4} \left(\frac{9m_{3/2}^3}{8\ M_P}\right)^{1/2} , \qquad (16)$$

with a typical lifetime of order 10^8 s. Using Lindley's analysis[24] of deuterium destruction, we obtain

$$T_{max} < \left(\frac{100 \text{ GeV}}{m_{3/2}}\right) \times \begin{cases} 2.1 \times 10^9 \text{ GeV for } \delta_B = 10^{-10} \\ 2.2 \times 10^{10} \text{ GeV for } \delta_B = 10^{-9} \end{cases} , \qquad (17)$$

where δ_B is the ratio of the baryon number to the photon number at the time of gravitino decay. If (17) is satisfied, there will not be any problem for thermalizing the decay photons.[10]

This is very bad news for baryogenesis in supergravity models. In the supersymmetric SU(5) model, the dimension five operators are not reconcilable with the bound (17). Therefore, the scenario of the 100 GeV gravitino is not permissible in cosmology.[25]

IV. CONCLUSION

We have discussed new particles arising from the theoretical requirement of naturalness. The existence of the axino makes the photino mass bound more flexible, but the current idea of the 100 GeV gravitino runs into cosmological trouble.

ACKNOWLEDGMENTS

I thank D. Lindley, R.J. Scherrer and, in particular, S. Sarkar for helpful discussions. I also thank K. Kang at Brown University, C.W. Kim at Johns Hopkins, P. Langacker and G. Segré at Penn University and E.W. Kolb and M.S. Turner at Fermilab for their kind hospitality.

REFERENCES

[1]A. Belavin et al., Phys. Lett. **95B**, 85 (1975); G. 't Hooft, Phys. Rev. **D14**, 3432, (1976); R. Jackiw and C. Rebbi, Phys. Rev. Lett. **37** 172, (1976); C.G. Callan, R.F. Dashen and D. Gross, Phys. Lett. **63B**, 334 (1976).

[2]V. Baluni, Phys. Rev. **D19**, 2227 (1979); R. Crewther, P. Di Vecchia, G. Veneziano and E. Witten, Phys. Lett. **89B**, 123 (1979).

[3]G. 't Hooft, in "Recent Developments in Gauge Theories," G. 't Hooft et al. eds., (Plenum, New York, 1980).

[4]R.D. Peccei and H.R. Quinn, Phys. Rev. Lett. **38**, 1440 (1977).

[5]S. Weinberg, Phys. Rev. Lett. **40**, 223 (1978); F. Wilczek, Phys. Rev. Lett. **40**, 279 (1978).

[6]J.E. Kim, Phys. Rev. Lett. **43**, 103 (1979); M.A. Shifman, A.I. Vainshtein and V.I. Zakharov, Nucl. Phys. **B166**, 493 (1980); M. Dine, W. Fischler and M. Srednicki, Phys. Lett. **104B**, 99 (1981); M.B. Wise, H. Georgi and S.L. Glashow, Phys. Rev. Lett. **47**, 402 (1981); H.P. Nilles and S. Raby, Nucl. Phys. **B198**, 102 (1982); R. Barbieri, R.N. Mohapatra, D.V. Nanopoulos and D. Wyler, Phys. Lett. **106B**, 303 (1981); J.E. Kim, Phys. Rev. **D24**, 3007 (1981).

[7]For an excellent review, see: H.P. Nilles, University of Geneva preprint UGVA-DPT 1983/12-412, to be published in Phys. Reports (1984); J. Polchinski, talk at this meeting; R. Barbieri and S. Ferrara, Surv. in HEP **4**, 33 (1983); J. Ellis, CERN preprint TH.3718 (1983); D.V. Nanopoulos, CERN preprint TH.3699 (1983).

[8]J.E. Kim, Phys. Lett. **136B**, 378 (1984).

[9]J.E. Kim, A. Masiero and D.V. Nanopoulos, Phys. Lett. **139B**, 346 (1984).

[10]J. Ellis, J.E. Kim and D.V. Nanopoulos, CERN preprint TH.3839 (1984).

[11]R. Cowsik and J. McClelland, Phys. Rev. Lett. **29**, 669 (1972).

[12]B.W. Lee and S. Weinberg, Phys. Rev. Lett. **39**, 165 (1977).

[13]H. Goldberg, Phys. Rev. Lett. **50**, 1419 (1983).

[14]J. Ellis, J.S. Hagelin, D.V. Nanopoulos, K. Olive and M. Srednicki, SLAC-PUB-3171 (1983).

[15]For unstable heavy neutrinos, see: D.A. Dicus, E.W. Kolb and V.L. Teplitz, Astro. Phys. J. **221**, 327 (1978).

[16]E.W. Kolb and R.J. Scherrer, Phys. Rev. **D25**, 1481 (1982).

[17]For small perturbations with bremsstrahlung, see: A.F. Illarianov and R.A. Sunyaev, Sov. Astron. **18**, 691 (1975); R.A. Sunyaev and Ya. B. Zel'dovich, Ann. Rev. Astron. Astrophys. **18**, 537 (1980); For large perturbations with double Compton scattering, see: S. Sarkar and A.M. Cooper, Proc. ESO/CERN Symposium, Geneva, November 1983.

[18]H. Pagels and J.R. Primack, Phys. Rev. Lett. **48**, 223 (1983).

[19]S. Weinberg, Phys. Rev. Lett. **48**, 1303 (1983).

[20]J. Ellis, A.D. Linde and D.V. Nanopoulos, Phys. Lett. **118B**, 59 (1982).

[21]S. Weinberg, unpublished; D.V. Nanopoulos, K.A. Olive and M. Srednicki, Phys. Lett. **127B**, 30 (1983); L.M. Krauss, Nucl. Phys. **B227**, 556 (1983).

[22]E. Cremmer et al., Nucl. Phys. **B212**, 413 (1983).
[23]P.J. Steinhardt, private communication (1984).
[24]D. Lindley, Mon. Not. R. Astr. Soc. **193**, 593 (1980).
[25]J.E. Kim, A. Masiero and D.V. Nanopoulos, CERN preprint TH.3866 (1984); J. Ellis et al., CERN preprint TH.3902 (1984).

The Recombination and Light-Inos

Pierre Salati

ABSTRACT

The recombination of neutral hydrogen is reinvestigated taking into account light neutral fermions, stable or radiatively unstable. When these are stable, their main effect is to increase the expansion rate of the Universe, and to increase the fossilized ionisation of matter. If the light neutral fermions decay radiatively, the emitted photon is in the UV-range and reionizes the neutral matter. Matter can be completely reionized at a redshift Z ~ 100 for radiative lifetimes in the range 10^{20}-10^{24} seconds.

If we admit that the standard Hot Big Bang Model[1] is essentially correct, the matter of the very early Universe was very hot and dense. When the temperature was over 10^4 °K, it was completely dissociated into electrons and ions like protons and ionized Helium for the most part. As the Universe expanded, it cooled and when its temperature reached over 4000 °K at a time of about 10^5 years the baryonic matter combined with electrons to form neutral matter. The recombination of the primeval plasma has been studied[2] and the details of the transition between ionized and neutral matter are of great interest:

1. Recombination plays a crucial role in galaxy formation[3] because the perturbations in matter density cannot develop or grow to form galaxies in an ionized medium, due to the Thomson friction of electrons upon the thermal radiation. So the recombination period is determinant for the evolution of perturbations which lead to gravitation-bound systems.

2. As radiation scatters onto the matter via the quasi-elastic Thomson diffusion the small scale angular inhomogeneities of the cosmic background radiation (CBR) depend straightly on the recombination features.

3. Electrons and protons catalyse the formation of molecular hydrogen.[4]

4. Finally, an intergalactic medium (IGM) may well be constituted by the relics of recombined matter which have escaped from gravitational collapse. So, observation of the IGM is a good way to test the theory and to go back into the past at the time matter recombined.

The main effect of stable light-inos is to increase the mass density of the Universe and its expansion rate. The recombination is speeded up and the residual ionization x_e of matter is increased. We have assumed that the Universe was quasi-flat:

$$\rho_{total} \sim \rho_{critic} = 2\times10^{-29}\,g/cm^3 \left(\frac{H_0}{100\ km/s/Mpc}\right)^2$$

and we have plotted[5] (fig. 1a and 1b) the behaviour of x_e versus the temperature for two values of the Hubble constant H_0 and various baryonic mass fractions:

$$\Omega_B = \frac{\rho_B}{\rho_{critic}}$$

Big Bang Nucleosynthesis provides density constraints on Ω_B and if light-inos are introduced to close the Universe, the ionized fraction x_e is always larger than 4×10^{-4}:

$$4\times10^{-4} \lesssim x_e \lesssim 2\times10^{-3}$$

instead of the previous baryon-dominated-Universe result

$$3\times10^{-5} \lesssim x \lesssim 3\times10^{-4}$$

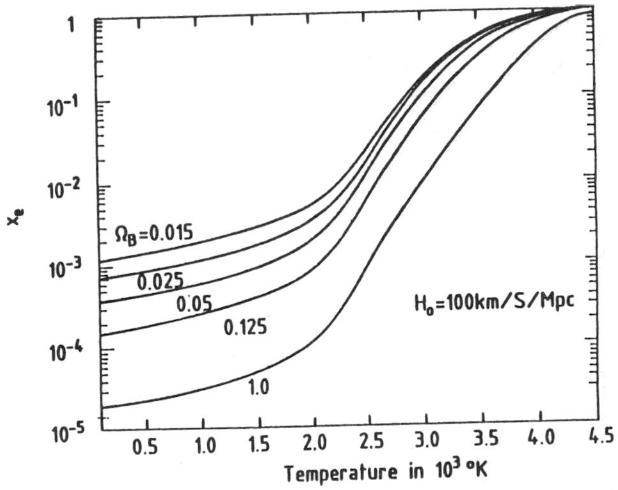

Fig. 1a

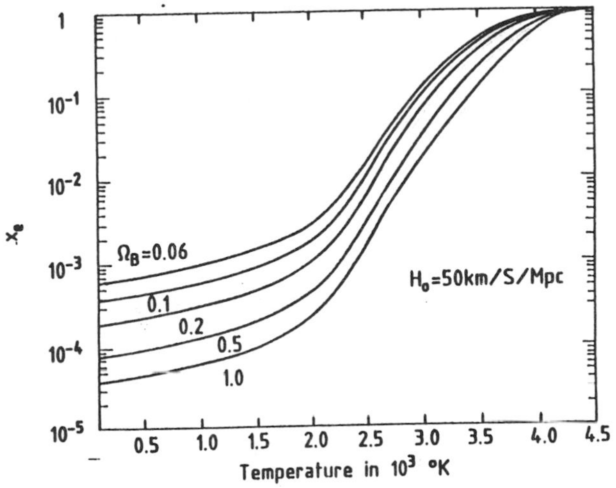

Fig. 1b

Finally, if a 30-100 eV -ino decays radiatively, the emitted photon is in the UV range. Figs. 2a and 2b show the reionization effect of such a radiative decay. A 10^{24}s. lifetime is still sufficient to induce a substantial increase of x_e at the time of galaxy formation. Supersymmetry provides us with such a -ino. The reaction

photino → photon + gravitino

produces a UV photon with a lifetime ranging between 10^{20} to 10^{25}s.

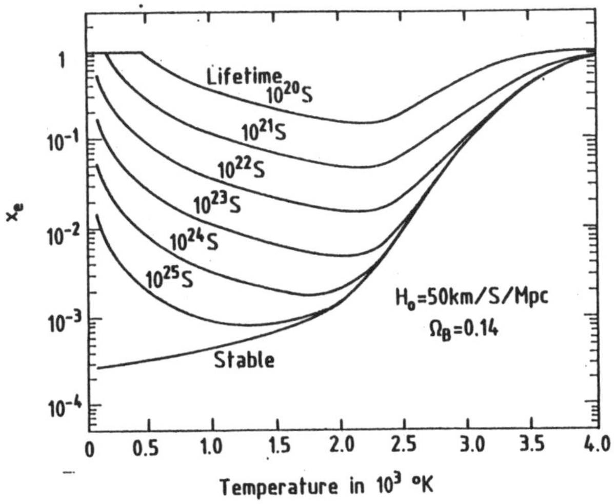

Fig. 2a

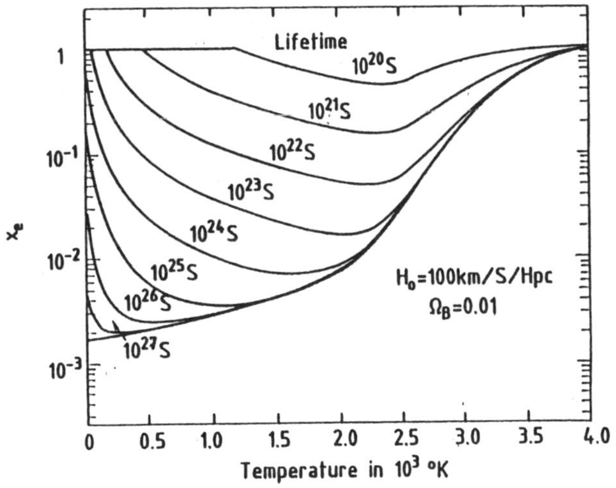

Fig. 2b

Supported in part by DRET under contract number 82/1252/DRET/DS/SR.

REFERENCE

[1]See for example:
a) S. Weinberg, Gravitation and Cosmology, John Wiley and Sons.
b) A. D. Dolgov and Ya B. Zeldovich, Reviews of Modern Phys. **53**, (1), 1 (1981).
[2]P. J. E. Peebles, Astrophys. J. **153**, 1 (1968); V. G. Kurt, R. A. Syunyaev and Ya B. Zeldovich, Sov. Phys., JETP **28**, 146 (1969).
[3]P. J. E. Peebles and J. T. Yu, Astrophys. J. **162**, 815 (1970).
[4]J. Silk, Formation and Evolution of Galaxies and Large Structures in the Universe, p. **253**, D. Reidel Publishing Company.
[5]Recombination of the primeval plasma and light-inos, P. Salati and J. C. Wallet, LAPP-TH-99, March 1984, to be published in Phys. Lett. B.

Mirror Fermions and Cosmology

Goran Senjanović

ABSTRACT

Extended supersymmetry, Kaluza-Klein theory and family unification all suggest the existence of mirror fermions, with the same quantum numbers but opposite helicities from ordinary fermions. The laboratory and especially cosmological implications of such particles are reviewed and summarized.

I. INTRODUCTION

One of the most striking features of the standard electroweak model is its left-right asymmetry in the fermionic spectrum f. Equally important is the repetitive family structure; both facts being just simply postulated ad hoc. Could it be that these phenomena have one and the same origin? This is the possibility I would like to discuss in my talk. Most of what will be presented here is based on the recent work in collaboration with Wilczek and Zee[1]; similar work has been simultaneously carried out by Bagger and Dimopoulos.[2]

The problem of family repetition persists in single grand unified theories, such as SU(5) and SO(10). In these theories families are xeroxed at free will, without any handle on their number. Furthermore, the fact that fermions have only left-handed weak interactions is taken for granted. It should be mentioned that SO(10) is the minimal one family unification theory, since all the fermions in one family form an irreducible $\underline{16}$ dimensional spinorial representation.

This immediately suggests larger orthogonal groups SO(10+2n) as ideal candidates for family unification.[3] Even more important, spinorial representations possess a unique property: <u>their decomposition on smaller subgroups contains only spinorial representations and nothing else.</u> For example, 2^{n+4} dimensional irreducible spinorial representation of SO(10+2n) contains 2^{n-1} $\underline{16}$ dimensional representations of SO(10) and 2^{n-1} $\overline{\underline{16}}$'s of SO(10). This leads to the vector-like low energy world: <u>for every ordinary family f there is an opposite helicity "mirror" family F!</u>[4-6]

Remarkably enough, completely independent theoretical ideas, such as extended supersymmetry (N>1) and Kaluza-Klein higher dimensional theories of gravity[7] require the existence of mirror fermions, too. Why have we not seen such particles yet? Why are they so weakly mixed with ordinary fermions? Are mirror fermions cosmologically stable or not? What about the increase in neutrino species that their existence leads to? These and similar questions must be answered before one has a consistent theory of mirror fermions.

To appreciate the problem, notice that the very existence of ordinary fermions in such theories appears to be a miracle. Namely, the mixing mass term $m_{fF} = M\bar{f}F$ is an SU(2) × U(1) × $SU(3)_C$ singlet and, according to the commonly accepted survival principle,[8] M should escape to the Planck mass or at least the unification scale. This means that <u>ordinary and mirror fermions should pair off to disappear from the low energy world</u>, so why do they not? Now, the survival principle sounds a lot like social Darwinism and I, for one, believe that the world may be much milder than such ideas suggest. The purpose of my talk is to demonstrate not only that fermions have natural ways to survive down to low energies, but also to try to convince you that the existence of mirror fermions has rather interesting, soon to be tested, laboratory and cosmological implications. In view of the nature of this conference, I will devote a substantial portion of my time to the cosmological issues.

As far as I know, the only natural way to forbid the bare mixing mass term requires larger orthogonal groups SO(10+2n). I start then, in the next section, by reviewing the basic properties of such theories and by presenting a realistic candidate for family unification based on SO(18). In Section III, we shall study the general low energy properties of mirror fermions, especially those of cosmological relevance. The paper ends with a brief summary in Section IV.

II. FAMILY UNIFICATION AND MIRROR FERMIONS

As I stressed in the Introduction, SO(10) is a perfect one family unified theory. Each family forms an irreducible $\underline{16}$ dimensional spinorial representation; anomalies are automatically cancelled.

Imagine a Clifford algebra

$$\{\gamma_i,\gamma_j\} = 2\,\delta_{ij} \qquad i,j = 1,\ldots,2N\ . \tag{2.1}$$

You can easily convince yourself that

$$T_{ij} \equiv \frac{1}{4i}\,[\gamma_i,\ \gamma_j]\ , \tag{2.2}$$

satisfy the usual commutation relations of SO(2N). Furthermore, $\gamma_{FIVE} = \gamma_1,\ldots\gamma_{2N}$ commutes with all the generators and the 2^{N-1} dimensional representation of (2.2) is characterized by a helicity: LEFT(RIGHT) = 1/2 (1 ± γ_{FIVE}). It is useful to use a basis in which $\hat{\varepsilon_i} \equiv 2T_{i-1,i}$ are diagonal operators; with eigenvalues $\varepsilon_i = \pm 1$. Therefore, each physical state can be characterized by the eigenvalues of $\hat{\varepsilon_i}$

$$|\varepsilon_i\ \ldots\ \varepsilon_N\rangle\ , \tag{2.3}$$

with $\varepsilon_i = +1$ or $\varepsilon_i = -1$. In this notation

$$\gamma_{FIVE} = \prod_{i=1}^{N} \varepsilon_i\ .$$

As I mentioned before, the 16 dimensional representation of SO(10) with

$$\prod_{i=1}^{5} \varepsilon_i = 1$$

contains a usual left-handed family.

SO(10+2n)

The irreducible representation is $2^{4+n} = 16 \times 2^n$ dimensional. Choose

$$\prod_{i=1}^{5+n} \varepsilon_i = 1 \, .$$

Then, either

$$\prod_{k=1}^{5} \varepsilon_k = 1 \text{ and } \prod_{k=6}^{5+n} \varepsilon_k = 1 \text{ or } \prod_{k=1}^{5} \varepsilon_k = -1 \text{ and } \prod_{k=6}^{5+n} \varepsilon_k = -1 \, .$$

In the first case we are dealing with ordinary families, i.e. 16's of SO(10), whereas in the latter case the helicities are switched -- we get mirror families ($\overline{16}$'s). In short, the theory contains 2^{n-1} standard families and 2^{n-1} mirror families.

The characteristics of the theory depend dramatically on whether n is even or odd. Namely, for n = 2k + 1 the representation is vectorlike (allows bare mixing mass term); for n = 2k the theory is chiral (no mixing mass terms). For n = 2k, gauge symmetry

$$U_M = e^{i(\pi/2)(\varepsilon_6+\ldots+\varepsilon_{2k})} = i^{2k}\varepsilon_6\ldots\varepsilon_{2k}, \tag{2.4}$$

becomes a discrete symmetry on families and mirror families

$$f \to i^{2k} f, \quad F \to -i^{2k} F \, . \tag{2.5}$$

Chirality of the theory is reflected in mirror symmetry being a gauge symmetry. Although we could make n = 2k + 1 theories work by adding extra symmetries, we stick here to physically more natural n = 2k theories.

SO(14)

Here n = 2; leading to 2f + 2F. This group is too small.

SO(18)

Here $n = 4$, so that we would get $8f + 8F$. This is the minimal natural unified theory of families. As opposed to SO(14), it appears too big. If all fermions are light, it is not an asymptotically free theory. Actually, asymptotic freedom allows at most 4 families and 4 mirror families. We must "kill" some families and mirror families, by pairing them off and decoupling them from low energies. Obviously, this requires breaking U_M at high energies, with the remnant of it staying unbroken down to M_W. If we stay in SO(18), it turns out that we always end up with either $4f + 4F$ or $6f + 6F$.

It turns out, however, that with the inclusion of Peccei-Quinn symmetry we can end up with precisely three families and three mirror families. I do not have time to discuss our work here,[1] so let me just summarize the main results[1]

(i) there are five light neutrinos
(ii) the mirror symmetry gets broken at M_W, with the lowest dimension breaking term being $d = 5$

$$H_{eff} = \frac{1}{M_X} \bar{f}F \, \phi_1^+ \phi_2 \, , \qquad (2.6)$$

where ϕ_1 and ϕ_2 are SU(2) doublets.

III. PHENOMENOLOGY AND COSMOLOGY OF MIRROR FERMIONS

The main message from the previous section, if for whatever reason you did not go through it, is that we can naturally keep both ordinary and mirror fermions light. This was absolutely necessary in order to live in the world with observed fermions. We still must answer the questions posed in the Introduction.

(1) Mixing and Masses

(a) Since F's get their mass at the SU(2) × U(1) level, we expect $m_F < 1$ TeV. If you further believe in unification, we must have $m_F \lesssim 250$ GeV to prevent Yukawa couplings from blowing up. We must see mirror fermions at the supercollider!

(b) Why is $m_f < m_F$? Could it be that separate Higgs doublets give them their masses, with $\langle\phi_F\rangle \neq \langle\phi_f\rangle$? But then, why is $\langle\phi_F\rangle > \langle\phi_f\rangle$? Bagger and Dimopoulos suggest that ordinary fermions get their masses radiatively, hence $m_f \ll m_F$.

(c) On the other hand, mirror fermions are naturally weakly coupled to ordinary fermions. Namely, there is no $d = 4$ renormalizable interaction of the type $fF\phi$, where ϕ is a SU(2) × U(1) Higgs doublet. To generate mixings, one needs the $d = 5$ effective interaction in (2.6) (two doublets are needed to construct SU(2) × U(1) singlet), implying a mixing

$$\theta_{fF} \simeq \frac{M_W}{M_X} \simeq 10^{-13}. \qquad (3.1)$$

The way to produce mirror fermions is to get them in pairs ($F\bar{F}$) in say e^+e^- or $p\bar{p}$.

(2) Cosmological Stability

Although practically stable for laboratory purposes, mirror fermions are cosmologically unstable. Actually, charged mirror leptons should be expected to decay very fast[2]: $E \to e\, N_R\, \bar{\nu}_L$. Expected lifetime

$$\tau_E \simeq G_F^{-2}\, m_E^{-5} \simeq 10^{-21} \text{ sec for } m_E \simeq 100 \text{ GeV} . \tag{3.2}$$

Neutral mirror leptons and mirror baryons are, on the other hand, long lived. From (2.6) one gets possible $Q \to qqq$ decay through the small mixing in (3.1). We estimate

$$\tau_Q \simeq \frac{M_x^2}{m_Q^3} \simeq 1 \text{ sec}$$

$$\text{for } M_x \simeq 10^{15} \text{ GeV}$$

$$m_Q \simeq 100 \text{ GeV}. \tag{3.3}$$

This explains why we see no mirror baryons; they have all decayed long ago.

(3) Neutrinos

Adding mirror neutrinos increases the number of light neutrino species. In our SO(18) model we predict $n_\nu = 5$. What about the usually quoted cosmological limit from observed helium abundance $n_\nu \leq 4$?

We could try to say that some of them, say N_R's, are heavy enough ($m \simeq$ 1-10 MeV), as to decay fast enough: $N_R \to e^+e^-\, \nu_L$. However, in our case the scale of superheavy singlet neutrinos is $M_{PQ} \simeq 10^9 - 10^{12}$ GeV, implying $m(N_R) \leq 10$ KeV (without fine tuning the parameters of the theory). To be honest, the natural prediction of our theory is five, effectively stable light neutrinos.

One possibility, as suggested by Bagger and Dimopoulos is to imagine a light SU(2) × U(1) Higgs triplet which gives mass to N_R's. This leads to various interesting phenomenological consequences, all to be tested. I do not find this alternative very appealing, since it violates the extended survival principle in the Higgs sector, in my opinion necessary to make predictive sense out of GUTs.

I would rather suggest that we ignore cosmological limits on n_ν at the moment. In due time, the precise measurements of the decay width of the Z boson will determine precisely n_ν. I suggest you have some patience before you reject the picture we suggested.

IV. SUMMARY

As I mentioned in the Introduction, various theoretical ideas that you may like (at least some of them) require the existence of mirror fermions. My aim in this talk was to try to convince you that you should not worry about this; but rather, take it as a blessing. If, however, I have failed in my task, I hope, at least, that our arguments which claim the consistency of the mirror fermion physics, appear convincing. If you find a mechanism which naturally forbids the large mixed mass terms for ordinary and mirror fermions (or accept what we suggested here), the rest follows naturally:

(a) f-F mixing is naturally small ($\sim M_W/M_X$)
(b) lightest Q is cosmologically unstable ($\tau_Q \sim 1$ sec)
(c) $m_F \lesssim 250$ GeV (in unified theories)
(d) there are more than three neutrinos (5?)

I hope you can live (if uneasily) with the last prediction and wait for the supercollider to pass the final verdict on the ideas presented here.

Let us look into mirrors!

ACKNOWLEDGMENTS

The work described here has been done in collaboration with Frank Wilczek and Tony Zee. I wish to thank Palash Pal for numerous discussions and a careful reading of the manuscript. Furthermore, thanks are due to the ITP, Santa Barbara, and International Atomic Energy Agency and UNESCO for hospitality at the International Centre for Theoretical Physics, Trieste, during various stages of this work. This work is supported in part by the U.S. Department of Energy under contract DE-AS05-80ER10713.

REFERENCES AND FOOTNOTES

[1]G. Senjanović, F. Wilczek and A. Zee, ITP preprint NSF-ITP-84-28 (1984) -- to appear in Phys Lett. B.
[2]J. Bagger and S. Dimopoulos SLAC preprint SLAC-PUB-3287 (1984). See also, H.M. Asatryan, Yerevan preprint (1982).
[3]M. Gell-Mann, P. Ramond and R. Slansky, in Supergravity, Eds. P. van Nieuwenhuizen and D.Z. Freedman (North-Holland, Amsterdam, 1979); F. Wilczek and A. Zee, Princeton report (1979) and Phys. Rev. **D25**, 553 (1982); R.N. Mohapatra and B. Sakita, Phys. Rev. **D21**, 1062 (1980).
[4]Mirror fermions have been suggested by J.C. Pati and A. Salam, Phys. Lett. **58B**, 333 (1975) and e.g. J.C. Pati, Proc. ICOBAN Conf., Bombay (1981) in order to achieve anomaly cancellation in their theory.
[5]For the physics and phenomenology of mirror fermions see also P. Ramond, University of Florida preprint IT-83-21 (1983). Mirror fermions have been studied extensively by the Helsinki group; see e.g. K. Enquist, K. Mursula and M. Roos, Nucl. Phys. **B226**, 121 (1983); J. Maalampi, Nucl. Phys. **B207**, 233 (1982); K. Enquist and J. Maalampi, Z. Phys. **C21**, 345 (1984). See also, K. Yamamoto, Phys. Lett. **120B**, 157 (1983); ibid. **117B**, 217 (1982) and references therein. For further references see Ref. 1 and Wilczek and Zee, Ref. 3.

[6]Mirror fermions have been discussed also in the context of composite models. See e.g. P. Senjanović, Fizika **15**, 126 (1979).
[7]See e.g. E. Witten, Nucl. Phys. **B186**, 412 (1981) and references therein.
[8]H. Georgi, Nucl. Phys. **B156**, 126 (1979).

Higgs Masses and Supersymmetry

Ricardo A. Flores and Marc Sher

The standard model (SM) of the strong, weak and electromagnetic interactions has a great deal of arbitrariness in it that has lead to the belief that there must be new physics beyond the SM. Supersymmetry (SUSY) models[1] have been among the most extensively studied extensions of the SM in recent years. SUSY implies the existence of SUSY partners of ordinary particles whose masses can in general be pushed to higher and higher values. However, it has been noted[2,3] that mass relations among ordinary Higgs scalars arise in a very wide class of SUSY models. If one only assumes that the effective low energy theory is such that

i) the gauge group is $G = SU(3) \times SU(2) \times U(1)$

ii) SUSY is either spontaneously or softly broken

iii) there are two Higgs doublets (the minimal number)

then, under these assumptions, the Higgs potential for two doublets Φ_1 and Φ_2 is $V = V_2 + V_4$, where (at tree level)

$$V_2 = m_1^2\Phi_1^\dagger\Phi_1 + m_2^2\Phi_2^\dagger\Phi_2 - m_3^2(\Phi_1\Phi_2 + \text{c.c.})$$

$$V_4 = g^2/2\ \Sigma\ |\Phi_1^\dagger\tau_i\Phi_1 + \Phi_2^\dagger\tau_i\Phi_2|^2 - g'^2/8\ |\Phi_1^\dagger\Phi_1 - \Phi^\dagger\Phi_2|^2$$

where the τ_i are the usual SU(2) generators and g and g' are the usual gauge couplings. V_2 is completely arbitrary, but contains only three free parameters, whereas the number of physical scalars is five: a charged scalar $\chi^\pm$, a pseudoscalar χ_0 and two neutrals ϕ and η; hence the mass relations. One then finds that

$$m_\chi^2 = m_{\chi_0}^2 + M_W^2$$

$$m_{\phi,\eta}^2 = 1/2\ \{M_Z^2 + m_{\chi_0}^2 + ((M_Z^2 + m_{\chi_0}^2)^2 - 4M_Z^2m_{\chi_0}^2\ (\frac{x^2-y^2}{x^2+y^2})^2)^{1/2}\}$$

where $x^2(y^2) = 2\Phi_1^\dagger\Phi_1(2\Phi_2^\dagger\Phi_2)$. Thus

$$m_\chi \geq M_W \quad \text{and} \quad m_\phi \leq M_Z\ .$$

The upper bound on m_ϕ is the most important result. It was shown in [4] that radiative corrections to the Higgs potential do not upset this bound by more than about 2 GeV. One then gets $m_\phi < 95$ GeV. A tough little theorem[3] shows that this result is independent of the number of doublets. It is worth emphasizing that assumptions i) - iii) are characteristic of most models. Extensions of the gauge group G to G × $\tilde{U}(1)$ have been considered in the literature[1], but the discovery of the Z at the right energy implies that $\tilde{g}$, the $\tilde{U}(1)$ coupling, should be very small and, thus, the upper bound is unchanged. Additional singlets eliminate the upper bound (unless they are very heavy), but generally make it very difficult for a gauge hierarchy to be preserved (see ref. [6] of [4]). This work was supported by NSF grants PHY-8305795 and PHY-815541.

REFERENCES

[1]For a review and references to the earlier literature see G. Kane and H. Haber, Michigan preprint UM HE TH 83-17.
[2]K. Inoue, A. Kakuto, H. Komatsu and S. Takeshita, Prog. Theor. Phys. **67** (1982) 1889; **68** (1982) 927.
[3]R. A. Flores and M. Sher, Ann. Phys. **148** (1983) 95.
[4]S. P. Li and M. Sher, Irvine preprint No. 84-7.

Recent Results in the Search for Fractional Charge at Stanford

William M. Fairbank and James D. Phillips

ABSTRACT

We have performed experiments to determine the residual charge of niobium spheres, which have consistently showed results of ±(1/3)e and 0e. We have carefully taken into account background forces as small as $0.01eE_A$. In a recent run, including a blind test with Luis Alvarez, inconsistencies and drifts appeared which were not present in the earlier data. Later measurements showed that the magnetic field was tilted by 12 milliradians with respect to gravity during that run as compared with a tilt of the order of 1 mr in the previous run. New calculations indicate that if the levitation magnetic field is not exactly vertical, there is a coupling to a magnetic force which can produce the inconsistencies observed. Spinning the ball about the vertical axis eliminates this effect, and we have found and demonstrated a method of spinning the ball. Since the previous results are narrowly distributed around 0 and ±(1/3)e, they form strong statistical evidence for fractional charge on matter of ±(1/3)e. However, it is essential to repeat these experiments with all background effects absent before we can be sure of the presence or absence of fractional charge on matter.

I. INTRODUCTION

The apparatus consists of a superconducting niobium sphere 1/4mm in diameter levitated between capacitor plates with a vertical resonant frequency of ~1 Hz. A 3kV/cm electric field is applied in phase with the ball's velocity, and the force which the field exerts on the ball is determined by the rate of increase of the ball's oscillation amplitude. The experiment is calibrated by measuring the force change due to single electron charge changes, and the fractional part of the force is determined. We can measure the alternating residual force on the ball, F^r_A, to within 0.01 eE_A. It is due to the residual charge plus background dipole forces and given by

$$F^r_A = [q_r E_A + P_F \cdot \nabla E_A + P_A \cdot \nabla E_F + \mu_A \cdot \nabla B + \ldots] \quad (1)$$

where q_r is the residual charge, P_F and P_A the ball's fixed and alternating electric dipole moment, E_F and E_A the fixed and alternating electric field, μ_A the ball's alternating magnetic dipole moment, and B the magnetic field. It was only after studying these effects in great detail that we published the data shown in Fig. 1.[1-6] The second term in Eq. 1, due to a fixed electric dipole moment, was found to be negligible. The third term, due to an induced electric dipole on the ball and a fixed gradient from the patch effect on the plates, although significant in size, could be accurately accounted for by measuring the position dependence of the residual force for several balls and taking the difference. The fourth term was calculated to be negligible with the electric, magnetic, and gravitational fields parallel.

Since the data in the figure was taken, we have had a run in which inconsistencies and drifts appeared which were not present in the earlier data. Some of the data in this run was very reproducible, while some was not. Even the reproducible data gave residual charge

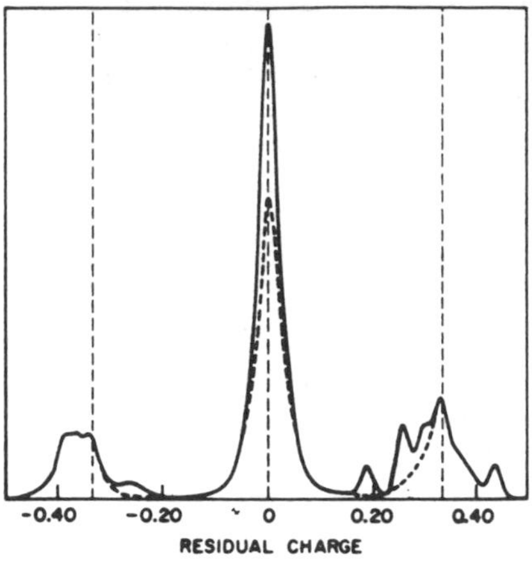

Figure 1 - Ideogram of residual charges measured to date. The solid curve shows all measurements made between August 1976 and the present for which the patch effect electric field gradient was constant (58 measurements). In the dashed curve, the 13 measurements of the run in which the blind test was done, during which the apparatus was inclined 12 milliradians with respect to the vertical, are excluded. In some runs the absolute charges are uncertain, but the relative charges are shown.

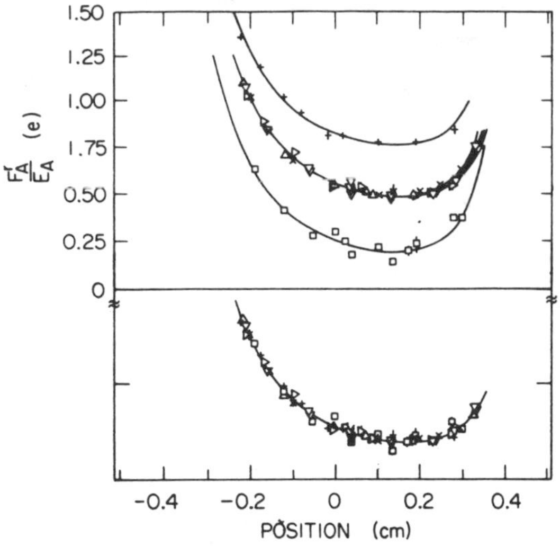

Figure 2 - Residual force per applied electric field vs. position for the second set of six levitations with constant patch effect gradient of the blind test run. The upper curves show the data as taken, and the lower shows the data superimposed to test the constancy of the background patch effect electric field gradient. Since the data all fall on the same curve, the gradient did not change. The statistical errors are smaller than the symbols, except where indicated.

∇=8T3.B,Δ=8T3.C,X=8T6.A,

□=2T2.A,+=8T6.B,and ▷=8T3.D.

differences between different balls which were not exactly 0e and ±(1/3)e, but (+0.296±0.005)e, (+0.259±0.005)e, and (-0.256±0.005)e (errors are statistical only), even though the patch effect field did not change (see Fig. 2). In the nonreproducible data, the apparent charge varied by several tenths of a charge over a period of minutes, so that no average could be obtained.

During this run we performed a blind test with Luis Alvarez in which a randomly chosen encoding number was added to the data so that we were able to determine the residual charge without knowing its significance. Although the patch effect field seems to have remained constant, the residual force at a single position of the capacitor plates with respect to the ball was irreproducible by $0.2eE_A$ (see Fig. 3). The apparent residual charge of the ball measured blind (after subtracting the encoding number) differed from that of a standard ball by 0.19e, and when the test was repeated without the blind procedure, the apparent charge difference was 0.24e. The discrepancy between these values is further evidence of nonreproducibility.

We investigated possible causes of these inconsistencies and found that the aparatus had been misaligned during the run in which the blind test was conducted, so that the magnetic field was 12±4 milliradians away from vertical, whereas it had been within ~1 mr of vertical in the previous run, in which there were 19 consecutive measurements showing ±(1/3)e and 0e, and had been within a few mr of vertical in the runs before that. (Since the run in which the blind test was done, the

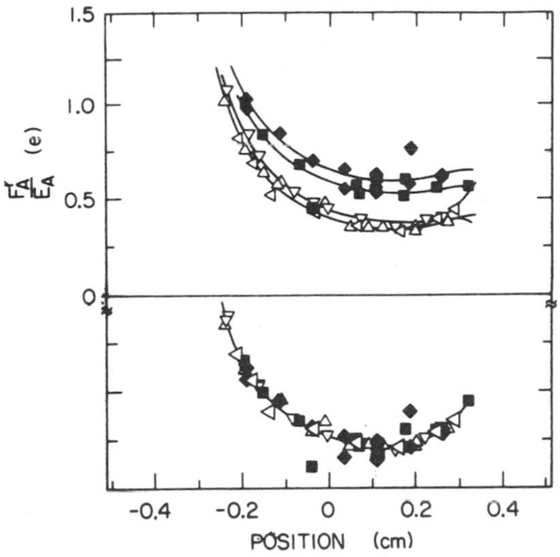

Figure 3 - Same as Fig. 2 for the fourth set of five levitations of the blind test run, including the ball measured blind. Since the superimposed data are all the same shape (lower curve), the patch effect gradient did not change. Note the points which fall $0.2eE_A$ off the superimposed curve (lower curve, solid diamond at upper right and solid square at lower left). As further evidence for nonreproducibility note that the data for the same ball relevitated (upper curves) do not follow the same curve. Statistical errors are again smaller than the symbols.

◁=8T3.E, ■=LT10.A, Δ=8T3.F,

◆=LT10.B, and ∇=8T3.I.

apparatus has been aligned to within <1mr with a laser technique before every run.) According to our previous calculations, a 12mr misalignment could not account for an error of the order of $0.2eE_A$ and certainly not for drifts in the residual force equivalent to several tenths of a charge. In a new analysis, we have found that the magnetic effect can be amplified by a rotation of the normal modes of the ball's translational oscillations by an angle 16 times the angle between the magnetic field and gravity.[6] The typical measured values of the ball's permanent electric dipole moment easily account for the inconsistencies and drifts observed in the run in which the apparatus was most tilted.

II. MAGNETIC BACKGROUND EFFECT DUE TO TILTED NORMAL MODES

The forces on the ball besides the charge force are the purely electric dipole terms and the magnetic term (Eq. 1). The purely electric term is well understood, and has most recently been described in Ref. 6. The primary magnetic force on the ball is the levitation force, which amounts to 10^7 times the force on one electron charge. Thus the smallest perturbation on the levitation force in phase with the applied electric field can cause a significant background effect. There is no significant source of alternating magnetic field gradient, but the ball's magnetic dipole moment varies with its orientation in the magnetic field, and its orientation varies slightly due to the torque exerted by the electric field on the ball's fixed electric dipole moment.

This alternating magnetic force has been known for some time, and acts at right angles to the magnetic field. It occurs in first order and has been observed in experiments in which the electric and magnetic fields are perpendicular.[7,8] In experiments such as this one in which the electric and magnetic fields are parallel, however, the force due to the tilting magnetic moment acts at right angles to the observation direction. Only a rotation of the measurement axis (the normal mode for translational oscillations) can cause the magnetic force to have an effect on the apparent residual charge.

Such a rotation can indeed occur if the apparatus is tilted with respect to gravity, and the angle of the rotation in this apparatus is 16 times the angle through which the apparatus is tilted. The factor 16 depends on higher derivatives of the magnetic field. The background force is given by

$$F=0.18(P_{Fx}/1x10^{-7}\text{ esu cm})(\Theta/0.001\text{ radian})eE_A, \quad (2)$$

where P_{Fx} is the component of the ball's electric dipole moment along a direction normal to the electric field and in the plane defined by the symmetry axis and gravity, and Θ is the angle between the apparatus' symmetry axis and gravity. Most of the dipole moments we have measured have been smaller than $1x10^{-7}$ esu cm, but in the run in which the blind measurements were made the apparatus was tilted by 12mr, so it is entirely possible that this background force was large enough that the true residual charge of the ball measured in the blind test could have been 0 or ±(1/3)e. Since then we have measured the effect of tilting the apparatus by 1.5mr with one ball and verified that the force changed by $0.04eE_A$.

We have long recognized that the magnetic force can be eliminated in experiments in which the electric and magnetic fields are parallel by spinning the ball about the field axis, although this is not effective in experiments in which the fields are perpendicular.[8] In fact, in one ferromagnetic levitation experiment the fields have been made parallel and the ball has been spun, and measurements on a number of balls have shown zero residual charge.[9]

In this experiment, in the absence of trapped flux, a magnetic field cannot exert a torque on the superconducting sphere, because the induced magnetic moment is exactly opposite to the applied field. However, we have realized that if the ball has even a small amount of trapped flux, it can be rotated if the spin-up field is applied at a frequency comparable with the resonant frequency for oscillations of the magnetic moment about the levitation magnetic field. This is because the ball's moment lags slightly behind the spin-up field.[10] (The required frequency is of the order of 60 Hz.) A ball has in fact been rotated in this way and the spin rate was observed visually to increase from zero to approximately 5 revolutions per second. This is easily fast enough to average the magnetic background effect to zero. The experiment is now being modified so that all the balls can be made to rotate during charge measurements and their rotation verified. Experiments performed henceforth will be done with the ball rotating and also with it not, to test the effect of the rotation.

Under these conditions we believe all background forces except the patch effect electric force will be negligible. The patch effect is accurately accounted for by measuring the position dependence of the residual force on several balls and taking differences.

III. SUMMARY

We first observed fractional charge in 1970, but did not publish until 1977, when we had not only statistical evidence but had demonstrated that the background forces were small. By 1981 we had made sufficiently definitive measurements to entitle our paper "Observation of Fractional Charge of (1/3)e on Matter". This was based on our determination by calculation and measurement that the background forces were negligible, as well as on the statistical evidence from the grouping of the results about 0e and ±(1/3)e. In light of the magnetic background effect which has appeared experimentally and which we understand theoretically, we would like to change the statement to "Evidence for the Existence of Fractional Charge on Matter" until we have repeated the measurements with the background effect absent. At present we do not believe that the magnetic effect can have accounted for the distribution of fractional charges in Fig. 1, but we can be completely confident in the results of the experiment only if, in addition to being statistically significant, they are obtained with the background forces completely taken into account. We have spun one ball, and are completing modifications to the apparatus to allow us to routinely spin all balls measured.

We are also proceeding with a room temperature ferromagnetic levitation experiment as an independent check on our results.[11] If fractional charges continue to be observed it can serve as a target for evaporated material in a mass spectrometer with a wide acceptance. In this way we can determine the e/m of the fractionally charged atom.

REFERENCES

[1] G. S. LaRue, W. M. Fairbank, and A. F. Hebard, "Evidence for the Existence of Fractional Charge on Matter", Phys. Rev. Lett. **38**, 1011 (1977).

[2] G. S. LaRue, W. M. Fairbank, and J. D. Phillips, "Further Evidence for Fractional Charge of (1/3)e on Matter", Phys. Rev. Lett. **42**, 142 (1979).

[3] G. S. LaRue, J. D. Phillips, and W. M. Fairbank, "Observation of Fractional Charge of (1/3)e on Matter", Phys. Rev. Lett. **46**, 967 (1981).

[4] A. F. Hebard, Ph. D. Dissertation, Stanford University (unpublished) (1970).

[5] G. S. LaRue, Ph. D. Dissertation, Stanford University (unpublished) (1978).

[6] J. D. Phillips, Ph. D. Dissertation, Stanford University (unpublished) (1983).

[7] M. Marinelli and G. Morpurgo, Phys. Lett. **94B**, 427 & 433 (1980).

[8] M. J. Buckingham and Conyers Herring, Phys. Lett. **98B**, 461 (1981).

[9] D. Liebowitz, M. Binder, and K. O. H. Ziock, Phys. Rev. Lett. **50**, 1640 (1983).

[10] Phillips, op. cit. pp. 228ff

[11] C. R. Fisel, Ph. D. Dissertation, Stanford University (in preparation)

Cosmic Ray Air Showers in a Fine Grained Calorimeter

Maury Goodman, and the E594 collaboration

A look at cosmic rays is particularly appropriate for a conference on "Inner Space/Outer Space". This presentation looks at the question of elemental composition in the energy range 10^{15} ev, using the E594 neutrino detector which consists of 608 1000 cell flash chambers. Its dimensions are 20 m. x 3 m. x 3 m. We studied extended cosmic ray air showers by using a trigger from four 0.6 m^2 scintillation counters on top of the detector.

An example of an event is shown in Figure 1. Each dot represents a flash chamber cell which was hit. A number of electrons at the top of the detector are seen, as well as a hadron near the center and a number of muons. In Figure 2 an event with 56 muons is shown.

A simulation of the showers in the atmosphere has been run to determine our response to different elemental abundances in the cosmic rays as a function of energy. Sensitivity to the elemental composition was found in the muon distribution. The proton simulation had events extending out to 15 in the number of muons traversing the detector. The iron simulation had a distribution starting at 12 and extending above 40.

The data after cuts was 457 events. Analyzing the muon component with a two component model, before trigger efficiency, the fraction of protons f(p) = 0.72 and iron f(Fe) = 0.28 with a statistical error of 0.03. Using an all particle spectrum with $dN/dE = c\ E^{-2.6}$, and including trigger efficiency, we conclude that f(Fe) >> f(p) above 10^{15} ev. This conclusion relies on the assumptions of the two component model, superposition so that an Fe nucleus acts like 56 nucleons with 1/56 th of the energy, pp physics at sqrt(s) = 10^{16} ev can be extrapolated from the collider parametrization with rising cross sections and scaling, and an all particle spectrum with an index of 2.6.

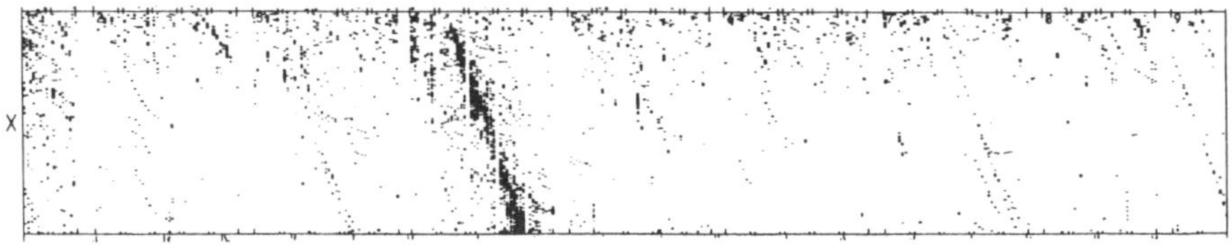

Figure 1 - Extended air shower event with electromagnetic, hadronic and muonic component.

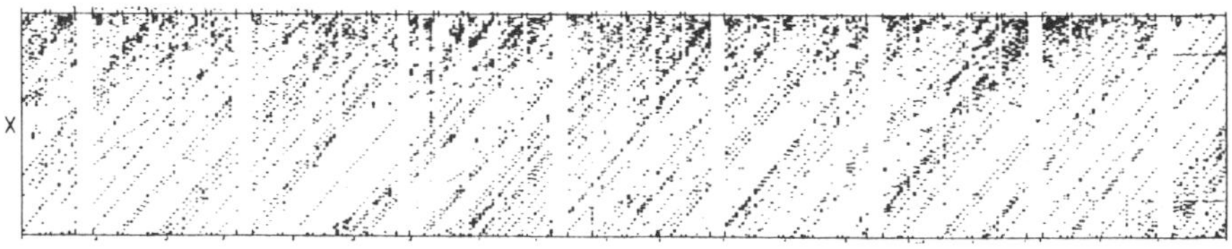

Figure 2 - Event with 56 muons.

PART VIII

KALUZA-KLEIN COSMOLOGY

Recent attempts to include gravity in the unification of forces and to search for a consistent quantum theory of gravity have led particle physicists to consider theories of gravity in more than four space-time dimensions. Various approaches - supergravity, superstrings, higher-dimensional gravity, etc. - all have the common assumption that there are additional spatial dimensions, unobservable today because they form a compact space with a very small volume. It is usually assumed that the dimensions of the internal space are close to the Planck length, 10^{-33} cm.

Although the internal dimensions are too small to be dynamically important in the present Universe, they may have played a crucial role in the evolution of the Universe at the Planck time, 10^{-43} seconds after the bang, when length scales of the dimensions we regard as large today were Planckian. Different higher-dimensional theories (here referred to generically as Kaluza-Klein theories) presumably result in different early histories for the Universe. The early Universe is the only site with enough energy to probe directly the extra dimensions, and perhaps cosmology can help constrain higher-dimensional theories.

In this chapter Weinberg reviews recent work on physics in higher dimensions. The evolution of the Universe in theories with extra dimensions is the subject of the remaining papers in the chapter. The chapter contains two papers on inflation with extra dimensions that are related to the material in Chapter IV. There is also a connection between extra dimensions and a certain class of N=8 supergravity theories covered in Chapter VI.

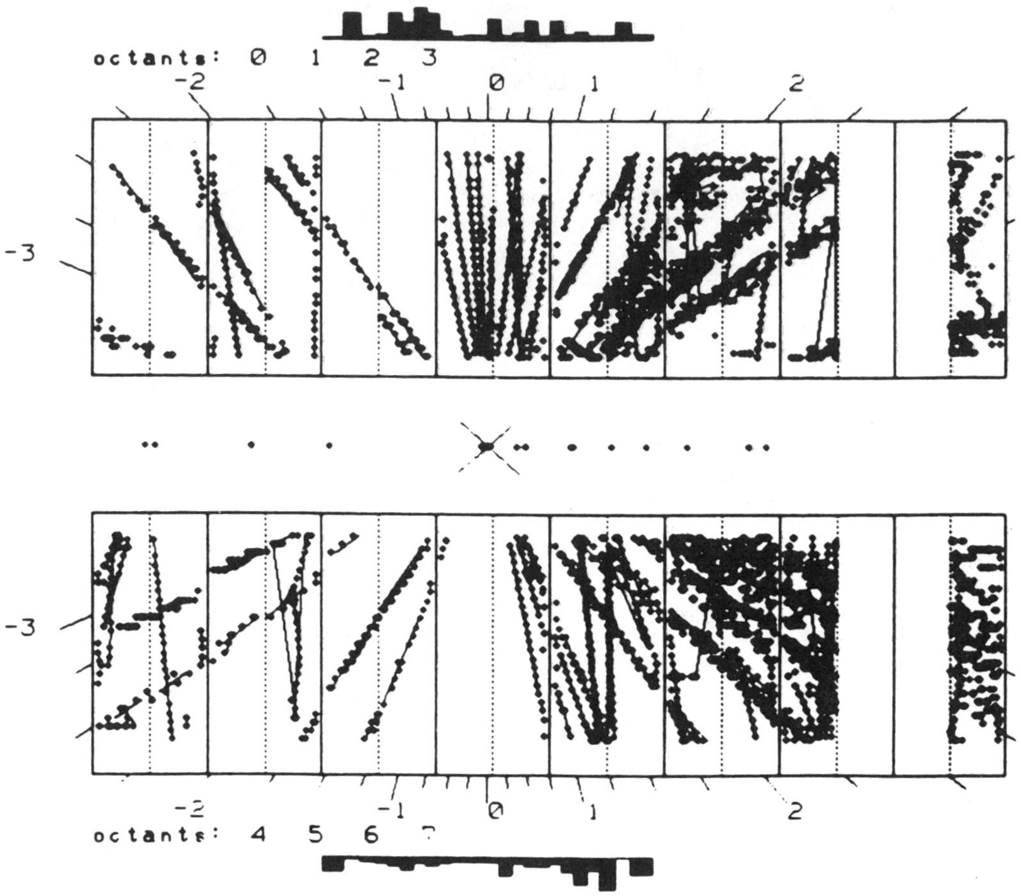

octants: 0 1 2 3
-2
-1
0
1
2
-3
-3
-2
-1
0
1
2
octants: 4 5 6 7

Physics in Higher Dimensions

Steven Weinberg

ABSTRACT

This is a lightly edited version of the transcript of a talk presented at the Inner Space/Outer Space Symposium at Fermilab on May 5, 1984. The talk is an elementary introduction to recent work on physical theories in higher dimensions, with a few accompanying historical remarks.

There are three different sorts of motivation that have led theorists to study physics in higher dimensions.

First, some physicists are interested in higher dimensions as an arena for mathematical tricks that will enable them to solve problems in 3+1 spacetime dimensions. In this category we may list the use of dimensional regularization by 't Hooft and Veltman and others; development by Scherk and Schwarz of extended supergravity theories by dimensional reduction of simple supergravity in higher dimensions; the use of dimensional continuation from $4-\varepsilon$ space dimensions down to our own 3 space dimensions to solve problems of critical phenomena by Wilson and Fisher, which has been outstandingly successful; and the use of dimensional continuation to do general relativity, continuing from $2+\varepsilon$ to 4 dimensions, by me, which has been totally unsuccessful. I recently also learned about the use of platonic solids in 4 spatial dimensions to make models of glasses in 3 dimensions, by Nelson and others.

A second reason for being interested in higher dimensions has to do with what we may call the "grand tour." You know that in the late 1800's and early 1900's the sons of the English gentry and nobility were sent to the continent of Europe so that when they came back they would understand that civilized life was only possible in England. In the same spirit, we sometimes study the physics of more than 4 spacetime dimensions because we would like to have some inkling of why it's impossible (if it is) for there to be more than four spacetime dimensions. In this connection, I'll just mention one example of something I've done recently and want to advertise. You know, it's very difficult to combine quantum mechanics and relativity in 3+1 spacetime dimensions. If you construct quantum fields out of the creation and annihilation operators of particles, it's not hard at all to make quantum fields that transform according to suitable representations of the Lorentz group, but the problem is to make them commute or anticommute with each other at spacelike separations, which is necessary not only because of intuitive ideas of causality, but in order to get a Lorentz invariant S-matrix. It's a standard topic in courses in quantum field theory that the only way to do this in familiar 3+1 dimensional theories is to combine the annihilation of particles with the creation of antiparticles, and furthermore that this only works if a whole host of conditions are met, included among which are for example the connection between spin and statistics, and the fact that antiparticles have the same mass and the same spin as the corresponding particles. In the spirit of the grand tour, I looked to see whether or not this construction of causal fields would work in more than 4 spacetime dimensions, hoping that it wouldn't, or at least that there would be some cases where it didn't, because that would be very instructive. Unfortunately it does work; you can construct causal quantum fields in any spacetime dimensionality you like.

(There is just one peculiarity here, which I found very surprising, that particles and antiparticles don't always have the same spin. By the "spin" of a particle I mean its transformation under the little group: for massive particles, the rotation group. Particles and antiparticles do always have the same spin in most dimensions, but not in 4N+5 spacetime dimensions; in 4N+5 dimensions some massive bosons and all massive fermions must transform differently under the rotation groups from their antiparticles.)

The third and most exciting reason for studying the physics of higher spacetime dimensionalities is the possibility that we really do live in more than 4 spacetime dimensions. Sitting here in this auditorium you can move to the left or right or you can move forward or backward, and with a little more difficulty you can move up or down. You know that the universe might possibly be closed, and if it is then you can only move a limited though large distance in these three spatial directions. But perhaps there are other directions into which one could move, for which we don't have names like up, down, north, south, east, west, but the space in those directions is very tightly curled up, and as soon as you start moving in one of those directions, you immediately come back to where you started. In that case you might very well live in more than 4 spacetime dimensions and never suspect it.

This can easily be put in the language of field theory. A free massless scalar field in 4+N dimensions might be expected to obey a wave equation of the form

$$0 = \Box_{4+N}\,\phi = \Box_4\phi + \nabla_N\phi$$

where $\Box_d$ and ∇_d denote the d'Alembertian and Laplacian in d dimensions. We can expand ϕ in eigenfunctions of the Laplacian on the N-dimensional curled-up manifold with coordinates y^a:

$$\phi(x,y) = \sum_\nu u_\nu(y)\phi_\nu(x)$$

with u_ν the ν-th eigenfunction of the Laplacian

$$\nabla_N u_\nu = \lambda_\nu u_\nu$$

and $\phi_\nu(x)$ a function of the ordinary spacetime coordinates x^μ, which, according to the 4+N-dimensional wave equation, must satisfy the wave equation in four dimensions

$$(\Box_4 + \lambda_\nu)\phi_\nu = 0$$

The eigenvalues λ_ν will typically be of the form

$$\lambda_\nu = -k_\nu^2/\ell^2$$

where ℓ is some circumference of the curled-up space, and k_ν is a dimensionless quantity, of order ν for large ν. A single massless particle in 4+N dimensions therefore appears in 4 dimensions as an infinite tower of particles, with masses

$$m_\nu = \frac{\hbar\sqrt{-\lambda_\nu}}{c} = \hbar\, k_\nu/\ell c$$

If ℓ is sufficiently small then all these masses can be so large that we would not expect to have seen any of these particles, except perhaps for a finite number for which $k_\nu = 0$. Similar remarks of course apply for fields of arbitrary spin and mass in 4+N dimensions.

But even if there might be extra space dimensions without our knowing it, why should anyone imagine such a weird idea? From the very beginning the interest in this possibility has arisen from the desire for a further degree of unification in physics; the hope is that a geometric theory in more than 4 spacetime dimensions can account not only for gravity but for all the other forces we observe in four dimensions.

A purely gravitational theory constructed along conventional lines in 4+N dimensions will involve a metric g_{AB}, whose coordinate indices run over values μ= 1,2,3,0 and a= 5,6,..., 4+N. Its components can be grouped into a four-tensor $g_{\mu\nu}$, N four-vectors $g_{\mu a} = g_{a\mu}$, and N(N+1)/2 four-scalars g_{ab}, each a function of the internal coordinates y^a as well as of the ordinary spacetime coordinates x^μ. Each of these gives rise in 4 dimensions to an infinite tower of particles of increasing mass and spin respectively 2, 1, and 0. It is the massless member of the spin 2 tower that plays the role of the ordinary graviton, and the massless members of the spin 1 towers that supposedly mediate the electroweak and strong forces. There may or may not also be massless particles in the spin 0 towers.

The simplest version of this picture was first described in 1921 by a rather obscure German mathematician, Theodore Kaluza, then at Königsberg. Kaluza took the indices A,B on g_{AB} to run over just 5 values, so that there was only one 4-vector $g_{\mu 5}$ in the metric, which he identified as the electromagnetic potential of Maxwell's theory. The title of Kaluza's paper shows his motivation. In those days everyone was talking about the possibility of unifying electricity with gravitation. (Well, not everyone.) Kaluza's paper was called, "On the unification problem." By the way, Kaluza is a rather interesting character, and if anyone wants to ask questions at the end, I'll be glad to tell you for example how he learned to swim.

The other name that's always coupled with Kaluza's is that of Oscar Klein, a very well known mathematician. Klein's contribution is contained in two papers written in 1926, while he was in Stockholm. One of them, the one in Zeitschrift für Physik, is the one that is always quoted in the present literature. However, this paper accomplished nothing but a clean-up of the mathematics in Kaluza's paper. The important paper by Oscar Klein is the one that's hardly ever quoted. It appeared in 1926 in Nature, and is called "The Atomicity of Electricity". In this paper for the first time it was suggested that there actually is a fifth dimension. Kaluza did not suggest this; he thought of the 5x5 array g_{AB} as a formal device for connecting electricity and gravitation. Klein in 1926 for the first time proposed that there really is a fifth dimension, and that the fifth dimension is curled up with some very small circumference which I have here called ℓ.

Klein, like Kaluza, considered the Einstein gravitational equations in 5 dimensions, with an energy-momentum tensor T^{AB} on the right-hand side of the equations arising from a single free particle. The field equations involving T^{AB} with $A=\mu=0,1,2,3$ and $B=5$ turned out to be just the inhomogeneous Maxwell equations, with the electric current four-vector identified as

$$J^{\mu} = \sqrt{16\pi G}\, T^{\mu 5}/c$$

Setting $\mu=0$ and integrating over 3-space then gave the charge q of this particle (in rationalized e.s.u.) as

$$q = \int d^3x\, J^0 = \sqrt{16\pi G}\, p^5/c$$

where p^5 is the component of the particle's momentum in the fifth direction. But Klein in 1926 knew (as Kaluza could not have known in 1921) that the momentum of a particle confined to a circle of circumference ℓ is quantized, with

$$p^5 = 2\pi\hbar\nu/\ell \qquad \nu = 0,\ 1,\ 2,\cdots$$

As we have already remarked, this leads to an effective mass in four dimensions, given by

$$m = |p^5|/c = 2\pi\hbar|\nu|/\ell c$$

and as Klein realized, it also leads to a quantization of electric charge

$$q = \frac{2\pi\hbar\nu\sqrt{16\pi G}}{\ell c}$$

This offered a possible explanation of the mysterious fact that the charges of all particles known in 1926 were integer multiples of the charge $-e$ of the electron. Identifying q for $\nu = -1$ with the electronic charge $-e$, Klein obtained the result that $e = 2\pi\hbar\sqrt{16\pi G}\ \ell c$. Unfortunately, he did not know ℓ, so this could not be used as a prediction of e, but turning the equation around he could use it to calculate the circumference of the 5-th dimension.

$$\ell = \frac{2\pi\hbar\sqrt{16\pi G}}{ec} = 0.8\cdot 10^{-30}\ \text{cm}$$

With a circumference this small, it's no wonder that the fifth dimension would not have become known to us. Equivalently, we can say that the mass of the lightest massive number of this tower of charged particles is

$$m = \frac{\hbar}{\ell c} = \frac{ec}{2\pi\sqrt{16\pi G}} = 2.5\cdot 10^{17}\ \text{GeV}$$

No particle this heavy could be produced in any laboratory experiment now forseeable.

It is now well understood how to generalize this original Kaluza-Klein model to more realistic theories in more than 5 spacetime dimensions. One supposes that the field equations somehow or other have a vacuum solution with a symmetry PxC: P is the group of ordinary Poincare transformations, acting on the first four coordinates z^μ (with $\mu=0,1,2,3$)

$$z^\mu \to \Lambda^\mu{}_\nu z^\nu + a^\mu \qquad \Lambda^T\eta\Lambda = \eta$$

$$y^a \to y^a$$

while C is a group of transformations of the remaining N coordinates y^a, of which the infinitesimal transformations take the form

$$y^a \to y^a + \sum_i \epsilon_i K_i^a(y), \qquad z^\mu \to z^\mu$$

$K^a{}_i$ is called the i-th Killing vector of the curled up space. As a consequence of this symmetry, the vacuum metric must take the form

$$g^{VAC}_{AB} = \begin{bmatrix} \eta_{\mu\nu} & 0 \\ 0 & g_{ab(y)} \end{bmatrix}$$

with $\eta_{\mu\nu}$ the usual Minkowski metric and $g_{ab}(y)$ the metric of the curled-up space. (If this space is not homogeneous, a y-independent factor may appear multiplying $\eta_{\mu\nu}$.) The perturbations to this vacuum metric appear in 4-dimensions as an infinite number of particle species, almost all of which are too heavy to have been observed.

For any piece of the perturbations to the 4+N-dimensional metric that appears in 4 dimensions as a massless gauge field $A_\mu(z)$, there must be a 4+N-dimensional coordinate transformation that subjects $A_\mu(z)$ to an infinitesimal gauge transformation

$$A_\mu(z) \to A_\mu(z) + \partial_\mu \epsilon(z)$$

Such a coordinate transformation is not a symmetry of the vacuum ($A_\mu = 0$ is not invariant) but it becomes one for the special case of constant ϵ. Therefore these gauge transformations must be identified with the symmetries of the curled-up space mentioned above, but with ϵ_i replaced with a function $\epsilon_i(z)$ of space-time coordinates

$$y^a \to y^a + \sum_i \epsilon_i(z) K_i{}^a(y) \qquad z^\mu \to z^\mu$$

Now, it is elementary to work out that the effects of such a coordinate transformation on the vacuum metric is to introduce off-diagonal components

$$g^{\mu a} \to \sum_i K^a{}_i(y)\eta^{\mu\nu}\partial_\nu \epsilon_i(z)$$

leaving the $g^{\mu\nu}$ and g^{ab} components unchanged. Thus in the perturbations δg_{AB} that appear in 4 dimensions as various fields of definite mass, there must appear one term of form

$$\delta g^{\mu a} = \sum_i K^a{}_i(y) A_i{}^\mu(z)$$

with $A^\mu{}_i(z)$ a gauge field that (for ε and A infinitesimal) is shifted under a gauge transformation by $\eta^{\mu\nu}\partial_\nu \varepsilon_i(z)$.

The moral of all this is that there is one gauge field for each symmetry of the curled up space; what we see in 4 dimensions as the gauge group can be identified with the symmetry group of the curled up space. This was, I believe, first recognized by B. deWitt, and emphasized more recently by Witten, who noted that the observed SU(3) × SU(2)×U(1) gauge group could arise in this way only from a curled up space with at least 7 dimensions, or 11 dimensions in all, because no smaller space could support an SU(3) × SU(2) × U(1) group of symmetries.

By inserting the formula for $\delta g^{\mu a}$ into the Einstein-Hilbert Lagrangian for gravity in 4+N dimensions and integrating over the curled up coordinates y^a, one obtains the Maxwell Lagrangian for the gauge field $A_i{}^\mu(z)$, but with a normalization that must be absorbed into the definition of $A_i{}^\mu$ and ε_i, leaving us with a definite value for the corresponding gauge couplings. Just as in grand-unified theories, each particle carries a "charge' with which it couples to each gauge field, equal to an integer multiple of a charge quantum e_i. Where the gauge field arises entirely from higher-dimensional gravity, the e_i are given by formulas just like Klein's;

$$e_i = \frac{2\pi\hbar\sqrt{16\pi G}}{\ell_i c}$$

where ℓ_i is the rms circumference of the curled up manifold, around circles taken in the direction of the Killing vector $K_i{}^a(y)$, and averaged over the manifold. This formula may need modifications if the 4 dimensional gauge field arises in part from gauge fields already present in 4+N dimensions. Also, there are important one-loop corrections if the 4+N-dimensional theory contains a large number of matter fields. Finally, these are not the gauge coupling constants that you can look up in the wallet card. These are the gauge coupling constants that would be measured at a characteristic energy scale of 10^{17} GeV, the Kaluza-Klein scale. In order to find out what the coupling constants are at ordinary energies one must integrate the renormalization group equations from 10^{17} GeV down to ordinary energies, in the way that by now has become thoroughly familiar in grand unified theories.

There has been much effort devoted to trying to calculate the parameters of the curled-up manifold by solving the classical field equations in 4+N dimensions, starting with papers of Cremmer and Scherk in 1976. A common feature of this work is that the dimensions of the compact manifold are not usually determined by the classical field equations, and can just be taken to have any value you like. Incidentally, this is also true of the original Kaluza-Klein theory; there is nothing in the field equations that tells us the value of ℓ, because there's no unit of length in these equations in terms of which ℓ could be determined if one doesn't bring $\hbar$ into the picture. These classical calculations are consequently not able to calculate or predict the gauge coupling constants.

Also, because there is a continuum of classical solutions of the field equations, such theories appear to involve massless scalars, which would be lovely because we need scalars of relatively low mass to explain SU(2) x U(1) breaking. (Any scalar with mass of order 1 TeV looks massless compared with the Kaluza-Klein scale of 10^{17} GeV.) Such massless scalars were always regarded in the early days of Kaluza-Klein theories as a terrible embarrassment; these days we would gladly welcome them as a possible solution to the "hierarchy" problem. Unfortunately there is an unpleasant aspect to the massless scalars of classical Kaluza-Klein theories. The vacuum expectation values of these scalars are free to drift as the universe expands, and as a result when you look back in time by looking at very distant quasars, for example comparing atomic spectra as they were 10 billion years ago with what they are now, you should see a different fine structure constant, because the compact manifold presumably has been expanding or contracting with the universe. And of course we don't see this; the fine structure splittings look the same relative to the Rydberg splittings in quasar spectra as they do here on earth.

However, it seems to me that the advantages and disadvantages of the massless scalars of classical Kaluza-Klein theories are somewhat beside the point, because there's no question that we have to consider quantum corrections, especially because where classical calculations don't give a unique answer, then it's got to be quantum corrections that do.

Suppose that the quantum corrections are produced by a certain number of species of massless particles, say f species. The one-loop quantum fluctuations in these fields produces a kind of Casimir pressure, except that instead of the boundary conditions being imposed by a pair of conducting plates, as in the Casimir effect, the boundary conditions here are just imposed by the "closedness" of the compact extra dimensions. With f species of massless particles, this "Casimir" vacuum energy density in 4+N dimensions must on dimensional grounds be of order f/ℓ^{4+N}. (From here on, we drop all factors $\hbar$ and c.) Such effects were first calculated in Kaluza-Klein theories by Appelquist and Chodos, but they studied the case d = 5, where the curled up space was a circle, and there was no curvature term to balance the Casimir pressure. For $d \geq 6$ the curvature terms on the left-hand side of the Einstein equation in any number of dimensions are typically of order $1/\ell^2$. Also, the gravitational constant $\bar{G}$ on the right-hand side of the field equations has dimensionality d^{-2+N}, so it must be of order ℓ^N times the Newton constant G of 4-dimensional gravity. (In fact, if the Einstein-Hilbert Lagrangian for gravitation in 4 dimensions is derived from a term of the same form in 4+N dimensions, then $G=\bar{G}/\Omega$, where Ω is the volume of the curled up space.) Putting this all together, the field equations become something like

$$\left(\frac{1}{\ell^2}\right) \approx (16\pi G\ell^N) \times \left(\frac{f}{\ell^{4+N}}\right)$$

so the circumference of ℓ is roughly

$$\ell \approx \sqrt{16\pi G f}$$

and the gauge couplings are of order

$$e \approx \frac{\sqrt{16\pi G}}{\ell} \approx \frac{1}{\sqrt{f}}$$

This is nice, because it offers an explanation of why dimensionless gauge couplings are all rather small at high energies; in these theories, this is just a consequence of the large number of matter fields.

Candelas and I have set up a general formalism for this sort of compactification, and have used it to calculate the circumferences and gauge couplings arising in 4+N dimensional theories with b massless boson and f massless fermion species, when the curled up space is a sphere S^N and the 4-dimensional gauge group is consequently O(N+1). For instance, for d=11 with f=1000 massless fermions, the O(8) gauge coupling at about 10^{17} GeV is calculated to be

$$g^2/8\pi^2 = 0.0987$$

It is a little disturbing that one needs so many matter fields to get a reasonably small gauge coupling, small enough to justify the use of perturbation theory. (Two-loop corrections are suppressed by an additional factor 1/f.) There are some indications that this is a special feature of compactification to a sphere. Also, it has been pointed out recently by Duff (in a private communication) and then in detailed calculations by Chodos and Myers, Ordoñez and Rubin and by Sarmeddi that (for reasons which so far are entirely obscure) a graviton in this business is worth about a million scalars or about 100,000 spin +1/2 fermions. It's certainly not the spin degeneracy of the graviton, but something about the one-graviton loop gives an enormously larger contribution to the Casimir pressure than matter loops of lesser spin. I don't know whether this can be made the basis of a systematic approximation, but if a one-graviton loop is the source of the vacuum energy-momentum tensor then one can explain immediately why the coupling constants are so small compared to one. But whether this can hold up in higher order perturbation theory I don't know.

Of course, it's not enough to get an equilibrium solution; you have to show it's a stable solution. In order to show it's a stable solution you have to consider all possible solutions, including those with no symmetry at all, that are infinitesimally close to the solution you're looking at. That's very difficult. It's not hard for example to just consider stability against a uniform dilation or contraction of the compact manifold, because then you keep the same symmetry, and you're just looking to see whether or not your potential has a local minimum as a function of the circumference. Candelas and I showed this is the case for many values of d, b, and f. As far as uniform dilation or contraction are concerned the quantum corrections do what the classical theories didn't do; they lock the radius of the compact manifold into a definite value, so you don't have to worry about cosmological drift. As far as I know there are no Kaluza-Klein theories at all in which when quantum corrections are included one finds massless scalars, which is not surprising because there is no reason why there should be any.

A theory with b massless scalar bosons and f massless spin 1/2 fermions in 11 dimensions will have a compactification to Minkowski space times S^7 that is stable against uniform dilation or contraction for a very wide range of values for b and f. What about other modes of instability? Candelas and I considered the possibility of exponentially growing Yang-Mills waves, and found that for stability we needed an upper limit on the number of bosons: the boson-fermion ratio couldn't be greater than 23.27. Gilbert, McClain and Rubin have recently re-worked this calculation in a time-dependent way, and they find that the effective action produced by the one-loop matter fluctuations contains terms involving time derivatives as well as the terms that Candelas and I found. It turns out if you have too few bosons then the kinetic term in the effective action has the wrong sign, so that a potential minimum means instability rather than stability: Gilbert, et al. show that this happens if the ratio of bosons to fermions is less than 1/137. So there is quite a large range of b/f for which the theory passes these stability tests, but I must admit that the full stability analysis has so far been beyond our mathematical powers.

Related to these stability calculations are the implications of these theories to cosmology. In cosmology you have to take into account quantum corrections at high temperatures, and you also have to consider everything as being time-dependent. This has not been done. There has been much discussion of Kaluza-Klein cosmology on the classical level, but I don't know what one can learn from classical Kaluza-Klein cosmology, if it is the quantum corrections that determine the size of the compact manifold. Gilbert and McClain have begun to set up this problem in its full generality. There is a suggestion in the paper by Candelas and me that there may be a critical temperature at which the compact manifold explodes, but as we emphasized in our paper, since we had not done the fully time-dependent temperature-dependent calculation we really didn't know whether or not the explosion actually occurs.

There is another interesting cosmological implication of Kaluza-Klein theories, that one can analyze without knowing too much about the dynamics. The heavy particles that Klein first realized had to appear in these theories might not be able to decay into ordinary particles. If they can't, then they also can't annihilate fast enough to be effectively eliminated - they're like monopoles in that respect - and they might still be around. (This has been studied extensively by Kolb and Slansky, and was also suggested but not written up by Fischler.) One might be able to rule out Kaluza-Klein theories on the grounds that these heavy particles would have such a large cosmic mass density that they would wreck the universe. Actually most theories do not have this problem. If SU(3) x SU(2) x U(1) is the only symmetry that survives down to accelerator energies, then the only kind of particle that would not be able to decay into ordinary quarks and leptons is one that belongs to a representation of SU(3) x SU(2) x U(1) that cannot be formed from direct products of the quark and lepton representations. Now, SU(3) x SU(2) x U(1) has the property that out of the representations of the group furnished by the ordinary quarks and leptons you can make any uncolored representation of SU(3) x SU(2) x U(1). With one qualification: The only kind of untrapped color-neutral particle which is not allowed by SU(3) x SU(2) x U(1) to decay into ordinary quarks and leptons would be a non-integer-charged particle. So the only thing you have to check is whether or not the Kaluza-Klein theory predicts that there should be a heavy non-integer-charged particle of zero color; if it doesn't then there is no cosmological problem with any heavy particles.

The dynamical compactification calculations are being continued, but no matter how they turn out we will be faced with a number of serious problems in all these theories. The first is the original unification problem of Kaluza. The whole idea of these theories is that everything is gravity, but in order to get a suitable theory we have had to add spin 0 and spin 1/2 particles by hand. This is not the most satisfying kind of unification.

The second problem is that of the light scalars. We'd like the theory to predict that at an energy around 300 GeV there should be some scalars left over from the compactification to provide the mechanism for breakdown of SU(2) x U(1). This is not an absolute requirement; there are other possibilities like technicolor that might serve this purpose, but it would be awfully nice if Kaluza-Klein theories predicted some kind of Higgs boson. They don't - the old hierarchy problem seems as bad in Kaluza-Klein theories as it is in grand unified theories.

Then there is a third problem particularly emphasized by Wetterich and Witten - the problem of the chiral fermions. If you put together all the left-handed fermions that we know about (including antiquarks and antileptons) you see that they form a complex representation of SU(3) x SU(2) x U(1); such fermions are called chiral. Furthermore, this property, that the left-handed fermions are in a complex representation of SU(3) x SU(2) x U(1), tells us also that the left-handed fermions must be in a complex representation of any big gauge group which is broken into SU(3) x SU(2) x U(1), because the left-handed fields of any massive fermions must be in a real representation of whatever symmetries are respected by these masses, so if all the left-handed fermions were in a real representation of some big gauge group that breaks into SU(3) x SU(2) x U(1) then the left-over light fermions, the ones which are still kept massless by SU(3) x SU(2) x U(1), would still be in a real representation of SU(3) x SU(2) x U(1). However, Kaluza-Klein theories generally predict that all left-handed fermions are in a real representation of whatever the isometry group is at low energy. It has been proven by Witten with great generality that as long as you don't add extra gauge fields to the theory, as long as you have pure Einsteinian gravity providing the mechanism for Yang-Mills fields at low energy, then no matter what fermions you add to the theory the left-handed ones will be a real representation of the low energy gauge group.

There are ways out of all these problems, and the leading candidate for a way out is supergravity (which of course has many things going for it in other respects). Supergravity provides a motivation for including a variety of fields besides the gravitational field, while preserving the beautiful geometrical flavor of pure gravity. There was a hope that supergravity would help with the hierarchy problem, but it hasn't so far. Finally, supergravity provides not only spin zero and spin 1/2 fields but in many cases (for instance, n=1 supergravity in 11 dimensions) there are extra tensor fields which can provide the kind of "twistedness" of compact manifolds which would allow chiral fermions to appear. Models of this type have been constructed by Cremmer, Horvath, Palla, and Scherk; Randjbar-Daemi, Salam, and Strathdee; Witten; Frampton and Yamamoto; and others.

Supergravity in eleven dimensions has been studied by a number of theorists: Cremmer and Julia; Freund and Rubin; Duff, Nilsson, and Pope; Castellani, D'Auria, Fré, and van Nieuwenhuizen; Englert and Nicolai;

Bars and MacDowell; Kallosh; and many others. A major problem with all supergravity theories that have been studied on the classical level (and almost all this work incidentally has been done on a purely classical level) is that when the compactification occurs you may get a compact 7-dimensional space or whatever it is you want for the extra dimensions, but you don't get a flat 4-dimensional spacetime. The 4 dimensional spacetime turns out to be something like an anti-DeSitter space, which might after all be the space of our real universe, but unfortunately the curvature of this anti-DeSitter space is about the same as the curvature of the extra compact dimensions, namely $(10^{-30}\text{cm})^{-2}$, and this doesn't correspond to reality as we know it.

The hope behind these models is I believe that when the dynamics is done in a better way, perhaps including quantum corrections, there will be found a correspondence between these results in anti-DeSitter space and realistic results in flat space. For example, it is hoped that particles that are found to satisfy conformally invariant field equations in anti-DeSitter space will ultimately correspond to massless particles when in flat space.

Candelas and Raines have studied the effect of the quantum one-loop fluctuations in higher dimensional supergravity and they show indeed that this hope is partly fulfilled. That is, when you include the one-loop matter fluctuations in the classical field equations (and remember, these are not a small correction) it is in fact possible to fine-tune the parameters so that Minkowski space is flat. The cosmological constant problem appears here as a fine tuning problem, but at least it's possible to do the fine tuning so that the spacetime is flat. But Candelas has very recently shown that unfortunately the hope that there would be a correspondence between results in flat space and previous results in anti-DeSitter space is not realized. Candelas has given very general arguments, not based on any specific dynamics, that if you have a supersymmetry theory in 11 dimensions which develops a 7-dimensional compact manifold of arbitrary shape but with a gauge group that is big enough to include subgroups like SU(3) and SU(2) (i.e., not all just U(1)'s) and if the other 4 coordinates form a flat Minkowski space, then no hint of supersymmetry remains, which does not correspond to what was found in purely classical models.

These problems have led me to consider recently the possibility that we may have been too timid, that even though we've taken the courageous step of going beyond 4 dimensions into higher dimensions, having done so we imagine that the theory we have in higher dimensions is just good old general relativity, that is, a theory based on metrics and Christoffel symbols, and all the rest of the apparatus we learn in nursery school, but with indices simply running from 1 to 11 or whatever instead of from 1 to 4.

You might ask, what other possibilities are there? There is at least one class of alternatives to conventional general relativity which preserve most of its geometric character and seems to me to be just as plausible as general relativity in more than 4 dimensions. General relativity can be regarded as based on two symmetries of the Lagrangian: invariance under general coordinate transformations and invariance under spacetime-dependent Lorentz transformations on the various tensor and spinor field indices. Why not keep general covariance, but replace the Lorentz group with some other arbitrary "tangent-space" group of 4+N-dimensional real matrices? An expanded account of work on this possibility will be published in the proceedings of the Fifth Workshop on Grand Unification. In brief, causality and the requirement of a Lorentz-invariant spacetime after compactification restricts the tangent space group to the form O(D-1,1) x G_T'[with O(D-1,1) acting on the first D coordinates, and G_T' on the remaining d-D coordinates] but this is just the sort of group we would expect in supergravity theories from an open-minded contraction of a simple graded Lie algebra. It is clear that these theories can solve the chiral fermion problem without extra gauge fields, but their dynamics has not yet been explored.

* * *

Question [source unidentified]: How did Kaluza learn to swim?

Answer: That is a very good question. First let me say a bit about his life. Kaluza was a contemporary of Einstein - born a few years later, died one year earlier. In 1921 when he wrote his famous paper he was starving as a privat dozent in Königsberg. As you may know, this was a very low academic rank, so low that the holder received no salary, but was just paid according to the number of students who attended his lectures. I say that Kaluza was starving, because in Jerusalem recently I met a Professor Sambursky, who recalls that Kaluza gave a course on tensor analysis in Königsberg in 1918 (a year before the British eclipse expedition made general relativity famous) and only one student took the course--namely, Sambursky. Einstein became interested in Kaluza's ideas a few years later, and helped him to get a professorship in Kiel, which doubtless saved his life. Kaluza never did anything so important again but he was famous as a master of many languages, and for never using notes in presenting lectures on mathematics. How did he learn to swim? According to the Dictionary of Scientific Biography, Kaluza learned to swim in the fashion of a true theorist: he read a book on how to swim, jumped in the water, and swam.

Kaluza-Klein Cosmology: Techniques for Quantum Computations

Mark A. Rubin

Why should we study the interplay of Kaluza-Klein (K-K) theory[1] and cosmology? And why must we conduct this study at the level of quantum physics, rather than at the purely classical level?

In answer to the first question: The length scales associated with the internal ("extra") dimensions in a K-K theory must be comparable to the Planck length ($\sim 10^{-33}$ cm.) in order for gauge couplings in the theory to be of their observed order of magnitude[2] (i.e., ~1). The energy scale of vibrational modes of the internal space is therefore the Planck energy ($\sim 10^{19}$ GeV). The only setting in which these modes can be excited above their vacuum state, and thus the only setting where we can hope to find direct evidence for the existence of the extra dimensions as physical reality, is the superheated environment of the very early universe.

The high energy scale associated with excitations of the internal space is also a reason for suspecting that quantum effects will play a significant role. On purely dimensional grounds we expect that quantum corrections to classical gravity should become important at Planckian energies and length scales. Indeed, the "evidence" obtained from one-loop ($O(\hbar)$) quantum-gravitational calculations in simple K-K models confirms our suspicion; the nature of solutions to the equations which determine the spacetime geometry is significantly affected when the lowest-order quantum fluctuations of the gravitational field or other fields are taken into account.[3]

In the context of a given K-K model for the local physics (i.e. a given set of fields, a classical action for these fields, and a given number of extra dimensons), the effective action method[4] is extremely useful in obtaining equations of motion, including quantum effects, for the spacetime geometry. In the one-loop ("semiclassical") approximation, the effective action Γ differs from the classical action S by the addition of a "quantum piece" Γ_Q:

$$\Gamma = S + \Gamma_Q \,, \tag{1}$$

where

$$\Gamma_Q \equiv (1/2) \log \det S_{2b} - \log \det S_{2f} \,. \tag{2}$$

$S_{2b(f)}$ is the second variational derivative of S with respect to bosonic (fermionic) fields. The equations of motion for the "background fields" (matrix elements of field operators between the initial and final vacuum states) are obtained by extremizing Γ with respect to variations of the background fields. In practice, we <u>assume</u> in advance the general form of the background fields (including the spacetime geometry) that we expect will extremize Γ, leaving only some small number of parameters to actually be determined by the condition that Γ be extremized. (We hope, of course, that our ansatz has been chosen with sufficient cleverness that Γ <u>is</u> extremized for <u>some</u> choice of parameters!)

An interesting static prototype of this kind of K-K quantum-cosmological calculation has been performed by Candelas and

Weinberg.[5] In their model, the background spacetime is the direct product of four-dimensional Minkowski spacetime M^4 with an N-dimensional sphere S^N whose radius is constant over M^4. In addition to the gravitational field, the model contains arbitrary numbers of massless matter fields, both scalars and Dirac spinors. These matter fields have vanishing background values and make no contribution to the classical action S. However, their contribution to Γ_Q yields field equations consistent with an internal space curved up into an N-sphere (the flatness of M^4 is obtained by fine-tuning a bare cosmological constant in the gravitational part of S). The contribution of the gravitational field to Γ_Q is neglected. This neglect can be justified by including a large number of matter fields in the model, provided that the contribution of a gravitational degree of freedom to the vacuum energy is comparable to that of a matter degree of freedom.

The first step in applying this kind of model to cosmology is to be able to compute Γ in a time-dependent background geometry. Such an analysis is also critical for understanding the stability of the original model against small perturbations. Recent work by Gilbert, McClain, and Rubin[6] has focused on stability with respect to spacetime-dependent fluctuations in the radius of the N-sphere. (Here and in what follows, "spacetime" refers to the four directly-observable dimensions, whose coordinates are denoted by "x".)

The effective action is computed on a background geometry which is perturbed away from $M^4 \times S^N$ by ordinary gravitational waves in spacetime as well as by fluctuations in the S^N radius r(x). Both the classical and quantum parts of Γ contribute to a "kinetic term" for r(x) viewed as a field on spacetime,

$$-c\, g^{\mu\nu}(x)\, \partial_\mu\left[\ell n\left(\frac{r(x)}{r_0}\right)\right] \partial_\nu\left[\ell n\left(\frac{r(x)}{r_0}\right)\right], \tag{3}$$

where r_0 is the spacetime-independent equilibrium radius of S^N. If the coefficient "c" is positive, then small perturbations in r(x) away from r_0 will give rise to oscillations about r_0; if negative, small perturbation will tend to grow exponentially, i.e., the model will be unstable against these modes.

The calculation is performed by means of the DeWitt-Schwinger method,[7] which yields an expansion in which succeeding terms contain increasing number of spacetime derivatives of r(x), or of the spacetime metric $g_{\mu\nu}(x)$. Gilbert and McClain[6] find that requiring the kinetic term for r(x) to have the "correct" sign places restrictions upon the ratio of the number of scalar fields to the number of spinor fields. These constraints are compatible with, though different from, the constraints imposed by the requirements of stability against exponentially-growing Yang-Mills and gravitational waves.[5,8] (The only modes whose stability has yet to be examined in this manner are modes of arbitrary spactime-dependent deformation of the internal space away from perfect sphericity. Analysis of these modes is currently in progress.[9])

The above analysis of perturbations about $M^4 \otimes S^N$ may be applied directly to cosmology at late times, since the curvature of spacetime is then exceedingly small compared to the curvature of the internal dimensions. However, we do not wish to simply assume that this was the case in the very early universe as well. It is of particular interest to see whether the observed spatial flatness of our present universe can arise "naturally" from a "Kaluza-Klein inflation."[10] It is also of vital importance to examine long-term changes in the internal geometry of a K-K model, since these changes give rise to changes in Yang-Mills coupling constants; changes in the electromagnetic coupling, at least, are subject to severe observational constraints.[11]

The technical problems which arise in computing Γ_Q on a "cosmological K-K" background (i.e., one of the form of a semi-direct product of a four-dimensional $k = \pm 1$ Friedmann-Robertson-Walker (FRW) spacetime with an S^N whose radius is determined dynamically) are essentially the same as those which arise in computing Γ_Q in the <u>static</u> spacetimes

$$R^1 \otimes S^3 \otimes S^N \qquad (k = +1) , \tag{4}$$

and

$$R^1 \otimes PS^3 \otimes S^N \qquad (k = -1) , \tag{5}$$

where PS^3 is the three-dimensional pseudosphere (3-space of constant negative curvature) and R^1 is the timelike direction. The $k = +1$ geometry is a special case of the more general spacetime

$$M^{n_0} \otimes S^{n_1} \otimes S^{n_2} . \tag{6}$$

For $n_0 = 4$, this coresponds to a static K-K model with the product gauge group $O(n_1+1) \otimes O(n_2+1)$; from the Γ for this theory, we can determine the ratios of coupling constants predicted by a K-K theory with such a product gauge group, and compare with prediction of a conventional grand unified theory. By considering the generalization (6) of the $k = +1$ case (4), we can thus kill two physical birds with a single mathematical stone.

Whatever the background geometry, the sums which arise in calculating Γ_Q are formally divergent:

$$\log \det S_2 = \sum_i \log \lambda_i , \tag{7}$$

where λ_i are the nonzero eigenvalues of S_2. One method of regularizing the sum (7) is the method of the generalized zeta function[12]:

$$\sum_i \log \lambda_i \equiv - \frac{d}{ds} \zeta(s) \Big|_{s=0} , \tag{8}$$

where

$$\zeta(s) \equiv \sum_i \lambda_i^{-s} . \tag{9}$$

The sum (9) will converge, provided it is evaluated for values of the parameter "s" which have a sufficiently large real part. The challenge is then to analytically continue $\zeta(s)$ to the physically-relevant region near $s = 0$.

For example: if S_2 arises from the action for scalars of mass m on $M^{n_0} \otimes S^{n_1} \otimes S^{n_2}$, $\zeta(s)$ (for Re $s >> 0$) is proportional to

$$\frac{\Gamma\left(s-\frac{n_0}{2}\right)}{\Gamma(s)} \sum_{j,k=0}^{\infty} \frac{(2j+n_1-1)}{j!} \frac{\Gamma(j+n_1-1)}{\Gamma(n_1)} \frac{(2k+n_2-1)}{k!} \frac{\Gamma(k+n_2-1)}{\Gamma(n_2)}$$

$$\cdot \left[\frac{\left(j+\frac{n_1-1}{2}\right)^2}{r_1^2} + \frac{\left(k+\frac{n_2-1}{2}\right)^2}{r_2^2} + \gamma^2 \right]^{-\left(s-\frac{n_0}{2}\right)} , \tag{10}$$

where

$$\gamma^2 \equiv m^2 - \left(\frac{n_1-1}{2}\right)^2 \frac{1}{r_1^2} - \left(\frac{n_2-1}{2}\right)^2 \frac{1}{r_2^2} , \tag{11}$$

and $r_1 (r_2)$ is the radius of $S^{n_1} (S^{n_2})$. For the corresponding fermion case, a method of analytic continuation has been devised by Candelas, Oppenheimer, Roth and Rubin. The numerical work involved in applying this method to the physical problems described above is currently in progress.[13]

Summations of a similar but more intricate form are involved in determining the contribution of gravitons to Γ_Q. The analytic continuation of these sums, as well as the extremization of the resulting Γ, has been performed by Ordóñez and Rubin.[14] We have discovered that, indeed, the graviton contribution to Γ_Q tends to be significantly larger than that of scalars or spinors, even after the increased number of degrees of freedom of the graviton is taken into account (see Table I). Furthermore, all the extrema of Γ with pure gravity, up to 13 extra dimensions, are unstable, as indicated by the presence of an imaginary part to Γ at these extrema. Since a model without fermionic matter is hardly realistic, this instability of "pure-gravity compactification" is not to be regarded as catastrophic. The corresponding analysis for suprgravity in the background $M^4 \otimes S^7$ is currently in progress.[15]

TABLE I. Contribution per degree of freedom to $-\Gamma_Q$ on $M^4 \times S^N$ in units of (radius of S^N)$^{-4}$.

N	Massless real scalar*	Massless Dirac spinor*	Graviton† real part	Graviton† imaginary part
3	7.6×10^{-5}	2.4×10^{-5}	-8.0×10^{-2}	1.4×10^{-1}
5	4.3×10^{-4}	-7.1×10^{-6}	1.9×10^{0}	2.3×10^{-1}
7	8.2×10^{-4}	1.9×10^{-6}	-2.5×10^{1}	1.9×10^{1}
9	1.1×10^{-3}	-4.7×10^{-7}	2.5×10^{2}	-1.5×10^{2}
11	1.4×10^{-3}	1.2×10^{-7}	-1.7×10^{3}	1.2×10^{3}
13	-8.8×10^{-3}	-2.8×10^{-8}	1.5×10^{4}	-8.8×10^{3}

*The results for scalars and spinors are from Ref. 5.

†In limit of radius and/or cosmologicial constant → 0.

ACKNOWLEDGMENT

I would like to thank Philip Candelas, Steven Weinberg, and all the participants in the Theory Group Lunch Seminars for many helpful discussions.

REFERENCES

[1]Th. Kaluza, Sitz. Preuss. Akad. Wiss. 966 (1921); O. Klein, Nature **118**, 516 (1926); for a pedagogical introduction, see A. Zee, in Proceedings of the Fourth Summer Institute on Grand Unified Theories and Related Topics, ed. by M. Kanuma and T. Maskawa (World Scientific Publishing Co., Singapore, 1981).
[2]P.G.O. Freund, Enrico Fermi Institute preprint EFI 82/83 (1982); S. Weinberg, Phys. Lett. **125B**, 265 (1983).
[3]T. Appelquist and A. Chodos, Phys. Rev. Lett. **50**, 141 (1983); Phys. Rev. **D28**, 772 (1983); T. Appelquist, A. Chodos and E. Myers, Phys. Lett. **127B**, 51 (1983); M.A. Rubin and B.D. Roth, Phys. Lett. **127B**, 55 (1983).
[4]See, e.g.; E.S. Abers and B.W. Lee, Phys. Rept. **9C**, 1 (1973); P. Ramond, Field Theory: A Modern Primer (The Benjamin/Cummings Publishing Company, Inc., Reading, Mass. 1981).
[5]P. Candelas and S. Weinberg, Nucl. Phys. **B237**, 397 (1984).
[6]G. Gilbert, B. McClain, and M.A. Rubin, Phys. Lett. **142B**, 28 (1984); G. Gilbert and B. McClain, Nucl. Phys. **B244**, 173 (1984).
[7]B. DeWitt, in Relativity, Groups and Topology (Gordon and Breach, Inc., New York and London, 1964).

[8]D.J. Toms, University of Wisconsin at Milwaukee preprint; M. Awada and D.J. Toms, University of Wisconsin preprint.
[9]G. Gilbert, B. McClain, C.R. Ordóñez, and M.A. Rubin, in progress.
[10]E. Alvarez and M.B. Gavela, Phys. Rev. Lett. **51**, 931 (1983).
[11]J.D. Beckenstein, Phys. Rev. **D25**, 1527 (1982) and references therein.
[12]P. Candelas and D.J. Raine, Phys. Rev. **D15**, 1494 (1977); S.W. Hawking, Commun. Math. Phys. **55**, 133 (1977).
[13]P. Candelas, S. Oppenheimer, B.D. Roth and M.A. Rubin, in progress; see also: K. Kikkawa, T. Kubota, S. Sawada and M. Yamasaki, Osaka University preprint OU-HET60; M. Kaku and J. Lykken, City College preprint CCNY-HEP-84-4; S. Ranjbar-Daemi, A. Salam and J. Strathdee, Phys. Lett. **135B**, 388 (1984).
[14]C.R. Ordóñez and M.A. Rubin, University of Texas Theory Group preprint; see also: A.Chodos and E.Myers, Yale University preprint YTP 84-09, and Ann. Phys. (N.Y.) **156** (1984); M.H. Sarmadi, ICTP preprint IC/84/3.
[15]C.A. Leutken, C.R. Ordóñez and M.A. Rubin, in progress.

High Temperature Quantum Effects in Kaluza Klein Cosmology

M. Yoshimura

One of the ultimate goals of the Kaluza-Klein cosmology is to understand dynamics of the compactification, namely the fact that the size of the extra space (of ~ 10^{-33}cm) << 10^{10} light years. A crucial key lies in the Planck era, or presumably the era slightly later than that. In the course of this exploratory study something unexpected will further be found.

When two sizes of the extra and the ordinary spaces are comparable and small, various quantum effects become important and modify the effective action. For instance, the ground state energy is changed by the presence of a finite size or a curvature of space: gravitational Casimir energy. This is only the first term of a systematic expansion in terms of the curvature. It is expected that the law of gravity is profoundly modified around the Planck epoch. I shall report some recent works of my own[1,2] that attempt to compute the effective action at finite temperatures in a curved spacetime. A basic underlying assumption is that gravity in higher dimensions is strong enough to maintain a thermal equilibrium. The equilibrium adiabatically changes as the universe slowly evolves, but we would like to estimate induced effects that depend on the derivative of scale factors, $\dot{a}_i$.

Use of a local equilibrium is justified only when the cosmological expansion rate is much smaller than the reaction rate of microscopic processes. We shall assume that this condition is satisfied, which typically means that $|\dot{a}/a| \lesssim G^2T^{2N-1}$ with N a dimension of the total space. In this hydrodynamic regime one can expand the background metric around a common time t_0, and analytically continue the relative time by $t=-i\tau(-\beta/2\leq\tau<\beta/2,\ \beta=1/T)$, to use the formalism of the Euclidean path integral,

$$a_i(t_0-i\tau) = a_i(t_0) - i\tau\dot{a}_i(t_0) - \frac{\tau^2}{2}\ddot{a}_i(t_0) + \dots \quad . \tag{1}$$

This eventually leads to a hydrodynamic equation with kinetic coefficients computed in equilibrium.

To illustrate the method, let us compute as an example the effective action caused by one loop quantum effects of a scalar field $\phi(x)$.[1] This is a prelude to the more complicated case[2] of a pure Kaluza-Klein theory. The background metric we take is the product of two spheres of dimensions, d_1 and d_2, characterized by two scale factors, $a_1(t)$ and $a_2(t)$. One loop effective action is the zero point quantum energy properly regularized and is formally given by a trace log of a Laplacian under the background metric,

$$W_q = \frac{1}{2}\,\mathrm{tr}\,\ell n\left[-\frac{d^2}{d\tau^2} - \frac{\Delta^{(1)}}{a_1^2} - \frac{\Delta^{(2)}}{a_2^2} + A\right]. \tag{2}$$

A contains derivatives of second order, whose precise form does not concern us. The eigenvalues and degeneracies of the Laplacian $\Delta^{(i)}$ on the unit sphere are well known and one only needs to evaluate the spectrum of a one-dimensional Schrödinger problem, with the Euclidean time τ identified as a position of a particle.

The method we use for the regularization is the ζ-function regularization. One first defines a covergent sum $\zeta(s)$ for a large, positive s,

$$\zeta(s) = \sum_n (\lambda_n)^{-s} = \Gamma(s)^{-1} \int_0^\infty dt\, t^{s-1} \sum_n e^{-\lambda_n t} . \tag{3}$$

The effective action W_q is then defined by its analytic continuation,

$$W_q = -\zeta'(0)/2 . \tag{4}$$

This method of regularization works only for $d_1 + d_2$ = even, corresponding to that at an odd dimension of spacetime there is no logarithmic divergence, thus one loop correction is finite in the dimensional regularization similar to the ζ-function method.

The spectrum is trivially found for the static spacetime

$$\lambda_{\ell mn} = (2\pi)^2 n^2 \beta^{-2} + \ell(\ell+d_1-1)\, a_1^{-2} + m(m+d_2-1)\, a_2^{-2} , \tag{5}$$

and the free energy, defined by the effective action times the temperature in this static case, is given by

$$F = -\sqrt{\pi}(2\beta)^{-1}\Gamma[(d_1+d_2+3)/2]^{-1} \int_0^\infty dt\, t^{-1/2} \left(\frac{d}{dt}\right)^{(d_1+d_2+2)/2}$$

$$\times \left[t^{(d_1+d_2+1)/2}\, h(t/\beta^2) f_1(t/a_1^2) f_2(t/a_2^2) \right] . \tag{6}$$

A nice feature of this integral representation is that temperature and scale dependence is factorized in the integrand. For instance,

$$h(x) = \sum_{n=-\infty}^{\infty} \exp(-4\pi^2 n^2 x) = \theta_3(0|i4\pi x) . \tag{7}$$

f_i is a little more complicated, but again is related[1] to the theta function.

The expression for the free energy (6) reduces to the known results in a few limiting cases. For instance, in the low temperature limit of $\beta \to \infty$, and in the case of $a_1 \gg a_2$, Eq. (6) gives the Casimir energy as computed by Candelas and Weinberg.[3] The high temperature limit of

$T \gg \max.(a_1^{-1},\ a_2^{-1})$ simply gives the free energy of (d_1+d_2) - dimensional ideal gas. Furthermore, Eq. (6) describes the free energy in all the region of the variables (T, a_1, a_2).

Time dependent part of the effective action can be obtained by performing a spectrum sum with $a_i(\tau)$ such as (1). In general, this is a difficult problem, but in the high temperature limit one can make a systematic expansion in powers of (derivative)/T; adiabatic expansion. Result of this calculation up to two powers is summarized as an effective Lagrangian of the form

$$T^{d_1+d_2-1}\,[c_{11}(\dot{a}_1/a_1)^2 + c_{22}(\dot{a}_2/a_2)^2 + 2\,c_{12}\dot{a}_1\dot{a}_2/a_1a_2]\ , \tag{8}$$

with c_{ij} a set of calculable constants. It is interesting to compare this with the original Einstein-Hilbert action,

$$(16\pi G)^{-1}\,[d_1(d_1-1)(\dot{a}_1/a_1)^2 + \ldots\ ]\ , \tag{9}$$

where a partial integration is performed to eliminate $\ddot{a}_1$. A remarkable result is that c_{ij} becomes negative for a large (d_1+d_2). For instance, with $d_3 = 3$, $c_{11} > 0$ when $d_2 = 1$, but $c_{11} < 0$ when $d_2 = 3,5,\ldots.$ It thus appears that there is a critical temperature T_c at which an effective gravitational constant vanishes, $T_c \sim G^{-1/(d_1+d_2-1)}$.

To analyze in more detail what may happen near T_c, let us take a model with a single scale a(t). When the dimension $d \geq 6$, the essential part of the Einstein equation is written as

$$\frac{1}{2}\,d(d-1)[1-(T/T_c)^{d-1}](\dot{a}/a)^2 + \frac{1}{2}\,d(d-1)\,a^{-2} = 8\pi G\rho\ . \tag{10}$$

The energy density can be taken as that of the ideal gas, $\rho \propto T^{d+1} \propto a^{-(d+1)}$. Near the critical size given by $a_0 = aT/T_c$, the solution behaves as

$$a - a_0 = \text{const.}\ |t - t_0|^{2/3}\ . \tag{11}$$

This may be taken as indicative of a possible bounce, but should not be accepted in a region very close to a_0 because the derivative, hence the curvature diverges and the approximation breaks down at a_0. The best one can hope at present is that the next correction of order R^2 will give a consistent bounce solution without any singular behavior.

Previous computation has been extended[2] to graviton loop in the pure Kaluza-Klein theory. Although this calculation is much more complicated technically, the result is rather simple. Dominant contribution at high temperatures and in a large dimension d gives an effective action of the magnitude, the scalar contribution times the number of degrees of freedom, $(d+1)(d-2)/2$. Thus, qualitative features of the previous solution are unchanged. Fermion loop gives a similar contribution, this time further modified by the factor, $(1-2^{-d})$, due to the difference between the statistics.

Although a definite answer on the bounce is yet to come, it would be interesting to speculate its implication on cosmology. It was long recognized[4] that the universe may not repeat itself after the bounce. There will always be some dissipative process which produces entropy in each cycle, and after many cycles a large amount of entropy may be accumulated. In the closed model the maximum size and the age in each cycle depends on the entropy Σ like $\Sigma^{(d+1)/d(d-1)}$. With accumulation of the entropy, the universe becomes more mature. The flatness problem may thus be solved. The horizon also enlarges because the light signal circumnavigates a small universe in the contracting phase. The recycling universe is thus an interesting alternative to the inflationary universe.

In the Kaluza-Klein cosmology one has to show how decoupling of the extra space takes place. Some interesting works on this problem are in progress. If this problem is solved, one might say that the extra dimension further sheds a new light on the outstanding problems of the cosmology.

REFERENCES

[1]M. Yoshimura, "Effective Action and Cosmological Evolution of Scale Factors in Higher Dimensional Curved Spacetime," Phys. Rev. **D30**, 344 (1984).

[2]M. Yoshimura, to appear in Prog. Theor. Phys. (1985, Feb. and April).

[3]P. Candelas and S. Weinberg, Nucl. Phys. **B237**, 397 (1984).

[4]R.H. Dicke and P.J.E. Peebles, in General Relativity: An Einstein Centenary Survey, ed. by S. Hawking and W. Isreal (Cambridge University Press, Cambridge, 1979).

More Dimensions - Less Inflation

David Lindley

Cosmologists and particle physicists alike are currently interested in Kaluza-Klein theories, but their viewpoints are rather different. To the cosmologist or relativist, extra dimensions are a priori on a par with the usual 3+1 dimensions of spacetime, and may evolve on a cosmological timescale. To the particle physicist, extra dimensions must be small, compact and stable, or one would see, for instance, time variation of gauge couplings. Perhaps the greatest difficulty with Kaluza-Klein theories is in reconciling some kind of extra dimensional general relativity, in which all dimensions are on an equal footing, with the observed division of the universe into 'large' and 'small' dimensions. (The easiest reconciliation is of course that there are no extra dimensions.)

Even if the extra dimensions are stable today, it is an exciting idea that at an earlier cosmological time they may have been dynamically important.[1,2] The major problem that arises is that somehow the extra dimensions must stop evolving, and settle into a stable, static configuration. I have nothing to say here about <u>how</u> this transition might occur, but I will attempt to pick out, using purely classical arguments, a definition of <u>when</u> it might occur. In our simple model, this allows quantitative estimates of the consequences of such an unusual period in the early history of the Universe.

To construct a dynamical model, we make the simple assumption that general relativity works in N+1 dimensions. A metric with signature (-++...+) yields an N+1 dimensional Ricci tensor, and for the stress energy tensor we take a perfect fluid, with pressure $p = \rho/N$, and a temperature such that $\rho \sim T^{N+1}$. (With these assumptions, this is the only temperature we can define. Any kind of 'effective 3-d temperature', obtained by integrating over the extra dimensions, is physically quite meaningless). An easy way of getting an anisotropic metric in N space dimensions is to put two Robertson-Walker metrics together in block diagonal form:[3]

$$g_{AB} = \mathrm{diag}(-1,\ R_d^2(t)g_{ab},\ R_D^2(t)g_{\alpha\beta}) \tag{1}$$

The co-ordinates x^a cover a d-dimensional space time with scale factor R_d and time-independent spatial metric $g_{ab}dx^a dx^b$ which may be of closed, flat or open form; similarly for the D-dimensional space in x^α. Since we want the D extra dimensions to be compact, we will take $g_{\alpha\beta}$ to be the metric of a unit D-sphere. For simplicity, we will assume the d 'ordinary' dimensions to be flat.

With a specific metric and a stress-energy tensor, we have enough to write down the dynamical equations of the system, which turn out to be, in effect, coupled Friedman equations,[2] together with a conservation law implying adiabaticity of the temperature with respect to the N spatial dimensions. Numerical solution of these equations is fairly straightforward; details are given elsewhere,[4] and we will quote results here. For each choice of d and D, there is a one parameter family of solutions, conveniently characterised by the maximum value, R_m, of the scale factor R_D of the closed dimensions. Figure 1 shows a typical solution, all of which behave in the same way: both scale factors expand

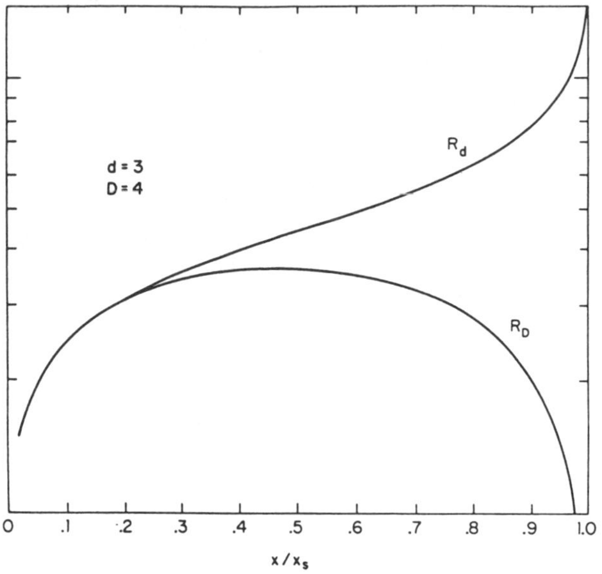

FIGURE 1

from an initial singularity, but the open dimensions continue to expand without limit while the closed dimensions contract back to zero. The solutions all finish at a second singularity, at $x=x_s$. Co-ordinate volumes, scaling like $R_d^{\,d}R_D^{\,D}$, start from zero, rise to a maximum, then go back to zero at the second singularity. Since the temperature evolves adiabatically, it goes to infinity at both singularities. This has the important consequence that, as the closed dimensions recontract, the temperature is rising at the same time that the open dimensions are expanding. This is what gives rise to the possibility that one might have 'inflationary' Kaluza-Klein cosmologies, because in the d open dimensions, the temperature, and therefore the entropy density, is rising at the same time as the causal horizon is growing. Accordingly, one may hope to get many more photons in a horizon volume than one would have at the same temperature in a conventional Universe.

It is obvious that we must stop this cosmological model from getting to the second singularity if we are to recover a more or less standard Universe from it. The strategy here is not to invent a model which accomplishes the transition from contracting to stable closed dimensions, but merely to produce a guess for when it occurs. One can imagine a number of things going wrong with our simple model as the second singularity is approached. The perfect fluid assumption will fail when the difference in expansion rate between the open and closed dimensions becomes too large for particle interactions to maintain an isotropic fluid. However, experience from studying anisotropy in conventional cosmology[5] indicates that, while such effects may alter the expansion rates, they will not prevent the singularity from being reached. A more important failure of the perfect fluid approximation will occur when the size of the closed dimensions becomes so small that one cannot imagine a classical continuum of energy states: instead, only excitations with wavelengths an integral fraction of R_D will contribute to the stress-energy tensor. This will occur roughly when $R_DT = 1$, and the simple assumption we employ is that the transition from contracting to 'frozen-out' extra dimensions happens at the same time as this classical-to-quantum transition in the stress-energy tensor.

In each solution to the field equations, specified by values of d, D and R_m, the point at which $R_D T = 1$ can be calculated, and this determines R_* and T_*, the scale-factor and temperature at freeze-out. If the transition to stable closed dimensions also occurs here, we know the size of the extra dimensions at stabilisation. Thus $R_* = R_{KK}$, the Kaluza-Klein lengthscale. Once the extra dimensions are stable, it makes sense to calculate the effective d-dimensional density, ρ_d, and temperature, T_d. Neglecting various factors of order one, these quantities can be calculated at stabilisation by factoring out the volume of the extra dimensions. This gives

$$T_d = \rho_d^{1/(d+1)} = (T_*^{N+1} R_{KK}^D)^{1/(d+1)} = T_* \qquad (2)$$

Because of the way in which we have chosen the stabilisation point we get a kind of 'second-order' transition, in which the N-dimensional temperature matches continuously onto the d-dimensional temperature.

We have now determined, for each solution, the Kaluza-Klein length and the temperature at which the extra dimensions stabilise. Given the solution for R_d, we can easily find the d-dimensional horizon length, and therefore the entropy, S_d, in a horizon volume in the open dimensions. This is what we want to make large to get a successful inflationary Kaluza-Klein cosmology. In the numerical calculations, we set d = 3 and chose different values of D. The results are easily summarised. By making stabilisation occur arbitrarily close to the second singularity (which we can do by suitably choosing R_m), the entropy S_3 can be made arbitrarily large. This looks promising, but we must also take note of the value of R_{KK} that emerges. As stabilisation occurs closer to the singularity, there is an asymptotic relation of the form $S_3 \propto (R_{KK}/R_{pl})^p$, where p is, for d = 3 and D > 1,

$$p = -(D+3)\left[\left(\frac{D(D+2)}{12}\right)^{1/2} - 1\right]^{-1} \qquad (3)$$

(For D = 1, p = 2.) The power p is negative for $D \geq 3$, which means that large values of entropy per horizon volume require small values of the Kaluza-Klein scale. In the limit D>>3, p has the asymptotic value $-2\sqrt{3}$, so that if we want $S_3 = 10^{88}$, then R_{KK}/R_{pl} must be 10^{-25}. This is quite implausible, since in Kaluza-Klein theories gauge couplings in the effective 3+1 dimensional theory look like R_{pl}/R_{KK}. On the other hand, choosing D = 1 would need $R_{KK}/R_{pl} = 10^{44}$, which means a Kaluza-Klein energy scale of $10^{-44} \times 10^{19}$GeV = 10^{-16}eV. The only likely possibilities are D = 2 or 3, which have p = 27 and -51 respectively; here one can have a large entropy in a horizon volume with a Kaluza-Klein scale not vastly different from the Planck scale.

Another way of looking at these results is to demand that $R_{KK} = R_{pl}$, and to plot the values of S_3 obtained with different D. This is done in figure 2, in which the entropy decreases with an increasing number of extra dimensions. (In this figure, R_* is the same as R_{KK}). Unfortunately, this conclusion depends strongly on the assumptions built into the model, especially those concerning the stabilisation of the extra dimensions. I believe the ideas proposed here deal economically with the unknown physics, but other schemes[6] may allow production of cosmologically interesting entropy. Nonetheless, in our model, Kaluza-Klein inflation works less well with more dimensions, whence the title of this paper.

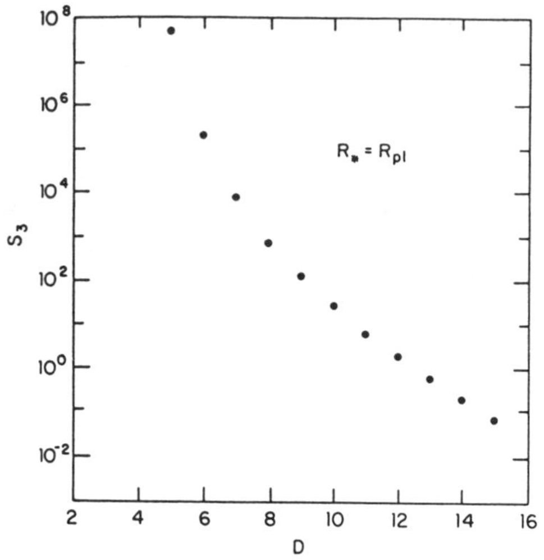

FIGURE 2

REFERENCES

[1]E. Alvarez and M. Belen-Gavela, Phys. Rev. Lett. **51**, 931 (1983); M. Bohm, J. L. Lúcio and A. Rosado, "Dynamical Compactification an Alternative to Inflation?" Preprint 1983; S. Barr and L. Brown, University of Washington Report 40048-01 P4 (1983).

[2]D. Sahdev, Phys. Lett. **137B**, 155 (1984).

[3]P. G. O. Freund, Nucl. Phys. **B209**, 146 (1982).

[4]E. W. Kolb, D. Lindley and D. Seckel, Fermilab-Pub-84/37-AST (1984, to appear in Phys. Rev. D).

[5]R. A. Matzner and C. W. Misner, Astrophys. J. **171**, 432 (1972).

[6]L. Abbott, S. Barr and S. Ellis, University of Washington Report 40048-03 P4 (1984); S. Barr, these proceedings.

Inflation From Extra Dimensions

S. M. Barr

I will be describing work done in collaboration with Richard Abbott and Steve Ellis. Similar work done by E. Kolb, D. Lindley and D. Seckel will be reported at a later session of this conference.

Recently there has been growing interest[1] in the possibility that the universe could have more than four dimensions. Aside from any light this may shed on problems in particle physics, if true it would undoubtedly have important implications for early cosmology. A rather speculative but very appealing possibility suggested by D. Sahdev and by E. Alvarez and B. Gavela is that the gravitational collapse of extra spatial dimensions could drive an inflation of ordinary space. This kind of inflationary cosmology would be quite different from the inflationary cosmologies now so intensively studied which are supposed to result from changes in vacuum energy during phase transitions in the early universe. In our work we examine the physics of these "Kaluza-Klein" inflationary cosmologies and come to three main conclusions. 1) It is desirable to have many extra dimensions, many being of order forty or fifty. 2) For models which give a realistically large inflation almost all of this inflation occurs in a period when quantum gravity is certainly important. This means that Einstein's equations cannot be used to calculate the details of this inflationary period. 3) Under plausible assumptions one may argue from the second law of thermodynamics that given appropriate initial conditions a large inflation will occur even when details of the inflationary phase cannot be calculated classically.

We consider a cosmology whose background metric has the form

$$g_{\mu\nu} = \begin{pmatrix} -1 & & \\ & r^2(t)\, g_{ij} & \\ & & R^2(t)\, g_{ab} \end{pmatrix}$$

where $r(t)$ and $R(t)$ are the scale factors of the ordinary $d(=3)$ spatial dimensions and the extra D compact spatial dimensions respectively. g_{ij} and g_{ab} are metrics of constant curvature surfaces. Let $n = d + D$. We will assume to begin with that we may use Einstein's equations and that the universe is radiation dominated, in equilibrium at temperature $T(t)$. Then the entropy S in a co-moving volume is constant. Einstein's equations are

$$d\ddot{r}/r + D\ddot{R}/R = -8\pi\bar{G}\rho \tag{1a}$$

$$k_d/r^2 + d/dt(\dot{r}/r) + (d\dot{r}/r + D\dot{R}/R)(\dot{r}/r) = 8\pi\bar{G}\rho/n \tag{1b}$$

$$k_D/R^2 + d/dt(\dot{R}/R) + (d\dot{r}/r + D\dot{R}/R)(\dot{R}/R) = 8\pi\bar{G}\rho/n \tag{1c}$$

$\bar{G}$ is the $n + 1$ dimensional gravitational constant[2]. If $k_D > 0$ and r and R start at $t = 0$ with a big bang type of expansion, two behaviors of the solution are possible. (See Fig. 1a and 1b.) For $k_d > 0$ one may get either behavior depending on initial conditions. For $k_d \leq 0$ one <u>always</u> has the situation in Fig. 1b. This latter assertion follows from the second of Einstein's equations. If we had the behavior in Fig. 1a there would be a turning point at which $\dot{r} = 0$ and $\ddot{r} \leq 0$. Then with $k_d \leq 0$ the left hand side of Eq. 1b is negative or zero at this point. But the right hand side of Eq. 1b is greater than zero.

It is also easy to see why r blows up at $t = t_o$. As $t \to t_o$ (for sensible behaviors of R) $\dot{R}/R \sim 1/t-t_o \to -\infty$. To keep the left-hand side of Eq. 1b positive r is driven to $+\infty$. This is the inflation which we desire. It will continue until R stops collapsing. We assume that some (quantum?) mechanism stops the collapse of R at its present size, which we call R_{KK}, the Kaluza-Klein radius. We call the time at which this occurs t_c. $t_c < t_o$. The inflation begins roughly speaking when R starts to collapse. This time, when R is at its maximum, we call t_m.

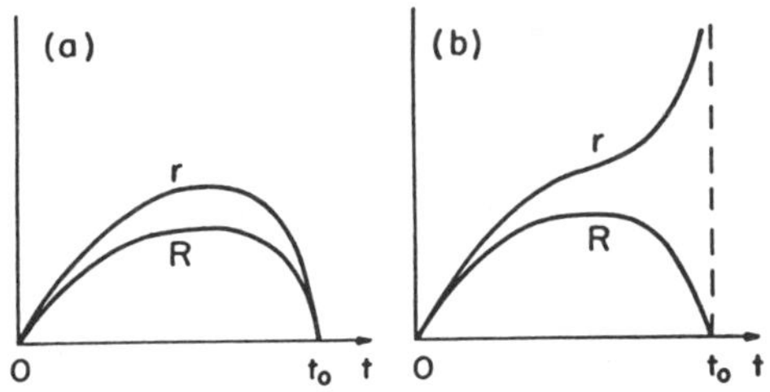

Figure 1: Two possible behaviors of the scale factors if they begin with a "big bang" at t=0. Fig. 1a can occur if $k_d>0$. Fig. 1b can also occur if $k_d>0$, but always occurs if $k_d \leq 0$.

For t near t = 0 the leading behavior of the scale factors is $r \sim t^{2/n+1}$, and $R \sim t^{2/n+1}$. For $\tau \equiv (t_o-t) \approx 0$ the leading behavior is $r \sim (t_o-t)^\eta = \tau^\eta$, and $R \sim (t_o-t)^\gamma = \tau^\gamma$, where $\eta = (1+((n-1)D/d)^{1/2})/n$ and $\gamma = (1+((n-1)d/D)^{1/2})/n$. For large D, $\eta \simeq -1/\sqrt{3}$ and $\gamma \simeq (1+\sqrt{3})/D$. We can see the situation more clearly perhaps if we plot $\ell n\, r$ versus $\ell n\, t$. (See Fig. 2). We normalize r so that it is equal to t at the present time (consider $k_d = 0$ for simplicity so that the scale of r is arbitrary). Before inflation $r \sim t^{2/n+1}$. After inflation is over and when the extra dimensions have "frozen out" $r \sim t^{1/2}$. Later still when matter dominates $r \sim t^{2/3}$. Between t_m and t_c one needs an inflation in r of 30 orders of magnitude or so. This is required if our presently observable universe is ever to have been within a horizon. We can crudely estimate how much inflation occurs between t_m and t_c. Call the inflation of the scale factor, r, during this interval I. Then

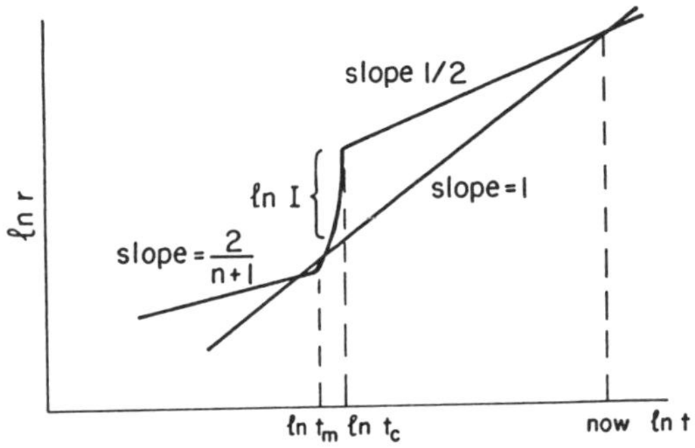

Figure 2: A plot of ℓn r versus ℓn t. Near t=0, $r \sim t^{2/n+1}$. When the universe is effectively four dimensional then $r \sim t^{1/2}$ in the radiation dominated era and $r \sim t^{2/3}$ in the matter dominated era. The line of slope 1 enables one to see when scales enter and leave the horizon.

$$I = \frac{r(t_c)}{r(t_m)} \simeq \left(\frac{\tau_c}{\tau_m}\right)^\eta \simeq \left(\frac{\tau_c}{\tau_m}\right)^{-1\sqrt{3}}$$

But

$$\frac{R_m}{R_{KK}} = \frac{R(t_m)}{R(t_c)} \simeq \left(\frac{\tau_m}{\tau_c}\right)^{\gamma} \simeq \left(\frac{\tau_c}{\tau_m}\right)^{-\left(\frac{1+\sqrt{3}}{D}\right)}$$

So

$$I \approx \left(\frac{R_m}{R_{KK}}\right)^{D/(3+\sqrt{3})} \gtrsim 10^{30}$$

We expect that the Kaluza-Klein radius R_{KK} is probably of order 1. Thus the quantity that must be large in the initial conditions is $(R_m)^D$ which is essentially the maximum volume attained by the extra space. This can be large because D is of order 1 and R_m is exponentially large. This is unappealing however as it does not explain in any sense the large cosmological numbers (first pointed out by Eddington and Dirac[3]). However if D is of order 40 or 50 then with R_m only of order 10 to 100 we can get enormous inflation. In some sense we can in this way explain the exponents of the Eddington-Dirac large numbers by relating them to the number of dimensions of space-time.

We must also, in any inflationary scenario, ensure that a huge entropy (corresponding to the 10^{88} or so photons within our present horizon) was present within a causal volume at some early time. At $t = t_m$ it is not hard to show that $R_m \sim t_m$. A causal region at that time will have volume of order $(t_m)^n \sim (R_m)^n$. The entropy of a gas of radiation at temperature T_m within this volume will be of order

$$S_m \sim (R_m)^n (T_m)^n$$

Again we see that the large number $(R_m)^D$ enters this expression. Thus the same initial conditions that lead to inflation will ensure that the "entropy problem" is solved.

We now are in a position to see why the classical Einstein equations are not valid during the later stages of inflation when $t \approx t_c$ (and when in fact most of the inflation occurs.) The time $\tau_c = (t_0 - t_c)$ is roughly

$$\tau_c \simeq \tau_m (R_m/R_{KK})^{-1/\gamma}$$

$$\simeq \tau_m I^{1/n} \lesssim \tau_m (10^{30})^{-\sqrt{3}}$$

Now for reasonable initial conditions τ_m is not very large ($\tau_m \sim R_m$ which we wish to be of or 10 or 100 as we have said). Thus τ_c is supersmall (compared, remember, to a Planck time[2]). Now the terms in the curvature of space-time which go (see Eq. 1) like $(\dot{R}/R)^2$ will thus be superlarge in Planck mass units. This means quantum effects are undoubtedly important and we are not justified in using Einstein's equations in the late stages of inflation.

Thus as R collapses and r begins to inflate a point will be reached when quantum effects become important. It is important to emphasize that we have no reason to suppose we know anything about the physics beyond the Planck scale. All we are assuming so far is that there are extra spatial dimensions and that at scales below the Planck mass classical relativity is a good effective theory. It is conceivable that quantum effects could cause all of the dimensions to collapse to a singularity. Let us suppose that this does not happen, but that the extra spatial dimensions continue their collapse while the ordinary three continue to expand. By how much will they expand? If we look at the universe at a later (but still extremely early) time when the extra dimensions have settled down to their final size and the temperature has dropped far enough that excitations in these directions have "frozen out" then we will find that at least as much inflation of r will have occurred as we would have found classically. This follows from the second law of thermodynamics. Comparing the entropy in a comoving volume before inflation (t_m) and at this later, post-inflationary time (t_f) we will find

$$S_m \sim (r_m^3 R_m^D) T_m^n \leq S_f \sim r_f^3 T_f^3$$

Where we have assumed that entropy increases. This leads to

$$(r_f/r_m) \gtrsim (T_m/T_f)(R_m^D T_m^D)^{1/3}$$

Note that the superlarge factor $(R_m)^D$ makes its appearance on the right hand side. (Temperature in these many dimensional cosmologies is a very gently varying function of time so that the factor T_m^D is not important compared to R_m^D.)

In conclusion we believe that an explanation of large cosmological numbers might be found in the number of space-time dimensions. By having a large number (40 or 50 or so) of such dimensions which expand and recollapse the universe can be driven to undergo an enormous inflation. The details of this inflation cannot however be calculated classically, but, rather, only if the physics beyond the Planck scale were better understood.

I have asserted a number of points, detailed justifications for which cannot be given in a talk of this length, but can be found in reference 4.

REFERENCES

[1]Sahdev, U. of Pennsylvania Preprint UPR-0238T (1983); E. Alvarez and M. Belen Gavela, Phys. Rev. Lett. **51**, 931 (1983). See Ref. 4 for a more complete set of references.

[2]Throughout we use natural units where lengths are in Planck lengths and energies and masses are in Planck masses.

[3]See, for example, J. D. Barrow, Q. Jl R. Astr. Soc. **22**, 388 (1981).

[4]R. B. Abbott, S. M. Barr, and S. D. Ellis, U. of Washington Preprint, 40048-03 P$ (1984).

Massless Fermions and Bound States in Kaluza-Klein Theories

Giovanni Venturi

We consider fermions coupled minimally to Einstein-Cartan gravity in a seven-dimensional background space-time[1] $M_4^0 \otimes S^3$ and use coordinates (x^μ, y^m), Greek indices go from 1 to 4 (M_4^0) and small Latin indices from 5 to 7 (S^3). We only exhibit the "internal" component of the contorsion K_{abc} which is the only one which can have a non-zero vacuum expectation value (v.e.v.) compatibly with Poincare invariance, further in the harmonic expansion for the contorsion we just retain the zero mode K^0_{abc} where

$$K^0_{abc} = \varepsilon_{abc}\ \chi\ (x) = \varepsilon_{abc}\ [\hat{\chi}(x) + \chi_0] \tag{1}$$

where the vacuum expectation value of $\chi(x)$ is $\langle\chi(x)\rangle = \chi_0$.

The fermions can also be harmonically expanded and each harmonic is associated with a physical (M_4^0) Dirac field $\psi_j(x)$ having mass eigenvalue M_j.[2] After integrating over S^3 we obtain for the sum of the truncated "internal" torsion and fermion actions

$$\bar{A}_\kappa + \bar{A}_D = \int e^1\wedge e^2\wedge e^3\wedge e^4 \left\{-\frac{3}{\kappa}(\hat{\chi}+\chi_0)^2 - \sum_j \bar{\psi}_j\ [\not\nabla^{(4)}(\overset{\circ}{\omega}) + M_j - \frac{3}{2}\hat{\chi}]\ \psi_j\right\} \tag{2}$$

where $\not\nabla^{(4)} = \gamma^\alpha e_\alpha - (1/4)\ \overset{\circ}{\omega}_{[\alpha\beta\gamma]}\gamma^{\alpha\beta\gamma}$, with $\gamma^{\alpha\beta\gamma}$ the skew symmetrized, normalized product of 3 Dirac γ matrices, $\overset{\circ}{\omega}_{[\alpha\beta\gamma]}$ the Levi-Civita connection and κ is Newton's constant. Further $e^a = e^a{}_m(y)\ dy^m$ and $e^\alpha = e^\alpha{}_\mu(x)\ dx^\mu = \delta^\alpha{}_\mu dx^\mu$ are the dreibein on S^3 and the vierbein on M_4^0. We shall later observe that with a suitable torsion $e^\alpha{}_\mu = \delta^\alpha{}_\mu$ is also a solution of the background Einstein-Cartan equations.

We now consider the functional integral

$$\bar{Z}\ (e^\alpha, e^a)\ \alpha \int D[\hat{\chi}]\ \prod_i D[\psi_i]\ D[\bar{\psi}_i]\ \exp\ i\ (\bar{A}_\kappa + \bar{A}_D) \tag{3}$$

and, in analogy with the evaluation of quantum corrections in superconductor models, integrate out the fermions obtaining an effective action for the torsion

$$\bar{Z}\ (e^\alpha, e^a) = \exp \mathrm{Tr}\ \ln\ (-\sum_j d_j S_j^{-1}) \int D[\hat{\chi}]\ \exp i\left\{-\frac{3}{\kappa}\int (\hat{\chi}+\chi_0)^2\ e^1\wedge..\wedge e^4 + \sum_{n=1}^{\infty} d_j\ \frac{i}{n}\ \mathrm{Tr}\ \left(\frac{3}{2}S_j\hat{\chi}\right)^n\right\} \tag{4}$$

where d_j is the multiplicity of the eigenvalue M_j and $S_j^{-1} = \nabla^{(4)} + M_j$. In Eq. (4) the terms quadratic and of higher order in $\hat{\chi}$ will correspond to mass, kinetic and potential terms for the torsion, which will then propagate as a quantum effect. The term linear in $\hat{\chi}$ vanishes since $\hat{\chi}$ by definition has zero v.e.v., hence we have the consistency condition

$$\chi_0 = i \frac{\kappa}{4} \sum_j d_j \, \mathrm{Tr}\, S_j \tag{5}$$

Further $M_4^0 \otimes S^3$ is a solution to the Einstein-Cartan equations if

$$L\,(M_0 + \frac{3}{2}\chi_0) = \frac{3}{4\sqrt{2}} = \frac{3}{2}\chi_0\, L \tag{6}$$

where L is the length of S^3.[2] Equation (6) corresponds, in analogy with superconductor models, to an anti-gap equation since as a consequence we have: a) zero physical mass for the lowest mode j = 0, b) vanishing of the Riemann-Cartan curvature on S^3.

On using the previous calculation of mass eigenvalues[2] M_j, introducing a cut-off $\Lambda^2 \geq P^2 + M_j^2$ and, in analogy with the usual calculation of Casimir energies, just retaining the leading contribution which depends on the background radius we obtain

$$1 = \frac{\kappa}{L^2} \frac{185}{128\pi^2} \ln(\Lambda L) \tag{7}$$

If the cut-off can be fixed (for example, by requiring that some torsion fields, ex. $K^{\alpha bc}$, have zero effective mass) one can compute L, then the Yang-Mills coupling constant is determined since it is given by $(1/L)\sqrt{\kappa}$ after dimensional reduction. Of course, in a finite theory, one expects a relation such as Eq. (7) but with a meaningful limit for $\Lambda \to \infty$.

Another possible application of torsion is to provide a binding mechanism between massive fermions:[3] in general the Dirac equation for a spinor $\Psi(x, y)$ in D = 4 + N dimensions is

$$(\Gamma^\mu \partial_\mu + \Gamma^a \nabla_a)\,\Psi(x,y) \equiv (\Gamma^\mu \partial_\mu + M_{OP})\,\Psi(x,y) =$$

$$\sum_{j,\rho} (\Gamma^\mu \partial_\mu + M_j)\,\Psi_{j\rho}(x) Y_{j\rho}(Y) = 0 \tag{8}$$

where Γ are the D-dimensional Dirac matrices and we have exhibited the harmonic expansion for fermions. For a zero mode

$$\nabla_a Y_{o\rho}(Y) = M_0 Y_{o\rho}(Y) = 0 \tag{9}$$

Otherwise the fermion has mass O(1/L) >> 1 GeV. On considering the outer products of fields

$$\Psi(x,y)\overline{\Psi}(x,y) = \sum_{j,k\neq o} \sum_{\rho,\xi} \psi_{j\rho}(x)Y_{j\rho}(y)Y^{+}_{k\xi}(y)\overline{\psi}_{k\xi}(x)+$$

$$\psi_{o\rho}(x)Y_{o\rho}(y)Y^{+}_{o\xi}(y)\overline{\psi}_{o\xi}(x) = \sum_{\rho,\xi} \sum_{j,k\neq o} \langle j,k|o,\overline{o}\rangle$$

$$Y_{o\rho}(y)Y^{+}_{o\xi}(y)\psi_{j\rho}(x)\overline{\psi}_{k\xi}(x)+\Sigma' \equiv \sum_{\rho,\xi} \pi_{\rho\xi}(x,y)+\Sigma' \qquad (10)$$

where we have separated the contribution to the product of zero mode representations from the rest. The resulting composite field will satisfy the Bargmann-Wigner equations

$$(\Gamma^{\alpha}\partial_{\alpha}\otimes 1 - 1\otimes\Gamma^{\beta}\partial_{\beta})\ \pi_{\rho\xi} = 0 \qquad (11)$$

corresponding to zero effective mass. Therefore if the internal space has an isometry group $G = H_L$ (left multiplication) $\otimes$ H_R (right multiplication) and if the background parallelizing torsion allows for zero modes in H_L but not in H_R one can identify H_R and H_L with the color and electroweak flavor groups respectively. Thus leptons and quarks would both be zero modes of H_L (thus having the same electroweak flavor) but the former would be color singlets (and therefore massless) whereas the latter belong to the lowest color representation and are therefore massive. The observed hadrons, however, are composite color (H_R) singlets, satisfy Eq. (11) and are therefore massless since quarks, although very massive, become massless when confined in a colorless hadron. Analogous statements are also valid for the gauge fields associated with H_R and H_L.

REFERENCES

[1]C. A. Orzalesi and G. Venturi, Phys. Lett. **139B**, 357 (1984).
[2]C. Destri, C. A. Orzalesi and P. Rossi, Ann. Phys. **147**, 321 (1983).
[3]S. Bergia, C. A. Orzalesi and G. Venturi, Phys. Lett. **123B**, 205 (1983).

The Reduced Gauge Group of a Symmetric Kaluza-Klein Space

R. Kantowski (work done with L. B. Hudson)

A symmetric Kaluza-Klein (K-K) space $[(\Omega,g_Q)$ with symmetry $(S,\Phi)]$ dimensionally reduces to a smaller K-K space (Ω^K,g_{QK}) with Higgs fields H.[1] The Einstein action of the original K-K metric g_Q automatically contains the desired kinetic and potential terms of H for spontaneous symmetry breaking. This approach amounts to a bundle analysis of previous work where it was shown that the bosonic sector of the electro-weak theory could be 'derived' by dimensionally reducing a symmetric gauge field (group G) on an extended space time M.[2] The gauge field is actually the 'off diagonal' part of g_Q. The main result here is that the reduced gauge group (N_K/K) is larger than previously thought (C_H) thus allowing room for the color group (SU_3) within the original symmetry group S.

A K-K space is a principal fiber bundle $\Omega = [Q,M,\pi,G,\Psi]$ with K-K metric on the bundle space Q, $g_Q = g_G o(\omega\otimes\omega) + \pi g_M$. g_M is the metric of extended spacetime M, ω the gauge field of the group G, and g_G the Jordan-Brans-Dicke scalar fields. A symmetric K-K space possesses a Lie Group of bundle automorphisms (S,Φ) which leave g_Q invariant, i.e., $\Phi_s:Q\to Q$ are diffeomorphisms, $\Phi_{s1}o\Phi_{s2} = \Phi_{s2s1}$, $\Phi_s o\Psi_g = \Psi_g o\Phi_s$, and $\Phi_s g_Q = g_Q$. The fiber preserving Φ action of S on Q induces an action ϕ of S on M which produces the dimensional reduction $\pi_\phi:M\to\bar{M}\equiv M/S$ of extended spacetime to spacetime. The Φ and Ψ actions of S and G on Q combine to define an action Σ of $S\times G$ on $Q(\Sigma(s,g) = \Phi_s o\Psi_g)$ as well as the associated projection $\pi_\Sigma:Q\to\bar{M}$. When the isotropy group K_q of q (the subgroup of $S\times G$ which leaves q fixed under the Σ action) is nontrivial, Q can be divided into subsets, Q^K, having the same isotropy group K, $Q^K = \{q\varepsilon Q|K_q = K\}$. Those symmetric bundles for which $\pi_\Sigma(Q^K\subseteq Q)=\bar{M}$, reduce to one of the equivalent principle bundles $\Omega^K\equiv[Q^K,\bar{M},\pi_{\bar{\Sigma}},N_K/K, \bar{\Sigma}]$, and fields on Ω reduce to related fields on Ω^K. In Ω^K, $\pi_{\bar{\Sigma}}$ is just the restriction of π_Σ to Q^K and N_K is the normalizer of K in $S\times G$, i.e., $N_K = \{n\in S\times G|nKn^{-1} = K\}$ and is the largest subgroup of $S\times G$ containing K as a normal subgroup. K has no effect on Q^K but N_K/K

acts freely and is appropriately called the reduced gauge group. $\bar{\Sigma}$ is the action of N_K/K on Q^K induced by the Σ action of N_K on Q^K. Specifying K is equivalent to giving two subgroups $R \equiv pr_1(K) \subset S$, $H \equiv pr_2(K) \subset G$ and a homomorphism $h: R \to H$ defined by $h = h(r) \leftrightarrow (r,h) \in K$. R is just the isotropy group for points in $\pi(Q^K)$ under the ϕ action of S. The part of N_K which acts vertically in Ω is (e, C_H) where $C_H = \{c \in G | chc^{-1} = h, \forall h \in H\}$ is the centralizer of H. This group sits inside N_K/K as a normal subgroup, $\tilde{C}_H \equiv \{(e,c)K | c \in C_H\}$, and was originally mistaken to be the reduced gauge group. $(N_K/K)/\tilde{C}_H \simeq pr_1(N_K)/R \subset N_R/R$ is not necessarily trivial, has the same algebra as N_R/R, and can, for example, be taken as the color part of the reduced gauge group.

The K-K metric $g_Q = g_G \circ (\omega \otimes \omega) \oplus \pi g_M$ on Q reduces to the K-K metric $g_{Q^K} = g_{N_K/K} \circ (\omega^K \otimes \omega^K) \oplus \pi_{\bar{\Sigma}} g_{\bar{M}}$ on Q^K plus other fields, e.g., ω reduces to the $\tilde{C}_H$ part of the reduced gauge field ω^K, plus the Higgs field H. g_M reduces to the metric on spacetime $g_{\bar{M}}$ plus the N_R/R part of ω^K. The original oversight in finding the reduced gauge group occurred because (1) a choice of g_M was made which produced a partially integrable ω^K (i.e., the N_R/R part) and (2) a local section (gauge) was used for which the corresponding potentials were zero.

REFERENCES

[1]L. B. Hudson and R. Kantowski, "Higgs Fields From Symmetric Connections - The Bundle Picture", to appear in J.M.P. (see complete references therein).
[2]P. Forgacs and N. S. Manton, Commun. Math. Phys. **72**, 15 (1980).

From Kaluza-Klein to Standard Cosmology

C.E. Vayonakis

One of the most appealing ways of unifying gauge field theories with gravity is through Kaluza-Klein type theories of higher dimensional gravity.[1] In discussing cosmology in these theories,[2-4] one associates a time dependent 'radius' $\tilde{R}(t)$ with the extra D dimensions in addition to the 'radius' R(t) associated with the 3 dimensions. Upon assuming that the compact D dimensional manifold is an Einstein space, the generalised Friedmann-Robertson-Walker ground state metric is

$$\bar{g}_{AB} = \mathrm{diag}\Big\{\underbrace{1,\ -R^2(t)\hat{g}_{ij}(x)}_{\leftarrow\ 4\ \mathrm{dim}\ \rightarrow}\ ,\ \underbrace{-\tilde{R}^2(t)\tilde{g}_{mn}(y)}_{\leftarrow\ D\ \mathrm{dim}\ \rightarrow}\Big\}. \tag{1}$$

Upon further assuming that the 4 + D energy momentum tensor is

$$\bar{T}_{\mu\nu} = T(\tilde{R})\bar{g}_{\mu\nu}$$

$$\bar{T}_{mn} = \tilde{T}(\tilde{R})\bar{g}_{mn} \tag{2}$$

where $T(\tilde{R})$ and $\tilde{T}(\tilde{R})$ are terms coming from whatever mechanism is responsible for compactification, one can show that the Einstein equations

$$\bar{\mathbf{R}}_{AB} - \frac{1}{2}(\bar{\mathbf{R}}+\bar{\Lambda})\bar{g}_{AB} = 8\pi\bar{G}\ \bar{T}_{AB} \tag{3}$$

where $\bar{\Lambda}$ and $\bar{G}$ are the 4 + D dimensional cosmological and gravitational constants, give the equations[4]

$$\frac{3}{R}\frac{d^2R}{dt^2} + \frac{D}{\tilde{R}}\frac{d^2\tilde{R}}{dt^2} = \frac{\bar{\Lambda}}{(D+2)} - \frac{8\pi\bar{G}[(D-2)T-D\tilde{T}]}{(D+2)} \tag{4}$$

$$\frac{1}{R}\frac{d^2R}{dt^2} + \frac{2}{R^2}\left(\frac{dR}{dt}\right)^2 + D\,\frac{1}{R}\frac{1}{\tilde{R}}\frac{dR}{dt}\frac{d\tilde{R}}{dt} + \frac{2K}{R^2}$$

$$= \frac{\bar{\Lambda}}{(D+2)} - \frac{8\pi\bar{G}[(D-2)T-D\tilde{T}]}{(D+2)} \tag{5}$$

$$\frac{1}{\tilde{R}}\frac{d^2\tilde{R}}{dt^2} + \frac{(D-1)}{\tilde{R}^2}\left(\frac{d\tilde{R}}{dt}\right)^2 + 3\,\frac{1}{\tilde{R}}\,\frac{1}{R}\,\frac{d\tilde{R}}{dt}\,\frac{dR}{dt} + \frac{2\tilde{K}}{\tilde{R}^2}$$

$$= \frac{\bar{\Lambda}}{(D+2)} + \frac{16\pi\bar{G}(2T-\tilde{T})}{(D+2)} \quad . \tag{6}$$

Now, there is a stringent constraint that there should be no significant time dependence of the 'radius' $\tilde{R}(t)$ at times later than the Planck time. Otherwise, variation of the gauge coupling constants and the four dimensional gravitational constant would be inconsistent with observation.[5] To have such solutions with $\tilde{R}(t) = \tilde{R}_0$ = const, we see from Eq. (6) that there must exist real solutions of

$$\frac{2\tilde{K}}{\tilde{R}_0^2} = \frac{\bar{\Lambda}}{(D+2)} + \frac{16\pi\bar{G}\left(2T(\tilde{R}_0)-\tilde{T}(\tilde{R}_0)\right)}{(D+2)} \tag{7}$$

This would not be the case if either there were a non-negligible R(t) dependence in T and $\tilde{T}$ or thermal terms were to be included on the righthand side of (6). Thus, compactification occurs at temperatures small on the scale $\bar{\Lambda}$, T and $\tilde{T}$. After that, the remaining equations are reduced automatically to standard cosmology

$$\frac{1}{R}\frac{d^2R}{dt^2} = \frac{\Lambda_4}{6} - \frac{4\pi G(\rho+3p)}{3} \tag{8}$$

$$\frac{1}{R^2}\left(\frac{dR}{dt}\right)^2 + \frac{K}{R^2} = \frac{\Lambda_4}{6} + \frac{8\pi G\rho}{3} \quad . \tag{9}$$

The four-dimensional cosmological constant is

$$\frac{\Lambda_4}{2} = \frac{\bar{\Lambda}}{(D+2)} - \frac{8\pi\bar{G}\left[(D-2)T(\tilde{R}_0)-D\tilde{T}(\tilde{R}_0)\right]}{(D+2)} \tag{10}$$

and we have now introduced the energy ρ and pressure p of radiation (for $R \gg \tilde{R}$)

$$\rho = 3\,p \propto T^4 \tag{11}$$

since, after compactification, these terms are now non-negligible ($\Lambda_4 \approx 0$ is of particular importance).

The form (2) of the energy momentum tensor happens in the most popular compactification mechanisms, i.e. compactification due to the presence of an antisymmetric tensor field strength (as in supergravity) à la Freund-Rubin,[6] or compactification by quantum fluctuations of towers of matter fields in four dimensions, arising from massless matter fields in 4 + D dimensions, à la Candelas and Weinberg.[7] In both cases[4]

$$T(\tilde{R}) = (\Omega_D \tilde{R}^D)^{-1} \phi(\tilde{R}) \tag{11}$$

$$\tilde{T}(\tilde{R}) = (D\ \Omega_D \tilde{R}^D)^{-1}\ \tilde{R}\ \frac{d\phi(\tilde{R})}{d\tilde{R}} \tag{12}$$

with

$$\phi(\tilde{R}) = C\ \tilde{R}^{-q},\ q > 0 \tag{13}$$

(q=D in Ref. (6) and q=4 in Ref. (7)). The actual case could be a combination of such terms. It is then easy[4] to establish the stability of the so obtained solutions for $\tilde{R}_0$ and R(t) against small perturbations. It is of particular importance that the criterion for such a stability[4] is automatically satisfied for zero four dimensional cosmological constant $\Lambda_4 \approx 0$. In that case, the approach of $\tilde{R}(t)$ to the equilibrium value $\tilde{R}_0$ is a damped oscillation, in which the damping rate of the oscillation amplitude is determined by the expansion of R(t) and the oscillation frequency is of order $\tilde{R}_0^{-1}$ (~Planck energy). This rapid oscillation can then produce[8] energetic particles, which in a realistic model include X, Y bosons, W, Z, photons, gluons, quarks, leptons etc.

REFERENCES

[1]Th. Kaluza, Sitz. Preuss. Akad. Wiss. Phys. Math. **K1**, 966 (1921); O. Klein, Z. Phys. **37**, 895 (1926); E. Witten, Nucl. Phys. **B186**, 412 (1981); A. Salam and J. Strathdee, Ann. Phys. (NY) **141**, 316 (1982).

[2]P.G.O. Freund, Nucl. Phys. **B209**, 146 (1982).

[3]S. Randjbar-Daemi, A. Salam and J. Strathdee, Phys. Lett. **135B**, 388 (1984).

[4]D. Bailin, A. Love, and C.E. Vayonakis, Phys. Lett. **142B**, 344 (1984).

[5]F.J. Dyson, 'Aspects of Quantum Theory', ed. by A. Salam and E.P. Wigner, (Cambridge Univ. Press, 1972).

[6]P.G.O. Freund and M.A. Rubin, Phys. Lett. **97B**, 233 (1980).

[7]P. Candelas and S. Weinberg, Nucl. Phys. **B237**, 397 (1984).

[8]T. Koikawa and M. Yoshimura, Phys. Lett. **150B**, 107 (1985).

Cosmology in Higher Dimensions

Roberto Bergamini

The metric tensor corresponding to an extended Kaluza-Klein "ansatz" can be written, in the local direct product basis, as

$$g_{\alpha\beta} = \left\{ \begin{array}{lll} \gamma_{\alpha\beta}+e^2K^2\gamma_{AB}M_a^AM_b^B & eKM_a^A\gamma_{AB} & (\alpha,\beta=0,\dots.N) \\ & & (a,b=0,\dots.3) \\ eK\gamma_{AB}M_b^B & \gamma_{AB} & (A,B=4,\dots.N) \end{array} \right\} \quad (1)$$

when the M_a^A can be identified as "gauge" potentials, e as the Yang Mills coupling constant and K as the typical length scale of the internal space,

$$\gamma_{ab} = \gamma_{ab}(\chi^c) \qquad \gamma_{AB} = \gamma_{AB}(\chi^\nu)$$

In cosmological situations, when compactification has been achieved (likely because of quantum effects), it becomes possible to introduce a purely classical energy-momentum tensor; for instance the perfect fluid one in which

$$T_{\alpha\beta} = (\rho + p)u_\alpha u_\beta + g_{\alpha\beta}p \quad (2)$$

ρ = density p = pressure

This energy-momentum tensor has to satisfy the Bianchi identities ($T^\alpha_\beta;\alpha = 0$) and this condition implies (putting p=o for simplicity) that

$$\rho(-g)^{1/2} = \text{constant} \qquad g = |g_{\alpha\beta}| \quad (3)$$

Because of 3) if the density of matter is assumed to depend only on time (as is the case for homogeneous cosmologies) g also will depend only on time. But g being the determinant of the complete metric the time evolution of the "visible" part of the Universe will reflect the presence of compactified dimensions[1].

Anyway if the length-scale of the compactified dimensions varies with time, even the gravitational constant and the fine-structure constant will vary. To avoid this effect Randjbar-Daemi, Salam and Strathdee[2] have introduced a peculiar state equation for the perfect fluid showing that, then, the hidden dimensions stay compactified during the cosmological evolution. Another way of obtaining the same result is to assume a dependence of g on the space coordinate so that the model of Universe will be in-homogeneous[3].

But, as it has been shown by Cho and Freund[4] the compactification produces a huge cosmological constant. To eliminate this "induced" cosmological constant, many suggestions have been advanced (see for instance Orzalesi-Venturi[5]). In the cosmological picture the N - dimensional Raychaudhuri equation

$$R^\alpha_\beta u_\alpha u^\beta = \dot{u}^\alpha;\alpha + 2(\omega^2 - \sigma^2) - \dot{\theta}^2 - \theta^2/N \quad (4)$$

where $\dot{u}^\alpha$ is the acceleration of the fluid, ω its "vorticity", σ the shear, and θ the volume expansion shows (as it is well known in the

4-dimensional case) that vorticity and shear have an opposite effect on the evolution of the Universe. Now the compactification of the hidden dimension implies that $\sigma \neq 0$ and that the expansion is not isotropic (in the N-dimensional space). It is this not-null shear that is ultimately responsible for the appearance of an "induced" cosmological constant. It seems thus natural to resort to the vorticity to counter balance the effect of the shear. Now introducing a reference frame and coordinates such that

$$g_{\alpha\beta} = \begin{pmatrix} -1 & 0 & g_{A0} \\ 0 & g_{ii} & \\ g_{0A} & & g_{AA} \end{pmatrix} \tag{5}$$

the vorticity $\omega_{\alpha\beta} \equiv u[\alpha;\beta] + \dot{u}[_{\alpha}u_{\beta}]$ can be shown to be given by two terms

$$\omega_{AB} = g_{0[A,B]} \tag{6}$$

$$\omega_{Ai} = g_{0[A,i]} \tag{7}$$

The term (7) will always $\neq 0$ because of the "ansatz" (1), while (6) will be $\neq 0$ only if some vorticity is assumed to exist in the compactified space. Thus in general ω will be different from nought.

By consequence it becomes possible to use the vorticity to induce a cosmological constant opposite to that connected to the shear. A detailed discussion of this suggestion will be found in (6).

REFERENCES

[1] R. Bergamini, C. A. Orzalesi, Phys. Lett. **135B**, 38 (1984).
[2] S. Randjbar-Daemi, A. Salam, J. Strathdee, Phys. Lett. **135B**, 388 ( ).
[3] R. Bergamini, R. Capovilla, submitted to Phys. Lett. (1984).
[4] Y. M. Cho, P. G. I. Freund, Phys. Rev. D **12**, 1711 (1975).
[5] S. Orzalesi, G. Venturi, Phys. Lett. **139B**, 357 (1984).
[6] R. Bergamini, G. Venturi, to be published.

PART IX

FUTURE DIRECTIONS AND CONNECTIONS IN PARTICLE PHYSICS AND COSMOLOGY

Although opinions are freely given in private, it is very difficult to find a physicist willing to discuss the future in print. In this concluding chapter we are fortunate to have Bjorken, an admitted futurist, discuss the future of particle physics at accelerators and possible future connections with astrophysics. It also proved difficult to find someone competent, willing, and able to write a concluding paper for this book. Since the competent were unwilling or unable, the editors (the incompetent) present their views of the future of the Inner Space/Outer Space connection.

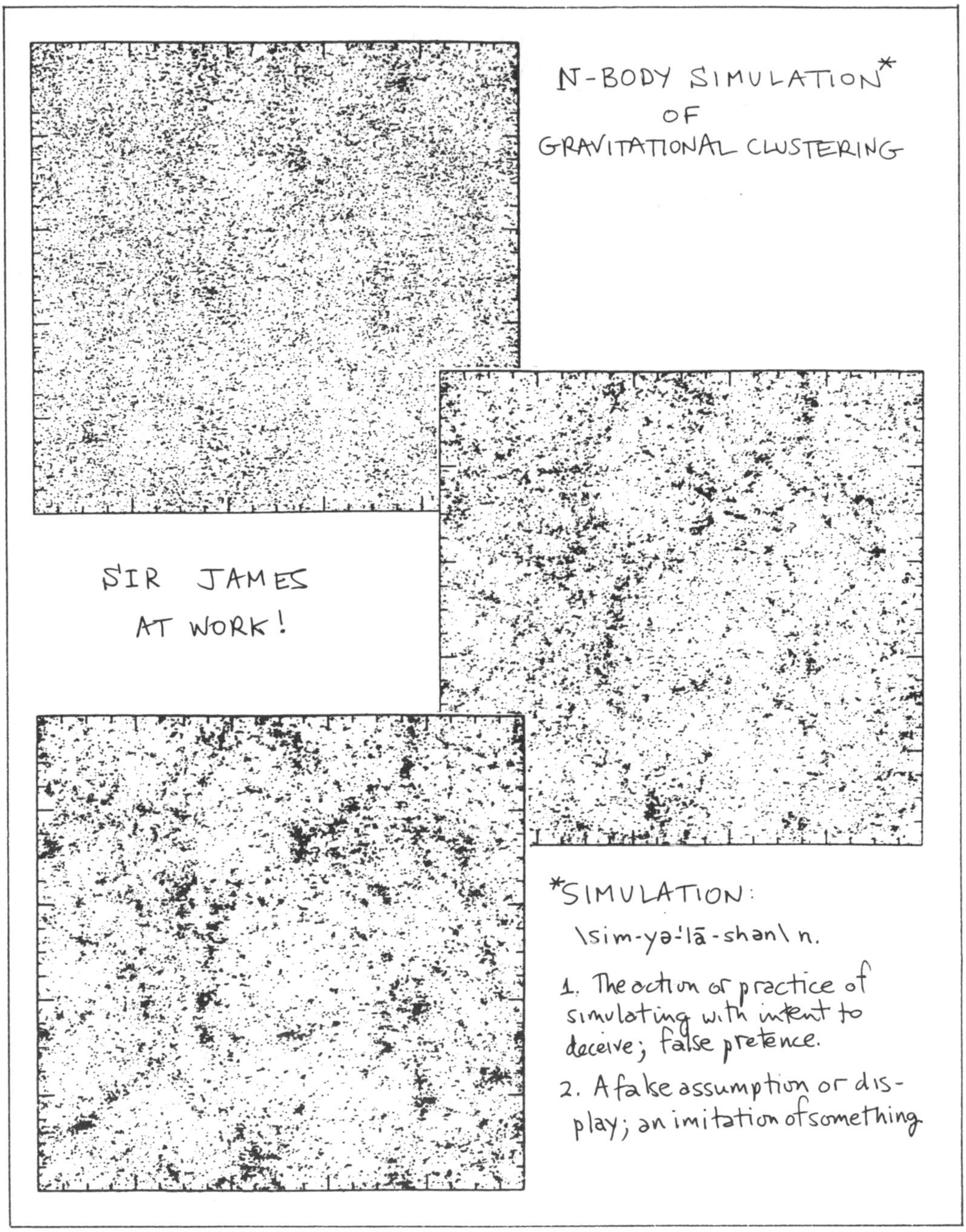
N-BODY SIMULATION* OF GRAVITATIONAL CLUSTERING
SIR JAMES AT WORK!
*SIMULATION:
\sim-yə-ˈlā-shən\ n.
1. The action or practice of simulating with intent to deceive; false pretence.
2. A false assumption or display; an imitation of something.

New and Old Accelerators: What Can They Do For Astrophysics?

J. D. Bjorken

ABSTRACT

The quantum numbers and energy spectrum of high energy accelerators and storage rings are described, along with some ways they may contribute to astrophysical issues. Some emphasis is given to the role of relativistic heavy-ion colliders in possibly providing laboratory samples of quark-gluon plasma.

I. INTRODUCTORY APOLOGIES

This talk, given by an amateur in astrophysical issues, must be regarded as an incomplete and subjective look at this subject. Emphasis will be given, not surprisingly, on topics of particular interest to this speaker. Little apology need be given for some of the omitted topics (e.g. monopoles, proton decay, neutrino masses and oscillations, and axions, standard and/or invisible), which are well covered in other talks[1] given at this workshop.

The material which will be discussed is organized from low energy to high (or in big-bang terms, from late to early times). We shall first, for those unfamiliar with the acronymics of high energy accelerators, briefly review the experimental facilities now available and planned for the future. In Section III we discuss phenomena at "late" times or low energies (<< 100MeV). Section IV concentrates on the hadronization epoch (100 MeV - 1 GeV), where study of relativistic heavy-ion collisions may be relevant. Section V deals with the bread-and-butter energy scales (1 GeV - 10 TeV) of the high energy physicist, with the concluding section mentioning the region beyond.

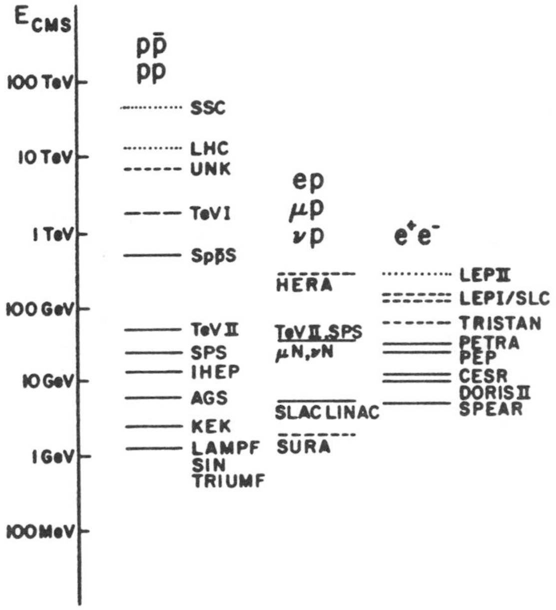

Figure 1: Energy-level diagram for particle accelerators.

II. THE SPECTRUM OF ACCELERATORS AND STORAGE RINGS

We here review the accelerators and storage rings existing, under construction, and planned for the future, which might contribute new information for the astrophysicist. These are shown in Fig. 1 in an energy-level diagram, as a function of their quantum numbers. One may collide hadrons with hadrons, hadrons with leptons, or leptons with leptons. The acronyms shown in Fig. 1 mean the following:

A) Hadron-Hadron Collisions:

LAMPF(Los Alamos), SIN(Zurich), TRIUMF(Vancouver): "Medium-energy" facilities with proton beams under 1 GeV. Especially useful for high energy physicists are the neutrino physics and rare decays of pion and (after proposed upgrades?) kaon.

KEK,AGS: 8 and 24 GeV proton synchrotrons in Tsukuba and Brookhaven, respectively.

IHEP: 70 GeV proton synchrotron in USSR.

SPS,TeV II: 400 and 1000 GeV synchrotrons operating at CERN (Geneva) and Fermilab respectively. Diversity of approach for investigations of collisions with center-of-mass energies of 20 - 40 GeV.

$Sp\bar{p}S$: 270 × 270 GeV proton-antiproton collider operating at CERN; site of $W^{\pm}$, Z^{0} weak gauge boson discovery.

Tev I: 1 TeV × 1 TeV proton-antiproton collider under construction at Fermilab, commissioning in 1986-1987.

UNK: Large proton synchrotron under construction in Serpukhov, USSR. 400 GeV protons onto fixed targets available no earlier than late 1980's; ultimately 3 TeV × 3 TeV proton-proton (or antiproton?) collisions might be provided at some uncertain date in the 1990's.

LHC: Possible proton-proton or proton-antiproton collider at CERN in LEP enclosure (see electron-positron listing); energy in the range of 5 × 5 TeV to 9 × 9 TeV; commissioning no earlier than 1994±2.

SSC: Possible 20 × 20 TeV proton-proton or proton-antiproton collider in US; highest priority of future US high-energy physics program.

B) Lepton-Hadron Collisions:

SURA: 4 GeV electron accelerator to be built in Newport News, Va.

SLAC: 20 GeV electron linac at Stanford, Ca.

SPS,TeV II: 200 - 600 GeV muon and neutrino secondary beams onto fixed targets.

HERA: 30 × 800 GeV electron-proton collider under construction at DESY (Hamburg, Germany); commissioning in 1989.

C) Electron-Positron Collisions:

SPEAR: 3 × 3 GeV electron-positron collider storage ring; site of Ψ and charmed quark discoveries; excellent machine for study of hadrons containing charmed quarks.

DORIS II, CESR: 5 × 5 GeV and 8 × 8 GeV electron-positron colliders at DESY and Cornell; excellent machines for study of hadrons containing bottom quarks.

PEP, PETRA: 14 × 14 GeV and 23 × 23 GeV electron-positron colliders.

TRISTAN: 30 × 30 GeV electron-positron collider under construction in Tsukuba, Japan; commissioning in 1986.

SLC, LEP I: 50 × 50 GeV electron-positron colliders under construction. SLC (at SLAC) uses a new "single-pass" technology employing the linear accelerator; LEP I (at CERN) is a "conventional" circular ring 26 km in circumference. SLC commissioning should occur in 1986; LEP I in 1988.

LEP II: Upgrade of the CERN collider to 70 - 100 GeV per beam.

In assessing the relative capability of the colliders, one must take into account the fact that at high energies a single proton should be regarded as a "beam" of its constituent quarks and gluons, and that seldom does more than one third of the proton momentum reside in a single constituent. Thus an electron-positron collider of center-of-mass-energy E_e should be compared with hadron-hadron colliders of energy $E_p \sim (3\text{-}5)E_e$.

III. LOW ENERGIES AND LATE TIMES

Neutrino physics and astrophysics are clearly intertwined but, as promised, little will be said here. Perhaps the next big step will be the resolution to the He^3 β-decay spectrum puzzle in the next year or two.

A direct contribution of accelerators will be the upcoming neutrino count from measurement of the width of the Z^0 at SLC and LEP II (with $Sp\bar{p}s$ and TeV I having an outside chance, too). The total width of Z^0 (of order 3 GeV) receives contributions from the decay $Z^0 \rightarrow T_n\bar{T}_n$. Assuming any new T_n to have "standard" weak coupling, the change in width $d\Gamma/dn$ with respect to the number n of neutrinos is 180 MeV per neutrino. The total width of ~3 GeV is measurable in e^+e^- annihilation by scanning the line shape, i.e. measuring the yield of Z^0's as function of beam energy. An alternative method uses radiative production, $e^+e^- \rightarrow Z^0+\gamma$, and compares $Z^0 \rightarrow \nu_n\bar{\nu}_n$ (in this case directly observable) with $Z^0 \rightarrow e^+e^-$ and/or $Z^0 \rightarrow \mu^+\mu^-$. At present, the limit from $Sp\bar{p}S$ on the number of neutrinos is somewhere around a dozen, scarcely of interest yet to astrophysicists.

One of the most interesting "late-time" astrophysical issues impacting particle-physics is the "dark-matter" problem. If mundane explanations do not suffice, some extension of the presently known low-mass particle spectrum seems to be required. Massive neutrinos are one option. Supersymmetric particles (photinos, sneutrinos?) are another. Axion-like entities are yet another. Accelerator searches can be (and have been) made in a variety of ways. Decay of mesons (e.g. $K^+ \to X^0+\pi^+$, $\psi \to X^0+\gamma$, $\Upsilon \to X^0+\gamma$) have been used to search for candidate bosons. Here low energy facilities, e.g. LAMPF, SIN, TRIUMPF, SPEAR, AGS, are especially valuable. Collider events with large missing transverse momentum (relative to the beam directions) may signal production of exotic light neutral particles if no conventional explanations suffice. For example, one search method for photinos $\tilde{\gamma}$ uses the process $e^+e^- \to \tilde{\gamma}\tilde{\gamma}+\gamma$, hopefully background-free when the photon has large transverse momentum and the center-of-mass energy is large. The $Sp\bar{p}S$ "zoo events", of which we speak later, are characterized by large missing transverse energy, and likewise invite speculations on new-particle production. Assuming, as appears reasonable, the detectors to be efficient in capturing all other forms of transverse energy, the balance must either be provided by neutrinos (e.g. by Z^0 production followed by $Z^0 \to \nu\bar{\nu}$) or else by some unknown neutral long-lived penetrating particle. Supersymmetry aficionados are now active in trying to interpret these events, but it is probably premature. The next $Sp\bar{p}S$ running period should clarify a presently murky experimental situation.

Beam-dump experiments are another useful way of looking for long-lived neutral penetrating particles. A variety of such experiments have been done with hadron beams at LAMPF, SIN, AGS, PS, SPS, and Fermilab. No convincing signals have been seen, and only limits on parameters for axions, heavy neutrinos, etc. exist. A somewhat typical experiment on these lines was done at SLAC. Since I was a participant in that one, I will dwell briefly on it as a prototype of such experiments. As shown in Fig. 2, 30 Coulombs of SLAC 20 GeV electrons were dumped in a large tank of water, just upstream of 200 meters of natural earth shielding. 400 meters downstream, an electromagnetic shower counter of good angular resolution was placed, where it could detect any decay products produced in

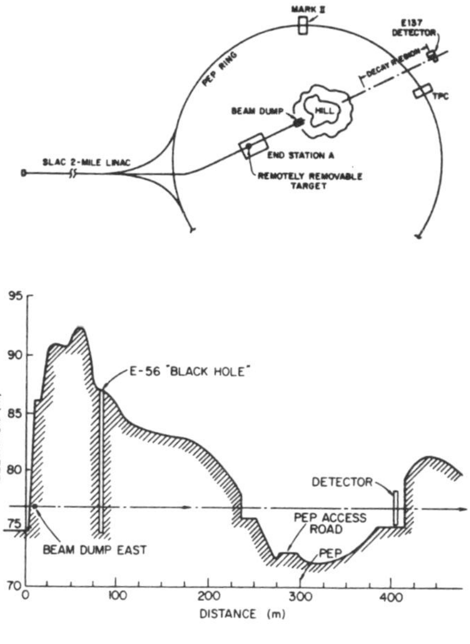

Figure 2: Beam dump experiment E137 carried out at SLAC.

the 200 meters of air upstream of the detector. For example, an axion-like entity X could be produced by the Primakoff mechanism (γ+ coulomb field → X) and then decay back into $\gamma\gamma$. Nothing was seen, and a preliminary lower limit (95% confidence level) for the product of mass and lifetime of 0.8 keV-sec was obtained[2].

IV. THE HADRONIZATION EPOCH

In the very early universe, the appropriate description must be in terms of a plasma of quarks, gluons, and leptons; the densities are simply too high to imagine individual hadrons as indentifiable. Much later this system must turn into hadrons (plus leptons), essentially a dilute gas of pions, with a small contamination of baryons. The transition between these two phases is expected to occur at a temperature of about 200 MeV. This is an issue of interest not only to astrophysicists, but also to lattice gauge theorists, who presume to have - or soon to have - calculational tools adequate to map out the equation of state of hadronic matter. Schematically this equation of state is shown in Fig. 3. At low and high temperature, Stefan and Boltzmann rule. However the large number of internal degrees of freedom (spin, color, etc.) for quarks and gluons outnumber the three for pions by an order of magnitude. Every indication from lattice calculations[3] is that the transition is abrupt, perhaps first-order with a latent heat $\gtrsim$ 1 GeV/fm^3. There may even be two phase transitions, one associated with deconfinement, the other with restoration of spontaneously broken chiral symmetry. This option is at present slightly disfavored, but remains an open question.

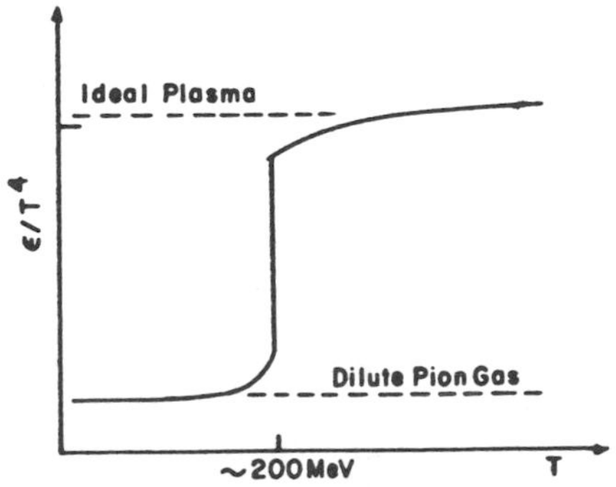

Figure 3: Equation of state of hadronic matter.

Location of the transition temperature experimentally, along with good theoretical calculations, would allow a very good measurement of the strong coupling constant α_s or, better, the QCD scale factor Λ. Thus, for this as well as many other reasons, there is growing interest in trying to observe experimentally the quark-gluon plasma. The opportunity may exist in collisions of relativistic heavy ions. At sufficiently high energy, as viewed in the center-of-mass frame, the incident nuclei (say U; 14 fm diameter) are Lorentz-contracted to a thickness much less than 1 fm. After they collide, the collision products, be they quarks, pions, gluons or something else, are (by causality) confined within the planes of outgoing excited nuclear matter, which in this frame recede at the speed of light. It is reasonable (but not at all proven) that these collision products equilibrate rapidly (ct $\lesssim$ 1 fm??) and even more reasonable that the initial energy density is $\sim$ 1-5 GeV/fm^3, corresponding to quark-gluon plasma above the phase transition (but not by an enormous amount; the temperature is ~ 200-300 MeV). Indeed a direct estimate from observed heavy-ion cosmic ray events tends to corroborate the theoretical estimates of initial energy density.

The geometry of the expansion phase appears to be quite simple. At sufficiently high collision energy the plasma within the receding planes undergoes (to reasonable approximation) a homogeneous one-dimensional longitudinal expansion[4]. That is, a fluid element in the midplane remains at rest, while one halfway between the midplane and the boundary plane moves at half the speed of light, etc. Thus, except for the one-dimensionality of the expansion, the conditions may indeed be quite similar to those which existed in the first three microseconds. In the case of heavy-ion collisions, this state of affairs cannot last very long. The news that the nuclei are finite is transmitted from the edge inward by a rarefaction front traveling at the velocity of sound in the plasma.

The central problem in any future experimental program in relativistic ion-ion collisions is to find connections between the properties of the 10^3 - 10^4 observed collision products, produced in general at late times, with the purported plasma extant (if at all) at early times. Proposed ideas include study of fluctuations and correlations in multiplicity density, composition, transverse momenta, etc., which might be correlated with bulk fluctuations (bubble formation? shocks?) occurring during the phase transition. Another attack is observation of direct photons and/or dileptons produced by the plasma during the initial hot phase. No signature looks supremely clean, and even with more work (sure to come) it will be a tough battle. My own prejudice is that if ion-ion collisions do turn out to reveal to us the existence and properties of quark-gluon plasma, we probably will not anticipate in advance the mechanism of revelation. It may be like neutron stars: theorists knew everything about them in advance - except how to find them. And, despite the fact that everyone is convinced that pulsars are indeed neutron stars, to this day the pulsar mechanism is not well understood.

The experimental prospects for a highly relativistic heavy-ion experimental program are reasonably bright. The first step will be acceleration of ions up to oxygen in the SPS at momenta of 200 GeV/nucleon. Brookhaven is building a transfer line from its Van de Graaf to the AGS, and with addition of a new booster ring ($\lesssim$ 4-5 years) will be able to accelerate the heaviest ions through the AGS. In the longer run, there is a rather detailed proposal to build a relativistic heavy-ion collider ($\lesssim$ 100 GeV/nucleon; $\lesssim$ 6 years) in the CBA tunnel. This facility would be of sufficient energy to attain the clean geometry for plasma production which we described.

V. HIGH ENERGY AND HIGH MASS SCALES/EARLY TIMES

The present theoretical perspective on dynamics above the strong interaction scale emphasizes the restoration of symmetries. As the energy scale increases, the symmetry of the vacuum state, broken at low energy scales, is expected to increase. The first such transition beyond the hadron-to-quark/gluon transition apparently occurs at the electroweak scale of ~100-200 GeV. The estimated[5] standard-model transition temperature is 425 GeV. At energies large compared to this scale, the SU(2) × U(1) symmetry of the electroweak theory is expected to be fully restored. The agreement of the measured W and Z masses with the predictions greatly strengthen confidence that this expectation is correct.

Beyond this scale lies speculation. Proposals for extra tiers of symmetry restoration at higher mass scales come and go with the seasons: "extended technicolor", restoration of global supersymmetry, restoration of left-right symmetry and/or CP invariance, and upward toward grand unification scales.

The phase transitions associated with these conjectured mechanisms may be of the most direct concern to the astrophysicist. To the particle physicist it is the extra particles as well as, perhaps, the breakdown of conservation laws, as in the case of proton decay. This feature might express itself in other ways (lepton nonconservation) and invites scrutiny of conservation laws at all energy scales.

The search phenomena high on the agenda for experimentalists includes the following incomplete list:

1. <u>The Standard Higgs Boson</u>:
Something like this particle <u>must</u> exist; the orthodox standard model predicts everything about it except its mass. The search technique depends a great deal on that parameter[6]. If the Higgs mass is less than about 40 GeV, the method of choice is the process $Z^0 \to e^+e^-h$ or $Z^0 \to \mu^+\mu^-h$, available at SLC and LEP I. For the 40 to 100 GeV range, LEP II may see it via the related process $e^+e^- \to Z^0h$, with $Z^0 \to e^+e^-$ or $Z^0 \to \mu^+\mu^-$. Between 170 GeV and 500 GeV, hadron-hadron colliders such as SSC are best. One uses the gluon component within the energetic proton and resonantly makes the Higgs h in a gluon-gluon "fusion": $gg \to h$; the Higgs boson then decays into W^+W^- or Z^0Z^0, which is a hopefully observable final state. The interval between 100 and 170 GeV is awkward; one needs luck or an e^+e^- collider of energy considerably beyond LEP II. The interval above 500 GeV is fraught with background problems; however the mass of h is bounded above by quite general arguments; the limit is around 1 TeV.

2. <u>Flavor Changing Decays</u>:
Rare K decays such as $K \to \mu e$, $K \to \pi\mu e$, etc. probe very high symmetry-restoration(?) scales, and limits on these are well worth pushing further. Searches for rare decays of hadrons containing c and b quarks should also be carried out. The b quark in particular seems to have a surprisingly long lifetime ($\sim 10^{-12}$ sec) and thus the branching fraction for crazy rare decays may be especially enhanced. The experiments can be done at e^+e^- machines such as SPEAR, CESR, and DORIS, as well as in the TeV II fixed target program. Also, one should not ignore the possibility of rare Z^0 decays (how about $Z^0 \to \gamma\mu e$?), given that SLC and LEP I may give us more than a million Z's per year.

3. <u>Right-handed Weak Currents</u>:
Here HERA will make important contributions. Positive evidence could well herald a phase transition associated with parity restoration.

4. Compositeness Test for Quarks And Leptons:
These tests (TeV I, SSC, LEP II) could indicate a phase transition to preons similar to the strong interaction transition from hadrons to quark-gluon plasma. Mass scales of up to 30 TeV appear attainable at the SSC.

5. Generic Axions:
Symmetry breaking at high mass scales might provide pseudo-Nambu-Goldstone bosons. A Fermilab experiment (E-635) proposes to search for $X \rightarrow \mu^+\mu^-$ in a range of parameter space which, while small, is exquisitely sensitive to mass scales up to 10^4 TeV (if m_x is $\lesssim$ 1 GeV).

These items have in part been listed because they exhibit sensitivity to phenomena at mass scales beyond what is directly available from the accelerator or storage ring. There will be of course the direct searches for generic new phenomena, such as discovery of new quarks, leptons, or gauge bosons. The big three concepts of theoretical speculation are[6]

Technicolor
Compositeness
Supersymmetry

They all imply proliferation of degrees of freedom - many new particles. The main technique of the future, just emergent now at PEP, PETRA, and $Sp\bar{p}S$, is multijet spectroscopy. A quark or gluon is now "seen" as a collimated jet of hadrons directed in the original direction of the parent quark or gluon. These are becoming "tracks" just about as well resolved as the cloud-chamber tracks of hadrons in the old days. The first such jets, back-to-back, were seen at SPEAR via the process $e^+e^- \rightarrow q\bar{q}$, followed by 3-jet $e^+e^- \rightarrow qqg$ events at PETRA. Spectacular 2-jet events $(q\bar{q},qg,gg) \rightarrow (q\bar{q},qg,gg)$ have been seen at the $Sp\bar{p}S$ with center-of-mass energies of over 200 GeV (out of the total 540 available). Multijet events are seen also. Very heavy particles which decay into quarks and/or gluons will contribute multijet final states. (For example $W,Z \rightarrow q\bar{q}$; regrettably these are swamped by backgrounds from the aforementioned scattering processes). An interesting class of "zoo" events from the $Sp\bar{p}S$ UA1 and UA2 experiments have been seen. They are characterized in part by large amounts of escaping transverse momentum, by a large amount of transverse energy (150-200 GeV) and often by presence of a charged lepton. As mentioned before, the missing transverse momentum, if not contributed by neutrinos, would be indicative of production of some kind of long lived neutral penetrating particle (photino? goldstino? sneutrino? axion?). Supersymmetry enthusiasts and other theorists are becoming active, but it is too early to conclude anything. Much more needs to be understood about multijet background processes, and larger samples need to be acquired. Nevertheless, at TeV I and higher energy hadron colliders, it seems certain that the reconstruction of multijet systems will become a routine technique and a powerful avenue toward new-particle discoveries in the 100-1000 GeV mass range.

VI. SUPERHIGH MASS SCALES/THE VERY EARLY UNIVERSE

Prospects for probing GUT mass scales ($10^{15\pm5}$GeV) clearly lie beyond accelerator physics. There is the nucleon decay program and the search for $n\bar{n}$ oscillations (much more speculative), searches for rare superheavy stable relics in terrestrial (or other) matter and the searches for monopoles. Were the monopole search successful, the use of GUT monopole-antimonopole annihilation would provide the ultimate high-energy physics experiment. If one had one SU(5) GUT monopole and one antimonopole and could bring them together to annihilate (nontrivial but thinkable), then the single annihilation could produce several SU(5) X and Y bosons, which could decay into observable, indeed lethal, hadron jets and leptons of energy 10^{14}-10^{15} GeV. A single event would be a radiation hazard. While far out, maybe this at least indicates that high energy experimental physics is a very long way from becoming a sterile discipline.

REFERENCES

[1]See for example, J. Preskill (monopoles), P. Langacker (proton decay), F. Scuilli (neutrinos), and P. Sikivie (axions); these proceedings.

[2]A preliminary report has been made by me at the Fourth Moriond Workshop on Massive Neutrinos, proceedings to be published; cf. Fermilab preprint FERMILAB-Conf-84/33-T.

[3]For a review, see the report of M. Jacob and J. Tran Than Van, Phys. Rep. **88** (1982).

[4]J. Bjorken, Phys. Rev. D**27**, 140 (1983).

[5]A. Linde, Rept. Prog. Phys. **42**, 389 (1979).

[6]E. Eichten, I. Hinchliffe, K. Lane, and C. Quigg, Revs. Mod. Phys. **56**, 579 (1984).

Future Directions at the Particle Physics-Cosmology Interface

Edward W. Kolb, Michael S. Turner, David Lindley,
Keith A. Olive, and David Seckel

In the concluding paper of this volume we give our personal views on the future of the particle physics - cosmology interface. Speculation about the future is often preceded by a review of the path to the present; by taking a few steps back it is possible to get a running start for the leap into the future. In the trip through the past toward the future, we will notice a singular change of attitude over the past few years. Cosmology in the 20th century has been hampered by a lack of confidence on the part of cosmologists, often leading to missed opportunities. Shortly after Einstein introduced the field equations in 1915, he attempted to use his equations to construct a cosmological model. Although he had revolutionized the concepts of space, time, and geometry, Einstein could not extricate himself from his 19th century view of a static Universe. For this reason he introduced the cosmological term to find static (but unstable) solutions. In so doing, Einstein (and others) missed one of the greatest opportunities of theoretical physics, the prediction of the expansion of the Universe. Some ten years later the expansion of the Universe was discovered by Hubble, and it was LeMaitre who first explicitly related the Hubble expansion to the time-varying solutions of Friedman.

This lack of confidence in the application of physical law to understand the Universe occurred again after the prediction of a relic background radiation by Gamow in the 1940's. It was not until 1965 that Penzias and Wilson accidently discovered the background radiation. It is somewhat ironic that this discovery came shortly before an experiment designed to look for the background would have found it. Nevertheless, there was a missed opportunity for many years. It seems again that a cosmological prediction, the existence of the microwave background, was ignored because cosmologists did not take their predictions seriously enough.

Soon after the discovery of the microwave background, several groups finally took seriously Gamow's suggestion that conditions in the Universe seconds after the bang were just right for the synthesis of light elements such as D, He, and Li. The successful predictions of primordial nucleosynthesis remain the most solid evidence that we have an accurate description of the Universe as early as 10^{-2} sec after the bang.

The standard big-bang model is not without its shortcomings. There are a number of cosmological facts that the standard model accommodates but does not explain. These facts include the observed predominance of matter over antimatter, the origin of the seed density inhomogeneities required for galaxy formation, the nature of the ubiquitous dark matter, the smallness of any cosmological constant, and the large-scale isotropy, homogeneity, and flatness of the Universe. In addition, the standard model predicts an initial singularity, which some people find distasteful.

If we somewhat arbitrarily date the beginning of modern cosmology at the discovery of the gravitational field equations in 1915, the first fifty years mark an era of hesitancy on the part of cosmologists, a hesitancy that led to missed opportunities. In the twenty years since the discovery of the microwave background a new confidence has emerged. Cosmologists in general now take the view that we do indeed have the basic model correct, and the above mentioned shortcomings will eventually be solved within the context of the model. In order to seek a solution to these problems it is necessary to turn to particle physics.

The importance of physics input to cosmology is well illustrated by a simple, although somewhat dated, example. In the late 60's and early 70's the success of primordial nucleosynthesis was somewhat dashed by the specter of a Universe with a limiting temperature of only a few 100 MeV, as models of hadronic physics at the time predicted an exponentially rising number of states. The introduction and eventual triumph of the quark/parton model in the 70's shattered that barrier, as the number of quark species appears to be small and their interactions are asymptotically free at very high energies.

The application of modern particle theory to cosmology has already begun to shed light on many of the above problems. It appears that the baryon number of the Universe evolved dynamically through interactions which operated in the very early Universe ($t = 10^{-34}$ sec) and which violated B, C, and CP conservation. Once we understand physics at the grand unification scale we should be able to calculate the baryon number of the Universe in the same way that we calculate the primordial abundances of the light elements. The new inflationary Universe scenario offers the possibility of resolving the isotropy, homogeneity, flatness, and inhomogeneity puzzles, although at the expense of once again introducing a cosmological constant -- albeit temporarily. It seems very attractive to believe that the dark matter consists of relic particles left over from the very early Universe, and particle physics has been generous in providing candidates. If relic particles are the solution, it is somewhat surprising that the present mass densities of ordinary matter and exotic relics are nearly comparable. An optimist would at present conclude that we are well on our way to answering many (if not all) of the most pressing cosmological questions. Even a less than optimistic person would conclude that the answers to many of the pressing cosmological questions lie in understanding the earliest history of the Universe.

That brings us to the future. First, on the theoretical side, it seems likely that inflation, or at least offshoots of the 'inflationary paradigm', will continue to be promising avenues to pursue. One of the most promising approaches to unification of all the forces seems to be through additional spatial dimensions. An area still in its infancy, cosmology with extra dimensions adds yet another puzzling, but perhaps not unrelated, fact to our list: why are all but three of the spatial dimensions so small? Superstring cosmology opens a Pandora's box of new problems -- whence came geometry, was there an initial singularity, does the Universe after all have a limiting temperature?

Structure formation seems to have been reduced once again to two competing pictures -- hot or cold dark matter, and unfortunately neither picture seems to be wholly satisfactory. The situation resembles the previous dilemma in the choicc of adiabatic vs. isothermal perturbations. Perhaps the logjam will be broken by an entirely new idea, e.g., cosmic strings, or an astrophysical role in structure formation, for instance through the energy released in the earliest generation of objects.

Of course, the ultimate test of theory is grounded in observation. Much of the present cosmological speculation is at very high redshift ($z = 10^{32}$ at the Planck era), while future observations are likely to be done at relatively low redshifts ($z \lesssim 3$). Hopefully these low z observations will bear in one way or another on the Universe at very high z. The Space Telescope should settle some uncertainties in the determination of the value of the Hubble constant, and perhaps even start to sort out the deceleration parameter. Infrared observations of the Universe, of both the diffuse background and individual sources, could prove to be very important. The cosmological importance of continued improvement of microwave anisotropy measurements cannot be overstated. If either the hot or the cold picture is correct, a detectable anisotropy is just around the corner. Continued scrutiny of the light element abundances is also very important, as they probe our last detailed test of the standard cosmology. It is clear that larger and more detailed surveys will be of great use in sorting out the story of structure formation.

Because of the Inner Space/Outer Space connection, experiments in particle physics will undoubtedly play an important role in the future of cosmology. A whole host of experiments -- accelerator searches for superpartners, neutrino mass experiments, etc., have the potential to discover a dark matter candidate, or at least to narrow the list. Discovering the light Higgs particle predicted by the Weinberg-Salam-Glashow theory would allow us to specify the scalar potential relevant for electroweak symmetry breaking and to understand the details of the electroweak phase transition in the early Universe. The electroweak interaction provides the only Higgs system accessible to experiment in the near future, and although it cannot be the scalar field driving inflation, it does provide a very useful toy model for understanding the dynamics of Higgs systems. At the very least, it would be nice to be reassured that the Higgs mechanism has something to do with Nature -- or to be rudely awakened. Numerical experiments on the deconfinement and chiral symmetry breaking transitions in QCD, and possibly direct experimental evidence from accelerators, should help to sort out the cosmological details of the QCD phase transition that should have occurred at a temperature of a few hundred MeV. The implications of the observation of proton decay for cosmology are enormous.

Then there are the non-accelerator experiments -- searches for exotic relic particles, such as monopoles, photinos, axions, etc., cosmic ray experiments, ...

Most certainly we have not included in our list the most important or exciting things that will occur in the next few years. Those almost by definition will be the surprises that await us. For example, definitive measurements of the Hubble constant and the age of the Universe might imply that their product is not equal to 2/3 (the value for an $\Omega = 1$ matter-dominated Universe); a value greater than 1 would once again force us to turn to a cosmological constant. A determination that the primordial He abundance is less than 22% would upset the excellent concordance which now exists between theory and observation and force us to rethink that relatively recent epoch. Lack of anisotropy in the microwave temperature at a level of even a few times 10^{-6} would again force us to rethink structure formation. And finally, at this point any evidence for spatial curvature in the Universe would be a real surprise.

Whatever future cosmologists write about cosmology in the 1980's, we can be certain that it will not be that the cosmologists of this era were afraid to take even their wildest ideas seriously. To what extent we will be rewarded for our boldness remains to be seen. We remain ever optimistic!

INNER SPACE/OUTER SPACE B-B-Q

BUFFALO CLASS* PICNIC

FRI 4 MAY 5-8 PM FINAL BARN

$5.00 Entertainment: Boy Keith

*or equivalent protein material VOID WHERE VALID

INNER SPACE/OUTER SPACE B-B-Q

BUFFALO CLASS* PICNIC

FRI MAY 4 5-8 PM FINAL BARN

$5.00 Lead Talk - J. Fonda*: "The Workout"

*or equivalent protein material VOID WHERE VALID

INNER SPACE/OUTER SPACE

AWARDS BANQUET

$30 THURS MAY 3 6-10 PM $30

- Best Paper Based Upon Original Work
- Best Paper Adapted From an Earlier Paper

Entertainment: Boy George*
Michael 'J.J.' Jackson*
The Amazing Leon*
(*or equivalent)

INNER SPACE/OUTER SPACE

AWARDS BANQUET

$30 THURS MAY 3 6-10 PM $30

- Best Paper Based Upon Original Work
- Best Paper Adapted From an Earlier Paper

Entertainment: Boy George*
Michael JJ Jackson*
The Amazing Leon*
(*or equivalent)

PROF. LAURENCE F. ABBOTT
PHYSICS DEPARTMENT
BRANDEIS UNIVERSITY
WALTHAM, MASSACHUSETTS 02254

FRANK ACCETTA
DEPT. OF PHYSICS AND ASTROPHYSICS
UNIVERSITY OF CHICAGO
5640 SOUTH ELLIS AVENUE
CHICAGO, ILLINOIS 60637

PROF. STEVEN P. AHLEN
PHYSICS DEPARTMENT
INDIANA UNIVERSITY
BLOOMINGTON, INDIANA 47405

DR. ANDREAS ALBRECHT
THEORY GROUP RLM 5.208
UNIVERSITY OF TEXAS
AUSTIN, TEXAS 78712

MS. VIRGINIA M. AYRES
DEPARTMENT OF PHYSICS
PURDUE UNIVERSITY
WEST LAFAYETTE, INDIANA 47907

DR. NETA BAHCALL
SPACE TELESCOPE SCIENCE INSTITUTE
JOHNS HOPKINS UNIVERSITY
HOMEWOOD CAMPUS
BALTIMORE, MARYLAND 21218

DR. JOHN N. BAHCALL
INSTITUTE FOR ADVANCED STUDY
SCHOOL OF NATURAL SCIENCES
BUILDING E, OLDEN LANE
PRINCETON, NEW JERSEY 08540

DR. WILLIAM A. BARDEEN
MS #106
FERMILAB
P.O. BOX 500
BATAVIA, ILLINOIS 60510

DR. JAMES M. BARDEEN
INSTITUTE FOR THEORETICAL PHYSICS
UNIVERSITY OF CALIFORNIA
SANTA BARBARA, CALIFORNIA 93106

DR. BARRY C. BARISH
256-48 LAURITSEN LABORATORY
CALIF. INSTITUTE OF TECHNOLOGY
PASADENA, CALIFORNIA 91125

DR. JOHN T. BARNETT
PHYSICS DEPARTMENT
UNIVERSITY OF NEVADA
LAS VEGAS, NEVADA 89154

DR. MADELEINE F. BARNOTHY
833 LINCOLN STREET
EVANSTON, ILLINOIS 60201

DR. JENO M. BARNOTHY
833 LINCOLN STREET
EVANSTON, ILLINOIS 60201

DR. STEPHEN M. BARR
DEPARTMENT OF PHYSICS, FM-15
UNIVERSITY OF WASHINGTON
SEATTLE, WASHINGTON 98195

MR. DAVID J. BATUSKI
DEPT. OF PHYSICS & ASTRONOMY
UNIVERSITY OF NEW MEXICO
ALBUQUERQUE, NEW MEXICO 87131

JAMES J. BEATTY
EFI/LASR
UNIVERSITY OF CHICAGO
933 E. 56TH STREET
CHICAGO, ILLINOIS 60637

DR. JOHN E. BECKMAN
DEPARTMENT OF PHYSICS
QUEEN MARY COLLEGE
MILE END ROAD
LONDON E1 4NS, ENGLAND

MS. SHARON BEGLEY
SCIENCE EDITOR
NEWSWEEK MAGAZINE
444 MADISON AVENUE
NEW YORK, NEW YORK 10022

DR. ROBERTO BERGAMINI
ISTITUTO DI RADIOASTRONOMIA
C. N. R.
VIA IRNERIO, 46
I-40126 BOLOGNA, ITALY

BEVERLY K. BERGER
DEPARTMENT OF ASTRONOMY
UNIVERSITY OF MICHIGAN
ANN ARBOR, MICHIGAN 48109

WILLIAM BERGLUND
UNIVERSITY OF MINNESOTA
SCHOOL OF ASTRONOMY
MINNEAPOLIS, MINNESOTA 55455

DR. STUART BERMON
IBM T.J. WATSON RESEARCH CENTER
P. O. BOX 218
YORKTOWN HEIGHTS, NEW YORK
10598

DR. PIERRE BINETROY
THEORETICAL PHYSICS BLDG 50A-3115
LAWRENCE BERKELEY LABORATORY
ONE CYCLOTRON ROAD
BERKELEY, CALIFORNIA 94720

DR. JAMES D. BJORKEN
MS #105
FERMILAB
P.O. BOX 500
BATAVIA, ILLINOIS 60510

DR. SIDNEY BLUDMAN
DEPARTMENT OF PHYSICS
UNIVERSITY OF PENNSYLVANIA
PHILADELPHIA, PENNSYLVANIA 19104

DR. ARNOLD R. BODMER
PHYSICS DIVISION, BLDG 203-B253
ARGONNE NATIONAL LABORATORY
9700 SOUTH CASS AVENUE
ARGONNE, ILLINOIS 60439

PROF. J. RICHARD BOND
PHYSICS DEPARTMENT
STANFORD UNIVERSITY
STANFORD, CALIFORNIA 94305

DR. STUART BOWYER
ASTRONOMY DEPARTMENT
UNIVERSITY OF CALIFORNIA
BERKELEY, CALIFORNIA 94720

DR. ROBERT H. BRANDENBERGER
INSTITUTE FOR THEORETICAL PHYSICS
UNIVERSITY OF CALIFORNIA
SANTA BARBARA, CALIFORNIA 93106

DR. RAYMOND L. BROCK
DEPT. OF PHYSICS AND ASTRONOMY
MICHIGAN STATE UNIVERSITY
EAST LANSING, MICHIGAN 48824

PROF. GEOFFREY R. BURBIDGE
DIRECTOR'S OFFICE
KITT PEAK NAT'L OBSERVATORY
P. O. BOX 26732
TUCSON, ARIZONA 85726

PROF. BLAS CABRERA
PHYSICS DEPARTMENT
STANFORD UNIVERSITY
VARIAN PHYSICS BLDG.
STANFORD, CALIFORNIA 94305

DR. RICHARD H. CAPPS
DEPARTMENT OF PHYSICS
PURDUE UNIVERSITY
WEST LAFAYETTE, INDIANA 47907

DR. ERIC D. CARLSON
ADLER PLANETARIUM
1300 SOUTH LAKE SHORE DRIVE
CHICAGO, ILLINOIS 60605

DR. BERNARD J. CARR
INSTITUTE OF ASTRONOMY
MADINGLEY ROAD
CAMBRIDGE CB3 0HA, ENGLAND

DR. RICHARD A. CARRIGAN
MS #208
FERMILAB
P.O. BOX 500
BATAVIA, ILLINOIS 60510

SEAN CASEY
AAC - F100
ASTRONOMY & ASTROPHYSICS CENTER
UNIVERSITY OF CHICAGO
CHICAGO, ILLINOIS 60637

DAVID W. CASPER
%RANDALL LABORATORY
UNIVERSITY OF MICHIGAN
ANN ARBOR, MICHIGAN 48109

DR. MICHAEL L. CHERRY
DEPARTMENT OF PHYSICS
UNIVERSITY OF PENNSYLVANIA
209 SOUTH 33RD STREET
PHILADELPHIA, PENNSYLVANIA 19104

DR. GEORGE B. COLLINS
PHYSICS DEPARTMENT
VIRGINIA POLYTECHNIC INSTITUTE
AND STATE UNIVERSITY
BLACKSBURG, VIRGINIA 24061

WILLIAM D. COLLINS
DEPT. OF PHYSICS AND ASTROPHYSICS
EFI-F100, UNIVERSITY OF CHICAGO
5640 SOUTH ELLIS AVENUE
CHICAGO, ILLINOIS 60637

MR. ARLIN CROTTS
DEPT. OF PHYSICS AND ASTROPHYSICS
UNIVERSITY OF CHICAGO
5640 SOUTH ELLIS AVENUE
CHICAGO, ILLINOIS 60637

DR. MARC DAVIS
DEPARTMENT OF ASTRONOMY
601 CAMPBELL
UNIVERSITY OF CALIFORNIA
BERKELEY, CALIFORNIA 94720

DR. AVISHAI DEKEL
INSTITUTE FOR THEORETICAL PHYSICS
UNIVERSITY OF CALIFORNIA
SANTA BARBARA, CALIFORNIA 93106

NIVETIEA DEO
DEPARTMENT OF PHYSICS
PURDUE UNIVERSITY
WEST LAFAYETTE, INDIANA 47907

MR. MARK DRAGOVAN
PHYSICS DEPARTMENT
UNIVERSITY OF CHICAGO
5630 SOUTH ELLIS AVENUE
CHICAGO, ILLINOIS 60637

DAVID J. EICHER
ASTRONOMY MAGAZINE
625 EAST ST. PAUL AVENUE
MILWAUKEE, WISCONSIN 53202

DR. STEPHEN D. ELLIS
DEPARTMENT OF PHYSICS, FM-15
UNIVERSITY OF WASHINGTON
SEATTLE, WASHINGTON 98195

MR. AUGUST E. EVRARD
PHYSICS DEPARTMENT
STATE UNIVERSITY OF NEW YORK
STONY BROOK, NEW YORK 11794

TIMOTHY FERRIS
NORTHSTAR ASSOCIATES
1741 NORTH IVAR, SUITE 217
LOS ANGELES, CALIFORNIA 90028

DR. PHILIPPE CRANE
EUROPEAN SOUTHERN OBSERVATORY
KARL-SCHWARZSCHILDSTRASSE 2
D-8046 GARCHING
WEST GERMANY

DR. DONALD C. CUNDY
EP DIVISION
CERN
CH-1211 GENEVE 23
SWITZERLAND

DAVID S. DE YOUNG
KITT PEAK NATIONAL OBSERVATORY
P. O. BOX 26732
TUCSON, ARIZONA 85726

DR. MAREK DEMIANSKI
DEPT. OF PHYSICS AND ASTROPHYSICS
WILLIAMS COLLEGE
WILLIAMSTOWN
MASSACHUSETTS 02167

DR. ROGER L. DIXON
MS #208
FERMILAB
P.O. BOX 500
BATAVIA, ILLINOIS 60510

PROF. KENNETH A. DUNN
DEPT. OF MATHEMATICS, STATISTICS
AND COMPUTING SYSTEMS
DALHOUSIE UNIVERSITY, HALIFAX
NOVA SCOTIA CANADA B3H 4H8

DR. MARTIN B. EINHORN
RANDALL LABORATORY OF PHYSICS
UNIVERSITY OF MICHIGAN
ANN ARBOR, MICHIGAN 48109

DR. ALLEN E. EVERETT
PHYSICS DEPARTMENT
TUFTS UNIVERSITY
MEDFORD, MASSACHUSETTS 02155

DR. WILLIAM M. FAIRBANK
PHYSICS DEPARTMENT
STANFORD UNIVERSITY
STANFORD, CALIFORNIA 94305

DR. GIOVANNI FIORENTINI
DEPARTIMENTO DI FISICA
UNIVERSITA DI PISA
PIAZZA TORRICELLI 2
I-56100 PISA, ITALY

DR. RICARDO A. FLORES
DEPARTMENT OF PHYSICS
UNIVERSITY OF CALIFORNIA
SANTA CRUZ, CALIFORNIA 95604

DR. STUART J. FREEDMAN
BLDG. 203, PHYSICS DIVISION
ARGONNE NATIONAL LABORATORY
9700 SOUTH CASS AVENUE
ARGONNE, ILLINOIS 60439

DR. KATHERINE FREESE
DEPT. OF ASTRONOMY & ASTROPHYSICS
UNIVERSITY OF CHICAGO
5640 SOUTH ELLIS AVENUE
CHICAGO, ILLINOIS 60637

DR. PETER G. O. FREUND
ENRICO FERMI INSTITUTE
5640 SOUTH ELLIS AVENUE
CHICAGO, ILLINOIS 60637

DR. JOSHUA A. FRIEMAN
DEPT. OF PHYSICS AND ASTROPHYSICS
UNIVERSITY OF CHICAGO
5640 SOUTH ELLIS AVENUE
CHICAGO, ILLINOIS 60637

PROF. HENRY J. FRISCH
HEP 320
ENRICO FERMI INSTITUTE
5640 SOUTH ELLIS AVENUE
CHICAGO, ILLINOIS 60637

DR. JAMES A. GAIDOS
PHYSICS DEPARTMENT
PURDUE UNIVERSITY
WEST LAFAYETTE, INDIANA 47907

DR. JOHN S. GALLAGHER III
KITT PEAK NATIONAL OBSERVATORY
BOX 26732
TUCSON, ARIZONA 85726

DR. NIKOS GIOKARIS
E-711, MS #122
FERMILAB
P.O. BOX 500
BATAVIA, ILLLINOIS 60510

HENRY GLASS
DEPARTMENT OF PHYSICS
STATE UNIVERSITY OF NEW YORK
STONY BROOK, NEW YORK 11794

DR. AUSTIN M. GLEESON
RLM 5.208
DEPARTMENT OF PHYSICS
UNIVERSITY OF TEXAS
AUSTIN, TEXAS 78712

DR. HAIM GOLDBERG
DEPARTMENT OF PHYSICS, 219 DA
NORTHEASTERN UNIVERSITY
360 HUNTINGTON AVENUE
BOSTON, MASSACHUSETTS 02115

DR. MAURY C. GOODMAN
MASS. INSTITUTE OF TECHNOLOGY
CAMBRIDGE, MASSACHUSETTS
02139

DR. J. RICHARD GOTT III
DEPT. OF ASTROPHYSICAL SCIENCE
PRINCETON UNIVERSITY
PRINCETON, NEW JERSEY 08544

DR. FRANK R. GRAZIANI
DEPARTMENT OF PHYSICS
UNIVERSITY OF COLORADO
BOULDER, COLORADO 80309

DR. MARCUS T. GRISARU
PHYSICS DEPARTMENT
BRANDEIS UNIVERSITY
WALTHAM, MASSACHUSETTS 02254

DR. DONALD E. GROOM
201 NORTH PHYSICS BUILDING
DEPARTMENT OF PHYSICS
UNIVERSITY OF UTAH
SALT LAKE CITY, UTAH 84112

DR. JAMES E. GUNN
PRINCETON UNIVERSITY OBSERVATORY
PEYTON HALL
PRINCETON, NEW JERSEY 08544

DR. ALAN H. GUTH
CENTER FOR THEORETICAL PHYSICS
MASS. INSTITUTE OF TECHNOLOGY
CAMBRIDGE, MASSACHUSETTS 02139

DR. JOHN S. HAGELIN
FACULTY
MAHARISHI INT'L UNIVERSITY
FAIRFIELD, IOWA 52556

ROBIN HANSON
5419 SOUTH ELLIS, #1E
CHICAGO, ILLINOIS 60615

MS. TINA HARRIOTT-MCCARTHY
22 WILLOWBIND COURT
HALIFAX, NOVA SCOTIA
CANADA B3M 3L3

DR. JAMES B. HARTLE
PHYSICS DEPARTMENT
UNIVERSITY OF CALIFORNIA
SANTA BARBARA, CALIFORNIA 93106

DR. JEFFREY A. HARVEY
JADWIN HALL
PRINCETON UNIVERSITY
PRINCETON, NEW JERSEY 08544

DR. MARK HAUGAN
DEPARTMENT OF PHYSICS
PURDUE UNIVERSITY
WEST LAFAYETTE, INDIANA 47907

DR. RICHARD W. HAYMAKER
DEPARTMENT OF PHYSICS & ASTRONOMY
LOUISIANA STATE UNIVERSITY
BATON ROUGE, LOUISIANA 70803

DR. DENNIS J. HEGYI
RANDALL LABORATORY
UNIVERSITY OF MICHIGAN
ANN ARBOR, MICHIGAN 48109

DR. CHRISTOPHER T. HILL
MS #106
FERMILAB
P.O. BOX 500
BATAVIA, ILLINOIS 60510

DR. LEWIS M. HOBBS
YERKES OBSERVATORY
P.O. BOX 258
WILLIAMS BAY, WISCONSIN 53191

DR. YEHUDA HOFFMAN
DEPARTMENT OF PHYSICS
UNIVERSITY OF PENNSYLVANIA
PHILADELPHIA, PENNSYLVANIA
19104

DR. CRAIG J. HOGAN
THEORETICAL ASTROPHYSICS
130-33
CALIFORNIA INST. OF TECHNOLOGY
PASADENA, CALIFORNIA 91125

DR. RICHARD HOLMAN
DEPARTMENT OF PHYSICS
WILLIAMSON HALL
UNIVERSITY OF FLORIDA
GAINESVILLE, FLORIDA 32611

PROF. BEI-LOK HU
DEPT. OF PHYSICS & ASTRONOMY
UNIVERSITY OF MARYLAND
COLLEGE PARK, MARYLAND 20742

MR. MARTIN HUBER
PHYSICS DEPARTMENT
STANFORD UNIVERSITY
STANFORD, CALIFORNIA 94305

DR. JOHN P. HUCHRA
HARVARD-SMITHSONIAN
CENTER FOR ASTROPHYSICS
60 GARDEN STREET
CAMBRIDGE, MASSACHUSETTS 02138

MR. JOE R. INCANDELA
UNIVERSITY OF CHICAGO
5640 SOUTH ELLIS AVENUE
CHICAGO, ILLINOIS 60637

DR. NAOKI IWAMOTO
PHYSICS DEPARTMENT
BOX 1105
WASHINGTON UNIVERSITY
ST. LOUIS, MISSOURI 63130

DR. KEITH JONES
"NUCLEAR PHYSICS B"
%NORDITA
BLEGDAMSVEJ 17
2100 COPENHAGEN 0, DENMARK

DR. DRASKO D. JOVANOVIC
MS #105
FERMILAB
P.O. BOX 500
BATAVIA, ILLINOIS 60510

DR. NICK KAISER
INST. FOR THEORETICAL PHYSICS
UNIVERSITY OF CALIFORNIA
SANTA BARBARA, CALIFORNIA 93106

DR. RONALD KANTOWSKI
DEPT. OF PHYSICS AND ASTRONOMY
440 WEST BROOKS
UNIVERSITY OF OKLAHOMA
NORMAN, OKLAHOMA 73019

DR. BORIS J. KAYSER
DIVISION OF PHYSICS
NATIONAL SCIENCE FOUNDATION
1800 G STREET N.W.
WASHINGTON, D.C. 20550

DR. THOMAS W. KEPHART
PHYSICS DEPARTMENT
PURDUE UNIVERSITY
WEST LAFAYETTE, INDIANA 47907

DR. CHUNG W. KIM
DEPARTMENT OF PHYSICS
JOHNS HOPKINS UNIVERSITY
HOMEWOOD CAMPUS
BALTIMORE, MARYLAND 21218

DR. JIHN E. KIM
CERN
CH-1211 GENEVE 23
SWITZERLAND

DR. SUNG K. KIM
DEPT. OF PHYSICS AND ASTROPHYSICS
MACALESTER COLLEGE
ST. PAUL, MINNESOTA 55105

DR. EDWARD W. KOLB, JR.
MS #209
FERMILAB
P. O. BOX 500
BATAVIA, ILLINOIS 60510

DR. DAVID S. KOLTICK
DEPARTMENT OF PHYSICS
PURDUE UNIVERSITY
WEST LAFAYETTE, INDIANA 47907

KENNETH L. KOWALSKI
DEPARTMENT OF PHYSICS
CASE WESTERN RESERVE UNIVERSITY
CLEVELAND, OHIO 44106

MR. STEVE KRISTOFF
DEPARTMENT OF MICROBIOLOGY
SCHOOL OF MEDICINE
INDIANA UNIVERSITY
INDIANAPOLIS, INDIANA 46223

MICHAEL KROUPA
DEPT. OF ASTRONOMY & ASTROPHYSICS
UNIVERSITY OF CHICAGO
5640 SOUTH ELLIS AVENUE
CHICAGO, ILLINOIS 60637

DR. NATHAN KRUMM
PHYSICS DEPARTMENT
UNIVERSITY OF CINCINNATI
CINCINNATI, OHIO 45221

DR. MOYSES KUCHNIR
MS #316
FERMILAB
P.O. BOX 500
BATAVIA, ILLINOIS 60510

MR. MAHMOUD K. KULLAB
SCHOOL OF ASTRONOMY
UNIVERSITY OF MINNESOTA
1173 RALEIGH STREET
ST. PAUL, MINNESOTA 55108

DR. GABOR KUNSTATTER
PHYSICS DEPARTMENT
UNIVERSITY OF TORONTO
TORONTO, ONTARIO
CANADA M5S 1A7

DR. PAUL G. LANGACKER
DEPARTMENT OF PHYSICS
UNIVERSITY OF PENNSYLVANIA
PHILADELPHIA, PENNSYLVANIA
19104

DR. ANTHONY N. LASENBY
NUFFIELD RADIO ASTRONOMY LABS
JODRELL BANK
ENGLAND

DR. LEON M. LEDERMAN, DIRECTOR
MS #105
FERMILAB
P.O. BOX 500
BATAVIA, ILLINOIS 60510

DR. DAVID LINDLEY
MS #209
FERMILAB
P. O. BOX 500
BATAVIA, ILLINOIS 60510

DR. JES MADSEN
INSTITUTE OF ASTRONOMY
UNIVERSITY OF AARHUS
DK-8000 AARHUS C
DENMARK

ROGER MALINA
SPACE SCIENCES LABORATORY
UNIVERSITY OF CALIFORNIA
BERKELEY, CALIFORNIA 94720

DR. ADRIAN C. MELISSINOS
DEPT. OF PHYSICS & ASTRONOMY
UNIVERSITY OF ROCHESTER
ROCHESTER, NEW YORK 14627

DR. ADRIAN L. MELOTT
ASTRONOMY AND ASTROPHYSICS
UNIVERSITY OF CHICAGO
5640 SOUTH ELLIS AVENUE
CHICAGO, ILLINOIS 60637

BRADLEY S. MEYER
DEPT. OF PHYSICS & ASTROPHYSICS
UNIVERSITY OF CHICAGO
5640 SOUTH ELLIS AVENUE
CHICAGO, ILLINOIS 60637

DR. SHOJI MIKAMO
MS #223
FERMILAB
P. O. BOX 500
BATAVIA, ILLINOIS 60510

DR. JOHN W. MOFFAT
DEPARTMENT OF PHYSICS
UNIVERSITY OF TORONTO
TORONTO M5S 1A7, ONTARIO
CANADA

PATRICK J. MOONEY
PHYSICS DEPARTMENT
UNIVERSITY OF NOTRE DAME
NOTRE DAME, INDIANA 46556

DR. DOMENICO NANNI
INFN FRASCATI
CASELLA POSTALE, 13
I-00044 FRASCATI (ROMA)
ITALY

JUAN NEGRET
PHYSICS DEPARTMENT
PURDUE UNIVERSITY
WEST LAFAYETTE, INDIANA 47907

MR. ROBERT NELSON
311 NORRIS COURT, #2N
MADISON, WISCONSIN 53703

DR. FRANK A. NEZRICK
MS #306
FERMILAB
P. O. BOX 500
BATAVIA, ILLINOIS 60510

MIROSLAV NIKOLIC
UNIVERSITY OF MINNESOTA
SCHOOL OF ASTRONOMY
MINNEAPOLIS, MINNESOTA 55455

DR. FRANCO OCCHIONERO
ISTITUTO ASTRONOMICO
UNIVERSITA DI ROMA
VIA F. M. LANCISI 29
I-00161 ROMA, ITALY

DR. KEITH A. OLIVE
MS #209
FERMILAB
P.O. BOX 500
BATAVIA, ILLINOIS 60510

DR. ARNE P. OLSON
APPLIED PHYSICS DIVISION
ARGONNE NATIONAL LABORATORY
9700 SOUTH CASS AVENUE
ARGONNE, ILLINOIS 60439

DAVID P. OLSON
LOOMIS LABORATORY OF PHYSICS
UNIVERSITY OF ILLINOIS
110 WEST GREEN STREET
URBANA, ILLINOIS 61801

DENNIS OVERBYE
DISCOVER MAGAZINE
TIME-LIFE BUILDING
ROCKEFELLER CENTER
NEW YORK, NEW YORK 10020

DR. B. E. J. PAGEL
ROYAL GREENWICH OBSERVATORY
HERSTMONCEAUX CASTLE
HAILSHAM
SUSSEX BN27 1RP, ENGLAND

DR. LEONARD E. PARKER
DEPARTMENT OF PHYSICS
UNIVERSITY OF WISCONSIN-MILWAUKEE
MILWAUKEE, WISCONSIN 53201

DR. STEPHEN J. PARKE
MS #106
FERMILAB
P. O. BOX 500
BATAVIA, ILLINOIS 60510

PROF. R. BRUCE PARTRIDGE
DEPARTMENT OF ASTRONOMY
HAVERFORD COLLEGE
HAVERFORD, PENNSYLVANIA 19041

MR. MALCOLM J. PERRY
DEPT. OF PHYSICS
PRINCETON UNIVERSITY
PRINCETON, NEW JERSEY 08544

DR. C. J. PETHICK
DEPARTMENT OF PHYSICS
UNIVERSITY OF ILLINOIS
1110 WEST GREEN STREET
URBANA, ILLINOIS 61801

DR. JAMES D. PHILLIPS
VARIAN PHYSICS BUILDING
STANFORD UNIVERSITY
STANFORD, CALIFORNIA 94305

DR. JOSEPH G. POLCHINSKI
DEPARTMENT OF PHYSICS
JEFFERSON LABORATORY
HARVARD UNIVERSITY
CAMBRIDGE, MASSACHUSETTS 02138

DR. JOHN W. PRESKILL
452-48
DEPARTMENT OF PHYSICS
CALIFORNIA INST. OF TECHNOLOGY
PASADENA, CALIFORNIA 91125

DR. CHRIS QUIGG
MS #106
FERMILAB
P.O. BOX 500
BATAVIA, ILLINOIS 60510

DR. JEFFREY M. RABIN
ENRICO FERMI INSTITUTE
5640 SOUTH ELLIS AVENUE
CHICAGO, ILLINOIS 60637

DR. JOHN P. RALSTON
BUILDING 362
ARGONNE NATIONAL LABORATORY
9700 SOUTH CASS AVENUE
ARGONNE, ILLINOIS 60439

MR. JAMES RANDI
51 LENNOX AVENUE
RUMSON, NEW JERSEY 07760

FRANCIS REDDY
ASTRONOMY MAGAZINE
625 EAST ST. PAUL AVENUE
MILWAUKEE, WISCONSIN 53202

DR. DAVID B. REISS
DEPARTMENT OF PHYSICS
UNIVERSITY OF MINNESOTA
116 CHURCH STREET S.E.
MINNEAPOLIS, MINNESOTA 55455

DR. PAUL L. RICHARDS
DEPARTMENT OF PHYSICS
366 LE CONTE HALL
UNIVERSITY OF CALIFORNIA
BERKELEY, CALIFORNIA 94720

DR. CARL ROSENZWEIG
DEPARTMENT OF PHYSICS
SYRACUSE UNIVERSITY
SYRACUSE, NEW YORK 13210

DR. MARK RUBIN
DEPARTMENT OF PHYSICS
UNIVERSITY OF TEXAS
AUSTIN, TEXAS 78712

DR. REMO RUFFINI
ISTITUTO DI FISICA
"G. MARCONI"
UNIVERSITA DI ROMA
I-00185 ROMA, ITALY

DR. MICHAEL SALAMON
PHYSICS DEPARTMENT
UNIVERSITY OF CALIFORNIA
BERKELEY, CALIFORNIA 94720

DR. PIERRE SALATI
LAPP
CHEMIN DE BELLEVUE, B.P. 909
F-74019 ANNECY-LE-VIEUX CEDEX
FRANCE

PROF. KATSUHIKO SATO
FACULTY OF SCIENCE
UNIVERSITY OF TOKYO
HONGO, BUNKYO-KU
TOKYO 113, JAPAN

MR. ROBERT J. SCHERRER
ASTRONOMY & ASTROPHYSICS CENTER
UNIVERSITY OF CHICAGO
5640 SOUTH ELLIS AVENUE
CHICAGO, ILLINOIS 60637

DR. DAVID N. SCHRAMM
DEPT. OF ASTRON. & ASTROPHYSICS
UNIVERSITY OF CHICAGO
5640 SOUTH ELLIS AVENUE
CHICAGO, ILLINOIS 60637

DR. FRANK SCIULLI
NEVIS LABORATORIES
P.O. BOX 137
IRVINGTON, NEW YORK 10533

DR. DAVID SECKEL
MS #209
FERMILAB
P.O. BOX 500
BATAVIA, ILLINOIS 60510

DAVID SEIBERT
DEPARTMENT OF PHYSICS
UNIVERSITY OF ILLINOIS
110 WEST GREEN STREET
URBANA, ILLINOIS 61801

SALLY SEIDEL
DEPT. OF PHYSICS, RANDALL LAB.
UNIVERSITY OF MICHIGAN
500 EAST UNIVERSITY STREET
ANN ARBOR, MICHIGAN 48109

MR. PHILIP E. SEIDEN
IBM RESEARCH CENTER
P. O. BOX 218
YORKTOWN HEIGHTS, NEW YORK
10598

DR. GORAN SENJANOVIC
DEPARTMENT OF PHYSICS
VIRGINIA POLYTECHNIC INSTITUTE
AND STATE UNIVERSITY
BLACKSBURG, VIRGINIA 24061

MR. DANIEL SHALIT
% DR. CHARLES RHODES
DEPARTMENT OF PHYSICS
UNIV. OF ILLINOIS, P. O. BOX 4348
CHICAGO, ILLINOIS 60680

DR. P. A. SHAVER
EUROPEAN SOUTHERN OBSERVATORY
KARL-SCHWARZSCHILD-STR. 2
D-8046 GARCHING BEI MUENCHEN
WEST GERMANY

MICHAEL J. SHEPKO
DEPARTMENT OF PHYSICS
TEXAS A & M UNIVERSITY
COLLEGE STATION, TEXAS 77843

DR. PIERRE SIKIVIE
215, WILLIAMSON HALL
PHYSICS DEPARTMENT
UNIVERSITY OF FLORIDA
GAINESVILLE, FLORIDA 32611

DR. JOSEPH I. SILK
ASTRONOMY DEPARTMENT
UNIVERSITY OF CALIFORNIA
BERKELEY, CALIFORNIA 94720

DR. LEE SMOLIN
ENRICO FERMI INSTITUTE
UNIVERSITY OF CHICAGO
5640 SOUTH ELLIS AVENUE
CHICAGO, ILLINOIS 60637

MR. SUNIL SOMALWAR
HEP #307
UNIVERSITY OF CHICAGO
5640 SOUTH ELLIS AVENUE
CHICAGO, ILLINOIS 60637

PROF. MARK SREDNICKI
PHYSICS DEPARTMENT
UNIVERSITY OF CALIFORNIA
SANTA BARBARA, CALIFORNIA 93106

DR. FLOYD W. STECKER
CODE 665
GODDARD SPACE FLIGHT CENTER
GREENBELT, MARYLAND 20771

DR. GARY STEIGMAN
BARTOL RESEARCH FOUNDATION
UNIVERSITY OF DELAWARE
NEWARK, DELAWARE 19711

DR. PAUL J. STEINHARDT
DEPARTMENT OF PHYSICS
UNIVERSITY OF PENNSYLVANIA
PHILADELPHIA, PENNSYLVANIA
19104

DR. JAMES L. STONE
RANDALL PHYSICS LABORATORY
UNIVERSITY OF MICHIGAN
ANN ARBOR, MICHIGAN 48109

DEAN M. SUMI
DEPARTMENT OF ASTRONOMY
UNIVERSITY OF ILLINOIS
1011 WEST SPRINGFIELD AVENUE
URBANA, ILLINOIS 61801

DR. ALEXANDER S. SZALAY
INSTITUTE OF THEORETICAL PHYSICS
UNIVERSITY OF CALIFORNIA
BERKELEY, CALIFORNIA 94720

MR. MICHAEL TABER
PHYSICS DEPARTMENT
STANFORD UNIVERSITY
STANFORD, CALIFORNIA 94305

DR. GUSTAV A. TAMMANN
ST. ALBAN RING 172
CH-4052 BASEL
SWITZERLAND

DR. TOMASZ R. TAYLOR
MS #106
FERMILAB
P.O. BOX 500
BATAVIA, ILLINOIS 60510

DR. CLAUDIA D. TESCHE
IBM WATSON RESEARCH CENTER
P. O. BOX 218
YORKTOWN HEIGHTS, NEW YORK 10598

DR. ROBERT L. THEWS
G-318, ER-221 GTN
U.S. DEPARTMENT OF ENERGY
WASHINGTON, D.C. 20545

MR. DIETRICK E. THOMSEN
SCIENCE NEWS MAGAZINE
1719 N STREET N.W.
WASHINGTON, D.C. 20036

ANDREW TOMASCH
DEPARTMENT OF PHYSICS
SWAIN HALL WEST 117
INDIANA UNIVERSITY
BLOOMINGTON, INDIANA 47405

DR. YASUNARI TOSA
THEORY DEVISION T-8, MS B285
LOS ALAMOS NATIONAL LABORATORY
LOS ALAMOS, NEW MEXICO 87545

MS. JENNIE TRASCHEN
HARVARD UNIVERSITY
141 OXFORD STREET
CAMBRIDGE, MASSACHUSETTS 02140

DR. W. PETER TROWER
PHYSICS DEPARTMENT
VIRGINIA POLYTECHNIC INSTITUTE
AND STATE UNIVERSITY
BLACKSBURG, VIRGINIA 24061

DR. JAMES W. TRURAN
DEPT. OF ASTRONOMY, ROOM 341
UNIVERSITY OF ILLINOIS
1011 WEST SPRINGFIELD
URBANA, ILLINOIS 61801

DR. EDWIN L. TURNER
123 PEYTON HALL
IVY LANE
PRINCETON, NEW JERSEY 08544

DR. MICHAEL S. TURNER
MS #209
FERMILAB
P. O. BOX 500
BATAVIA, ILLINOIS 60510

DR. JACK D. ULLMAN
DEPT. OF PHYSICS AND ASTRONOMY
HERBERT LEHMAN COLLEGE
BEDFORD PARK BOULEVARD WEST
BRONX, NEW YORK 10464

DR. JUAN M. USON
JOSEPH HENRY LABORATORIES
PHYSICS DEPARTMENT, JADWIN HALL
PRINCETON UNIVERSITY, P.O.BOX 708
PRINCETON, NEW JERSEY 08544

DR. C. E. VAYONAKIS
INTERNATIONAL CENTRE FOR
THEORETICAL PHYSICS, P.O.B. 586
MIRAMARE, STRADA COSTIERA II
I-34100 TRIESTE, ITALY

MR. ROBERTO VEGA
DEPARTMENT OF PHYSICS
UNIVERSITY OF TEXAS
AUSTIN, TEXAS 78712

DR. GIOVANNI VENTURI
DIPARTIMENTO DI FISICA
UNIVERSITA DI BOLOGNA
VIA IRNERIO 46
40126 BOLOGNA, ITALY

DR. AGOSTINO VIGNATO
LAB. NAZIONAL DI FRASCATI
C. P. 13
00044 FRASCATI (ROMA) ITALY

DR. ALEXANDER VILENKIN
DEPARTMENT OF PHYSICS
TUFTS UNIVERSITY
MEDFORD, MASSACHUSETTS 02155

DR. ETHAN T. VISHNIAC
ASTRONOMY DEPARTMENT
UNIVERSITY OF TEXAS
AUSTIN, TEXAS 78712

DR. NICOLA VITTORIO
ASTRONOMY DEPARTMENT
UNIVERSITY OF CALIFORNIA
BERKELEY, CALIFORNIA 94720

DR. DAVID E. WAGONER
%E 705, MS #219
FERMILAB
P.O. BOX 500
BATAVIA, ILLINOIS 60510

DR. ROBERT M. WALD
ENRICO FERMI INSTITUTE
UNIVERSITY OF CHICAGO
5640 SOUTH ELLIS AVENUE
CHICAGO, ILLINOIS 60637

M. MITCHELL WALDROP
SCIENCE MAGAZINE
1515 MASSACHUSETTS AVE., N.W.
WASHINGTON, D. C. 20005

RONALD A. WALTON
1671 VIA RANCHO
SAN LARENZO, CALIFORNIA 94580

DR. THOMAS J. WEILER
PHYSICS DEPARTMENT, B-019
UNIVERSITY OF CALIFORNIA
LA JOLLA, CALIFORNIA 92093

DR. STEVEN WEINBERG
DEPARTMENT OF PHYSICS
R.L. MOORE HALL, 5-208
UNIVERSITY OF TEXAS
AUSTIN, TEXAS 78712

DR. ERICK J. WEINBERG
PHYSICS DEPARTMENT
COLUMBIA UNIVERSITY
P.O. BOX 133
NEW YORK, NEW YORK 10027

DR. SIMON D. M. WHITE
DEPARTMENT OF ASTRONOMY
UNIVERSITY OF CALIFORNIA
SANTA BARBARA, CALIFORNIA 93106

MR. PAUL L. WHITEHOUSE
DEPARTMENT OF PHYSICS
NORTHWESTERN UNIVERSITY
EVANSTON, ILLINOIS 60201

DR. DAVID T. WILKINSON
JOSEPH HENRY LABORATORIES
PRINCETON UNIVERSITY
P. O. BOX 708, JADWIN HALL
PRINCETON, NEW JERSEY 08544

DR. CLIFFORD M. WILL
DEPARTMENT OF PHYSICS
WASHINGTON UNIVERSITY
ST. LOUIS, MISSOURI 63130

DR. JAMES P. WRIGHT
DIV. OF ASTRONOMICAL SCIENCES
NATIONAL SCIENCE FOUNDATION
1800 G STREET, N.W.
WASHINGTON, D.C. 20550

DR. AMOS YAHIL
ASTRONOMY PROGRAM
STATE UNIVERSITY OF NEW YORK
STONY BROOK, NEW YORK 11794

WILOX YANG
MS #335
FERMILAB
P.O. BOX 500
BATAVIA, ILLINOIS 60510

DR. MOTOHIKO YOSHIMURA
NAT'L LAB FOR HIGH ENERGY PHYSICS
OHO-MACHI, TSUKUBA-GUN
IBARAKI-KEN 305, JAPAN

DR. LEONARD I. ZANE
DEPT. OF PHYSICS
UNIVERSITY OF NEVADA
LAS VEGAS, NEVADA 89154

DR. DANIELA ZANON
40 FERNALD DRIVE, #11
CAMBRIDGE, MASSACHUSETTS
02138

DAVID JEFFERY ZOLLER
DEPT. OF PHYSICS & ASTROPHYSICS
UNIVERSITY OF CHICAGO
5640 SOUTH ELLIS AVENUE
CHICAGO, ILLINOIS 60637

Battle Cry of Freedom
McPherson